Functional Mesoporous Carbon-Based Film Devices for Energy Systems

Biao Kong · Hongbin Xu · Lei Xie · Shan Zhou

Functional Mesoporous Carbon-Based Film Devices for Energy Systems

上海交通大學出版社
Shanghai Jiao Tong University Press

Springer

Biao Kong
Fudan University
Shanghai, China

Hongbin Xu
Massachusetts Institute of Technology
Cambridge, MA, USA

Lei Xie
Xi'an Jiaotong University
Xi'an City, China

Shan Zhou
Qingdao University
Qingdao, China

ISBN 978-981-99-7497-9 ISBN 978-981-99-7498-6 (eBook)
https://doi.org/10.1007/978-981-99-7498-6

Jointly published with Shanghai Jiao Tong University Press
The print edition is not for sale in China (Mainland). Customers from China (Mainland) please order the print book from: Shanghai Jiao Tong University Press.

This Springer imprint is published by the registered company Springer Nature Singapore Pte Ltd.
The registered company address is: 152 Beach Road, #21-01/04 Gateway East, Singapore 189721, Singapore

Paper in this product is recyclable.

Contents

Chapter 1
Introduction of Energy Materials

With the development of science and technology, human beings are increasingly demanding energy. Various energy materials were developed for the application of energy harvest. In the following context, diverse materials that favor for the energy generation were introduced.

1.1 Introduction

Currently, the energy system is crucial for the economies of most countries as it facilitates the exploration of energy sources and their conversion into various usable forms to support industrial manufacturing, transportation, and personal lifestyles. Consequently, energy holds significant importance in our world today.

In today's world, the primary energy sources encompass solar power, nuclear energy, hydroelectricity, wind power, geothermal energy, natural gas, coal, biomass, and petroleum. The 2015 Quadrennial Technology Review (QTR 2015) published by the U.S. Department of Energy (DOE) depicted the flow of energy throughout the U.S. energy system in 2014 [1]. These initial energy sources can be transformed into fuels or electricity, yet some energy is inevitably lost during the conversion and utilization processes. Upon analyzing the aforementioned figure, it becomes evident that fossil fuels account for approximately 82% of the primary energy employed in the United States. Electricity assumes a critical role in both industrial production and daily life, and it can be generated through diverse means utilizing all the aforementioned energy sources. Unfortunately, a significant portion of electricity is ultimately wasted due to various factors, which is a regrettable circumstance.

In contrast to previous decades, the global energy landscape has been gradually transforming, shifting away from heavy reliance on fossil fuels towards the emergence and advancement of diverse new energy sources. This necessitates the development of various energy materials to effectively adapt to and drive this change

B. Kong et al., *Functional Mesoporous Carbon-Based Film Devices for Energy Systems*,
https://doi.org/10.1007/978-981-99-7498-6_1

[2–5]. In a broad sense, energy materials encompass all materials utilized in the energy industry and energy technology. However, within the realm of new materials, energy materials typically refer to materials that are being developed and have the potential to support the establishment of new energy systems, catering to the specific demands of various new energy and energy-saving technologies.

Currently, some significant new energy materials include: (1) Fission reactor materials: Examples include uranium and plutonium. (2) Fusion reactor materials: This category encompasses thermonuclear fusion fuel, structural materials, and more. (3) High-energy propellant: It comprises liquid propellant and solid propellant. (4) Battery materials: These include battery electrode materials, electrolytes, and related components. (5) Hydrogen energy materials: Primarily solid hydrogen storage materials and associated application technologies. (6) Superconducting materials: This involves traditional superconducting materials, high-temperature superconducting materials, and their applications in energy conservation and storage. (7) Solar cell materials: Materials specifically designed for solar cell technology. (8) Other new energy materials: Such as those required for wind energy, geothermal energy, magnetohydrodynamic power generation, and related technologies.

In the case of a lithium-ion battery, which is widely used in our daily lives, the anode material is graphite. However, the cathode material, which is equally essential, varies. The most commonly employed cathode materials in lithium-ion batteries are lithium cobalt oxide and lithium iron phosphate. These materials were invented by Professor Goodenough, the recipient of the 2019 Nobel Prize in chemistry for his significant contributions to the development of lithium-ion batteries. This serves as just one example among numerous energy materials, yet it is this material that has made a tremendous impact on our lives. Thus, the importance of advancing energy materials becomes inherently evident.

Since the twenty-first century, governments around the world have heavily invested in the advancement of new energy technologies due to the increasingly prominent clash between energy demands and environmental concerns. They consider the development of energy technology as a crucial element for future economic growth. Consequently, the research and industrialization of energy materials have experienced rapid progress within this context.

1.2 The Challenge About Energy

Over the past few decades, significant progress has been made in the rapid development of energy technology. However, due to the rapid growth of society and the economy, the environmental and energy contradiction has become increasingly prominent. There are several challenges we face in the field of energy:

(1) Depletion of traditional fossil fuels [6, 7]. Current energy sources heavily rely on non-renewable options such as petroleum, coal, and natural gas. While these resources are not expected to deplete in the immediate future, our heavy

dependence on them necessitates a prompt transition towards renewable energy sources like solar, wind, hydro, and nuclear energy to ensure a better future for the generations to come.

(2) Environmental pollution from traditional energy sources [8, 9]. Coal mining leads to surface subsidence, ecological deterioration, and the generation of large amounts of wastewater that harm crops and water sources. Burning fossil fuels emits NOx, a significant contributor to ozone layer destruction, increasing the risk of skin cancer. Additionally, SO2 and NOx from coal burning cause acid rain, which damages ecosystems. Automobile and industrial emissions also result in air pollution, including harmful particulate matter like PM2.5. Addressing these issues requires the development of clean energy solutions to minimize environmental pollution.

(3) Greenhouse gas emissions and the greenhouse effect [10, 11]. The massive combustion of fossil fuels releases greenhouse gases, predominantly carbon dioxide, which contributes to the greenhouse effect. This phenomenon leads to global temperature rise, glacier melting, sea level rise, and species extinction. Urgent actions are needed to alleviate the greenhouse effect through the use of clean energy sources like carbon dioxide-free fuel cells and the capture and conversion of carbon dioxide into useful chemicals.

(4) Increasing demand for energy [12, 13]. The growing reliance on energy, especially electricity, poses challenges. There is a need for extended battery life for electronic devices such as mobile phones, computers, and iPads. The rise of electric vehicles and drones demands high-energy density, lightweight, and affordable batteries. Enhancing the efficiency of solar cells for efficient photoelectric conversion is also a desired goal.

(5) Security of energy technologies [14, 15]. As various new energy technologies are developed, security becomes a critical concern. Proper selection of materials is essential to ensure the safety of systems, especially in the case of nuclear energy, given past incidents like Chernobyl and Fukushima. Issues such as lithium dendrite formation and short circuits in lithium–ion batteries require attention, as safety concerns arise, including reports of electric vehicle fires. Security must be a priority when developing new energy technologies.

In summary, to achieve sustainable development, numerous challenges must be overcome. These challenges represent opportunities for researchers to conquer. Energy materials play a crucial role in the development of various new energy technologies and systems. The design, synthesis, selection, and application of energy materials will significantly influence the future development of the energy field and contribute to overall human progress.

1.3 Research Contents and Methods of Energy Materials

The modern world economy has greatly benefited from the use of fossil fuels due to their abundance and global availability. However, as discussed earlier, these resources are finite and contribute to environmental pollution, posing challenges to human survival. In response to these challenges, we have no alternative but to develop and utilize new energy systems, embracing the path of sustainable development. The twenty-first century is witnessing rapid advancements in new energy technologies, and the realization of these technologies relies not only on the utilization of new energy systems but also on the development and application of corresponding energy materials. This integration is vital to realize the potential of new energy systems, improve efficiency, and reduce costs.

Solar energy, as a clean and environmentally friendly renewable energy source, holds tremendous potential for development and application [16, 17]. It can be harnessed through various methods such as photothermal conversion, photoelectric conversion, and photochemical conversion. Among these, solar cells have garnered significant attention. Semiconductor materials play a crucial role in solar cells. When sunlight falls on the semiconductor PN junction, electron–hole pairs are generated. The internal electric field facilitates the movement of photogenerated holes to the P region and photogenerated electrons to the N region, resulting in the generation of electric current. Through continuous research and development efforts, the photoelectric conversion efficiency of solar cells has been steadily improved, and the manufacturing costs associated with them have also been progressively reduced.

In the twenty-first century, hydrogen energy is poised to play a crucial role on the global energy stage, and the production, storage, transportation, and application of hydrogen will garner significant attention [18, 19]. Hydrogen possesses a high calorific value, approximately three times that of gasoline, 3.9 times that of alcohol, and 4.5 times that of coke. When hydrogen is combusted, the only byproduct is water, making it one of the cleanest sources of energy worldwide. Moreover, hydrogen is abundant and offers sustainable development prospects. Photochemical hydrogen production involves the use of specific photocatalysts or electrode materials to absorb solar energy, generating photocarriers that split water into hydrogen and oxygen. This method, which directly converts solar energy into usable hydrogen, is considered the ultimate goal of hydrogen production. Ongoing research focuses on designing and preparing efficient catalytic materials, improving conversion efficiency, and ensuring the stability of water splitting processes. Electrocatalytic hydrogen production, on the other hand, utilizes electricity to split pure water into hydrogen and oxygen. The cost of electricity represents a significant portion of the production cost. Therefore, reducing energy consumption (overpotential) and ensuring the stability of catalytic materials are critical research directions in this area. Additionally, excess electricity generated from sources like solar power, wind power, and hydropower can be stored and used to electrolyze water for hydrogen production, presenting a promising approach. The development of hydrogen energy and the advancement of hydrogen production technologies hold great potential for addressing the challenges posed by traditional energy sources. With ongoing research and innovation, hydrogen has the opportunity to become a key player in the global energy landscape, promoting a cleaner and more sustainable future.

Hydrogen storage is a critical technology that represents a bottleneck in the widespread utilization of hydrogen at scale [20, 21]. The challenge of hydrogen storage spans across the entire hydrogen production, transportation, and final application processes. Without a solution to the hydrogen storage problem, the widespread adoption of hydrogen energy becomes difficult. Unlike solid coal or liquid petroleum, hydrogen is a gas, making its transportation and storage more challenging. Various forms of hydrogen storage are being explored, including gas storage, liquid storage, and storage through compounds. High-pressure gas storage involves compressing hydrogen gas to high pressures, while low-temperature liquid storage involves cooling hydrogen to extremely low temperatures to achieve a liquid state. Additionally, hydrogen storage materials are being investigated as a promising approach. Many studies have been conducted on metal compound hydrogen storage materials, which have shown promise. However, in recent years, nano-hydrogen storage materials, such as carbon nanotubes and metal–organic frameworks (MOFs), have garnered increased attention. The stability of these materials, the mechanisms of hydrogen storage, and the quest for more efficient hydrogen storage methods require further study and exploration. Efficient and effective hydrogen storage is crucial for realizing the full potential of hydrogen energy. Researchers continue to investigate and develop storage technologies to overcome the challenges associated with hydrogen storage, ensuring its viability and practicality for a wide range of applications.

A fuel cell is a device that directly converts the chemical energy of a fuel into electrical energy [22, 23]. It is considered a fourth-generation power generation technology, following hydropower, thermal power, and nuclear power. Fuel cells offer several advantages, including high power generation efficiency, minimal environmental pollution, higher energy density, low noise levels, and the ability to utilize a wide range of fuels. Hydrogen fuel cells, in particular, have garnered significant attention. Hydrogen fuel cells operate by combining hydrogen and oxygen to produce water, without emitting pollutants such as hydrocarbons, carbon monoxide, nitrogen oxides, or carbon dioxide. They are environmentally friendly and exhibit high power generation efficiency. Hydrogen fuel cells have been successfully used in space applications since the 1960s, with the Apollo spacecraft featuring these compact, high-capacity devices. Fuel cells also hold promise for use in future transportation, including vehicles. Fuel cell technology is rapidly advancing, with ongoing research focused on developing electrolyte membranes and electrode materials to improve efficiency and reduce costs. Efforts are being made to enhance the performance and durability of fuel cells, making them more viable for widespread commercial and industrial applications. Continued progress in fuel cell technology holds the potential to transform energy generation, providing a clean and efficient source of power for various sectors.

In addition to fuel cell vehicles, the electric vehicle industry is experiencing significant growth, exemplified by Tesla's Model S. The coming years are expected to be a golden period for the development of new energy vehicles. Many countries are implementing policies to promote the adoption of new energy vehicles, with some even planning to phase out the sale of traditional fuel vehicles within a designated timeframe. Within the field of power batteries, lithium–ion batteries are widely recognized as the ideal energy storage devices [24, 25]. Extensive research is being conducted on various aspects of lithium-ion batteries, including positive and negative electrode materials, membranes, electrolytes, and more. Each component warrants further study and exploration to enhance battery performance. Additionally, ensuring the safety of lithium-ion batteries is a critical consideration for the application of new energy vehicles. While lithium-ion batteries dominate the energy storage landscape, other battery systems such as lithium-sulfur batteries and other emerging technologies are also worth exploring. Research and development efforts are focused on improving battery performance, energy density, lifespan, and safety, aiming to advance the capabilities of energy storage systems for a variety of applications, including electric vehicles.

Catalysts play an indispensable role in various chemical reactions and industrial systems, such as the industrial catalytic synthesis of ammonia and electrocatalytic hydrogen production [26, 27]. The advent of these catalysts has brought about significant transformations in production and daily life. For instance, in electrocatalytic reactions, catalysts effectively reduce overpotential and accelerate reaction rates, thereby conserving energy and reducing costs. This aligns with the principles of low-carbon economy and green development. It is no exaggeration to state that energy catalytic materials have emerged as a vital branch of energy materials. The development and application of these catalysts enable the realization of numerous reactions and contribute to the advancement of new energy systems and technologies.

The research on energy materials can be approached from two perspectives: the material perspective and the energy perspective. In the field of material science and engineering, the study of energy materials focuses on the preparation techniques, composition, structure, and application performance of the materials. In the field of energy, the focus is on the energy storage and conversion capabilities, micro-mechanisms, safety considerations, and other relevant aspects. Several specific areas of study can be identified:

(1) Research and development of new energy materials: Researchers explore the design and synthesis of new energy materials to enhance their performance, stability, and safety. This process often leads to the discovery of new material structures, synthesis methods, and theories that can be applied in other fields. For example, the study of organic metal frameworks has opened up new possibilities in various disciplines.
(2) Investigation of the microscopic composition, electronic structure, and surface morphology: When synthesizing new energy materials, it is important to analyze their microscopic composition, crystal structure, electronic properties, and surface morphology to establish the structure–property relationship. Advanced characterization techniques enable a deeper understanding of these materials,

shedding light on their working mechanisms. For example, synchrotron radiation can be used to analyze the microstructure of single-atom catalysts, providing insights into the catalytic process.

(3) Theoretical calculations for guidance and explanation: The rapid advancement of computer technology has made theoretical calculations an integral part of energy material research. Theoretical calculations aid in material design and the interpretation of material properties. They allow researchers to build computational platforms for materials, reducing the need for trial and error. For instance, in the context of the oxygen evolution reaction, preliminary calculations can help identify suitable catalyst materials, guiding subsequent experimental efforts. Theoretical calculations also provide insights into the energy and electronic structure changes during reactions, explaining material performance.

(4) Safety and reliability of energy materials: Safety is a crucial aspect from the development to the final application of energy materials. For example, lithium-ion batteries, while exhibiting excellent performance, require continuous attention to safety concerns. Instances of short circuits, battery explosions, and electric vehicle fires highlight the importance of developing safer and more efficient lithium-ion batteries. Safety considerations are also relevant in the development and utilization of nuclear and hydrogen energy.

(5) Industrial production, environmental impact, and cost: After materials are developed in the laboratory, their large-scale production and environmental impact need to be addressed. The conditions for laboratory-scale and industrial-scale production often differ, requiring adjustments to maintain the desired material properties during large-scale manufacturing. Furthermore, the environmental impact of materials during and after their use should be considered. Energy materials should be designed with sustainability in mind to address environmental concerns. Finally, the cost of materials is a crucial factor in determining their widespread applicability. High-cost materials, such as platinum as a catalyst, may hinder large-scale adoption and limit their impact on society.

Currently, there are numerous popular energy materials, and it is beyond the scope of this response to provide an exhaustive list. Readers are encouraged to explore and study energy materials based on their own interests and research directions. In conclusion, the development and application of energy materials aim to improve performance, ensure safety, reduce costs, and minimize pollution. Energy materials are expected to contribute significantly to economic development, energy structure transformation, environmental protection, and other aspects both in China and globally.

References

1. Department of Energy (United States of America) (2015) An assessment of energy technologies and research opportunities. Quadrenn Technol Rev

2. Larcher D, Tarascon JM (2015) Towards greener and more sustainable batteries for electrical energy storage. Nature Chem 7:19
3. Gallezot P (2012) Conversion of biomass to selected chemical products. Chem Soc Rev 41:1538
4. Habisreutinger SN, Schmidt-Mende L, Stolarczyk JK (2013) Photocatalytic reduction of CO_2 on TiO_2 and other semiconductors. Angew Chem Int Ed 52:7372
5. Nugent P, Belmabkhout Y, Burd SD et al (2013) Porous materials with optimal adsorption thermodynamics and kinetics for CO_2 separation. Nature 495:80
6. Ma S, Zhou HC (2010) Gas storage in porous metal–organic frameworks for clean energy applications. Chem Commun 46:44
7. Li X, Yu J, Low J et al (2015) Engineering heterogeneous semiconductors for solar water splitting. J. Mater. Chem. A 3:2485
8. Huang RJ, Zhang Y, Bozzetti C et al (2014) High secondary aerosol contribution to particulate pollution during haze events in China. Nature 514:218
9. Davis SJ, Caldeira K (2010) Consumption-based accounting of CO_2 emissions. Proc Natl Acad Sci 107:5687
10. Paracchino A, Laporte V, Sivula K, Grätzel M, Thimsen E (2011) Highly active oxide photocathode for photoelectrochemical water reduction. Nature Mater 10:456
11. Kirschke S, Bousquet P, Ciais P, Saunois M et al (2013) Three decades of global methane sources and sinks. Nature Geosci 6:813
12. Brennan L, Owende P et al (2010) Biofuels from microalgae—a review of technologies for production, processing, and extractions of biofuels and co-products. Renew Sust Energ Rev 14:557
13. Moniz SJA, Shevlin SA, Martin DJ et al (2015) Visible-light driven heterojunction photocatalysts for water splitting – a critical review. Energy Environ Sci 8:731
14. Zou H, Gratz E, Apelian D et al (2013) A novel method to recycle mixed cathode materials for lithium ion batteries. Green Chem 15:1183
15. Chen W, Lei T, Wu C et al (2018) Designing safe electrolyte systems for a high-stability lithium-sulfur battery. Adv Energy Mater 8:1702348
16. Burschka J, Pellet N, Moon SJ et al (2013) Sequential deposition as a route to high-performance perovskite-sensitized solar cells. Nature 499:316
17. Atwater HA, Polman A (2010) Plasmonics for improved photovoltaic devices. Nat Mater 9:205
18. Chhowalla M, Shin HS, Eda G et al (2013) The chemistry of two-dimensional layered transition metal dichalcogenide nanosheets. Nat Chem 5:263
19. Li Y, Wang H, Xie L et al (2011) MoS_2 nanoparticles grown on graphene: an advanced catalyst for the hydrogen evolution reaction. J Am Chem Soc 133:7296
20. Kreno LE, Leong K, Farha OK et al (2012) Metal–organic framework materials as chemical sensors. Chem Rev 112:1105
21. Liu C, Li F, Ma LP et al (2010) Advanced materials for energy storage. Adv Mater 22:E28
22. Steele BCH, Heinzel A (2001) Materials for fuel-cell technologies. Nature 414:8
23. Winter M, Brodd RJ (2004) What are batteries fuel cells, and supercapacitors? Chem Rev 104:4245
24. Bruce PG, Scrosati B, Tarascon JM (2008) Nanomaterials for rechargeable lithium batteries. Angew Chem Int Ed 47:2930
25. Goodenough JB, Park KS (2013) The Li-ion rechargeable battery: a perspective. J Am Chem Soc 135:1167
26. Lai L, Potts JR, Zhan D et al (2012) Exploration of the active center structure of nitrogen-doped graphene-based catalysts for oxygen reduction reaction. Energy Environ Sci 5:7936
27. Bao D, Zhang Q, Meng FL et al (2017) Electrochemical reduction of N_2 under ambient conditions for artificial n_2 fixation and renewable energy storage using N_2/NH_3 cycle. Adv Mater 29:1604799

Chapter 2
Mesoporous Materials

Mesoporous materials are defined as a kind of porous materials with a pore size of 2–50 nm, they exhibit great potential in various applications.

2.1 Introduction of Mesoporous Materials

In the following text, the detailed description and application of mesoporous materials were introduced.

2.1.1 Introduction

Pore can be classified as microporous (<2 nm), mesoporous (2 nm < pore diameters < 50 nm) and macroporous (>50 nm) according to pore diameters [1–3]. Mesoporous materials in particular refer to a class of materials with pore sizes ranging from 2 to 50 nm [4] Since the classic family of M41S was reported by the researchers at mobil research and development corporation in 1992 [5, 6]. The mesoporous materials can be classified according to their shape, a large variety of geometries such as slit-shaped, cylindrical, spherical, conical, ink-bottle or interstitial mesoporous materials can be synthesized by different methods. Mesoporous materials, as a huge family of porous materials, exhibit remarkable physicochemical properties, including the hydrophobicity of their surfaces, high corrosion resistance, excellent thermal stability, high surface area, large pore volume, outstanding mechanical stability, easy be handling and low cost of manufacture. Therefore, mesoporous materials are ubiquitous and indispensable in many scientific applications, such as water and air purification, shape-selective catalysts, gas hosts, templates and components of electrodes for electrochemical double-layer capacitors. Besides, mesoporous materials are of great

B. Kong et al., *Functional Mesoporous Carbon-Based Film Devices for Energy Systems*,
https://doi.org/10.1007/978-981-99-7498-6_2

significance for fundamental research and practical applications. Firstly, they can be developed into high-performance electrode materials for batteries, fuel cells, and supercapacitors; they can also be used as sorbents for separation processes and gas storage, and as supports for many important catalytic processes. The diverse applications of mesoporous materials benefit from their superior physical and chemical properties, such as electric conductivity, thermal conductivity, chemical stability, and low density. In addition, their wide availability is also contributed to their wide huge application valuable. Therefore, in regard to the fundamental investigation and practical application, it is important to develop new mesoporous materials with controlled surface properties and structural ordering.

Mesoporous carbon, as an important member of the mesoporous family, has drawn great attention in the research field. Traditional porous carbon materials, such as activated carbon and carbon molecular sieves, are synthesized by pyrolysis and physical or chemical activation of organic precursors, such as coal, wood, fruit shell, or polymers, at elevated temperatures. Generally, the synthesis of porous carbon materials have two methods, namely the hard-templating and soft-templating methods [7, 8]. The hard-templating method involves nano-casting by using pre-synthesized mesoporous solids or colloidal crystals as a sacrificial mould [9]. Taking the preparation of mesoporous carbon material by hard-templating strategy as an example, the specific steps include the synthesis of an ordered mesoporous silica, the filling of mesopores by organic precursors, carbonization at high temperature, and template removal in acidic/basic solutions. Several ordered mesoporous silica (e.g., MCM-48, SBA-1, SBA-15, SBA-16, KIT-6 and KIT-5 [10, 11]) and even mesoporous metal oxides have been employed as hard templates for OMC replicas [12] (Recent advances in functionalized micro and mesoporous carbon materials: synthesis and applications). The stability of the hard-templating scaffold allows the preparation of mesoporous nanomaterials with a broad range of compositions. However, the method is costly, time-consuming and difficult in the tuning of the mesopore parameters (for example, mesostructured and pore size). By contrast, the soft-templating method synthesizes mesoporous materials from single micelles which typically comprise amphiphilic surfactants or block copolymers. Then, these micelles assemble with organic or inorganic oligomers to obtain surfactant–oligomers micelles, which further assemble into functional mesoporous materials [6, 13]. The mesostructures can be regulated by adjusting the assembly of the soft micelles, in addition, the method is convenient, low cost and, hence, soft-templating is widely used for the preparation of mesoporous materials [1]. The mesoporous carbon materials with multi-functions exhibit great potential in the practical fields of sensing, drug delivery, carbon capture, energy storage and conversion and adsorption and separation.

2.1.2 Applications of Mesoporous Carbons

Mesoporous carbons synthesized via hard- and soft-templating approaches exhibit extremely high specific surface area, well-developed ordered porous structure, and high thermal and chemical stabilities which make them available for various practical applications such as adsorption, gas purification and separation, electrode materials for energy storage and conversion, in sensors and as catalyst support for industrially important reactions. Here, we briefly describe some of the important applications of mesoporous carbon molecular sieves.

Adsorption

Adsorption is a key industrially relevant process that has been extensively used for removal of pollutants. Mesoporous carbons are potential candidate adsorbent for adsorbing hydrocarbons, amino acids, hydrogen, methane, carbon dioxide, heavy metals, phenols, dyes, oil, etc. benefiting from their high surface area, large pore volume, chemical inertness, and high mechanical strength. An ordered pore system and well-defined pore size distribution favor easy diffusion properties, which play a crucial role in a material's performance as an adsorbent [14]. Study demonstrates that the mesoporous carbons with highly ordered pores present excellent adsorption capacity for bulky dye molecules such as rhodamine B, brilliant yellow, methyl orange, Sudan G, and the adsorption capacity is almost twice that of conventional activated carbons [15]. Importantly, the adsorption capacity of mesoporous carbons can be further improved by functional modification. For example, the adsorption properties of FDU-15 carbon can be enhanced after modified with ammonium peroxysulfate for trapping heavy metal ions, such as Fet3, Cu 2t, Cd2t, Pb2t, and efficient removal of organic dye Fuschin was also achieved. Importantly, the functionalized mesoporous carbon also shows a high loading capacity for the very commonly used medicinal drug ibuprofen [16], which extends its practical application value.

Batteries and Supercapacitors

Besides, mesoporous carbons are also potential electrode materials for the sulfur-carbon batteries [17–19]. Recently, CMK-3/sulfur composite has been used as electrode materials for Li-sulfur battery, an extremely high reversible capacities of up to 1320 mAh/g was realized by this electrode material. Further, a sulfur-loaded highly ordered mesoporous carbon with bimodal pore structure has been synthesized by soft-templating approach and applied as an active material for Li–S battery, and a higher discharge capacities of 1675 mAh/g with good cycling stability at high current rates were obtained. Importantly, mesoporous carbons have also been used as electrode materials for supercapacitor application and which have shown a reasonably high capacitance of 180 F/g with good cyclability and a high power density of more than 90% even after 700 cycles at a current density of 3.0 A/g [20]. Nanocomposites of CMK-3/polyaniline nanowhiskers have also exhibited a very high capacitance of 470 F/g, with 94% of capacity retention even after 1000 cycles repetition at a current density of 1 A/g [21].

Oxygen Reduction Reactions Catalytic Activity

Ordered mesoporous carbon has been considered as cheap and efficient electro catalyst for oxygen reduction reactions due to its inherent electrochemical and mass transport properties. The diffusion of air through the connected pores of mesoporous carbon would be more effective than the commonly expensive noble metals being used as electrocatalyst in the ORR catalytic activity. The surface modification for the mesoporous carbons has been suggested to achieve the significant ORR performance. The metal oxide/mesoporous composites or heteroatom doping into the mesoporous structures can greatly enhance the performance at higher current density. However, metal oxide incorporation decrease the surface area of mesoporous carbon and it makes difficult for oxygen to pass through the walls of mesoporous carbon. In contrast to this, heteroatom doping changes the chemisorption mode of oxygen and also reduces the ORR potential by inducing the charge/spin redistributions on mesoporous carbon electrode materials. Recently, ammonia treated ordered mesoporous carbon has been used as an electrocatalyst for ORR and these materials exhibited higher current density and significant ORR catalytic activity. There are several investigations on the utilization of mesoporous carbon to improve the ORR catalytic activity by modifying the physico-chemical properties of mesoporous carbon materials, using various heteroatoms and polymers.

2.2 Advantages of Mesoporous Materials

Mesoporous materials are appealing materials in many energy applications owing to their unique pore structure. According to the International Union of Pure and Applied Chemistry (IUPAC) definition, porous materials are classified into three categories according to their pore sizes: microporous (<2 nm), mesoporous (2–50 nm) or macroporous (>50 nm). Since the first report of mesoporous silica in the 1990s [5, 22], the variety of mesoporous materials available has rapidly expanded, encompassing a very broad range of compositions. Mesoporous materials have attracted great attention worldwide and opened up broad spaces for their applications in various fields, such as catalysis, adsorption and separation, drug storage and delivery, nanofabrication, etc. (1) Due to outstanding properties of mesoporous materials such as extraordinarily large specific surface area and pore volume, well-defined ordered mesostructure, tunable pore size, varieties of the framework. (2) The ultrahigh surface areas, large pore volumes, controllable pore sizes and shapes. (3) The materials bearing redox-active moieties can be promising to induce intra-silica electron transfer chains or to act as electron shuttles or mediators, with promising applications in electrocatalysis. (4) The possibility for nanobioen capsulation in such functionalized and mesostructured reactors could result in the development of integrated systems combining molecular recognition, catalysis and signal transduction, with applications in the field of electrochemical biosensors. These features are particularly advantageous for applications in energy conversion and storage. In principle, high specific surface areas should provide a large number of reaction or interaction sites for surface or interface-related processes such as adsorption, separation, catalysis and energy storage.

2.3 Future Development of Mesoporous Carbon Materials

However, it is still a great challenge to develop mesoporous materials with high-efficient storage capacity and conversion yields for energy applications. Although the various application of mesoporous materials extensively studied, but has yet to gain practical commercial products. For the practical applications, the stability of mesoporous carbons is quite an important question needs to be answered. Another issue preventing the practical application of mesoporous materials is the insufficient understanding about the influence of the textural parameters of mesoporous materials, such as the and pore volume, wall thickness, pore geometry and size, on the performance of the fabricated electrochemical devices. Therefore, the relationship between the textural parameters of mesoporous materials and electrochemical performance should be constructed by further researching. Besides, the composition of mesoporous framework also has great influence on the application of energy conversion. Hence, it is also a great research focus on the development of the mesoporous multicomponent systems.

The presence of mesopores in bulk materials can considerably improve their performance in energy conversion and storage applications in terms of energy and power density, lifetime and stability. The energy-related applications of mesoporous materials include, but are not limited to, solar cells, solar fuel production, supercapacitors, rechargeable batteries and fuel cells. To realize the full potential of these materials in energy conversion and storage, several challenges need to be overcome.

Firstly, the primary challenge for the future development of mesoporous carbon materials is production. The methods of simple, template-free, reproducible, scalable and cost-effective remain avidly sought after. In an energy conversion and storage device, it is also very important to control the overall particle size, surface structure, morphology and packing density, among other features. Meanwhile, the favourable structural parameters vary for different applications. However, methods to systematically control and optimize such parameters for specific applications have not yet emerged. It is becoming increasingly evident that the required functionality for energy applications cannot be attained with a single material or structure. Complex materials with several components or with particular, desired crystalline structures and structural ordering exhibit a broader application prospect. In addition, high purity is desirable, and properties such as high conductivity, electrochemical performance, catalytic activity and stability are also essential. It is anticipated that the fabrication of hierarchical porous scaffolds will enable promising synergies between mechanical and transport properties, and provide enhanced accessibility to the active sites. For example, hierarchically ordered large-pore mesoporous zeolite crystals have been regarded as ideal candidates for improving the mass transfer and catalytic conversion of bulky molecules, which may revolutionize the application of conventional microporous zeolites in oil refining and the petrochemical industry.

There are drawbacks of mesoporous materials that may hinder their practical applications. (1) Although high surface area and porosity are desirable for enhanced activity, these features provide more opportunities for undesired side reactions. This is particularly true in batteries, in which the high surface area, together with defects on the surface, increase the reactivity towards the electrolyte and cause uncontrolled solid electrolyte interphase reactions. These conditions also apply to solar energy conversion applications, in which the high surface area and defects increase the trap states and charge recombination, thereby lowering the power conversion efficiency. (2) The low packing density may be another drawback of many mesoporous materials in terms of the volumetric energy and power density (reflecting how much and how fast energy can be stored in a unit volume of a packed device) in supercapacitors and batteries. (3) Moreover, mesoporous materials have a nanometre-sized framework and high surface energy, which generally lead to low thermal stability, thereby compromising the catalytic performance under high temperatures. With this in mind, it will be challenging to achieve a balance among the different characteristics of mesoporous materials and hence optimize devices for energy applications.

As the research work goes further, it is certain that in the near future, optimized mesoporous carbon materials with superior performance will become a promising and prominent selection in energy-storage market.

References

1. Zhao T, Elzatahry A, Li X et al (2019) Single-micelle-directed synthesis of mesoporous materials. Nat Rev Mater 4:775–791
2. Zhang L, Jin L, Liu B et al (2019) Templated growth of crystalline mesoporous materials: from soft/hard templates to colloidal templates. Front Chem 7:22
3. Nemanashi M, Noh JH, Meijboom R (2018) Dendrimers as alternative templates and pore-directing agents for the synthesis of micro-and mesoporous materials. J Mater Sci 53:12663−12678
4. Kueasook R, Rattanachueskul N, Chanlek N et al (2020) Green and facile synthesis of hierarchically porous carbon monoliths via surface self-assembly on sugarcane bagasse scaffold: influence of mesoporosity on efficiency of dye adsorption. Micropor Mesopor Mater 296:110005
5. Kresge CT, Leonowicz ME, Roth WJ et al (1992) Ordered mesoporous molecular sieves synthesized by a liquid-crystal template mechanism. Nature 359:710–712
6. Beck JS, Vartuli JC, Roth WJ et al (1992) A new family of mesoporous molecular sieves prepared with liquid crystal templates. J Am Chem Soc 114:10834–10843
7. Corma A (1997) From microporous to mesoporous molecular sieve materials and their use in catalysis. ChemInform 28:2373–2420
8. Schuth F, Schmidt W (2002) Microporous and mesoporous materials. Adv Mater 14:629–638
9. Schüth F (2003) Endo- and exotemplating to create high-surface-area inorganic materials. Angew Chem Int Ed 42:3604–3622
10. Zhai YP, Dou YQ, Zhao D et al (2011) Carbon materials for chemical capacitive energy storage. Adv Mater 23:4828–4850
11. Sayari A, Yang Y (2005) SBA-15 templated mesoporous carbon: new insights into the sba-15 pore structure. Chem Mater 17:6108–6113
12. Jun S, Joo SH, Ryoo R et al (2000) Synthesis of new, nanoporous carbon with hexagonally ordered mesostructure. J Am Chem Soc 122:10712–10713
13. Yu K, SmarslyC B, Brinker J (2003) Self-assembly and characterization of mesostructured silica films with a 3d arrangement of isolated spherical mesopores. Adv Mater Opt Electr 13:47–52

14. Xin W, Song Y (2015) Mesoporous carbons: recent advances in synthesis and typical applications. ChemInform 5:83239–83285
15. Zhuang X, Wan Y, Feng C et al (2009) Highly efficient adsorption of bulky dye molecules in wastewater on ordered mesoporous carbons. Chem Mater 21:706–716
16. Wu Z, Webley PA, Zhao D (2010) Comprehensive study of pore evolution, mesostructural stability, and simultaneous surface functionalization of ordered mesoporous carbon (FDU-15) by wet oxidation as a promising adsorbent. Langmuir Acs J SurfS & Colloids 26:10277
17. Ji X, LeeL KT, Nazar F (2009) A highly ordered nanostructured carbon–sulphur cathode for lithium–sulphur batteries. Nat Mater 8:500–506
18. He G, Ji X, Nazar L (2011) High "c" rate Li-S cathodes: sulfur imbibed bimodal porous carbons. Energy & Environ Envir Sci 4:2878–2883
19. Liang C, Dudney NJ, Howe JY (2009) Hierarchically structured sulfur/carbon nanocomposite material for high-energy lithium battery. Chem Mater 21:4724–4730
20. Lei Z, Christov N, Zhang LL et al (2011) Mesoporous carbon nanospheres with an excellent electrocapacitive performance. J Mater Chem 21:2274–2281
21. Yan YF, Cheng QL, Wang GC et al (2011) Growth of polyaniline nanowhiskers on mesoporous carbon for supercapacitor application. J Power Sources 196:7835–7840
22. Zhao D, Feng et al (1998) Triblock copolymer syntheses of mesoporous silica with periodic 50 to 300 angstrom pores. Science 279:548−552

Chapter 3
Synthesis Methods of Mesoporous Carbon-Based Materials

Various methods for synthesizing mesoporous materials are shown in Fig. 3.1, each with its own advantages and disadvantages (Table 3.1). These methods can be categorized as soft-templating, hard-templating, or template-free approaches.

In the soft-templating approach (Fig. 3.1a), surfactant molecules and guest species co-assemble into ordered mesostructured composites. Upon removal of the template, mesoporous materials with open pores and ordered structures are obtained. The interaction between the guest species and surfactant molecules plays a critical role in the formation of these ordered mesostructures. Various synthetic routes based on soft templating have been employed to prepare mesoporous materials with diverse morphologies and tunable pore architectures. These routes include aqueous, non-aqueous (also known as the evaporation-induced self-assembly (EISA) process), and hydrothermal processes. The integration of aqueous and EISA processes has led to the development of a versatile method called the solvent evaporation-induced aggregating assembly (EIAA) method. This approach is more tolerant to water and allows for the creation of mesoporous materials by controlling solvent evaporation.

In the hard-templating route (Fig. 3.1b), also known as nanocasting, mesostructures are formed using preformed hard templates such as mesoporous silica, carbon, or aggregates of nanoparticles. The use of hard templates eliminates the need to control the hydrolysis and condensation of guest species, as well as their assembly with surfactants. Furthermore, hard templating ensures complete filling of the mesochannels. Consequently, this method allows for the preparation of a wide range of materials, including highly crystalline or even single-crystal materials, due to the protection provided by rigid templates at high temperatures. However, the hard-templating route has limitations, such as the availability of fewer hard templates compared to soft templates and the complexity and time-consuming nature of the procedure.

The combination of soft- and hard-templating routes allows for the synthesis of hierarchical porous materials, which find applications in batteries and supercapacitors (Fig. 3.1c). These materials possess macropores within the electrode structure that

B. Kong et al., *Functional Mesoporous Carbon-Based Film Devices for Energy Systems*,
https://doi.org/10.1007/978-981-99-7498-6_3

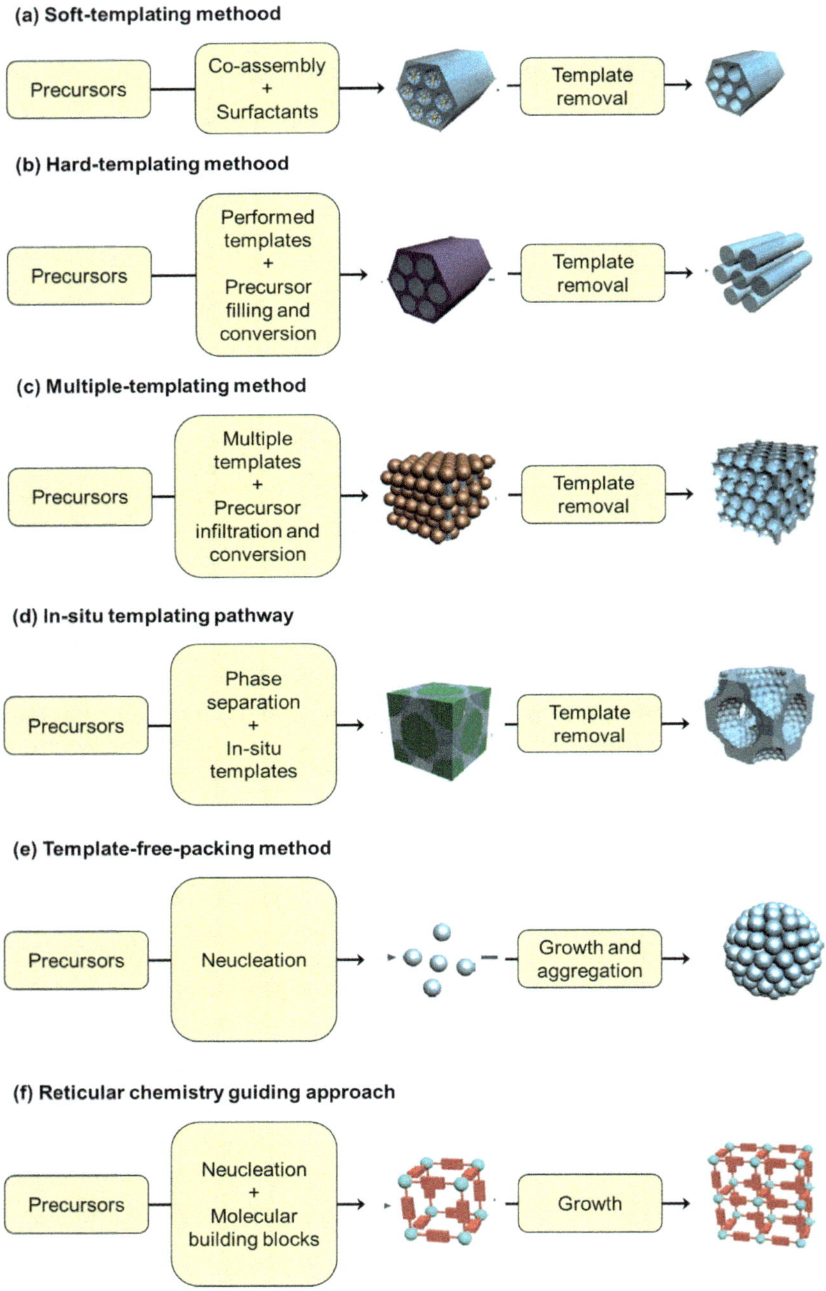

Fig. 3.1 The principal methods for synthesizing mesoporous materials. **a** Soft-templating method. **b** Hard-templating method. **c** Multiple-templating method. **d** In-situ templating pathway. **e** Template-free packing method. **f** Reticular chemistry guiding approach

Table 3.1. A comparison of synthetic methods for mesoporous materials

Methods	Examples of mesoporous materials	Advantages	Disadvantages
Soft-templating method	• Silica • Alumino silicates • Organo silica • Carbon • Metaloxides • Metals • Polymers • Phosphates	• Controllable mesostructures and pore sizes • Tunable morphologies and easily processable • High quality product • Large scale production	• Uses surfactants • Mesostructure formation is highly sensitive to the reaction conditions • Relatively low crystallinity
Hard-templating method	• Carbon • Metal oxides • Metals • Metal sulfides • Metal nitrides • Metal carbides • Zeolites • Non-oxide ceramics	• Low sensitivity to ther eaction conditions • High quality product • Highly crystallineproduct • Ordered nanoarraysstructure	• Uses preformed hard templates • High cost • Time consuming
Multiple-templating method	• Silica • Carbon • Metal oxides • Zeolites	Hierarchically porous structure	• Requires multiple templates • High cost • Time consuming
In-situ templating pathway	• Silica • Carbon • Metal oxides • Metals • Polymers	• Simple method • No preformed templatesor surfactants required • Low cost	• Low quality • Hard to obtain ordered structures
Template-free packing method	• Carbon • Metal oxides • Metal sulfides	• Simple method • Easily processable • No templates required • Highly crystalline product	• Hard to obtain ordered structures
Reticular chemistry guiding approach	• Metal-organic frameworks • Covalent organic frameworks • Zeolitic imidazol ate frameworks	• Controllable mesostructures and pore sizes • No templates • Highly crystalline product	• Low stability • Mesostructure formation is highly sensitive to the reaction conditions • Not easily processable

enhance electrolyte transport properties and serve as reservoirs for electrochemically active ions, resulting in improved rate capability and cycling performance. A wide range of macroscopic templates, including biological materials, colloidal crystals, and foams, have been utilized in the design and synthesis of hierarchical porous structures. It is worth noting that hard templates do not necessarily need to be preformed;

chemical species in the precursor solution or even the solvent itself can undergo phase separation and transform into templates in situ (Fig. 3.1d). However, the resulting mesopores in these cases are typically disordered and randomly distributed. The precise design and control of the template, including its quantity, location, and removal process, are crucial for achieving high-quality mesoporous materials using the in situ templating route.

In addition to templating routes, template-free methods are widely employed in the synthesis of mesoporous materials. In most template-free approaches, the mesopore voids arise from the aggregation of nanoscale building blocks (Fig. 3.1e). Recent advancements in nanomaterial synthesis have expanded the scope of mesoporous materials. However, it remains challenging to predict the resulting pore architectures, except in the case of closed-packing structures. In contrast to these synthetic strategies, reticular chemistry utilizes molecular building blocks, such as organic molecules, inorganic clusters, and complexes, to create extensive metal–organic frameworks (MOFs) and covalent organic frameworks. These frameworks enable the preparation of mesoporous materials with crystalline pore walls [1] (Fig. 3.1f). For instance, an isoreticular series of MOF 74 structures with pore sizes ranging from 1.4 to 9.8 nm has been synthesized by expanding the linking group in the original MOF structure, incorporating up to 11 phenylene rings.

3.1 Hard Template Methods

The hard-templating method, also known as nanocasting, is a widely used approach for synthesizing crystalline mesoporous materials, particularly mesoporous carbon materials (Fig. 3.2). This technique utilizes mesoporous inorganic materials such as mesoporous silica and zeolites as hard templates to introduce mesoporous structures into carbon materials. By employing different porous templates, various mesoporous carbons with different textural properties can be prepared using this simple technique.

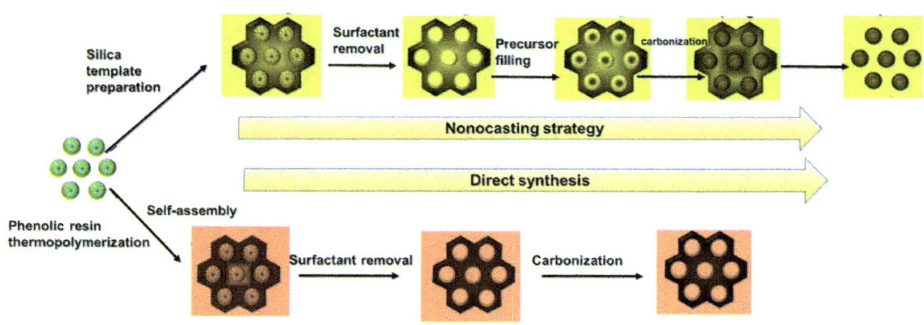

Fig. 3.2 Two typical methods for the preparation of ordered mesoporous carbon materials: the nanocasting strategy from mesoporous silica hard templates and the direct synthesis from block copolymer soft templates

The control of the structure and morphology of the materials has garnered significant attention due to its correlation with the final textural parameters. Porous silica materials, including MCM-41 [2], MCM-48 [3], SBA-1 [4], SBA-2 [5], SBA-11 [6], SBA-12 [7], SBA-15 [8], SBA-16 [6], FDU-1, FDU-12 [9], KIT-5 [10], and KIT-6 [11], are commonly used as hard templates in the typical hard-templating approach. The synthesis process involves several steps: (a) Preparation of a mesoporous silica matrix with controlled architecture. (b) Introduction of a suitable carbon precursor into the mesopores. (c) Pore infiltration with a suitable carbon precursor using wet impregnation, chemical vapor deposition (CVD), or a combination of both. (d) Removal of the porous silica template through etching with aqueous HF or NH_4F solution [12].

The synthesis of mesoporous materials, particularly in the context of hard template synthesis of mesoporous carbons with well-defined mesoporous structures, typically involves the following steps: (a) Preparation of silica gel with controlled pore structure: Silica gel is prepared with a specific pore structure using appropriate methods, such as sol–gel synthesis or hydrothermal treatment. (b) Impregnation/infiltration of the silica template: The prepared silica gel template is impregnated or infiltrated with monomer or polymer precursors. These precursors will later undergo cross-linking and carbonization to form the carbon material. (c) Cross-linking and carbonization of the organic precursors: The impregnated precursors are subjected to cross-linking reactions to form a stable polymer network. This network is then carbonized under controlled conditions, such as high temperature or chemical treatment, to convert the organic precursors into carbon. (d) Dissolution of the silica template: The final step involves the removal of the silica template. This is typically achieved by using an etching agent, such as an acid or base, which selectively dissolves the silica while leaving behind the carbon framework. The space once occupied by the silica template is transferred into the resulting mesoporous carbon material, with the carbon framework forming the continuous structure within the pores. Overall, this synthesis approach enables the replication of the mesoporous structure of the silica template into the resulting carbon material, providing a well-defined and ordered mesoporous carbon structure.

3.1.1 Ordered Mesoporous Carbon Materials

The interest in synthesizing ordered mesoporous carbon materials gained traction in the late 1990s following the successful development of mesoporous molecular sieves with uniform pore size and ordered structures using ionic surfactants, block copolymers, and neutral amines as structure-directing agents [3, 6]. The porous structure of mesoporous silica serves as a template for synthesizing carbon materials with well-defined morphologies, leveraging the three-dimensional structure of mesoporous silica.

This breakthrough highlighted the potential of utilizing mesoporous silica as a template for synthesizing ordered carbon materials and spurred further research and development in the field. Since then, various methods and techniques have been explored to synthesize ordered mesoporous carbon materials with controlled pore sizes and structures for diverse applications.

In 1999, Ryoo et al. reported the synthesis of the first self-supported highly ordered mesoporous carbon material, CMK-1, using the ordered aluminosilicate MCM-48 as a hard template. This work demonstrated the feasibility of using hard templates to achieve ordered structures in mesoporous carbon materials. Hyeon et al. synthesized another ordered mesoporous carbon material named SNU-1, also using MCM-48 as the template and phenolic resin as the carbon precursor. The performance of SNU-1 in an electrochemical double-layer capacitor was evaluated, showcasing its potential for energy storage applications. In an earlier study, Schuster et al. reported the synthesis of spherical ordered mesoporous carbon nanoparticles with a bimodal pore size distribution. They utilized a phenol and formaldehyde mixture precursor and spherical silica as a hard template via the nano templating approach. The resulting material exhibited a high specific surface area of 2445 $m^2\ g^{-1}$ and significant inner pore volumes up to 2.32 $cm^3 g^{-1}$, making it promising for various applications [13]. Seo et al. demonstrated the preparation of ordered mesoporous carbon using furfuryl alcohol as a carbon source and acidified mesoporous silica impregnated with phosphoric acid or sulfuric acid as a precursor. The resulting ordered mesoporous carbon showed a lower surface area compared to unmodified mesoporous silicas, highlighting the influence of precursor and acid impregnation on the final material properties [14]. These studies demonstrate the diverse approaches and techniques employed to synthesize ordered mesoporous carbon materials, opening up opportunities for their application in various fields.

Riccardo Brandiele et al. introduced a novel approach for the hard template synthesis of nitrogen-doped carbons using a propylamine silica template [15]. The propylamine silica template refers to a mesoporous silica material that has been functionalized with propylamine groups (Fig. 3.3). In their study, sucrose was employed as the carbon source, and the propylamine functionalized silica acted as both the nitrogen source and the templating agent. Through this innovative hard template method, nitrogen-doped carbon materials with mesopores ranging from 4 to 8 nm in width were successfully synthesized. These materials exhibited a high surface area of approximately 1100 m2 g^{-1} [15]. This approach offers a promising strategy for the controlled synthesis of nitrogen-doped carbon materials with well-defined mesoporous structures. The utilization of propylamine silica as a functionalized template provides a means to introduce nitrogen functionalities into the resulting carbon materials, which can enhance their performance for various applications such as energy storage, catalysis, and sensing.

In their work, Rumeli Banerjee and co-workers achieved the synthesis of Au/OMC (gold nanoparticle-intercalated mesoporous carbon) through a one-step hard template method (Fig. 3.4). They utilized mesoporous silica with gold nanoparticles as the template to fabricate the catalyst. This approach allowed them to investigate the catalytic properties of Au/OMC in two different aspects. The synthesized Au/OMC catalyst was then applied for the spectrophotometric detection and quantification of GSH (glutathione) with high sensitivity. The limit of detection (LOD) value obtained

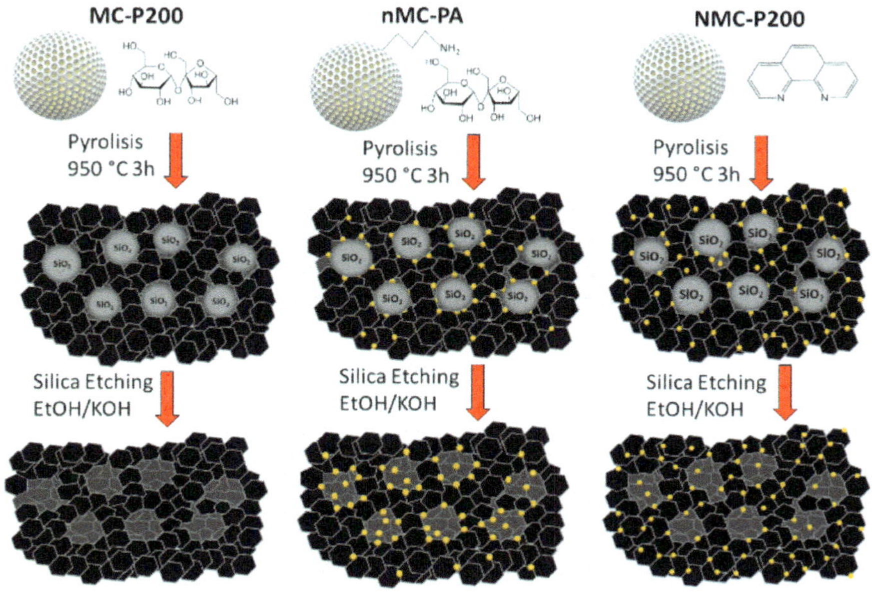

Fig. 3.3 Sketch of the procedure employed for the synthesis of doped mesoporous carbons. Copyright 2020 Wiley–VCH, Ref. [15]

was 0.604 nM, indicating the high sensitivity of the catalyst for GSH detection. This study demonstrates the potential of Au/OMC as a catalyst with diverse catalytic properties and highlights its applicability in sensitive detection methods, such as for quantifying GSH. The successful synthesis of Au/OMC through the one-step hard template method offers a convenient and effective approach for the preparation of functionalized mesoporous carbon materials for various catalytic and sensing applications [16].

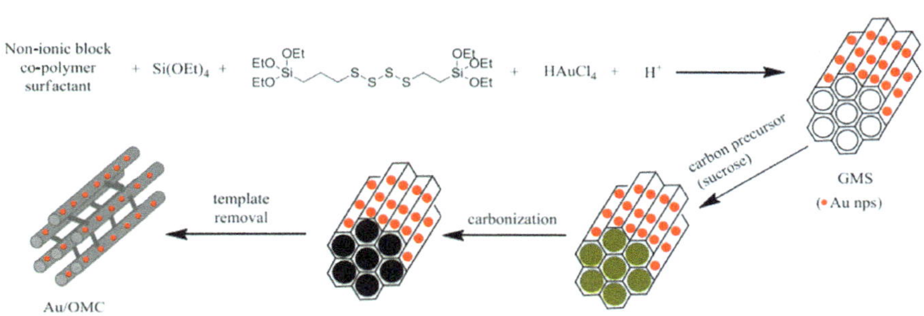

Fig. 3.4 Synthesis of Au/OMC using AuNPs intercalated mesoporous silica (GMS) as a hard template. Copyright 2019 American Chemical Society, Ref. [16]

Liu et al. synthesized N, S-codoped ordered mesoporous carbon materials using SBA-15 as a template (Fig. 3.5). They utilized phenanthroline and KSCN as nitrogen and sulfur precursors, respectively, and varied the pyrolysis temperature to obtain different catalyst samples. Among the synthesized catalysts, the catalyst calcined at 900 °C demonstrated the highest catalytic performance for the selective oxidation of ethylbenzene. This catalyst exhibited high specific surface areas and uniform pore sizes, contributing to its superior performance. The unique architectures and synergistic effect of pyridinic-N, graphitic-N, and thiophene-S dopants were attributed to the enhanced catalytic activity of the catalyst. Overall, this study showcases the potential of N, S-codoped ordered mesoporous carbon materials as efficient catalysts for selective oxidation reactions [17].

Ryoo and his colleagues recognized the potential and demonstrated a groundbreaking synthesis process for microporous 3D graphene-like carbon materials using lanthanum-catalyzed zeolite templating. They utilized ethylene as a carbon precursor, resulting in carbon materials with twice the conductivity of CMK-3. The remarkable aspect of this synthesis process is its energy efficiency, requiring a carbonization temperature of only 600 °C, compared to the conventional range of 900–1000 °C for microporous graphene synthesis. Additionally, the deposition of microporous carbon in lanthanum-modified zeolite Y occurs more than 20 times faster than in acidic HY zeolite. Furthermore, modified zeolites enable the use of acetylene gas instead of ethylene, allowing carbon framework construction at even lower temperatures, as low as 340 °C, due to the higher reactivity of acetylene. In the case of poorly ordered porous structures, heat treatment at 900 °C can transform them into highly ordered materials. The synthesis of microporous carbon materials with controlled pore size and ordered structure has been achieved using ionic surfactants, block copolymers, and neutral amines as structure-directing agents.

Ryoo's group conducted the synthesis of CMK-1 using a method involving impregnation of sucrose and sulfuric acid solutions into the mesopores of the cubically structured aluminosilicate MCM-48 (Fig. 3.6). The presence of sulfuric acid

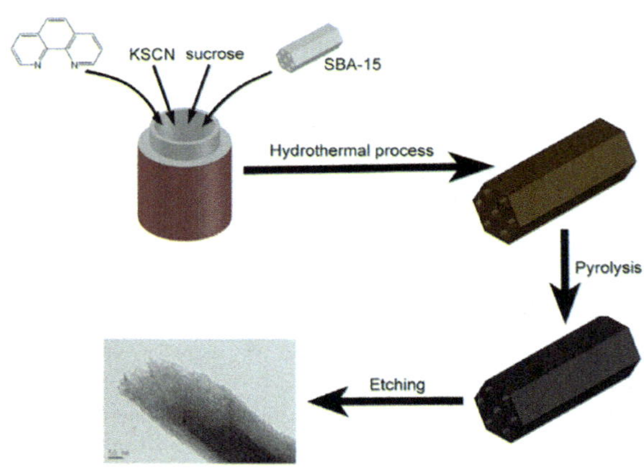

Fig. 3.5 Schematic illustration of the preparation of N, S-codoped catalysts [17]

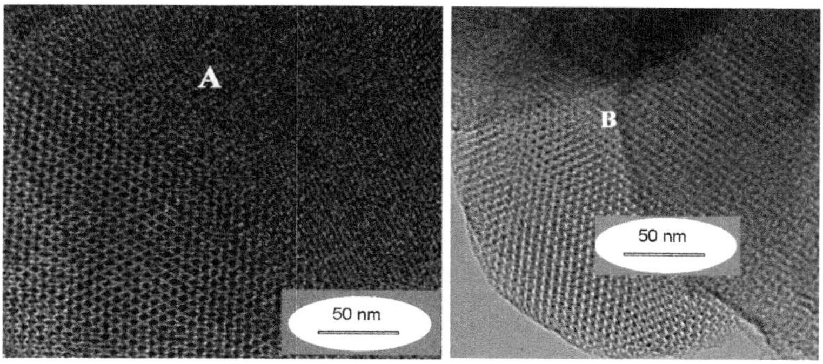

Fig. 3.6 Typical HRTEM images of ordered mesoporous carbon molecular sieves [21]

acted as a catalyst during the carbonization process, which was carried out at temperatures ranging from 1073 to 1373 °C. Subsequently, the silica framework was removed by treating the material with a hot ethanolic solution of NaOH. This synthetic procedure yielded CMK-1, a mesoporous carbon material with a cubic structure and an average pore size of approximately 3 nm. The adsorption isotherms of nitrogen or argon on CMK-1 exhibited a distinct adsorption step, attributed to the capillary condensation of adsorbate molecules within the uniform mesopores [18].

Several months later, Hyeon and his colleagues successfully synthesized an ordered mesoporous carbon material called SNU-1. They utilized MCM-48 as the template and phenolic resin as the carbon precursor in the synthesis process. The performance of SNU-1 was also evaluated in an electrochemical double-layer capacitor [19].

In a separate study, Ryoo, Terasaki, and others reported the synthesis of CMK-4 using chemical vapor deposition (CVD) method, which exhibited similar X-ray diffraction (XRD) patterns to the MCM-48 silica template used [20]. Vix-Guterl et al. demonstrated the feasibility of preparing ordered mesoporous carbon sieves through chemical vapor infiltration (CVI) method (Fig. 3.6). In this approach, MCM-48 silica materials were used as a porous solid template and infiltrated with an appropriate carbon precursor such as sucrose. This CVI process offered several advantages over liquid impregnation method, including higher carbon content achieved in a single step, elimination of additional drying and heat-treatment steps, milder experimental conditions resulting in reduced thermal shrinkage of the silica matrix, and a lower lattice contraction. Chemical analysis of the carbon materials indicated a higher carbon-to-oxygen (C/O) atomic ratio for the carbon prepared by CVI compared to that obtained through liquid impregnation method [21].

In an interesting study, Jeong Yeon Kim and Jong-Sung Yu developed a novel synthesis method for a nanostructured silica material with an ordered arrangement of uniform mesopores. The process involved the nanocasting of a highly ordered mesoporous carbon template. First, a newly synthesized MCM-48 template was calcined to remove the surfactants, resulting in CM48T. Mesoporous carbon was then synthesized by infiltrating the template mesopores with divinylbenzene (DVB) and the free

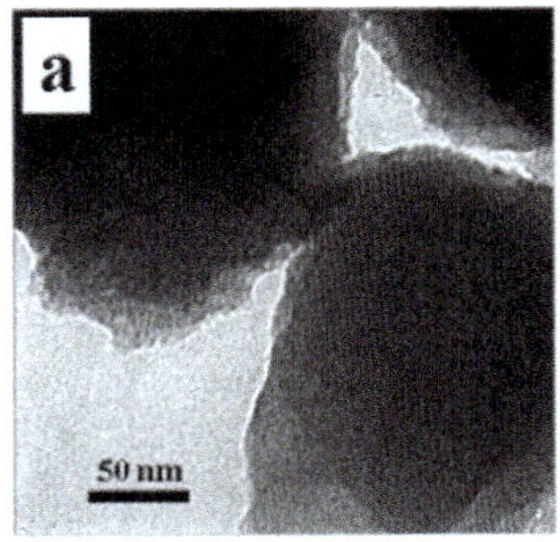

Fig. 3.7 Transmission electron microscopy (TEM) images of CM48T-C [22]

radical initiator azobisisobutyronitrile (AIBN), followed by in situ polymerization and carbonization. After dissolving the template, the resulting mesoporous carbon material was denoted as CM48T-C. To create a regenerated silica replica, the pores of the carbon template were infiltrated with a silica precursor solution (TEOS). The carbon/silica composite was exposed to HCl vapor, which induced hydrolysis and condensation reactions. Through this process, a nanostructured silica material with periodic ordering, not achievable through conventional surfactant assembly methods used for typical mesoporous silicas, was successfully produced. This innovative approach utilizing MCM-48 as a starting template provides a promising avenue for the synthesis of new silica materials with unique nanostructured order (Fig. 3.7) [22].

In a recent study by the Sheng Dai group, ordered mesoporous carbons C15 and C48 were synthesized using ordered silica templates SBA-15 and MCM-48, respectively, along with mesophase pitch AR (Mitsubishi) as the precursor. The mesophase pitch used in the synthesis was obtained from naphthalene with the HF/BF_3 catalyst. In the synthesis process, 3.0 g of SBA-15 or MCM-48 ordered silicas were mixed with 3.0 g and 3.6 g of finely powdered mesophase pitch, respectively, in ethanol. The mixture was then subjected to gradual evaporation of ethanol. Subsequently, the mixture was placed in a tube furnace under nitrogen protection at 360 °C for 2 h, followed by the stabilization of the pitch structure at 220 °C in air for 10 h. Carbonization was carried out at 800 °C in a nitrogen atmosphere. Finally, the silica template was removed by treatment with hot 3 M sodium hydroxide solution (Fig. 3.8) [23].

The pitch-based carbon material exhibits high strength and conductivity, allowing for the acquisition of the first high-resolution scanning electron microscopy (SEM) image of the material (Fig. 3.9). The SEM image reveals that the framework of the C48 carbon network is comprised of interconnected carbon nanorods, measuring approximately 1.5–2 nm in length, in a three-dimensional arrangement. In contrast, the MCM-48 template contains a system of interconnected mesopores with a diameter of around 2.5 nm, also arranged in three dimensions [23].

Tian et al. have introduced two innovative methods for the facile synthesis of carbon nanowire arrays utilizing MCM-41 as the hard template. The MCM-41 materials were customized in terms of mesoscale texture and macroscale morphology. Two representative synthesis approaches are described as follows.

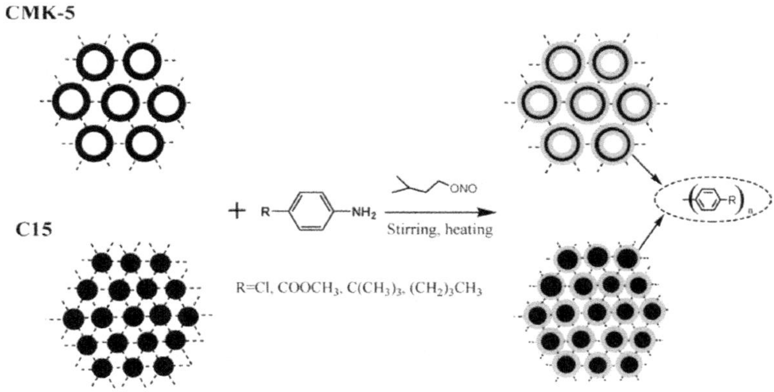

Fig. 3.8 Schematic illustration of the surface functionalization reaction for ordered carbons CMK-5 and C15 and its effects on the pore dimensions [23]

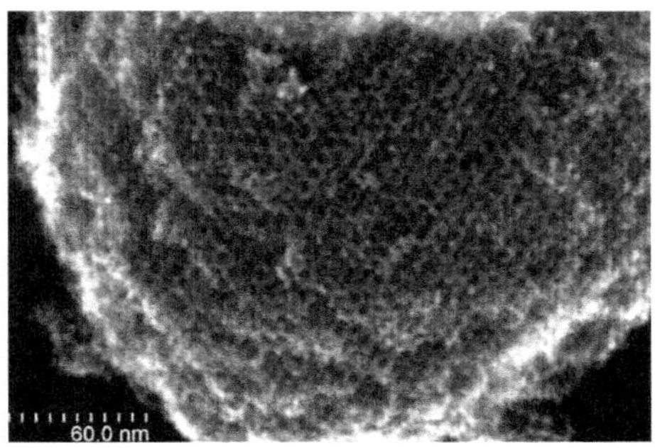

Fig. 3.9 High-resolution SEM image of C48 templated with MCM-48 ordered silica and using mesophase pitch as the carbon precursor [23]

In the first method, conventional as-made MCM-41 was prepared under basic conditions, and the organic amphiphiles were completely removed via microwave digestion. The as-prepared MCM-41 samples (0.5 g) were dispersed in water (1 g). Subsequently, sucrose (0.5–0.6 g) and H_2SO_4 (0.05 g) were added to the mixture, followed by stirring for 0.5 h. The resulting mixture was dried at 353 K for 2–10 h, and then further dried at 433 K for 8–24 h. This impregnation and drying process was repeated once with an additional 0.3–0.4 g of sucrose. The obtained black samples were carbonized at 1173 K under nitrogen or argon flow for 6 h. Finally, the silica framework was etched using HF (5–10 wt.%) to remove it. For comparison, mesoporous carbon materials (C10) templated by calcined MCM-41 were also synthesized and characterized.

In the second method, MCM-41 with a distinctive leaf-like morphology was prepared using docosyltriethylammonium bromide as a surfactant and tetraethoxysilicon (TEOS) as the silica source. The reactant mixture was subjected to specific temperature and time conditions, including treatment at 373 K for 16 h, adjustment of the H_2O/SiO_2 ratio to about 50, and pH adjustment to 10. The resulting mixture was

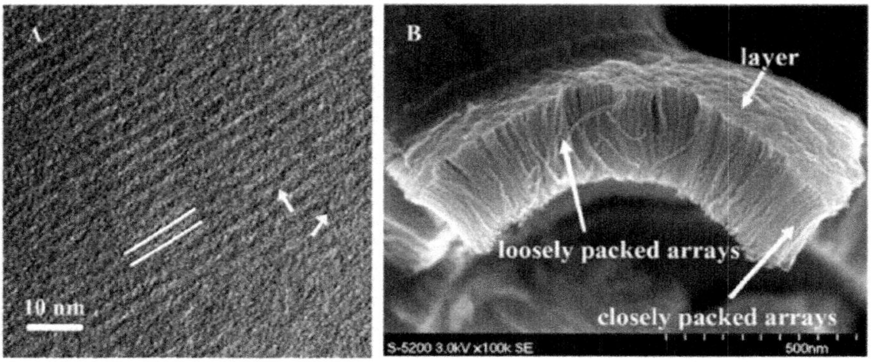

Fig. 3.10 TEM and FESEM images of **a** C1 and **b** C2, respectively [24]

then heated at 413 K for 4 days. Calcination in air at 923 K was performed to remove the surfactant. The templated carbon arrays (C2) were synthesized by utilizing furfuryl alcohol (FA, $C_5H_6O_2$) as the carbon precursor. The pore volume of MCM-41 was filled with FA using the incipient-wetness technique. Without evacuation, the FA-filled aluminosilicate MCM-41 was dried and carbonized following procedures similar to the first method. High-resolution transmission electron microscopy (HRTEM) images reveal that the diameter of the carbon nanowires is approximately 2.0–2.4 nm (Fig. 3.10) [24].

Ryoo's group has reported the synthesis of CMK-3, an ordered nanoporous carbon material, utilizing the ordered mesoporous silica molecular sieve SBA-15 as the template instead of MCM-48. A high-quality SBA-15 sample was prepared using the surfactant Pluronic P123 from BASF and tetraethyl orthosilicate (TEOS) as the silica source. The calcined SBA-15 was impregnated with an aqueous solution of sucrose containing sulfuric acid, following a similar procedure as the synthesis of CMK-18, with adjustments in the amounts of sucrose and H_2SO_4 [25, 26].

CMK-3 is composed of carbon nanorods arranged in a hexagonal pattern with connecting bridges between them. The pore wall structure of CMK-3 contains a significant number of complementary micropores, resulting in a much higher specific surface area than predicted based on the assumption of nonporous pore walls in CMK-3. The synthesis of CMK-3 using SBA-15 as the template confirmed the presence of interconnecting micropores and small mesopores among the primary hexagonal pores in the SBA-15 silica template. High-resolution transmission electron microscopy (TEM) images reveal that the carbon nanorods in CMK-3 have a diameter of 7 nm, with centers of adjacent rods being 10 nm apart and a surface-to-surface distance of 3 nm. The carbon nanorods are interconnected by spacers composed of carbon that fills the channel-interconnecting micropores within the SBA-15 wall. TEM images in Fig. 1 demonstrate that CMK-3 particles exhibit the same morphology as SBA-15 silica particles (Fig. 3.11) [25].

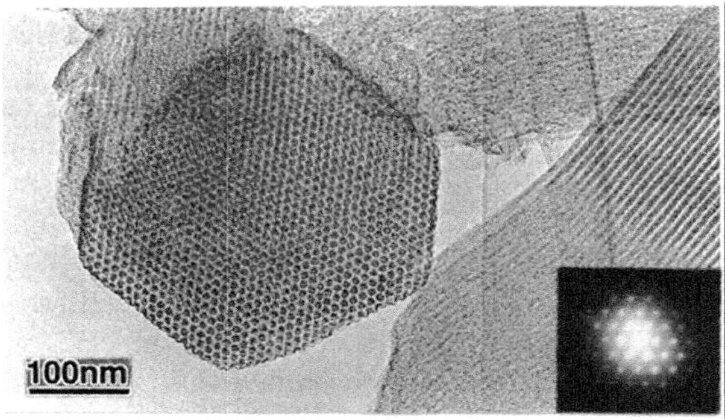

Fig. 3.11 Typical TEM images of the ordered mesoporous carbon molecular sieve, CMK-3. This carbon was synthesized using sucrose as a carbon source and SBA-15 silica as a template. The TEM images were obtained with a JEM-4000 EX operated at 400 kV from the carbon material after silica template was completely removed with HF solution [25]

Dongyuan Zhao et al. successfully synthesized a rod-like, hexagonally ordered mesoporous carbon material, similar to CMK-3, by utilizing a rod-shaped SBA-15 template. This unique template morphology was achieved by incorporating the inorganic salt KCl into the synthesis process of SBA-15 silica. In a separate study, Kim and Pinnavaia reported the synthesis of an ordered mesoporous carbon material with pore sizes of approximately 4 nm, referred to as C-MSU-H. This material was prepared using MSU-H silica as the template, which is structurally similar to SBA-15 but is synthesized under near-neutral conditions and employs sodium silicate as the silica source.

Efforts have also been directed towards the direct conversion of the triblock copolymer present in silica/triblock copolymer nanocomposites into ordered carbon materials through carbonization and template removal [27]. Jaeyun Kim reported a simple direct synthesis method for ordered mesoporous carbons, wherein sulfuric-acid-treated silica/triblock copolymer nanocomposites were carbonized. This process resulted in the formation of mesocellular carbon foams and an ordered carbon nanorod array, exhibiting similarities to CMK-3 carbon material (Fig. 3.12) [28].

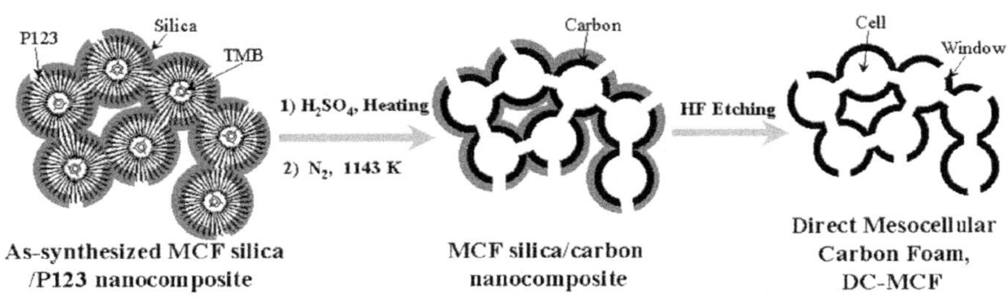

Fig. 3.12 Schematic diagram of the simple direct synthesis of mesocellular carbon materials from the carbonization of as-synthesized MCF/P123 nanocomposite [28]

Hyeon's group employed sulfuric acid treatment on the as-synthesized SBA-15 silica/triblock copolymer nanocomposite to synthesize ordered mesoporous carbon materials with narrow pore-size distributions centered around 3 nm. Sulfuric acid serves as a crucial catalyst in the dehydration and carbonization processes of the triblock copolymer surfactant. Without sulfuric acid treatment, it is not possible to directly obtain ordered mesoporous carbon materials from the as-synthesized SBA-15-type silica.

Chemical vapor deposition (CVD) has also been utilized for the synthesis of ordered mesoporous carbons. Ryoo's group reported the CVD approach for the synthesis of ordered mesoporous CMK-4 carbon. In this method, the as-synthesized MCM-48 silica was first transformed into an aluminosilicate form with a Si/Al ratio of 20. The silica template was then subjected to washing with a mixture of ethanol and HCl to remove the surfactant. The resulting mesoporous aluminosilicate template was heated in a fused quartz reactor to 1073 K while acetylene gas flowed through at a rate of 200 cm^3/min for 30 min. After ceasing the acetylene gas flow, the temperature was increased to 1173 K and held for 2 h under vacuum. The carbon/template composite obtained was subsequently washed with HF or NaOH solution to remove the template (Fig. 3.13) [29].

Zhang et al. conducted a study where they successfully synthesized CMK-5-type hexagonal arrays of carbon nanopipes using cobalt-incorporated SBA-15 as the template through chemical vapor deposition (CCVD). The researchers found that an adequate deposition time was crucial for the successful synthesis of high-quality ordered arrays of carbon nanopipes.

Liu, Zhao et al. achieved the synthesis of large-pore 3D bicontinuous mesoporous silica designated as FDU-5 at room temperature. They utilized triblock poly(ethylene oxide)-b-poly(propylene oxide)-b-poly(ethylene oxide) copolymer ($EO_{20}PO_{70}EO_{20}$, P123) as the template, tetraethyl orthosilicate (TEOS) as the silica source, and introduced small amounts of 3-mercaptopropyltrimethoxysilane (MPTS), benzene, or benzene derivatives (such as methyl-, ethyl-, dimethyl-, or trimethylbenzene) as

Fig. 3.13 HREM images of CMK-4. Incident electrons are parallel of [100] and [110][29]

additives. Notably, this was the first successful preparation of *Ia*3*d* mesostructured material at room temperature under acidic conditions. The resulting FDU-5 products exhibited uniform large pores with diameters ranging from 4.5 to 9.5 nm. These mesoporous materials with large 3D bicontinuous pores hold potential applications in sorption and transport, particularly for large molecules (Fig. 3.14) [30, 31].

Zhao et al. conducted a synthesis of a cubic mesoporous silica with large cavities, denoted as FDU-12, in the presence of block copolymer templates. The size of the entrances to the cavities could be adjusted within the range of 4–9 nm, as confirmed by nitrogen-sorption studies and the examination of negative replicas of gold and carbon. The synthesis of mesoporous FDU-12 samples involved using an acidic solution containing a nonionic block copolymer, $EO_{106}PO_{70}EO_{106}$ (where EO represents poly(ethylene oxide) and PO represents poly(propylene oxide)), as a template. Additives such as 1,3,5-trimethylbenzene (TMB) and inorganic salts, including KCl, were also used, along with tetraethyl orthosilicate (TEOS) as the silica source. The resulting silica-surfactant composite precipitates were subjected to hydrothermal treatment at different temperatures (denoted as FDU-12-X, with X representing the temperature of hydrothermal treatment). The synthesized samples

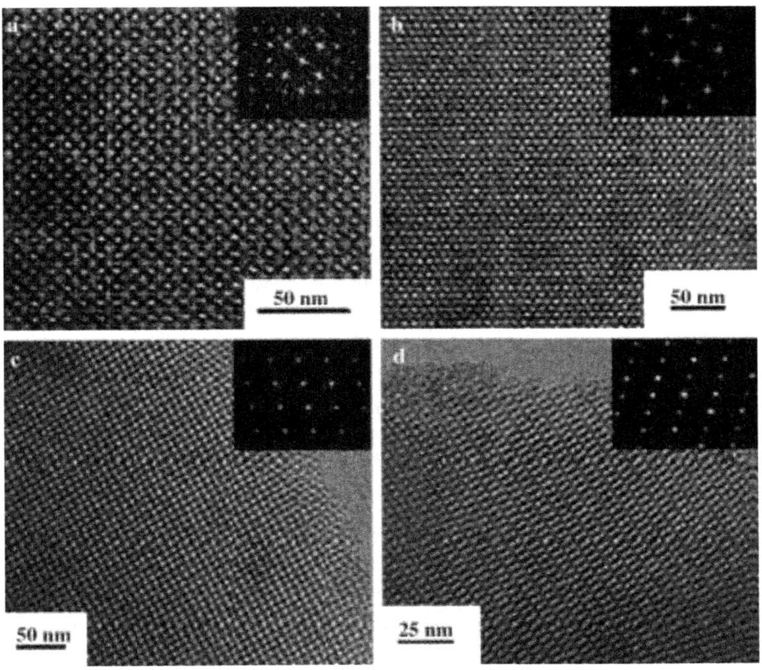

Fig. 3.14 TEM images and corresponding Fourier diffractograms of calcined FDU-5 prepared at room temperature under acidic conditions with triblock copolymer P123 as template and MPTS as additive. The images were recorded along the directions [100] (**a**), [111] (**b**), [110] (**c**), and [331] (**d**) [30]

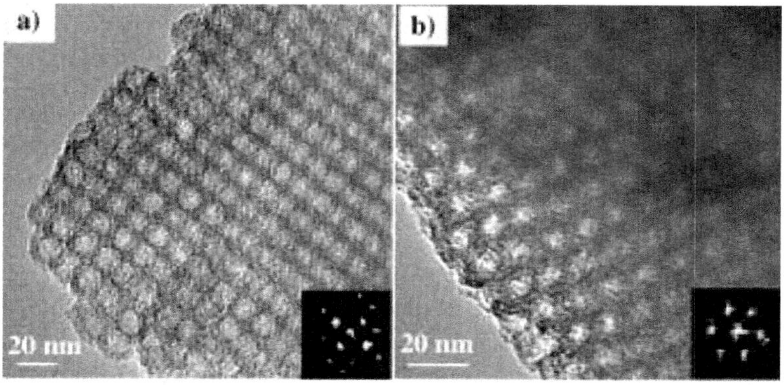

Fig. 3.15 TEM images **a** and **b** of FDU-12-373 recorded along the [100] and [111] directions, respectively [9]

were further calcined at 823 K for 6 h to obtain mesoporous silica materials (Fig. 3.15) [9].

3.1.2 Ordered Mesoporous Carbon Materials with Graphitic Pore- Wall Structure

Recently, there has been a growing interest in the synthesis of ordered mesoporous carbon materials with a graphitic pore wall structure due to their potential applications in fields such as fuel cells and double-layer capacitors. Unlike carbon materials with amorphous pore walls, those with a graphitic structure exhibit improved properties. Three main approaches have been employed to synthesize ordered mesoporous carbon with a graphitic structure. The first approach involves using carbon precursors with fused aromatic structures. The second approach is the synthesis through chemical vapor deposition (CVD) at high temperatures (>900 °C). Lastly, the third approach involves treating carbon materials with amorphous pore walls at temperatures above 2000 °C to induce graphitization.

Tzong-Horng Liou conducted a study to prepare a new adsorbent called GO/CMK-3 composite by utilizing rice husk ash (RHA) as a source of silica template. The process involved extracting sodium silicate solution from RHA waste, synthesizing the SBA-15 template using the sodium silicate solution in the presence of surfactant micelles, and subsequently mixing graphene oxide (GO) and SBA-15 with a carbon precursor to obtain the GO/CMK-3 composite. This approach aimed to utilize RHA as a sustainable source of silica template for the synthesis of mesoporous carbon materials [19].

Kim, Ryoo, and their colleagues developed a method to synthesize nanoporous carbon materials with graphitic framework structures through the in situ conversion of aromatic compounds to mesophase pitch inside silica templates. The synthesis process involves several steps. Firstly, the mesoporous silica template and the carbon source are placed in an autoclave. The aluminum sites on the silica walls act as catalysts for the formation of in situ mesophase pitch within the pores of the silica template at a relatively low pyrolysis temperature of 400 °C. Subsequently, the temperature of the autoclave is increased to 750 °C to carbonize the mesophase pitch within the template. After cooling the autoclave, the product is further heated to 900 °C under vacuum in a fused-quartz reactor to ensure complete carbonization of the carbon source. Finally, the carbon product is obtained by removing the silica template using an aqueous solution of HF or NaOH. This synthesis method has been successfully applied to various mesoporous silica templates with different structures, such as MCM-48 (cubic Ia3d), SBA-1 (cubic Pm3n), and SBA-15 (2-dimensional hexagonal p6mm) mesoporous silicas. The resulting carbon materials, synthesized using acenaphthene as the carbon source and these silica templates, are referred to as CMK-1G, CMK-2G, and CMK-3G, respectively (Fig. 3.16) [32].

Dongyuan Zhao and his team developed a simple melt impregnation method to synthesize ordered mesoporous carbon materials with ordered graphitized pore walls at low temperatures. This straightforward procedure involves melting inexpensive mesophase pitches (MPs) at 140 °C and impregnating the molten material into hexagonal or bicontinuous cubic mesoporous silica templates. After the impregnation process, the silica templates are removed using an HF solution, resulting in ordered mesoporous carbon materials with replica structures exhibiting either 2-D hexagonal p6mm or cubic *Ia*3*hd* symmetry. The resulting carbon materials retain the ordered pore structure and possess graphitic pore walls (Fig. 3.17) [33].

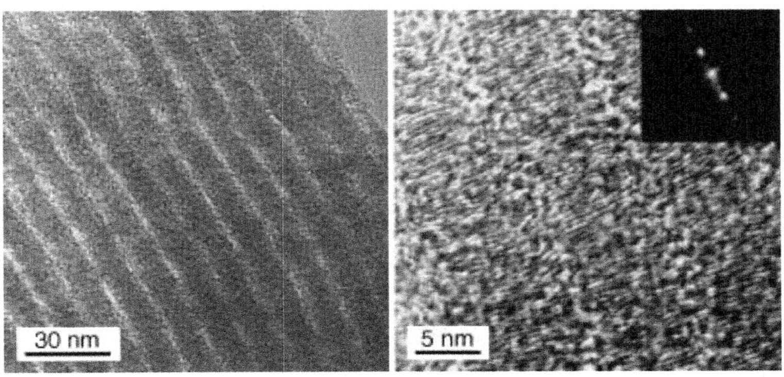

Fig. 3.16 TEM image of CMK-3G (left) and its photomagnification (right) and the corresponding electron diffraction pattern (inset)[32]

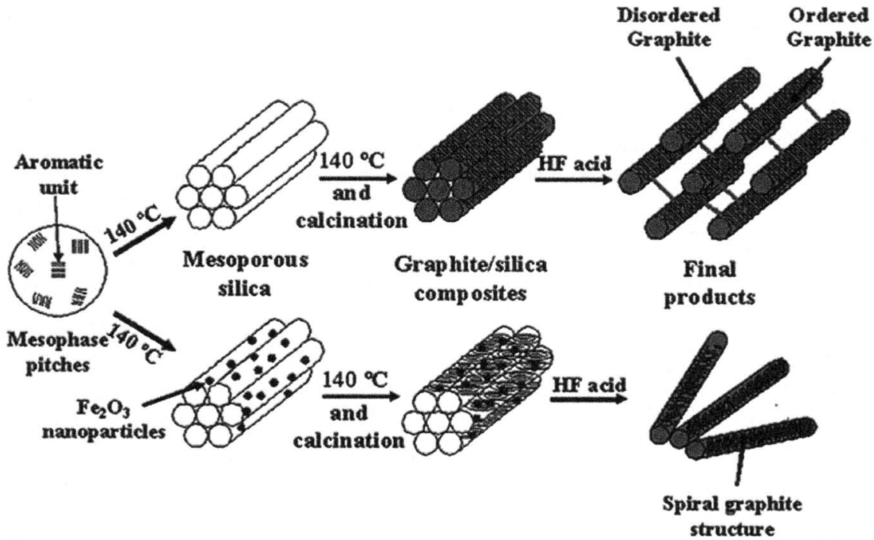

Fig. 3.17 The simple melt impregnation procedure for preparation of ordered mesoporous carbon materials and carbon nanofiber bundles with graphitized structure [33]

3.1.3 Disordered Mesoporous Carbon Materials with Uniform Mesopores

In addition to the rapid development in synthesizing ordered mesoporous carbon materials using ordered silica templates, various approaches have been developed for the synthesis of disordered mesoporous carbon materials with uniform pores. These approaches include templating with silica nanoparticles or crystals [34–37], anodic alumina [38, 39], silica gel, or alumina-silica composites [40, 41], as well as copolymerization of carbon precursors with inorganic precursors like tetraethylorthosilicate (TEOS) [1, 42]. By utilizing these methods, mesoporous carbon materials with diverse hierarchical pore structures and intriguing morphologies can be synthesized.

3.1.3.1 Templated with Colloidal Silica Particles

Han, Sohn, and Hyeon conducted a synthesis of nanoporous carbons using silica nanoparticles as templates. They utilized Ludox HS-40, a 40 wt.% aqueous colloidal silica sol solution with an average particle size of 12 nm, which was purchased from Aldrich Chemicals Co. The synthesis involved polymerizing resorcinol (99%, ACS reagent) and formaldehyde (37 wt.% aqueous solution, ACS reagent) in the presence of silica sol particles to generate silica-resorcinol–formaldehyde (RF) gel composites. Carbonization followed by HF etching of the composites resulted in porous carbons.

In the typical synthesis process, a mixture of resorcinol and formaldehyde in a 1:2 ratio was added to the Ludox HS-40 silica sol solution (Fig. 3.18). The pH of the reaction mixture was adjusted, and the mixture was then aged at 85 °C for 1 week to obtain a composite of silica and RF gel. To remove residual water, the silica-RF gel composites were heated at 85 °C for 2 days in an open-air environment. Carbonization was performed by heating the composite under a nitrogen atmosphere from room temperature to 850 °C at a heating rate of 5 °C/min, and holding it at that temperature for 3 h. Subsequently, the resulting silica–carbon composite was stirred in a 48 wt.% aqueous HF solution (J. T. Baker) for 12 h. The carbon materials were collected by filtration and washed thoroughly with deionized water until the pH of the filtrate reached 7. The carbon materials obtained from this process were designated as SMC1 (silica-sol-mediated synthesized carbon one) [43].

Jang and Lim also reported a technique to prepare porous carbon using surfactant-modified silica and constrained polymerization of divinylbenezene on the silica surface (Fig. 3.19) [44].

Hampsey, Lu, et al. introduced a novel method for the synthesis of mesoporous carbon particles using sucrose and silica as building blocks through a direct one-step aerosol process. In this process, aerosol droplets containing sucrose and silica are subjected to continuous solvent evaporation at the air/liquid interface. This leads to the concentration of sucrose and silica within the droplets, resulting in the formation of spherical silica/sucrose nanocomposite particles. These particles are then collected on a membrane filter using a press.

Subsequent carbonization of the sucrose component, followed by the removal of silica, leads to the formation of spherical porous carbon particles. This one-step aerosol process allows for the synthesis of mesoporous carbon particles with a well-defined structure.

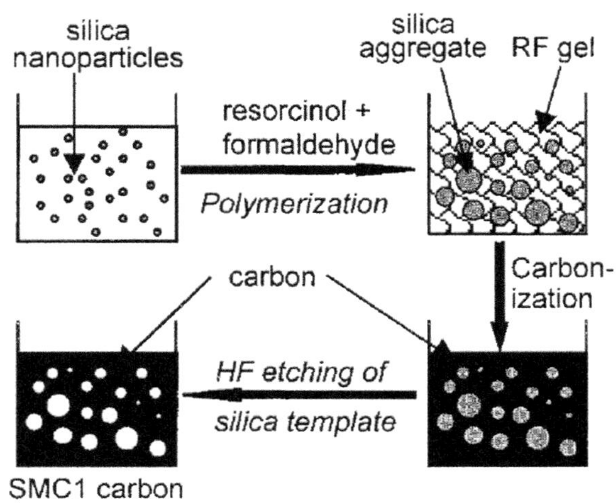

Fig. 3.18 Schematics for the synthesis of nanoporous SMC1 carbons: (1) polymerization of resorcinol and formaldehyde (RF) in the presence of silica nanoparticles; (2) carbonization of RF gel-silica composite at 850 °C to get a carbon–silica composite; (3) HF etching of silica aggregates to get nanoporous SMC1 carbons [43]

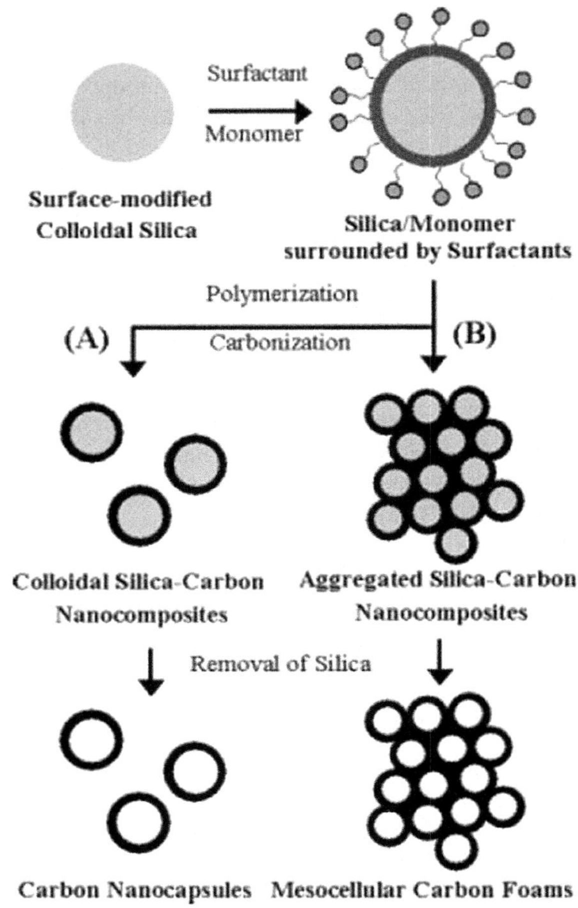

Fig. 3.19 Schematic diagram for the selective fabrication of carbon nanocapsules and mesocellular foams [44]

Gierszal and Jaroniec conducted a study in which they synthesized ordered uniform porous carbon frameworks using removable colloidal silica crystalline templates. The pore sizes of these frameworks ranged from 10 to 1000 nm. The carbonization process involved the use of phenol and formaldehyde as carbon precursors. The synthesized porous carbon materials were then utilized as supports for a Pt (50)-Ru (50) alloy catalyst in a direct methanol fuel cell (DMFC). The researchers investigated the effect of these porous carbons as catalyst supports on the anodic performance of the fuel cell. The results showed that the use of the ordered uniform porous carbons significantly improved the catalytic activity for methanol oxidation in the fuel cell. This improvement can be attributed to several factors, including the high surface areas, large pore volumes, and three-dimensionally interconnected uniform pore structures of the carbon frameworks. These characteristics facilitate a higher degree of catalyst dispersion and efficient diffusion of reagents within the fuel cell system (Fig. 3.20) [45].

Li and Jaroniec developed a colloidal-imprinting technique for synthesizing mesoporous carbon materials with uniform mesopores (Fig. 3.21). They used mesophase pitch as the carbon precursor and commercial colloidal silicas as templates. The resulting carbon materials exhibited uniform mesopore sizes of 13 and 24 nm, respectively, which were determined by the dissolution of the corresponding colloidal spheres. The BET surface area of the carbon materials agreed well with calculations based on a model of nonmicroporous pore wall structure [34].

On the other hand, Fuertes reported the synthesis of mesoporous carbon materials using mesostructured silica materials as templates. These silica templates were synthesized with the assistance of surfactants as structure-directing agents. The resulting porous carbons maintained the size and shape of the template, allowing for the selection of primary particle sizes in a wide range from 10 nm to 10 μm by using appropriate silica templates. These carbons exhibited large pore volumes (>1.4 $cm^3\ g^{-1}$), high BET surface areas (>1800 $m^2\ g^{-1}$), and a porosity consisting of mesopores in the range of sizes. The micrometric carbon materials exhibited a spherical morphology and a unimodal pore system consisting of structural mesopores with a diameter of approximately 3 nm [46].

Mesoporous/macroporous carbon materials with uniform pore sizes were synthesized using colloidal silica crystals as templates through impregnation [47, 48] and chemical vapor deposition (CVD) methods [49]. Hyeon et al. developed a low-cost and environmentally friendly route to prepare bimodal mesoporous silica materials. These bimodal mesoporous silica materials were then used as templates for synthesizing bimodal mesoporous carbon materials. The synthesized silica and carbon materials with bimodal mesopores were denoted as Meso-nano-S and Meso-nano-C, respectively. The typical synthetic procedure for Meso-nano-S involved dissolving

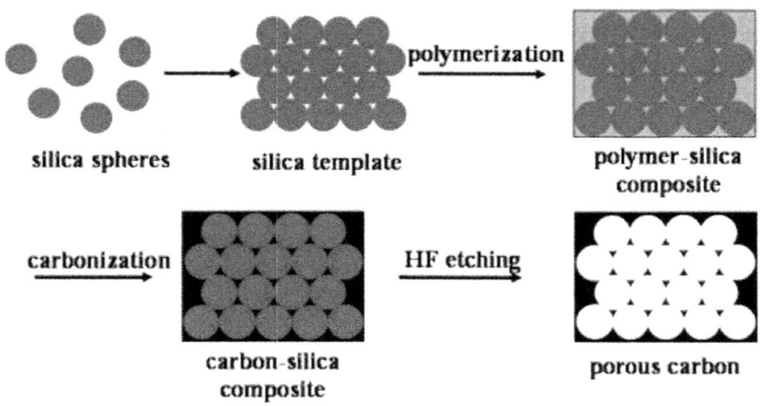

Fig. 3.20 Synthetic procedure for uniform porous carbons of tunable pore sizes through colloidal crystal template approach [45]

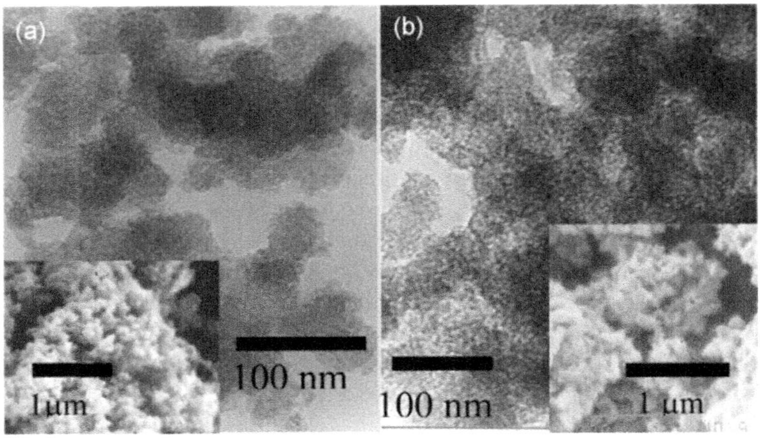

Fig. 3.21 Transmission electron micrograph (TEM) of (a) Meso-nano-S and (b) Meso-nano-C Inset: Scanning electron microscope (SEM) of Meso-nano-S silica and Meso-nano-C [50]

8.6 g of Triton X-114 in 200 ml of water containing 4.43 ml of concentrated acetic acid. Then, 16 ml of sodium silicate solution (Aldrich, 27% SiO_2, 14% NaOH) diluted with 200 ml of water was added to the solution with vigorous stirring. The resulting mixture was heated at 45 °C for 20 h. Afterward, the mixture was transferred to a polypropylene bottle and aged for an additional 20 h at the same temperature under static conditions. The white precipitate obtained was isolated by filtration, dried in air, and calcined at 550 °C for 4 h to obtain the final Meso-nano-S material (Fig. 3.21a). Using the bimodal mesoporous silica (Meso-nano-S) as a template, the researchers synthesized bimodal mesoporous carbon materials with framework mesopores of approximately 4 nm in size and textural pores of around 30 nm in size (Fig. 3.21b) [50].

Gierszal and Jaroniec developed a cost-effective and straightforward method for synthesizing bimodal carbons derived from pitch. These carbons exhibit both ordered mesopores, which are an inverse replica of the SBA-15 ordered mesoporous silica, and spherical mesopores formed after the dissolution of uniform silica colloids [51] .The synthesis involved a mixture of ordered SBA-15 and colloidal silica spheres as templates. By adjusting the temperature and duration of the thermal treatment, the researchers were able to control the surface area and total pore volume of the resulting carbons. Notably, this study demonstrated a simple approach to accurately control the size of larger pores by utilizing commercially available silica colloids, distinguishing it from previous reports.

3.1.3.2 Templated with Silica/Aluminosilicate Gels

Lee, Sohn, and Hyeon developed a method to synthesize monolithic mesoporous carbon materials using a monolithic silica or colloidal silica as a template. The process involved partially impregnating a mesocellular aluminosilicate foam with a mixture of phenol and formaldehyde, followed by carbonization and template removal. The addition of aluminum during the impregnation step introduced strong acid catalytic sites within the silicate framework, facilitating the polymerization of phenol and formaldehyde. Different mesocellular aluminosilicate foams with varying cell and window dimensions were used as templates. The resulting mesocellular carbon foams exhibited uniform mesopores and showed potential applications as electrode materials in supercapacitors, catalyst supports, and adsorbents for large-sized water pollutants (Fig. 3.22) [40].

Oda, Tatsumi et al. successfully synthesized mesoporous carbon in a foam structure using mesocellular foam (MCF) silica as a template. The carbon material consisted of spherical hollows with closed walls and mesopores with a diameter of approximately 4 nm. The synthesis involved a two-step impregnation process, where sucrose was first impregnated into the MCF silica template, followed by carbonization under flowing argon at 900 °C. Feng and co-workers developed a templating method to synthesize a carbon monolith with a co-continuous structure and trimodal pores using a novel silica monolith as the template. The resulting carbon monolith exhibited high surface area, micropores, uniform mesopores, and macroporosity, making it suitable for applications in adsorption, separation, purification, and other areas.

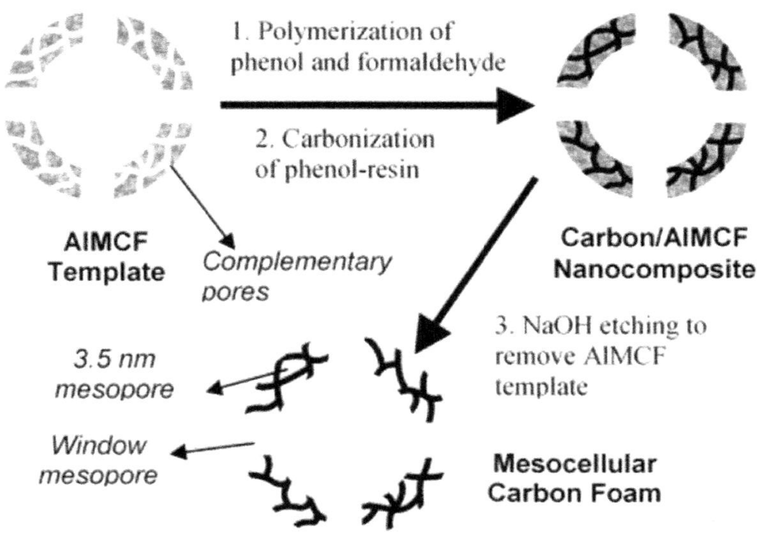

Fig. 3.22 Schematic Illustration for the Synthesis of Mesocellular Carbon Foam [40]

Lee, Hyeon et al. reported the synthesis of mesoporous carbon material called SNU-2 using hexagonal mesoporous aluminosilicate (Al-HMS) as the template and phenolic resin as the carbon precursor [52]. The synthesis involved treating the Al-HMS template with AlCl3 to introduce strong acid catalytic sites for the polymerization of phenol and formaldehyde. The resulting Al-HMS/phenol resin composite was carbonized under a constant stream of nitrogen, followed by the dissolution of Al-HMS using hydrofluoric acid to obtain the mesoporous carbon material with a uniform pore size of 2 nm [53].

Xia and Mokaya conducted a study on the synthesis of hollow spherical mesoporous carbon material using mesoporous silica SBA-15 as a template through a chemical vapor deposition (CVD) route. This research was the first to demonstrate that mesoporous carbons, when nanocast using conventional mesoporous silica, can adopt a particle morphology that is different from that of the templating silica [54].

Knox et al. developed a method for synthesizing rigid mesoporous carbon using a phenol-hexamine mixture as the carbon precursor. The process involved polymerization and carbonization of the phenolic resin inside the pores of spherical silica gel, followed by dissolution of the silica template. The resulting carbon material had a rigid structure. Additional treatment at high temperatures (2500 °C) in an inert atmosphere was performed to remove complementary micropores through graphitization. The final specific surface area of the carbon material was approximately 150 $m^2\ g^{-1}$. This carbon material was later commercialized under the name "Hypercarb," and its unique characteristics, such as graphitized mesoporous carbon spheres, have shown excellent retention properties in liquid chromatographic separations.

3.2 Soft Template Methods

Except the hard template method, the soft template method is another effective method for the preparation of mesoporous materials. In the following text, some cases of preparing mesoporous materials by soft template method are presented.

3.2.1 Soft-Templating Approach

3.2.1.1 Methodologies

In the soft-templating approach, various methods are employed to synthesize ordered mesoporous carbon using different types of amphiphilic block copolymers and surfactants. These surfactants and polymers consist of a hydrophobic unit and a hydrophilic unit, which self-assemble with carbon sources such as polymeric resins, sucrose, or furfuryl alcohol. Upon appropriate thermal treatment, ordered mesoporous carbon materials are formed. Several common routes have been successfully used for the synthesis of mesoporous carbon, including:(1) Solvent Evaporation

Induced Self-Assembly (EISA): This method involves the self-assembly of block copolymers and resols (such as phenol, resorcinol, or phloroglucinol/formaldehyde) induced by solvent evaporation. The resulting mesoporous carbon structure is formed when the solvent evaporates, leaving behind ordered mesopores. (2) Hydrothermal Autoclaving: In this process, the carbon precursor and surfactant are mixed together in an aqueous solution and subjected to hydrothermal treatment under high pressure and temperature. The hydrothermal conditions promote the self-assembly of the surfactant and carbon precursor, leading to the formation of ordered mesoporous carbon. (3) Dilute Aqueous Route: This method involves the dispersion of the carbon precursor and surfactant in a dilute aqueous solution. The mixture is then subjected to a thermal treatment, which triggers the self-assembly of the surfactant and carbon precursor into an ordered mesoporous structure. (4) Macroscopic Phase Separation: In this route, the carbon precursor and surfactant are mixed together to form a macroscopic phase-separated system, such as an emulsion or gel. The subsequent thermal treatment induces the formation of ordered mesoporous carbon by phase separation and carbonization of the precursor material. These methods provide versatile approaches for synthesizing ordered mesoporous carbon materials with controlled pore structures. The choice of method depends on the specific requirements and desired properties of the final carbon material.

3.2.1.2 Evaporation Induced Self-Assembly

In the EISA method, the preferential evaporation of the solvent plays a crucial role in inducing the organization of precursors around the template and the formation of voids, which leads to the development of mesoporous structures. One unique aspect of the EISA method is that the cross-linking of precursor molecules and the solvent evaporation occur simultaneously, allowing for the formation of the desired mesoporous structures. The solvent evaporation process can be carried out using different techniques depending on the desired application and geometry of the material. It can be as simple as evaporating the solvent in an open-mouthed dish, or more sophisticated techniques such as spin coating or dip coating can be employed to achieve more controlled and uniform mesoporous films. The EISA method has been widely utilized in the synthesis of mesoporous carbon materials with various mesostructures, including 2D hexagonal (p6mm), 3D bi-continuous (*Ia*3*d*), body-centered (Im3m), and lamellar mesostructures. These mesostructured materials have been predominantly used for the fabrication of mesostructured films and mesoporous oxide particles, enabling a wide range of applications in areas such as catalysis, energy storage, and separation (Fig. 3.23) [55].

3.2.1.3 Dilute Aqueous Method

Indeed, the dilute aqueous route offers several advantages over the EISA method, particularly in terms of scalability for large-scale production. The method relies on

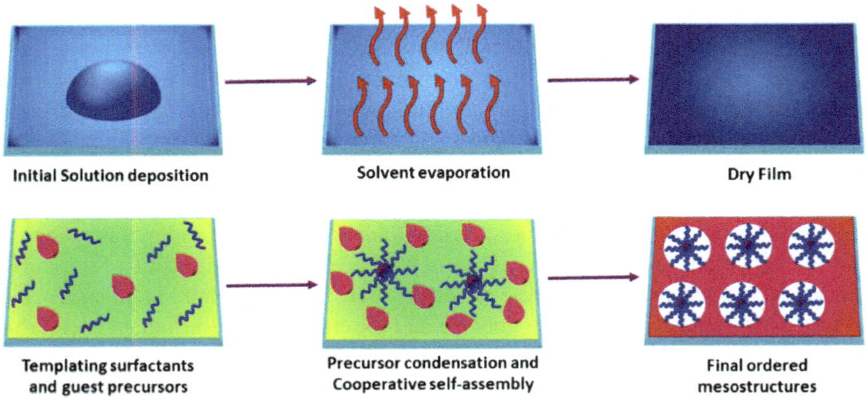

Fig. 3.23 The evaporation induced self-assembly [55]

the cooperative self-assembly process, where phenolic groups of the resol and ether groups of the PEO block copolymers interact through monolayer hydrogen bonding. This interaction leads to the formation of a complex mesostructure, which undergoes polymerization to generate a composite of phenolic resin and Pluronic.

Subsequently, the block copolymer is removed, and the resulting composite is carbonized at an appropriate temperature to yield ordered mesoporous carbon. The success of the dilute aqueous route is highly influenced by the reaction conditions, especially the pH level. Polymerization occurs more rapidly under basic conditions, around pH 9, while it proceeds extremely slowly under neutral or weakly acidic conditions.

By controlling the pH during the synthesis process, it is possible to regulate the rate of polymerization and achieve the desired mesoporous carbon structure. This method provides flexibility in tuning the reaction conditions to optimize the synthesis process and obtain ordered mesoporous carbon materials with tailored properties. Additionally, the scalability of the dilute aqueous route makes it advantageous for large-scale production of mesoporous carbon materials.

3.2.1.4 Macroscopic Phase Separation Method

The dilute aqueous route is highly sensitive to the synthetic conditions, particularly the pH of the solution, aging time, and quantity of carbon precursor. When using resorcinol/formaldehyde (R/F) as a carbon source under strongly acidic conditions, a clear phase separation can occur. The solvent, such as water or ethanol, forms a top layer, while the composite of resol/block-polymer settles in the lower layer.

To obtain ordered mesoporous carbon, the lower layer containing the composite needs to be carefully isolated. It is then subjected to aging, curing, and carbonization processes. By controlling the synthetic conditions, such as pH and the ratio of carbon precursor, the formation of phase separation can be achieved, which is essential for the successful synthesis of ordered mesoporous carbon materials.

A similar phase separation phenomenon can also be observed when using a mixture of phloroglucinol with formaldehyde under strong acidic conditions. The understanding and control of these phase separation processes are crucial for obtaining the desired mesostructure and properties in the resulting ordered mesoporous carbon materials.

3.2.1.5 Hydrothermal Autoclaving Method

Indeed, the hydrothermal autoclaving route is widely used in the synthesis of mesoporous materials, including mesoporous carbon. It is a commonly adopted method due to its effectiveness, efficiency, and suitability for large-scale production.

In the hydrothermal autoclaving route, the synthesis mixture, typically containing the carbon precursor and a template (such as block copolymers or surfactants), is sealed in an autoclave and subjected to high temperature and pressure conditions. The autoclave is then heated and maintained at a specific temperature for a defined period of time.

The hydrothermal conditions promote the self-assembly and organization of the precursor molecules around the template, leading to the formation of ordered mesostructures. The high temperature and pressure facilitate the chemical reactions, such as polymerization or condensation, to occur and result in the formation of the desired mesoporous materials.

Compared to the dilute aqueous route, the hydrothermal autoclaving method offers faster reaction kinetics and is more energy-efficient. It allows for the synthesis of a wide range of mesoporous materials, including mesoporous carbons, with controlled pore sizes, surface areas, and pore structures.

Overall, the hydrothermal autoclaving route is a versatile and effective method for the large-scale production of mesoporous materials, making it highly valuable in the field of materials synthesis.

3.2.2 Recent Developments in Soft Template Synthesis

The soft-templating approach in the synthesis of mesoporous carbon involves the use of carbon precursors that can cross-link and form a stable structure, along with a template that drives the micelle formation. Strong interactions between the templating block copolymer and the precursor resin are necessary to prevent macrophase separation. Additionally, the cross-linked precursor resins should have higher thermal stability than the templates to avoid structural collapse during template removal.

Bähr and co-workers utilized a soft-templating method to synthesize ordered mesoporous carbon materials with tunable porosity, surface area, and degree of graphitization. The synthesis process involved dissolving phenolic resin in tetrahydrofuran (THF) and stirring it for 30 min. In a separate step, PEO-b-PS block copolymer was dissolved in THF using ultrasonication. The PEO-b-PS solution was slowly added to the phenolic resin solution under stirring, followed by an additional 2 h of stirring. The solution was then divided into crystallization bowls to allow solvent evaporation and subjected to two subsequent heating steps at 50 and 100 °C for 24 h each. The resulting polymer/resin film was scraped off the bowls and subjected to pyrolysis under an argon atmosphere. Pyrolysis was performed at temperatures of 600 °C, 800 °C, or 1000 °C for 5 h, with a heating rate of 5 K/min. The final ordered mesoporous carbon materials were labeled as OMeCX, with X representing the pyrolysis temperature. Through this soft-templating approach, the researchers achieved the synthesis of ordered mesoporous carbon materials with controlled properties, allowing for tunability of porosity, surface area, and degree of graphitization (Fig. 3.24) [56].

Haibo Tan et al. reported the synthesis of two-dimensional (2D) sandwich-structured phosphorus- and nitrogen-doped mesoporous carbon nanosheets. In their method, graphene oxide (GO) was used as the substrate, triblock copolymer F127 micelles served as the soft template, and a combination of resin and phytic acid acted as the organic precursor. This synthesis approach allowed for the incorporation of phosphorus and nitrogen dopants into the carbon nanosheets, resulting in the formation of mesoporous structures. The use of GO as the substrate provided a stable and conductive platform for the formation of the nanosheets, while the F127 micelles and organic precursors facilitated the templating and doping processes (Fig. 3.25) [57].

Dipendu Saha employed a soft-templating approach to synthesize mesoporous carbon using phloroglucinol as the carbon precursor and pluronic F127 as the porogen. Scanning electron microscopy (SEM) analysis of the resulting mesoporous carbon particles revealed an average particle size in the range of 400–600 μm. The particle surface exhibited larger macropores, which were clearly visible and ranged in size from 1 to 5 μm. This indicates the presence of a hierarchical pore structure in the synthesized mesoporous carbon material, with both mesopores and macropores contributing to its porosity (Fig. 3.26) [58].

Ratchadaporn Kueasook synthesized hierarchically macro- and mesoporous carbon materials (HPCs) using environmentally friendly phloroglucinol/glyoxylic acid precursors with the soft-template F127. Sugarcane bagasse was used as a scaffold via self-assembly surface coating. The resulting monoliths obtained through this method exhibited a stable monolithic feature with a hierarchically porous structure. The HPCs had high BET surface areas ranging from approximately 500 to 645 m^2/g, total pore volumes of 0.2–0.4 cm^2/g, and mesoporosity contribution as large as 50–60% volume. The mesopore sizes were around 5 nm (Fig. 3.27) [59].

Guangzhi Hu synthesized ordered mesoporous carbon (OMC) using a soft template method. The synthesis process is illustrated in Fig. 35. The procedure involved dispersing 2 g of sodium dodecyl sulfate (SDS) and 1.27 g of resorcinol in a mixture of 7 mL of benzaldehyde and 15 mL of deionized water. The pH of the solution was adjusted to between 8.5 and 9, and the mixture was stirred at 60 °C in

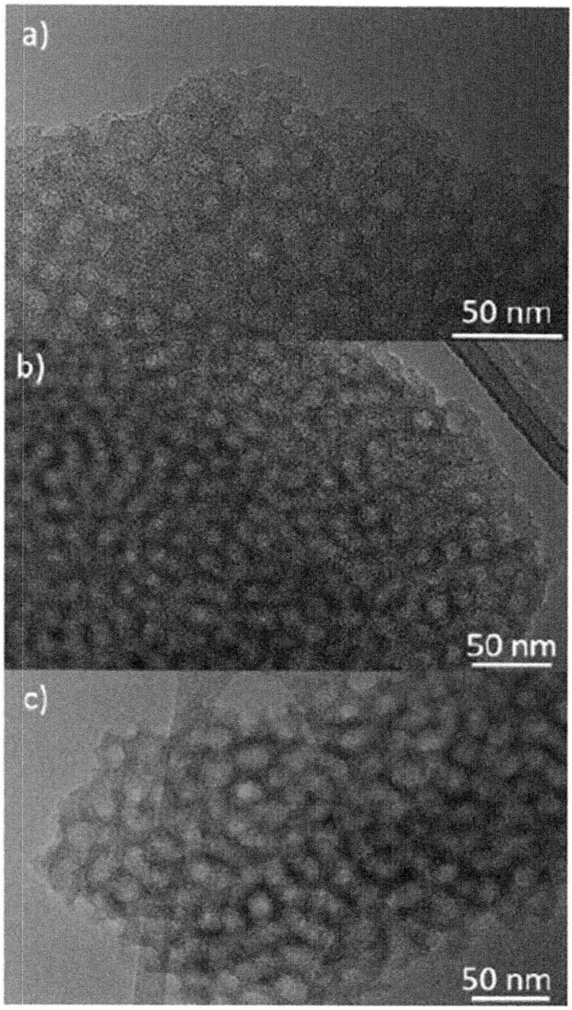

Fig. 3.24 TEM micrographs of the samples **a** OMeC600, **b** OMeC800 and **c** OMeC1000 [56]

a water bath for 2 h. The water bath was then removed, and the mixture was stirred overnight to form a colloidal substance. The colloidal substance was freeze-dried for 36 h and subsequently carbonized at 700 °C for 2 h under a nitrogen atmosphere. The final OMC material was obtained by washing it with deionized water and drying at 120 °C for 5 h (Fig. 3.28) [60].

Weizheng Li developed a synthesis method for nitrogen-doped mesoporous hollow carbon spheres (NMHCSs) that is simple, efficient, and tunable. The synthesis is a one-pot process using the sol–gel method. It involves the use of 3-aminophenol-formaldehyde resin as the carbon source (organic precursor), tetraethyl orthosilicate

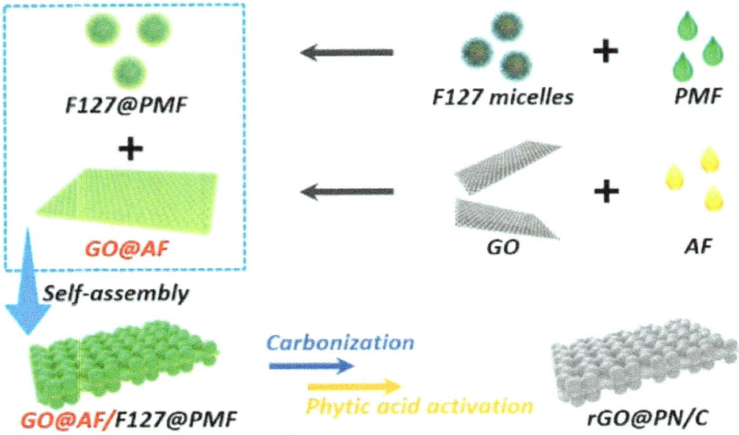

Fig. 3.25 Illustration of the Synthetic Processes for Nitrogen-Doped Carbon (rGO@N/C) and Phosphorusand Nitrogen-Doped Carbon (rGO@PN/C) [57]

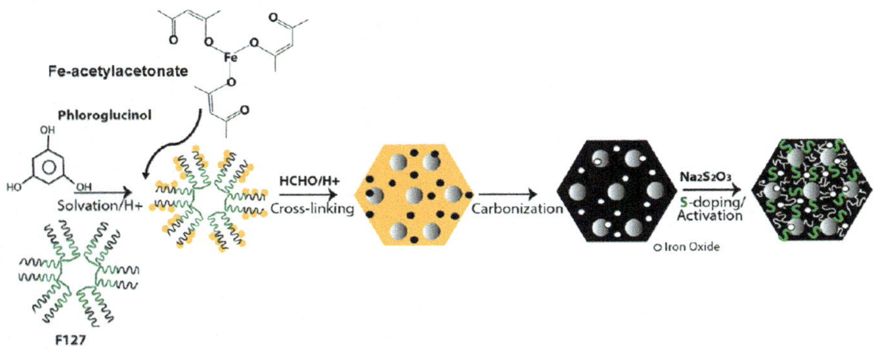

Fig. 3.26 Schematic of synthesis of iron and sulfur dual-doped mesoporous carbon [58]

(TEOS) as the silicon source (inorganic precursor), and cationic Gemini surfactant pentane-1,5-bis(dimethylcetyl ammonium bromide) as the soft template. The resulting NMHCSs possess a unique hollow structure with mesoporous walls and are doped with nitrogen. This structure allows for shortened transmission paths for electrolyte ions, increased utilization of the specific surface area, and reduced diffusion resistance for ions. These characteristics make the NMHCSs promising for applications in energy conversion and storage (Fig. 3.29) [61].

Alex M. Volosin synthesized high-surface area mesoporous carbons using hydrous alumina gel as the soft template. The synthesis method involved the formation of a three-dimensional continuous network of mesopores in the alumina gel, which served as a negative replica in the resulting mesoporous carbon product. The synthesis process included the controlled and sequential formation of the alumina gel network

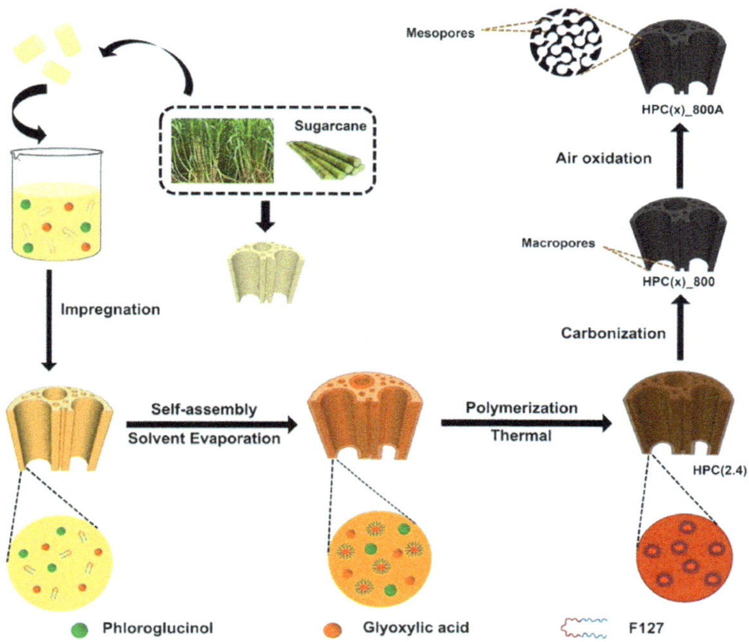

Fig. 3.27 Schematic illustration of the preparation steps for all HPCs by self-assembly methods [59]

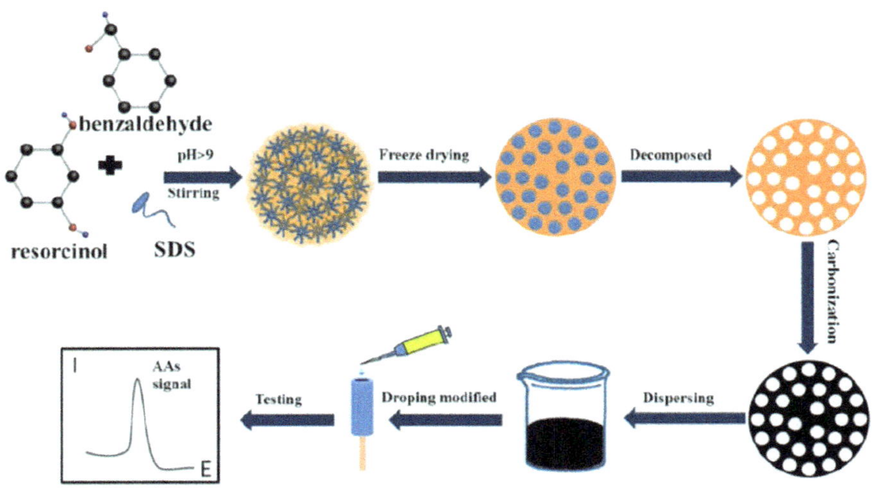

Fig. 3.28 Synthetic route for OMC[60]

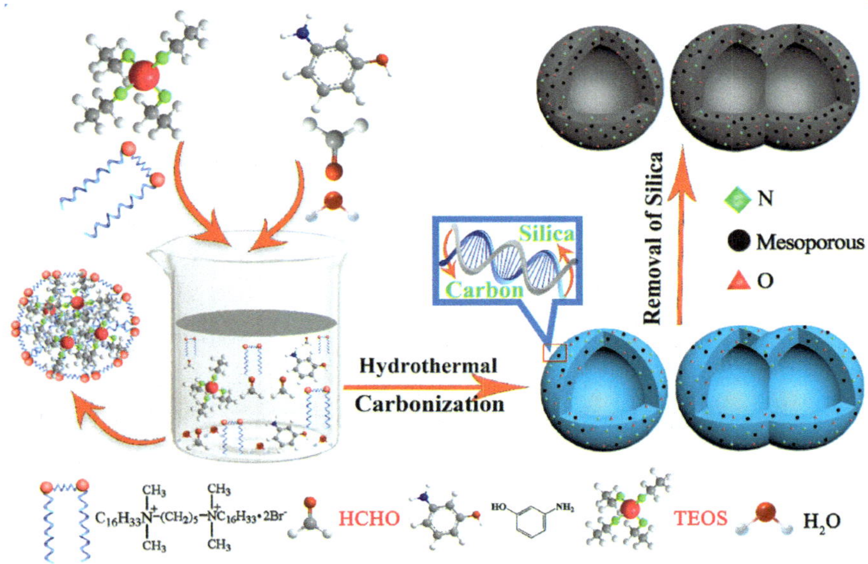

Fig. 3.29 Synthesis of Nitrogen-doped polymer-silica and carbon–silica composite nanospheres with cross-interlocking mesostructures [61]

and the polymer network. This ensured the interpenetration of the two gel networks, which is crucial for the formation of open and connected mesopores in the final mesoporous carbon material. The inorganic alumina gel was formed first and then catalyzed the polymerization of a carbon precursor polymer in a one-pot preparation. This method allows for the synthesis of mesoporous carbons with high surface areas, providing a framework of interconnected mesopores. These materials have potential applications in various fields, including catalysis, adsorption, and energy storage (Fig. 3.30) [62].

Inayat Ali Khan developed a simple and effective strategy for synthesizing carbon nanospheres (CNSs) using a metal–organic framework (MOF-5) as a soft template. This approach involves the careful selection of a carbon precursor (sucrose), the MOF-5 template, and the carbonization temperature. By utilizing this method, well-defined and spherical carbon nanostructures in the form of CNSs were obtained. These CNSs exhibited superior electrochemical performance compared to fragmented porous carbon obtained through the direct thermal conversion of MOF-5. The synthesis route of carbon nanospheres via soft-template carbonization offers a new approach to designing and constructing well-defined carbon nanostructures. This method has potential applications in various fields, such as energy storage and conversion (Fig. 3.31) [63].

Pengyuan Yang conducted a study where they synthesized a composite material consisting of a 3D floor-like ordered mesoporous carbon functionalized graphene (FLOMC-GO) using a combination of the soft template method and chemically

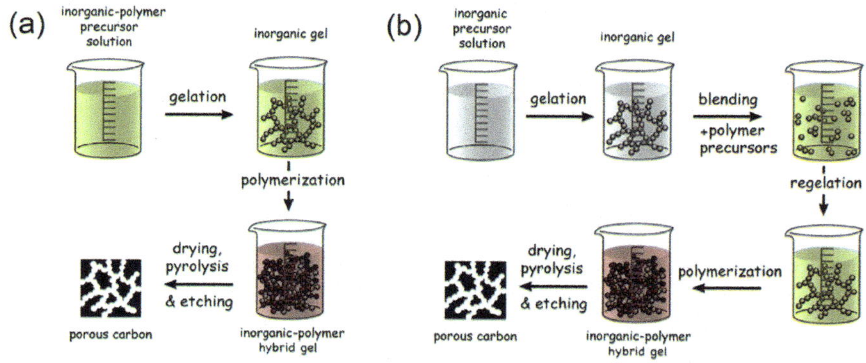

Fig. 3.30 Synthetic schemes of **a** single mix route and **b** thixotropic mixing route [62]

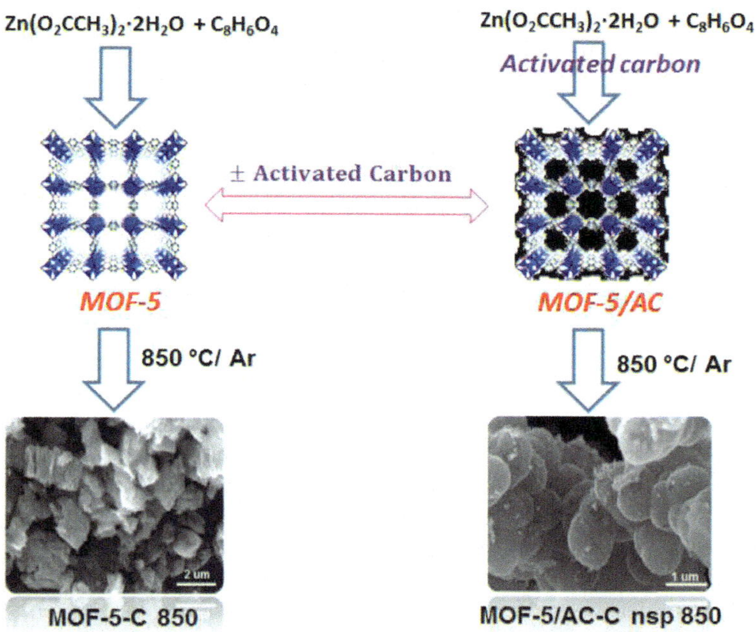

Fig. 3.31 Schematic representation of thermal conversion of MOF-5 to fragmented porous carbon and MOF-5/AC to carbon nanospheres [63]

derived graphene (CDG). The synthesis process involved using cetyltrimethylammonium bromide (CTAB), a cationic surfactant, as a structure-directing agent during the formation of a mesoporous silica template. Sucrose was then utilized as the carbon source to form the support for the carbonized floors within the mesoporous template. The resulting composite material exhibited a high content of graphitized carbon, along with highly ordered mesoporous structures and large surface areas. The unique properties of the FLOMC-GO composite make it suitable for various applications. In the study, the composites were specifically utilized for enriching N-linked glycans from complex bio-samples, showcasing high sensitivity, selectivity, and efficiency in this application. Overall, the synthesized composite material offers significant potential for use in diverse fields due to its desirable characteristics and performance.

Gang Zhao conducted a study in which they synthesized ordered mesoporous carbon FDU-15 using resol and Pluronic F127 as the carbon precursor and soft template, respectively. The addition of tetraethyl orthosilicate (TEOS) and decane served as pore-expanding agents to achieve FDU-15 with varying pore sizes. In the synthesis process, the resol solution was prepared as the carbon precursor for FDU-15. This involved mixing phenol, NaOH (20 wt.%), and formaldehyde solution (37 wt.%) at 75 °C for 1 h. The pH of the mixture was adjusted to 7.0 using HCl solution (1 M) after cooling to room temperature. Water was then removed through vacuum distillation, and ethanol was added to form a 20 wt.% resol solution. For the preparation of FDU-15, Pluronic F127 and resol were dissolved and mixed in ethanol, followed by transferring the mixture to a dish for evaporation at room temperature for 8 h. The resulting mixture was further heated at 100 °C for 24 h for thermopolymerization. The obtained product was then calcined at 800 °C for 3 h in an argon atmosphere with a heating rate of 1 °C/min. Finally, the resulting material was ground into powder and designated as FDU-15. This synthetic method allowed for the creation of ordered mesoporous carbon material FDU-15, with its specific properties and characteristics (Fig. 3.32) [64].

Dai et al. developed a stepwise assembly approach for the fabrication of highly ordered nanoporous carbon films. The process involves several steps to achieve the desired structure and properties. Firstly, the resorcinol monomers are preorganized into a well-ordered nanostructured film with the assistance of self-assembly of polystyrene-block-poly(4-vinylpyridine) (PS-P4VP) and solvent-induced structural annealing. This step helps to create a template for the subsequent formation of the nanoporous carbon film. Next, solvent annealing is carried out to further enhance the organization and stability of the nanostructured film. This step helps to remove any defects and optimize the arrangement of the resorcinol monomers. The polymerization of the resorcinol monomers takes place between the P4VP regions using gaseous formaldehyde vapor. This leads to the formation of a highly crosslinked phenolic resin, which serves as the precursor for the carbon film. Finally, the thermal treatment is conducted at 800 °C under a nitrogen atmosphere. During this step, the block copolymer template decomposes, releasing gaseous species, while the phenolic resins undergo carbonization. This transformation results in the formation of a highly ordered nanoporous carbon film.The self-assembly of the PS-P4VP/resorcinol mixture is attributed to the hydrogen bonding interaction between the resorcinol and P4VP blocks, which contributes to the overall organization and stability of the nanostructured film. Overall, this stepwise assembly approach allows for the fabrication of highly ordered nanoporous carbon films with controlled structure and properties, offering potential applications in various fields (Fig. 3.33) [65].

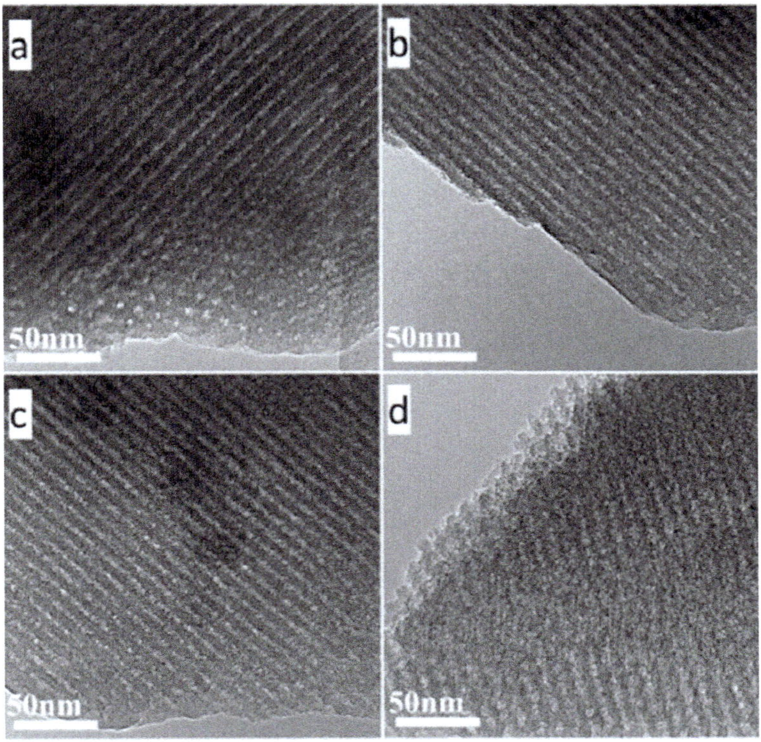

Fig. 3.32 TEM images of as-prepared FDU-15-1 (**a**), FDU-15-2 (**b**), FDU-15-3 (**c**), and FDU-15-4 (**d**) [64]

Tanaka et al. reported a method for synthesizing mesoporous carbons with ordered channel structure (COU-1) through the direct carbonization of an organic-organic nanocomposite. The synthesis procedure involved the use of resorcinol/formaldehyde resins as the carbon precursor, triethylorthoacetate (EOA) as a co-carbon source, and Pluronic block copolymer F127 as the surfactant. To prepare the mesoporous carbon, a solution was prepared by mixing the surfactant, carbon precursor, and co-carbon source. This solution was then spin-coated onto a silicon substrate by spinning at a maximum speed of 1000 rpm for approximately 2 min. The coated substrate was then subjected to the polymerization of resorcinol formaldehyde resins at 90 °C for 5 h in an air environment. After the polymerization step, a brown product was obtained and subsequently subjected to carbonization at different temperatures: 400, 600, and 800 °C for 3 h. The resulting materials were designated as COU-1. The COU-1 materials obtained through this synthesis method exhibit mesoporous structures with ordered channel arrangements. The choice of carbon precursor, co-carbon source, and surfactant allows for the control of the mesoporous structure and properties of the final carbon material [66].

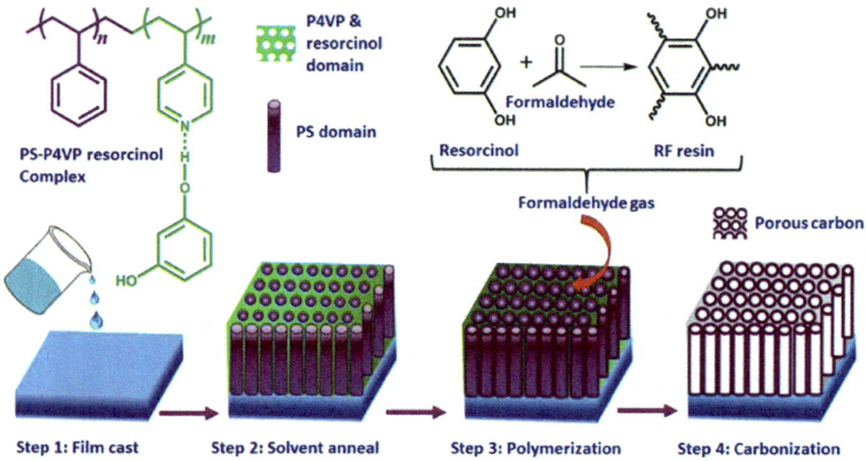

Fig. 3.33 Schematic representation of the synthesis process. Step 1: film casting of PS-b-P4VP/resorcinol supramolecular assembly on Si wafer. Step 2: microphase separation during solvent annealing at 80 °C in DMF–benzene vapor. Resorcinol organizes between the P4VP domains. Step 3: polymerization of resorcinol exposing the film to formaldehyde gas. Step 4: pyrolysis in N_2 resulting in arrays of hexagonally arranged carbon channels [65].

Zhao's group has successfully synthesized a family of highly ordered mesoporous carbon materials known as C-FDU-x [31, 67–69], where x represents different symmetries such as 2D hexagonal p6m, 3D bi-continuous Ia3d, and body-centered cubic Im3m. These materials were synthesized using the solvent evaporation-induced self-assembly (EISA) method. The templates used in this process were amphiphilic triblock copolymers (PEO-PPO-PEO), and the carbon precursors were low-molecular-weight phenolic resols obtained from the reaction of phenol and formaldehyde under basic conditions.

Yan Meng and colleagues have also reported a simple and reproducible synthesis of highly ordered and stable mesoporous polymers using an amphiphilic surfactant-templating approach, inspired by the assembly of mesoporous silica. In their study, they utilized resol (phenol/formaldehyde) as an affordable precursor that can form three-connected covalently bonded zeolite-like frameworks through thermopolymerization. By employing F127 ($EO_{106}PO_{70}EO_{106}$) as a template, they obtained a 2D hexagonal (p6m) structure, while a 3D cubic (*Im*3*m*) structure was achieved using P123 ($EO_{20}PO_{70}EO_{20}$) as the template. In addition, they were able to prepare a lamellar mesostructure using P123 as well. The resulting ordered mesoporous carbon frameworks retained the same structures as the parent polymers and exhibited high surface areas and uniform pores. The carbon framework walls were amorphous, relatively thick (approximately 7 nm), and highly stable even at temperatures exceeding 1400 °C (Fig. 3.34) [67].

Yan Meng conducted a systematic study on the synthesis of highly ordered mesoporous polymer resin and carbon structures using copolymers PEOPPO-PEO

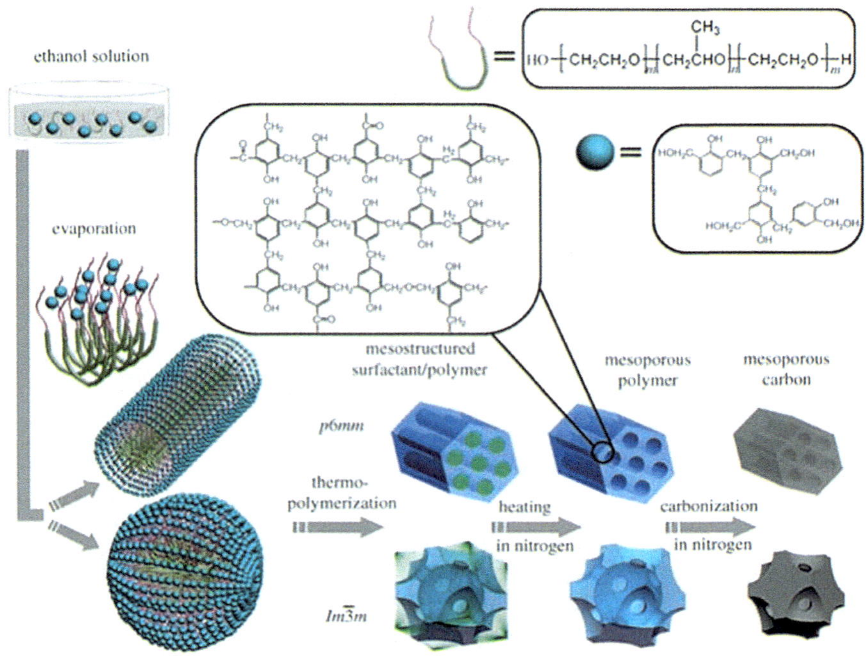

Fig. 3.34 Schematic representation of the procedure used to prepare mesoporous polymers and carbon frameworks [67]

and resols through an EISA (Evaporation-Induced Self-Assembly) strategy. They proposed a five-step procedure to synthesize C-FDU-x mesoporous carbon, as depicted in Fig. 3.35. The steps involved in the synthesis are as follows: Synthesis of resol: Phenol is polymerized with formaldehyde under basic conditions to produce the resol, which serves as the carbon precursor. Formation of surfactant/resol complex and assembly of mesostructures: The resol is mixed with the amphiphilic copolymer, forming a complex. This complex undergoes self-assembly, guided by the copolymer, to form ordered mesostructures. Curing of resol: The resol is cured through thermopolymerization, which involves heating the mixture to a specific temperature to promote cross-linking and solidification of the polymer structure. Removal of the template: The templating copolymer is selectively removed, typically by solvent extraction or calcination, leaving behind a porous resin structure.

Carbonization: The porous resin is subjected to high-temperature carbonization, where it undergoes a thermal treatment process in an inert atmosphere to convert the resin into mesoporous carbon. Through this multi-step process, highly ordered mesoporous carbon structures with uniform pore sizes and high surface areas can be obtained [68].

Dongyuan Zhao conducted a study on the synthesis of 3D face-centered cubic ($Fd\overline{3}m$) mesostructures, known as FDU-17, using a phenolic resol precursor and

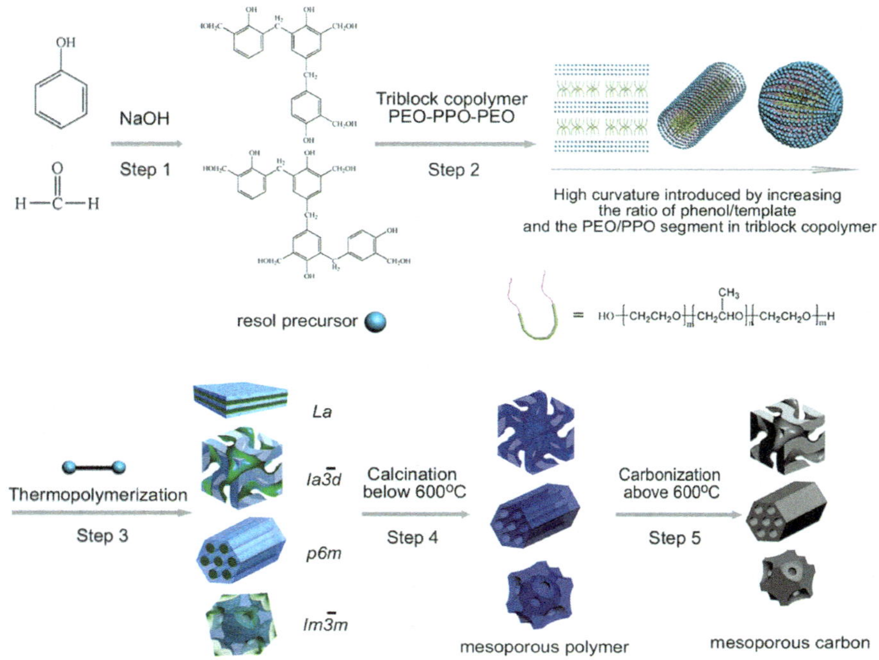

Fig. 3.35 Scheme for the preparations of the ordered mesoporous polymer resins and carbon frameworks [68]

a reverse amphiphilic triblock copolymer ($PO_{53}EO_{136}PO_{53}$) as a template through the evaporation-induced self-assembly (EISA) method. The resulting polymeric and carbon materials exhibited a stable and highly ordered mesostructure. In this study, a phenolic resol precursor was used as the carbon source, and the reverse amphiphilic triblock copolymer ($PO_{53}EO_{136}PO_{53}$) served as a template. The EISA method was applied, involving solvent evaporation to induce the self-assembly of the template and carbon precursor. The resulting mesostructures, named FDU-17, displayed a 3D face-centered cubic ($Fd\bar{3}m$) arrangement. These materials possessed open polymer and carbon frameworks, demonstrating stability and a highly ordered mesostructure. This synthesis approach demonstrates the ability to create stable and well-ordered mesostructures using a phenolic resol precursor and reverse amphiphilic triblock copolymer as a template, showing promise for various applications (Fig. 3.36) [69].

An-Hui Lu developed a simple and efficient one-pot synthesis approach for the fabrication of ordered mesoporous carbon through self-assembly in an aqueous phase under mild reaction conditions. The synthesis method involved using F127 as the structure directing agent, glutamic acid as the catalyst, and resorcinol and formaldehyde as carbon precursors. The pH of the reaction solution was controlled at 3, and phase separation occurred within 10 h at 60 °C. Following further aging at 90 °C for

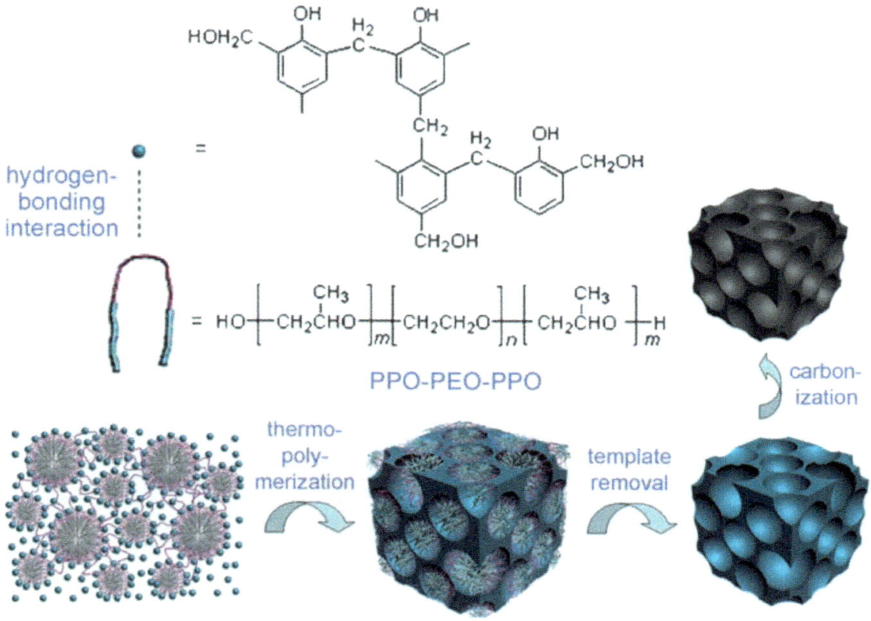

Fig. 3.36 Organic–organic self-assembly process of the triblock copolymer PPO-PEO-PPO with a phenolic resol precursor to form 3D facecentered-cubic (Fd $\overline{3}$ m) polymers and carbon mesostructures with bimodal architectural pores [69]

48 h, a glass-like bright red solid was formed. Carbonization of the initial hydrogel resulted in the formation of ordered mesoporous carbon (Fig. 3.37) [70].

The study revealed that the degree of mesostructural ordering in the synthesized mesoporous carbon was influenced by the amount of glutamic acid used. The obtained mesoporous carbon exhibited a BET surface area of approximately 720 m^2/g and was found to be free from any halogen impurities.

Guang-Ping Hao conducted a study on the rapid and scalable synthesis of crack-free and nitrogen-doped carbon monoliths with fully interconnected macropores and an ordered mesostructure using the soft-template method (Fig. 3.38). The carbon precursor utilized was resorcinol and formaldehyde-based polymers, while the structural directing agent was the triblock copolymer Pluronic F127. Additionally, the organic base lysine served as both the polymerization catalyst and mesostructure assembly promoter. The resulting carbon monoliths exhibited high surface area and macropore volume, reaching values of up to 600 m^2/g and 3.52 cm^3/g, respectively. Furthermore, subjecting the carbon monoliths to steam activation led to a significant increase in surface area, reaching 2422 m^2/g, while retaining the ordered mesostructure. The nitrogen-doped hierarchical carbon monoliths with their multi-level continuous pore system, high surface area, and nitrogen doping show potential for applications in CO_2 capture, separation, capacitors, and catalysis [71].

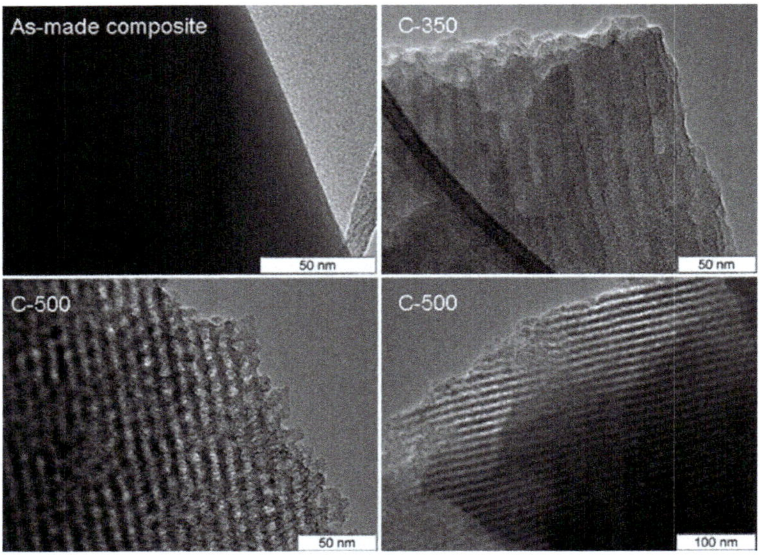

Fig. 3.37 TEM images of as-made composite and samples pyrolyzed at 350 and 500 °C [70].

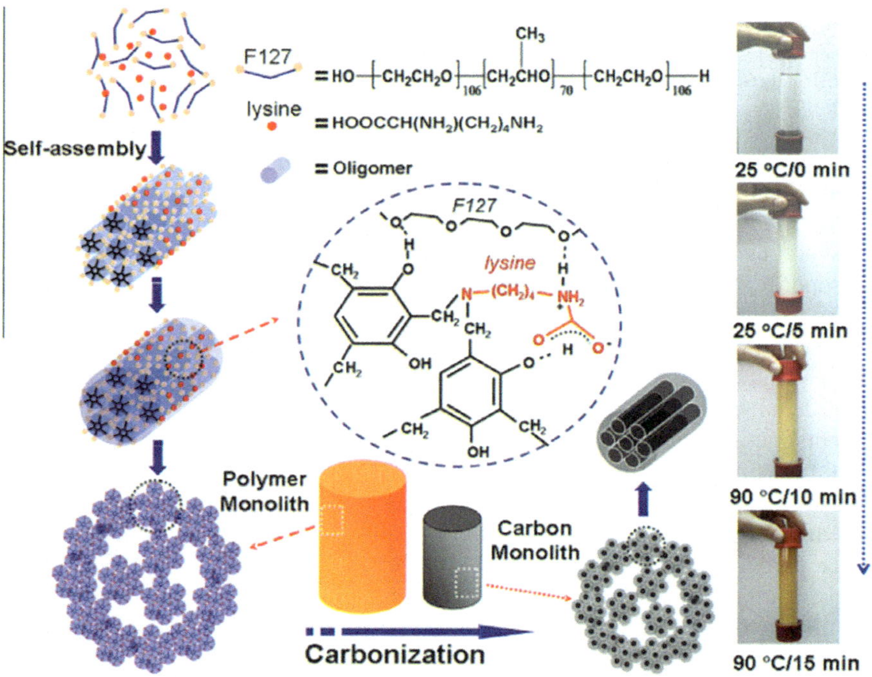

Fig. 3.38 Schematic of the rapid synthesis of N-doped a hierarchically porous carbon monolith with an ordered mesostructured [71]

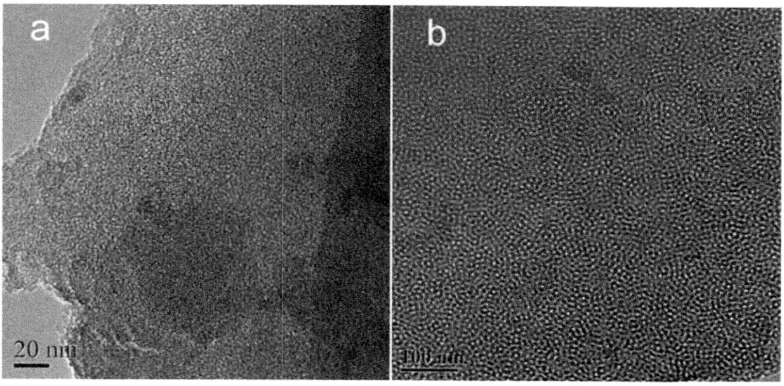

Fig. 3.39 TEM images of mesoporous carbons synthesized at F127/RF of 0.2 (**a**) and 0.3 (**b**) [72]

Donghui Long conducted a study on the synthesis of mesoporous carbons using resorcinol-furfural oligomers as carbon precursors and the triblock copolymer F127 as templates (Fig. 3.39). The quality of interactions between F127 and the oligomers, which is influenced by the physicochemical properties of the oligomers, plays a crucial role in determining the resulting mesostructures. It was observed that, at a constant F127/RF ratio, there is a transition in the mesostructure from Im3m to p6m to disorder as the degree of polymerization of the oligomers increases. This transition is attributed to a decreased thermodynamic driving force for the oligomers to mix with the poly(ethylene oxide) (PEO) blocks of F127. Furthermore, decreasing the F/R ratio, or the ratio of F127 to resorcinol-furfural oligomers, was found to improve the hydrophilic properties of the oligomers. This enhancement in hydrophilicity favored the formation of highly ordered mesostructures in the resulting mesoporous carbons [72].

Qian et al. proposed a simple method for the rational design of ordered mesoporous carbon structures based on the molar ratio of hydrophilic and hydrophobic sections (nH/nL). By introducing PPO-PEO-PPO-type or PEO-PPO-PEO-type copolymers with low PEO segments content into the templating synthesis process using PEO-PPO-PEO copolymers with high-content PEO blocks, highly ordered cubic and hexagonal carbon mesostructures can be achieved (Fig. 3.40). Upon carbonization, the resulting ordered mesoporous carbon products exhibit high surface areas (495–777 m^2/g), large pore volumes (0.32–0.58 cm^3/g), and uniform pore sizes (2.5–5.1 nm) [73].

Kubo et al. employed a direct sustainable hydrothermal carbonization (HTC) and soft templating approach to synthesize ordered porous carbonaceous materials. The soft template F127 and a small amount of trimethyl benzene as a swelling agent were used during the HTC process of D-fructose at 130 °C. The removal of the template was carried out at 550 °C under N_2 atmosphere, resulting in the

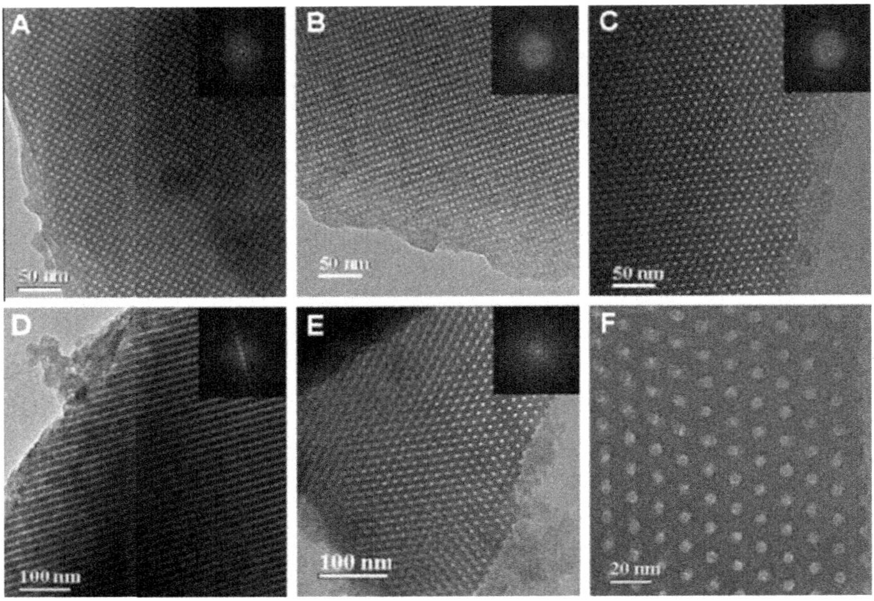

Fig. 3.40 TEM images of mesoporous carbon R-F127–1.2 viewed along the **a** [1 0 0], **b** [1 1 0] and **c** [1 1 1] directions and mesoporous carbon R-F127–9.1 viewed along the **d** [1 1 0] and **e**, **f** [0 0 1] directions. Insets are corresponding FFT (fast Fourier transform) diffractograms [73]

formation of highly ordered mesoporous carbon. High-resolution transmission electron microscopy (HRTEM) images revealed the presence of a long-range regularly ordered pore structure. Small-angle X-ray scattering (SSAXS) analysis of the obtained material showed the retention of the near perfect cubic Im3m symmetry observed in the parent composite. N_2 sorption analysis indicated a nonreversible microporous type I isotherm, suggesting possible structural changes in pore wall dimensions during sorption processes. The pore size distribution exhibited a distinct peak at 0.9 nm and a broader shoulder centered around 2 nm (Fig. 3.41) [74].

The synthesis of mesoporous carbon materials using the self-assembly process via a soft-templating approach has made significant advancements in the field of nanomaterials. These materials possess unique and distinctive characteristics that make them suitable for various applications, such as adsorption, catalysis, sensing, and energy storage and conversion. The discovery of mesoporous carbon through a simple nano hard-templating approach and self-assembly process has ushered in a new era for the synthesis of advanced carbon nanostructures. This strategy holds great potential for generating a new series of advanced nanostructures, opening up exciting opportunities in areas such as petrochemicals, adsorption and separation, and energy storage and conversion.

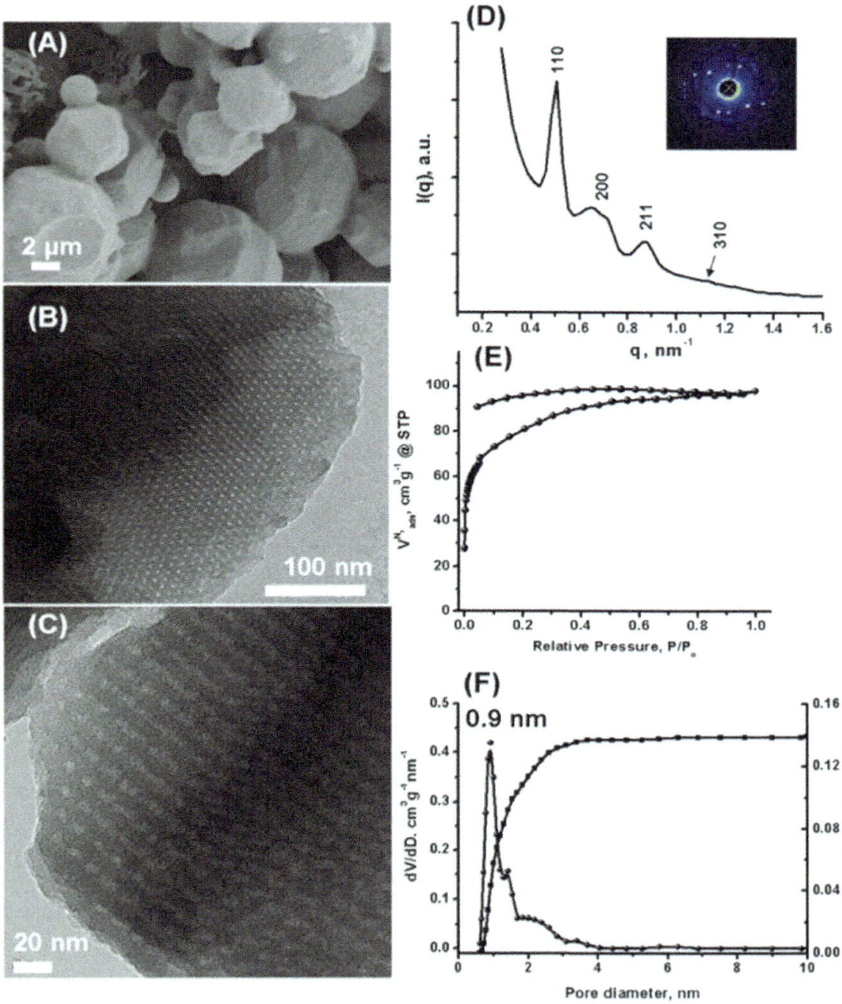

Fig. 3.41 **a** SEM, **b** TEM, and **c** HRTEM micrographs, **d** synchrotron SAXS pattern (inset; 2D scattering pattern), **e** N_2 sorption isotherm, and **f** QSDFT pore size distribution of assynthesized C-MPG1-meso [74]

3.3 Multiple Template Methods

Before delving into the investigation of morphologies achieved through multiple templating, it is important to highlight a few key points. Firstly, the overall synthesis scheme for multiple templating closely resembles that of templating with a single hard template. However, depending on the chosen soft templating approach, additional steps may be necessary. For instance, in the case of evaporation-induced self-assembly (EISA), solvent evaporation is required after infiltration to facilitate the formation of a liquid crystalline phase. Secondly, the combination of soft and hard templating can lead to significant alterations in the resulting templated mesostructured.

The subsequent sections are organized according to the type of hard template employed in the multiple templating synthesis. These include macroscopic templates such as biological materials, colloidal crystals, three-dimensionally ordered macroporous (3DOM) structures, foams, and anodized aluminum oxide (AAO) membranes. In situ templates and discrete particles are also covered. Each subsection highlights the various compositions that can be achieved through multiple templating. For certain sections, particularly those where confinement effects play a significant role (such as colloidal crystals and AAO membranes), emphasis is placed on the factors that influence the morphology of the mesopores. Additionally, when relevant, the applications of multiple templated materials are discussed to provide a broader understanding of the motivations behind their synthesis.

Ruili Liu reported a novel approach for the synthesis of unconventional freestanding ordered mesoporous carbon (OMC) sheets, referred to as OMCSs, using a combination of soft and hard templates. In this method, MgAl-layered double hydroxide (MgAl-LDH) was employed as the hard template, while triblock copolymer F127 served as the soft template, and phenolic resols acted as the carbon sources. The interactions between MgAl-LDH and the resol-F127 monomicelles were found to significantly influence the orientation of the mesopores in the OMCSs, enabling the fabrication of OMCSs with vertically or horizontally aligned mesopore arrays. This structural control subsequently impacted the electrochemical energy storage properties of the OMCSs in supercapacitors with different configurations. In particular, OMCSs with vertically aligned mesopore arrays exhibited outstanding energy density (28.9 Wh kg^{-1}) and power density (288 kW kg^{-1}) in an all-solid-state supercapacitor (ASSS) with two face-to-face electrodes, comparable to state-of-the-art supercapacitors based on OMCs (Fig. 3.42) [75].

Hu et al. proposed a method that combines the advantages of hard-template and soft-template approaches in the synthesis of ordered mesoporous carbons (OMCs). This hybrid method was chosen because the soft-templated frameworks offer improved stability during nanoparticle loading and electrochemical cycling processes. The researchers synthesized three types of OMCs, namely FDU-15, CMK-8, and OMCNW, using the soft-template, hard-template, and soft-hard dual-template methods, respectively. In addition, they successfully prepared OMC/Fe_2O_3 nanowires using this combined method. The resulting material exhibited a high pore volume, excellent structural stability, and enhanced electrical conductivity. This approach holds promise for the development of advanced materials for various applications (Fig. 3.43) [76].

Zhang et al. reported the synthesis of a three-dimensionally ordered macro-/mesoporous carbon (3DOM-mC) as a sulfur supporter for rechargeable lithium/sulfur (Li/S) batteries. The 3DOM-mC material was prepared using a dual-templating method, where a silica crystal served as the hard template, triblock copolymer $EO_{20}PO_{70}EO_{20}$ acted as the soft template, and resol (phenol/formaldehyde) was used

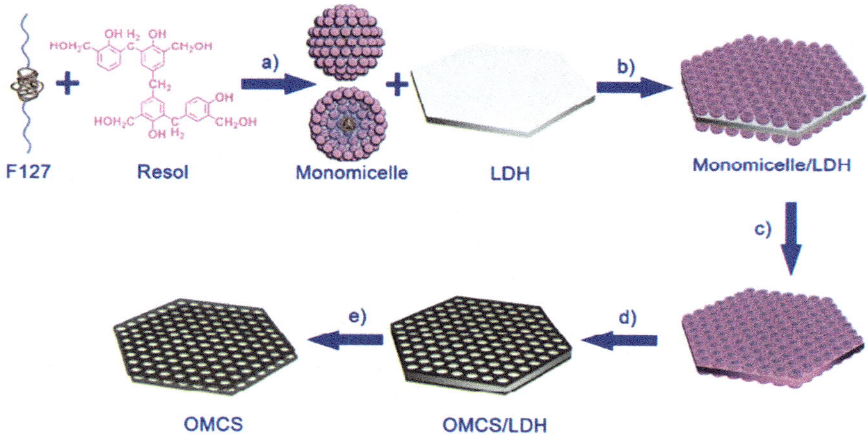

Fig. 3.42 Schematic illustration of the fabrication process for the OMCSs [75]

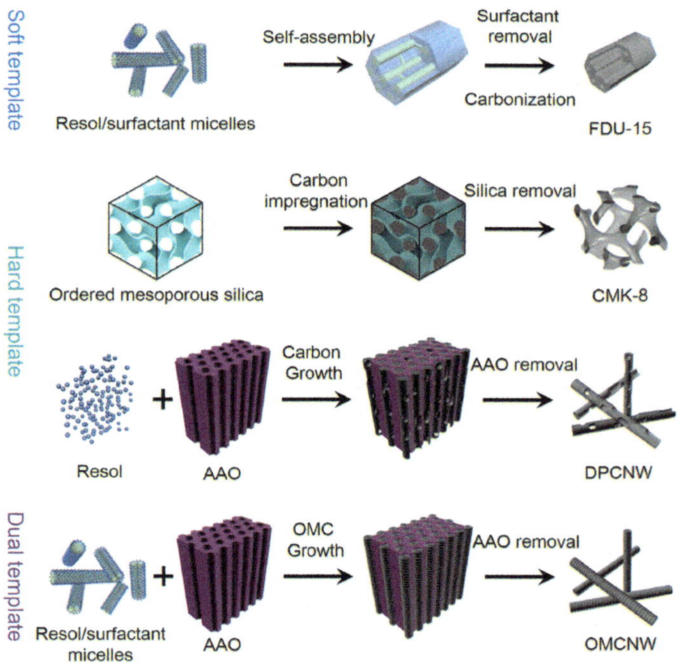

Fig. 3.43 Schematic illustrations of the proposed dual-template method in comparison with soft- or hard-template methods for the controlled synthesis of porous carbon frameworks [76]

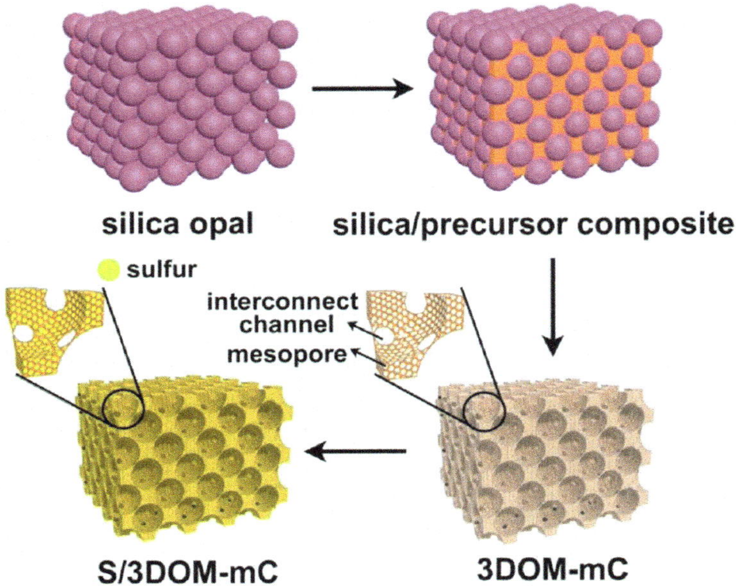

Fig. 3.44 Schematic of the preparation of the S/3DOM-mC [77]

as the carbon source. The resulting 3DOM-mC exhibited a large BET surface area of 818.5 $m^2\ g^{-1}$ and possessed ordered interconnected macropores with ordered mesopores on the macropore walls. This unique structure makes it a promising candidate for applications in Li/S batteries, where it can enhance the performance and stability of the sulfur cathode (Fig. 3.44) [77].

Aibing Chen conducted the synthesis of macro-mesoporous carbon materials and hollow core/mesoporous shell (HCMS) carbon spheres using an aqueous route with monodisperse silica spheres as the hard template and F127 as the soft template. By adjusting the reaction conditions and the SiO_2/F127 ratio, they were able to fabricate ordered macro-mesoporous carbon materials, disordered macro-mesoporous carbon materials, and HCMS carbon spheres. The resulting HCMS carbon spheres exhibited notable characteristics, including a high specific surface area of 972 $m^2\ g^{-1}$, a large pore volume of 1.27 $cm^3\ g^{-1}$, and a thin shell thickness of 5 nm. These unique hollow structures make them highly promising for applications in supercapacitors, as they offer a high capacitance and excellent performance (Fig. 3.45) [78].

Q. L. Kong introduced a novel approach known as the "soft to hard templating" strategy for the synthesis of N-doped mesoporous carbon nanospheres (MCNs) with excellent dispersion. This approach involves the copolymerization of a carbon source (dopamine) and a silica source (tetraethyl orthosilicate). The resulting MCNs possess desirable properties that make them highly suitable for electrochemical supercapacitor applications. In particular, they exhibit a capacitance of 223 and 140 F/g at current densities of 0.5 and 10 A/g, respectively, in a 1 mol/L KOH electrolyte.

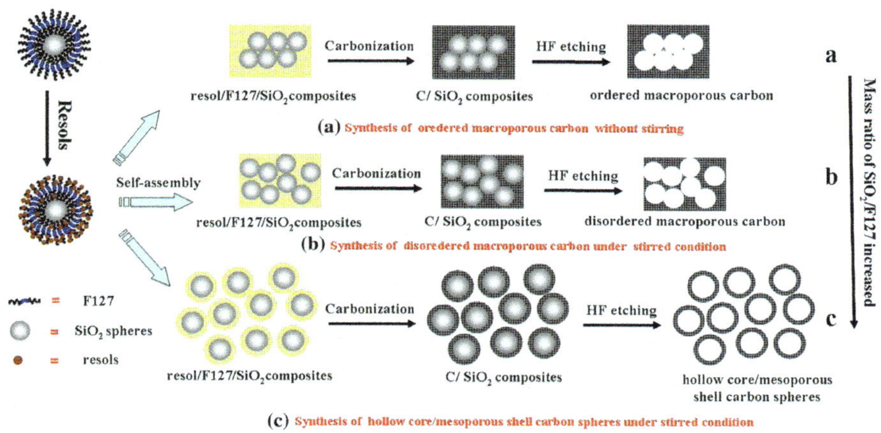

Fig. 3.45 Schematic illustration for the synthesis of carbon materials [78]

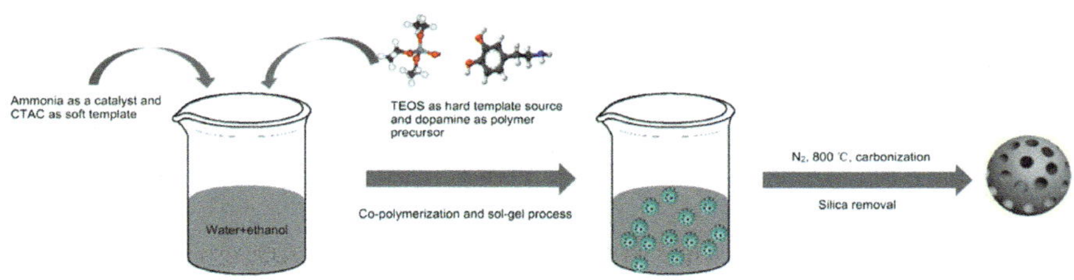

Fig. 3.46 Schematic illustration of the synthesis of MCNs [79]

These promising results demonstrate the potential of the nitrogen-doped MCNs as high-performance electrode materials for supercapacitors. Additionally, the unique characteristics of the MCNs open up possibilities for their application in various fields (Fig. 3.46) [79].

Aibing Chen developed a synthesis method to prepare mesoporous tubular carbon using Halloysite as a cost-effective and abundant hard template. In this approach, resin was used as the carbon precursor, and F127 served as the soft template. The resulting mesoporous tubular carbon material exhibited a replicated tubular structure similar to that of Halloysite. The incorporation of F127 not only introduced abundant pore structure but also significantly enhanced the surface area of the carbon material. This synthesis strategy offers a promising route to fabricate mesoporous tubular carbon materials with improved properties and potential applications in various fields (Fig. 3.47) [80].

N. W. Li conducted a study on the synthesis of hierarchically ordered macro-/mesoporous polymer resins and macro-/mesoporous carbon monoliths using a combination of hard and soft templating techniques (Fig. 3.48). In this method, SiO_2 opal was used as the hard template to create the macropores, while the amphiphilic

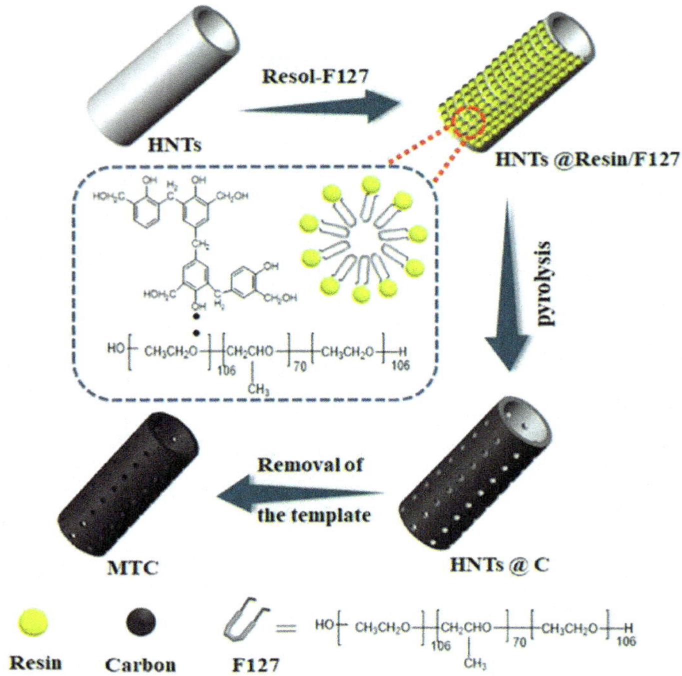

Fig. 3.47 The formation mechanism of the mesoporous tubular carbon [80]

triblock copolymer PEO-PPO-PEO served as the soft template for the mesopores. Phenolic resin was used as the precursor material for both the polymer and carbon phases. The resulting hierarchical macro-/mesoporous frameworks exhibited highly regular arrays of uniform macropores surrounded by walls containing mesoporous structures. The size of the mesopores increased as the heat-treatment temperature rose, up to 700 °C. This study demonstrated the successful restriction of framework shrinkage during the thermosetting and carbonization process by utilizing the SiO_2 hard template [81].

Tong Xue conducted research on the fabrication of Pt-loaded mesoporous carbon nanowires (Pt/MCNWs) using a confined soft template approach within the cylindrical pores of porous anodic alumina (PAA) membranes (Fig. 3.49). The synthesis involved the evaporation-induced self-assembly method, where the reduction of Pt and formation of MCNWs occurred simultaneously in a one-step process. The Pt/MCNWs exhibited a higher electrochemical active surface area and better catalytic performance compared to a commercial Pt/Vulcan XC-72 catalyst. This improved performance can be attributed to the excellent dispersion of Pt particles and the unique morphology of MCNWs. The Pt/MCNWs synthesized using the dual template method show great potential as electrocatalysts [82].

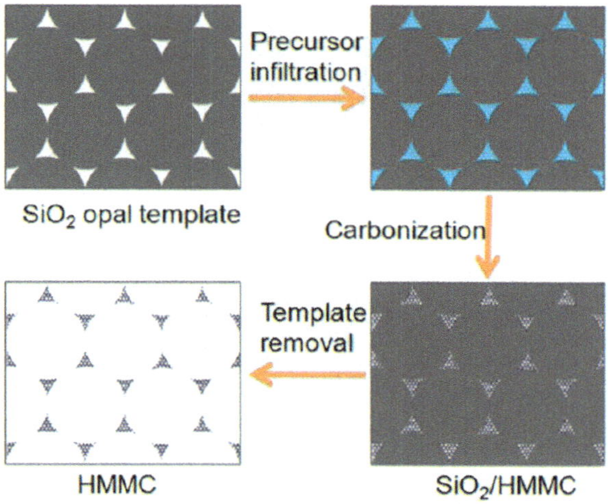

Fig. 3.48 Schematic Illustration for the Preparation of hierarchical macro-/mesoporous carbon (HMMC) [81]

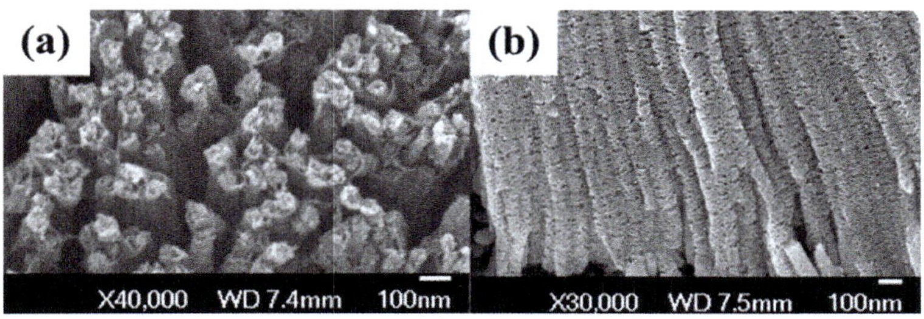

Fig. 3.49 e SEM images of 10 wt.% Pt/MCNW nanowires: **a** top view and **b** side view [82]

Shuwei Chen conducted research on the synthesis of mesoporous carbons using resorcinol and formaldehyde as carbon precursors. A poly(ethylene oxide)-poly(propylene oxide)-poly(ethylene oxide) triblock copolymer was employed as a soft template, while tetraethylortho silicate-generated silica and colloidal silica served as hard templates (Fig. 3.50). The resulting mesoporous carbons exhibited a spherical morphology with average particle diameters ranging from 2 to 7 μm, and the mesopores had a size of approximately 2 nm. Characterization of the materials was carried out using nitrogen adsorption–desorption isotherms and electron microscopy techniques (SEM and TEM). The mesoporous carbons were further tested as electrode materials for aqueous electric double-layer capacitors. They demonstrated a high specific surface area of 1,000 m^2/g and a mesopore volume of 0.86 cm^3/g. The specific capacitance achieved was 130 F/g, as determined through galvanostatic charging/discharging and cycle voltammetry techniques [83].

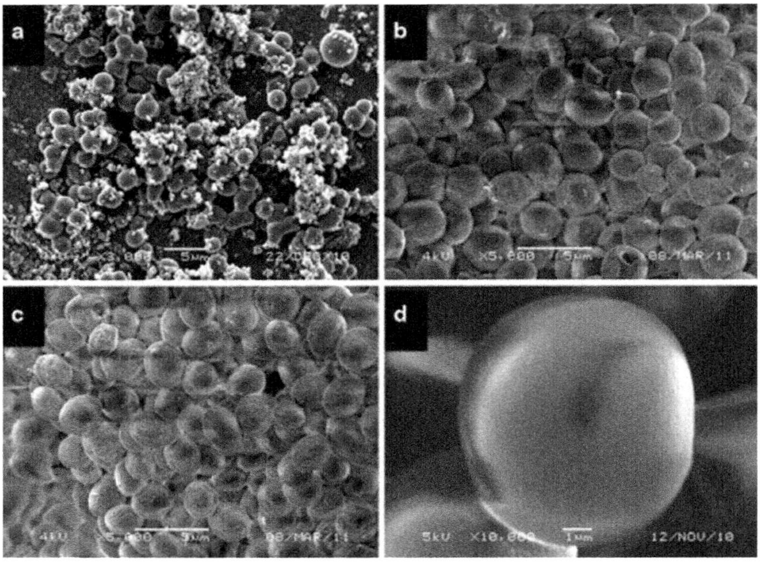

Fig. 3.50 The SEM images of spherical mesoporous carbon. MA (**a**), MCT (**b**) and MAT (**c, d**) [83]

Qiang Li conducted research on the synthesis of mesoporous carbon spheres with a hierarchical pore structure as an efficient electrode material for electrochemical double-layer capacitors (EDLCs). A dual-template strategy was employed, utilizing resol as the carbon source, and Pluronic F127 and Si-MCF spheres as the templates (Fig. 3.51). The resulting mesoporous carbons exhibited a hierarchical pore size range of approximately 3.5–60 nm, high specific surface areas of up to 1321 m^2/g, thin pore walls of around 2 nm, and an oxygen-enriched carbon surface. Electrochemical measurements demonstrated that the hierarchically mesoporous carbon materials exhibited a high specific capacitance of 208 F/g at 0.5 A/g in a 2 M H_2SO_4 aqueous solution. The materials also showed excellent rate capability and long cyclic stability, indicating their potential for use as electrode materials in energy storage devices [84].

Guiwang Zhao conducted a study on the direct preparation of mesoporous carbon nanofibers using a combination of hard and soft templates. Anodic aluminum oxide (AAO) membrane served as the hard template, while the block copolymer surfactant F127 (PEO_{106}-PPO_{70}-PEO_{106}) acted as the soft template (Fig. 3.52). The AAO membrane allowed for precise control of the length-to-diameter ratio of the resulting nanofibers. The AAO template was immersed in a nanocomposite sol, which was mixed and self-assembled with the low molecular weight resol in the presence of the surfactant. Subsequent carbonization under a nitrogen atmosphere yielded ordered mesoporous carbon nanofibers. Platinum (Pt) particles were loaded onto the mesoporous carbon nanofibers to create Pt/MCNFs catalysts. The Pt/MCNFs catalysts exhibited a significantly higher electrochemical active surface area for Pt particles in sulfuric acid solution compared to Pt/OMCs and Pt/C(E-TEK), reaching a peak value

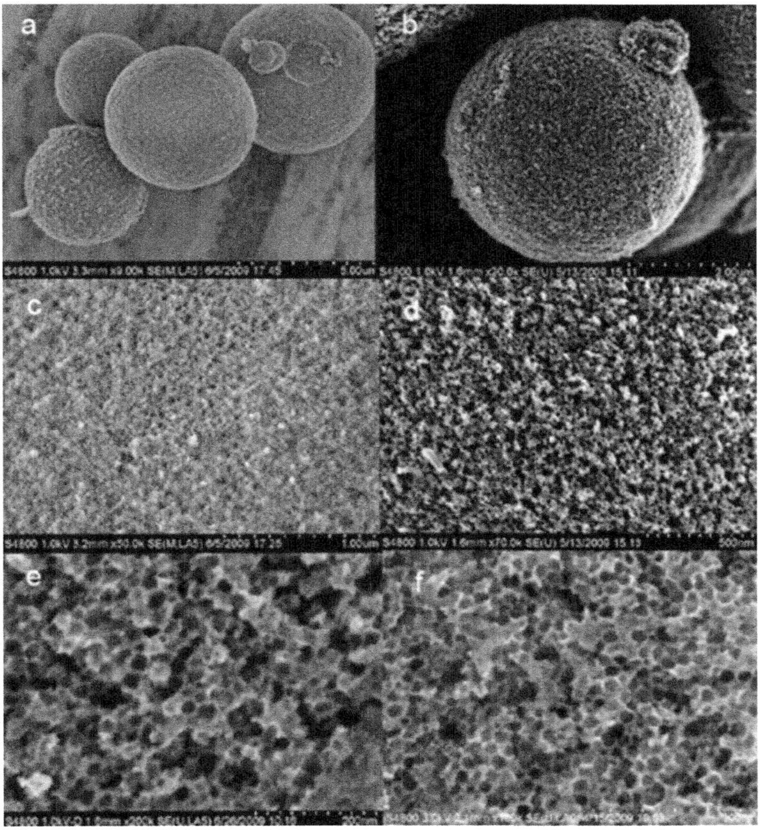

Fig. 3.51 FE-SEM images of **a**, **c** the silica template (Si-MCFs), **b**, **d**, **e** hierarchical MC-1 material synthesized by dual-template method, and **f** MC-2 material just replicated by the silica template [84]

of 235.2 m^2/g. Furthermore, the Pt/MCNFs catalysts displayed higher activity for methanol electrooxidation, demonstrated by their higher current density and higher ia/ib values in methanol solution, compared to the other two catalysts [85].

Kaixue Wang conducted a study on the preparation of well-aligned free-standing mesoporous carbon nanofiber arrays on silicon wafers using a confined self-assembly process within the pores of anodic aluminum oxide (AAO) membranes (Fig. 3.53). A soluble and low molecular weight phenolic resin derived from the reaction of phenol with formaldehyde was employed as the carbon source. The triblock copolymer, Pluronic F127 ($EO_{106}PO_{70}EO_{106}$, Mav) 12,600), served as the structural directing agent for creating the mesoporous phases. Transmission electron microscopy (TEM) analysis revealed the presence of hexagonal-arranged circular mesochannels at the edges of the carbon nanofibers, while columnar-oriented mesochannels, enveloped by the circular mesochannels, formed in the centers of certain nanofibers. These

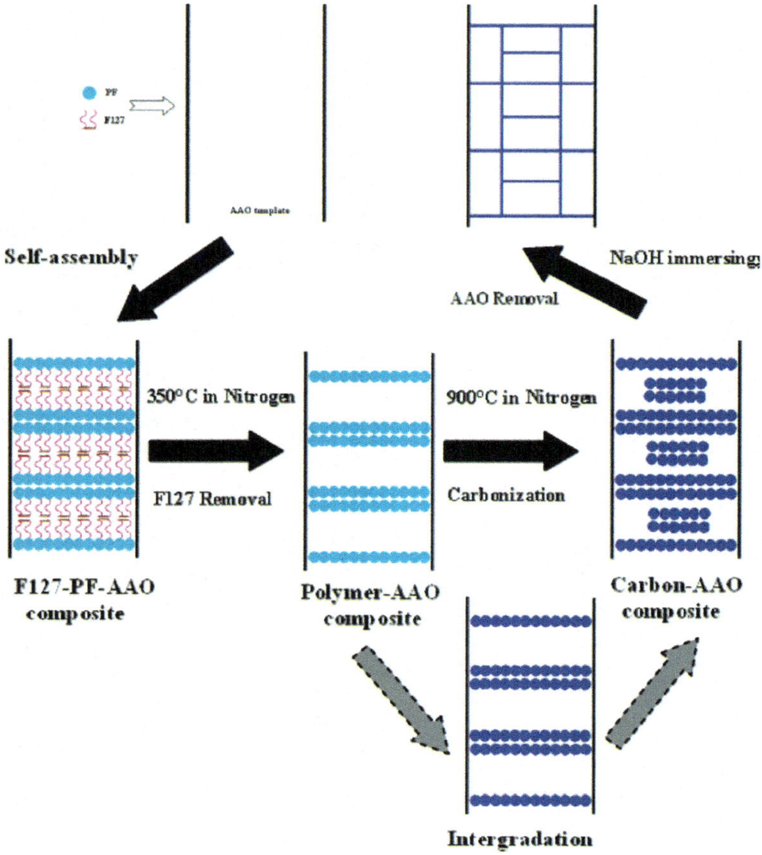

Fig. 3.52 Synthesis process of mesoporous nanofibers frameworks [85]

ordered arrays of carbon nanofibers fabricated on silicon substrates hold the potential for various nanoscale applications, including interconnects, sensors, and electrodes. The study is currently underway to measure the electrical properties of these nanofiber arrays [86].

3.4 Template Free Methods

Traditional methods for synthesizing mesoporous carbon involve the use of hard templates, such as metal oxides, metal salts, and mesoporous silica, or soft templates, such as surfactants and block copolymers. However, these methods have limitations in terms of complexity, cost, and scalability. The hard template method requires the removal of the template, often using harsh conditions like HF etching [87, 88],

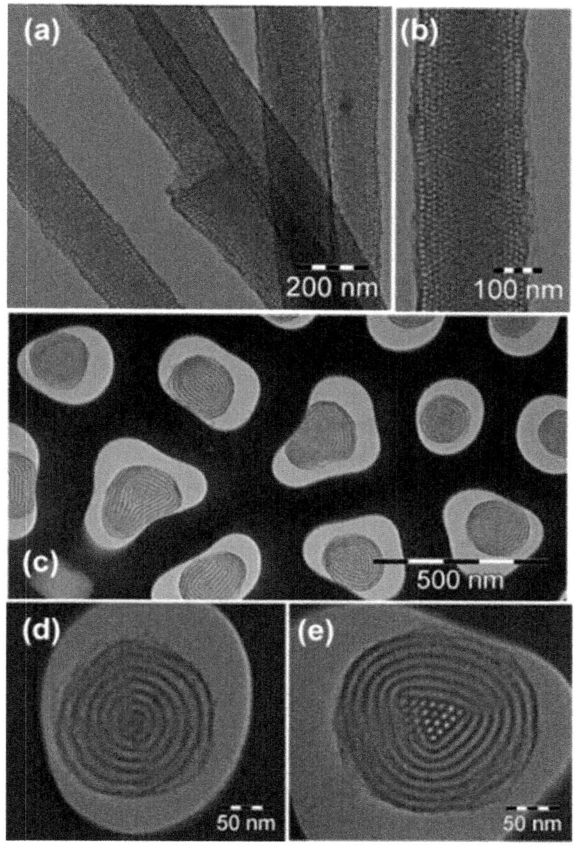

Fig. 3.53 Transmission electron microscopy images of mesoporous carbon nanofibers after being calcined at 600 °C for 3 h (**a**) and (**b**) side views of the carbon nanofibers showing the circular hexagonal-ordered mesoporous structuresl; (**c**–**e**) plan views of the mesoporous carb [86]

which can be expensive and environmentally stressful in large-scale production. On the other hand, the soft template method utilizes organic templates within a polymer network, which influences the formation of mesoporous structure during pyrolysis or carbonization. Template-free synthesis focuses on the influence of chain structure and organic functional groups on the porous structure of mesoporous carbon. However, achieving the desired porous structure from readily available thermoplastic polymer precursors can be challenging due to the complex nature of pyrolysis. Chemical modification of polymer chains through simple reactions offers a way to achieve the desired porous structure [89].

In a study by Li Sun, a novel and versatile precursor called protic ionic liquid ([Megl][HSO_4]) was used in combination with a double soft-template and solvent-free self-assembly method to synthesize nitrogen/sulfur-codoped hierarchically porous carbon materials (N/S-HPC). The N/S-HPC materials possessed an interconnected structure with micropores, mesopores, and macropores, high electrical conductivity, large specific surface area (SBET), and abundant nitrogen and sulfur

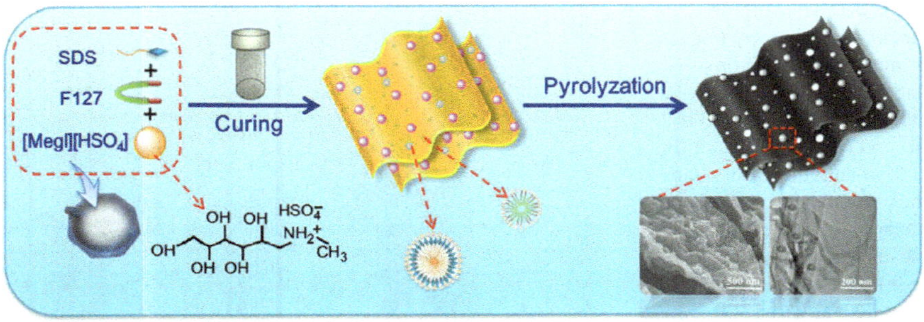

Fig. 3.54 The synthesis process of N/S-HPC [90]

content, making them suitable for high-performance supercapacitors. The precursor, [Megl][HSO_4], served as the carbon source, heteroatom source, and microporous forming agent, while F127 acted as the mesoporous soft-template and sodium dodecyl sulfate served as the macroporous soft-template. Through self-assembly and pyrolysis, the N/S-HPC materials were obtained. The materials exhibited stable cycling at a large scan rate, excellent rate capability at a high current density, and temperature resistance and flexibility across a wide range (Fig. 3.54) [90].

Ming-bo Wu developed a simple and template-free method to prepare mesoporous carbon (MC) from rice husk (RH). The method involved a combination of H_3PO_4 activation, pretreatment of the RH with a NaOH solution, and pre-oxidation in air (Fig. 3.55). The NaOH pretreatment served to remove silicon and disrupt the crystal structure of cellulose in the RH, which contributed to the formation of MC with a high surface area and mesoporosity. The resulting MC exhibited a specific surface area of 2009 m^2/g and a mesoporosity of 90.8%. It demonstrated excellent electrochemical performance, with a specific capacitance of 176 F/g at a current density of 50 mA/g, and a retention of 126 F/g at 1000 mA/g, indicating remarkable rate capability. Furthermore, the MC electrode maintained a stable specific capacitance of approximately 150 F/g at 200 mA/g, with no significant capacitance decay observed after 1000 cycles, demonstrating excellent electrochemical stability. Overall, the template-free synthesis of MC from rice husk via the described method offers a promising approach for producing high-performance carbon materials with desirable properties for energy storage applications [91].

Yanzhi Sun proposed a new green and recyclable synthesis method for porous carbon (Fig. 3.56). Aluminum hydroxide (Al(OH)3) and trimesic acid (BTC) were used as raw materials to obtain aluminum trimesate (referred to as Al-BTC) through their covalent reaction. Subsequently, the porous carbon was obtained by carbonization and dissolution processes, with the removal of aluminum oxide (Al_2O_3), which was recovered using the Bayer method for subsequent batches. Scanning electron microscopy (SEM) images revealed that the porous carbon exhibited a rugby-like morphology with a width of 400 nm and a length of 1000 nm, indicating that the porous carbon with a c/a ratio of 2.5 provided the largest specific surface area. The

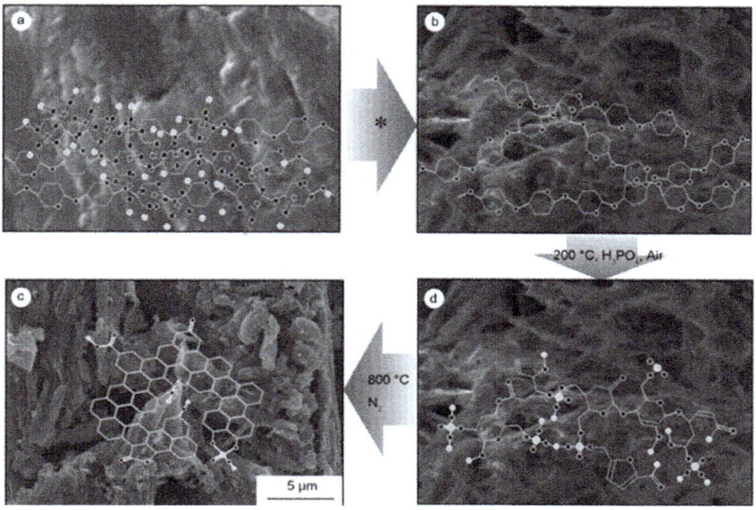

Fig. 3.55 A schematic illustration of MC synthesis from rice husk [91]

porous carbon demonstrated excellent durability with a capacitance decline of only 5.05% after 50,000 cycles at 50 A g − 1. Moreover, the porous carbon exhibited ultrafast charge acceptance, requiring only 4.4 s for a single charging process, making it promising for electric vehicle applications [52]

Fig. 3.56 **a** Diagram from rugby-like Al-BTC to rugby-like PC and **b** diagram of preparation route of PC[52]

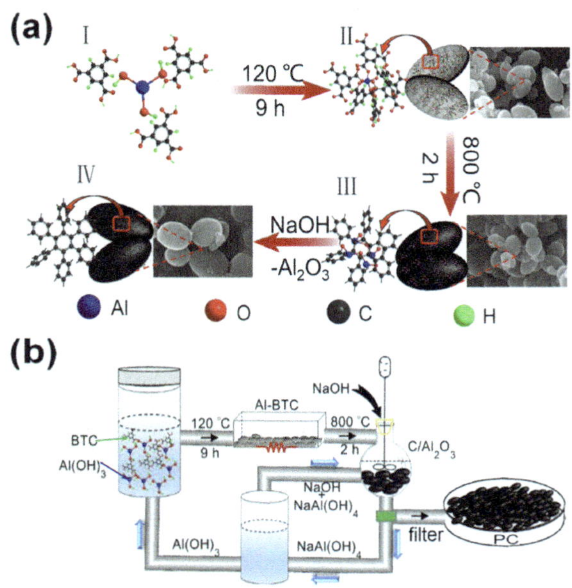

Mojtaba Jahanbakhshi synthesized mesoporous carbon foam (MCF) using the Pechini method, which is a facile and template-free approach (Fig. 3.57). This method represents an improvement over the traditional Pechini method, with significant changes made to the metal precursor and final stage of the process. The resulting carbon nanocomposite serves as a suitable matrix for immobilizing Mb (myoglobin). Furthermore, dispersing MCF in salep enhances its biocompatibility and greatly enhances the bio-electrochemical capabilities of MCF. The developed biosensor exhibits good sensitivity, excellent stability, and acceptable repeatability and reproducibility, making it a promising platform for various applications [92] [132].

Yu Wang developed a method for fabricating 3D metal–organic chelate foams using gas blowing and coordination assembling under solvent- and template-free conditions (Fig. 3.58). In this process, zinc nitrate hexahydrate was decomposed, leading to the assembly of metal clusters and organic ligands. Simultaneously, the release of gas during the decomposition caused the formation of macropores in the chelate network, resulting in the formation of 3D foam-like structures. Subsequently, the 3D chelate foams were directly pyrolyzed to produce 3D nitrogen-doped carbon

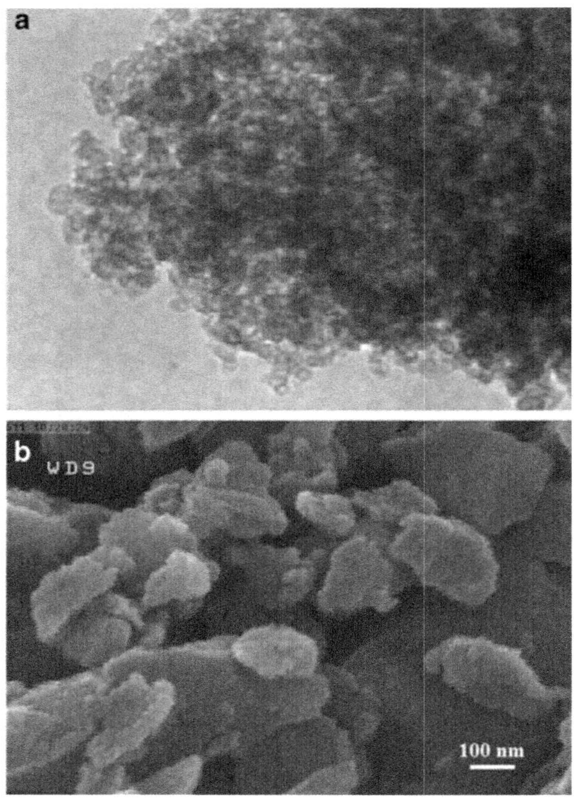

Fig. 3.57 **a** TEM image of the mesoporous carbon foam (MCF). **b** FE-SEM image of the MCF [92]

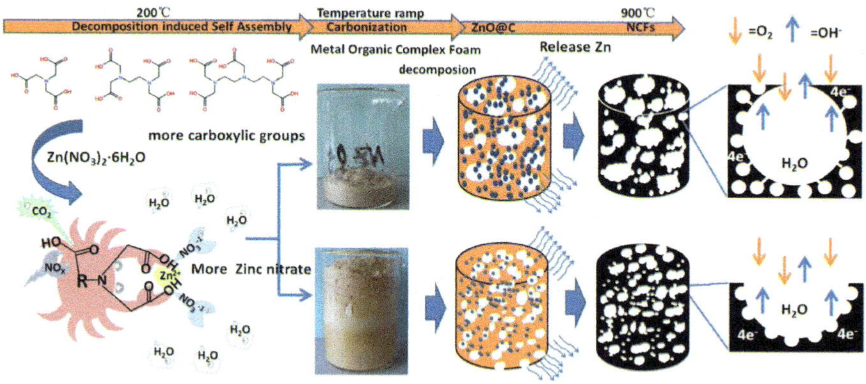

Fig. 3.58 Schematic illustration for the formation process of NCFs [93]

foams (NCFs). During pyrolysis, the ZnO present in the chelate foams acted as a porogen, generating mesopores as it was reduced to volatile metal Zn and released from the system at a high calcination temperature of 900 °C. Additionally, the Zn vapor further activated the carbon framework, leading to the development of additional porosity. The resulting 3D nitrogen-doped carbon foams exhibited excellent performance for oxygen reduction reactions (ORR) in alkaline media, with an onset potential that was 50 mV more positive compared to a commercial 20 wt.% Pt/C catalyst tested under the same conditions [93].

Heliang Zhou developed a template-free method to prepare mesoporous carbon spheres (MCS) with tunable porosity. The method involves a spray-drying process using chitosan as the carbon precursor and ethanol as the porosity tuning agent (Fig. 3.59). By adjusting the volume ratio of ethanol in the solvent for chitosan, MCSs with different porosity levels can be obtained. The resulting MCSs exhibit controllable surface areas, pore volumes, and bimodal pore size distribution. These MCSs were then used as substrates for the fabrication of sulfur/carbon composite cathode materials for lithium-sulfur batteries. The electrochemical performance of the MCSs with 50 wt.% sulfur loading was investigated [94].

Mingming Chen presented a straightforward and bottom-up approach for synthesizing mesoporous carbon without the need for additional templates (Fig. 3.60). This strategy involves the sodium-assisted carbonization of bromobenzene and consists

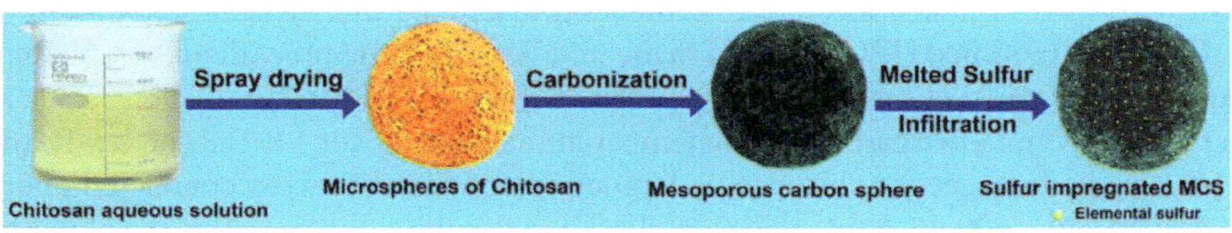

Fig. 3.59 Illustration for the formation process of the S/MCS composite spheres [94]

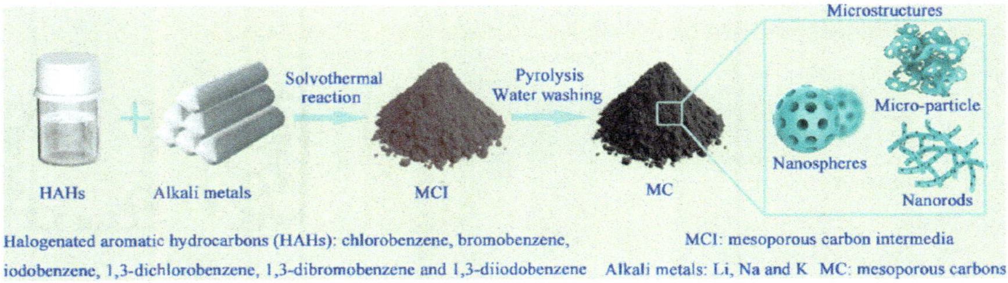

Fig. 3.60 The synthesis process of MCs prepared by using alkali metals (Li, Na and K)-assisted carbonization of six kinds of HAHs [95]

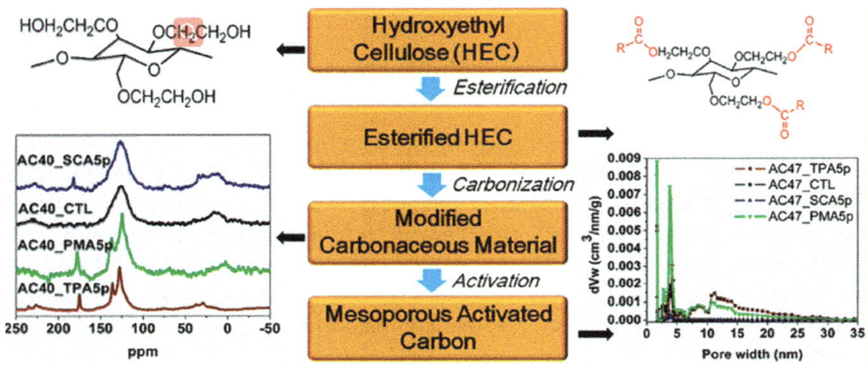

Fig. 3.61 The synthesis process of mesoporous activated carbon [89]

of three steps: solvothermal reaction, pyrolysis, and water washing. Through this process, bromobenzene is directly converted into porous carbon materials. The obtained material, characterized by nanorod structures, exhibits a high surface area of 1902 $m^2\ g^{-1}$, a large pore volume of 1.21 $cm^3\ g^{-1}$, and a high degree of graphitization. The pore formation mechanism is attributed to the "in situ templating" effect of NaBr. The resulting material, when pyrolyzed at 800 °C, demonstrates remarkable rate performance, long lifespan, and high power density [95].

Ming Sun investigates a template-free method for synthesizing a mesoporous activated carbon (AC) matrix by creating interchain bonding in its precursor (Fig. 3.61). Hydroxyethyl cellulose (HEC) is used as the precursor, and its polymer chains are covalently cross-linked through esterification reactions. In a typical preparation of AC, 10 g of modified HEC is placed in the middle of a quartz tube (diameter: 50 mm, length: 1200 mm). The sample is then purged with Ar gas (500 mL/min) and heated at a rate of 6 °C/min to a carbonization temperature of 400 °C for 1 h. Subsequently, the resulting carbonaceous substance undergoes an activation process conducted under a flow of CO_2 (500 mL/min) at 700 °C for 2 h. After the activation step, the obtained AC is cooled down under an Ar atmosphere until it reaches room temperature [89].

References

1. Hyeon T, Han S, Sung YE et al (2003) High-performance direct methanol fuel cell electrodes using solid-phase-synthesized carbon nanocoils. J Ind Eng Chem 42:4352–4356
2. Kresge CT, Leonowicz ME, Roth WJ et al (1992) Ordered mesoporous molecular sieves synthesized by a liquid-crystal template mechanism. Nature 359:710–712
3. Beck JS, Vartuli JC, Roth WJ et al (1992) A new family of mesoporous molecular sieves prepared with liquid crystal templates. J Am Chem Soc 114:10834–10843
4. Huo Q, MargoleseG DI, Stucky D (1996) Surfactant control of phases in the synthesis of mesoporous silica-based materials. Chem Mater 8:1147–1160
5. Zhou W, Hunter HMA, Wright PA et al (1998) Imaging the pore structure and polytypic intergrowths in mesoporous silica. J Phy Chem B 102:6933–6936
6. Zhao D, Huo Q, Feng J et al (1998) Nonionic triblock and star diblock copolymer and oligomeric surfactant syntheses of highly ordered, hydrothermally stable, mesoporous silica structures. J Am Chem Soc 120:6024–6036
7. Sakamoto Y, Díaz I, Terasaki O et al (2002) Three-dimensional cubic mesoporous structures of sba-12 and related materials by electron crystallography. J Phys Chem B 106:3118–3123
8. Zhao D, Feng J, Huo Q et al (1998) Triblock copolymer syntheses of mesoporous silica with periodic 50 to 300 angstrom pores. Science 279:548–552
9. Fan J, Yu C, Gao F et al (2003) Cubic mesoporous silica with large controllable entrance sizes and advanced adsorption properties. Angew Chem Int Ed Engl 42:3146–3150
10. Liang C, LiS Z, Dai (2008) Mesoporous carbon materials: synthesis and modification. Angew Chem Int Ed Engl 47:3696–3717
11. Kleitz F, ChoiR SH, Ryoo R (2003) Cubic Ia3d large mesoporous silica: synthesis and replication to platinum nanowires, carbon nanorods and carbon nanotubes. Chem Commun, 2136–2137
12. Ma TY, Lei L et al (2013) Direct synthesis of ordered mesoporous carbons. Chem Soc Rev 42:3977–4003
13. Schuster J, He G, Mandlmeier B et al (2012) Spherical ordered mesoporous carbon nanoparticles with high porosity for lithium–sulfur batteries. Angew Chem Int Ed 51:3591–3595
14. Seo Y, Kim K, Jung Y et al (2015) Synthesis of mesoporous carbons using silica templates impregnated with mineral acids. Micropor Mesopor Mater 207:156–162
15. Brandiele R, Poli F, Picelli L et al (2020) Nitrogen-doped mesoporous carbon electrodes prepared from templating propylamine-functionalized silica. ChemElectroChem 7:1914–1921
16. Banerjee R, Ghosh D, Satra J et al (2019) One step synthesis of a gold/ordered mesoporous carbon composite using a hard template method for electrocatalytic oxidation of methanol and colorimetric determination of glutathione. ACS Omega 4:16360–16371
17. Li K, Jie S, Li Y et al (2018) The synthesis of n, s-codoped ordered mesoporous carbon as an efficient metal-free catalyst for selective oxidation of arylalkanes. Catal Commun 112:39–42
18. Ryong R, Sang HJJ (1999) Synthesis of highly ordered carbon molecular sieves via template-mediated structural transformation. J Phys Chem B 103:7743–7746
19. Jinwoo L, Yoon S, Hyeon T et al (1999) Synthesis of a new mesoporous carbon and its application to electrochemical double-layer capacitors. Chem Commun, 2177–2178
20. Mizue K, Toshikazu T, Carlsson A et al (2002) Structural study of mesoporous mcm-48 and carbon networks synthesized in the spaces of mcm-48 by electron crystallography. J Phys Chem B 106:1256–1266
21. Vix-Guterl C, Boulard S, Parmentier J et al (2002) Formation of ordered mesoporous carbon material from a silica template by a one-step chemical vapour infiltration process. Chem Lett 2002:1062–1063
22. Kim JY, YoonJ SB, Yu S (2003) Template synthesis of a new mesostructured silica from highly ordered mesoporous carbon molecular sieves. Chem Mater 15:1932–1934
23. Li Z, Yan W, Dai S (2005) Surface functionalization of ordered mesoporous carbons–a comparative study. Langmuir 21:11999

24. Tian B, Che S, Liu Z et al (2003) Novel approaches to synthesize self-supported ultrathin carbon nanowire arrays templated by MCM-41electronic. Chem Commun, 2726–2727
25. Jun S, Joo SH, Ryoo R et al (2000) Synthesis of new, nanoporous carbon with hexagonally ordered mesostructure. J Am Chem Soc 122:10712–10713
26. Shin H J, Ryoo R, Kruk M et al (2001) Modification of SBA-15 pore connectivity by high-temperature calcination investigated by carbon inverse replication. Chem Commun, 349–350
27. Kim SS, Lee DK, Shah J et al (2003) Nanocasting of carbon nanotubes: in-situ graphitization of a low-cost mesostructured silica templated by non-ionic surfactant micelles. Chem Commun, 1436–1437
28. Kim J, Lee J, Hyeon T (2004) Direct synthesis of uniform mesoporous carbons from the carbonization of as-synthesized silica/triblock copolymer nanocomposites. Carbon 42:2711–2719
29. Kaneda M, Tsubakiyama T, Carlsson A et al (2002) Structural study of mesoporous MCM-48 and carbon networks synthesized in the spaces of MCM -48 by electron crystallography. J Phys Chem B 106:1256–1266
30. Liu X, Tian B, Yu C et al (2002) Room-temperature synthesis in acidic media of large-pore three-dimensional bicontinuous mesoporous silica with Ia3d symmetry. Angew Chem Int Ed 114:4032–4034
31. Suryavanshi UB, Ijima T, Hayashi A et al (2011) Simple methods for tuning the pore diameter of mesoporous carbon. Chem Commun 47:10758–10760
32. Kim TW, ParkR IS, Ryoo R (2003) A synthetic route to ordered mesoporous carbon materials with graphitic pore walls. Angew Chem Int Ed 42:4375–4379
33. Yang H, Yan LY et al (2004) A simple melt impregnation method to synthesize ordered mesoporous carbon and carbon nanofiber bundles with graphitized structure from pitches. J Phys Chem B 108:17320–17328
34. Li Z, Jaroniec M (2001) Colloidal imprinting: a novel approach to the synthesis of mesoporous carbons. J Am Chem Soc 123:9208–9209
35. Li Z, Jaroniec M (2003) Synthesis and adsorption properties of colloid-imprinted carbons with surface and volume mesoporosity. Chem Mater 15:1327–1333
36. Jang J, LimM B, Choi M (2005) A simple synthesis of mesoporous carbons with tunable mesopores using a colloidal template-mediated vapor deposition polymerization. Chem Commun, 4214–4216
37. Hampsey JE, Hu Q, Rice L et al (2005) A general approach towards hierarchical porous carbon particles. Chem Commun 28:3606–3608
38. Kyotani T, TsaiA L, Tomita A (1995) Formation of ultrafine carbon tubes by using an anodic aluminum oxide film as a template. Cheminform 7:1427–1428
39. Li J, Papadopoulos C, Xu JM et al (1999) Highly-ordered carbon nanotube arrays for electronics applications. Appl Phys Lett 75:367–369
40. Lee J, Sohn K, Hyeon T (2001) Fabrication of novel mesocellular carbon foams with uniform ultralarge mesopores. J Am Chem Soc 123:5146–5153
41. Oda Y, Fukuyama K, Nishikawa K et al (2004) Mesocellular foam carbons: aggregates of hollow carbon spheres with open and closed wall structures. Chem Mater 16:3860–3866
42. Hu Q, Pang J, Jiang N et al (2005) Direct synthesis of palladium-containing mesoporous carbon. Micropor Mesopor Mater 81:149–154
43. Han S, Sohn K, Hyeon T (2000) Fabrication of new nanoporous carbons through silica templates and their application to the adsorption of bulky dyes. Chem Mater 12:3337–3341
44. Jang J, Lim B (2002) Selective fabrication of carbon nanocapsules and mesocellular foams by surface-modified colloidal silica templating. Adv Mater 14:1390–1393
45. Chai GS, Yoon SB, Yu JS et al (2004) Ordered porous carbons with tunable pore sizes as catalyst supports in direct methanol fuel cell. J Phys Chem B 108:7074–7079
46. Fuertes AB (2003) Template synthesis of mesoporous carbons with a controlled particle size. J Mater Chem 13:3085–3088
47. Yu JS, YoonG SB, Chai S (2001) Ordered uniform porous carbon by carbonization of sugars. Carbon 39:1442–1446

48. Kang S, Yu JS, Kruk M et al (2002) Synthesis of an ordered macroporous carbon with 62 nm spherical pores that exhibit unique gas adsorption properties. Chem Commun, 1670–1671
49. Zakhidov AA, Baughman RH, Lqbal Z et al (1998) In: Diamond opal-replica photonic crystals and graphitic metallic photonic band gap structures: fabrication and properties, Aps March Meeting
50. Lee J, Kim J, Hyeon T (2003) A facile synthesis of bimodal mesoporous silica and its replication for bimodal mesoporous carbon. Chem Commun 9:1138–1139
51. Gierszal K, Jaroniec PM (2004) Novel pitch-based carbons with bimodal distribution of uniform mesopores. Chem Commun, 2576–2577
52. Sun Y, Guo S, Li W et al (2018) A green and template-free synthesis process of superior carbon material with ellipsoidal structure as enhanced material for supercapacitors. J Power Sour 405:80–88
53. Lee J, Yoon S, Oh SM (2000) Development of a new mesoporous carbon using an HMS aluminosilicate template. Adv Mater 12:359–362
54. Xia Y, Mokaya DR (2004) Ordered mesoporous carbon hollow spheres nanocast using mesoporous silica via chemical vapor deposition. Adv Mater 16:886–891
55. Florent M, Xue C, Zhao D et al (2012) Formation mechanism of cubic mesoporous carbon monolith synthesized by evaporation-induced self-assembly. Chem Mater 24:383
56. Bähr A, Diedenhoven J, Tüysüz H (2020) Cl2 adsorption and desorption over ordered mesoporous carbon materials as an indicator for catalytic phosgene formation. Chem Ing Tech 92:1508–1513
57. Tan H, Zhao Y, Xia W et al (2020) Phosphorus- and nitrogen-doped carbon nanosheets constructed with monolayered mesoporous architectures. Chem Mater 32:4248–4256
58. Saha D, Richards CP, Haines RG et al (2020) Soft-templating of sulfur and iron dual-doped mesoporous carbons: lead adsorption in mixtures. Molecules 25:403
59. Kueasook R, Rattanachueskul N, Chanlek N et al (2020) Green and facile synthesis of hierarchically porous carbon monoliths via surface self-assembly on sugarcane bagasse scaffold: influence of mesoporosity on efficiency of dye adsorption. Micropor Mesopor Mater 296:110005
60. Wang Y, Qiao M, Baikeli Y et al (2020) Soft-templated mesoporous carbon-modified glassy carbon electrode for sensitive and selective detection of aristolochic acids. J Hazard Mater 385:121550
61. Li W, Li B, Shen M et al (2020) Use of gemini surfactant as emulsion interface microreactor for the synthesis of nitrogen-doped hollow carbon spheres for high-performance supercapacitors. Chem Eng J 384:123309
62. Volosin AM, ChenD S, Seo K (2020) High-surface area mesoporous carbons from gel templating and inorganic-organic hybrid gel formation. J Solid State Chem 281:121040
63. Khan IA, Badshah A, Khan I et al (2017) Soft-template carbonization approach of mof-5 to mesoporous carbon nanospheres as excellent electrode materials for supercapacitor. Micropor Mesopor Mater 253:169–176
64. Zhao G, Zhao TS, Xu J et al (2017) Impact of pore size of ordered mesoporous carbon fdu-15-supported platinum catalysts on oxygen reduction reaction. Int J Hydrogen Energ 42:3325–3334
65. Liang C, Hong K, Guiochon GA et al (2004) Synthesis of a large-scale highly ordered porous carbon film by self-assembly of block copolymers. Angew Chem 116:5909–5913
66. Tanaka S, Nishiyama N, Egashira Y et al (2005) Synthesis of ordered mesoporous carbons with channel structure from an organic–organic nanocomposite. Chem Commun, 2125–2127
67. Meng Y, Gu D, Zhang F et al (2005) Ordered mesoporous polymers and homologous carbon frameworks: amphiphilic surfactant templating and direct transformation. Angew Chem 117:7215–7221
68. Meng Y, Gu D, Zhang F et al (2006) A family of highly ordered mesoporous polymer resin and carbon structures from organic-organic self-assembly. Chem Mater 18:4447–4464
69. Huang Y, Huaqiang C et al (2007) Formation of mesoporous carbon with a face-centered-cubicfdm structure and bimodal architectural pores from the reverse amphiphilic triblock copolymer PPO-PEO-PPO. Angew Chem 119:1107–1111

70. Lu A, Spliethoff B, Schuth F (2008) Aqueous synthesis of ordered mesoporous carbon via self-assembly catalyzed by amino acid. Chem Mater 20:5314–5319
71. Hao GP, Li WC, Wang S et al (2011) Lysine-assisted rapid synthesis of crack-free hierarchical carbon monoliths with a hexagonal array of mesopores. Carbon 49:3762–3772
72. Long D, Qiao W, Zhan L et al (2009) Effect of template and precursor chemistry on pore architectures of triblock copolymer-templated mesoporous carbons. Micropor Mesopor Mater 121:58–66
73. Qian X, Li H, Wan Y (2011) Structure design of mesoporous carbons by blending PEO–PPO–PEO-type and PPO–PEO–PPO-type amphiphilic block copolymers in organic–organic self-assembly. Micropor Mesopor Mater 141:26–37
74. Kubo S, White RJ, Yoshizawa N et al (2011) Ordered carbohydrate-derived porous carbons. Chem Mater 23:4882–4885
75. Xin D, Han L et al (2018) Highly uniform carbon sheets with orientation-adjustable ordered mesopores. ACS Nano 12:5436–5444
76. Hu J, Sun CF, Gillette E et al (2016) Dual-template ordered mesoporous carbon/Fe_2O_3 nanowires as lithium-ion battery anodes. Nanoscale 8:12958
77. Zhang C, Zheng Z et al (2017) Three-dimensionally ordered macro-/mesoporous carbon loading sulfur as high-performance cathodes for lithium/sulfur batteries. J Alloy Compound 714:126–132
78. Chen A, Yu Y, Li Y et al (2016) Synthesis of macro-mesoporous carbon materials and hollow core/mesoporous shell carbon spheres as supercapacitors. J Mater Sci 51:4601–4608
79. Kong Q, Zhang L, Wang M et al (2016) Soft-to-hard templating to well-dispersed N-doped mesoporous carbon nanospheres via one-pot carbon/silica source copolymerization. Sci Bulletin 61:1195–1201
80. Zhang L, Yu Y, Liu B et al (2018) Synthesis of mesoporous tubular carbon using natural tubular halloysite as template for supercapacitor. J Mater Sci Mater Electron 29:1–8
81. Li N, Zheng M, Feng S et al (2013) Fabrication of hierarchical macroporous/mesoporous carbons via the dual-template method and the restriction effect of hard template on shrinkage of mesoporous polymers. J Phys Chem C 117:8784–8792
82. Xue T, Sun ZP, Wei L et al (2013) One-step dual template synthesis of platinum on mesoporous carbon nanowires for electrocatalysts. Int J Hydrogen Energ 38:2754–2759
83. Chen S, Shen W et al (2011) Synthesis of spherical mesoporous carbon for electric double-layer capacitors. J Sol–Gel Sci Techn 60:131–136
84. Li Q, Jiang R, Dou Y et al (2011) Synthesis of mesoporous carbon spheres with a hierarchical pore structure for the electrochemical double-layer capacitor. Carbon 49:1248–1257
85. Zhao G, He J, Zhang C et al (2008) Highly dispersed pt nanoparticles on mesoporous carbon nanofibers prepared by two templates. J Phys Chem C 112:1028–1033
86. Wang K, Zhang W, Phelan R et al (2007) Direct fabrication of well-aligned free-standing mesoporous carbon nanofiber arrays on silicon substrates. J Am Chem Soc 129:13388–13389
87. Dai W, Zheng M, Zhao Y et al (2010) Template synthesis of three-dimensional cubic ordered mesoporous carbon with tunable pore sizes. Nanoscale Res Lett 5:103–107
88. Gao R, Yin L, Wang C et al (2009) High-yield synthesis of boron nitride nanosheets with strong ultraviolet cathodoluminescence emission. J Phys Chem C 113:15160–15165
89. Ming S, Liang H et al (2011) Mesoporous activated carbon structure originated from cross-linking hydroxyethyl cellulose precursor by carboxylic acids. J Phys Chem C 116:352–360
90. Li S, Ying Y, Yanmei Z et al (2018) Solvent-free synthesis of N/S-codoped hierarchically porous carbon materials from protic ionic liquids for temperature-resistant, flexible supercapacitors. ACS Sustain Chem Eng 6:13494–13503
91. Wu MB, Li LY, Liu J et al (2015) Template-free preparation of mesoporous carbon from rice husks for use in supercapacitors. New Carbon Mater 30:471–475
92. Jahanbakhshi M (2018) Myoglobin immobilized on mesoporous carbon foam in a hydrogel (selep) dispersant for voltammetric sensing of hydrogen peroxide. Mikrochim Acta 185:121
93. Wang Y, Pan Y, Zhu LK et al (2018) Fabrication of 3D heteroatom-doped porous carbons from self-assembly of chelate foams via a solid state method. Inorg Chem Front 5:656–664

94. Zhou HL, Wang DX, Fu AP et al (2018) Mesoporous carbon spheres with tunable porosity prepared by a template-free method for advanced lithium-sulfur batteries. Mater Sci Eng B 227:9–15
95. Liu K, Wang J, Yang T et al (2020) An "in situ templating" strategy towards mesoporous carbon for high-rate supercapacitor and high adsorption capacity on dye macromolecules. Carbon 164:19–27

Chapter 4
Nano Materials Self-assembly

The self-assembly of nano materials can obtain nanomaterials with orderd structure, leading to the generation of regulated mesostructured. In the following text, various self-assembly strategies were introducted.

4.1 Introduction of Nano Materials Self-assembly

Nano materials have attracted significant attention due to their remarkable properties and vast potential for applications in the field of nanotechnology. To achieve diverse functionalities, it is essential to assemble nano materials in orderly patterns on surfaces and interfaces. The assembly of nano materials gives rise to novel nanostructures, exhibiting unexpected collective and inherent physical properties. These properties can be exploited for multipurpose applications in nanoelectronics, spintronics, sensors, etc. Quantum size effects (QSE), stemming from the quantization of the electronic states of nanoparticles, are the driving forces behind the current intense study of synthesis and applications of nanoparticles. In general, the physicochemical properties of nanoparticles are size- and shape-dependent [1, 2]. These properties undergo significant alterations at the nanometer scale, rendering nanoparticles excellent contenders for applications in energy stoage [3], single-electron devices [4]/nanoelectronics [5], sensing [6], biodiagnostics [7], catalysis [8], and so forth. In recent years, the synthesis of diverse nanoparticles is well developed [9]. Nevertheless, the organization of nanoparticles at the nanoscale remains the most crucial and challenging aspect.

Nanoparticle self-assembly refers to a technology wherein the fundamental building blocks of nanostructures autonomously arrange themselves within a specific range. During the self-assembly process, these basic structural units spontaneously organize or cluster into a relatively stable structure, exhibiting a distinct geometric configuration under the influence of non-covalent bonds [10–12]. The self-assembly

B. Kong et al., *Functional Mesoporous Carbon-Based Film Devices for Energy Systems*,
https://doi.org/10.1007/978-981-99-7498-6_4

system of particles with different scales and configurations is constructed [13], and the orderly assembly and precise control of nanoparticles can obtain some properties beyond nanoparticles [14], so it has always been a hotspot for scientists. Relying on the interaction of non-covalent bonds between nanoparticles (such as occupied volume effect, surface force, van der Waals force, electrostatic effect, dipole interaction, hydrophobic effect, or conjugation) [15], the nano unit can be spontaneous organize the collection to form an assembly structure with a certain geometric appearance. At present, nanoparticles are classified by morphology, which can be divided into isotropic spherical nanoparticles, anisotropic zero-dimensional non-spherical nanoparticles [16], one-dimensional nanotubes, one-dimensional nanowires, one-dimensional nanofibers, two-dimensional nanomaterials such as nanosheets [17, 18], the shape control and application of nanoparticles themselves are the foundation for the development of nanotechnology [19].

The morphology of nanoparticles plays a significant role in their assembly, influencing the orientation and arrangement of particles during the process. Moreover, different nanoparticle structures require distinct approaches and techniques for assembly control. Spherical nanoparticles, being a prevalent form, are widely encountered in nanoparticle assemblies.

The multi-dimensional assembly of nanoparticles with controlled morphology in highly ordered arrays is important for realizing their novel applications [20, 21]. Organized or self-assembled nanostructures show remarkable collective properties, useful for engineering nanoarchitectures. It has been demonstrated that the arrangement of nanocrystals in multi-dimensional superlattices alters their properties compared to when they are isolated. It is widely acknowledged that the key to successful applications of nanoparticle-based devices lies in the creation of well-defined, defect-free ordered nanostructures. As a result, the generation of periodic assembled nanostructures through novel and innovative procedures has garnered significant attention in recent times. In general, nano-organization can be achieved through "top-down" approaches such as lithography and patterning, or "bottom-up" approaches based on self-assembly. Self-assembly serves as a crucial tool for producing integrated nanoscale devices. Through self-assembly, robust structures with exceptional mechanical stability are formed, driven by favorable free-energy changes or kinetics. Specific non-covalent interactions play a fundamental role in organizing atomic, molecular, and other nanoentities, leading to the spontaneous formation of hierarchical and complex architectures. The importance of self-assembly in technology is pervasive and far-reaching. Interesting analogies to nature's art of putting things together are sought after by the current nanotechnologists. In this context, exploitation of the spontaneous and directed assembly of nano-objects with supramolecular structures [22], polymers [23], dendrimers [24], biological molecules [25] or inorganic nanoparticles [26, 27] is well investigated. Nanoparticle self-assembly is very interesting as one has the possibility of exploiting spatial or temporal attributes of nanoparticles to tailor assembly formation. By altering the substituents of the assembling building blocks, desirable changes in the assembly process are achieved.

Moreover, the desired properties and structural effects in assembled materials can be achieved by employing appropriate templates. In various nanoparticle assembly methods, supramolecular chemistry and surface properties play a crucial role. Non-covalent interactions, including hydrogen bonding, electrostatic interactions, van der Waals forces, interfacial surface wettability, surface free energy, chemical conjugations, templation, and patterning, are essential factors influencing spontaneous or directed assembly [28–31]. The interactions between nanoparticles are most likely dominated by van der Waals forces. However, the assembly process is also influenced by factors such as the size and shape of the particles, as well as the thickness of the surface passivation layer. Stabilized colloidal nanoparticles are generally well-suited for controlled assembly processes. The coupling interactions between nanoparticles and their surroundings play a crucial role in determining the physicochemical properties of the nanoparticles. Furthermore, the proximity of nanoparticles and subsequent dipole–dipole interactions lead to significant changes in their electron-transport properties, thereby affecting their overall characteristics. For instance, altering the length of the protective ligands can impact quantum transitions and electronic properties [32]. Surfaces of nanoparticles can be chemically modified using different, well-developed chemical strategies. The surface passivation of nanoparticles with ligands serves to prevent their agglomeration and offers the opportunity for tailoring functional groups through chemical reactions. Additionally, chemical patterning and surface electrostatic interactions are employed to create precisely patterned nanoparticle assemblies. At interfaces, the reduction in interfacial energies drives the spontaneous assembly process [33].

In the last decade, numerous nanoparticle assembly methods were investigated. Different approaches of nano-ordering are separately discussed, for example, polymer-mediated nanoparticle assembly [34], biological assembly [35], magnetic nanoparticle superstructures [36], three-dimensional superlattices [37], and one- or two-dimensional nanocrystal arrays [38]. In addition, some specific technique-based overview articles are available such as drying-mediated assembly [39], Langmuir–Blodgett approach [40], Layer-by-Layer (LbL) assembly [41], and so forth.

Interfacial assembly is an elegant approach to assemble different nanoparticles [42]. This is an easy and fast route for large-scale applications. A well-known example is of Pickering emulsions, in which large particles (>1 mm) effectively stabilize emulsions by adsorbing onto the liquid–liquid interface [43]. A similar dynamic approach is being applied for creating nanoparticle assemblies at the fluid interfaces. During this process, thermal fluctuations and interfacial energy play an important role. One can achieve size-dependent nanoparticle assembly at the interfaces. Lin et al. have shed light on the mechanism involved in this process [44]. Usually, the assembly process is driven by the minimization of the Helmholtz free-energy. The low energy associated with the positioning of nanoparticles at the interfaces leads to a thermally activated release of particles from the interface. Furthermore, Binks et al. developed an in-depth theory of nanoparticle self-assembly at liquid–liquid interfaces [45]. This theory proposes that the interfacial energy between the oil and water interface decreases during nanoparticle assembly, leading to the formation of a monolayer of nanoparticles at the interface. Nanoparticles tend to assemble as disordered, but densely packed monolayers. Additionally, the interfacial energy and absorption of nanoparticles at the interface exhibit a clear dependence on particle size, with smaller nanoparticles absorbing more weakly than larger ones. This sometimes results in size-dependent particle exchange, where larger particles displace smaller ones based on their adsorption energies. The fluorescence photobleaching method was utilized to investigate the in-plane dynamics of nanoparticles, taking advantage of their inherent fluorescence emission. The assembly process is primarily driven by dipole moment, small positive charge, and directional hydrophobic attractions. The assembly progression is highly dynamic and involves error corrections. Two-dimensional arrays of nanoparticles can be produced at various two-phase interfaces, such as gas–solid and liquid–liquid interfaces, which serve as templates. Furthermore, nanoparticles can also be assembled at polymeric surfaces, such as the interface of immiscible block copolymer micelles. In this case, spin-coating of micelles results in an ultrathin film that concentrates the nanoparticles at the interface, forming a hexagonal pattern between the core and corona.

Likewise, the Langmuir–Blodgett technique is also a type of the interfacial assembly process [46]. Typically, surface-protected nanoparticles are made to float over water surface, using hydrophobic interactions. At the water–air interface, nanoparticles assemble into a monolayer that undergoes gradual compression. This compressed monolayer can then be transferred to substrates, such as silicon, using either horizontal or vertical lift-off methods. By utilizing this approach, films composed of nanoparticles with ordered cores, reaching several millimeters in size, can be engineered.

4.2 Interfacial Assemblies of Mesoporous Materials

With recent advancements in modern nanoscience and nanotechnology, ordered mesoporous materials have emerged as a highly sought-after research topic across various scientific disciplines, including chemistry, materials science, physics, and biology [47–49]. That is because ordered mesoporous materials possess fascinating properties including regular, uniform and interpenetrating mesopores, tunable pore sizes, high surface areas as well as abundant framework compositions. Compared with their bulk counterparts, they can interact with atoms, ions, molecules or even larger guest species not only at the external surface, but also through the whole internal pore system. As a result, ordered mesoporous materials exhibit substantial performance boosts in numerous applications such as adsorption [50], separation [51], catalysis [52], sensors [53], drug delivery [54], energy conversion and storage [55], and so on. Since the exciting discovery of this new kind of materials based on the supramolecular assembly chemistry in the early nineties [56–58], considerable work has been done to synthesize ordered mesoporous materials with diverse compositions, morphologies and pore symmetries; meanwhile, tremendous effort has been devoted to elucidate the mechanism of mesostructure formation and explore their applications.

The construction of mesoporous materials primarily focuses on creating monodisperse mesosized (2–50 nm) pore voids and arranging them in a well-organized, long-range ordered array [59, 60]. In generally, two kinds of templates are used to produce the mesopores: supramolecular aggregates such as surfactant micelle arrays, and rigid preformed mesoporous solids such as ordered mesoporous silica, carbon, and colloidal crystals [61, 62]. The synthesis pathways for these mesoporous materials are commonly referred to in literature as soft-templating and hard-templating (nano-casting) methods. It is important to note that, in addition to the template, the interface also plays a central role in the processing as it provides a critical space for the assembly and construction of mesostructures. In general, there are two types of interfaces involved in the synthetic system. The first type is between the surfactant templates and guest species, which has been extensively studied by various research groups [63–66]. It suggests that the effective interaction of surfactants-guest species is critical to govern the soft-templating route for synthesis of ordered mesoporous materials [59]. Although significant progress has been made in the cooperative assembly of mesostructures in an aqueous phase system, there are several inherent drawbacks to consider. Firstly, the resulting products are typically powders with poorly defined morphology, which limits their applicability in thin films or other shape-based technologies. Secondly, the preparation of non-siliceous mesoporous materials is more challenging due to the difficulty in controlling the hydrolysis and condensation of non-siliceous precursors, such as metal alkoxides. Thirdly, the arrangement patterns and sizes of mesopores are often limited in their versatility. Lastly, achieving multifunctional mesoporous materials through a one-pot cooperative assembly remains a significant challenge.

Another important interface is the two-phase (solid, liquid and gas) one in the synthetic system, including liquid–solid, gas–liquid, liquid–liquid, gas–solid, and solid–solid interface, which has been well developed for synthesis of mesoporous materials (Fig. 4.1). Compared to one-phase synthesis, which involves homogeneous nucleation and growth, the inclusion of a two-phase interface in the system alters the growth behavior of mesoporous materials and enables the formation of molded or multifunctional mesoporous materials. This approach can help overcome the aforementioned drawbacks associated with one-phase synthesis. For instance, mesoporous thin films or membranes have been successfully fabricated on substrates using the evaporation-induced self-assembly (EISA) method [67, 68]. Multifunctional core–shell structured mesoporous materials can be obtained by rationally depositing mesoporous shells on well-designed cores [69, 70]. Furthermore, the widely recognized hard-templating method for producing mesoporous materials is also an exemplary interface reaction. In this approach, a fluid precursor (liquid or even gas) is initially infused into the nanometer-sized pore channels of solid templates. Subsequently, it undergoes a conversion process within the nanostructured confinement, resulting in the formation of the desired nanomaterial [62, 71–75]. This interfacial casting strategy effectively eliminates the need to control the cooperative assembly between surfactants and guest species, as well as the sol–gel process of guest species. As a result, it has proven to be highly successful in the synthesis of numerous mesoporous materials.

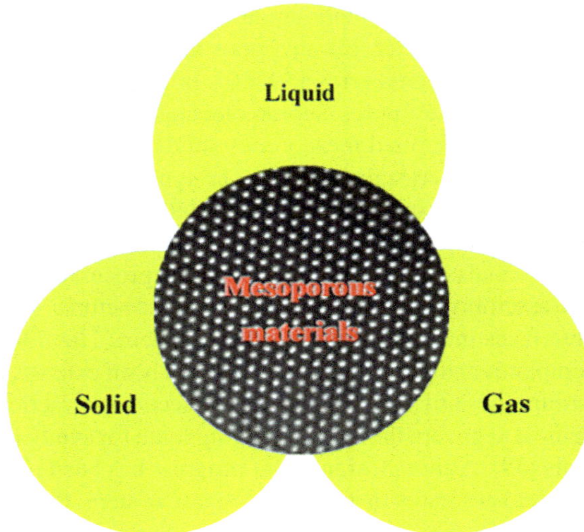

Fig. 4.1 The two-phase interfaces for the synthesis of mesoporous materials, including liquid–solid, gas–liquid, liquid–liquid, gas–solid, and solid–solid interface

Wan and Zhao have provided a comprehensive overview of the fundamental interactions between surfactant-guest species at the interface [59, 76, 77]. Therefore, a detailed explanation will not be provided here. In this context, we aim to review the synthesis of ordered mesoporous materials based on the interfacial assembly and engineering, unless otherwise specified, which refer to the two-phase interface. Overall, the discussion of interfacial assembly and synthesis of ordered mesoporous materials will be classified into five categories, including liquid–solid interfacial assembly, gas–liquid interfacial assembly, liquid–liquid interfacial assembly, gas–solid interfacial synthesis, and solid–solid interfacial synthesis.

4.3 Self-assembly of Mesoporous Film Functional Materials

EISA Process

EISA process is a mature method to prepare ordered mesoporous materials especially for mesoporous thin films, membranes and monoliths [67, 68, 78]. It was first used by Brinker and coworkers in the preparation of mesoporous silica thin films [79]. Figure 4.2 illustrates the typical EISA process for mesoporous thin films. Firstly, a homogeneous solution containing dissolved templating surfactants and guest precursors is prepared. Subsequently, the solution is deposited onto a substrate through chemical solution deposition. The self-assembly process is initiated as volatile solvents gradually evaporate, resulting in the formation of a metastable phase comprising template-guest species with well-ordered mesostructures. Following a treatment step to enhance the stability of mesostructured composites and boost porosity by removing the template, ordered mesoporous materials are achieved. Extensive research investigating this strategy has been conducted by Sanchez, Brinker, and their collaborators [79, 80]. Indeed, the EISA (Evaporation-Induced Self-Assembly) process is inherently intricate, comprising a minimum of three dynamic steps that are influenced by distinct parameters. These steps encompass the chemical aspects associated with the initial solution, the processes related to the layer-deposition technique, and the treatment aimed at eliminating the template while ensuring the stability of the mineral networks, thereby preventing any significant pore collapses. Detailed studies of such process have been conducted for various mesostructured films (eg., SiO_2–CTAB [80, 81], SiO_2–Pluronics [82], TiO_2–Pluronics [83, 84]) by using various techniques, including in situtime-resolved small-angle X-ray scattering (SAXS), grazing incidence SAXS (GI-SAXS), ellipsometric fourier transform infrared spectroscopy (FTIR) and conventional FTIR, and so on. The ultimate mesostructures are influenced by a multitude of factors, including the type and concentration of surfactants, inorganic precursors, the final treatment process, and even seemingly insignificant parameters (eg., water concentration, processing humidity and evaporation temperatures). It has been observed that the formation of ordered mesostructures typically occurs during the final stages of solvent evaporation, and even during the aging stage [83]. The interaction among the guest species themselves is of utmost importance in determining the resulting ordered mesostructures.

General Factors

To achieve uniform and homogeneous thin films through the EISA process, it is crucial that the solvent effectively wets the substrate and is volatile. Typically, solvents with low polarity, such as ethanol, methanol, and tetrahydrofuran (THF),

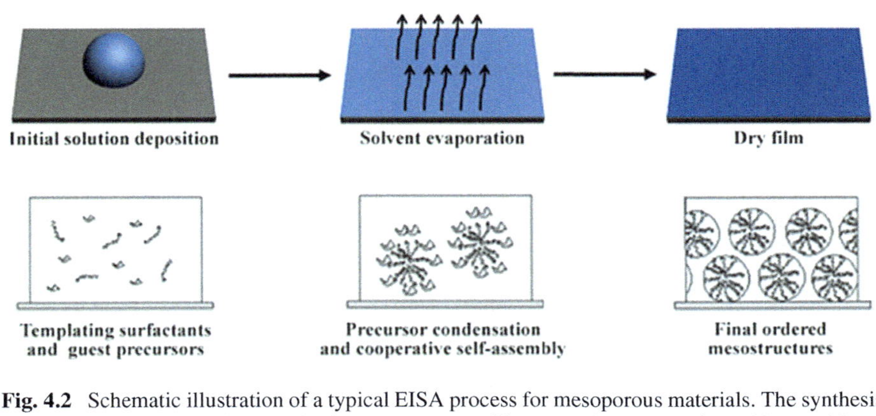

Fig. 4.2 Schematic illustration of a typical EISA process for mesoporous materials. The synthesis procedure is shown on the top, the bottom sequence illustrates the cooperative self-assembly of mesostructures

are preferred in the EISA process. In these solvents, the hydrophilic/hydrophobic properties of the surfactant templates are diminished, as both the hydrophilic and hydrophobic segments can interact with these solvents. Consequently, the self-assembly of surfactants in the initial solution may be hindered, and the assembly process is primarily induced during solvent evaporation. On the other hand, nonpolar and oily solvents are rarely used. For instance, silica nanowires with adjustable diameters have been synthesized using the EISA approach with Pluronic P123 and F127 in toluene or xylene solutions. The formation of such arrays corresponds to the reverse mesophases of surfactants in oily solvents [85].

The EISA process predominantly occurs on the substrate, which serves as a vital and diverse interface for the self-assembly of mesostructures. The characteristics of the substrate significantly influence the morphology and orientation of the resulting mesoporous materials. Hence, the selection of a suitable substrate involves various considerations. Firstly, it is essential to ensure that the substrate exhibits strong affinity towards the template and guest species to achieve a uniform and homogeneous film. Secondly, the chemical composition and nanostructure of the substrate are crucial factors as they directly impact the alignment of mesopores. For example, Hara et al. demonstrated the vertical alignment of silica mesochannels by utilizing the π–π interaction between the organic template molecule of a planar discotic liquid crystalline and two-dimensional (2D) π-plane of graphite [86]. Furthermore, it is imperative for the substrate to exhibit excellent physical and chemical stabilities. This ensures that chemical reactions with the film and self-degradation of the substrate are effectively prevented during the solidification of the framework and template elimination processes. Additionally, special consideration should be given to the thermoshrinkage of the substrate when preparing ordered mesoporous materials with significant framework shrinkage during thermal processing. This helps accommodate the stresses that may arise [87]. Undoubtedly, the morphology of the resulting mesoporous materials is closely tied to the choice of substrate. Typically, planar supports like glass substrates and silicon wafers are commonly employed for producing mesoporous films. However, other materials that offer abundant interfaces can also serve as suitable substrates, such as polyether polyol-based polyurethane (PU) foams [88], hierarchical biological materials [89], colloidal crystals [90], and so on.

The typical techniques for casting initial solution on the substrate include dip coating [79], spray coating [91], and spin coating [92], etc. The dip coating method involves withdrawing the substrate from the coating solution at a controlled rate to ensure the formation of a liquid layer on the substrate surface. In spray coating, the solution is sprayed onto the substrate surface using an aerosol generator or an atomizer. On the other hand, spin coating works by uniformly dispersing the coating solution at the center of a spinning substrate and then spreading the solution droplet through the centrifugal force generated by high-speed rotation. This process simultaneously forms a film while subjecting it to a shearing force between the interfaces of the drying solution layer. Among these techniques, dip coating and spin coating are the most commonly used due to their low equipment cost and ease of operation.

Typical Examples

The preparation of continuous mesoporous silica film is a most successful case of the EISA strategy [93]. In a typical process, silica precursors (e.g., tetraethyl orthosilicate, TEOS) dissolved in organic solvent (normally ethanol) are prehydrolyzed with stoichiometric quantity of water (catalyzed by acids, such as HCl) at a temperature of 25–70 °C. Upon solvent evaporation, the silicate species further polymerize and condense. At the final stage of solvent evaporation, high concentrated surfactants such as cetyltrimethylammoniumboromide (CTAB)or polyethylene-oxide based block copolymers (e.g., Pluronic P123) form liquid–crystal phases in the presence of inorganic oligomers. Simultaneously, the low hydrolysis and crosslinkage degree of silicate species improve the assembly on organic/inorganic interfaces, leading to the formation of ordered mesostructures. The process is incredibly fast, taking only a few seconds to complete. To date, researchers have successfully achieved mesoporous silica thin films with a wide range of mesostructures, compositions, film thicknesses, and adjustable porosities. Yang et al. first extended the EISA process to the synthesis of mesoporous metal oxides [94]. In this non-aqueous sol–gel process, metal halides were used as the inorganic precursors and amphiphilic block copolymers as templates to generate several large-pore mesoporous metal oxides, including TiO_2 ZrO_2 Nb_2O_5, Ta_2O_5, Al_2O_3, SiO_2, SnO_2, WO_3, HfO_2, and mixed oxides $SiAlO_y$, Al_2TiO_y, $ZrTiO_y$, $SiTiO_y$ and $ZrWO_y$. The resultant mesoporous metal oxides show relatively thermally stability (about 400 °C), narrow pore size distributions, but low surface areas (<200 m^2/g), as well as possessed semicrystalline frameworks. The most important feature is alcoholysis of inorganic salts generates $M(OEt)_nCl_m$ species with a low polymerization rate, which can slowly react with water in air via hydrolysis and crosslinking to form mesostructures. It is widely recognized that metal alkoxides exhibit significantly higher reactivity compared to silicon alkoxides. Consequently, controlled hydrolysis and condensation of inorganic species play a vital role in the formation of mesophases for most non-siliceous oxides. This is necessary due to their inherent propensity to precipitate and crystallize directly into bulk oxide phases when exposed to aqueous media. Various strategies have been developed to control the reactions, including the optimization of pH, water content in precursor solutions, relative humility [83]. For example, under strongly acidic condition, the condensation reaction is effectively hindered by protonation of MOH nucleophilic species, in which small oligomers can retain during the self-assembly process of surfactants [95, 96]. The use of additives such as acetic acid and acetylacetone is also effective for controls of the hydrolysis and condensation [97, 98].

In order to facilitate the selection of precursors for the fabrication of mesoporous metal oxides with different components using the EISA process, a novel "acid–base pair" concept was proposed by Zhao and coworkers [99]. This concept takes a different perspective to address the influence of inorganic–inorganic interplay on the synthesis of mesoporous materials, in which the inorganic precursors are divided into "acid" and "base" according to their alcoholysis (here, alcohol is used as a solvent) behaviors. Inorganic metallic or nonmetallic chlorides are considered as strong "acids" since a large amount of acid is generated during the alcoholysis process. Hydrate metallic salts and inorganic acids (Brønsted acids) are attributed as a medium acidic precursor. Metallic alkoxides and nonmetallic alkoxides (e.g. phosphatides) are assigned as bases because acid substances are seldom generated. In this process, two or more inorganic species are used to self-generate a reaction medium with the correct acidity to favor the sol–gel process of solvated products, and give a self-adjusted sol–gel synthesis, and promote the assembly of mesostructured materials. Generally, the "acid–base" pairs formed from strong "acid" and strong "base", or strong "acid (base)" and medium "base (acid)" in non-aqueous media are required. Guided by this concept, a wide variety of well-ordered, large-pore, homogeneous, multicomponent, mesostructured solids have been synthesized, including metal phosphates, silicoaluminophosphates, metal borates, as well as various metal oxides and mixed metal oxides. For example, using phosphorus trichloride and zirconium propoxide as inorganic precursors, and triblock-copolymers as templates, ordered mesoporous zirconium phosphates have been prepared with surface areas between 78 and 177 $m^2\ g^{-1}$ and controllable pore sizes between 2 and 4 nm. This concept, together with the increased understanding on EISA strategy, sol–gel chemistry and organic–inorganic interaction, which are interdependent of each other, could pave the way for preparing ordered mesoporous nonsiliceous oxides [100–102].

Another advantage of the EISA process is its ability to produce ordered mesoporous polymers and carbons. It was first demonstrated by Zhao's group [103]. Highly ordered mesoporous carbons, denoted as FDU-15 and FDU-16 (Fig. 4.3), were presented by using commercial triblock copolymers as templates, and water-soluble and low-molecular-weight phenolic resins (resols) as precursors, which are polymerized by phenol and formaldehyde under alkaline conditions. Figure 4.4 illustrates the formation process of the ordered mesoporous carbons. The initial homogeneous solution is prepared by dissolving triblock copolymers and resols in ethanol. The choice of resol as a precursor and triblock copolymer as a template is essential for the successful organization of organic–organic mesostructures because resol has a 3D network structure and plenty of hydroxyl groups (–OH), can interact strongly with the PEO blocks of triblock copolymer templates via hydrogen bonds. The assembly of resols and copolymer templates occurs readily to form ordered mesostructures without macrophase separation. The preferential evaporation of ethanol progressively enriches the concentration of copolymers and drives the organization of resol-copolymer composites into an ordered liquid–crystal mesophase. Furthermore, curing the resol-copolymer composites at 100 °C with a long time (>24 h) to solidify the polymeric framework yields a rigid hydrocarbon network with three-connected benzene rings through the formation of covalent bonds, the same as silicates in zeolites or/and mesoporoussilicate molecular sieves. This step holds significant importance in ensuring the stability of mesoporous products. It is worth noting that the cross-linking and polymerization processes of phenolic resin frameworks are distinct from the assembly with surfactants, which is a key feature of the EISA strategy. This differs from a cooperative formation assembly mechanism, where surfactant-directed assembly and polymerization of inorganic oligomers occur simultaneously. Due to the disparity in chemical and thermal stability between the resin and Pluronic copolymers, the templates can be easily removed at low temperatures without compromising

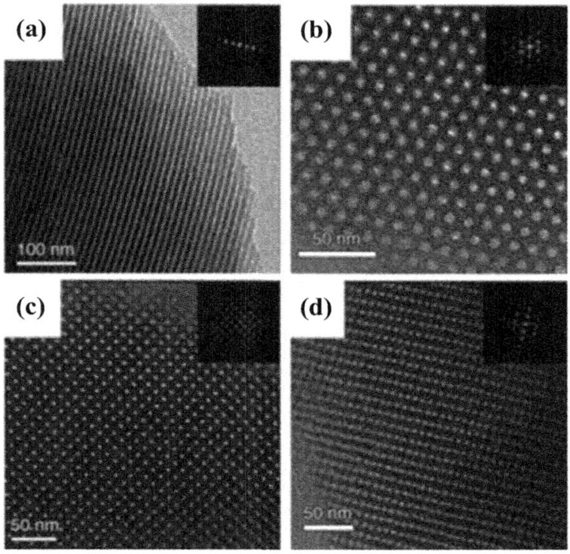

Fig. 4.3 TEM images of mesoporous carbons FDU-15 (**a** and **b**) and FDU-16 (**c** and **d**)

the resin framework. The continuous hydrocarbon frameworks of mesoporous resins confer exceptional stability, enabling their direct transformation into mesoporous carbon frameworks by heating to temperatures ranging from 600 to 1400 m °C under an inert atmosphere. From the viewpoint of synthesis, these mesoporous polymers and carbons are excellent examples to expatiate the EISA process. A series of ordered mesoporous carbons with pore sizes ranging from 2 to 23nm can be obtained from the well-established self-assembly [104–109].

Assembly in a Confined Space

In addition to using flat substrates, one can also exploit physically confined environments like porous anodic aluminum oxide (AAO) membranes to facilitate the EISA process in preparing mesoporous materials with unique mesostructures and morphologies [110]. In their synthetic approach (Fig. 4.5a), when the initial depositing solutions enter the porous matrices of AAO and undergo drying until the solvents are completely evaporated, the strong interfacial forces can lead to the formation of mesostructures with thermodynamically stable arrangements [111]. When small molecular CTAB was used as the structure-directing agent (SDA), the long axis of the mesochannels aligns parallel to the long axis of AAO channels, exhibiting columnar orientations (Fig. 4.5b) [111, 112]. However, in most cases, the mesochannels are oriented perpendicular to the AAO channels and circularly packed like stacked donuts when nonionic surfactants, such as block copolymers P123, are used as the SDA (Fig. 4.5c and d) [111]. Moreover, different mesostructures were discovered, which depended on the confinement conditions imposed by the varying diameters of AAO nanochannels. These ranged from individual chains of spherical mesopores to concentric or chiral helical mesopores [113, 114]. Following a rapid

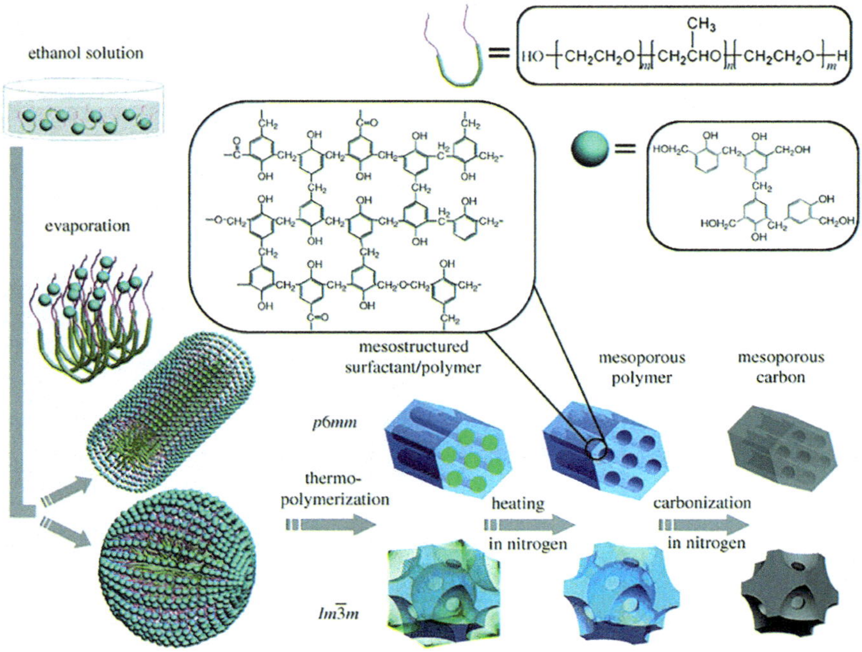

Fig. 4.4 Schematic representation of the EISA procedure used to prepare mesoporous polymers and carbon frameworks

evolution of synthetic techniques, a great number of various mesoporous materials (e.g., titania, and carbon) with highly regular structures can now be prepared within these membranes [115, 116]. Another common used confined space is 3D colloidal crystals [117–119]. By utilizing the initial depositing solutions within these matrices, it becomes possible to synthesize hierarchical materials featuring an interconnected, face-centered cubic macropore and a mesopore network. A noteworthy example is the work done by Deng et al., who employed monodispersed silica colloidal crystals as a substrate to facilitate the self-assembly of triblock copolymers and soluble resols at the interface [120]. After removing the copolymers and silica scaffolds, hierarchically porous carbons with highly ordered face-centered cubic structure were obtained. The macro-/meso-porous carbon products have tunable macropore sizes of 230–430 nm and interconnected windows with a size of 30–65 nm, a high surface area (up to 760 m^2/g), a large pore volume ($\sim$1.25 cm^3/g) and a mesopore size ($\sim$11 nm). Interestingly, Schuster et al. used the inverse opal as a template for a triconstituent precursor solution containing resol as a carbon precursor, TEOS as a silica precursor and Pluronic F127 as a SDA. After carbonization, the silica template and silica component within the carbon/silica nanocomposites were etched, leading to the creation of well-organized mesoporous carbon nanospheres [121]. The resultant carbon nanospheres possessed a bimodal pore size distribution of large and small mesopores of ~6 and 3.1 nm, a high specific surface area of ~2445 $m^2\ g^{-1}$, and a high reversible charge capacity of up to 1200 mA h g^{-1} and a good cycling stability when applied in lithium–sulfur batteries.

Recently, Zhao and coworkers developed a kilogram-scale synthesis of mesoporous materials based on the EISA process by using commercially available PU foam as a substrate [122–125]. It challenges the conventional analysis-scale production scenario by employing the EISA process on a standard 2D substrate. Because PU foam as a substrate possess several advances for the EISA process including: (i) the 3D opened macrostructure facilitating the evaporation of the solvent; (ii) the abundantly porous structure, rich-surface hydrophilic groups and low density providing a plenty of idea interfaces for the EISA process, which can save a large number of space; (iii) the nature of decomposition in an inert atmosphere at 300–400 °C and low-cost making the removal of PU foamsimple and industrial, and which can occur with the surfactant elimination, thus avoiding the excessive introduction of impurities

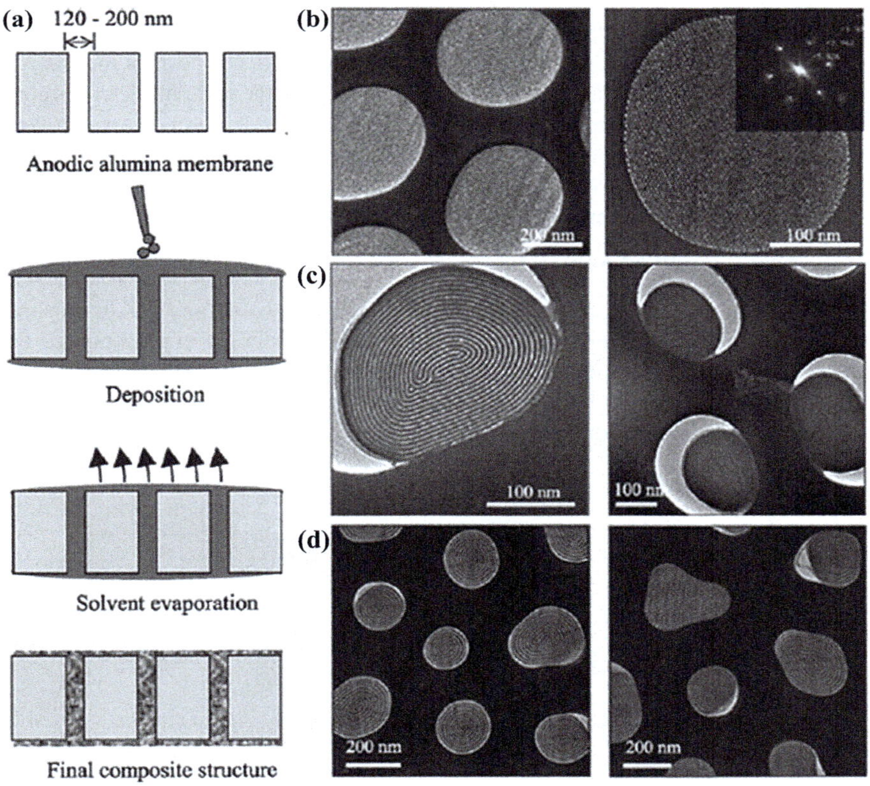

Fig. 4.5 **a** Schematic representation of the EISA process proceeding in AAO membranes. Plan-view TEM images of mesoporous silica: **b** templated with CTAB, **c** templated with Brij 56, and **d** templated with Pluronic P123

and treatment process. For instance, by applying the initial triconstituent precursor solution in PU foam, one can obtain ordered mesoporous carbon–silica composite monoliths with a wide range of macroporous architectures (Fig. 4.6a) [122]. After impregnation, the initial sol solution can be infused into the interconnecting 3D networks and large macropore voids by capillary and wetting driving forces. During the solvent evaporation, the precursors can coat onto the struts of the foams because of the strong interaction between the sol precursors and the surface of the PU foam and the low concentration of the sol. With a further increase of the concentration, the resol precursor and crosslinked silica species can assemble with Pluronic F127 to form a uniform layer of the ordered mesostructured resol–silica composite on the 3D interconnecting skeletons of PU foams. Once ethanol has completely evaporated, the expanded PU-foam scaffold regains its original shape and size. The thickness of the layer is determined by the concentration of sol precursors. To achieve an ordered mesostructure, it is preferable for the layer thickness to be less than 20 mm. For a uniform coating, a continuous rotary evaporation process can be employed, preferably in a well-ventilated area. The calcination process involved a gradual increase in temperature (1 °C min^{-1}) from 100 °C to 350 °C to ensure uniform shrinkage. Wang et al. later introduced an additional post-cure process for the synthesis of mesoporous carbons on a kilogram scale. (Fig. 4.6b). It was found that the resin frameworks could be greatly enhanced by the post-cure process and thus resist the framework shrinkage during the carbonization. As a result, high-quality mesoporous carbon with a high specific surface area (~690 m^2 g^{-1}), a large pore volume (~0.45 cm^3 g^{-1}) and a uniform pore size (~4.5 nm) can be obtained on kilogram-scale [125].

Assembly on Modified Substrates

As mentioned earlier, the chemical compositions and nanostructures of the substrate play a crucial role in determining the alignment of mesopores in mesoporous thin films. Consequently, significant efforts have been dedicated to adjusting the orientation of mesopores due to their significant impact on practical applications, particularly in controlling the direction of substance transmission within films [126]. In

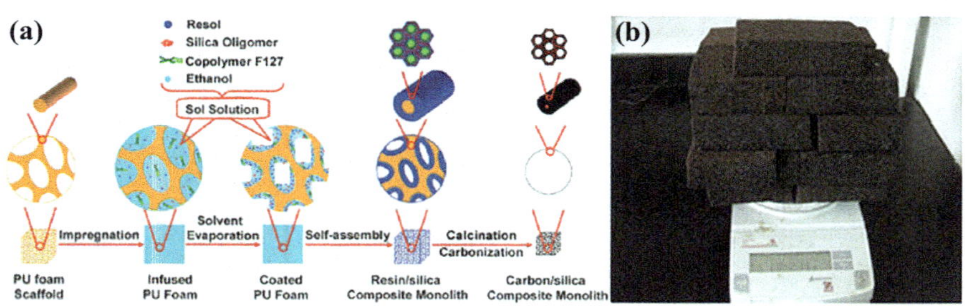

Fig. 4.6 **a** The scheme of the EISA process by using PU foams as the substrate. **b** The photograph of the kilogram-scaled mesoporous carbons

recent years, there has been a growing interest in engineering the nature of interfaces to achieve control over 1D straight mesochannels that align parallel or perpendicular to the substrate. For instance, Yamauchi et al. devised a method involving the design and preparation of AAO substrates with conical holes. By spin-coating a Pluronic P123-TEOS solution onto these substrates, they successfully produced standing mesochannels [127]. They observed that mesochannels oriented perpendicular to the substrate could be achieved within the conical holes with a low aspect ratio of 1–1.6 (Fig. 4.7a). However, as the aspect ratio increased to 2.45–3.25 (Fig. 4.7b), parallel-oriented mesochannels were formed. The formation of standing mesochannels in the low aspect ratio regime is believed to be influenced by the presence of sponge-like structures formed within the conical holes, resulting in an "oriented growth" mechanism. On the other hand, in the high aspect ratio regime, the stripes originating from the mesochannels in the conical holes were oriented parallel to the substrate, leading to parallel orientations of the mesochannels in the continuous film region. In another study, Tolbert and colleagues utilized cubic mesoporous films as a substrate to vertically orient hexagonal mesoporous TiO_2 films [128], in which, the cubic phase generally results in a close-packed array of spheres at the interface, and many cubic templating materials exhibit a hexagonal pore array on their top surface, a result that perfectly matches nanometre-scale epitaxy (Fig. 4.7c). It was discovered that when a good lattice-match was achieved, resulting in a mismatch of less than 7%, it led to the epitaxial growth of vertically oriented mesochannels (Fig. 4.7d). Conversely, with a mismatch of approximately 15–18%, epitaxial interactions were only strong in a limited range close to the substrate. Additionally, Miyata et al. demonstrated the preferential control of mesochannels in mesoporous silica films prepared on a substrate coated with rubbing-treated polyimide film [129, 130]. The formation of uniaxially-aligned mesochannels was achieved through hydrophobic interactions between the alkyl chains of surfactants and elongated polyimide chains. Rankin and colleagues successfully prepared mesoporous silica films with perpendicularly aligned mesochannels using Pluronic P123 as a SDA [131]. In their synthesis, substrates modified with PEO–PPO random copolymers or Pluronic P123 were employed as chemically neutral interfaces, facilitating the formation of perpendicularly aligned mesochannels. This method can also be applied to prepare mesoporous TiO_2 films.

Another crucial aspect of interface engineering is the fabrication of free-standing mesoporous thin films. This innovative strategy was initially demonstrated by Zhao and colleagues for synthesizing free-standing mesoporous carbon films [132]. A groundbreaking coating-etching mechanism has been proposed to synthesize versatile compositions and structures of free-standing mesoporous thin films. This mechanism involves the selective removal of interlayers, such as SiO_2, resulting in the exfoliation of mesoporous films from the substrate. Through this innovative approach, researchers successfully achieved the synthesis of intact free-standing mesoporous carbon films, reaching sizes as large as square centimeters and exhibiting variable thicknesses ranging from hundreds of nanometers to several micrometers.

These obtained free-standing carbon films possess highly-ordered orthorhombic mesostructure, remarkable porosity with a high specific surface area (~700 m^2/

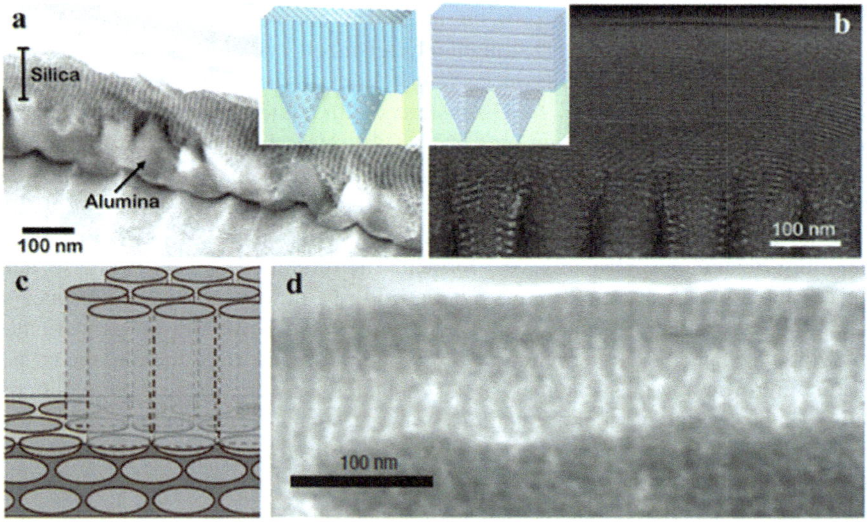

Fig. 4.7 SEM and schematic (inset) images of mesoporous silica film on the AAO substrate with an aspect ratio of 1.60 (**a**) and 3.25 (**b**). **c** Schematic diagram of the nanometre-scale epitaxy method. **d** FESEM image of a hexagonal silica fi lm with vertical pores that was formed by epitaxial growth on a cubic titania substrate

g), uniform pore size (~4.3 nm), tunable thicknesses (90–3000 nm), and maintain their intact morphology over several square centimeters. Additionally, these films can consist of multiple layers, each with distinct mesostructures. To facilitate their application, a polymer-layer assisted transfer method was developed, enabling the easy transfer of these free-standing carbon films onto substrates, even those with large curvatures, for novel device applications.

The researchers further demonstrated the potential of these free-standing carbon thin films by directly fabricating them into supercapacitor devices and nanofilter devices with size-selective permeability for cytochrome c and bovine serum albumin.

Aqueous-Phase Process

Aqueous solution synthesis is a widely used method for the production of mesoporous materials. One notable example is the synthesis of the renowned MCM-type mesoporous silicas by researchers at Mobil. This approach involves the assembly of surfactant templates and silica precursors in aqueous solutions [57, 58]. After two decades of development, this pioneering work has been significantly expanded to enable the production of a wide range of mesoporous materials in aqueous solutions. These materials exhibit tunable framework compositions, mesostructures, and porosities [133–135]. Despite the significant achievements, the resulting materials from aqueous-phase synthesis typically exist in powder form and only exhibit their inherent functionality. To achieve precise control over the functionality and structure of mesoporous materials through aqueous-phase synthesis, several strategies have been employed. These strategies involve adjusting the assembly process by utilizing specifically designed substrates with varying properties and geometries.

Assembly on Hydrophilic Substrates

In 1996, substrates (e.g., mica) were first introduced into the silicate/surfactant acidic aqueous system, thus the liquid-solid interface immediately form between the mica and acidic solution. Consequently, stable and supported mesoporous silica films were obtained [136]. In this scenario, the crucial factor for successfully preparing the films lies in achieving rapid and selective heterogeneous nucleation of silicate-surfactant mesostructures on the substrate. This should be followed by uniform growth of the mesophase without any additional homogeneous nucleation [69, 137]. This desirable heterogeneous nucleation process is directly influenced by the surface chemistry of the substrate and is controlled by charge- and structure-matching at the interface. Additionally, effective separation from the homogeneous nucleation is achieved through careful control of the reaction kinetics [70, 136–138]. When nanostructures are utilized as substrates instead of 2D planes, it becomes possible to obtain uniform mesoporous materials in the form of shelled core–shell nanostructures. In 1998, Büchel et al. were the first to report this innovative core@mesoporous SiO_2 shell nanostructure with controlled mesoporosity [139]. They initially produced monodispersed solid silica spheres as cores and then, in a subsequent aqueous step, coated them with a mesostructured silica layer. This layer was formed by introducing n-octadecyltrimethoxysilane (C18TMS) as both a mesophase-directing agent and co-precursor. During a subsequent calcination step, the alkyl chain could be eliminated, resulting in the formation of disordered mesopores in the outer shells. The successful deposition of the mesoporous silica layer at the interface is primarily attributed to the strong interactions between the silicate species, ensuring heterogeneous nucleation and growth. This strategy has been widely employed for the preparation of various core@mesoporous SiO_2 nanostructures. However, the presence of disordered mesopores in the shell may limit their application performance [140–142].

In 2007, Yoon et al. first reported the synthesis of ordered mesoporous core–shell nanostructures with perpendicularly oriented mesochannels to the surface of silica spheres by using alkyltrimethylammonium bromide (CnTAB) as a SDA [143]. Later, Deng et al. further developed this method for the synthesis of magnetic mesoporous silica microspheres [144]. In this case, the surface of magnetite nanoparticles was initially modified with a nonporous silica layer. This modification facilitated the growth of silica–surfactant mesostructures, resulting in the formation of a well-defined ordered mesoporous core–shell nanostructure. The core–shell nanostructure resembled that of a silica sphere with a core (Fig. 4.8a).

The favorable formation of ordered mesoporous silica on the interface is directly influenced by the surface chemistry of the cores. This includes their affinity towards the template molecules and deposition species, such as silicate. In basic conditions, cationic surfactants are typically effective in synthesizing ordered mesoporous silica, as the silicate anions can directly interact with the surfactant cations through Coulomb forces. Therefore, a crucial factor for the interfacial deposition of mesoporous silica is that the surface of the target substrate is negatively charged in a basic

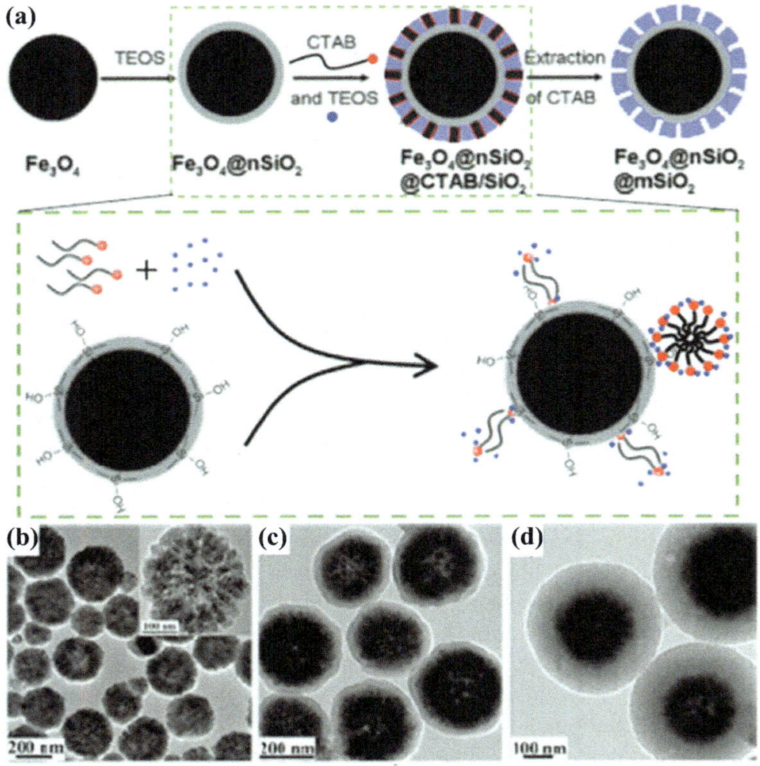

Fig. 4.8 **a** Schematic illustration of the formation process for Fe_3O_4@$nSiO_2$@$mSiO_2$ microspheres. TEM images of **b** Fe_3O_4 particles, **c** Fe_3O_4@$nSiO_2$ and **d** Fe_3O_4@$nSiO_2$ @$mSiO_2$ microspheres

aqueous solution. This allows for the effective adsorption of surfactant cations and enables the self-assembly of micelles on the interface through electrostatic interaction, facilitated by the inherent electronegativity of silicate species in the mesopore frameworks. Additionally, the coating of ordered mesoporous silica benefits from large diameter and smooth surface of the core, as they effectively reduce interfacial curvature energy. The unique microstructure of the obtained sandwich-like mesoporous materials could be very useful for many applications (Fig. 4.8c, d). Firstly, the intermediate nonporous silica layer serves to protect the core from harsh application conditions. Secondly, the mesoporous silica shells not only provide a high surface area for anchoring various functional groups but also offer a large accessible pore volume for the adsorption and encapsulation of guest molecules. Furthermore, the unique perpendicular orientation of the mesopore channels in the microspheres allows for easy access, promoting the adsorption and release of large species triggered by external stimuli. As a result, this versatile method has garnered significant global attention for the synthesis of multifunctional ordered mesoporous SiO_2 core–shell nanostructures, such as Fe_2O_3@SiO_2@mSiO_2 [145, 146], Ag@SiO_2@mSiO_2 [147] ZSM-5@mSiO_2 [148] and doped-NaYF4@SiO_2@mSiO_2 [149, 150]. In addition to the silica layer, the presence of negatively charged molecules (such as citric acid, polyacrylic acid, and rich-oxygen functional groups) on the surface of the core can also facilitate the interfacial assembly of mesoporous silica [151–153].

Taking inspiration from the vertical mesopore channels reaching the surface, Teng et al. successfully expanded this strategy to create large and uniform mesoporous silica thin films. These films exhibited continuous, ordered, and perpendicular mesopore channels with significant domain sizes on the interface of substrates (Fig. 4.9) [154]. In this sol–gel process, the surfactant cations (CTA^+) are first strongly adsorbed on a negatively charged substrate and formed spherical CTAB micelles assemblies on the surface. When TEOS molecules are added, the resulted negatively charged oligomeric silicate species approach the spherical micelle surface through electrostatic interaction.

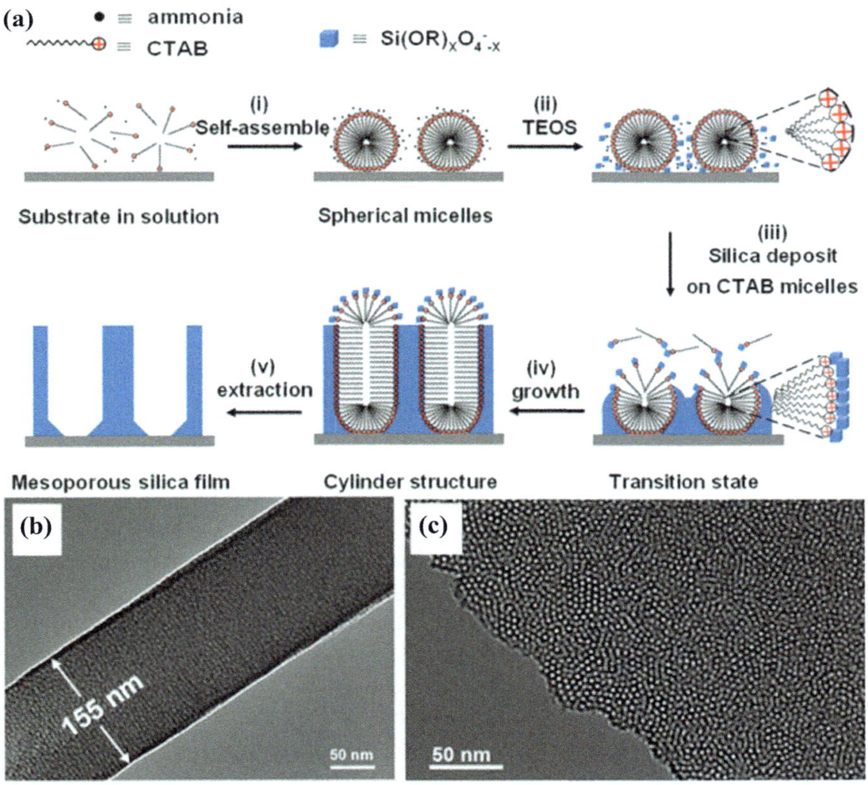

Fig. 4.9 **a** Schematic illustration of the formation process of the ordered mesoporous silica films with perpendicular mesochannels on the substrate via an aqueous process. Crosssection **b** and top-view **c** TEM images of the mesoporous silica films after vertical cleaving

At low concentrations, the deposition of free silicate species primarily occurs at the junction between the spherical micelles and the substrate. This deposition serves to mitigate the electrostatic repulsion among adjacent head groups within the micelles. Additionally, the diffusion of ethanol into the CTAB micelles reduces the interaction of the alkyl tails, thereby increasing the hydrophobic volume of the micelles and reducing their curvature. Furthermore, the presence of ammonia facilitates the formation of hydrogen bonds between adjacent CTAB micelles and silicate oligomers, promoting the formation of parallel mesochannels and minimizing the curvature energy of the micelles.

These three factors collectively drive the structural transformation of CTAB–silicate composites from spherical to cylindrical micelles. Through the continuous diffusion and re-assembly of CTAB molecules, newly hydrolyzed silicate oligomers are adsorbed and cross-linked onto the charged micelle head groups at the solution-silica film interface. This process results in the continuous growth of a large-domain film along the longitudinal direction of the perpendicularly oriented cylindrical micelles on the substrate.

Furthermore, this method can also be extended to fabricate mesoporous silica materials on 3D hierarchical interfaces. For example, Wu et al. successfully integrated a uniform mesoporous silica layer within 3D hierarchical interconnected macroporous graphene-based frameworks [155]. The resultant 3D graphene@mSiO_2 materials possess macro- and meso-porous structures with a high specific surface area,importantly which can enable the construction of other 3D graphene-based materials (eg.,mesoporous carbons and metal oxides) with hierarchical porous features via a simple nanocasting method.

One of the advantages of core–shell structures superior to single-component is the possible interface engineering between the core and shell materials. Recently, we reported the preparation of multifunctional mesoporous core–shell microspheres by incorporation of metal nanoparticles between the magnetic core and mesoporous silica shell via a combining sol–gel and interfacial deposition process [156]. As illustrated in Fig. 4.10a, these samples were prepared by following a procedure analogous to that for the aforementioned Fe_3O_4@nSiO_2@mSiO_2 microspheres except for an additional procedure for introduction of metal nanoparticles. Fe_3O_4@SiO_2 microspheres were first modified with 3-aminopropyl triethylsilane (APTS), and gold nanoparticles prepared by citrate reduction of $HAuCl_4$ were immobilized on the interface (Fig. 4.10b).Then, the resulting Fe_3O_4@SiO_2/Au microspheres were coated with a mesoporous silica layer, leading to a unique nanostructure (Fig. 4.10c) with multicomponents and multifunctionalities. The resultant microspheres have a high magnetization (~18.6 emu/g), a high surface area (~236 m^2/g),highly opened mesopores (~2.2 nm), and stably confined but accessible Au nanoparticles and, as a result, show high performances in catalytic reduction of 4-nitrophenol (with conversion of 95% in 12 min), styrene epoxidation with a high conversion (72%) and selectivity (80%),especially convenient magnetic separability, long life and good reusability. Following this concept, a wide variety of multifunctional mesoporous core–shell nanostructures have been fabricated [157–159]. Interesting, Song and coworkers

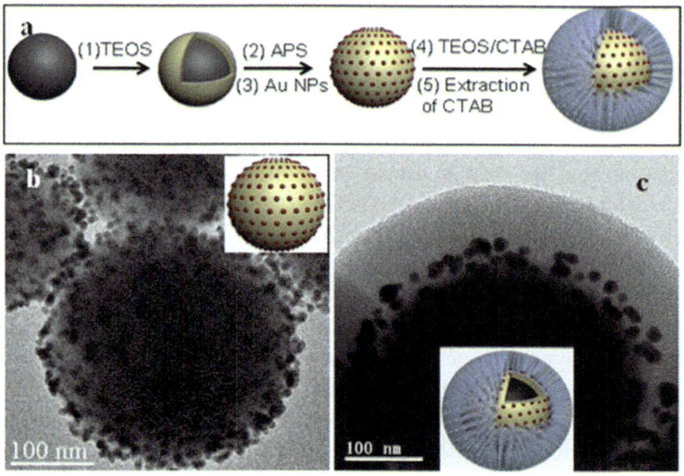

Fig. 4.10. **a** Schematic illustration of the interface engineering for Fe_3O_4@$nSiO_2$–Au@$mSiO_2$ microspheres. TEM and schematic (inset) images of **b** Au nanoparticle decorated Fe_3O_4@$nSiO_2$ microspheres and **c** Fe_3O_4@$nSiO_2$–Au@$mSiO_2$ microspheres

selectively removed the core, leading to a unique nanoreactor with mesoporous silica shell and Pd nanoparticles residing inside the hollow spheres [160].

While mesoporous silica shells have been extensively developed, they are typically utilized as surface protectors or supports rather than active centers. This is because colloidal silicas lack specific functionalities. In contrast, mesoporous metal oxides are expected to exhibit unique properties, such as titanium dioxide [152], a photoactive semiconductor, tin oxide [161], being used as an anode material in lithium ion batteries, and zirconium oxides [162], a famous catalyst supports.

These metal oxides can also be successfully synthesized on the surface of a substrate using the sol–gel coating method. The crucial factors for achieving well-defined metal oxide shells are still the compatibility between the templates and guests, as well as preferential heterogeneous nucleation [69]. High quality core@TiO_2 structures have been prepared via the sol–gel method by several groups [163–165]. Li et al. have, for the first time, demonstrated a typical example of uniformly constructing mesoporous TiO_2 at the interface through a versatile kinetics-controlled coating process [166]. A modified Stöber system of ethanol/ammonia was used with titanium butoxide (TBOT) as a TiO_2 precursor. The ammonia content in the system was found to play a crucial role in controlling the hydrolysis and condensation rates of mesoporous TiO_2. The kinetics controlled mechanism involves preferentially heterogeneous nucleation and growth of mesoporous TiO_2 on a core, resulting in uniform core–shell nanostructures within a narrow range of ammonia content. The thickness of the mesoporous shell can be adjusted between 5–70 nm. Using spindle-like α-Fe_2O_3 as a model substrate, mesoporous α-Fe_2O_3@TiO_2 microellipsoids with a uniform core–shell structure (Fig. 4.11a) were obtained. These microellipsoids have a high surface area (~404 m^2/g), which increases to ~105 m^2/g after calcination at 500 °C to transform into anatase phase (Fig. 4.11b). They also have a narrow pore size of 2.5–4.3 nm and exhibit excellent performance in lithium-ion storage and catalytic reactions. Importantly, this strategy is versatile and can be applied to various cores such as silica spheres, GO sheets, and magnetic nanospheres. In other cases, the outer layer of mesoporous TiO_2 can be transformed into unique nanostructures as needed. For instance, Zhao and colleagues prepared Fe_3O_4@titanate yolk shell microspheres with ultrathin nanosheets using a simple interfacial "hydrothermal etching assisted crystallization" strategy (Fig. 4.11c and d) [167]. The process involved obtaining a typical sandwich structure (Fe_3O_4@SiO_2@TiO_2), followed by hydrothermal treatment in an alkaline solution. During this process, the SiO_2 interlayer was etched, and then the TiO_2 layer was etched while epitaxial titanate nanosheets grew simultaneously. The SiO_2 layer promoted the coating of TiO_2 and acted as the template for the first cavity. The TiO_2 layer provided interfaces for nucleation and discrete space for self-assembled nanosheets to grow along opposite directions, serving as the template for the second cavity. This resulted in a yolk–shell structure with nanosheet-assembled double shells. The resulting Fe_3O_4@titanate microspheres had a uniform size (~560 nm), tailored shell structure, good structural stability, versatile ion-exchange capability, high surface area (~150 m^2/g), large magnetization (~17.7 emu/g), and enhanced acid catalysis performance for Friedel Crafts alkylation. Importantly, the titanate shell could be easily crystallized into anatase phase to generate magnetic mesoporous titania for photocatalysis.

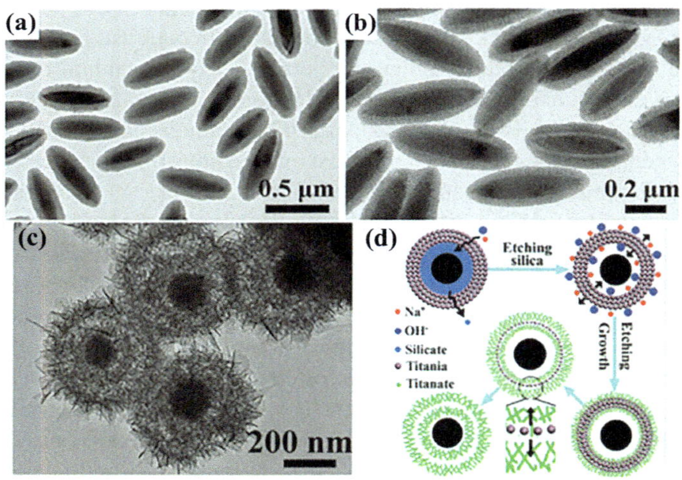

Fig. 4.11. TEM images of mesoporous core–shell α-Fe_2O_3@TiO_2 microellipsoids before **a** and after **b** calcinations at 500 °C in air. **c** TEM image and **d** schematic illustration of the formation mechanism of the Fe_3O_4@titanate yolk–shell microspheres

Hydrophobic Substrates

Another promising interfacial engineering method for constructing mesoporous materials in aqueous phase is first developed by Hyeon's group (Fig. 4.12a) [168]. In their approach, hydrophobic ligand-capped nanocrystals are mainly used as the substrates, and first converted into hydrophilic nanoparticles by the adsorption of CTAB molecules through hydrophobic–hydrophobic interaction. Then, the bimolecular layer stabilizes hydrophobic nanoparticles, which can act as nucleation seeds for the growth of a mesostructured composite layer. For instance, uniform magnetic mesoporous silica nanospheres with core–shell structures can be prepared by first treating hydrophobic oleic-acid stabilized magnetite nanoparticles with CTAB and then coating them with a mesoporous silica shell in the sol–gel aqueous process. The resulting spheres are uniform in size, with an average particle diameter of ~150 nm. Each silica sphere contains several monodisperse magnetite nanocrystals (Fig. 4.12b and c). The authors later improved the synthesis and achieved well-controlled magnetic mesoporous silica nanospheres by constructing mesoporous silica on the interface of individual nanocrystals (Fig. 4.12d) [169]. After being further modified with hydrophilic PEG and fluorescence isothiocyanate (FITC), these magnetic mesoporous nanospheres were successfully employed for multimodal imaging and drug delivery. This method has been extensively utilized for the fabrication of various other functional mesoporous materials [170, 171].

Assembly in a Confined Space

The porous AAO matrices are also utilized as confinement spaces for fabricating 1D mesoporous materials through the aqueous sol–gel method. In a typical process, AAO membranes are immersed in a synthesis solution during the gelation process.

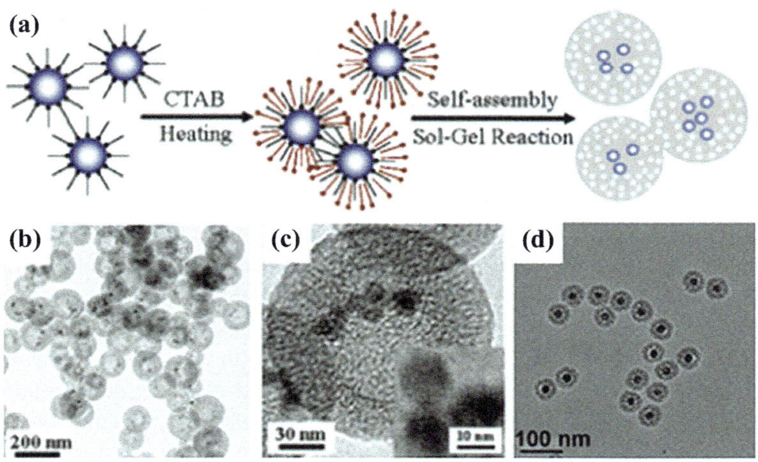

Fig. 4.12. Synthetic procedure (**a**) and (**b–d**) TEM images of hydrophobic magnetite nanocrystals embedded in mesoporous silica spheres

These solutions typically contain low amounts of solvent and high concentrations of surfactants. The gelation process usually requires a prolonged induction period, ranging from several hours to days. After complete infiltration, the AAO membranes are removed and subjected to further aging treatment. The surfactant is eliminated through calcination, and the alumina is removed with hydrochloric acid, resulting in the formation of uniform arrays of 1D mesoporous materials. Yang et al. discovered that mesoporous silica nanotubes could be formed in unmodified AAO, while nanofibers were obtained when hydrophobized membranes were used as substrates [172]. Lu et al .reported the synthesis of SBA-15 nanorod arrays inside AAO membranes [173]. In this case, the formed SBA-15 exhibits vertical mesochannels that are parallel to the channels of the AAO membrane. In contrast, silica fibers with a phase mixture of circular and columnar orientations were produced using the same recipe but with different immersion and aging times [174].

Later, Lawrence and his colleagues conducted extensive research to examine the impact of synthesis conditions on the resulting silica mesostructures. Their findings revealed that the aging environment played a crucial role in determining whether nanotubes or nanowires could be achieved. Specifically, aging the sol in the presence of additional water vapor resulted in silica rods with a circular hexagonal arrangement of mesopores. On the other hand, aging without additional water vapor led to the formation of silica rods with a columnar hexagonal mesostructured [175].

Assembly at Asymmetric Interfaces

While a conformal coating is typically preferred in most cases, there are certain special instances where a nonuniform coating, such as an anisotropic coating, is required to achieve unique structures and properties [176, 177]. For example, Feyen et al. synthesized PSD-SiO_2 Janus particles by first asymmetrically coating poly(styrene-co-divinylbenzene) (PSD) on functionalized iron oxide by free-radical polymerization and then coating silica using a sol–gel process with spatial confinement, in which the part was not covered by the polymer serves as the nucleation site because of its affinity toward silicate species [178]. Furthermore, the introduction of CTAB as porogens during the synthesis process effectively enables the creation of mesoporosity within the silica shell. Additionally, Wiesner and his colleagues successfully prepared multicompartment mesoporous silica nanoparticles, which feature a core exhibiting a cage-like cubic mesoporous morphology and up to four branches with hexagonally packed cylindrical mesopores [179]. The extent of cylindrical mesostructure growth can be well controlled via a single additive of ethyl acetate in the system. It is suggested that the nucleation and growth of hexagonally mesostructured compartments on the cubic mesoporous silica cores can occur on multiple facets of the cubic core, agreeing with models in which hexagonal branches grow in the [139] direction off of the core, consistent with the epitaxial relationship [179]. This result would suggest a path toward high levels of architectural complexity in locally amorphous, mesostructured nanoparticles, and enable tuning of different pore environments of the same particle for specific chemistries in catalysis or drug delivery.

Liquid-Phase Impregnation Nanocasting

Hard template nanocasting has been widely developed in the past decade as another approach to the synthesis of ordered mesoporous materials with new chemical compositions or pore structure topologies. For example, Ryoo's group first reported the synthesis of mesoporous carbons (CMK-1) by using aluminosilicate MCM-48 as a hard template [180]. Later, mesoporous Pt networks were fabricated in a similar way [181]. In a nanocasting process, three major steps are involved: (i) fluid precursor (typically liquid phase) infiltration inside mesochannels of the template; (ii) conversion of the precursor into target product in the mesochannels; (iii) removal of the mesoporous template. Some excellent and exhaustive reviews have been published covering various aspects of this technique for mesoporous materials, reflecting the tremendous advances in the past [71–75]. In the following, we will focus on the interface engineering of liquid precursors and rigid templates for the creation of well-defined mesoporous materials.

In addition to the nanoscale, nano-casting differs from conventional casting in that capillary condensation rather than mechanical penetration acts as a unique driving force, which can transport fluid into the nanoscale voids. Therefore, improving the capillary force of the system is a key problem [182]. It is well known that surface modification of pore channels can effectively change the wettability and improve the capillary force, thus promoting the penetration of precursors. In addition, the surface properties also have a great influence on the interaction between the precursor and the pore surface, the migration and aggregation of the guest material. Ion, molecule). If ions migrate and accumulate randomly in the middle channel, they will block the channel, resulting in low infiltration. This is because the interactions between the pore wall and the precursor are complex, including hydrogen and coordination bonds, Coulomb and van der Waals forces [183]. Typically, the precursors used for nanocasting are hydrophilic. Therefore, the hydrophilicity between the pore wall and the precursor can significantly improve the impregnation capacity and has a high degree of filling, reducing the tendency of the precursor to stick outside the mesopore or in the mesopore. Therefore, various methods have been developed to improve the hydrophilicity of templates. Of course, the hydroxyl group (Si–OH) on the silica pore wall is very much needed. However, in the process of removing surfactants by traditional calcination, the number of surface silanol (Si–OH) groups is dramatically reduced. Gently removing the surfactant from mesoporous silicates by solvent extraction may be a possible solution, but the resulting mesoporous silica is not suitable for the nanocasting process [184]. Zhao and co-workers developed a novel microwave digestion (MWD) method to remove surfactants but maintain abundant silanols on the silica pore walls [185]. Almost 100% of the surfactants in mesopores can be removed and the obtained mesoporous silica maintains a high density of silanol groups on the surface, which were confirmed by IR spectra (Fig. 4.13a), 1 H-NMR (nuclear magnetic resonance) spectroscopy, proton titration and chemical analysis. Framework shrinkage during calcination is avoided, besides the pore volume and surface area of the resultant mesoporous silica are larger than those of the extracted products. Due to abundant silanols, the interaction between the precursors and silicate walls is enhanced, facilitating the precursor infiltration with a high degree. As a consequence, MWD-mesoporous silica is an excellent candidate for nanocasting synthesis of ordered mesoporous replicas (Fig. 4.13b) [186–189]. For example, Tian et al.demonstrated the production of various high-surface-area ordered crystallized metal oxides (including, Cr_2O_3, Mn_2O_3, Fe_2O_3, Co_3O_4, NiO, In_2O_3 and CeO_2) by using such MWD-mesoporous silica as the powerful hosts, which does not require the functionalization process of mesochanneals [187]. The functionalization of the mesoporous silica templates by certain organic groups (e.g., $-NH_2$, $-CH_2{=}CH_2$) can also enhance the incorporation of the precursors. The silanols on the surface are active and easily modified, thus which can vary the surface property of the pore walls, for example, from hydrophilic to hydrophobic, from weak acidic to basic, finally leading to an obviously enhanced adsorption of a precursor into mesopores due to the strong interaction between the precursors and walls. For example, using amino functionalized SBA-15 or vinyl-functionalized KIT-6 as a hard template, a number of replica mesostructures of V_2O_5, Cr_2O_3, Fe_2O_3, WO_3 and Co_3O_4 have been successfully obtained [190–193].

Mesoporous carbon, as an alternative to preformed silica, is also being explored as a hard template for the preparation of mesoporous materials. However, the inertness and hydrophobicity of the original mesoporous carbon material make it less wettability and dispersibility in polar solvents (especially water), so it is unfavorable to the nano-casting process [194–196]. In 2003, Ryoo and coworkers first reported the surface modification of CMK-1 and CMK-5 by oxidation with concentrated HNO_3 solution at 110 °C for a very short period of 15 min in order to introduce surface oxygen-containing groups [197]. Zhao and coworkers have conducted a comprehensive study of pore evolution, mesostructural oxidation resistance, and simultaneous surface functionalization of ordered mesoporous carbon FDU-15 under various oxidation conditions by acidic $(NH_4)_2S_2O_8$, HNO_3,and H_2O_2 solutions [198]. The contents and properties of the oxygen-containing groups on the carbon surface are extensively studied by FTIR, thermogravimetric analysis (TG), elemental analyses, and water and ammonia adsorption techniques. After oxidation, the surface of FDU-15 has a high concentration of oxygen-containing groups, which makes the resultant carbons more hydrophilic (Fig. 4.13c). In addition, they found that the modified FDU-15 showed excellent mesostructural stability under strong oxidation conditions, compared with the mesostructural analogue CMK-3 carbon. Therefore, the functionalized mesoporous carbons can be utilized as a promising stable supports or hard templates for various functional mesoporous materials. Later on, the same authors first reported the controllable synthesis of ordered mesoporous aluminas with variable pore architectures and high mesoporosity, as well as crystalline framework by using such functionalized bimodal mesoporous carbon as the hard template [199]. When the pristine mesoporous carbon was adopted as a hard template, it was almost impossible to get any ordered mesostructure because of their hydrophobic nature, resulting in only partial filling and disordered aluminas. Therefore, the surface functionalization of the carbon framework is a critical issue. Meanwhile, such a method can also be extended to synthesize mesoporous MgO and ZnO, which are difficult

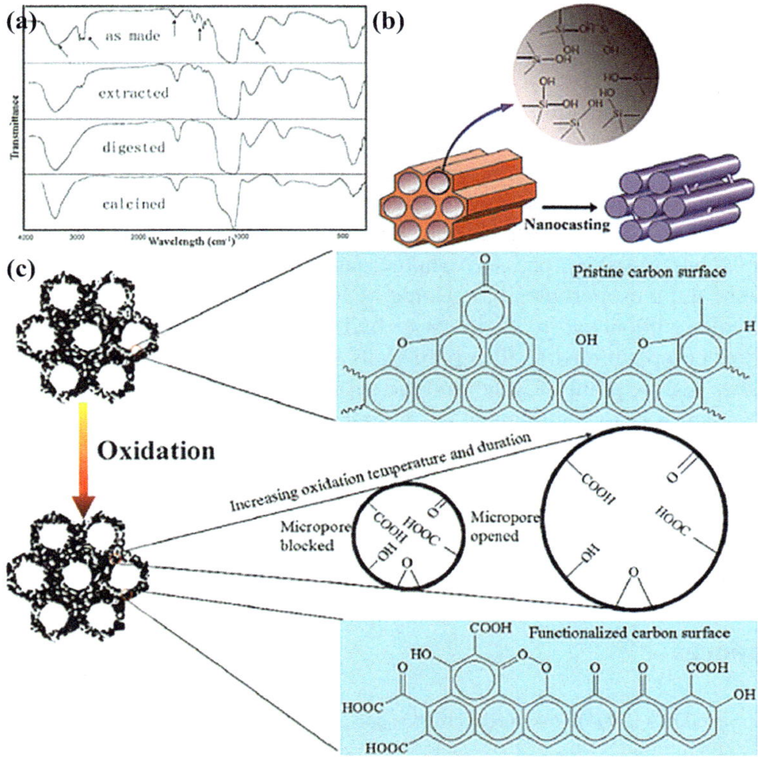

Fig. 4.13. **a** IR spectra of as-made, ethanol extracted, microwave digested and calcined SBA-15 mesoporous materials. **b** The scheme of nanocasting method templated with microwave digested SBA-15 mesoporous materials. **c** Possible models for surface structures of pristine mesoporous carbon (**a**) and the oxidized surface (**b**)

to be prepared by using mesoporous silica as a hard template. In addition, with a simple impregnation step, the surface-oxidized bimodal mesoporous carbon matrix can be easily infiltrated by metal nitrate precursors with a high loading level [200]. Upon the pre-hydrolysis step under ammonia atmosphere, oxide nanoparticles with high concentration (>52 wt%) homogeneously and exclusively dispersed in predefined mesopores. By combining such fascinating features, the mesoporous Fe_2O_3@C encapsulates show excellent performance for arsenic removal, with high adsorption capacities (up to ~29.4 mg g^{-1}), fast adsorption rate with pseudo-second-order kinetics, ready magnetic separation, and excellent cyclic stability.

Since most precursors require a specific solvent as the carrying medium, liquid is another important factor at the interface that should be considered during the process. The solubility of the precursor is related to the solvent. In general, the higher the solubility, the better the precursor filling and the higher the penetration efficiency. In addition, the solvent affects the surface wettability and thus the capillary force [204].

Solvation of the inorganic precursors shows influence on migration and aggregation. In general, solvents having low boiling points, such as ethanol, THF or water, is often used in nanocasting. High boiling point solvents are not recommended because later on they are difficult to remove by evaporation. Ethanol serves as one of the best ideal solvents in the nanocasting process. The advantages include that: (i) ethanol has a lower boiling point and is more volatile than water; (ii) most inorganic precursors are soluble in it; (iii) it has a mild interaction with precursors, which is weaker than water, facilitating facile migration and aggregation of precursors in the nanochannels during the nanocasting process; (iv) numerous hydroxyl groups are presented on the surface of the mesoporous silica. Domains with hydroxyl groups are hydrophilic, and domains without hydroxyl groups are hydrophobic. Ethanol also has amphiphilic properties compatible with silica pore walls, enhancing capillary forces. However, the higher boiling point of water and the stronger hydration effect make it difficult for the precursors to accumulate and migrate on the surface of silica. Since most precursors are low in solubility in tetrahydrofuran, and have no hydroxyl group and are low in wettability, tetrahydrofuran is rarely used. However, due to the high solubility of polymer precursors, it is a good solvent for organic polymers as carbon precursors to synthesize ordered mesoporous carbon.

References

1. El-Sayed MA (2004) Small is different: shape-, size-, and composition-dependent properties of some colloidal semiconductor nanocrystals. Acc Chem Res 37:326–333
2. Burda C, Chen X, Narayanan R et al (2005) Chemistry and properties of nanocrystals of different shapes. Chem Rev 105:1025–1102
3. Li D, Liang K, Cheng W et al (2019) Liquid-solid interfacial assemblies of soft materials for functional freestanding layered membrane-based devices toward electrochemical energy systems. Adv Energy Mater 9:1804005
4. Liu X, Hersam MC (2019) 2D materials for quantum information science. Nat Rev Mater 4:669–684
5. Yu B, Meyyappan M (2006) Nanotechnology: role in emerging nanoelectronics. Solid State Electron 50:536–544
6. Gao M, Zeng J, Liang K et al (2020) Interfacial assembly of mesoporous silica-based optical heterostructures for sensing applications. Adv Funct Mater 30:1906950
7. Rosi NL, Mirkin CA (2005) Nanostructures in biodiagnostics. Chem Rev 105:1547–1562
8. Zhang ZX, Liu Y, Meng WJ et al (2017) One-pot synthesis of Ni nanoparticle/ordered mesoporous carbon composite electrode materials for electrocatalytic reduction of aromatic ketones. Nanoscale 9:17807–17813
9. Loza K, Heggen Mepple M (2020) Synthesis, structure, properties, and applications of bimetallic nanoparticles of noble metals. Adv Funct Mater 30:1909260
10. Arciniegas MP, Kim MR, De Graaf J et al (2014) Self-assembly of octapod-shaped colloidal nanocrystals into a hexagonal ballerina network embedded in a thin polymer film. Nano Lett 14:1056–1063
11. Baker JL, Widmer-Cooper A, Toney MF et al (2010) Device-scale perpendicular alignment of colloidal nanorods. Nano Lett 10:195–201
12. Boal AK, Ilhan F, DeRouchey JE et al (2000) Self-assembly of nanoparticles into structured spherical and network aggregates. Nature 404:746–748

13. Fava D, Nie Z, Winnik MA et al (2008) Evolution of self-assembled structures of polymer-terminated gold nanorods in selective solvents. Adv Mater 20:4318–4322
14. Ferrier RC, Lee H-S, Hore MJA et al (2014) Gold Nanorod linking to control plasmonic properties in solution and polymer nanocomposites. Langmuir 30:1906–1914
15. Boles MA, Engel M, Talapin DV (2016) Self-assembly of colloidal nanocrystals: from intricate structures to functional materials. Chem Rev 116:11220–11289
16. Xu ZC, Shen CM, Xiao CW et al (2006) Fabrication of gold nanorod self-assemblies from rod and sphere mixtures via shape self-selective behavior. Chem Phys Lett 432:222–225
17. Thorkelsson K, Bai PXuT (2015) Self-assembly and applications of anisotropic nanomaterials: a review. Nano Today 10:48–66
18. Dessombz A, Chiche D, Davidson P et al (2007) Design of liquid-crystalline aqueous suspensions of rutile nanorods: evidence of anisotropic photocatalytic properties. J Am Chem Soc 129:5904–5909
19. Ye X, Chen J, Engel M et al (2013) Competition of shape and interaction patchiness for self-assembling nanoplates. Nat Chem 5:466–473
20. Ji Q, Acharya S, Hill JP et al (2008) Multi-dimensional control of surfactant-guided assemblies of quantum gold particles. Adv Mater 20:4027–4032
21. Li Z, Liang R, Xu S et al (2016) Multi-dimensional, light-controlled switch of fluorescence resonance energy transfer based on orderly assembly of 0D dye@micro-micelles and 2D ultrathin-layered nanosheets. Nano Res 9:3828–3838
22. Reinhoudt DN, Crego-Calama M (2002) Synthesis beyond the molecule. Science 295:2403
23. Ikkala Oten Brinke G (2002) Functional materials based on self-assembly of polymeric supramolecules. Science 295:2407
24. Chen LJ, Yang HB (2018) Construction of stimuli-responsive functional materials via hierarchical self-assembly involving coordination interactions. Acc Chem Res 51:2699–2710
25. Jiao T, Wu G, Zhang Y et al (2020) Self-assembly in water with N-substituted imines. Angew Chem Int Ed 59:18350–18367
26. Lan K, Xia Y, Wang R et al (2019) Confined interfacial monomicelle assembly for precisely controlled coating of single-layered Titania Mesopores. Matter 1:527–538
27. Zhang X, Liang H, Li H et al (2020) Sequential chemistry toward core-shell structured metal sulfides as stable and highly efficient visible-light photocatalysts. Angew Chem Int Ed 59:3287–3293
28. Lu C, Wu N, Jiao X et al (2003) Micropatterns constructed from Au nanoparticles. Chem Commun 1056-1057
29. Zhang H, Wang C, Li M et al (2005) Fluorescent nanocrystal−polymer composites from aqueous nanocrystals: methods without ligand exchange. Chem Mater 17:4783–4788
30. Dubertret B, Skourides P, Norris DJ et al (2002) In vivo imaging of quantum dots encapsulated in phospholipid micelles. Science 298:1759
31. Mulder WJM, Koole R, Brandwijk RJ et al (2006) Quantum dots with a paramagnetic coating as a bimodal molecular imaging probe. Nano Lett 6:1–6
32. Daniel M-CAstruc D (2004) Gold nanoparticles: assembly, supramolecular chemistry, quantum-size-related properties, and applications toward biology, catalysis, and nanotechnology. Chem Rev 104: 293–346
33. Lin Y, Böker A, Skaff H et al (2005) Nanoparticle assembly at fluid interfaces: structure and dynamics. Langmuir 21:191–194
34. Wang X, Fei Y, Wang W et al (2019) Polymer-mediated self-assembly of amorphous metal-organic complexes toward fabrication of three-dimensional graphene supported cop nanoparticle-embedded N-doped carbon as a superior hydrogen evolution catalyst. ACS Appl Energy Mater 2:8851–8861
35. Hanpanich OMaruyama A (2020) Artificial chaperones: From materials designs to applications. Biomaterials 254: 120150
36. Sun S, Murray CB, Weller D et al (2000) Monodisperse FePt nanoparticles and ferromagnetic FePt nanocrystal superlattices. Science 287:1989

37. Derrien TL, Hamada S, Zhou M et al (2020) Three-dimensional nanoparticle assemblies with tunable plasmonics via a layer-by-layer process. Nano Today 30:100823
38. Zhang H, Kinnear C, Mulvaney P (2020) Fabrication of single-nanocrystal arrays. Adv Mater 32:1904551
39. Sollami Delekta S, Laurila M-M, Mäntysalo M et al (2020) Drying-mediated self-assembly of graphene for inkjet printing of high-rate micro-supercapacitors. Nano-Micro Lett 12:40
40. Chen H, Zhang W, Li M et al (2020) Interface engineering in organic field-effect transistors: principles, applications, and perspectives. Chem Rev 120:2879–2949
41. Viswanathan P, Kim YJ, Hong JD (2020) Nanoporous Silver submicrocubes layer by layer encapsulated with polyelectrolyte films: nonenzymatic catalysis for glucose monitoring. Langmuir 36:3452–3460
42. Li W, Yue Q, Deng Y et al (2013) Ordered mesoporous materials based on interfacial assembly and engineering. Adv Mater 25(5129–5152):5128
43. Kralchevsky PA, Ivanov IB, Ananthapadmanabhan KP et al (2005) On the thermodynamics of particle-stabilized emulsions: curvature effects and catastrophic phase inversion. Langmuir 21:50–63
44. Lin Y, Skaff H, Emrick T et al (2003) Nanoparticle assembly and transport at liquid-liquid interfaces. Science 299:226
45. Binks BP, Clint JH (2002) Solid wettability from surface energy components: relevance to pickering emulsions. Langmuir 18:1270–1273
46. Yin Z, Tian B, Zhu Q et al (2019) Characterization and application of PVDF and its copolymer films prepared by spin-coating and Langmuir-Blodgett Method. Polymers 11:2033
47. Lebeau B, Galarneau ALinden M (2013) Introduction for 20 years of research on ordered mesoporous materials. Chem Soc Rev 42:3661–3662
48. Kresge CT, Roth WJ (2013) The discovery of mesoporous molecular sieves from the twenty year perspective. Chem Soc Rev 42:3663–3670
49. Li W, Zhao D (2013) An overview of the synthesis of ordered mesoporous materials. Chem Commun 49:943–946
50. Coasne B, Galarneau A, Pellenq RJM et al (2013) Adsorption, intrusion and freezing in porous silica: the view from the nanoscale. Chem Soc Rev 42:4141–4171
51. Kärger J, Valiullin R (2013) Mass transfer in mesoporous materials: the benefit of microscopic diffusion measurement. Chem Soc Rev 42:4172–4197
52. Ivanova II, Knyazeva EE (2013) Micro–mesoporous materials obtained by zeolite recrystallization: synthesis, characterization and catalytic applications. Chem Soc Rev 42:3671–3688
53. Wagner T, Haffer S, Weinberger C et al (2013) Mesoporous materials as gas sensors. Chem Soc Rev 42:4036–4053
54. Lai CY, Trewyn BG, Jeftinija DM et al (2003) A Mesoporous silica nanosphere-based carrier system with chemically removable CdS nanoparticle caps for stimuli-responsive controlled release of neurotransmitters and drug molecules. J Am Chem Soc 125:4451–4459
55. Zhai Y, Dou Y, Zhao D et al (2011) Carbon materials for chemical capacitive energy storage. Adv Mater 23:4828–4850
56. Yanagisawa T, Shimizu T, Kuroda K et al (1990) The Preparation of alkyltriinethylaininonium-kaneinite complexes and their conversion to microporous materials. B Chem Soc Jap 63:988–992
57. Kresge CT, Leonowicz ME, Roth WJ et al (1992) Ordered mesoporous molecular sieves synthesized by a liquid-crystal template mechanism. Nature 359:710–712
58. Ruiz JP, Dharia JR, Reynolds JR et al (1992) Repeat unit symmetry effects on the physical and electronic properties of processable, electrically conducting, substituted poly[1,4-bis(2-thienyl)phenylenes]. Macromolecules 25:849–860
59. Wan Y, Zhao N (2007) On the controllable soft-templating approach to mesoporous silicates. Chem Rev 107:2821–2860
60. Deng Y, Wei J, Sun Z et al (2013) Large-pore ordered mesoporous materials templated from non-pluronic amphiphilic block copolymers. Chem Soc Rev 42:4054–4070

61. Wan Y, Shi Y, Zhao D (2008) Supramolecular aggregates as templates: ordered mesoporous polymers and carbons. Chem Mater 20:932–945
62. Schüth F (2003) Endo- and exotemplating to create high-surface-area inorganic materials. Angew Chem Int Ed 42:3604–3622
63. Huo Q, Margolese DI, Ciesla U et al (1994) Generalized synthesis of periodic surfactant/inorganic composite materials. Nature 368:317–321
64. Huo Q, Margolese DI, Ciesla U et al (1994) Organization of organic molecules with inorganic molecular species into nanocomposite biphase arrays. Chem Mater 6:1176–1191
65. Yuan Z, Zhou W (2001) A novel morphology of mesoporous molecular sieve MCM-41. Chem Phys Lett 333:427–431
66. Ruthstein S, Schmidt J, Kesselman E et al (2006) Resolving intermediate solution structures during the formation of mesoporous sba-15. J Am Chem Soc 128:3366–3374
67. Sanchez C, Boissière C, Grosso D et al (2008) Design, synthesis, and properties of inorganic and hybrid thin films having periodically organized nanoporosity. Chem Mater 20:682–737
68. Sanchez C, Julián B, Belleville P et al (2005) Applications of hybrid organic–inorganic nanocomposites. J Mater Chem 15:3559–3592
69. Li W, Zhao D (2013) Extension of the Stöber method to construct mesoporous SiO_2 and TiO_2 shells for uniform multifunctional core–shell structures. Adv Mater 25:142–149
70. Deng Y, Cai Y, Sun Z et al (2011) Magnetically responsive ordered mesoporous materials: a burgeoning family of functional composite nanomaterials. Chem Phys Lett 510:1–13
71. Lu A, Schüth HF (2006) Nanocasting: a versatile strategy for creating nanostructured porous materials. Adv Mater 18:1793–1805
72. Lee J, Kim J, Hyeon T (2006) Recent progress in the synthesis of porous carbon materials. Adv Mater 18:2073–2094
73. Shi Y, Wan Y, Zhao D (2011) Ordered mesoporous non-oxide materials. Chem Soc Rev 40:3854–3878
74. Ren Y, Ma Z, Bruce PG (2012) Ordered mesoporous metal oxides: synthesis and applications. Chem Soc Rev 41:4909–4927
75. Petkovich N, Stein DA (2013) Controlling macro- and mesostructures with hierarchical porosity through combined hard and soft templating. Chem Soc Rev 42:3721–3739
76. Shi Y, Wan Y, Tu B et al (2008) Nanocasting synthesis of ordered mesoporous silicon nitrides with a high nitrogen content. J Phys Chem C 112:112–116
77. Wan Y, Yang H, Zhao D (2006) "Host−guest" chemistry in the synthesis of ordered nonsiliceous mesoporous materials. Acc Chem Res 39:423–432
78. Brinker CJ, Lu Y, Sellinger A et al (1999) Evaporation-induced self-assembly: nanostructures made easy. Adv Mater 11:579–585
79. Lu Y, Ganguli R, Drewien CA et al (1997) Continuous formation of supported cubic and hexagonal mesoporous films by sol–gel dip-coating. Nature 389:364–368
80. Grosso D, Cagnol F, Soler-Illia GJdAA et al (2004) Fundamentals of mesostructuring through evaporation-induced self-assembly. Adv Funct Mater 14:309–322
81. Grosso D, Babonneau F, Albouy A et al (2002) An in situ study of mesostructured ctab−silica film formation during dip coating using time-resolved saxs and interferometry measurements. Chem Mater 14:931–939
82. Grosso D, Balkenende AR, Albouy PA et al (2001) Two-dimensional hexagonal mesoporous silica thin films prepared from block copolymers: detailed characterization and formation mechanism. Chem Mater 13:1848–1856
83. Crepaldi EL, Soler-Illia GJdAA, Grosso D et al (2003) Controlled formation of highly organized mesoporous titania thin films: from mesostructured hybrids to mesoporous nanoanatase tio2. J Am Chem Soc 125:9770–9786
84. Soler-Illia GJdAA, Crepaldi EL, Grosso D et al (2003) Block copolymer-templated mesoporous oxides. Curr Opin Colloid In 8:109–126
85. Yu C, Fan J, Tian B et al (2003) Synthesis of mesoporous silica from commercial poly(ethylene oxide)/poly(butylene oxide) copolymers: toward the rational design of ordered mesoporous materials. J Phys Chem B 107:13368–13375

86. Hara M, NaganoT Seki S (2010) $\Pi-\pi$ interaction-induced vertical alignment of silica mesochannels templated by a discotic lyotropic liquid crystal. J Am Chem Soc 132:13654–13656
87. Vogt BD, Chavez VL, Dai M et al (2011) Impact of film thickness on the morphology of mesoporous carbon films using organic–organic self-assembly. Langmuir 27:5607–5615
88. Wang J, Xue C, Wu Z et al (2012) Hollow micro-mesoporous carbon polyhedra produced by selective removal of skeletal scaffolds. Carbon 50:2546–2555
89. Liu H-J, Wang X-M, Cui W-J et al (2010) Highly ordered mesoporous carbon nanofiber arrays from a crab shell biological template and its application in supercapacitors and fuel cells. J Mater Chem 20:4223–4230
90. Stein A, Li F, Denny NR (2008) Morphological control in colloidal crystal templating of inverse opals, hierarchical structures, and shaped particles. Chem Mater 20:649–666
91. Lu Y, McCaughey BF, Wang D et al (2003) Aerosol-assisted formation of mesostructured thin films. Adv Mater 15:1733–1736
92. Petkov N, Mintova S, Jean B et al (2003) Stable mesostructured silicate films containing nanosized zeolite. Chem Mater 15:2240–2246
93. Che S, Garcia-Bennett AE, Liu X et al (2003) Synthesis of large-pore *Iad* mesoporous silica and its tubelike carbon replica. Angew Chem Int Ed 42:3930–3934
94. Yang P, Zhao D, Margolese DI et al (1998) Generalized syntheses of large-pore mesoporous metal oxides with semicrystalline frameworks. Nature 396:152–155
95. Yuan Q, Yin A-X, Luo C et al (2008) Facile synthesis for ordered mesoporous γ-aluminas with high thermal stability. J Am Chem Soc 130:3465–3472
96. Yuan Q, Duan H-H, Li L-L et al (2010) Homogeneously dispersed ceria nanocatalyst stabilized with ordered mesoporous alumina. Adv Mater 22:1475–1478
97. Fan J, Boettcher SW, Stucky GD (2006) Nanoparticle assembly of ordered multicomponent mesostructured metal oxides via a versatile sol–gel process. Chem Mater 18:6391–6396
98. Zhang J, Deng Y, Gu D et al (2011) Ligand-assisted assembly approach to synthesize large-pore ordered mesoporous titania with thermally stable and crystalline framework. Adv Energy Mater 1:241–248
99. Tian B, Liu X, Tu B et al (2003) Self-adjusted synthesis of ordered stable mesoporous minerals by acid–base pairs. Nat Mater 2:159–163
100. Tian B, Yang H, Liu X et al (2002) Fast preparation of highly ordered nonsiliceous mesoporous materials via mixed inorganic precursors. Chem Commun, 1824–1825
101. Xu X, Tian BZ, Kong JL et al (2003) Ordered mesoporous niobium oxide film: a novel matrix for assembling functional proteins for bioelectrochemical applications. Adv Mater 15:1932–1936
102. Yuan Q, Liu Q, Song W-G et al (2007) Ordered mesoporous $Ce_{1-x}Zr_xO_2$ solid solutions with crystalline walls. J Am Chem Soc 129:6698–6699
103. Meng Y, Gu D, Zhang F et al (2005) Ordered mesoporous polymers and homologous carbon frameworks: Amphiphilic surfactant templating and direct transformation. Angew Chem Int Ed 44:7053–7059
104. Liu R, Shi Y, Wan Y et al (2006) Triconstituent co-assembly to ordered mesostructured polymer–silica and carbon–silica nanocomposites and large-pore mesoporous carbons with high surface areas. J Am Chem Soc 128:11652–11662
105. Meng Y, Gu D, Zhang F et al (2006) A family of highly ordered mesoporous polymer resin and carbon structures from organic–organic self-assembly. Chem Mater 18:4447–4464
106. Huang Y, Cai H, Yu T et al (2007) Formation of mesoporous carbon with a face-centered-cubic fd$\bar 3$m structure and bimodal architectural pores from the reverse amphiphilic triblock copolymer ppo-peo-ppo. Angew Chem Int Ed 46:1089–1093
107. Wan Y, Qian X, Jia N et al (2008) Direct triblock-copolymer-templating synthesis of highly ordered fluorinated mesoporous carbon. Chem Mater 20:1012–1018
108. Deng Y, Yu T, Wan Y et al (2007) Ordered mesoporous silicas and carbons with large accessible pores templated from amphiphilic diblock copolymer poly(ethylene oxide)-b-polystyrene. J Am Chem Soc 129:1690–1697

109. Florent M, Xue C, Zhao D et al (2012) Formation mechanism of cubic mesoporous carbon monolith synthesized by evaporation-induced self-assembly. Chem Mater 24:383–392
110. Platschek B, Keilbach A, Bein T (2011) Mesoporous structures confined in anodic alumina membranes. Adv Mater 23:2395–2412
111. Platschek B, Petkov N, Bein T (2006) Tuning the structure and orientation of hexagonally ordered mesoporous channels in anodic alumina membrane hosts: A 2d small-angle x-ray scattering study. Angew Chem Int Ed 45:1134–1138
112. Yamaguchi A, Uejo F, Yoda T et al (2004) Self-assembly of a silica–surfactant nanocomposite in a porous alumina membrane. Nat Mater 3:337–341
113. Wu Y, Cheng G, Katsov K et al (2004) Composite mesostructures by nano-confinement. Nat Mater 3:816–822
114. Yamaguchi A, Kaneda H, Fu W et al (2008) Structural control of surfactant-templated mesoporous silica formed inside columnar alumina pores. Adv Mater 20:1034–1037
115. Schuster J, Köhn R, Keilbach A et al (2009) Two-dimensional-hexagonal periodic mesoporous polymer resin thin films by soft templating. Chem Mater 21:5754–5762
116. Wang K, Wei M, Morris MA et al (2007) Mesoporous titania nanotubes: Their preparation and application as electrode materials for rechargeable lithium batteries. Adv Mater 19:3016–3020
117. Sun Z, Deng Y, Wei J et al (2011) Hierarchically ordered macro-/mesoporous silica monolith: Tuning macropore entrance size for size-selective adsorption of proteins. Chem Mater 23:2176–2184
118. Sen T, Tiddy GJT, Casci JL et al (2003) One-pot synthesis of hierarchically ordered porous-silica materials with three orders of length scale. Angew Chem Int Ed 42:4649–4653
119. Holland BT, Blanford CF, Do T et al (1999) Synthesis of highly ordered, three-dimensional, macroporous structures of amorphous or crystalline inorganic oxides, phosphates, and hybrid composites. Chem Mater 11:795–805
120. Deng Y, Liu C, Yu T et al (2007) Facile synthesis of hierarchically porous carbons from dual colloidal crystal/block copolymer template approach. Chem Mater 19:3271–3277
121. Schuster J, He G, Mandlmeier B et al (2012) Spherical ordered mesoporous carbon nanoparticles with high porosity for lithium–sulfur batteries. Angew Chem Int Ed 51:3591–3595
122. Xue C, Tu B, Zhao D (2008) Evaporation-induced coating and self-assembly of ordered mesoporous carbon-silica composite monoliths with macroporous architecture on polyurethane foams. Adv Funct Mater 18:3914–3921
123. Xue C, Zhao XQ, Bai FW (2010) Effect of the size of yeast flocs and zinc supplementation on continuous ethanol fermentation performance and metabolic flux distribution under very high concentration conditions. Biotechnol Bioeng 105:935–944
124. Xue C, Wang J, Tu B et al (2010) Hierarchically porous silica with ordered mesostructure from confinement self-assembly in skeleton scaffolds. Chem Mater 22:494–503
125. Wang J, Xue C, Lv Y et al (2011) Kilogram-scale synthesis of ordered mesoporous carbons and their electrochemical performance. Carbon 49:4580–4588
126. Wu KCW, Jiang X, Yamauchi Y (2011) New trend on mesoporous films: Precise controls of one-dimensional (1d) mesochannels toward innovative applications. J Mater Chem 21:8934–8939
127. Yamauchi Y, Nagaura T, Ishikawa A et al (2008) Evolution of standing mesochannels on porous anodic alumina substrates with designed conical holes. J Am Chem Soc 130:10165–10170
128. Richman EK, Brezesinski TTSH (2008) Vertically oriented hexagonal mesoporous films formed through nanometre-scale epitaxy. Nat Mater 7:712–717
129. Miyata H, Noma T, Watanabe M et al (2002) Preparation of mesoporous silica films with fully aligned large mesochannels using nonionic surfactants. Chem Mater 14:766–772
130. Miyata H, Kuroda K (2000) Formation of a continuous mesoporous silica film with fully aligned mesochannels on a glass substrate. Chem Mater 12:49–54
131. Koganti VR, Rankin SE (2005) Synthesis of surfactant-templated silica films with orthogonally aligned hexagonal mesophase. J Phys Chem B 109:3279–3283

132. Feng D, Lv Y, Wu Z et al (2011) Free-standing mesoporous carbon thin films with highly ordered pore architectures for nanodevices. J Am Chem Soc 133:15148–15156
133. Zhao D, Feng J, Huo Q et al (1998) Triblock copolymer syntheses of mesoporous silica with periodic 50 to 300 angstrom pores. Science 279:548
134. Asefa T, MacLachlan MJ, Coombs N et al (1999) Periodic mesoporous organosilicas with organic groups inside the channel walls. Nature 402:867–871
135. Che S, Garcia-Bennett AE, Yokoi T et al (2003) A novel anionic surfactant templating route for synthesizing mesoporous silica with unique structure. Nat Mater 2:801–805
136. Yang H, Kuperman A, Coombs N et al (1996) Synthesis of oriented films of mesoporous silica on mica. Nature 379:703–705
137. Ogawa M (1996) A simple sol–gel route for the preparation of silica–surfactant mesostructured materials. Chem Commun 10:1149–1150
138. Aksay IA, Trau M, Manne S et al (1996) Biomimetic pathways for assembling inorganic thin films. Science 273:892
139. Büchel G, Unger KK, Matsumoto A et al (1998) A novel pathway for synthesis of submicrometer-size solid core/mesoporous shell silica spheres. Adv Mater 10:1036–1038
140. Zhao W, Gu J, Zhang L et al (2005) Fabrication of uniform magnetic nanocomposite spheres with a magnetic core/mesoporous silica shell structure. J Am Chem Soc 127:8916–8917
141. Chen Y, Chen H, Zeng D et al (2010) Core/shell structured hollow mesoporous nanocapsules: A potential platform for simultaneous cell imaging and anticancer drug delivery. ACS Nano 4:6001–6013
142. Chen Y, Chen H, Guo L et al (2010) Hollow/rattle-type mesoporous nanostructures by a structural difference-based selective etching strategy. ACS Nano 4: 529-539
143. Yoon SB, Kim J-Y, Kim JH et al (2007) Synthesis of monodisperse spherical silica particles with solid core and mesoporous shell: Mesopore channels perpendicular to the surface. J Mater Chem 17:1758–1761
144. Deng Y, Qi D, Deng C et al (2008) Superparamagnetic high-magnetization microspheres with an Fe_3O_4@SiO_2 core and perpendicularly aligned mesoporous SiO_2 shell for removal of microcystins. J Am Chem Soc 130:28–29
145. Guo X, Deng Y, Gu D et al (2009) Synthesis and microwave absorption of uniform hematite nanoparticles and their core-shell mesoporous silica nanocomposites. J Mater Chem 19:6706–6712
146. Teng Z, Zhu X, Zheng G et al (2012) Ligand exchange triggered controlled-release targeted drug delivery system based on core–shell superparamagnetic mesoporous microspheres capped with nanoparticles. J Mater Chem 22:17677–17684
147. Yang J, Zhang F, Chen Y et al (2011) Core-shell Ag@SiO_2@mSiO_2 mesoporous nanocarriers for metal-enhanced fluorescence. Chem Commun 47:11618–11620
148. Qian XF, Li B, Hu YY et al (2012) Exploring meso-/microporous composite molecular sieves with core–shell structures. Chem Eur J 18:931–939
149. Yang P, Quan Z, Hou Z et al (2009) A magnetic, luminescent and mesoporous core–shell structured composite material as drug carrier. Biomaterials 30:4786–4795
150. Garcia JV, Yang J, Shen D et al (2012) Nir-triggered release of caged nitric oxide using upconverting nanostructured materials. Small 8:3800–3805
151. Qian X, Lv Y, Li W et al (2011) Multiwall carbon nanotube@mesoporous carbon with core-shell configuration: A well-designed composite-structure toward electrochemical capacitor application. J Mater Chem 21:13025–13031
152. Zhang Z, Wang L, Wang J et al (2012) Mesoporous silica-coated gold nanorods as a light-mediated multifunctional theranostic platform for cancer treatment. Adv Mater 24:1418–1423
153. Yang S, Feng X, Wang L et al (2010) Graphene-based nanosheets with a sandwich structure. Angew Chem Int Ed 49:4795–4799
154. Teng Z, Zheng G, Dou Y et al (2012) Highly ordered mesoporous silica films with perpendicular mesochannels by a simple stöber-solution growth approach. Angew Chem Int Ed 51:2173–2177

155. Wu Z-S, Sun Y, Tan Y-Z et al (2012) Three-dimensional graphene-based macro- and mesoporous frameworks for high-performance electrochemical capacitive energy storage. J Am Chem Soc 134:19532–19535
156. Deng Y, Cai Y, Sun Z et al (2010) Multifunctional mesoporous composite microspheres with well-designed nanostructure: a highly integrated catalyst system. J Am Chem Soc 132:8466–8473
157. Wang Y, Biradar AV, Duncan CT et al (2010) Silica nanosphere-supported shaped pd nanoparticles encapsulated with nanoporous silica shell: Efficient and recyclable nanocatalysts. J Mater Chem 20:7834–7841
158. Yang H, Chong Y, Li X et al (2012) Encapsulation of a catalytically active core with a nanoporous shell: a new strategy for designing size-selective catalysts. J Mater Chem 22:9069–9076
159. Okada S, Ikurumi S, Kamegawa T et al (2012) Structural design of Pd/SiO_2@Ti-containing mesoporous silica core–shell catalyst for efficient one-pot oxidation using in situ produced H_2O_2. J Phys Chem C 116:14360–14367
160. Chen Z, Cui Z-M, Niu F et al (2010) Pd nanoparticles in silica hollow spheres with mesoporous walls: a nanoreactor with extremely high activity. Chem Commun 46:6524–6526
161. Lou XW, Yuan C, Archer LA (2007) Shell-by-shell synthesis of tin oxide hollow colloids with nanoarchitectured walls: cavity size tuning and functionalization. Small 3:261–265
162. Arnal PM, Comotti M, Schüth F (2006) High-temperature-stable catalysts by hollow sphere encapsulation. Angew Chem Int Ed 45:8224–8227
163. Sakai H, Kanda T, Shibata H et al (2006) Preparation of highly dispersed core/shell-type titania nanocapsules containing a single ag nanoparticle. J Am Chem Soc 128:4944–4945
164. Demirörs AF, van Blaaderen A, Imhof A (2010) A general method to coat colloidal particles with titania. Langmuir 26:9297–9303
165. Ye M, Zhang Q, Hu Y et al (2010) Magnetically recoverable core–shell nanocomposites with enhanced photocatalytic activity. Chem Eur J 16:6243–6250
166. Li W, Yang J, Wu Z et al (2012) A versatile kinetics-controlled coating method to construct uniform porous TiO_2 shells for multifunctional core–shell structures. J Am Chem Soc 134:11864–11867
167. Li W, Deng Y, Wu Z et al (2011) Hydrothermal etching assisted crystallization: a facile route to functional yolk-shell titanate microspheres with ultrathin nanosheets-assembled double shells. J Am Chem Soc 133:15830–15833
168. Kim J, Lee JE, Lee J et al (2006) Magnetic fluorescent delivery vehicle using uniform mesoporous silica spheres embedded with monodisperse magnetic and semiconductor nanocrystals. J Am Chem Soc 128:688–689
169. Kim J, Kim HS, Lee N et al (2008) Multifunctional uniform nanoparticles composed of a magnetite nanocrystal core and a mesoporous silica shell for magnetic resonance and fluorescence imaging and for drug delivery. Angew Chem Int Ed 47:8438–8441
170. Kim T, Momin E, Choi J et al (2011) Mesoporous silica-coated hollow manganese oxide nanoparticles as positive T1 contrast agents for labeling and mri tracking of adipose-derived mesenchymal stem cells. J Am Chem Soc 133:2955–2961
171. Chen F, Zhang S, Bu W et al (2012) A uniform sub-50 nm-sized magnetic/upconversion fluorescent bimodal imaging agent capable of generating singlet oxygen by using a 980 nm laser. Chem Eur J 18:7082–7090
172. Yang Z, Niu Z, Cao X et al (2003) Template synthesis of uniform 1d mesostructured silica materials and their arrays in anodic alumina membranes. Angew Chem Int Ed 42:4201–4203
173. Lu Q, Gao F, Komarneni S et al (2004) Ordered sba-15 nanorod arrays inside a porous alumina membrane. J Am Chem Soc 126:8650–8651
174. Wang D, Kou R, Yang Z et al (2005) Hierachical mesoporous silica wires by confined assembly. Chem Commun, 166–167
175. Yao B, Fleming D, Morris MA et al (2004) Structural control of mesoporous silica nanowire arrays in porous alumina membranes. Chem Mater 16:4851–4855

176. Li J, Wei Y, Li W et al (2012) Magnetic spherical cores partly coated with periodic mesoporous organosilica single crystals. Nanoscale 4:1647–1651
177. Zhao N, Gao M (2009) Magnetic janus particles prepared by a flame synthetic approach: synthesis, characterizations and properties. Adv Mater 21:184–187
178. Feyen M, Weidenthaler C, Schüth F et al (2010) Regioselectively controlled synthesis of colloidal mushroom nanostructures and their hollow derivatives. J Am Chem Soc 132:6791–6799
179. Suteewong T, Sai H, Hovden R et al (2013) Multicompartment mesoporous silica nanoparticles with branched shapes: an epitaxial growth mechanism. Science 340:337
180. Ryoo R, Joo SH, Jun S (1999) Synthesis of highly ordered carbon molecular sieves via template-mediated structural transformation. J Phys Chem B 103:7743–7746
181. Shin HJ, Ryoo R, Liu Z et al (2001) Template synthesis of asymmetrically mesostructured platinum networks. J Am Chem Soc 123:1246–1247
182. Scott BJ, Wirnsberger G, Stucky GD (2001) Mesoporous and mesostructured materials for optical applications. Chem Mater 13:3140–3150
183. Yang H, Zhao D (2005) Synthesis of replica mesostructures by the nanocasting strategy. J Mater Chem 15:1217–1231
184. Gao F, Lu Q, Liu X et al (2001) Controlled synthesis of semiconductor pbs nanocrystals and nanowires inside mesoporous silica SBA-15 phase. Nano Lett 1:743–748
185. Tian B, Liu X, Yu C et al (2002) Microwave assisted template removal of siliceous porous materials. Chem Commun, 1186–1187
186. Tian B, Che S, Liu Z et al (2003) Novel approaches to synthesize self-supported ultrathin carbon nanowire arrays templated by MCM-41. Chem Commun, 2726–2727
187. Tian B, Liu X, Yang H et al (2003) General synthesis of ordered crystallized metal oxide nanoarrays replicated by microwave-digested mesoporous silica. Adv Mater 15:1370–1374
188. Tian B, Liu X, Solovyov LA et al (2004) Facile synthesis and characterization of novel mesoporous and mesorelief oxides with gyroidal structures. J Am Chem Soc 126:865–875
189. Liu X, Tian B, Yu C et al (2003) Ordered nanowire arrays of metal sulfides templated by mesoporous silica sba-15 via a simple impregnation reaction. Chem Lett 32:824–825
190. Zhu K, Yue B, Zhou W et al (2003) Preparation of three-dimensional chromium oxide porous single crystals templated by SBA-15. Chem Commun, 98–99
191. Jiao F, Yue B, Zhu K et al (2003) A-Fe_2O_3 nanowires. Confined synthesis and catalytic hydroxylation of phenol. Chem Lett 32:770–771
192. Zhu K, He H, Xie S et al (2003) Crystalline WO_3 nanowires synthesized by templating method. Chem Phys Lett 377:317–321
193. Wang Y, Yang CM, Schmidt W et al (2005) Weakly ferromagnetic ordered mesoporous Co_3O_4 synthesized by nanocasting from vinyl-functionalized cubic ia3d mesoporous silica. Adv Mater 17:53–56
194. Stein A, Wang Z, Fierke MA (2009) Functionalization of porous carbon materials with designed pore architecture. Adv Mater 21:265–293
195. Lu A-H, Li W-C, Muratova N et al (2005) Evidence for C–C bond cleavage by H_2O_2 in a mesoporous CMK-5 type carbon at room temperature. Chem Commun, 5184–5186
196. Bazuła PA, Lu A-H, Nitz J-J et al (2008) Surface and pore structure modification of ordered mesoporous carbons via a chemical oxidation approach. Micropor Mesopor Mat 108:266–275
197. Jun S, Choi M, Ryu S et al (2003) Ordered mesoporous carbon molecular sieves with functionalized surfaces. Stud Surf Sci Catal 146:37–40
198. Wu Z, Webley PA, Zhao D (2010) Comprehensive study of pore evolution, mesostructural stability, and simultaneous surface functionalization of ordered mesoporous carbon (fdu-15) by wet oxidation as a promising adsorbent. Langmuir 26:10277–10286
199. Wu Z, Li Q, Feng D et al (2010) Ordered mesoporous crystalline γ-Al_2O_3 with variable architecture and porosity from a single hard template. J Am Chem Soc 132:12042–12050
200. Wu Z, Li W, Webley PA et al (2012) General and controllable synthesis of novel mesoporous magnetic iron oxide@carbon encapsulates for efficient arsenic removal. Adv Mater 24:485–491

Chapter 5
Interfacial Assemblies for Film Devices

The interfacial assemblies are developed for the preparation of different mesoporous membrane devices. In the following context, diverse assembly strategies are introduced.

5.1 Liquid–Solid Interfacial Assembly

Liquid–solid interfacial assembly is a quite versatile interface in nature, which is widely used to construct and assemble various functional film devices. For example, as a typical example of liquid–solid interfacial synthesis, Langmuir–Blodgett (LB) technique with diverse advantages of high-throughput, low-cost and easily integrated was universally applied to fabricate both closely packed super-structures and well-defined patterns with low density through assembling amphiphilic molecules or nano-sized building blocks [1, 2]. In addition, liquid–solid interface has also been widely employed to produce a wide range of supported catalysts in the industry [3]. Specially, in this section, we will describe the interfacial synthesis of mesoporous materials which takes advantage of the liquid–solid interface.

5.1.1 Flow-Directed Interfacial Assembly

Carbon-based materials such as graphene oxide (GO) sheets, graphene, and carbon nanotubes (CNTs) can indeed be engineered into a thin film device under a directional flow [4]. A film can be constructed by vacuum filtration of colloidal dispersions with the assistance of membrane filter. The detailed operations are as follows: (1) Firstly, the powdered precursor is evenly dispersed in aqueous or organic solvent either by sonication or by mechanical stirring to form a suspension with comparatively low

B. Kong et al., *Functional Mesoporous Carbon-Based Film Devices for Energy Systems*,
https://doi.org/10.1007/978-981-99-7498-6_5

solid content. (2) The suspension is then filtered through a porous support (e.g., polycarbonate or mixed cellulose ester membrane). During the filtration process, the flow speed is enhanced with the aid of a vacuum pump. Consequently, the precursors are filtered on the membrane and form paper-like materials. (3) Finally, the film device is then peeled from the porous substrate and dried. Notably, the controllable film thickness can be achieved simply by adjusting the filtered volume and the concentration of the suspension. The flow-directed interfacial assembly offers a strategy for building film device with the advantages of high reproducibility and precise control in the film thickness.

This method is generally used to construct 2D film devices. Figure 5.1a illustrates the synthesis of the functional bilayer membrane device using poly(ionic liquid) (PIL)-grafted GO nanosheets (GO-PIL) on a porous filter paper [5]. GO nanosheets can be stabilized or chemically reduced by the strong interactions between the charged GO and PIL chains in a dispersion solution. In this work, the stabilization effect is achieved concurrently by radical polymerization of the PIL monomer, where the as synthesized PIL polymer chains are absorbed onto the GO surface with the aid of electrostatic interaction. After filtration through a filter paper, the micrometer-sized GO-PIL nanosheets can be placed onto the top of the filter paper in a flat orientation to form a thin coating layer. The freestanding film device (such as dense packed GO films [6, 7], rGO films [8, 9]) can be obtained after being peeled off from the filter membrane (Fig. 5.1b and c). In addition, the freestanding film can be transferred onto other substrates with diverse surface properties, including metal, conductive glass, and plastic substrates (Fig. 5.1d) [10], to construct functional asymmetric film devices. Besides GO, graphene films can also be deposited onto a membrane filter by vacuum filtration. Wallace et al. fabricated chemically converted graphene (CCG) paper which is flexible bendable and exhibits a shiny metallic lustre (Fig. 5.1e) [11]. Similar to GO and graphene sheets, CNTs can also be arranged and assembled into a membrane material under the directional flow. Recently, Chen et al. designed and built an editable supercapacitor capable of tailor-made stretchability from mechanically strengthened ultralong MnO_2 nanowires/CNT films synthesized via vacuum filtration method (Fig. 5.1f) [12]. Thereinto, the honeycomb-like structural supercapacitor presents a specific capacitance of 227.2 mF cm^{-2}. Notably, the MnO_2 nanowires/CNT films can still maintain excellent performance without any decline in electrochemical performance even be stretched up to 500%, placing it among one of the best stretchable supercapacitors. Therefore, the flow-directed interfacial assembly is an important strategy to construct high-performance and functional film devices, which show great potential in practical application.

5.1.2 Superlattice Interfacial Assembly

Superlattice interfacial assembly is a unique approach to fabricate highly ordered nanoparticle superlattice nano-membranes [13]. The nanoparticle superlattices have

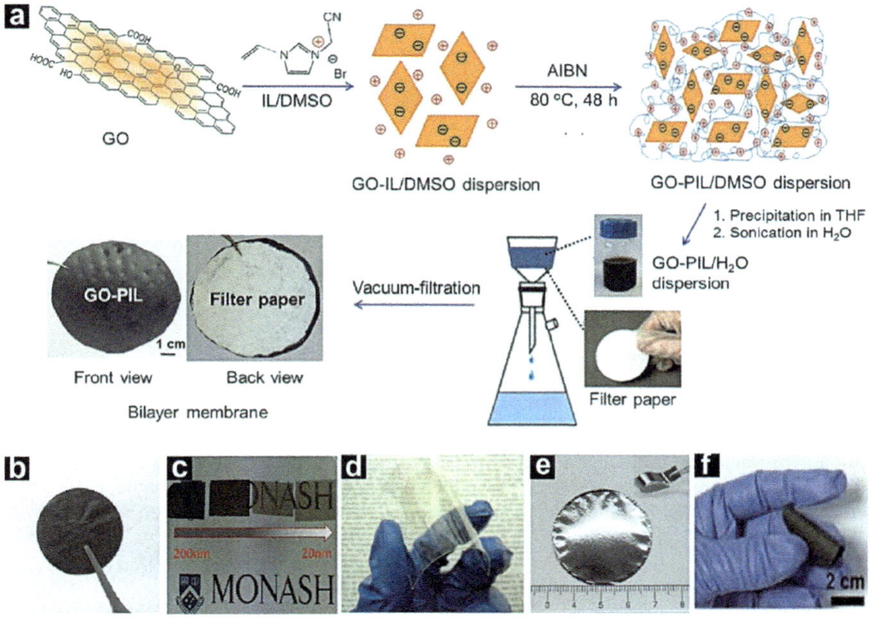

Fig. 5.1 **a** Schematic illustration of the preparation for GO-PIL/filter paper bilayer membrane. Reproduced with permission [5]. Copyright 2016, Wiley–VCH. **b-f** Corresponding digital camera images of **b** GO film. Reproduced with permission [6]. Copyright 2016, nature publishing group; **c** rGO films with the thicknesses of 200, 100, 40, and 20 nm (from left to right). Reproduced with permission [9]. Copyright 2015, Wiley–VCH; **d** GO film on plastic substrate. Reproduced with permission [10]. Copyright 2008, Nature publishing group; **e** CCG film and paper. Reproduced with permission [11]. Copyright 2008, Nature publishing group; **f** the MnO_2 nanowires/CNT composite film being bended. Reproduced with permission [12]. Copyright 2018, Wiley–VCH

been universally assembled onto solid and liquid surfaces as well as in bulk solutions [14], although post-processing these superlattices into separate structures while maintaining their order remains a challenge. It is known that organic linkers such as DNA can guide the nanoparticle assembly into superstructures in aqueous buffers [15]. These linkers can be designed with stimuli-responsive properties upon specific binding events or environmental changes. For instance, superlattice sheets were obtained by employing DNA as a 'dry ligand' together with microhole-confined crystallization (Fig. 5.2a) [16]. Free volumes of DNA-capped gold nanoparticles are restricted with water evaporating, leading to controlled crystallization at a particular concentration threshold. The ordered nanoparticle structures are preserved by the interactions between soft DNA coronas on the nanoparticle surface. The shapes of discrete freestanding superlattice sheets can be adjusted, meanwhile the sheet exhibits unique structural and functional properties (such as discrete, regularly sized, free-standing ones), which can be rationally tuned by adjusting DNA lengths (Fig. 5.2b).

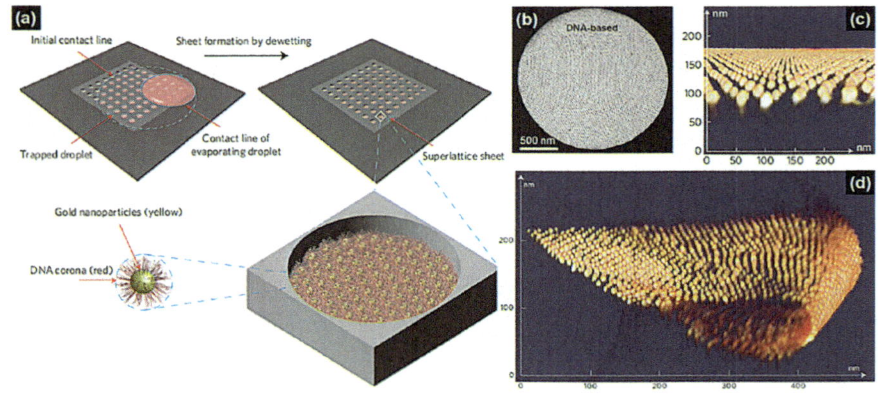

Fig. 5.2 Freestanding Au nanoparticle superlattice nano-membranes. **a** Schematic diagram of fabrication of freestanding Au nanoparticle superlattice sheets. **b** TEM micrograph of a DNA-based freestanding, monolayered superlattice suspended over a 2 μm-diameter microhole. **c** 3D STEM tomography. **d** 3D STEM tomography reconstruction of a folded sheet. Reproduced with permission [16]. Copyright 2009, Nature publishing group

Three-dimensional (3D) scanning transmission electron microscopy (STEM) characterizations confirmed the ultraflat orientation of the fully attached superlattice nano-membranes (Fig. 5.2c). Furthermore, the partially attached nano-membrane exhibits paper-like morphology (Fig. 5.2d). This fabrication strategy for freestanding superlattice demonstrates unique DNA-based method for self-assembled superlattice materials. Further, this strategy overcomes the stability concerns for biologically inspired assemblies in aqueous environments, which may greatly expands the potentials of biomaterials in solid-state devices. In addition, this approach is highly modular and versatile towards the fabrication of advanced systems beyond just plasmonic superlattice sheets.

5.1.3 Solvent-Casting Interfacial Assembly

Casting interfacial assembly is a common method for functional thin film preparation [17]. During this process, the as-prepared solution is casted on a substrate by an external force, followed by drying to obtain of a film. The thin film devices prepared by this method exhibit high-throughput, low-cost, easily integrated in the industry.

Typically, freestanding graphene-based film can also be acquired by a solvent-casting interfacial assembly process. Gao et al. proposed a protocol to synthesize collapsed GO and graphene papers from GO gels (Fig. 5.3a) [18]. A jelly-like film was prepared by blade-casting of GO/DMF liquid-crystalline dispersions on polytetrafluoroethylene (PTFE) substrate. Then, the collapsing process was promoted by soaking the films in the ethyl acetate (EA) pool. Subsequently, a freestanding gel film with a degree of shrinkage was prepared after gelation in EA. Finally, a flexible rubber-like GO papers with the width of 6 cm was obtained when the freestanding gel films are hanged and dried at 70 °C (Fig. 5.3b), and a novel air-stable and freestanding lithium alloy/graphene film is fabricated by solvent-casting interfacial assembly [19]. Besides, this method can be used to prepare composite film devices by mixing various functional materials. *Cui* et al. reported a strategy that can be mass-produced to prepare a Li_xM/graphene foil (M = Si, Sn, or Al). Figure 5.3c depicts the detailed film building process of the Li_xM/graphene foils. Generally, Li_xM nanoparticles are firstly mixed with graphene sheets and poly (styrene–butadiene–styrene) (SBS) rubber to form an even solution, and then cast on a release substrate made of polyethylene terephthalate (PET). After drying, it can be further removed as a freestanding film of Li_xM/graphene, which appears excellent flexibility and shows great potential for scaled up fabrication of densely packed reactive LixM nanoparticles in large graphene sheets (Fig. 5.3d).

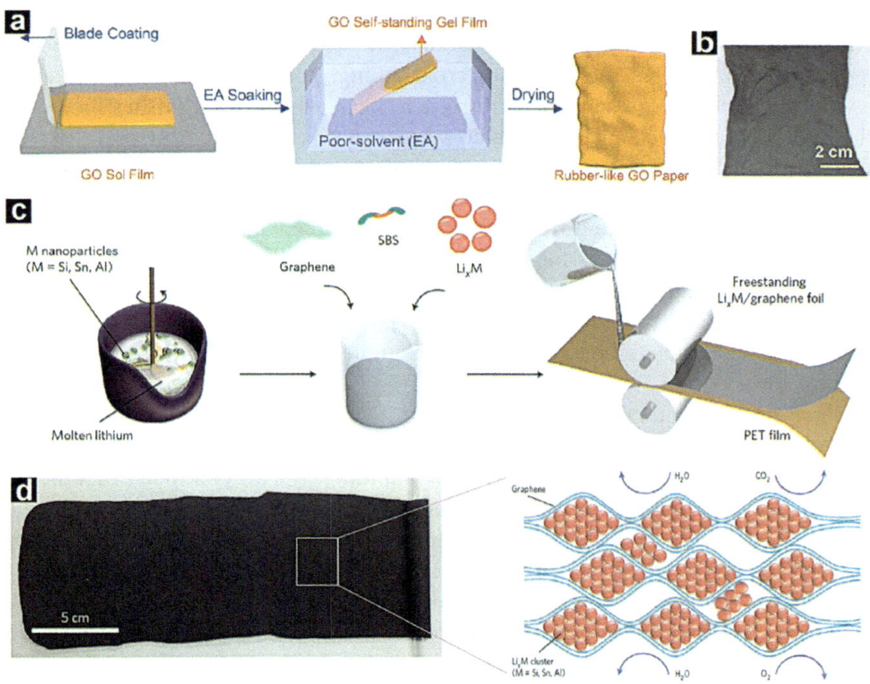

Fig. 5.3 **a** Schematic of the preparation process for collapsed GO paper. **b** The digital camera image of the drying of GO film. Reproduced with permission [18]. Copyright 2017, American chemical society. **c** Schematic of the fabrication process for LixM/graphene foils. **d** The digital camera image and the schematic microstructure of a large LixSi/graphene foil. Reproduced with permission [19]. Copyright 2017, Nature publishing group

5.1.4 Evaporation-Induced Self-Assembly (EISA)

Brinker and Ozin firstly reported EISA process to prepare mesoporous silica thin films [20]. The EISA strategy is usually designed for mesostructured thin films, where the self-assembly is induced with the successive evaporation of volatile solvents, leading to the formation of template-guest metastable phase with ordered mesostructure [21, 22]. In the field of functional film devices prepared by EISA, the most successful case is the preparation of well-ordered mesoporous polymers and carbon films [21]. Generally, the system of EISA consists of precursor solution, soft template (such as block copolymer) and volatile solvent. The interfacial assembly of precursor and block copolymer occurs rapidly results in ordered mesostructures without macrophase separation. The preferential evaporation of ethanol causes the copolymer to accumulate in space gradually, and then the precursors and copolymer composites form ordered liquid crystal mesophase. In addition, thermal curing of precursors and copolymers to form rigid hydrocarbon networks connected by covalent bonds, similar to zeolite or mesoporous silicate zeolite, is an important step to ensure the stability of mesoporous products. It should be noted that an important aspect of EISA is that the precursor crosslinking, polymerization, and surfactant assembly processes are separated from each other, which distinguishes them from the co-assembly mechanism, in which surfactant-directed assembly and the polymerization of inorganic oligomers occur simultaneously. Considering the difference in chemical and thermal stability between the precursor and copolymers, the soft templates can be easily removed at mild temperatures without damaging the resin framework. The continuous mesoporous hydrocarbon skeleton resin shows highly stable properties and can be directly converted into mesoporous carbon skeleton by annealing in an inert atmosphere. From a synthetic perspective, these mesoporous polymers and carbons films are excellent examples to extend the potential of EISA process.

Freestanding metal oxide-based films can also be synthesized via EISA process [23]. Zhao et al. synthesized hierarchically ordered macro-/mesoporous TiO_2 films (H-TiO_2) with high thermal stability and crystallinity by a the confined EISA method (Fig. 5.4a) [24]. Triblock copolymer Pluronic P123 is selected as a soft template to manufacture the mesopores, while the 3D periodic colloidal crystals made of polystyrene spheres (PS) are chosen as a hard template to induce the formation of macropores. The hierarchically ordered macro-/mesostructures can be visualized by Scanning electron microscopy (SEM). The thickness of H-TiO_2 film is measured to be 3 μm, and the size of the hexagonally arrayed macropores is ~255 nm (Fig. 5.4b, c). It should be noted that a bi-block copolymer can also be selected as a soft template, but this copolymer tends to aggregate and surface phase separation on the substrate due to the strong interface interaction. Therefore, acetylacetone (AcAc) coordinated titanium complex has served as an excellent titanium precursor for attaining ordered TiO_2 mesostructures (Fig. 5.4d) [25]. The effect of AcAc complexes controlling the hydrolysis and condensation of titanium precursors minimizes the effects of temperature and humidity and leads to controlled co-assembly with surfactant micelles, which ensures highly ordered mesopores with large specific surface area. SEM and TEM analysis of the pristine mesoporous TiO_2 films revealed the well-ordered mesostructure with ~12 nm in pore size after pyrolysis in air (Fig. 5.4e and f). The resulted large porosity gives access for efficient post-deposition of other semiconductor nanomaterials such as cadmium sulfide quantum dots (CdS QDs) inside the primary mesopores. Additionally, other transition metal ion dopants can be readily inserted into the TiO_2 frameworks by a co-assembly process during EISA, rendering it functions. Our group successfully incorporated ultrasmall (<5 nm) graphitic pencil nanodots (PNDs) into the ordered mesoporous frameworks (TiO_2, SiO_2, carbon, and carbon-SiO_2) by super-assembly, a process driven by hydrogen bonding between the PNDs and the frameworks' precursors during the EISA process. The synthesis involves site-inserted super-assembly of ultrasmall PNDs synthesized from commercial graphite pencils using an electrochemical tailoring process with mesoporous material precursors and surfactants as the template (Fig. 5.4g). The SEM and TEM images indicate that ultrasmall PNDs have successfully inserted into well-ordered mesostructured frameworks (Fig. 5.4 h, i). This strategy can be applied to a wide range of mesoporous materials, allowing the incorporation of guest molecules/materials in a well-dispersed fashion. In addition, this method offers a route to control the spatial distribution of nanodots within the mesoporous framework matrix.

5.1.5 Dip-Coating Assembly

Dip coating assembly, sometimes called immersion assembly, is usually performed by mechanically dipping the substrate into a solution containing the desired material at a specified rate to control the formation of a liquid layer on the surface of the substrate, and then washing to remove the unbound material [27]. Compared with gas deposition and nucleation deposition, dip coating assembly allows for a more uniform film, allowing this assembly method to be widely used for the formation of layered films. The assembly process is based on various interactions such as electrostatic interaction, covalent bond, hydrogen bond, van der Waals interaction or affinity interaction at the interface between solution and substrate [28, 29]. Importantly, the property of the substrate plays an important role on preparing film devices in terms of the morphology and mesochannels orientations [30]. Generally, planar substrates such as glass or silicon wafers are commonly used as film supports [50]. Other substrates with desirable interfaces can also be employed, such as polyether polyol-based polyurethane (PU) foams [21], anodic aluminum oxide (AAO) [31], Ni foam [32], and so on. For example, a 2D ordered mesoporous carbon nanosheets was prepared by a controlled low concentration monomolecular micelles close-packing dip assembly approach with the assistance of AAO substrate [31]. The surface of the synthesized carbon nanosheets has only one layer of ordered mesoporous pores, while the internal AAO pores can be further transformed into mesoporous graphene nanosheets by carbonization (Fig. 5.5a and b). The construction of mesoporous graphene nanosheets is achieved by controlling the heterogeneous

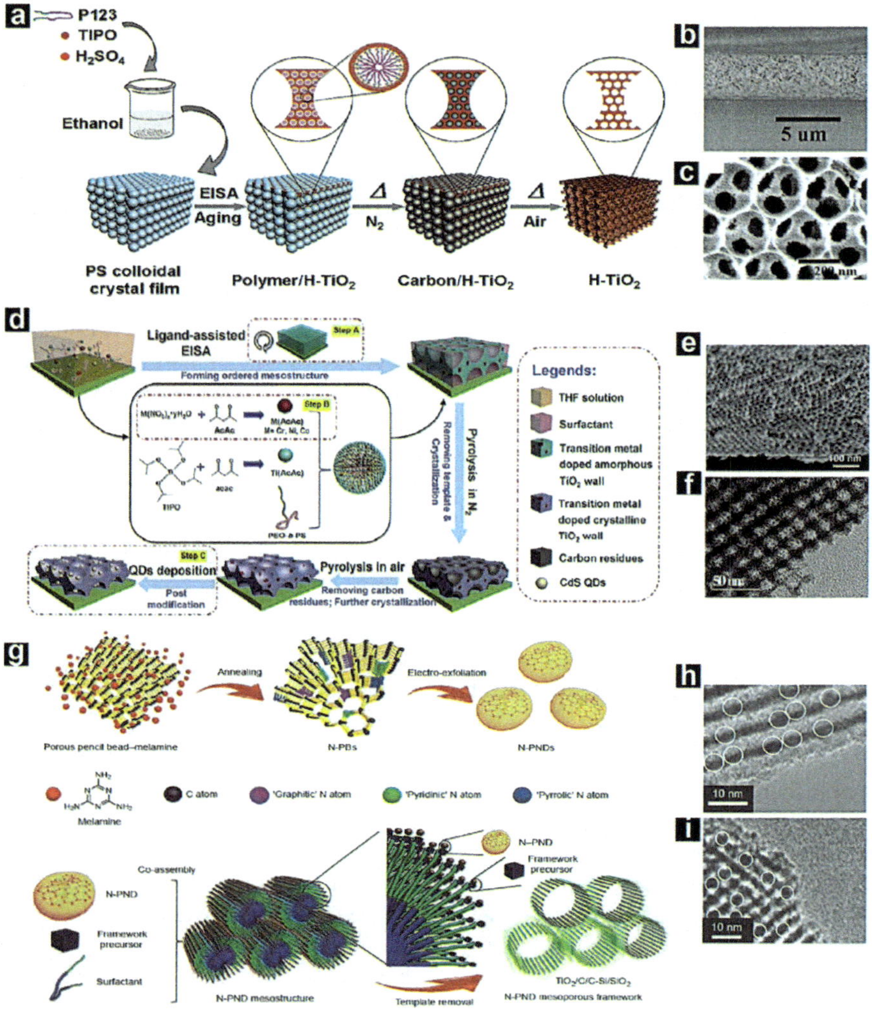

Fig. 5.4 **a** Schematic of the preparation process for H-TiO_2 films. **b**, **c** SEM images of H-TiO_2 films. Reproduced with permission [24]. Copyright 2014, Wiley–VCH. **d** Schematic of the preparation process for freestanding ordered mesoporous TiO_2 films, transition metal (Cr, Ni, Co) doping and CdS QD deposition. **e** SEM image and **f** TEM image of the mesoporous TiO2 films calcined in air at 450 °C. Reproduced with permission [25]. Copyright 2013, Royal society of chemistry. **g** Schematic of the preparation process for a series of PND mesoporous frameworks (TiO_2, C, C-Si, C-SiO_2). **h** TEM image of PND-inserted mesoporous SiO_2. (i) TEM image of PND-inserted mesoporous C-SiO2. Reproduced with permission [26]. Copyright 2016, Nature publishing group

nucleation and growth of single molecule micelles on the substrate surface (Fig. 5.5c). Firstly, due to the strong interaction between the single-molecule micelles and the substrate, the synthesized sol can be assembled with the Pluronic triblock copolymer template to form spherical single-molecule micelles and deposited on the substrate surface. The surface of single-molecule micelles is rich in phenolic hydroxyl groups, which can adhere to the substrate surface by electrostatic attraction because of their strong negative charges under alkaline conditions. Secondly, continuous deposition, assembly, and then slow transformation of the monomolecular micelles subunits into an ordered micellular array was achieved by a close packing route to form the lowest energy state. In the process of array formation, the heterogeneous nucleation of single molecule micelles on the substrate surface through electrostatic interaction is more likely than spontaneous nucleation and accumulation, which prevents the multi-layer deposition of single molecule micelles to form 3D microstructure. Consequently, these micelle arrays can be polymerized and assembled tightly to form bowl-like 2D planar mesostructure structures after hydrothermal treatment. Finally, single-layer bowl-like mesoporous graphene nanosheets are obtained by high temperature calcination. The unique 2D mesoporous structure is favorable for enhanced lithium-ion storage, while the flat mesopore-graphene layers offer high specific surface area for lithium ion adsorption/intercalation, and the perpendicular mesopore-graphene layers enable high-efficient ion transport and volume expansion flexibility.

Dip-coating assembly on 3D substrates can also be used to controllably synthesis 3D composite electrodes. Multilayer films fabricated by Ni foam supported poly(ethyleneimine) (PEI)/reduced graphene oxide (RGO) presented tunable oxygen evolution reaction activity and stability by this method [32]. The corresponding fabrication process is illustrated in Fig. 5.5d. In the assembly step, PEI is self-assembled on the Ni foam by coordination interaction and hydrogen bonding with in-situ formed $Ni(OH)_2$ on its surface, while GO nanosheets are self-assembled on the Ni foam by electrostatic interaction and covalent bonding with the PEI. The step is repeated for predetermined times to obtain multilayers with controllable numbers of bilayers.

5.1.6 Spinning Assembly

Spinning-coating is another general method to synthesize diverse freestanding membranes [33]. Spinning assembly uses rotating substrates to deposit layers and remove excess material. The solution of desired material was first deposited at the center of a spinning substrate, and the excess solution can be spread out by the centrifugal force under the high-speed rotation. Then, a film can be formed at the same time of being subjected to a shearing force at the interface of the drying solution layer and the substrate. Lee et al. demonstrated for the first time that spin assembly is caused by a range of forces, including electrostatic interactions that lead to adsorption and rearrangement of polymer chains, centrifugation, air shear, and viscous forces that lead to desorption of weakly bound soft materials and dehydration of membranes [34]. This significant difference in adsorption capacity is caused by

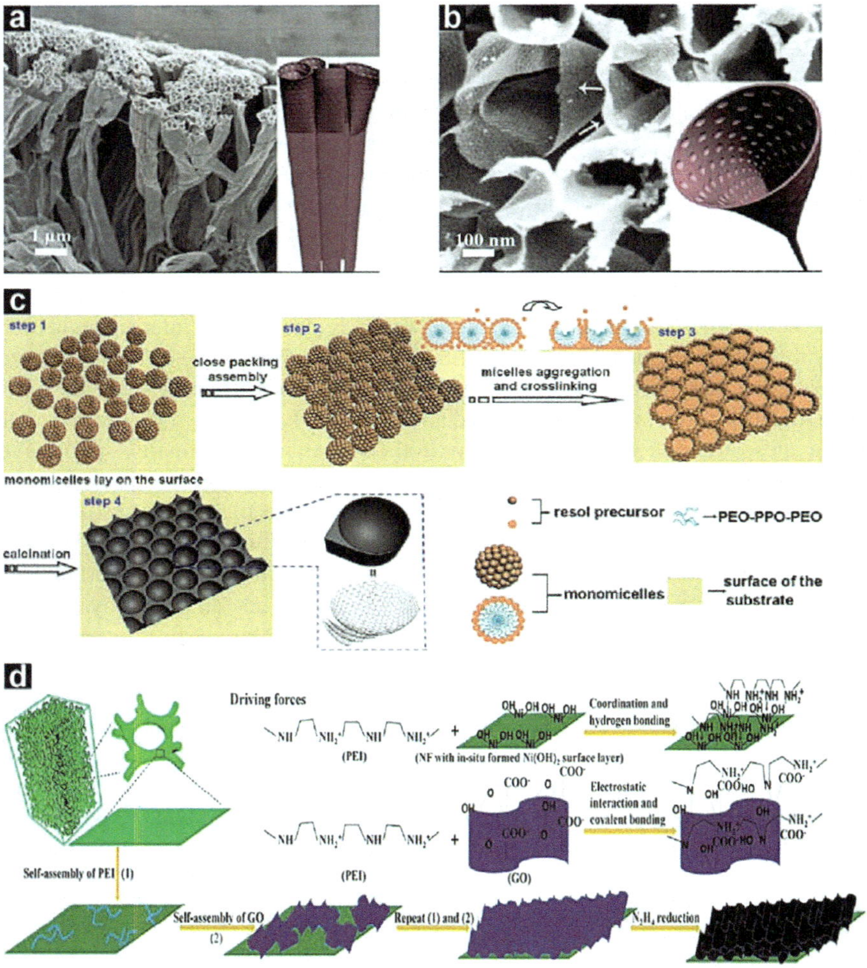

Fig. 5.5 **a** SEM image and schematic (inset) of the mesoporous graphene nanosheets. **b** SEM images and schematic (inset) of the mesoporous graphene nanosheets observed along the nanosheets. **c** Schematic of the formation process for the 2D mesoporous graphene nanosheets. Reproduced with permission [31]. Copyright 2012, American chemical society. **d** Schematic illustration of the fabrication process of NF supported PEI/RGO multilayers. Reproduced with permission [32]. Copyright 2017, Royal society of chemistry

different adsorption mechanisms. Compared with the membranes prepared by dip-coating assembly, whereas spin assembly produces thinner, smoother films. A significant number of recent reports employed spinning assembly to prepare films, such as mesoporous carbon film (Fig. 5.6a) [35], mesoporous silica, mesoporous carbon–silica composite films (Fig. 5.6b) [36], Au nanowires-based flexible film (Fig. 5.6c) [37], and perovskite thin films (Fig. 5.6d) [38]. Kunitake et al. reported a simple spin-coating procedure for the facile synthesis of freestanding metal-oxide-gel films with controllable nanometer thicknesses macroscopic lateral dimensions [39]. In this study, the inorganic network was brittle after removing the sacrificial layer. To overcome the lack of flexibility and robustness of inorganic substrates, the group later described an improved spin coating method for creating ultra-thin organic–inorganic hybrid films with interpenetrating network (IPN) structures (Fig. 5.6e) [40]. In their work, poly(vinylphenol) of 100 nm thickness as a 'sacrificial' layer is first spin-coated onto a silicon substrate, followed by the deposition of a 5-nm-thick layer of poly(vinyl alcohol) (PVA) layer. Afterwards, the solution comprising of both organic and inorganic precursors is spin-coated within a nitrogen atmosphere under the influence of ultraviolet (UV) light. The UV radiation triggers the degradation of the photo-initiator, subsequently initiating radical polymerization. Simultaneously, the presence of residual humidity in the PVA under layer induces a sol–gel process, while the UV photo polymerization takes place. This unique combination enables the simultaneous growth of two distinct networks. As a result, the film can be separated from the substrate by delicately scratching the sample's edges with a needle. By immersing the substrate in ethanol, the sacrificial underlayer is dissolved, resulting in the freestanding film being of comparable size to the substrate without any observed cracks. Recently, Bao et al. introduced a process that enables the high-yield and uniform production of various intrinsically stretchable electronic polymers. (Fig. 5.6f) [41]. In this study, a water-soluble sacrificial layer is initially spin-coated onto a silicon wafer. Subsequently, a stretchable dielectric is deposited onto the sacrificial layer and photo-patterned. Following that, a top-contact structure is created by consecutively depositing a stretchable semiconductor and a stretchable conductor to serve as source/drain electrodes, along with the first layer of interconnect. The fabrication of the device involves laminating the stretchable substrate, followed by immersion in water to release the flexible device from the rigid substrate. Finally, the gate electrodes (along with the second layer of interconnect) are deposited and patterned on the dielectric layer, resulting in the final transistor structure.

5.1.7 Electrospinning Assembly

Initially developed as a technique for synthesizing nanofibers from polymer solutions, electrospinning assembly utilizes the electrostatic repulsion between surface charges to continuously generate nanofibers from a viscoelastic fluid. Through the manipulation of their alignment, stacking, and/or folding, the electrospun nanofibers can be organized into freestanding films featuring ordered arrays or hierarchical structures. To date, over one hundred different types of polymers have been employed in electrospinning, consisting of polyacrylonitrile (PAN), PU [42], polycarbonate (PC) [43], polyvinyl alcohol (PVA) [44], polylactic acid (PLA) [45], polymethacrylate (PMMA) [46], polyethylene oxide (PEO) [47], PS [48], and so on. The integration of electrospinning and sol–gel chemistry enables the production of diverse layered devices, encompassing small molecules, carbon composites, and their combinations.

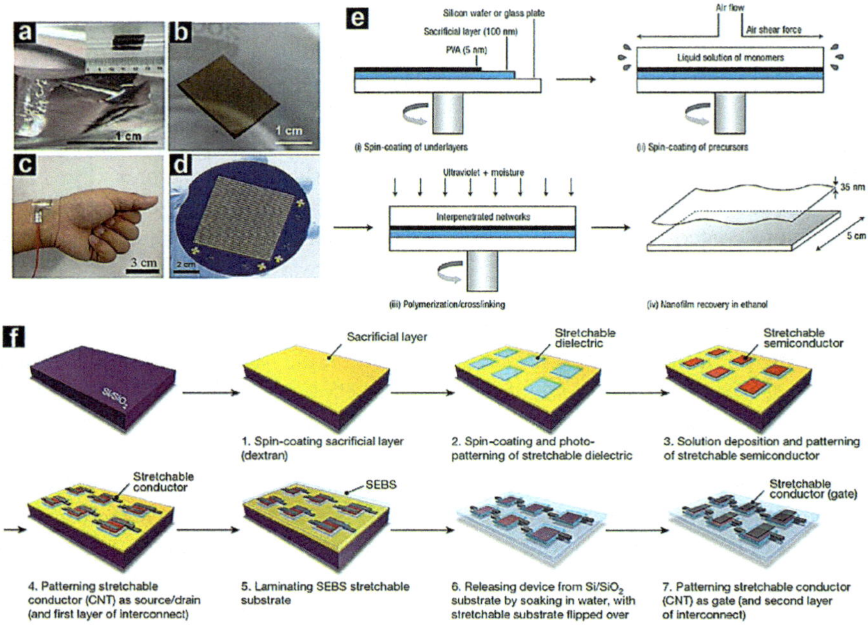

Fig. 5.6 (a-d) Photograph of films prepared by spin-coating assembly: **a** mesoporous carbon film floating in water (the inset shows the freestanding carbon film transferred onto a cylindrical substrate). Reproduced with permission [35]. Copyright 2011, American chemical society; **b** mesoporous carbon–silica composite thin film floating on water. Reproduced with permission [54]. Copyright 2013, Royal society of chemistry; **c** Au nanowires-based flexible film. Reproduced with permission [37]. Copyright 2014, Nature publishing group; **d** perovskite thin films. Reproduced with permission [38]. Copyright 2017, Wiley–VCH. **e** Schematic of the synthesis process for freestanding hybrid IPN nanofilm. Reproduced with permission [40]. Copyright 2006, Nature publishing group. **f** Schematic illustration of the fabrication process for stretchable transistor arrays. Reproduced with permission [41]. Copyright 2018, Nature publishing group

As an illustration, in a preliminary investigation, a composite nanofiber comprising of polyvinylpyrrolidone (PVP) and amorphous TiO2 was effortlessly fabricated by electrospinning an alcoholic solution containing PVP and titanium tetraisopropoxide. Typically, solid nanofibers exhibit a smooth surface. Nevertheless, electrospinning assembly has been employed to generate nanofibers with diverse secondary structures, containing porous [49], hollow [50], or core-sheath structures [51]. Many studies have also prepared mesoporous carbon-fiber (CNF) film devices via electrospun assembly [52]. When PS is incorporated into the PAN solution, it is observed that a micro emulsion can form and transform into nanoscale wires within the PAN fibers during the electrospinning process. Additionally, nanochannels can be generated through decomposition during the subsequent pyrolysis process. Lou et al. developed a freestanding pie-like paper electrode by electrospinning PAN/PS precursors.

The inner fillings of this electrode consist of 3D interconnected lotus root-like multichannel carbon (LRC) nanofibers, while the outer crust is a thin layer of ethylenediamine (EDA)-functionalized reduced graphene oxide (EFG) [53]. As depicted in Fig. 5.7a, the electrospinning of a mixed polymer precursor of PAN/PS allows for the formation of 3D networks comprising multichannel LRC nanofibers with controllable channel structures within each fiber. Upon PS removal through carbonization, parallel channels are formed within the carbonized PAN shells. The introduction of sulfur into the LRC channels is achieved by heating commercial sulfur powder with the LRC paper. Subsequently, the freestanding LRC/S electrode can be divided into smaller square pieces and dip-assembled with a layer of interconnected EFG nanosheets on their surfaces. Completing the process involves thorough drying to obtain the freestanding LRC/S@EFG structure. Furthermore, during the electrospinning process, functional materials can be added to further enhance the properties of the nanofibers. The synthesis strategy involves a two-step process: (1) Co-solvation of the metal salt and polymer to obtain a homogeneous solution, and (2) electrospinning the solution mixture to form composite nanofibers consisting of the polymer and metal salt. Using this approach, Lou et al. successfully designed and synthesized a freestanding lotus root-like carbon fiber network decorated with CoS_2 nanoparticles (CoS_2@LRC) to serve as the host material for enhancing the performance of Li-SeS_2 batteries [48].

Cui et al. demonstrated a high throughput method that significantly reduces the run time required for the common electrospinning assembly strategy. Their approach involves the fast transfer of electrospun nanofiber films from roughed metal foil to a mesh substrate [54]. In contrast to directly electrospinning the nanofiber film onto a conductive mesh (Fig. 5.7b i), the transferred nanofiber film exhibits a much more uniform fiber distribution under an optical microscope (Fig. 5.7b ii). Additionally, the transfer method is 10 times faster due to the enhanced uniformity of the transferred nanofiber film. The novel transfer process of electrospun nanofibers involves the following steps: (1) Using the same polymers for surface coating of both the plastic mesh and the electrospun fibers, ensuring strong adhesion between the fibers and the mesh. (2) Utilizing rough copper foil as the electrospinning substrate, which is crucial for achieving a clean and complete transfer. (3) Applying uniform pressure to laminate the plastic mesh onto the copper foil, effectively sandwiching the nanofibers in between. (4) Transferring the nanofiber film to the plastic mesh by removing the plastic mesh from the copper foil, taking advantage of the stronger adhesion of the nanofibers to the plastic mesh. By further transferring onto facial masks, a freestanding round nanofiber network with a diameter of 15 cm can be obtained, similar to the case of mesh transfer (Fig. 5.7b iii). Based on this method, the authors have successfully achieved roll-to-roll production of freestanding nanofiber films for commercial products. Although electrospinning is considered the most promising and reliable method for commercial nanofiber fabrication, several challenges still need to be addressed, including the risk associated with high operating voltages, solvent compatibility in terms of dielectric constant, and complex operating conditions [55]. Taking all of these factors into consideration, Cui et al. have successfully developed a new blow spin methodology that not only eliminates the need for high voltages

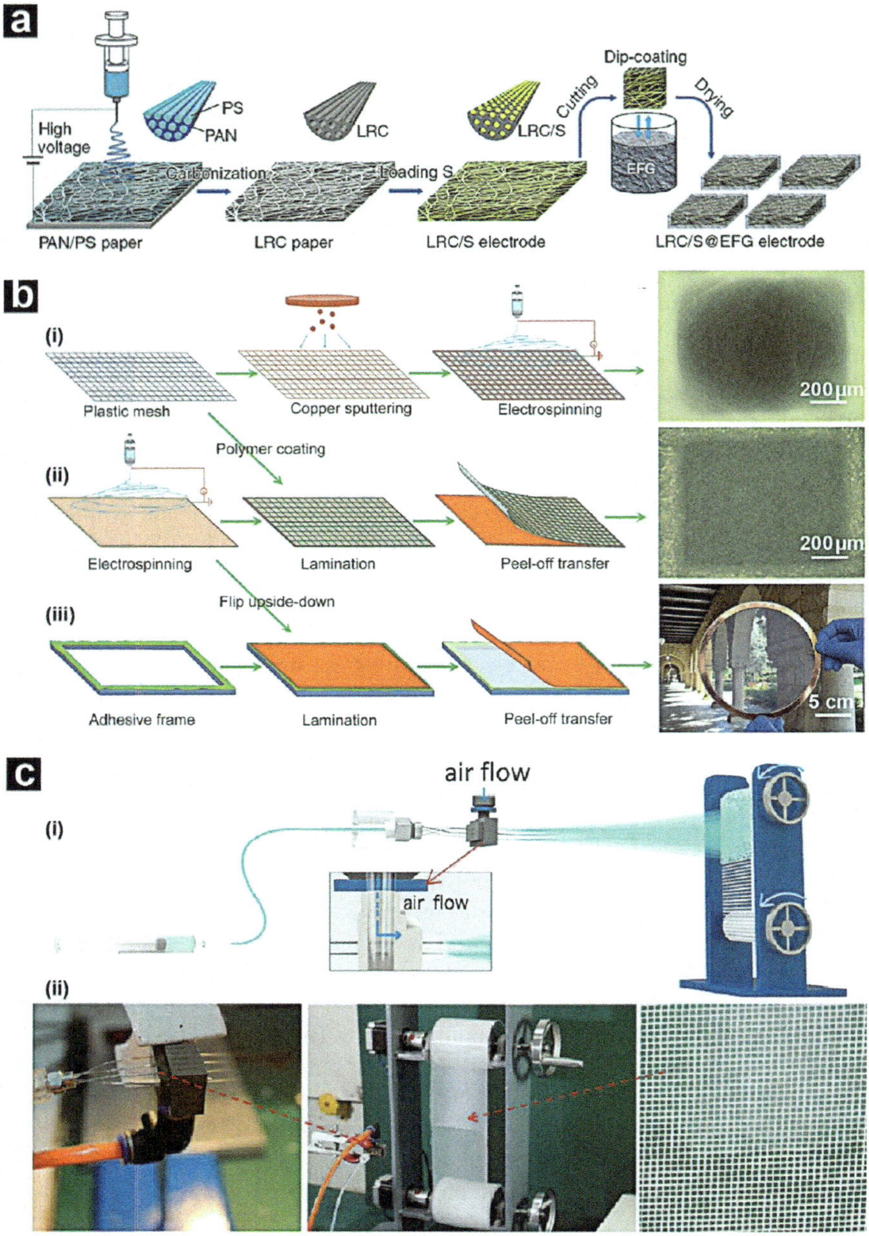
a
High voltage
PS
PAN
LRC
LRC/S
Cutting
Dip-coating
Drying
EFG
Carbonization
Loading S
PAN/PS paper
LRC paper
LRC/S electrode
LRC/S@EFG electrode
b
(i)
Plastic mesh
Copper sputtering
Electrospinning
200 μm
Polymer coating
(ii)
Electrospinning
Lamination
Peel-off transfer
200 μm
Flip upside-down
(iii)
Adhesive frame
Lamination
Peel-off transfer
5 cm
c
(i)
air flow
air flow
(ii)

◀**Fig. 5.7** **a** Schematic illustration of the fabrication process for the LRC/S@EFG electrode. Reproduced with permission [56]. Copyright 2015, Nature publishing group. **b** Schematic illustration and optical microscopy image of a serial of transparent air filter. (i) Transparent air filter by direct-spinning on a conductive mesh. (ii) Transparent air filter by transferring electrospun nanofiber film onto a plastic mesh. (iii) transparent air filter by transferring electrospun nanofiber film onto a facial mask. Reproduced with permission [54] Copyright 2016, American chemical society. **c** Roll-to-roll production of nanofibers. (i) Experimental arrangement for roll-to-roll deposition of nanofibers. (ii) Sequential multineedle blow spinning on rolling polymeric mesh of larger size and successful deposition of fibers. Reproduced with permission [55]. Copyright 2017, American chemical society

but also offers a facile, high yield, and easy-to-operate fabrication process. The experimental setup of the entire roll-to-roll process is depicted in Fig. 5.7c. In this setup, a syringe is connected to a syringe pump, and a curved tip is positioned in front of a continuous air blowing hollow pipe. Figure 5.7c ii illustrates the sequential multineedle blow spinning process onto a rolling polymeric mesh. By adjusting the applied pressure, pump rate, blow-spinning duration, and needle-to-collector distance, the researchers are able to control the nanofiber diameter and packing density.

5.1.8 Other Liquid–Solid Interfacial Assembly

In addition to the aforementioned liquid–solid assembly techniques, various other commonly employed techniques including extrusion [57], induced polymerization [58], and electrodeposition [59], have also been exploited for the preparation of layered films. Cui et al. successfully achieved the production of mass spools of nanoporous polyethylene (nanoPE) microfibers with a soft, cotton-like texture through extrusion interfacial assembly (Fig. 5.8a) [60]. In this process, polyethylene (PE) is dissolved in paraffin oil, resulting in a highly viscous and homogeneous solution that can be extruded to produce microscale fibers. The PE fibers rapidly cool and solidify, allowing them to be separated from the paraffin oil. During the phase separation step, nanoscale oil domains become intermixed with the PE, creating nanoscale pores. The paraffin oil is then selectively removed, resulting in the formation of nanoporous polyethylene (nanoPE) microfibers. These microfibers exhibit a milky white color and possess a soft, cotton-like texture due to the presence of contiguous nanoscale cavities (Fig. 5.8b and c). Figure 5.8d showcases a piece of woven nanoPE fabric measuring 1.8×0.5 m. This extrusion process significantly accelerates the assembly process, enabling the rapid and continuous preparation of nanoPE microfibers. As a result, nanoPE fabrics can be efficiently produced using industrial knitting/weaving techniques. Jiang and his colleagues have also successfully designed and fabricated various polymer heterogeneous membranes with different micro-/nanoscale pore networks on a large scale using the liquid–solid self-assembly approach [61, 62]. As an example, hydrogel layers of poly (acrylamide-co-acrylic acid) (P(AAm-co-AA)) were fabricated through UV-induced free radical polymerization (Fig. 5.8e) [62]. In their study, the researchers dropped the hydrogel precursor solution onto the FTO glass surface using a tape as a spacer, and then covered it with a cover glass. After 30 min of UV irradiation, the P(AAm-co-AA) hydrogel film was successfully synthesized. Additionally, they electrodeposited a nanoporous conducting polymer polypyrrole (PPy) layer with a positive charge on top of the negatively charged P(AAm-co-AA) substrate to create the P(AAm-co-AA)/PPy heterogeneous films (Fig. 5.8f). The integration of the two films with a clearly defined boundary can be observed in Fig. 5.8g and h. This fabrication process can be easily scaled up for large-scale production, as both the photopolymerization and electrochemical deposition methods can be easily expanded. Our research group has also utilized a combination of electrochemical deposition and dip-coating techniques to design a multifunctional biointerface using layered graphene-artificial peroxidase (AP)-extracellular matrix protein nanostructures. This biointerface demonstrates excellent cell proliferation and enables in-situ selective and quantitative detection of H_2O_2 [59].

5.2 Gas–Liquid Interfacical Assembly

Gas–liquid interfacical assembly is another import strategy for the preparation of self-supporting mesoporous films. In the following text, detailed cases about this strategy are presented.

5.2.1 Assembly at Air–Water Interface

Gas–liquid interface is an effective strategy for the growth of freestanding and self-supporting ordered mesoporous films. The interface serves to concentrate the different reactive precursors and templating agents. Yang et al. firstly reported a freestanding mesoporous silica film formed at the air-solution interface from acidic solutions in 1996. In their methodology, the availability of the resultant mesoporous membranes may be caused by that the initially adsorbed surfactant layer at the air-solution interface above the CMC formed a surface bound layer of hemi-micelles, thus effectively attracted silicate species causing the local condensation and polymerization, onto which more micelles could adsorb, then facilitating a hexagonal mesophase to occur in a gradual manner from the surface into the subphase. The resultant films exhibited a hexagonal mesostructure with the channels in an orientation parallel to the interface and a continuous morphology with the resilience to withstand significant bending as well as flexibility to be transferred onto substrates of different shapes. Later on, the majority of work in regard of the synthesis of mesoporous silica films at the air solution interface concern films formed from acidic solutions using an alkoxysilane (eg.,TEOS, tetramethoxysilane, TMOS)as a silica source and a cationic surfactant (eg.,C_nTAB, C_nTAC) as a template in terms of the growth time, concentration, film thickness and mesostructures [63–67]. Until 2004,

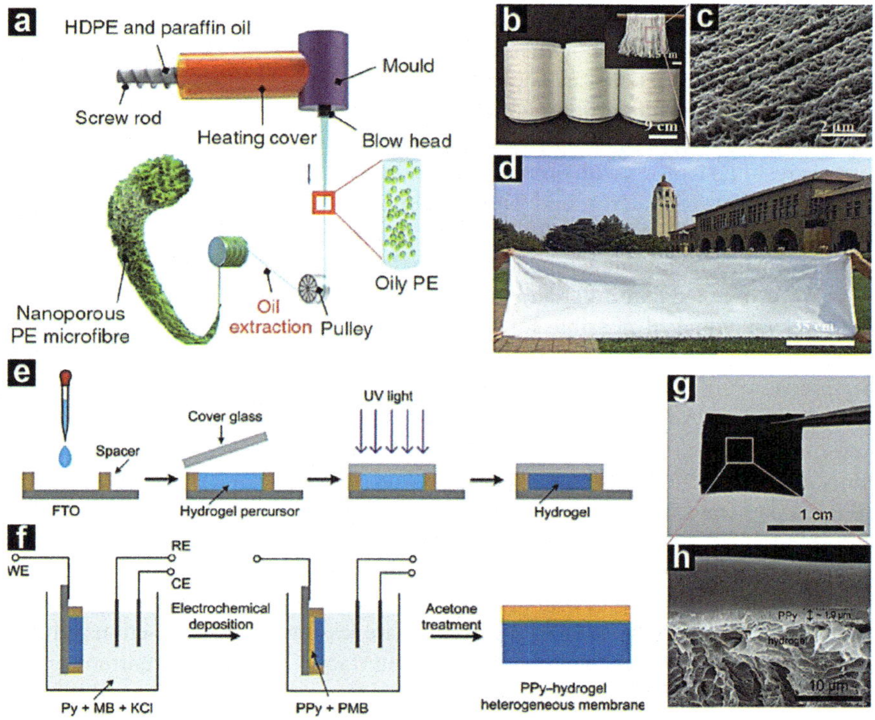

Fig. 5.8 **a** Schematic illustration of the manufacturing process for the nanoPE microfibre. **b** Photograph of three spools of nanoPE microfibres produced by industrial extrusion. **c** SEM image of the surface of a nanoPE microfibre. **d** A photograph of a large woven nanoPE fabric. Reproduced with permission [60]. Copyright 2018, Nature publishing group. **e** Schematic illustration of fabrication for the P(AAm-co-AA) hydrogel layer by UV-induced free radical polymerization. **f** Schematic illustration of fabrication for porous P(AAm-co-AA)/PPy film by electrodeposition. **g** Photograph of the P(AAm-co-AA)/PPy film. **h** Cross-sectional SEM image of P(AAm-co-AA)/PPy film. Reproduced with permission [62]. Copyright 2017, Wiley–VCH

Park et al. first developed an air–water interfacial synthesis of oriented periodic mesoporous organosilica (PMO) films from alkaline solutions by using binuclear alkoxysilane ($(R'O)_3Si–R–Si(OR')_3$ where R = ethane) as a silica precursor and CTAB as a template at 95 °C in a closed polypropylene bottle [68]. The resultant films contained a 2D hexagonal mesostructure with the pores aligned with the long axis parallel to the surface, and the pore size was tunable from 2.4 to 3.3 nm with the varied chain lengths of C_nTAB surfactants similar to that observed in silica films from acidic synthesis [69]. Apart from these organosilica films, film forming surfactant (CTAB)-polyelectrolyte (polyethylenimine, PEI) mixtures from alkaline solutions are another possible templates to grow mesoporous silica films [70]. In this case, the electrostatic interactions between silicate species and CTAB/PEI as well as the dipole-cationic interaction between CTAB and PEI allowed their co-assembly and resulted in homogeneous dispersion of the polymer and silicate species around CTAB micelles. The conbination of surfactants and polymers (either with or without silica) made the complex increasingly hydrophobic, so it moved to the solution surface. Migration to the interface and rearrangement of these CTAB/PEI/silicate complexes at the solution surface resulted in films with ordered mesostructures. In addition, mesoporous metal oxides (such as TiO_2, ZrO_2 and TiO_2-ZrO_2) can also be prepared at the air–water interface from both alkaline solution (using a cationic template C_{14}TAB) and acidic solution (using an anionic template, sodium dodecylsulphate,SDS) [71–73].

It is worth noted that Edler and Yang have recently proposed a more comprehensive understanding of possible formation mechanisms of mesoporous films at the air–liquid interface, as shown in Fig. 5.9 [74]. Routes A and B are concentration dependent with route (A) involving micelles coated with inorganic species occurring at high inorganic/surfactant ratios while particle formation in the sub-phase, leading to film formation; route (B) occurs at intermediate silica/surfactant ratios where the films are formed fastest [74]. Evaporation from the solution surface takes place in all three pathways, but it can also give rise to route (C), where a concentrated phase is formed at the interface and organizes into a mesostructured film, even in cases where there is minimal interaction between the species in the subphase [74]. There are several requirements for film formation, including the presence of an evaporating component, and the formation of species that are less favorable interactions with the solvent, leading to their accumulation at the interface. Once the concentration at the interface of reactive and structure directing species was sufficiently high, film formation would occur. It was found that several systems could promote this interfacial aggregation behavior, including the use of more hydrophobic inorganic precursors, the addition of soluble polymers with a strong interaction with the inorganic materials and manipulation of evaporation rates.

5.2.2 *Aerosol-Assisted Assembly*

The aerosol method, being a representative gas-phase reaction, offers an industrial-scale approach for producing materials due to its continuous nature. Thus far, it has been extensively utilized for the fabrication of various nanostructures with controlled size, shape, and architecture [75]. In 1999, Brinker and co-workers first designed the synthesis of mesoporous materials through a rapid, aerosol-based process, which is a special form of the EISA route proceeding within a droplet based on the gas–liquid interface [76]. The process is schematically illustrated in Fig. 5.10a. The feed solution is dispersed by the atomizer, transforming into uniform and size-controllable droplets. These droplets are then guided through a drying zone, facilitating the self-assembly of mesostructures by a hot carrier gas. The process of Evaporation-Induced Self-Assembly (EISA) is initiated based on the formed droplets. The increased concentration within the droplets induces the formation of a liquid-crystalline mesophase, and the final products are formed around the assembled

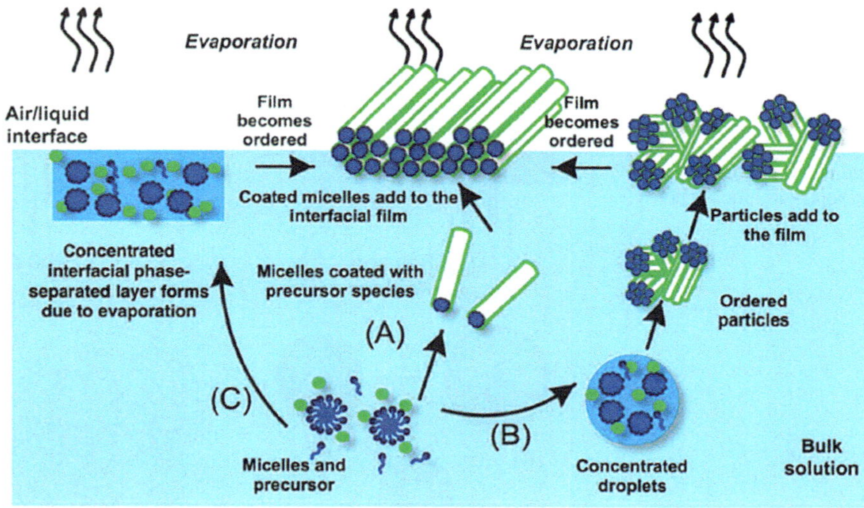

Fig. 5.9 The scheme of the formation mechanism for mesoporous films at the air–water interface

mesophase. Subsequent removal of surfactant templates in the heating zone results in the formation of mesoporous spheres. Unlike the typical EISA process for film formation, where composition gradients develop along the film or through its thickness, this approach focuses on the formation of mesoporous spheres. The process is confined to spherical aerosol droplets and results in radial concentration gradients, and the self-assembly is accordingly directed from the outer gas–liquid interface toward theinterior (Fig. 5.10b and c) [77]. By changing reagent ratios and drying conditions in the aerosol process, SiO_2 microspheres with internal hexagonal, cubic, and vesicular mesostructures can be obtained [76]. Zhao and coworkers used this technique for the preparation of ordered mesoporus carbonaceous spheres at scale in which industrial triblock copolymers were used as templates and soluble phenol resols as carbon precursors [78]. The obtained mesoporous carbons exhibit polydispersed spherical morphology with diameters ranging from 10 0 nm to 5 μm, and possess a high surface area of $\sim 530 m^2 g^{-1}$. Furthermore, utilizing a similar process combined with an acid-mediated sol–gel chemistry system, it is possible to fabricate mesoporous metal oxide microspheres with highly crystalline frameworks. This involves dissolving metal alkoxides in a mixture of acetic acid, hydrochloric acid, and ethanol solutions (denoted as AcHE method) [79]. In the AcHE system, acetate ligands are bonded to metal cores of metal alkoxides in a bi-dentate and bridging fashion to form stable, nanometer-sized metal oxo-acetate particles. These preformed precursor particles serve as inorganic building blocks, and are rapidly co-organized with amphiphilic block copolymers into ordered mesostructures during the evaporation process. The hydrolysis and condensation behaviors controlled by acetic acid are even able to enhance the framework crystallinity at relatively low calcination temperatures while maintaining high surface areas. Guided by this concept, a wide variety

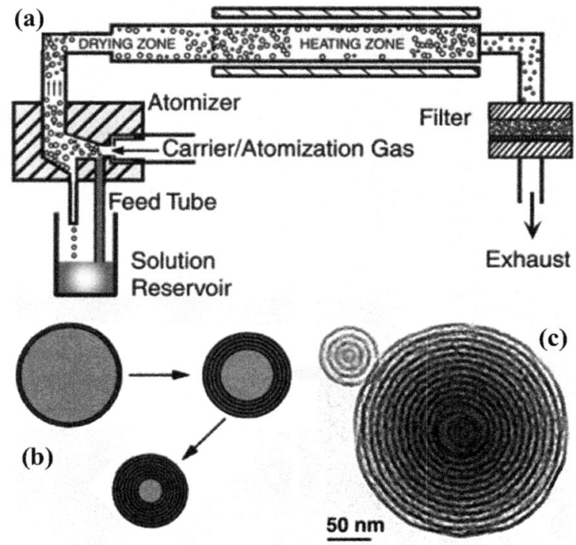

Fig. 5.10 Schematic diagrams of **a** the aerosol process for mesoporous materials and **b** the cooperative self-assembly of mesostructures in a droplet. and **c** TEM images of the mesoporous silica spheres

of mesoporous spheres with a high specific surface area and framework crystallinity can be fabricated, including single-component TiO_2, ZrO_2 and Al_2O_3 and multicomponent CuO-TiO_2, Al-Si–O and $BaTiO_3$ [79]. Moreover, Jiang et al. extended this process to precipitate NaCl crystals within mesoporous SiO_2 microspheres; the salt could be subsequently leached through the shells to produce hollow microspheres [80]. This architecture relies on the development of radial concentration gradients within evaporating droplets, which is facilitated by high ionic diffusivity and differential ionic solubility in a binary solvent system. Moreover, the aerosol process has been extensively utilized for synthesizing multifunctional mesoporous materials by dispersing the corresponding functional components into precursor solutions prior to nebulization [81–85].

5.3 Liquid–Liquid Interfacial Assembly

Liquid–liquid interface (eg., oil–water interface) synthesis is a widely adopted and versatile method. A typical example is the microemulsions. Generally, they are macroscopically homogeneous mixtures of aqueous phase (water), organic phase (oil), and surfactants. On a microscopic scale, these systems consist of heterogeneous domains where water and oil are separated by a surfactant monolayer. Importantly, these domains are intercommunicated with each other [86]. Therefore, microemulsion has been widely used as confined reaction media for synthesizing nanostructures with diverse sizes, shapes, and functions [87]. In this section, we focus on the synthesis of mesoporous materials with ordered pore structure.

In 1996, Schacht et al. first reported the synthesis of various mesoporous materials with macroscale structures by well-control of the interface on water–oil (1, 3, 5-trimethylbenzene,TMB). The ordered mesoporous silica with *P6mm*, *Pm3n* and *P63/mmc* structures were obtained, their morphologies were in the form of porous fibers with 50–1000 μm in length, hollow spheres with diameter of 1–100 μm, and thin sheets up to 10 cm in diameter and about 10 to 500 μm in thickness [88]. Regev and co-workers further studied the effect of interfacial synthesis conditions on the formation of free-standing mesoporous silica film through the separation of the ingredients into two stratified phases: aqueous phase containing the surfactant molecules and acid or base as a catalyst, and oil phase (heptanes) as a reservoir of the silica precursor [89]. They found that the concentration and diffusion rate of reactants were critical for the film formation. The processing time for this technique is generally longer compared to EISA, but it allows for the formation of freestanding and self-supporting films. This is particularly useful for studying the mechanisms involved in surfactant templating growth. Additionally, a significant advantage of this technique is the ability to grow thick mesoporous films without cracks at liquid/liquid or liquid/air interfaces.

Later, Zhou et al. employed 1, 3, 5-triisopropylbenzene (TIPB) as an oil phase, which can effectively tune the packing parameter of Pluronic P123 so as to transform hexagonal arrays of cylindrical micelles into multilamellar vesicles (MLVs) [90, 91] resulting in high-quality multi-lamellar vesicle-like silica or onion-like silica with large uniform inter shell spaces. Recently, Zhao and coworkers first extended this method to prepare onion-like mesoporous carbon or carbon–silica MLVs with uniform lamellar structure (Fig. 5.11a) by using TMB as a cosolvent [92]. The key issue for the successful synthesis of hybrid mesoporous carbon–silica MLVs is the aqueous co-assembly process on the emulsion oil/water interface under a hydrothermal condition. Actually, the co-condensation and formation of 3D interpenetrated frameworks of carbon and silica pose significant challenges due to the substantial disparity between the carbon precursor (a soluble resol polymer with low molecular weight) and silicate polymerizations. In their approach, the authors separated the condensation and co-assembly processes. First, the condensation of silicate species occurs at low temperature (40 °C) accompanied by the co-assembly process. After the mesostructure is formed, a hydrothermal treatment (at 90 °C) is adopted to in situ polymerize resol precursors and form rigid organic–inorganic hybrid frameworks. At last, spherical MLVs with particle sizes of 100–250 nm were obtained. After calcinations at 900 °C, the mesoporous carbon–silica nanocomposites with MLV structure could be yielded (Fig. 5.11b). Etching silica led to mesoporous carbon vesicles with 3–9 layers (Fig. 5.11c); combustion or digestion to remove organic components resulted in mesoporous silica vesicles (Fig. 5.11d). These hollow vesicles show a bimodal pore in the range of 4–23and 66-82 nm and a large pore volume ($\sim$1.44 $cm^3 g^{-1}$). More recently, Wei et al. developed a novel selective solvent evaporation induced aggregating assembly (EIAA) method to synthesize highly ordered mesoporous silica in acidic THF/H_2O mixture by using poly(ethylene oxide)-b-poly(methyl methacrylate)(PEO-b-PMMA) as a template and TEOS as a silica precursor, in which the self-assembly of the mesostructures occurs at the liquid–liquid interface (Fig. 5.12) [93]. In this case, the water insoluble block copolymer PEO-b-PMMA can be dissolved in an acidic THF-H_2O solution (v/v 8:2). As the template molecule's good solvent, THF, evaporates, the solvency power of the mixed solvent diminishes considerably, resulting in the formation of spherical micelles with a PMMA core enveloped by a PEO shell. Due to the contrasting solvency properties, water is a subpar solvent for PMMA segments but highly suitable for PEO segments. Simultaneously, silicate oligomers hydrolyzed from TEOS in the acidic solution can associate with PEO moieties through hydrogen bonding, giving rise to the individual hybrid micelles (Fig. 5.12, Step1). With the evaporation of THF proceeding, the ever-increasing concentration of the tiny hybrid micelles enables the self-assembly, and finally aggregation and packing into irregular particles with an ordered fcc mesostructured to decrease the interface energy (Fig. 5.12, Step2).The increasing acidity of the solution accelerates the condensation of silicate oligomers, thus fixing the ordered PEO-b-PMMA/SiO_2 mesostructure in a short time, and large crystal-like particles are formed, precipitated from the solution phase (Fig. 5.12, Step3). The resultant mesoporous silica possesses a large tunable pore size of ~37.0 nm and window size of ~8.7 nm, high specific surface area of ~508m^2/g and pore volume of 1.46cm^3/g. The EIAA process is different from that of the common EISA method, where only a stoichiometric amount of water is allowed to be used in the initial solution, and the ordered mesostructure is formed at the interface of the solid (substrate)-liquid (solution) when the good solvent for the template evaporates off. The EIAA method is more tolerant to water, and the self-assembly occurs at the liquid–liquid interface. Such EIAA method can also be applied to prepare ordered dual mesoporous silica with two sets of mesopores and independently controlled pore size by using PEO-b-PMMA block copolymers and C_nTAB surfactants as contemplates [94]. It is suggested that, during the evaporation of THF under static conditions, block copolymers with associated silicate oligomers begin to assemble into spherical composite micelles with a PMMA core and a PEO/silicate shell. At this stage, small C_nTAB molecules can tend to approach the PEOb-PMMA/silicate composite micelles because of the high density of silicate species on their surface, and then organize into small rod-like micelles around the large micelles, assisted by the silicate species. The resultant "alloyed" bi-micellar spherical nanocomposites further assemble into large particles of several micrometers with an fcc structured mesophase. Finally, the removal of PEO-b-PMMA and C_nTAB templates in air results in ordered dual-mesoporous silica. Moreover, this EIAA method can also be extended to other block copolymers such as poly(ethylene oxide)-b-polystyrene (PEO-b-PS), for synthesis of large-pore mesoporous silica spheres [95].

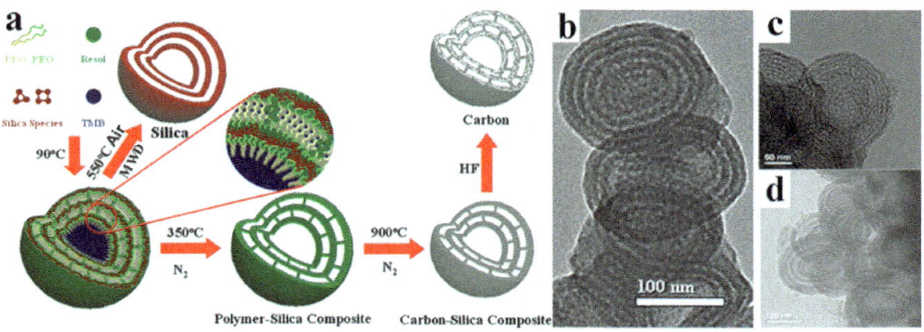

Fig. 5.11 **a** The scheme for the formation process of the mesoporous carbon and carbon–silica MLVs by multiconstituent co-assembly approach in an aqueous emulsion solution under a low-temperature hydrothermal treatment. TEM images of **b** mesoporous carbon–silica composite MLV calcinated at 900 °C, **c** mesoporous carbon MLV and **d** mesoporous silica MLV

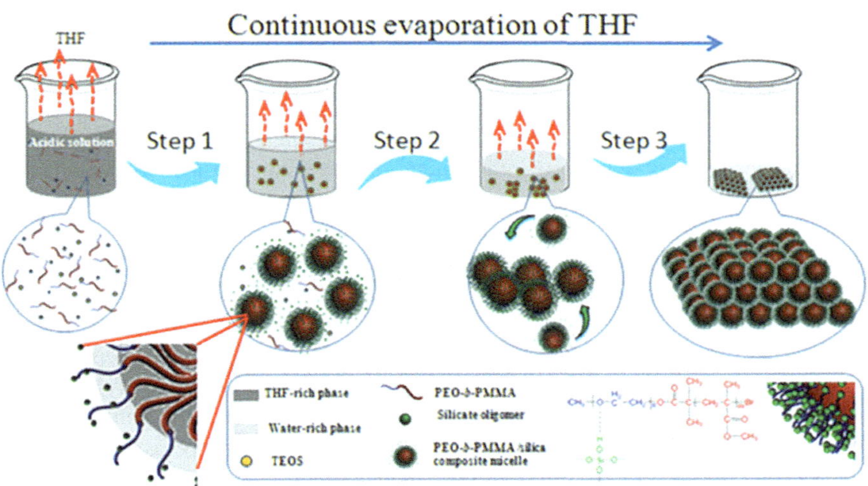

Fig. 5.12 Formation mechanism of the ordered mesoporous silica through the solvent evaporation induced aggregating (EIAA) method

5.4 Gas–Solid Interfacial Synthesis

The gas–solid phase reaction holds significant prominence in the chemical industry, with approximately 90% of catalytic reactions occurring at the gas–solid interface, as per statistical data. Numerous crucial reactions in industrial processes, such as carbon monoxide conversion in synthetic ammonia production and ammonia synthesis, involve gas–solid catalytic reactions. In this context, our focus lies on expanding the gas–solid interface for the synthesis of mesoporous materials.

5.4.1 Vapor-Phase Synthesis

Vapor-phase synthetic technique is an alternative method for fabricating ordered mesoporous films. This method typically involves two steps: first, a pure template (eg., surfactant, latex beads, etc.) layer is coated onto a substrate surface. This template forms the ordered mesoscopic solid phase through micro-phase separation. Then, the coated layer is placed in a vessel with saturated precursor steam. The volatile precursor molecules diffuse and are absorbed into the ordered template layer. They then undergo hydrolysis/condensation and crosslink in the presence of a catalyst, forming a stable framework around the template. Once the templates are removed, the porosity is released [96–98]. Nishiyama et al. first devised this novel route to prepare ordered mesoporous silica films. In that case, a CTAB/ethanol/water solution is first deposited onto a silicon substrate by spin coating, thus an ordered CTAB (template) layer is formed after the solvent evaporation. Then, the silicon wafer is placed vertically in a closed vessel and exposed to the vapor mixture of TEOS or TMOS and a catalyst (ammonia or HCl vapor). After a heat treatment at 120 °C for 2 h and calcination at 400 °C in air for 5 h, mesoporous silica thin films are obtained. Obviously, this technique provides experimental simplicity, facile accessibility and mass production, and high reproducibility. These features make it particularly interesting and promising for synthesizing mesoporous silica films [99–101]. Besides, uniform mesoporous hybrids or other metal oxides films can also be achieved by infusing the copolymer layer with the corresponding reactive vapors, subjecting it to thermal annealing. For example, mesoporous SiO_2-P_2O_5 films were synthesized from TEOS vapor phase onto a silicon substrate, on which a CTABH$_3$ PO$_4$ composite layer was pre-loaded [102]. The molar ratio of P/Si in the resultant SiO_2-P_2O_5 film was as high as 0.43, which are difficult to be prepared via conventional sol–gel methods. Recently, Parsons and coworkers demonstrated the preparation of mesoporous alumina or titania by the vapor phase infiltration of trimethylaluminum or titanium tetrachloride into the preformed films of ordered surfactant molecules (Pluronic F127,CTAB and SDS) films [103]. Using water vapor as a co-reactant, mesoporous film products with high specific surface area ($\sim$1000 m^2/g) and large pores (3–10 nm) were obtained.

Interestingly, Lee et al. demonstrated that the mesostructured silica nanofibers could be produced in channels of the AAO membrane using the vapor-phase synthetic technique [104]. In their process, the surfactant was first deposited into the channel of AAO membranes. As surface charge of AAO membranes is normally negative, the head of the cationic surfactant (CTAB) would be preferentially attached onto the pore wall of AAO channels by charge-charge interaction. The surfactant could construct a lamellar mesostructure on the surface. Then, similar to the preparation for mesoporous films, the AAO membrane was placed in a closed vessel, and vapor-phase synthesis was performed at 100 °C for 2 h. The authors found that the parallel alignment of mesopores in the fibers was changed into a tilted alignment when the alkyl chain surfactant length was decreased from C_{16} to C_{10} owing to the enhanced surface tension. Although significant process has been made, it is crucial to ensure that the copolymer layer remains stable and does not dissolve under such conditions. Additionally, it is important to ensure that the precursors effectively polymerize around the micelle arrays.

5.4.2 Chemical Vapor Deposition

Chemical vapor deposition (CVD) is a typical gas–solid interfacial reaction initiated by the thermal, laser, or plasma for the synthesis of nanomaterials with uniform structures, high purity and low energy-consumption. It can be initiated by thermal, laser, or plasma sources. CVD has been well established as a reliable technique for preparing carbon molecular sieves and carbon nanotubes [105]. As an alternative process of nanocasting method for ordered mesoporous carbons, CVD has also been widely developed via using a low-carbon hydrocarbon gas or vapor as a precursor for carbon decomposition on the performed mesoporous templates under thermal treatment (Fig. 5.13). In 2002, Kaneda et al. first reported a CVD approach to prepare ordered mesoporous carbon CMK-4 at 800 °C using the $AlCl_3$ -impregnanted MCM-48 as a template and acetylene as a carbon precursor [106]. The gas precursor can be directly adsorbed onto the surface of the template's pores, and subsequently decomposed in situ to form carbon within the channels. Li and colleagues were the pioneers in developing a catalytic chemical vapor deposition (CVD) technique for synthesizing hexagonal arrays of mesoporous carbon nanopipes, known as CMK-5. In that case, the pore walls of SBA-15 template is first modified with cobalt ions, which is then used as the template for carbon deposition under an ethylene/Ar gas flow at 700 °C. Additionally, they discovered that a sufficient deposition time is crucial for the successful synthesis of high-quality, well-ordered arrays of carbon nanopipes [107].

The utilization of a mesoporous silica template without the need for catalyst loading on the pore surface holds practical significance as it eliminates the catalyst loading step. A notable example of this approach was described by Vix-Guterl et al., who utilized pure silica MCM-48 as the template to fabricate ordered mesoporous carbons. These carbons exhibited a mesostructure similar to that of well-known carbon replicas [108]. In 2006, Feng and coworkers first proposed 3D carbon decomposition approach for the synthesis of hierarchical porous carbon monoliths starting from mesoporous silica powders by integration of gel casting and CVD technique [109]. The carbon monoliths exhibit a combination of macroporosity, well-ordered mesoporosity with either hexagonal or bicontinuous cubic symmetry, and controlled microporosity. They also possess excellent mechanical strength properties, even under pressures as high as 51 psi. In recent years, significant progress has been made in the field of gas–solid interface engineering using the CVD method to prepare mesoporous carbons. This progress has been achieved by adjusting various reaction conditions (such as temperature, deposition time, and precursor) and modifying surface properties (such as catalysts). Additionally, considerable effort has

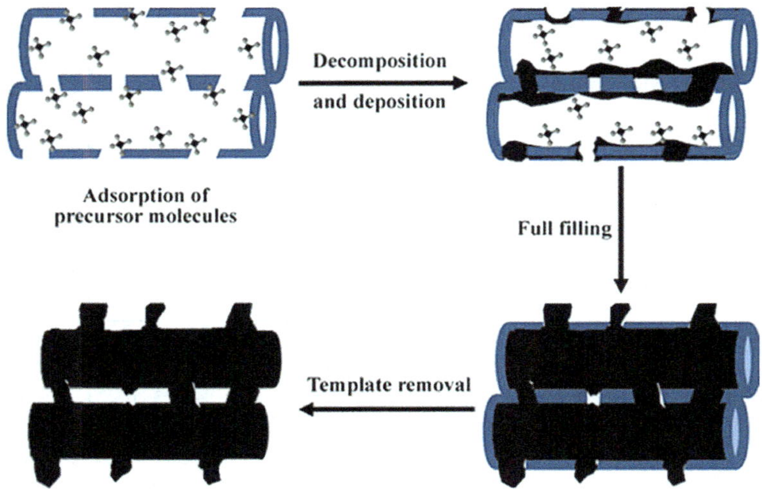

Fig. 5.13 Schematic illustration of the chemical vapor deposition process for mesoporous carbons

been dedicated to investigating the formation mechanisms, fine-tuning the three-dimensional nanostructures, and controlling the morphology of the resulting mesoporous carbon materials [110–114]. Notably, this technique offers several advantages over the liquid-phase impregnation method. These include a high degree of pore filling, time and cost efficiency, ease of control, the ability to form graphitic pore walls, and the avoidance of additional microporosity. Furthermore, due to the ease of introducing heteroatoms (such as nitrogen) into hydrocarbons, another advantage is the simple production of highly graphitized mesoporous carbons with abundant hetero-aryl atoms in their skeleton [115, 116]. For example, Xia et al. first reported the synthesis of N-doped mesoporous carbons using SBA-15 as a hard template and acetonitrile as a carbon precursor at a temperature above 900 °C and preferably in the range of 950–1100 °C [117]. The resultant mesoporous carbons possess high graphitic pore walls and high nitrogen content of ~8.5%, which mainly present as highly coordinated (quaternary) and pyridine-like N atoms in pore wall.

5.4.3 *Post Transformation*

In a nanocasting process, the majority of precursors are typically transformed into desired materials through a straightforward thermal treatment conducted under either air or an inert gas (e.g., Ar). This method finds widespread application in the synthesis of various materials, including metal oxides, mixed metal oxides, and carbon-based materials, etc. However, it is still a challenge to get a suitable and available precursor for nitrides, sulfides, and other non-oxide materials via such direct conversion process. Recently, gas–solid reactions have been extensively developed and employed to effectively convert preformed mesoporous materials into targeted products with unique framework compositions. In this case, a component of the preformed framework (e.g., normally O element) is substituted or eliminated by other elements (eg.,C,N,H) during a high temperature treatment. This transformation results in the formation of mesoporous products with novel compositions (eg., SiN_xO_y,CoN,WS_2) that cannot be achieved through alternative methods. A typical example is to prepare mesoporous silicon oxynitrides (SiN_xO_y) via treating mesoporous silica under an ammonia gas flow [118]. For example, Masuda et al. first reported the synthesis of mesoporous silicon oxynitrides (SiN_xO_y) by directly treating as-made silica MCM-41 with ammonia gas at 950 °C [119]. The resultant silicon oxynitrides retain ordered mesostructures with high specific surface areas of 880–1100 m^2g^{-1} and a narrow pore size of ~2 nm, as well as nitrogen contents as high as 9.7–12.9 wt%. In this prcoess, the oxygen atoms on pore walls of the MCM-41 can be substituted by nitrogen atoms step by step through a reaction: ($SiO_2 + NH_3 \rightarrow Si_3N_4 + H_2O + N_2 + H_2$). Subsequent research efforts have been directed towards augmenting the nitrogen content, as it is considered crucial for enhancing the alkalinity of silicon oxynitrides. The attempts mainly included the increase of nitridation temperature, change of silica precursors and improvement of the technology. Vinu et al. utilized this strategy for synthesis of mesoporous boron nitride (BN) and boron carbonitride (BCN) by using a well-ordered hexagonal mesoporous carbon (CMK-3) as a hard template. In which the carbon element on the framework can be replaced by boron and/or nitrogen, at a higher temperature such as 1750 °C leading to mesoporous BNs while the mesoporous BCN can be obtained at considerably lower temperatures (1450–1550 °C) with short reaction times [120].

Nanostructured transition metal nitrides with significant specific surface areas and pore volumes are highly anticipated because the confined-space effects inside the nanospace may improve their physical and chemical behaviors to meet the demands of advanced materials [121]. Nevertheless, the synthesis of metal nitride nanoparticles has typically involved their formation within the confined spaces of mesoporous materials [122, 123]. The predominance of nanoparticle formation, rather than nanowires, within the mesopores is primarily attributed to the significantly low volume conversion of most metal nitride precursors. Recently, Shi et al. have successfully demonstrated an ammonia nitridation approach utilizing gas–solid reactions to convert mesostructured metal oxides into their corresponding ordered mesoporous metal nitrides [124]. Two synthetic transformation pathways are designed as shown in Fig. 5.14a. One approach involves the direct nitridation of mesoporous metal oxide nanowire replicas to produce metal nitrides. By this method, highly ordered mesoporous cobalt nitrides (CoN) could be obtained by the transformation of Co_3O_4 nanowire replica under ammonia atmosphere within the temperature range of 275 to 350 °C (Fig. 5.14b). Remarkably, this transformation occurred without a discernible loss of the mesostructural regularity. This phenomenon could be attributed to the relatively insignificant change in volume (approximately 9.7%) observed after nitridation. Subjecting the samples to temperatures exceeding 375 °C resulted in the formation of metallic cobalt and the subsequent collapse of the mesostructures due to significant volume shrinkage. An alternative approach involves the conversion of mesostructured metal oxide/silica composites into nitride/silica composites by subjecting them to ammonia at temperatures ranging from 750 to 1000 °C. Following the removal of the silica template through NaOH erosion, we successfully obtained ordered mesoporous CrN nanowire arrays (Fig. 5.14c). A slowly temperature program-decrease process could effectively reduce the influence of silica nitridation and improve the purity of final CrN product. The resultant mesoporous metal nitrides have high specific surface areas (70–90 m^2g^{-1}) and large pore volumes ($\sim$0.2 cm^3g^{-1}). The sectionalized introduction of metal and nitrogen precursors offers multiple advantages, including increased filling capacity for metal precursors, enhanced volume conversion to the desired metal nitrides, and avoidance of the limitations associated with the low volume conversion directly from most precursors to metal nitrides. These f characteristics enable the faithful replication of the silica hard template mesostructures and the successful synthesis of highly ordered, self-supported mesoporous metal nitride nanowire arrays, eliminating the need for a silica matrix as a support. This strategy can also be extended for in-situ transformation and mesostructure inheritance from α-Fe_2O_3 to Fe_3O_4 and γ-Fe_2O_3 materials [125]; Mn_2O_3 to Mn_3O_4 [126]; Co_3O_4 to CoO [127]; and Ta_2O_3 to Ta_3N_5 [128].

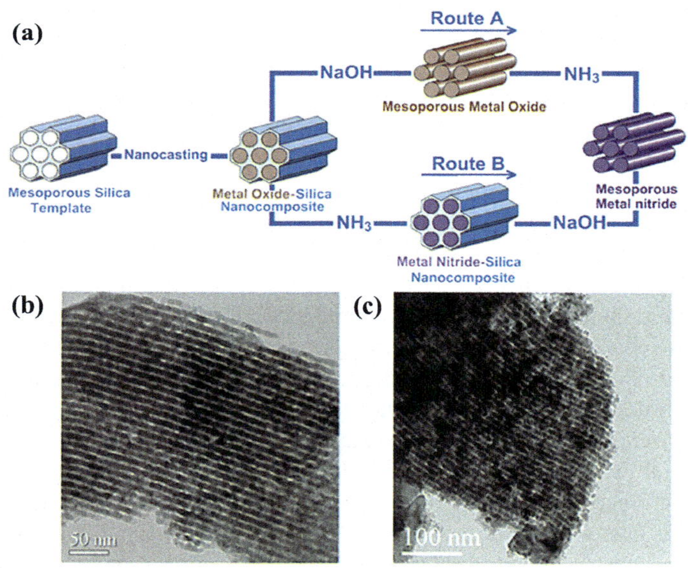

Fig. 5.14 (**a**) The scheme of two pathways of the gas–solid reaction for synthesis of selfsupported mesoporous cobalt nitrides CoN (A) and chromium nitrides CrN (B). TEM images of CoN (**b**) and CrN (**c**)

5.5 Solid–Solid Interfacial Synthesis

Solid–solid reaction is a significant synthetic process for high-performance solid light emitting and ceramic materials in the industry. For example, high temperature silicon carbide ceramics can be directly synthesized by reacting silicon with carbon in an inert atmosphere at high temperatures (>1200°C, $Si + C \rightarrow SiC$). However, mesoporous SiC materials with high specific surface area, which are desirable for catalyst carriers, are seldom reported. The major reason lies in the facts of the lack of appropriate controllable sol–gel process for SiC precursors and their air- and/or watersensitive feature. Carbothermal reduction, which is the most common way to fabricate SiC materials, offers chances for the synthesis of mesoporous SiC materials [129] In this case, uniform mesoporous silica/carbon composites are first prepared. Followed by a thermal treatment, the component of silica can be in situ reduced, leading to the formation of SiC; simultaneous, the ordered mesostructures can be replicated and retained. For example, Parmentier et al. demonstrated a carbothermal reduction method based on the solid–solid reaction to fabricate mesoporous SiC [130]. In their synthesis, an ordered mesostructured SiO_2 /C nanocomposite was first prepared via chemical vapor infiltration of the carbon (propylene as a precursor) into the pore channels of MCM-48 hard template. The nanocomposite was then treated in an inert atmosphere at high temperatures (1200–1450 °C) to trigger the carbothermal reduction reaction. Yang and Lu et al. also fabricated mesoporous SiCs through the similar synthesis strategy but with different carbon sources i.e. sucrose and furfuryl alcohol [131], However, the resultant SiC materials normally possess low surface areas around 120–190 m^2g^{-1} with a poor mesostructural regularity, suggesting that the ordered mesostructure collapse after exquisite reaction. These phenomena can be explained by the reaction mechanism of carbothermal reduction (The overall reaction is: $SiO_2 + 3C \rightarrow SiC + 2CO$). Actually, this is not a pseudomorphic transformation process due to the intermediate formation of gas phase SiO and CO, so that the obtained SiC products possess different structures than their SiO_2 precursors [132]. The SiC products are mostly in the form of nanoparticles, nanorods, and nanotubes because of the crystal growth tendencies of SiC [132].

Recently, Stucky and coworkers have reported a low temperature synthesis approach for directly converting SiO_2/C composites to the corresponding SiC materials without losing their nanostructures via a so-called magnesiothermic reduction at 600~900 °C [133]. The overall reaction can be described as $SiO_2 + C + 2\,Mg \rightarrow SiC + 2MgO$. In this case, silica was first reduced to silicon via the magnesiothermic reduction ($SiO_2 + 2\,Mg \rightarrow Si + 2MgO$), in which Mg acted as a reductant [134]. It was found that silica was not reduced by carbon via the carbothermal reduction under this condition. Then, the obtained silicon reacted with carbon to form crystalline SiC at a temperature as low as 600 °C, only approximately half of that applied in the earlier carbothermal reduction. Although Mg was not a reagent in the second step, it was still indispensable because SiC can only be formed above 1200 °C from carbon and silicon if no metal catalyst is present. Therefore, in condition, Mg played a critical catalytic role in the carbothermal reduction. The uniform distribution of silica and carbon in the SiO_2/C precursor enables it possible for the silicon intermediate to directly react with nearby carbon species, making the SiO_2/C to SiC conversion a feasible macroscale pseudomorphic transformation. The resultant.

SiC retains the ordered macroscale structure and possesses a high BET surface area of ~410 m^2/g and a pore size distribution of 2.0 to 7.0 nm. Most importantly, neither a high temperature special furnace nor a high purity protection inert gas is required, which greatly reduces the fabrication costs and saves energy. This pseudomorphic transformation can be regarded as a general synthesis method for different kinds of SiC nanostructures, and it can also be readily extended to other metal carbide materials like TiC and WC.

References

1. Tao AR, Huang J, Yang P (2008) Langmuir−blodgettry of nanocrystals and nanowires. Acc Chem Res 41:1662–1673
2. Acharya S, Hill JP, Ariga K (2009) Soft langmuir–blodgett technique for hard nanomaterials. Adv Mater 21:2959–2981
3. Bourikas K, Kordulis C, Lycourghiotis A (2007) The role of the liquid—solid interface in the preparation of supported catalysts. Catal Rev 48:363–444
4. Cao W-T, Chen F-F, Zhu Y-J et al (2018) Binary strengthening and toughening of mxene/ cellulose nanofiber composite paper with nacre-inspired structure and superior electromagnetic interference shielding properties. ACS Nano 12:4583–4593
5. Song H, Lin H, Antonietti M et al (2016) From filter paper to functional actuator by poly(ionic liquid)-modified graphene oxide. Adv Mater Interfaces 3:1500743
6. Lin D, Liu Y, Liang Z et al (2016) Layered reduced graphene oxide with nanoscale interlayer gaps as a stable host for lithium metal anodes. Nat Nanotechnol 11:626–632
7. Zhu X, Zhou Y, Hao J et al (2017) A charge-density-tunable three/two-dimensional polymer/ graphene oxide heterogeneous nanoporous membrane for ion transport. ACS Nano 11:10816–10824
8. Li Y, Lu Q, Liu H et al (2015) Antibody-modified reduced graphene oxide films with extreme sensitivity to circulating tumor cells. Adv Mater 27:6848–6854
9. Liu H, Wang H, Zhang X (2015) Facile fabrication of freestanding ultrathin reduced graphene oxide membranes for water purification. Adv Mater 27:249–254
10. Eda G, Fanchini G, Chhowalla M (2008) Large-area ultrathin films of reduced graphene oxide as a transparent and flexible electronic material. Nat Nanotechnol 3:270–274
11. Li D, Müller MB, Gilje S et al (2008) Processable aqueous dispersions of graphene nanosheets. Nat Nanotechnol 3:101–105
12. Lv Z, Luo Y, Tang Y et al (2018) Editable supercapacitors with customizable stretchability based on mechanically strengthened ultralong MnO_2 nanowire composite. Adv Mater 30:1704531
13. Cheng W (2017) Free-standing nanoparticle superlattice sheets: from design to applications. EPL (Eur Lett) 119:48004
14. Meder F, Thomas SS, Bollhorst T et al (2018) Ordered surface structuring of spherical colloids with binary nanoparticle superlattices. Nano Lett 18:2511–2518
15. Lin Q-Y, Mason JA, Li Z et al (2018) Building superlattices from individual nanoparticles via template-confined DNA-mediated assembly. Science 359:669–672
16. Cheng W, Campolongo MJ, Cha JJ et al (2009) Free-standing nanoparticle superlattice sheets controlled by DNA. Nat Mater 8:519–525

17. Trotochaud L, Ranney JK, Williams KN et al (2012) Solution-cast metal oxide thin film electrocatalysts for oxygen evolution. J Am Chem Soc 134:17253–17261
18. Xiao Y, Xu Z, Liu Y et al (2017) Sheet collapsing approach for rubber-like graphene papers. ACS Nano 11:8092–8102
19. Zhao J, Zhou GM, Yan K et al (2017) Air-stable and freestanding lithium alloy/graphene foil as an alternative to lithium metal anodes. Nat Nanotechnol 12:993–999
20. Yang H, Coombs N, Sokolov I et al (1996) Free-standing and oriented mesoporous silica films grown at the air–water interface. Nature 381:589–592
21. Xue C, Tu B, Zhao D (2008) Evaporation-induced coating and self-assembly of ordered mesoporous carbon-silica composite monoliths with macroporous architecture on polyurethane foams. Adv Func Mater 18:3914–3921
22. Brinker CJ (2004) Evaporation-induced self-assembly: functional nanostructures made easy. MRS Bull 29:631–640
23. Zhu Y, Zhao Y, Ma J et al (2017) Mesoporous tungsten oxides with crystalline framework for highly sensitive and selective detection of foodborne pathogens. J Am Chem Soc 139:10365–10373
24. Zhang R, Shen D, Xu M et al (2014) Ordered macro-/mesoporous anatase films with high thermal stability and crystallinity for photoelectrocatalytic water-splitting. Adv Energy Mater 4:1301725
25. Feng D, Luo W, Zhang J et al (2013) Multi-layered mesoporous tio2 thin films with large pores and highly crystalline frameworks for efficient photoelectrochemical conversion. J Mater Chem A 1:1591–1599
26. Kong B, Tang J, Zhang Y et al (2016) Incorporation of well-dispersed sub-5-nm graphitic pencil nanodots into ordered mesoporous frameworks. Nat Chem 8:171–178
27. Lee D, Rubner MF, Cohen RE (2006) All-nanoparticle thin-film coatings. Nano Lett 6:2305–2312
28. Guo CX, Ng SR, Khoo SY et al (2012) Rgd-peptide functionalized graphene biomimetic live-cell sensor for real-time detection of nitric oxide molecules. ACS Nano 6:6944–6951
29. Ma X, Hu W, Guo C et al (2014) DNA-templated biomimetic enzyme sheets on carbon nanotubes to sensitively in situ detect superoxide anions released from cells.
30. Fang Y, Lv Y, Tang J et al (2015) Growth of single-layered two-dimensional mesoporous polymer/carbon films by self-assembly of monomicelles at the interfaces of various substrates. Angew Chem Int Ed 54:8425–8429
31. Fang Y, Lv Y, Che R et al (2013) Two-dimensional mesoporous carbon nanosheets and their derived graphene nanosheets: synthesis and efficient lithium ion storage. J Am Chem Soc 135:1524–1530
32. Zhao M, Yuan W, Li CM (2017) Controlled self-assembly of ni foam supported poly(ethyleneimine)/reduced graphene oxide three-dimensional composite electrodes with remarkable synergistic effects for efficient oxygen evolution. J Mater Chem A 5:1201–1210
33. Kotsuki K, Tanaka H, Obata S et al (2014) The importance of spinning speed in fabrication of spin-coated organic thin film transistors: film morphology and field effect mobility. Appl Phys Lett 104:233306
34. Cho J, Char K, Hong JD et al (2001) Fabrication of highly ordered multilayer films using a spin self-assembly method. Adv Mater 13:1076–1078
35. Si M, Feng D, Qiu L et al (2013) Free-standing highly ordered mesoporous carbon–silica composite thin films. J Mater Chem A 1:13490–13495
36. Feng D, Lv Y, Wu Z et al (2011) Free-standing mesoporous carbon thin films with highly ordered pore architectures for nanodevices. J Am Chem Soc 133:15148–15156
37. Gong S, Schwalb W, Wang Y et al (2014) A wearable and highly sensitive pressure sensor with ultrathin gold nanowires. Nat Commun 5:3132
38. Lee W, Lee J, Yun H et al (2017) High-resolution spin-on-patterning of perovskite thin films for a multiplexed image sensor array. Adv Mater 29:1702902
39. Hashizume M, Kunitake T (2003) Preparation of self-supporting ultrathin films of titania by spin coating. Langmuir 19:10172–10178

40. Vendamme R, Onoue S-Y, Nakao A et al (2006) Robust free-standing nanomembranes of organic/inorganic interpenetrating networks. Nat Mater 5:494–501
41. Wang S, Xu J, Wang W et al (2018) Skin electronics from scalable fabrication of an intrinsically stretchable transistor array. Nature 555:83–88
42. Zhao J, Li Y, Sheng J et al (2017) Environmentally friendly and breathable fluorinated polyurethane fibrous membranes exhibiting robust waterproof performance. ACS Appl Mater Interfaces 9:29302–29310
43. Bagheri H, Aghakhani A (2011) Novel nanofiber coatings prepared by electrospinning technique for headspace solid-phase microextraction of chlorobenzenes from environmental samples. Anal Methods 3:1284–1289
44. Destaye AG, Lin C-K, Lee C-K (2013) Glutaraldehyde vapor cross-linked nanofibrous pva mat with in situ formed silver nanoparticles. ACS Appl Mater Interfaces 5:4745–4752
45. Xu T, Yang H, Yang D et al (2017) Polylactic acid nanofiber scaffold decorated with chitosan islandlike topography for bone tissue engineering. ACS Appl Mater Interfaces 9:21094–21104
46. Mo J, Xu N, Xiao C et al (2014) Structure and property of electrospun fibrous mat based on polymethacrylate. J Mater Sci 49:4816–4824
47. Freitag KM, Walke P, Nilges T et al (2018) Electrospun-sodiumtetrafluoroborate-polyethylene oxide membranes for solvent-free sodium ion transport in solid state sodium ion batteries. J Power Sources 378:610–617
48. Zhang J, Li Z, Lou XW (2017) A freestanding selenium disulfide cathode based on cobalt disulfide-decorated multichannel carbon fibers with enhanced lithium storage performance. Angew Chem Int Ed 56:14107–14112
49. Koo JY, Hwang S, Ahn M et al (2016) Controlling the diameter of electrospun yttria-stabilized zirconia nanofibers. J Am Ceram Soc 99:3146–3150
50. Li L, Peng S, Lee JKY et al (2017) Electrospun hollow nanofibers for advanced secondary batteries. Nano Energy 39:111–139
51. Shi Q, Hou J, Zhao C et al (2016) A smart core–sheath nanofiber that captures and releases red blood cells from the blood. Nanoscale 8:2022–2029
52. Wang H, Zhang C, Chen Z et al (2015) Large-scale synthesis of ordered mesoporous carbon fiber and its application as cathode material for lithium–sulfur batteries. Carbon 81:782–787
53. Li Z, Zhang JT, Chen YM et al (2015) Pie-like electrode design for high-energy density lithium–sulfur batteries. Nat Commun 6:8850
54. Xu J, Liu C, Hsu P-C et al (2016) Roll-to-roll transfer of electrospun nanofiber film for high-efficiency transparent air filter. Nano Lett 16:1270–1275
55. Khalid B, Bai X, Wei H et al (2017) Direct blow-spinning of nanofibers on a window screen for highly efficient $PM_{2.5}$ removal. Nano Lett 17:1140–1148
56. Li Z, Zhang JT, Chen YM et al (2015) Pie-like electrode design for high-energy density lithium-sulfur batteries. Nat Commun 6:8850
57. Liu Y, Zhang B, Xu Q et al (2018) Supercapacitors: development of graphene oxide/ polyaniline inks for high performance flexible microsupercapacitors via extrusion printing (adv. Funct. Mater. 21/2018). Adv Funct Mater 28:1870142
58. Ding Y, Cai M, Cui Z et al (2018) Synthesis of low-dimensional polyion complex nanomaterials via polymerization-induced electrostatic self-assembly. Angew Chem Int Ed 57:1053–1056
59. Guo CX, Zheng XT, Lu ZS et al (2010) Biointerface by cell growth on layered graphene–artificial peroxidase–protein nanostructure for in situ quantitative molecular detection. Adv Mater 22:5164–5167
60. Peng Y, Chen J, Song AY et al (2018) Nanoporous polyethylene microfibres for large-scale radiative cooling fabric. Nat Sustain 1:105–112
61. Zhang Z, Kong X-Y, Xiao K et al (2016) A bioinspired multifunctional heterogeneous membrane with ultrahigh ionic rectification and highly efficient selective ionic gating. Adv Mater 28:144–150
62. Bao B, Hao J, Bian X et al (2017) 3D porous hydrogel/conducting polymer heterogeneous membranes with electro-/pH-modulated ionic rectification. Adv Mater 29:1702926

63. Brown AS, Holt SA, Dam T et al (1997) Mesoporous silicate film growth at the air–water interface direct observation by X-ray reflectivity. Langmuir 13:6363–6365
64. Edler KJ, Goldar A, Hughes AV et al (2001) Structural studies on surfactant-templated silica films grown at the air/water interface. Microporous Mesoporous Mater 44–45:661–670
65. Brennan T, Hughes AV, Roser SJ et al (2002) Concentration-dependent formation mechanisms in mesophase silica–surfactant films. Langmuir 18:9838–9844
66. Edler KJ, Brennan T, Roser SJ et al (2003) Formation of CTAB-templated mesophase silicate films from acidic solutions. Microporous Mesoporous Mater 62:165–175
67. Edler KJ, Brennan T, Roser SJ (2006) Formation of mesostructured thin films at the air/water interface. Thin Solid Films 495:2–10
68. Park SS, Ha C-S (2004) High-quality free-standing and oriented periodic mesoporous organosilica films grown without a solid substrate at the air–water interface. Chem Commun 1986–1987
69. Park SS, Ha C-S (2005) Free-standing and oriented periodic mesoporous organosilica films with variable pore size at the air–water interface. Chem Mater 17:3519–3523
70. Yang B, Edler KJ (2009) Free-standing ordered mesoporous silica films synthesized with surfactant–polyelectrolyte complexes at the air/water interface. Chem Mater 21:1221–1231
71. Henderson MJ, King D, White JW (2004) Time-dependent changes in the formation of titania-based films at the air–water interface. Langmuir 20:2305–2308
72. Ji H, Liu X, Wang X et al (2011) Influence of counter-ions on the self-assembly of ZrO_2 nanodisks. J Colloid Interface Sci 353:356–362
73. Kan C, Liu X, Duan G et al (2007) Synthesis and characterization of the air–water interfacial TiO2/ZrO2 binary oxide film. J Colloid Interface Sci 310:643–647
74. Edler KJ, Yang B (2013) Formation of mesostructured thin films at the air–liquid interface. Chem Soc Rev 42:3765–3776
75. Buesser B, Gröhn AJ (2012) Multiscale aspects of modeling gas-phase nanoparticle synthesis. Chem Eng Technol 35:1133–1143
76. Lu Y, Fan H, Stump A et al (1999) Aerosol-assisted self-assembly of mesostructured spherical nanoparticles. Nature 398:223–226
77. Brinker CJ, Lu Y, Sellinger A et al (1999) Evaporation-induced self-assembly: nanostructures made easy. Adv Mater 11:579–585
78. Yan Y, Zhang F, Meng Y et al (2007) One-step synthesis of ordered mesoporous carbonaceous spheres by an aerosol-assisted self-assembly. Chem Commun 2867–2869
79. Tsung C-K, Fan J, Zheng N et al (2008) A general route to diverse mesoporous metal oxide submicrospheres with highly crystalline frameworks. Angew Chem Int Ed 47:8682–8686
80. Jiang X, Brinker CJ (2006) Aerosol-assisted self-assembly of single-crystal core/nanoporous shell particles as model controlled release capsules. J Am Chem Soc 128:4512–4513
81. Julián-López B, Boissière C, Chanéac C et al (2007) Mesoporous maghemite–organosilica microspheres: a promising route towards multifunctional platforms for smart diagnosis and therapy. J Mater Chem 17:1563–1569
82. Li L, Tsung C-K, Ming T et al (2008) Multifunctional mesostructured silica microspheres from an ultrasonic aerosol spray. Adv Funct Mater 18:2956–2962
83. Bore MT, Marzke RF, Ward TL et al (2005) Aerosol synthesized mesoporous silica containing high loading of alumina and zirconia. J Mater Chem 15:5022–5028
84. Jin Z, Wang F, Wang F et al (2013) Metal nanocrystal-embedded hollow mesoporous TiO_2 and ZrO_2 microspheres prepared with polystyrene nanospheres as carriers and templates. Adv Funct Mater 23:2137–2144
85. Wang W-N, Kim S-G, Wuled Lenggoro I et al (2007) Polymer-assisted annealing of spray-pyrolyzed powders for formation of luminescent particles with submicrometer and nanometer sizes. J Am Chem Soc 90:425–432
86. Guerrero-Martínez A, Pérez-Juste J, Liz-Marzán LM (2010) Recent progress on silica coating of nanoparticles and related nanomaterials. Adv Mater 22:1182–1195
87. Holmberg K (2003) Organic reactions in microemulsions. Curr Opin Colloid Interface Sci 8:187–196

88. Schacht S, Huo Q, Voigt-Martin IG et al (1996) Oil-water interface templating of mesoporous macroscale structures. Science 273:768
89. Faget L, Berman A, Regev O (2001) Synthesis of unsupported mesoporous silica interphasic films at the oil–water boundary. Thin Solid Films 386:6−13
90. Zhou G, Chen Y, Yang J et al (2007) From cylindrical-channel mesoporous silica to vesicle-like silica with well-defined multilamella shells and large inter-shell mesopores. J Mater Chem 17:2839–2844
91. Schmidt-Winkel P, Glinka CJ, Stucky GD (2000) Microemulsion templates for mesoporous silica. Langmuir 16:356−361
92. Gu D, Bongard H, Deng Y et al (2010) An aqueous emulsion route to synthesize mesoporous carbon vesicles and their nanocomposites. Adv Mater 22:833–837
93. Wei J, Wang H, Deng Y et al (2011) Solvent evaporation induced aggregating assembly approach to three-dimensional ordered mesoporous silica with ultralarge accessible mesopores. J Am Chem Soc 133:20369–20377
94. Wei J, Yue Q, Sun Z et al (2012) Synthesis of dual-mesoporous silica using non-ionic diblock copolymer and cationic surfactant as co-templates. Angew Chem Int Ed 51:6149–6153
95. Yue Q, Wang M, Wei J et al (2012) A template carbonization strategy to synthesize ordered mesoporous silica microspheres with trapped sulfonated carbon nanoparticles for efficient catalysis. Angew Chem Int Ed 51:10368–10372
96. Sanchez C, Boissière C, Grosso D et al (2008) Design, synthesis, and properties of inorganic and hybrid thin films having periodically organized nanoporosity. Chem Mater 20:682–737
97. Tanaka S, Nishiyama N, Oku Y et al (2004) Nano-architectural silica thin films with two-dimensionally connected cagelike pores synthesized from vapor phase. J Am Chem Soc 126:4854–4858
98. Li X, Song L, Vogt BD (2008) Tuning mechanical properties of mesoporous silicas using associating homopolymers/block copolymer blends as templates. J Phys Chem C 112:53−60
99. Tanaka S, Tate MP, Nishiyama N et al (2006) Structure of mesoporous silica thin films prepared by contacting PEO106−PPO70−PEO106 films with vaporized TEOS. Chem Mater 18:5461–5466
100. Li X, Vogt BD (2008) Carbon dioxide mediated synthesis of mesoporous silica films: tuning properties using pressure. Chem Mater 20:3229−3238
101. Tanaka S, Maruo T, Nishiyama N et al (2005) Formation of ordered mesostructured silica by vapor infiltration of tetraethoxysilane into hexagonally arranged surfactant-catalyst nanocomposites. Chem Lett 34:1148–1149
102. Nishiyama N, Kaihara J, Nishiyama Y et al (2007) Vapor-phase synthesis of mesoporous SiO2−P2O5 thin films. Langmuir 23:4746–4748
103. Zhang Z, Wang L, Wang J et al (2012) Mesoporous silica-coated gold nanorods as a light-mediated multifunctional theranostic platform for cancer treatment. Adv Mater 24:1418–1423
104. Lee KJ, Min SH, Jang J (2008) Vapor-phase synthesis of mesostructured silica nanofibers inside porous alumina membranes. Small 4:1945−1949
105. Yang Z, Xia Y, Mokaya R (2007) Enhanced hydrogen storage capacity of high surface area zeolite-like carbon materials. J Am Chem Soc 129:1673−1679
106. Kaneda M, Tsubakiyama T, Carlsson A et al (2002) Structural study of mesoporous MCM-48 and carbon networks synthesized in the spaces of MCM-48 by electron crystallography. J Phys Chem B 106:1256–1266
107. Zhang WH, Liang C, Sun H et al (2002) Synthesis of ordered mesoporous carbons composed of nanotubes via catalytic chemical vapor deposition. Adv Mater 14:1776–1778
108. Vix-Guterl C, Boulard S, Parmentier J et al (2002) Formation of ordered mesoporous carbon material from a silica template by a one-step chemical vapour infiltration process. Chem Lett 31:1062–1063
109. Wang X, Bozhilov KN, Feng P (2006) Facile preparation of hierarchically porous carbon monoliths with well-ordered mesostructures. Chem Mater 18:6373−6381
110. Wu Z, Li W, Xia Y et al (2012) Ordered mesoporous graphitized pyrolytic carbon materials: synthesis, graphitization, and electrochemical properties. J Mater Chem 22:8835–8845

111. Xia YD, Mokaya R (2004) Ordered mesoporous carbon hollow spheres nanocast using mesoporous silica via chemical vapor deposition. Adv Mater 16:886–891
112. Xia Y, Yang Z, Mokaya R (2006) Simultaneous control of morphology and porosity in nanoporous carbon: graphitic mesoporous carbon nanorods and nanotubules with tunable pore size. Chem Mater 18:140–148
113. Xia Y, Mokaya R (2008) Mesoporous MCM-48 aluminosilica oxynitrides: synthesis and characterization of bifunctional solid acid–base materials. J Phys Chem C 112:1455–1462
114. Su F, Zeng J, Bao X et al (2005) Preparation and characterization of highly ordered graphitic mesoporous carbon as a Pt catalyst support for direct methanol fuel cells. Chem Mater 17:3960–3967
115. Xia Y, Mokaya R (2005) Generalized and facile synthesis approach to N-doped highly graphitic mesoporous carbon materials. Chem Mater 17:1553–1560
116. Wang B, Ang TP, Borgna A (2012) A rapid hard template method for the synthesis of N-doped mesoporous carbon replicated from TUD-1. Microporous Mesoporous Mater 158:99–107
117. Xia Y, Mokaya R (2004) Synthesis of ordered mesoporous carbon and nitrogen-doped carbon materials with graphitic pore walls via a simple chemical vapor deposition method. Adv Mater 16:1553–1558
118. Asefa T, Kruk M, Coombs N et al (2003) Novel route to periodic mesoporous aminosilicas, PMAs: ammonolysis of periodic mesoporous organosilicas. J Am Chem Soc 125:11662–11673
119. Masuda H, Asoh H, Watanabe M et al (2001) Square and triangular nanohole array architectures in anodic alumina. Adv Mater 13:189–192
120. Vinu A, Terrones M, Golberg D et al (2005) Synthesis of mesoporous BN and BCN exhibiting large surface areas via templating methods. Chem Mater 17:5887–5890
121. Shi Y, Wan Y, Zhao D (2011) Ordered mesoporous non-oxide materials. Chem Soc Rev 40:3854–3878
122. Winkler H, Birkner A, Hagen V et al (1999) Quantum-confined gallium nitride in MCM-41. Adv Mater 11:1444–1448
123. Fischer A, Antonietti M, Thomas A (2007) Growth confined by the nitrogen source: synthesis of pure metal nitride nanoparticles in mesoporous graphitic carbon nitride. Adv Mater 19:264–267
124. Shi Y, Wan Y, Zhang R et al (2008) Synthesis of self-supported ordered mesoporous cobalt and chromium nitrides. Adv Funct Mater 18:2436–2443
125. Jiao F, Jumas J-C, Womes M et al (2006) Synthesis of ordered mesoporous Fe_3O_4 and γ-Fe_2O_3 with crystalline walls using post-template reduction/oxidation. J Am Chem Soc 128:12905–12909
126. Jiao F, Harrison A, Hill AH et al (2007) Mesoporous Mn_2O_3 and Mn_3O_4 with crystalline walls. Adv Mater 19:4063–4066
127. Tüysüz H, Liu Y, Weidenthaler C et al (2008) Pseudomorphic transformation of highly ordered mesoporous Co_3O_4 to CoO via reduction with glycerol. J Am Chem Soc 130:14108–14110
128. Hisatomi T, Otani M, Nakajima K et al (2010) Preparation of crystallized mesoporous Ta_3N_5 assisted by chemical vapor deposition of tetramethyl orthosilicate. Chem Mater 22:3854–3861
129. Jin G-Q, Guo X-Y (2003) Synthesis and characterization of mesoporous silicon carbide. Microporous Mesoporous Mater 60:207–212
130. Parmentier J, Patarin J, Dentzer J et al (2002) Formation of SiC via carbothermal reduction of a carbon-containing mesoporous MCM-48 silica phase: a new route to produce high surface area SiC. Ceram Int 28:1–7
131. Yang Z, Xia Y, Mokaya R (2004) High surface area silicon carbide whiskers and nanotubes nanocast using mesoporous silica. Chem Mater 16:3877–3884
132. Yao J, Wang H, Zhang X et al (2007) Role of pores in the carbothermal reduction of carbon–silica nanocomposites into Silicon carbide nanostructures. J Phys Chem C 111:636–641

133. Shi Y, Zhang F, Hu Y-S et al (2010) Low-temperature pseudomorphic transformation of ordered hierarchical macro-mesoporous SiO2/C nanocomposite to SiC via magnesiothermic reduction. J Am Chem Soc 132:5552–5553
134. Richman EK, Kang CB, Brezesinski T et al (2008) Ordered mesoporous silicon through magnesium reduction of polymer templated silica thin films. Nano Lett 8:3075–3079

Chapter 6
Applications for Energy Storage

Efficient utilization of green energies requires the development of rechargeable electrical energy storage devices with high energy–density and high power-density. So far, researchers have reported various rechargeable electrical energy storage devices. Among these systems, electrodes are crucial components that significantly impact performance, especially in terms of energy density and power density. The latest advancements in layered devices made from interfacial assembly of soft materials have specifically targeted applications in supercapacitors, lithium-ion batteries (LIBs), sodium-ion batteries (NIBs), lithium-sulfur batteries, lithium-air batteries, and lithium-metal batteries.

6.1 Supercapacitor

Supercapacitors, developed after over a century of capacitor advancements (Fig. 6.1), surpass the power delivery capabilities of conventional capacitors, bridging the gap between rechargeable batteries and capacitors. They play a vital role in meeting the growing energy demands, especially for high-power applications like electric vehicles [1–3]. Layered-based supercapacitors, compared to traditional ones, are typically constructed using thin-film electrodes with sandwich structures, fiber-shaped electrodes with core–shell structures, or arrays of in-plane microelectrode fingers with microscale dimensions. The layered design enhances accessibility to the active electrode materials, exposing the edges to the electrolyte. This enables the achievement of ultrahigh power densities, surpassing those of batteries and conventional supercapacitors, particularly when utilizing two-dimensional (2D) layered materials. Carbon materials, commonly used as electrode materials in EDLC (double-layer capacitor), exemplify this class with their ability to store charges electrostatically at the electrode–electrolyte interface, exhibiting high power density and extended cycle life. Pseudo-capacitors, on the other hand, store electrical energy through reversible redox

B. Kong et al., *Functional Mesoporous Carbon-Based Film Devices for Energy Systems*,
https://doi.org/10.1007/978-981-99-7498-6_6

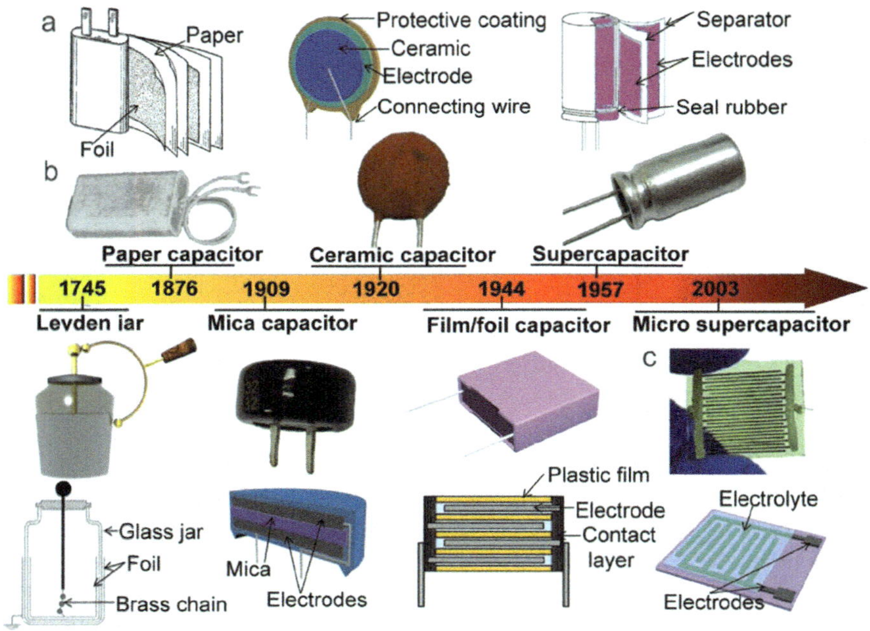

Fig. 6.1 Development history of capacitors. **a** Reproduced with permission Ref. [1]. **b** Reproduced with permission Ref. [2]. **c** Reproduced with permission Ref. [3]

reactions, representing a different form of capacitance beyond the electrostatic nature of EDLC. They rely on conducting polymers and metal oxides and are not part of the layered-carbon-based supercapacitor category. Hybrid supercapacitors combine both EDLC and pseudocapacitance mechanisms. This section will focus on carbon-based film devices, encompassing EDLC and hybrid supercapacitors.

6.1.1 Double-Layer Capacitance

EDLCs rely on the capacitance formed by the charging and discharging of ions at the electrode interfaces. These interfaces exhibit remarkable recyclability, with the ability to undergo this process approximately 105–106 times. As the charge-storage mechanism is non-Faradaic, there is ideally no electron transfer occurring across the electrode interface during the storage of electric charge and energy. Carbon-based materials, known for their excellent electrical properties, stable performance, and large surface area, are widely employed as EDLC materials. These include active carbon [4], carbon nanotubes (CNTs), carbon nanofibers, graphene, and carbon derived from metal carbide.

Durou et al. utilized inkjet-printing technology to fabricate a supercapacitor with electrode fingers made of active carbons (Fig. 6.2a). To transform active carbon powders into thin-film electrode fingers, polymer binders like PTFE and PVDF were incorporated. The micro-device exhibited excellent capacitive behavior across a wide potential range of 2.5 V, achieving a cell capacitance of 2.1 mF/cm^2. Kim et al. introduced atomic layer deposition (ALD) encapsulation of activated carbons for stable supercapacitors at high voltages [5]. Figure 6.2b illustrates the fabrication process, where ALD replaced the conventional carbon activation steps of thermal annealing and vacuum drying. This simplification reduced production costs and minimized time delays. Creating novel structures with enhanced electrolyte ion diffusivity is an effective approach for performance improvement. Onion-like carbon, despite having a moderate surface area of 500 m^2/g compared to activated carbons (1500 m^2/g), offers full accessibility for electrolyte-ion adsorption due to the absence of internal porous networks within the particles. The electrophoretic deposition of nanostructured carbon onions as a layer on micro-electrode fingers resulted in a supercapacitor with capacitance four orders of magnitude higher, an energy per volume one order of magnitude higher, and a discharge rate three orders of magnitude higher than conventional supercapacitors (Fig. 6.2c) [6]. Graphene quantum dots (GQDs), ultrafine lamellar graphene derivatives with sizes less than 10 nm, possess a π-conjugated core structure with abundant edge sites and functional groups. Fan et al. developed a synthesis method for ultramicroporous carbons (CoDCs) assembled using GQDs as building units through chemical welding and in situ activation (Fig. 6.2d) [7]. The resulting CoDCs exhibited uniform diameters of 2.96 nm (Fig. 6.2e) and homogeneous distribution of ultramicropores (marked by yellow circles in Fig. 6.2f) due to inside-out activation utilizing potassium hydroxide (KOH). The high-specific-surface-area (1730 m^2/g) and high-packing-density (0.97 g/cm^3) porous carbon structure ensured remarkable capacitive and rate performances, resulting in high gravimetric capacitance (270 F/g) and volumetric capacitance (262 F/cm^3 at 1 A/g), along with excellent rate capability and cyclic stability (Fig. 6.2g).

Graphene has garnered significant attention due to its extremely high specific surface area (2600 m^2/g), high electron mobility (15,000 cm^2/V s), and theoretical double-layer capacitance (550 F/g). Various porous graphene microstructures, including activated graphene, mesoporous graphene, and macroporous graphene foams, have been synthesized. These structures offer unique characteristics such as low volumetric capacitances and poor rate performance for activated graphene due to ultrasmall micropores, and fast rate capability for mesoporous graphene and microporous graphene foams due to large empty spaces facilitating electrolyte flooding. The preparation methods for porous graphene-based supercapacitors can be divided into two categories. The first category utilizes graphene produced by chemical vapor deposition (CVD) as electrode materials, although scaling-up using CVD graphene is costly. The second category involves solution processes using graphene oxide (GO) dispersion as a starting material, followed by controllable reduction to obtain porous reduced graphene oxide (RGO) films. Methods in this category include light scribe patterning, hydrothermal or chemical reduction of GO dispersions, template-etching,

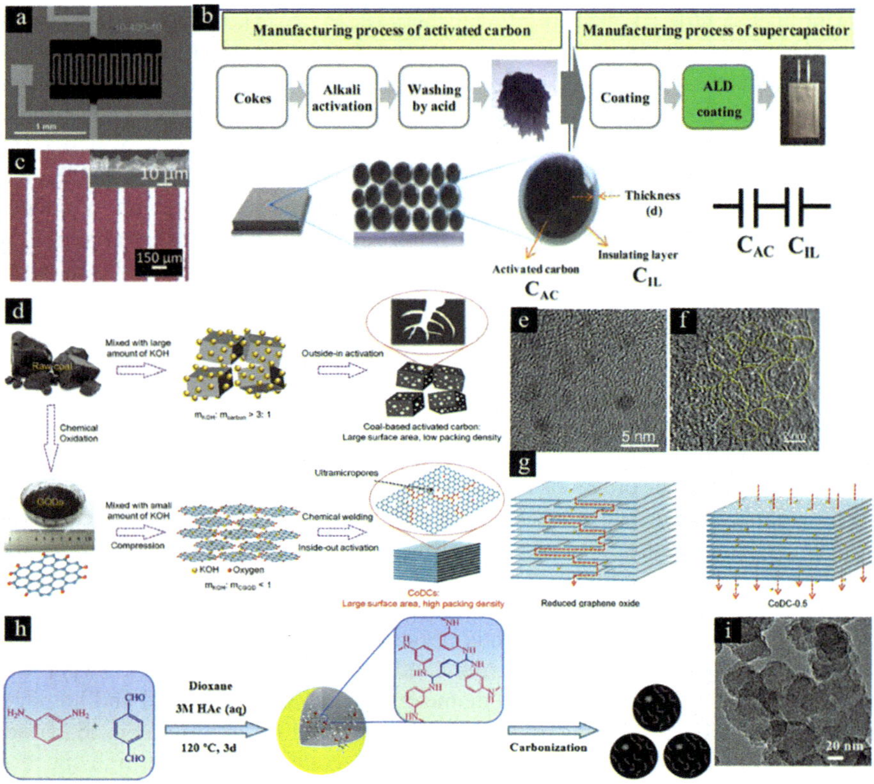

Fig. 6.2 Schematic of zero-dimensional structure for EDLC and the corresponding materials of: **a** active carbon. Reproduced with permission Ref. [4]. Copyright 2010, Elsevier. **b** Al_2O_3 deposited activated carbon. Reproduced with permission Ref. [5]. Copyright 2015, American Chemical Society. **c** Onion like carbon. Reproduced with permission Ref. [6]. Copyright 2010, Nature Publishing Group. **d** Ultramicroporous carbons (CoDCs). **e** HRTEM images of the GQDs. **f** HRTEM images of the CoDC-0.5. **g** Scheme of the ion transport processes in RGO and CoDC-0.5 electrodes with the same high mass loadings. Reproduced with permission Ref. [7]. Copyright 2018, Wiley-VCH. **h** nitrogen-functionalized microporous carbon nanoparticles. Reproduced with permission Ref. [8]. Copyright 2014, Elsevier

and breath-figure methods. However, these techniques may be complex or time-consuming for practical applications. Wang et al. developed a flexible supercapacitor composed of large-area hierarchical porous graphene films using GO hydrogel as the initial feedstock (Fig. 6.3a) [9]. The excellent processability and structure stability of GO hydrogel make it suitable for fabricating large-area RGO films. The resulting RGO film displayed a shiny metallic luster with good flexibility and could be rolled into a multilayer cylinder (Fig. 6.3c). Inspired by the structure of natural vein-textured leaves, Lee et al. reported a scalable synthesis of nanochanneled and PDDA-mediated RGO (nc-PDDA-Gr) film with high packing density and efficient 2D ion transport pathways [10]. The film was obtained by chemically reducing graphene oxide and removing nanowires, resulting in vein-like-textured nanochannels that allowed ion diffusion parallel to the graphene plane. Interdigitated electrodes fabricated using this film facilitated ion diffusion parallel to the film, maintaining high rate capability regardless of the film thickness.

Carbon nanodots (CNDs) with graphitic/graphenic cores, considered smaller counterparts of graphene, have been employed as precursors for larger extended π-systems. Kaner et al. employed a combination of thermolysis and infrared laser treatment to convert biomolecule-based CNDs into high-surface-area turbostratic graphene networks (Fig. 6.3g) [11]. The resulting 3D turbostratic graphene networks exhibited a hierarchical porous structure, high surface area, and excellent electrochemical properties. The volumetric capacitance reached 27.5 mF/L at a current density of 560 A/L, corresponding to an energy density of 24.1 mWh/L at a power density of 711 W/L. Remarkably, the charge–discharge cycling rate was extremely fast, with a time constant of 3.44 ms.

6.1.2 Hybrid Supercapacitors

EDLCs are known for their rapid charge storage but relatively low capacitance, while pseudocapacitors offer high capacitance but often suffer from poor rate capability and cycle stability. Hybrid supercapacitors, which combine the advantages of both EDL capacitors and pseudocapacitors, have emerged as an appealing choice for achieving high energy and power densities in a single device. Various carbon-based materials are currently being explored for the fabrication of hybrid supercapacitors, including graphene paper, CNT paper, and carbon paper integrated with active components. However, carbon materials often have an inert surface due to high-temperature carbonization, making it challenging for the electrolyte to penetrate the inner layers. Nitrogen functionalization of carbon materials has shown promising results in improving surface wettability, electrical conductivity, and capacitance properties. For instance, Liu et al. synthesized nitrogen-rich porous carbon shells that exhibited an electrochemical capacitance of 155 F/g at 10 A/g. Gan et al. prepared nitrogen-functionalized microporous carbon nanoparticles through direct carbonization of a novel polymer derived from the Schiff base reaction of terephthalaldehyde and m-phenylenediamine (Fig. 6.2h). The electrode based on these nanoparticles demonstrated outstanding capacitive behavior (391 F/g at 0.1 A/g), ultrahigh-rate performance (145 F/g at 100 A/g), and excellent long-term cycling stability (98% retention over 5000 cycles at 2 A/g) in a 6 M KOH aqueous electrolyte.

One-dimensional (1D) carbon nanoarchitectures, such as nanowires, nanorods, and nanofibers (CNFs), have demonstrated superior kinetic properties compared to carbon matrices with particle morphology. Figure 6.4a showcases the fabrication of a freestanding film composed of hierarchically porous CNFs doped with boron, phosphorus, nitrogen, and oxygen (B/P/N/O) using a facile electrospinning and one-step carbonization/activation method with a polyacrylonitrile solution containing

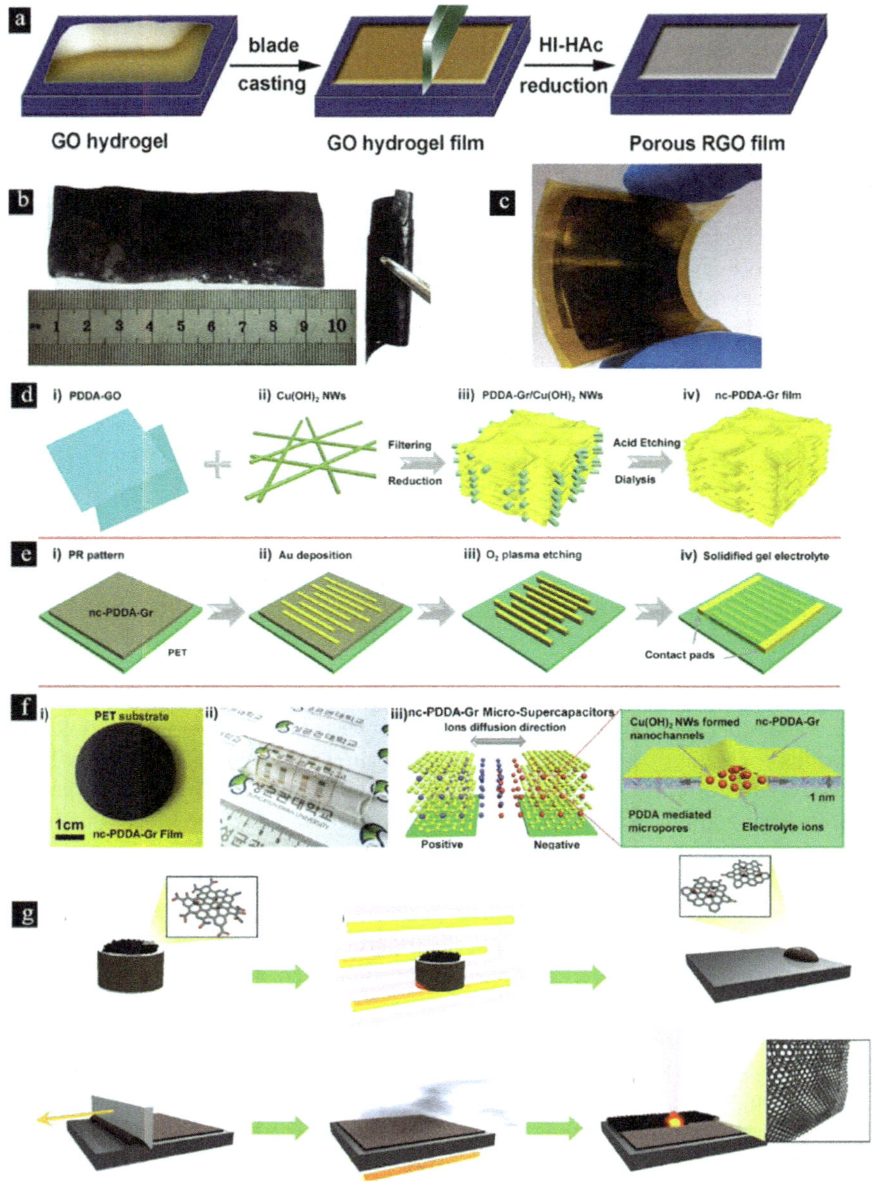

Fig. 6.3 **a** Schematic illustration of the fabrication process of the porous RGO film. **b, c** Photographs of the RGO film and the corresponding flexible supercapacitor. Reproduced with permission Ref. [9]. Copyright 2015, Wiley–VCH. **d** Synthesis procedure of the nc-PDDA-Gr film. **e** MSCs fabrication process. **f** As-prepared MSCs and ion transport pathway. Reproduced with permission Ref. [10]. Copyright 2015, Wiley–VCH. **g** Illustration of the conversion process of CND300 into 3D-ts-graphene. Reproduced with permission Ref.[11]. Copyright 2018, Wiley–VCH

boric acid and phosphoric acid [12]. The optimized film exhibited excellent bending performance, significant volumetric capacitance, and high gravimetric capacitance. The outstanding capacitance performance resulted from the suitable combination of pseudocapacitance behavior from B/P/N/O heteroatom co-doping, particularly active heteroatom species co-doping, and a dense self-standing carbon nanofiber film structure with an appropriate surface area and pore structure characterized by a high meso-/macropore ratio. These advantageous features contributed to remarkable energy storage performance, with a notable volumetric capacitance of 395 F/cm^3 at 1 A/g and a high gravimetric capacitance of 332 F/g. On the other hand, CNTs can be easily fabricated into thin films and possess excellent mechanical properties due to their high length-to-diameter ratio. Consequently, CNTs are promising candidates for electrode materials in supercapacitors. Long et al. designed an all-carbon-based supercapacitor that employed CNTs as the core and high-surface-area microporous carbon as the shell. This structure was achieved through in-situ resorcinol–formaldehyde resin coating of CNTs, followed by carbonization and controlled KOH activation (Fig. 6.4b) [13]. The unique core–shell configuration allowed the CNT core to address the issue of low electronic conductivity in microporous carbons. Additionally, the 1D tubular structure with short pore-pathway micropores and a 3D entangled network enhanced the utilization of overall porosity and improved electrode kinetics. As a result, these CNT/microporous carbon core–shell nanocomposites exhibited high specific capacitance of 237 F/g, excellent rate performance with 75% retention from 0.1 to 50 A/g, and exceptional cyclability.

The use of active materials in powder form necessitates the casting of planar electrodes using binders and conductive carbon materials for the final supercapacitor. However, a challenge arises as the active material can easily detach from the current collector when the supercapacitors are subjected to bending, twisting, or stretching. A freestanding configuration is highly desirable as it offers high flexibility and facile fabrication methods. To address this, Chan et al. developed a solid-state asymmetric fiber-shaped supercapacitor utilizing carbon fiber thread@polyaniline and functionalized carbon fiber thread electrodes with a high operating voltage of 1.6 V (Fig. 6.5a) [14]. The carbon fiber thread served as the current collector and scaffold, while the conducting PANI nanowire networks were deposited on the carbon fiber thread to act as both the active material and electron transport path due to their good conductivity, high energy storage capability, long-term stability, and flexibility. In order to test the flexibility of the device in electronic textiles, a 10-cm length device was sewn into a glove. Even under different bending angles (Fig. 6.5b), no significant change in the CV curves was observed (Fig. 6.5c), and the capacitance ratio remained nearly constant (Fig. 6.5d). Moreover, the prepared device exhibited high stretchability and good mechanical stability under various stretching conditions (Fig. 6.5e). The excellent performance of the device can be attributed to several factors: (1) the asymmetric system expands the operating voltage, thereby significantly increasing the volumetric energy density and power density; (2) the porous channels of CFT@PANI and FCFT electrodes facilitate ion accessibility, resulting in good rate capability of the fiber-shaped asymmetric supercapacitor (FASC); (3) the high flexibility of CFT and PANI contributes to the device's performance during bending tests. Li et al. reported a

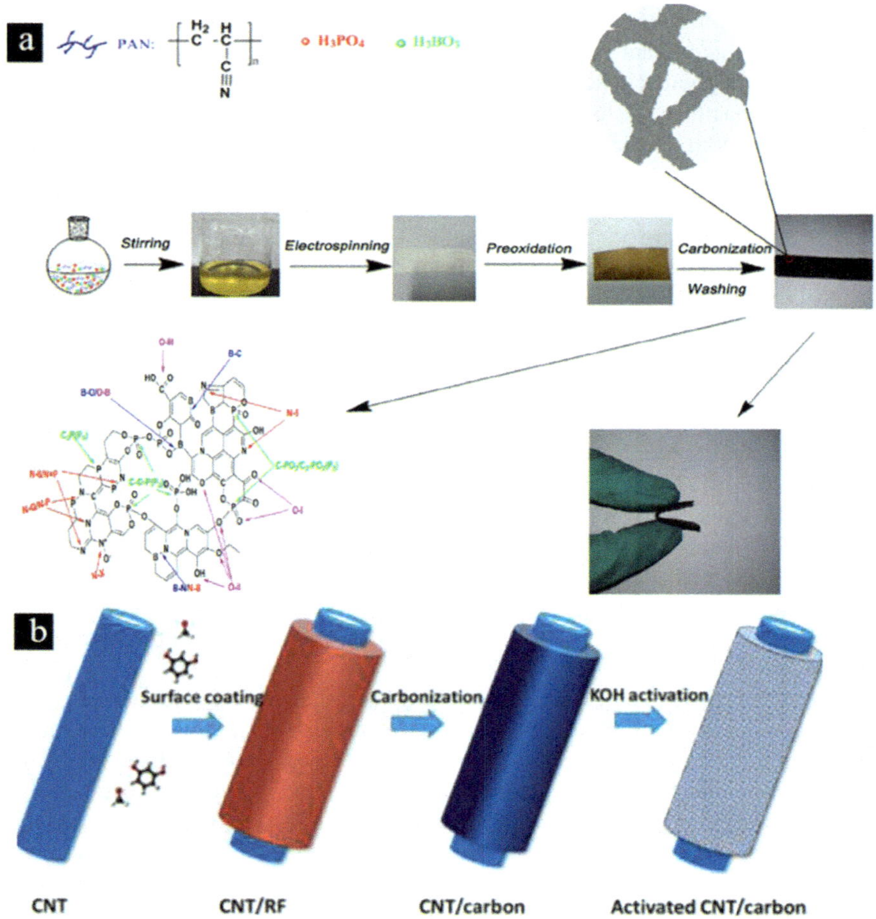

Fig. 6.4 Schematic of one-dimensional structure for EDLC and the corresponding materials of: **a** B/P/N/O co-doped CNF film. Reproduced with permission Ref. [12]. Copyright 2020, Elsevier. **b** CNT/microporous carbon core–shell nanocomposites. Reproduced with permission Ref. [13]. Copyright 2015, American Chemical Society

substrate-free, self-stretchable CNT/PPy yarn supercapacitor that utilizes the helical loop structure incorporated into the CNT yarns (Fig. 6.5f) [15]. Two individual helical yarns were used as symmetric electrodes, coated with a thin layer of H3PO4-PVA gel electrolyte, and then twisted into a freestanding double-helix structure. The high stretchability is achieved by the separation of the loops within the helical yarn under tension, similar to a stretched spring. Figure 6.5g shows photos of the freestanding double-helix supercapacitor in its original state and after being stretched to 2.5 times its initial length. The deformation of the CNT yarns resembles that of stretched telephone wires, which also have a helical structure.

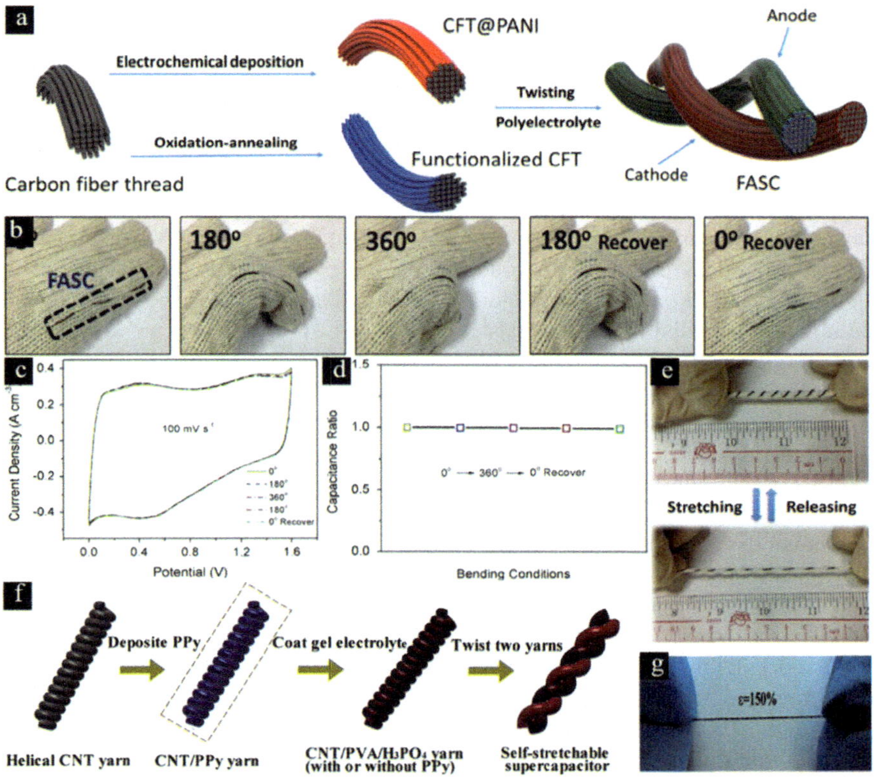

Fig. 6.5 **a** Schematic illustration of the fabrication process of FASC. **b** Digital images of a FASC woven into a glove at different bending states. **c** CV curves of solid-state FASC at different bending states with a scan rate of 100 mV/s. **d** Capacitance ratio for the FASC with different bending angles. **e** Digital images of the stretchable FASC before and after stretching. Reproduced with permission Ref. [14]. Copyright 2015, Elsevier. **f** Synthesis procedure of the self-stretchable CNT yarn supercapacitor. **g** Snapshots of a double-helix CNT yarn supercapacitor manually stretched to 1.5 times of original length. Reproduced with permission Ref. [15]. Copyright 2015, Elsevier

6.2 Li-Ion Battery

Lithium-ion batteries (LIBs) are highly promising rechargeable energy storage devices widely used in electric vehicles (EVs) and hybrid electric vehicles (HEVs). The anode material plays a crucial role in determining the energy density, safety, and cycling life of LIBs. While graphite has been the commercialized choice for its excellent cycling stability and minimal volume change during lithiation/delithiation (only 12%), it offers a low theoretical capacity (LiC6, 372 mAh/g) and a low delithiation potential (0.05 V vs. Li^+/Li). Therefore, graphite is not suitable for next-generation LIBs required for smart electrical grid systems and wearable electronic devices. Novel anode materials with higher Li-storage capability and operational safety are needed. Alloy-type materials are the primary candidates due to their reversible alloying and dealloying with lithium, offering high specific capacity and safety characteristics. Table 6.1 illustrates the high gravimetric capacities of group IV (Ge, Sn) and V (Sb, Bi) metal-based materials, which reach 1625, 994, 660, and 385 mAh/g, respectively (Table 6.1) [16–20]. Additionally, these anode materials exhibit superior volumetric capacities, such as 8645, 7216, 4420, and 3800 mAh/cm^3, far exceeding that of graphite (837 mAh/cm^3). Moreover, these metal-based anodes demonstrate favorable average delithiation voltage of approximately 0.4–0.6 V versus Li/Li$^+$, contributing to increased cut-off voltage to avoid Li dendrite formation and enhance safety. Compared to zero-strain anode materials like $Li_4Ti_5O_{12}$, the relatively low delithiation potential of these metal-based anodes results in higher working voltage and energy density. However, the major obstacle preventing the practical use of these materials is the significant volume change they undergo during charge–discharge cycles in a battery. Ge, Sn, Sb, and Bi phases experience substantial volume variations estimated at 370%, 260%, 200%, and 215%, respectively. Such dramatic expansion and contraction lead to electrode pulverization, cracking, loss of electrode contact, and detachment of electrode materials from the current collector, consequently affecting battery capacity and cyclic stability.

Considerable efforts have been focused on reducing the sizes of metal-based nanomaterials to micro- or nano-scales to mitigate volume change and enhance Li$^+$/electron transport by increasing surface area-to-volume ratios and reducing diffusion lengths [55–57]. Although these nanomaterials show improved cycling performance compared to their bulk counterparts, their practical application in next-generation LIBs is hindered by challenges such as volume change, severe aggregation, complex preparation, excessive SEI formation, low tap/packing density, and limited areal/volumetric capacities. To address these issues, metal-based nanomaterials are often transformed into metal–carbon frameworks (MCFs) by incorporating them into carbon matrices. Numerous studies have demonstrated that MCF composites exhibit enhanced electrochemical performance compared to unsupported metal or carbon alone, benefiting from the combination of high metal capacity and carbon stability. Nanostructured materials offer several advantages, including larger

Table 6.1 Comparison of C and Ge, Sn, Sb, Bi anode materials

Sample	Theoretical gravimetric capacity (mAh/g)	Theoretical volumetric capacity (mAh/cm^3)	Potential vs Li/Li$^+$ (V)	Volume variation (%)	Li-ion transport rate (cm^2/s)	References
C	372	837	0.05	12	10^{-7}~10^{-6}	[16]
Ge	1625	8645	0.5	370	10^{-12}~10^{-10}	[17]
Sn	994	7216	0.6	260	10^{-16}~10^{-13}	[18]
Sb	660	4420	0.99	200	–	[19]
Bi	385	3800	0.85	215	~10^{-13}	[20]

specific surface areas to reduce capacity loss during charge/discharge, shorter diffusion lengths for faster rate capability, and improved cycling life by mitigating the stress of lithium insertion/extraction. The carbon frameworks serve as buffers to accommodate volume expansion of active materials, maintain structural integrity during cycling, and prevent metal particle aggregation. Depending on the carbon component, the nanocomposites can be either active-active (with carbon also storing lithium) or active-inactive (with carbon solely providing an electron migration path). In this section, metal-based anode materials are broadly classified as nanomaterials or composites without explicitly discussing the underlying anode construction principles. The aim is to provide readers with a clear understanding of developing MCF materials from a new perspective. The review covers structural design principles, fabrication methods, morphological features, and cycling performance of several representative MCF anode materials to promote the practical application of MCF anodes in next-generation LIBs.

6.2.1 Metal–Carbon Compounds

To enhance the performance of metal anodes, researchers have explored the use of various metal–carbon framework (MCF) composites. Carbon materials with different shapes and dimensions, such as carbon nanospheres, carbon nanowires, carbon rods, carbon nanotubes (CNTs), graphene, and three-dimensional (3D) carbon matrices, have been employed to prepare carbon-based nanocomposites for use as LIB anodes. The inclusion of a carbon layer in these composites serves multiple purposes. Firstly, it acts as a buffering matrix, accommodating the volume changes and preventing metal agglomeration, thereby enhancing cycling stability. Secondly, the carbon layer improves the electrical conductivity of the composites, leading to enhanced rate capability. In this section, we provide a detailed overview of several typical MCFs, including Sn, Ge, Sb, Bi, and two-component composites with various carbon nanomaterials, for their application in rechargeable LIBs.

6.2.1.1 Sn-Carbon-Based Frameworks

Table 6.2 presents a comparison of the electrochemical performance of Sn-carbon-based frameworks using different carbon nanomaterials as anode materials for LIBs. Wang et al. utilized the aerosol spray pyrolysis technique to synthesize nano Sn–C composites, where Sn nanoparticles (10 nm) were uniformly dispersed within a spherical carbon matrix (Fig. 6.6a) [21]. These composites demonstrated enhanced lithium-storage performance, including high reversible capacity (710 mAh/g at 0.25 C), excellent cycling stability (100% retention after 130 cycles at 0.25 C), and remarkable rate capability (750 mAh/g at 0.1 C and 600 mAh/g at 20 C) (Fig. 6.6b). The carbon matrix played a crucial role in improving lithium storage performance by providing mechanical support to accommodate the volume change of Sn nanoparticles and reducing lithium ion diffusion and electronic transport distances. Carbon nanotubes (CNTs), being hollow, robust, and flexible, are particularly favorable for lithium-ion storage. Hence, significant research has focused on developing electrochemically active CNT-based anode materials. Figure 6.6c illustrates the growth process of yolk-shell Sn-CNT composites with a four-layer structure prepared using a SnO_2-CNT-mediated method [29]. The nanoparticles were supported by bilayer CNTs, benefiting from large void spaces. The Sn-CNT composite exhibited a large reversible capacity of 611 mAh/g after 200 cycles at 0.5 C (495 mA/g) (Fig. 6.6d).

Graphene has garnered significant attention due to its exceptionally high specific surface area of 2600 m^2/g and flexibility, making it an excellent candidate for buffering volume changes. Consequently, various Sn-graphene hybrids with different structures have been developed, including directly decorated Sn-graphene [30], Sn-CNT-graphene, and sandwich-like graphene-supported hybrids. These Sn-graphene hybrid anodes have demonstrated improved electrochemical performance. However, their rate capabilities and cycling stability at high rates have not consistently outperformed other Sn-carbon-based frameworks. To address these challenges, Li et al. devised unique Sn-graphene hybrids with structures that prevent the aggregation of Sn nanostructures and graphene nanosheets, while maintaining high conductivity, activity, and integration during high-rate charge/discharge processes (Fig. 6.6e) [30]. The resulting 3D hybrid anode exhibits exceptional rate performance (1022 mAh/g at 0.2 C, 865 mAh/g at 0.5 C, 780 mAh/g at 1 C, 652 mAh/g at 2 C, 459 mAh/g at 5 C, and 270 mAh/g at 10 C) and remarkable cycling stability, even at high rates (achieving a capacity of 682 mAh/g at 2 A/g and maintaining approximately 96.3% capacity after 1000 cycles) (Fig. 6.6f).

To enhance the performance of MCFs, nitrogen (N) or boron (B) doping, as well as N/B-containing functional groups, have been utilized. These additions improve electrical conductivity for fast electron transport and provide defect structures for enhanced lithium storage. Lu et al. developed an electrode composed of continuous porous N-doped carbon encapsulating a high content of Sn nanoparticles (88.7 wt.%). This structure was derived from binary metal oxides using the outward growth of ZIF-8 approach (Fig. 6.7a) [31]. The resulting anode exhibited an initial discharge capacity of 1321 mAh/g with a superior coulombic efficiency of 80.1% at 0.2 A/g. Moreover, the reversible capacity remained at 96% of its initial discharge capacity after 150 cycles (Fig. 6.7b). The remarkable electrochemical performance was attributed to the synergistic effect of well-dispersed Sn nanoparticles and the interconnected N-doped porous carbon network. In a different approach from previous methods involving the mixing of Sn and carbon precursors, Chen et al. achieved the preparation of ultrasmall Sn nanoparticles finely embedded in an N-doped porous carbon network by carbonizing a single precursor, a divalent Sn complex. The overall fabrication process is illustrated in Fig. 6.7b [32]. During the pyrolysis process, the Sn(salen) ligand transformed into a porous carbon network while the metal cation underwent in-situ reduction to form uniform Sn nanoparticles. At the same time, a portion of the N species in the precursor was retained in the composite, resulting in the final Sn/

Table 6.2 Comparisons of electrochemical performance of Sn-carbon-based frameworks with different carbon nanomaterials as anode materials for LIBs

Sample	Method	Initial capacity (mAh/g)	Coulombic efficiency	Cycling stability	References
Sn@spherical carbon	Aerosol spray pyrolysis	710 (0.25 C)	100% (10 cycles)	710 mAh/g (130 cycles, 0.25 C)	[21]
F-G/Sn@C	Hydrogen assisted thermal reduction method	727 (0.1 A/g)	–	645 mAh/g (100 cycles, 0.1 A/g)	[22]
Sn@CNT@Sn@CNT	SnO_2-CNT-mediated method	751 (1 C)	–	611 mAh/g (200 cycles, 0.5 C)	[23]
TiO_2-Sn@carbon nanofibers	Electrospinning and atomic layer deposition	1644 (0.1 A/g)	66.8% (1st cycle)	643 mAh/g (1100 cycles, 0.2 A/g)	[24]
Sn-graphene	Microwave hydrothermal and hydrogen reduction	2136 (0.1 C)	65.9% (1st cycle)	1100 mAh/g (30 cycles, 0.1 C)	[25]
Sn@CNT-Graphene nanosheets	Microwave hydrothermal and CVD	1160 (0.1 A/g)	70.3% (1st cycle)	982 mAh/g (100 cycles, 0.1 A/g)	[26]
RGO-Sn@C	Hydrolysis and CVD	1351 (0.1 A/g)	66.9% (1st cycle)	630 mAh/g (50 cycles, 0.1 A/g)	[27]
Sn@porous CNFs	Electrospinning	1880 (0.8 A/g)	61.8% (1st cycle)	774 mAh/g (200 cycles, 0.8 A/g)	[28]

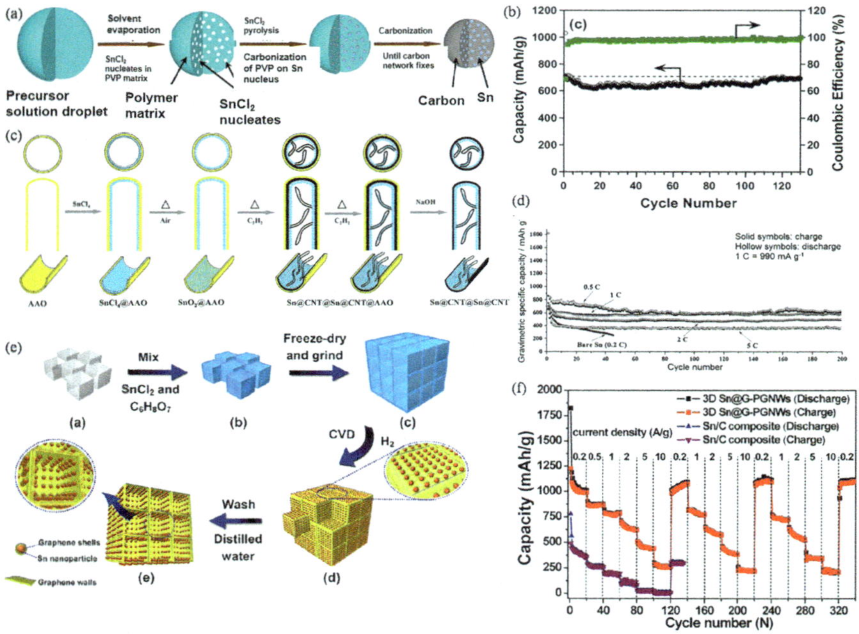

Fig. 6.6 **a** Schematic formation processes of ultrasmall nano-Sn dispersed carbon spheres. **b** Cycling performance at a current density of 0.20 A/g in a potential window of 0.02–3 V versus Li/Li^+ of the nano-Sn/C composite (Reproduced with permission Ref. [21]. Copyright 2013 American Chemical Society). **c** Schematic illustration of the growth process of the four-layer yolk-shell Sn–C nanotube arrays. **d** The cycling performance of the four-layer yolk-shell Sn–C nanotubes in a potential window of 5 mV–3 V versus Li/Li^+ (Reproduced with permission Ref. [29]. Copyright 2014 Wiley–VCH). **e** Schematic fabrication illustration of a 3D porous graphene network anchored with graphene-encapsulated Sn NPs (Sn@G-PGNW). **f** Rate capabilities of the Sn@G-PGNW electrode in a potential window of 5 mV–3 V versus Li/Li^+ (Reproduced with permission Ref. [30]. Copyright 2014 American Chemical Society)

N-doped porous carbon network. This unique structure contributed to the excellent electrochemical performance for LIBs.

Carbon nanostructures, including nanowires, nanorods, and nanofibers (CNFs), have demonstrated superior kinetic properties compared to carbon matrixes with particle morphology. To further enhance performance, a novel design utilizing one-dimensional (1D) carbon nanostructures and Sn quantum dots (QDs) was developed. As depicted in Fig. 6.7c, N-doped CNFs embedded with uniformly distributed Sn QDs were synthesized using the electrospinning technique followed by nitrogen annealing [33]. This unique combination of ultrasmall Sn QDs and CNFs resulted in extraordinary performance in LIBs, including exceptional reversible capacity, excellent rate capability, and superior cyclic stability.

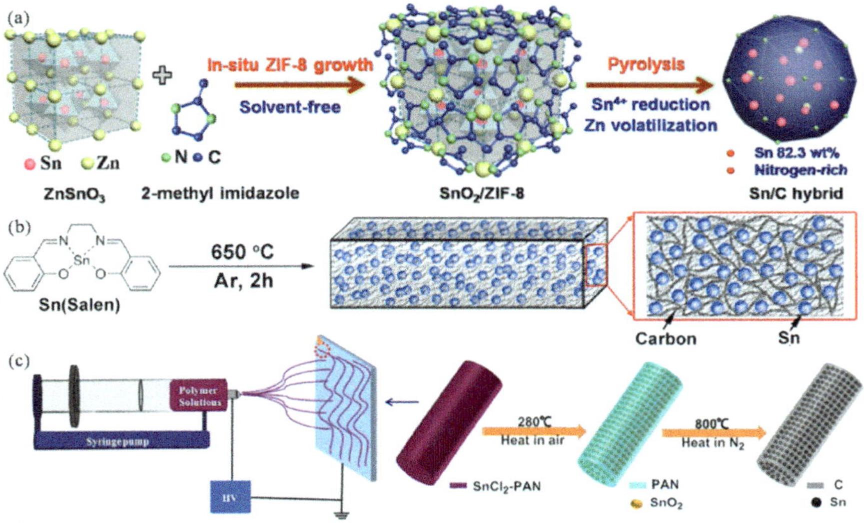

Fig. 6.7 Schematic formation processes of **a** Sn/C hybrid (Reproduced with permission Ref. [31]. Copyright 2016, Elsevier), **b** Sn NPs dispersed in N-doped porous carbon matrix (Reproduced with permission Ref. [32]. Copyright 2013, American Chemical Society), **c** Sn QDs@CNFs (Reproduced with permission [33]. Copyright 2014, Elsevier)

6.2.1.2 Ge-Carbon-Based Frameworks

Despite its high cost, Ge has gained increased attention due to its superior Li diffusivity and theoretical gravimetric capacity compared to Sn [59, 100, 101]. To address the challenges associated with Ge, researchers have developed innovative approaches and unique structures. For instance, Guo et al. designed a core–shell structure where Ge nanoparticles were encapsulated within robust carbon boxes by layering carbon onto cubic GeO_2, followed by reduction to Ge particles within the carbon boxes (Fig. 6.8a) [34]. This synthesis design offered multiple advantages, leading to excellent lithium storage performance. Similarly, Yu et al. designed a flexible and self-supported electrode consisting of Ge nanoparticles encapsulated in carbon nanofibers (CNFs) using an electrospinning technique (Fig. 6.8b) [35]. The improved electrochemical performance was attributed to the synergistic effects of 0D Ge nanoparticles, 1D CNFs, and a 3D interconnected CNF framework, providing effective accommodation of the volume change of Ge nanoparticles and maintaining excellent electrical conductivity. In another approach, a Ge/reduced graphene oxide (RGO)/carbon nanotube (CNT) composite electrode was constructed through a simple process involving ultrasonication and H_2 reduction (Fig. 6.8c) [36]. This unique structure, incorporating graphene sheets, enhances electrical conductivity and buffers volume changes, while the CNT mechanically stabilizes the electrodes and maintains a conductive network for the active Ge nanoparticles, resulting in improved cycling performance.

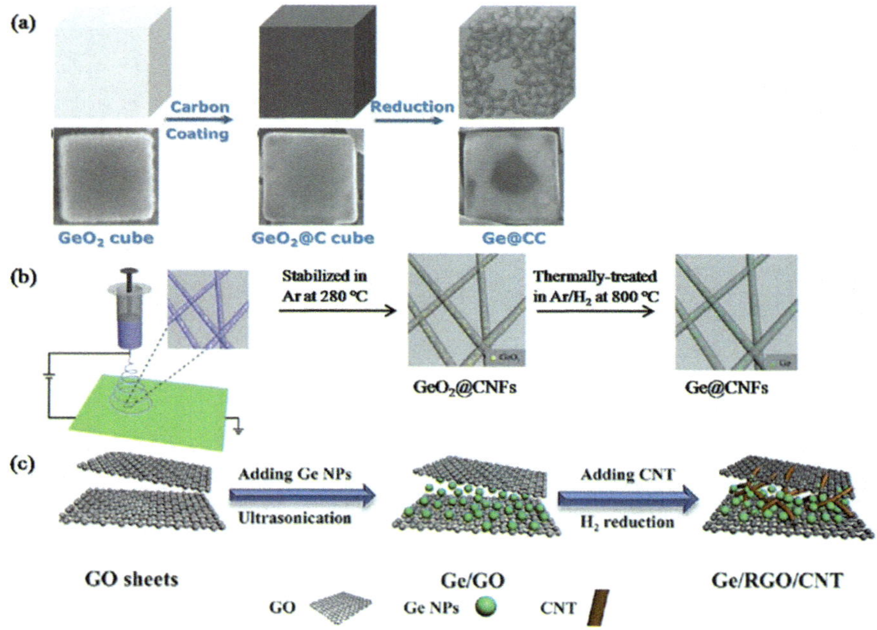

Fig. 6.8 Schematic formation processes of **a** Ge/carbon cubes (Reproduced with permission Ref. [34]. Copyright 2016, Wiley–VCH), **b** Ge@CNF electrode (Reproduced with permission Ref. [35]. Copyright 2014, The Royal Society of Chemistry), **c** Ge/RGO/CNT nanocomposite (Reproduced with permission Ref. [36]. Copyright 2015, The Royal Society of Chemistry)

Similar to Sn/heteroatom-doped carbon nanomaterials, Ge/N-doped carbon composites have been extensively studied for their battery performance. Table 6.3 provides a summary of the electrochemical performance of various Ge/N-doped carbon composite electrodes. It is evident from the table that the morphologies and structures significantly influence the performance. For instance, Yang et al. employed a facile approach to prepare a 3D interconnected porous nitrogen-doped graphene foam (NGF) with encapsulated Ge quantum dot@N-doped graphene yolk-shell nanoarchitecture (Ge-QD@NG/NGF) (Fig. 6.9a, b) [37]. The internal void space in this architecture effectively mitigated volume expansion during lithiation, while the N-doped graphene outer shell minimized pulverization, exfoliation, and aggregation of Ge. The Ge-QD@NG/NGF exhibited high specific reversible capacity (1220 mAh/g), long cycling capability (over 96% capacity retention from the second to 1,000 cycles), and ultra-high rate capability (over 800 mAh/g at 40 C). Wu et al. developed a novel olive-like N-doped carbon material with embedded high-level Ge nanoparticles using dopamine as a carbon and nitrogen source (Fig. 6.9c, d) [38]. The unique olive-like structure, along with the porous carbon framework, facilitated lithiation/delithiation reactions and accommodated volume expansion during cyclic alloying/dealloying. Nanocrystalline Ge/N-doped carbon (Ge/NC) composites with

Table 6.3 Electrochemical performance of various Ge/N-doped carbon composite electrodes

Sample	Feature	Electrochemical performance			References
		Current density	Cycle number	Capacity retention (mAh/g)	
Ge/NC	Ge NPs (~10 nm) distributed in the N-doped carbon spheres (300–600 nm)	0.2 A/g	100	684	[7]
Ge@N-CNTs	Ge cores encapsulated by the N-CNTs shell	0.1 A/g	200	892	[40]
Ge/NC	Olive-like with a minor axis size of ~250 nm	0.5C	2000	1042	[41]
Ge-QD@NG/NGF	3D interconnected porous graphene network Ge-QD (~5 nm)	1C	1000	1200	[42]
Ge/CN_x	Peapods-like	0.5C	1200	1080	[43]

controlled nitrogen content were fabricated through condensation, hydrolysis, and annealing under nitrogen (Fig. 6.9e, f) [39]. The enhanced performance of the Ge/NC electrode can be attributed to the increased electrical conductivity for efficient charge transfer and the presence of defect sites for enhanced lithium insertion.

6.2.1.3 Sb-Carbon-Based Frameworks

Sb anode research distinguishes itself from Sn and Ge anodes by focusing primarily on applications in LIBs, highlighting notable Na-storage capacity, rate capability, and cyclability. Figure 6.10a illustrates the fabrication process of Sb nanoparticles embedded uniformly in carbon nanofibers using a single-nozzle electrospinning technique followed by carbothermal reaction [44]. Electrochemical experiments demonstrate that the Sb/C nanofiber electrode exhibits a large reversible capacity (528 mAh/g at C/6, 337 mAh/g at 5 C) (Fig. 6.10b). Notably, the Sb/C electrode displays a remarkable recovery in capacity, reaching 514 mAh/g when the current decreases to C/6, indicating a high tolerance for rapid Na ion insertion and extraction. This superior electrochemical performance is attributed to the unique nanofiber structure and uniform distribution of Sb nanoparticles in the carbon matrix. These features provide a conductive and buffering matrix that effectively alleviates mechanical stress caused by Na ion insertion/extraction and prevents the aggregation of Sb nanoparticles. Several studies have reported the successful application of these anode materials in LIBs.

Yi et al. utilized pyrolytic ZIF-67 as both a template and reducing agent to enable the replacement reaction from Sb3 + to Sb, while serving as a carbon matrix for assembling reduced Sb nanoparticles (Fig. 6.10c) [45]. The rate capability of the Sb/

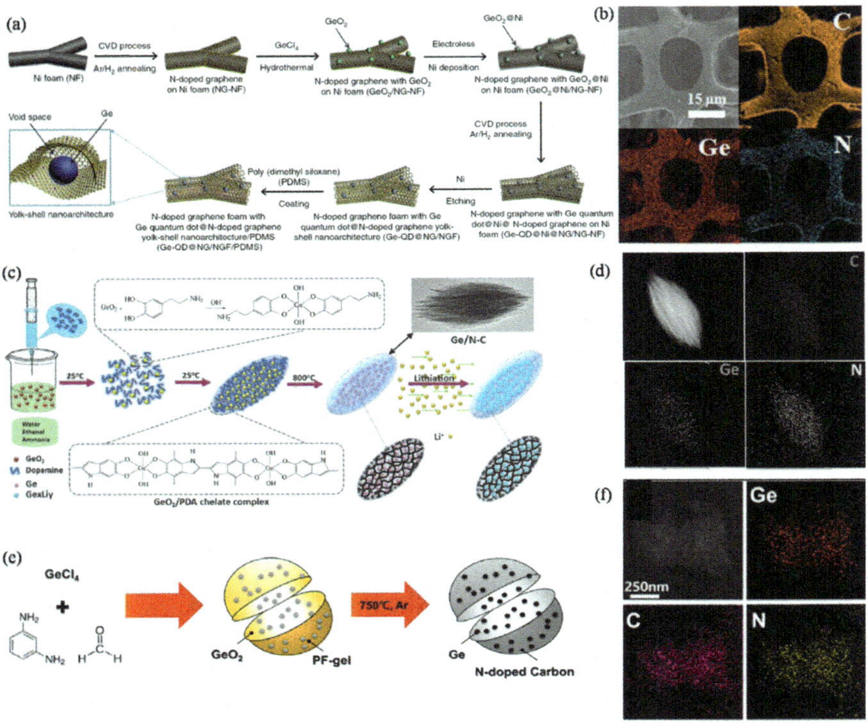

Fig. 6.9 Schematic formation processes of **a**, **b** Ge-QD@NG/NGF (Reproduced with permission Ref. [37]. Copyright 2016, Nature), **c** Ge/N–C composites (Reproduced with permission Ref. [38]. Copyright 2017, Wiley–VCH), **e** Ge/NC (Reproduced with permission Ref. [39]. Copyright 2016, American Chemical Society). **b**, **d**, **f** EDS elemental maps of sample in (**a**, **c**, **e**), respectively

C composite electrode was investigated at various current densities ranging from 0.2 to 5 A/g. As shown in Fig. 6.10d, the Sb/C composite electrode exhibited decent capacity retention, delivering average discharge capacities of 315.4 mAh/g even at the high current density of 5 A/g. The skillful packaging of Sb nanoparticles within the carbon matrix provided an effective buffer to mitigate volume changes during lithium insertion/extraction.

In another study, Wang et al. employed Ni foam as a template and reducing agent for the replacement reaction from Sb^{3+} to Sb, followed by deposition of an Sb thin film onto the substrate (Fig. 6.10e) [46]. To further enhance the cycling performance, a reduced graphene oxide (RGO) thin film was deposited onto the surface of Sb/Ni using a facile pulsed electrophoretic deposition route. Figure 6.10f illustrates the rate capability of RGO/Sb/Ni at various current densities. The improved performances can be attributed to the novel hierarchical sandwich laminated structure, which promotes excellent electrical contact, and the synergistic effect between

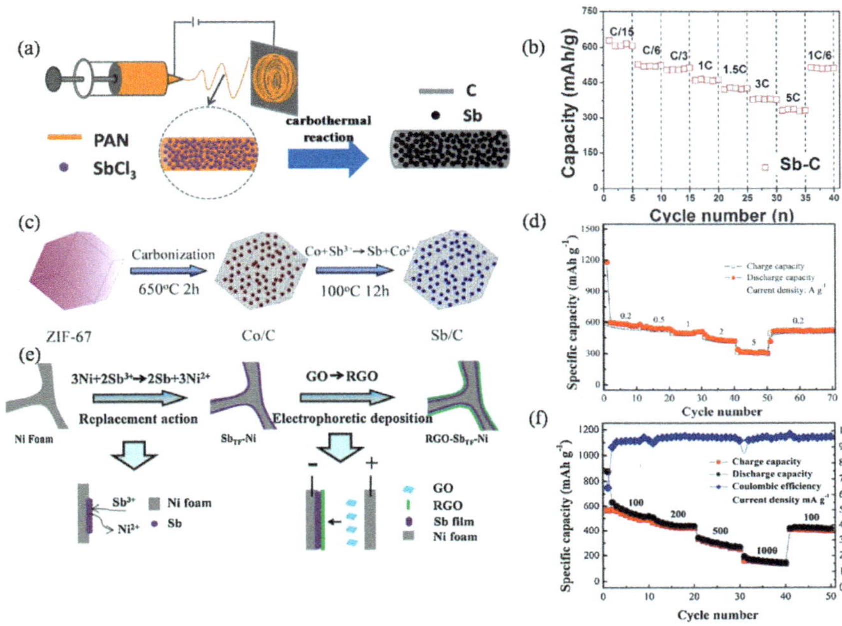

Fig. 6.10 **a** Schematic illustration of the preparation process for the Sb/C nanofibers. **b** Rate capability of the Sb/C nanofibers at various current rates from C/15 to 5 C. (Reproduced with permission Ref. [44]. Copyright 2014 The Royal Society of Chemistry). **c** Schematic illustration of the growth process of Sb/C composite. **d** Rate capability of the Sb/C electrode from 0.2 to 5 A/g (Reproduced with permission Ref. [45]. Copyright 2016, Wiley–VCH). **e** Schematic fabrication illustration of RGO/Sb/Ni. **f** Rate capability of RGO/Sb/Ni at various current densities from 100 to 1000 mA/g (Reproduced with permission Ref. [46]. Copyright 2016 Elsevier)

the high-specific-capacity Sb thin film and the flexible RGO, providing a buffering effect.

6.2.1.4 Bi-Carbon-Based Frameworks

Bi possesses intrinsic advantages as a LIB anode, including relatively higher stability in moisture/air and an oxygen atmosphere at room temperature, as well as easier preparation compared to Sn and Sb due to its lower melting point [47]. Recent studies have highlighted the potential of Bi-based materials as anodes for LIBs [48–50]. For instance, Yang et al. successfully prepared sphere-like Bi/C nanocomposites using the aerosol spray pyrolysis technique [48]. These microspheres featured uniformly distributed Bi nanoparticles with a diameter of approximately 15 nm within the carbon matrix (Fig. 6.11a). The small size of the Bi nanoparticles facilitated Li^+/electron diffusion. Additionally, the dispersion of Bi nanoparticles in the

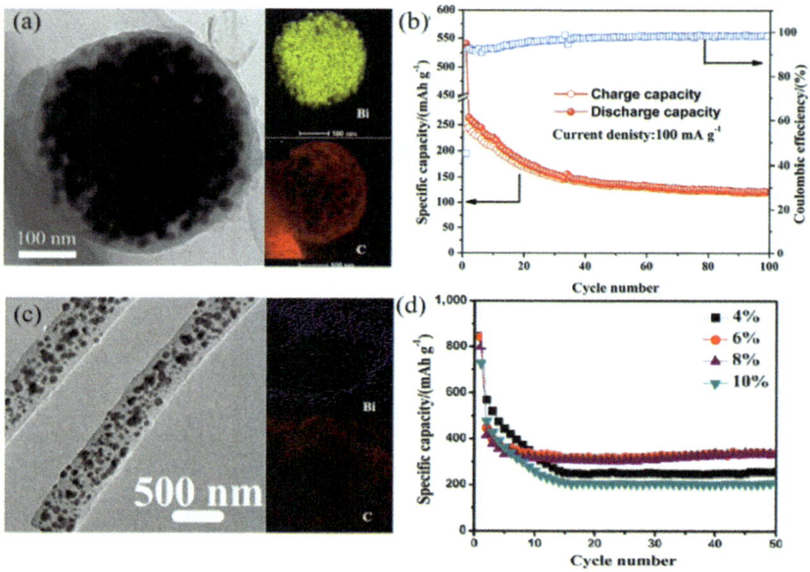

Fig. 6.11 TEM images, corresponding Bi and carbon elemental mapping of **a** Bi/C microspheres. Reproduced with permission Ref. [48]. Copyright 2016, Wiley–VCH. **c** Bi/C nanofibers (Reproduced with permission Ref. [49]. Copyright 2016, Springer. **b**, **d** Cycle performance of sample in (**a**, **c**) at 100 mA/g, respectively

carbon matrix during charge/discharge processes allowed for accommodating volumetric changes and preventing aggregation, resulting in enhanced cycling stability (Fig. 6.11b). Furthermore, the carbon matrix improved the utilization of the active material (Bi) by enhancing the electrode's conductivity.

Furthermore, the incorporation of nanosized Bi into 1D carbon nanofibers offers a promising solution to address issues related to volume expansion, electron transport, and cycle stability. Li et al. employed a carbothermal reduction method, involving the high-temperature pyrolysis of polyacrylonitrile (PAN) and deoxidization of an appropriate bismuthate, to fabricate Bi/C nanofibers [49]. The transmission electron microscopy (TEM) image in Fig. 6.11c reveals the homogeneous embedding of Bi nanoparticles with diameters of approximately 20 nm within the carbon nanofibers, with a few Bi nanoparticles adhering to the fiber surface. The Bi/C nanofiber anodes with varying Bi contents demonstrated different rates of capacity decay in LIBs (Fig. 6.11d). Although capacity decay was observed across all four Bi contents, Bi-6 and Bi-8 exhibited higher capacities of up to 339.3 and 334.8 mAh/g, respectively, at a cycling rate of 100 mA/g.

6.2.1.5 Alloy-Carbon-Based Frameworks

Alloying is a preferred approach when utilizing group IV and V elements for anode materials. Table 6.4 illustrates that certain intermetallic compounds based on alloys can produce less active metal after the initial lithiation process, thereby serving as a buffer to accommodate expansion during subsequent lithiation stages.

6.2.2 Designed Architectures

Over the years, various strategies have been employed to enhance the performance of anodes through innovative structure design techniques. Composite materials have

Table 6.4 Electrochemical performance of various alloy-carbon composite electrodes

Sample	Feature	Electrochemical performance			References
		Current density	Cycle number	Capacity retention (mAh/g)	
$Sn_{0.92}Sb_{0.08}/C$	Sn0.92Sb0.08 nanoparticles encapsulated hollow carbon nanofibers	0.05 A/g	150	637	[51]
$CoSb_2/C$	Disproportionated CoSb nanocrystalline dispersed within amorphous-C matrix	0.1 A/g	100	490	[52]
SnTe/C	SnTe nanocrystallites (4–10 nm) dispersed within amorphous-C matrix	0.1 A/g	100	647	[53]
$Fe_{0.74}Sn_5/RGO$	FeSn nanospheres (30–50 nm) dispersed on RGO nanosheets	0.05 A/g	100	674	[54]
NiSbC	NiSb alloy embedded in carbon hollow spheres	1 A/g	1000	309	[55]
SnCo/NC	SnCo nanoparticles (20 nm) uniformly embedded in porous N-doped carbon microboxes (2 μm)	2 A/g	500	500	[56]

emerged as a promising solution by combining the exceptional electrical conductivity, tensile strength, and flexibility of carbon-based materials with the reactivity and high theoretical capacity of active components. The rational design and engineering of these composites are crucial in achieving superior overall performance and desired structural characteristics. These composites have evolved from simple carbon coatings to more complex and advanced hybrid structures, incorporating multiple carbon-based materials within a single composite. Examples of such designs include anchored structures, layered/sandwich configurations, yolk-shell architectures, core–shell morphologies, and porous structures that provide void spaces to accommodate volume expansion.

6.2.2.1 Anchored Structures

MCF-anchored structures exhibit enhanced cycling performance compared to isolated metal-based particles, thanks to their unique formation process. This microstructure is achieved through rapid heating of precursor droplets, facilitating the quick formation of metal nanograins within a carbon framework. The short residence time and rapid cooling allow for the stabilization of nonaggregated and uniformly sized metal grains firmly anchored in the carbon matrix. Techniques like spray pyrolysis and electrospinning of mixtures containing metallic precursors offer effective means to fabricate micro-sized composites, where the active metal materials are completely encapsulated or embedded within conductive carbon matrices. This results in increased conductivity and effective suppression of SEI formation. Joo et al. successfully optimized the anchored structure of Sn/C composites by carefully controlling pressure equilibrium reactions between electrospun carbon fibers and gaseous decomposition products. Additionally, embedding the metal-based particles into micro-sized composites partially addresses the low tap/packing density challenge associated with nanomaterials.

Various solid-phase and gas-phase methods are utilized to prepare metal-based composites with metal particles fully embedded within matrices. For instance, Chen et al. developed a novel hierarchical Sn/C structure where nanosized Sn particles were anchored on the tips of carbon nanotubes (CNTs). These CNTs were rooted on the surfaces of microsized hollow carbon cubes, with additional Sn nanoparticles encapsulated within the cubes (Fig. 6.12a). This stable hierarchical structure provided ample voids for Sn expansion, resulting in excellent cyclic performance. The carbon component in the Sn/C composite not only contributed to its electrical conductivity, enabling an excellent rate capability, but also acted as an active component for capacity. In a different approach, Wen et al. demonstrated the uniform anchoring of core–shell structured Ge@C onto reduced graphene oxide (rGO) nanosheets using dopamine's strong adhesion. Ge particles were initially prepared through a reduction reaction of GeO_2 and Mg powders, followed by an etching process with HCl solution. The Ge particles were then coated with polydopamine (PDA) films to form Ge@PDA composites, which were uniformly anchored on GO through the adhesion of dopamine. Subsequent calcination resulted in the formation of Ge@C/rGO

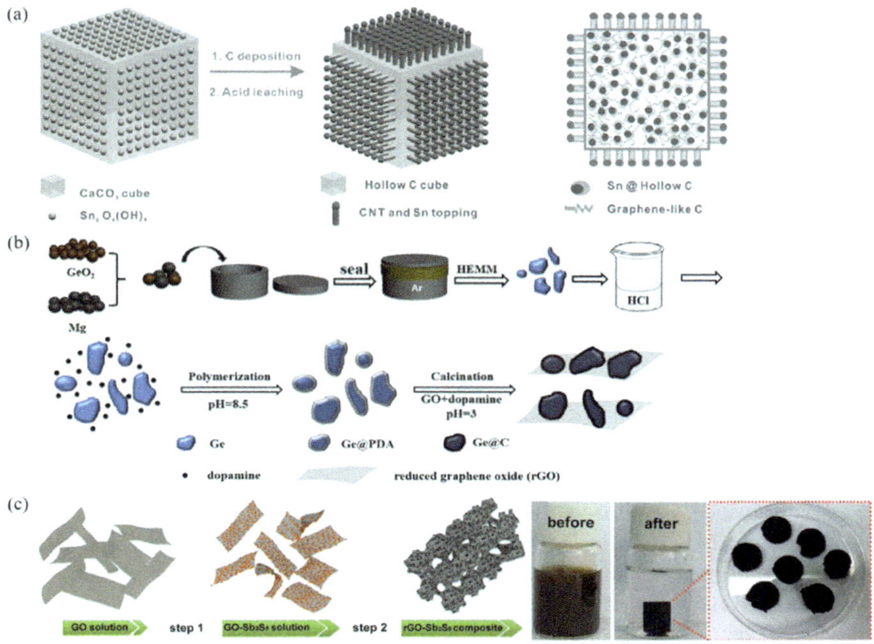

Fig. 6.12 Schematic formation processes of **a** the Sn/C composites (Reproduced with permission Ref. [57]. Copyright 2014, Wiley–VCH), **b** Ge@C/rGO composites (Reproduced with permission Ref. [58]. Copyright 2017, Elsevier), **c** Sb_2S_5/rGO composites (Reproduced with permission Ref. [59]. Copyright 2017, American Chemical Society)

composites (Fig. 6.12b). Additionally, Sb2S5 nanoparticles were uniformly encapsulated in a 3D porous graphene foam (GF) through a facile hydrothermal coassembly strategy (Fig. 6.12c). The obtained composite, where most of the Sb2S5 nanoparticles and reduced rGO were conjugated, could be directly used as electrodes without the need for a binder or current collector. It exhibited outstanding electrochemical performance.

However, the use of anchored metal-based active materials presents challenges, as they may become detached from the multi-functional supports and reaggregate during long-term cycling. This can result in diminished electronic conductivity, capacity loss, self-discharge, and potential electrode failure. Additionally, the direct exposure of metal-based particle surfaces to the electrolyte leads to the continuous formation of unstable and thick solid-electrolyte interphase (SEI) films.

6.2.2.2 Layered/Sandwich Structures

Layered/sandwich structures have emerged as a promising approach to enhance the cycling performance of LIBs by incorporating hybrid films. These films consist of active material layers, composed of metal-based materials with high Li storage capability, and electrically conductive layers made of carbon, which act as pliable buffer spaces to accommodate the significant volume changes resulting from Li alloying/dealloying reactions. Unlike the anchored structure, in the layered/sandwich structure, both sides of the metal-based active materials are effectively confined by the adjacent multifunctional matrices, thereby improving the overall electrode integrity and stability during cycling. Additionally, the void spaces created during the fabrication process play a dual role by mitigating the volume change during lithiation/delithiation and shortening the diffusion length of Li^+ [18, 60, 61].

A common approach to achieve a layered structure involves the incorporation of metal-based nanoparticles (NPs) into graphene sheets. In this regard, a microwave-assisted solvothermal reaction followed by carbon coating and thermal reduction was employed to prepare a unique sandwich-structured C/Ge/graphene composite, with Ge nanoparticles trapped between graphene sheets [62]. Initially, a GeO_2/graphene composite was synthesized through a microwave-assisted solvothermal reaction, as depicted in Fig. 6.13a. Subsequently, a carbon layer was uniformly coated on the precursor surface, resulting in the formation of C/Ge/GeO_2/graphene by decomposing and partially reducing acetylene gas. After reduction treatment with hydrogen gas, a distinct material structure was obtained, consisting of metallic Ge nanoparticles situated between graphene sheets and carbon layers. TEM analysis of GeO_2/graphene (Fig. 6.13b) revealed clusters of small particles with sizes ranging from 20 to 30 nm. TEM images of C/Ge/graphene (Fig. 6.13c) exhibited a carbon shell with a similar shape to that of the GeO_2 precursor, but with shrunken particles inside the shell and the presence of voids. These observations resulted from the reduction of GeO_2 to metallic Ge, accompanied by the release of oxygen. The schematic illustration in Fig. 6.13d demonstrates the mechanically robust and flexible matrix composed of graphene sheets and carbon coating during lithiation/delithiation. The improved electrochemical performance of C/Ge/graphene, compared to Ge/graphene, can be attributed to the prevention of Ge detachment or LixGe formation on the graphene sheets during prolonged charge–discharge cycles. The unique sandwich structure effectively buffers volume changes during lithium ion intercalation/de-intercalation processes, enabling the composites to exhibit excellent lithium storage performance, including high reversible capacity, excellent rate capability, and superior cycling retention.

6.2.2.3 Yolk-Shell Structures

Yolk-shell structures have gained significant attention in the realm of LIBs due to their distinctive structural characteristics. These structures feature a void space that acts as a buffer zone, allowing for the expansion of active materials without damaging the outer protective shells. The outer shells, composed of materials like carbon or conductive polymers, serve multiple functions. Firstly, they enhance the electrical conductivity within the electrode. Additionally, they prevent the agglomeration of active materials and help maintain the stability of SEI films.

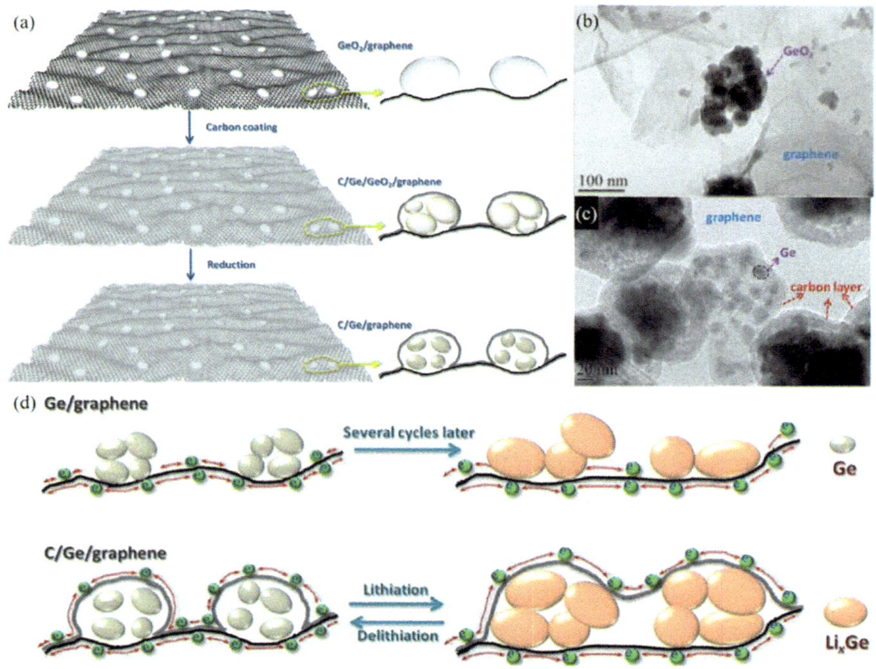

Fig. 6.13 **a** Schematic formation processes of C/Ge/graphene composites. **b** TEM image of GeO_2/graphene composites. **c** TEM image of C/Ge/graphene composites. **d** The schematic illustration of the Ge/graphene and C/Ge/graphene during lithiation/delithiation (Reproduced with permission Ref. [62]. Copyright 2013, The Royal Society of Chemistry)

Controlling the composition, shape, thickness, and uniformity of the shell is crucial in yolk-shell structures, as they greatly impact lithium ion diffusion and the ability to withstand mechanical stress caused by volume changes in metal-based materials. Yu et al. developed a controllable yolk-shell Sn@C nanobox composite using a simple method [63]. In the process, $ZnSnO_3$ nanocubes were coated with a uniform layer of PDA, forming a $ZnSnO_3$@PDA core–shell precursor (Fig. 6.14a). Subsequent thermal treatment under an H_2/N_2 atmosphere resulted in the reduction of $ZnSnO_3$ to metallic Zn/Sn and the conversion of PDA to amorphous carbon. The final product, yolk-shell Sn@C, was obtained by removing Zn and oxygen. The thickness of the carbon shell played a significant role in the nanostructures and electrochemical performance. Optimizing the carbon shell thickness at 25 nm, the Sn@C composite exhibited excellent rate capability and a high reversible capacity of 810 mAh/g after 500 cycles, corresponding to 90% retention (Fig. 6.14b, c). The exceptional electrochemical properties of Sn@C can be attributed to several factors: (1) the carbon nanobox shell effectively prevented the agglomeration of Sn particles; (2) the designed void space provided sufficient room for the volume change of Sn particles without compromising the integrity of the carbon shells, ensuring stable

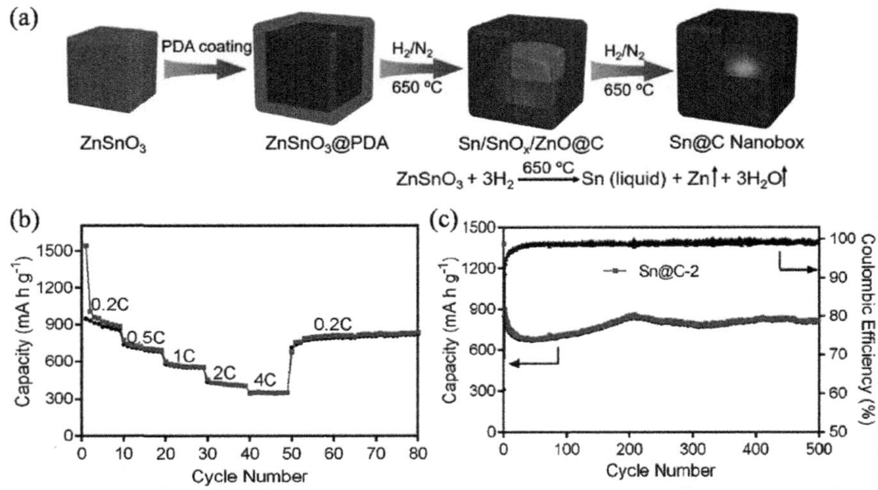

Fig. 6.14 **a** Schematic illustration of the preparation for yolk-shell Sn@C nanobox, **b** rate performance, **c** cycling performance at a current density of 200 mA/g (Reproduced with permission Ref. [63]. Copyright 2017, Wiley–VCH)

SEI films; (3) the relatively high active material content contributed significantly to the composite's specific capacity; and (4) the extensive contact area between Sn particles and the carbon nanoboxes facilitated fast ion and electron transfer in the unique Sn@C yolk-shell structure.

Li et al. employed PDA as a carbon source to fabricate a yolk-shell structured Sb@C nanocomposite [64]. In the process, Sb_2O_3 nanoparticles were coated with PDA, forming a core–shell structure of Sb_2O_3@PDA. The composite was then annealed in Ar/H_2, leading to partial reduction of Sb_2O_3 and the formation of a mixed-phase of Sb_2O_3/Sb. Simultaneously, the carbon shell was formed through polymer carbonization. Finally, the Sb@C yolk-shell structure was obtained by etching away the Sb_2O_3 with HCl. The 3D volume expansion of Sb was examined using in-situ TEM. The sodiation process of a representative yolk-shell structure containing multiple Sb particles within the carbon shells. It is observed that, at the end of the sodiation process, the encapsulated Sb nanoparticles experienced approximately 119.7% and 66.9% expansion in 2D projected area for the small and large Sb yolks, respectively (). This indicates that the yolk-shell structure as a whole did not exhibit significant expansion during sodiation. Moreover, the yolk-shell structure aids in mitigating the pulverization effect of Sb particles, as Sb nanoparticles tend to fracture due to their large volume change during sodiation [64].

Zhu et al. developed a novel Sb@C nanosphere anode with a biomimetic yolk-shell structure using a nanoconfined galvanic replacement approach [65]. This synthetic route enabled the production of hollow and yolk-shell particles with different compositions, which can also serve as nanoscale reactors in catalytic applications. The fabrication process of the hollow Sb@C yolk-shell spheres is depicted in Fig. 6.15a.

Initially, uniform hollow SnO_2 nanospheres were synthesized through a hydrothermal route, utilizing the Ostwald ripening process to create the hollow structure. Subsequently, these hollow SnO_2 spheres were coated and annealed in an H_2 atmosphere. The hollow cores of the SnO_2 spheres provided the necessary volume for the formation of a yolk-shell structure in the Sn@C composite. In the subsequent galvanic replacement process, the encapsulated solid Sn yolks were chemically transformed into hollow Sb. The SEM image (Fig. 6.15b) and element mapping (Fig. 6.15c–e) confirmed the achievement of fully-encapsulated hollow Sb yolks with thin carbon nanocages through spatially confined galvanic replacement. The electrochemical stability of the Sb@C anode at 1000 mA/g was demonstrated in Fig. 6.15f, where a substantial capacity of 405 mAh/g was maintained after 300 cycles, indicating its superior performance. Both SEM and in-situ TEM measurements revealed that the framework of the Sb@C particles remained intact during lithiation, with minimal microstructural damage upon cycling. Additionally, a stable SEI layer could be efficiently formed on the outer carbon shell, avoiding continual rupturing and reforming. The outer carbon layer also exhibited excellent electronic conductivity, facilitating superior transport kinetics and rate capability. Furthermore, the inner voids within the Sb yolk and the gaps between the Sb yolk and carbon shell accommodated the swelling of electroactive Sb during alloying, thereby preserving the integrity of the electrode (Fig. 6.15g).

6.2.2.4 Porous Structures

Yolk-shell structures provide ample hollow spaces to accommodate the volume changes of the internal metal-based materials during cycling. Similarly, the transformation of metal-based materials into porous or hollow structures can also serve as a buffer against volume expansion/contraction during lithiation/delithiation reactions, thanks to the presence of interior void spaces/channels. Furthermore, these porous/hollow structures facilitate fast electrochemical reactions by offering short diffusion pathways for lithium ions and easy access of the liquid electrolyte into the bulk of electrode materials. Park et al. developed a simple method to prepare 3D nanoarchitectures of Ge coated with a carbon layer (3D-Ge/C) through the reaction of nano GeO_2 with carbon during the carbonization of PVP in an inert gas atmosphere at 775 °C (Fig. 6.16a) [66]. The resulting porous 3D-Ge/C structure consisted primarily of coalesced Ge/C composite particles with a size of approximately 100 nm (Fig. 6.16b, c). Each coalesced Ge/C composite particle comprised finer individual Ge nanoparticles, approximately 20 nm in diameter, coated with carbon. The carbon coating effectively prevented excessive agglomeration of Ge particles. As a result, the 3D-Ge/C electrodes exhibited exceptional capacity and rate capability, as demonstrated in Fig. 6.16d.

A surfactant-assisted assembly method was employed to fabricate 3D porous graphene network-encapsulated Sn-based nanoarchitectures. This bottom-up assembly approach involved two steps, as depicted in Fig. 6.17a [67]. Initially, a precursor solution was prepared, consisting of graphene oxide (GO), polystyrene (PS)

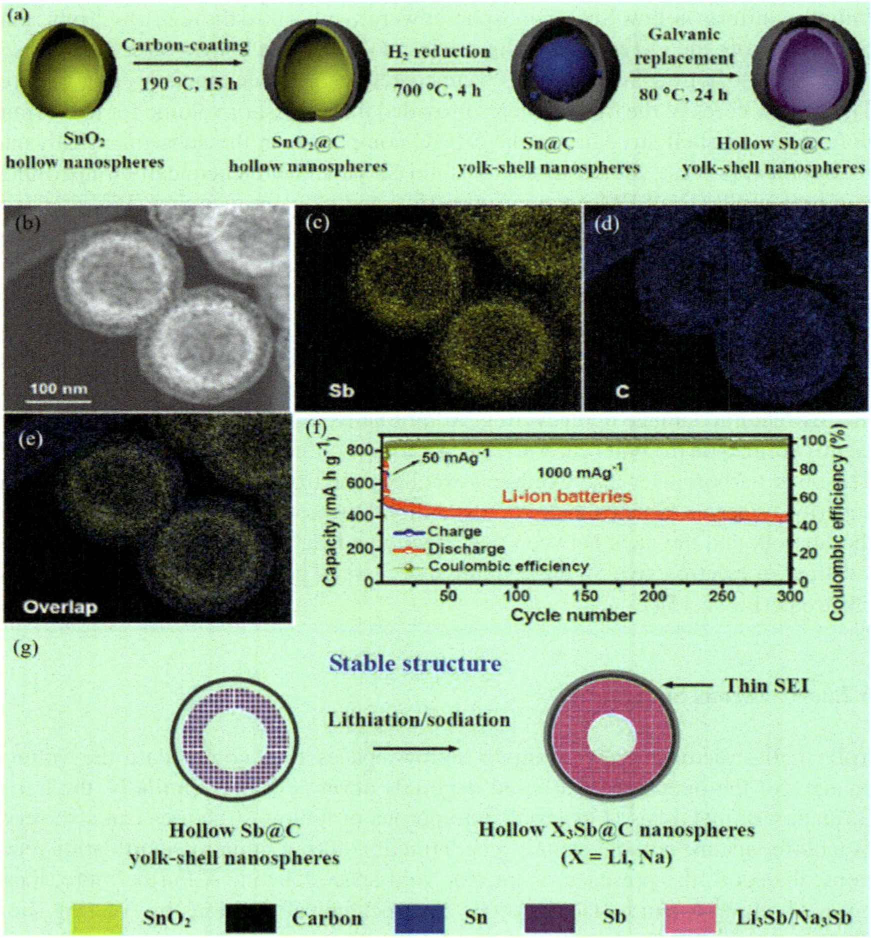

Fig. 6.15 **a** Schematic evolution of hollow Sb@C yolk-shell nanospheres. **b** Dark field TEM, **c–e** EFTEM element mapping images of hollow Sb@C yolk-shell spheres. **f** Long cycling performances of hollow Sb@C yolk-shell anode at 1000 mA/g. **g** Schematic structure transformation of hollow Sb@C yolk-shell nanospheres during the lithiation/sodiation process (produced with permission Ref. [65]. Copyright 2017, American Chemical Society)

latex particles, and $CoSn(OH)_6$ nanoparticles. Subsequently, cetyltrimethyl ammonium bromide (CTAB) solution was added dropwise to the mixture, leading to coagulation under stirring. GO, with its abundant oxygen-containing groups, prevented the negatively charged surface in the aqueous solution, enabling the capture of cationic CTAB molecules and transforming the GO surface from hydrophilic to hydrophobic. This hydrophobic interaction among GO sheets facilitated assembly. Finally, the coagulum was annealed under different atmospheres to obtain the desired 3D porous

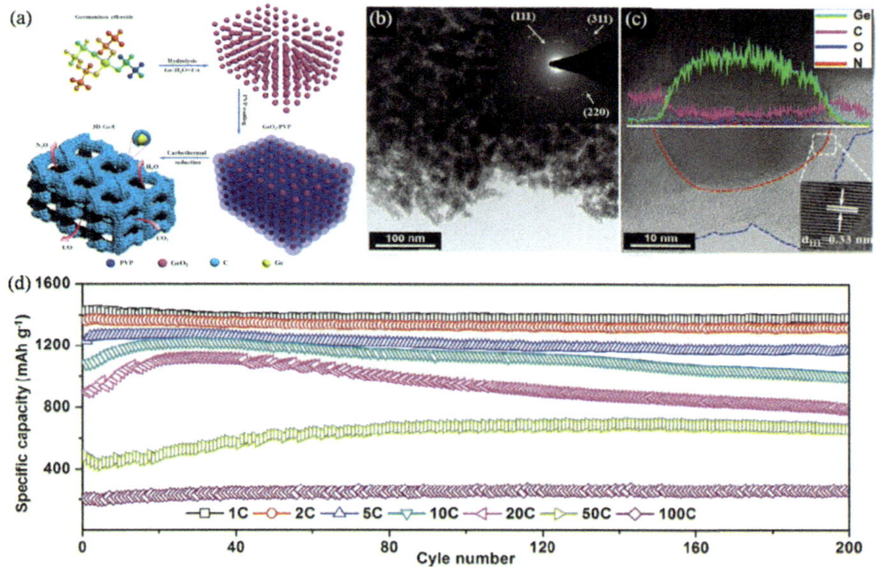

Fig. 6.16 **a** Schematic evolution of 3D-Ge/C. **b** TEM images and SAED pattern (inset). **c** HR-TEM image and EDS line scan profile. **d** Long cycling performances under various charge/discharge rates from 1 to 100 C. Reproduced with permission Ref. [66]. Copyright 2015, The Royal Society of Chemistry

composites. The resulting porous 3D-Ge/C structure mainly comprised coalesced Ge/C composite particles with a size of approximately 100 nm (Fig. 6.17b). Each coalesced Ge/C composite particle consisted of individual Ge nanoparticles coated with carbon, effectively preventing excessive agglomeration. Figure 6.17c illustrates the advantages of this novel structure design: (i) the nanoscale Sn-based particles suppress mechanical pulverization and enable short diffusion lengths; (ii) the interconnected nanopores provide space for expansion of Sn-based particles; and (iii) the microsized 3D conductive graphene framework counteracts nanoparticle self-aggregation, facilitates electron transfer, and stabilizes SEI formation. As a result, the 3D-Ge/C electrodes exhibit exceptional capacity and rate capability.

6.3 NIBs

Concerns have arisen regarding the limited reserves of lithium on Earth, which may significantly increase the cost of Li-ion batteries in the near future. Sodium-ion batteries (SIBs) have emerged as a promising alternative for energy storage due to the similar electrochemistry and redox potential between sodium and lithium ions, as well as the abundant availability of sodium resources in nature. However,

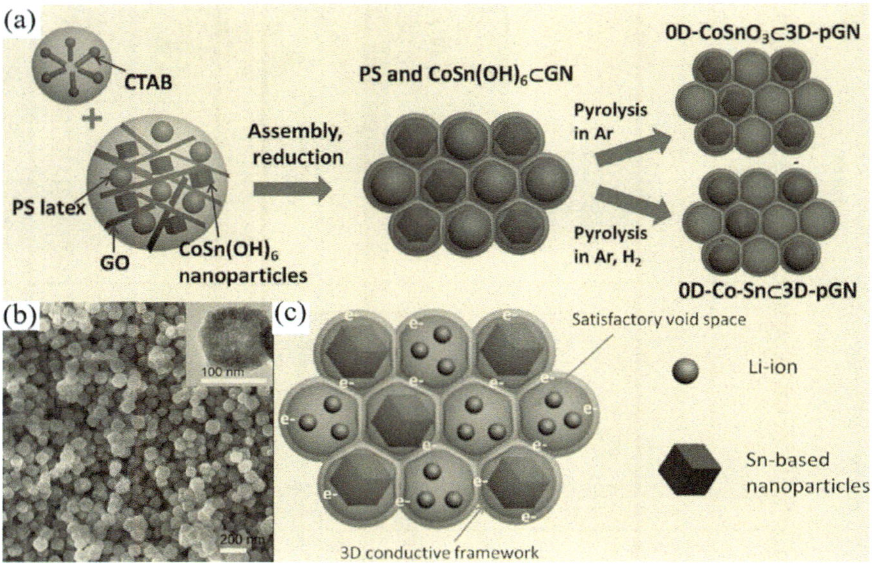

Fig. 6.17 **a** Fabrication of 3D porous graphene network-encapsualted Sn-based nanocomposites. **b** SEM image of $CoSnO_3$ nanoparticles, and inset showing TEM image of a single nanoparticle. **c** Schematic of 3D porous graphene network encapsulating Sn-based nanoparticles. Reproduced with permission Ref. [67]. Copyright 2015, Wiley–VCH

sodium ions have a larger ionic radius (approximately 1.5 times) and higher ionic mass (approximately 3.3 times) than lithium ions. Consequently, SIBs face higher energy barriers for sodium-ion intercalation/extraction in electrodes with the same crystal structure and ionic channels, resulting in inherently slower sodium transport kinetics. To achieve high energy density in SIBs, one approach is to design freestanding and binder-free electrodes where all the materials actively participate in sodium storage. Various polymers, such as PAN, PAN/refined lignin, and PAN/Pluronic F127, can be used to fabricate flexible electrodes for SIBs. Lou et al. utilized poly(amic acid) (PAA) as the polymer precursor to fabricate freestanding and flexible nitrogen-doped carbon nanofibers (N-CNFs) (Fig. 6.18a) [68]. Upon calcination, the PAA nanofibers transformed into N-CNFs (Fig. 6.18b, c). As shown in Fig. 6.18d and e, the flexible electrode exhibits a reversible capacity of 210 mAh/g after 7000 cycles at 5 A/g, with a capacity retention of 99%. The key concept behind this method lies in the selection of a nitrogen-rich and thermally stable polyimide (PI), which can be converted into carbon nanofibers with high nitrogen content, exceptional structural stability, and good mechanical flexibility. Thus far, numerous active materials, including Sn nanodots, Sb/Bi/SbBi nanodots, and MoS_2 nanosheets, have been embedded in flexible carbon films.

Our research group developed a range of freestanding metal–carbon frameworks (MCF), including Sb/C, Bi/C, and SbBi/C composites, using a space-confined

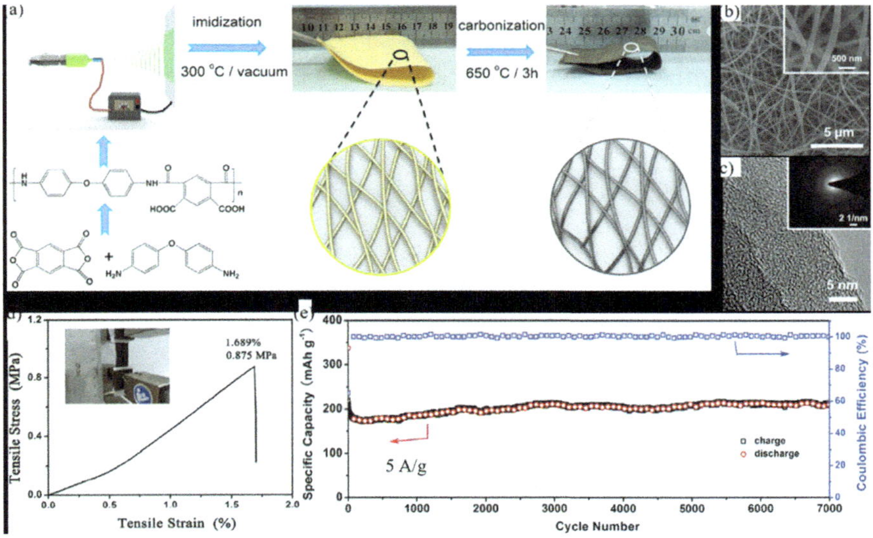

Fig. 6.18 **a** Schematic illustration of the fabrication for highly flexible N-CNFs and digital photos. **b** SEM and **c** TEM images of N-CNFs. **d** Stress–strain curve from tensile test. **e** The cycling performance at the current density of 5 A g^{-1}. Reproduced with permission [68]. Copyright 2016, Wiley–VCH

superassembly strategy (Fig. 6.19a) [69]. These MCF composites consist of ultra-small metal nanodots distributed within the gaps of stacked graphene nanosheets. Specifically, the Sb/C framework, characterized by large mesopores (~21 nm) and macropores (~60–100 nm), exhibited exceptional performance, including a high specific capacity of 246 mAh/g, long cycle life (5000 cycles), and remarkable capacity retention (nearly 100%) even at a high rate of 7.5 C (Fig. 6.19b). The unique framework structure provides sufficient space to accommodate the volume changes of Sb nanodots and non-deformable channels for stable and rapid diffusion of Na^+ ions. Furthermore, the highly dispersed Sb nanodots effectively alleviate volume expansion and shorten the diffusion length for Na^+ ions. Additionally, the reversible crystalline-phase transformation contributes to the fast and durable sodium storage (Fig. 6.19c). This synthetic strategy can be extended to other metal or alloy systems by simply modifying the precursor molecules, showcasing its versatility and applicability.

To address the challenges related to volume pulverization and achieve high capacity in sodium-ion batteries (SIBs), metal sulfides or oxides have emerged as promising anode candidates due to their favorable kinetics and weak metal-sulfur bonds. Among them, SnS_2 and SnS have gained significant attention for their high theoretical capacities (1123, 1022 mAh/g) and low volume expansion (324, 242%). However, these materials suffer from limited electron transfer and ion diffusion properties as they are semiconductors. To enhance their electrochemical behavior

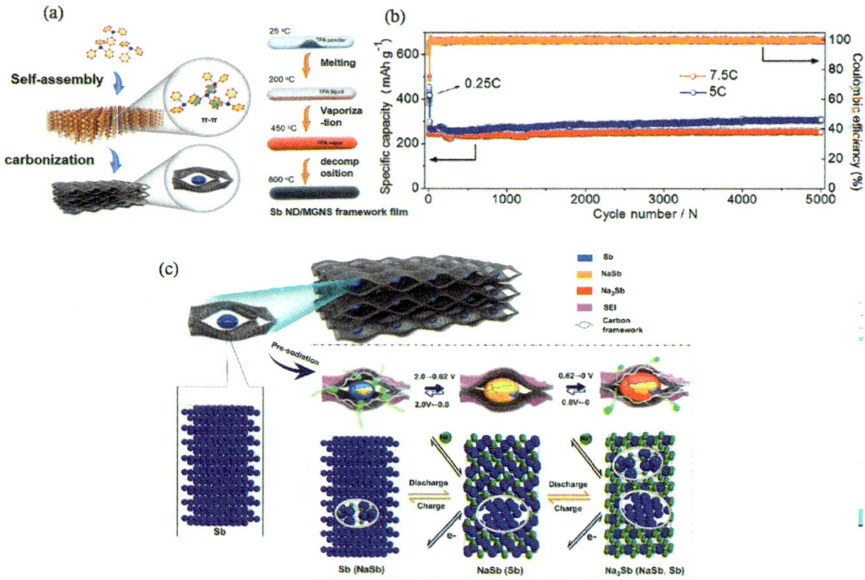

Fig. 6.19 **a** Schematic formation processes of the Sb/C framework films. **b** Cycle performance of Sb/C framework films electrode at high current densities (5 C and 7.5 C). **c** Schematic illustration of the proposed discharge–charge mechanism for the Sb/C framework film anodes as the high-performance sodium battery with unusual reversible crystalline-phase transformation (Reproduced with permission [69], copyright 2016 American Chemical Society)

and stability, carbonaceous materials have been incorporated with SnS_2 or SnS, resulting in improved cycle stability. For instance, SnS/C microspheres prepared via spray pyrolysis exhibited a reversible capacity of 433 mAh/g after 50 cycles at 0.5 A/g. Similarly, SnS nanotubes coated with carbonaceous materials showed a retention of 96% after 100 cycles at 200 mAh/g. Nonetheless, the capacity was limited to 440.4 mAh/g, and the material stability remained a challenge. In order to address these issues, Lee et al. designed a hydrothermal-treated SnS@CNT film, where SnS nanoparticles were well-dispersed and anchored on a porous CNT network (Fig. 6.20a) [70]. The inherent chemical bonds between SnS and CNT surface significantly increased the contact surface area between the active material and sodium ions. This freestanding film exhibited a high capacity of up to 762 mAh/g, attributed to improved conductivity and enlarged surface area. Another approach involved the fabrication of a 3D porous graphene nanosheet/SnS_2 (3D-GNS/SnS_2) film as a high-performance SIB electrode [71]. The fabrication process involved hydrothermal synthesis of GNS/SnS_2 nanosheets, their assembly into porous GNS/SnS_2 microblocks, and the formation of a 3D-GNS/SnS_2 film using these microblocks and secondary GNS. The resulting hybrid film featured a 3D porous and interconnected framework, facilitated fast electron/ion transport, and provided satisfactory

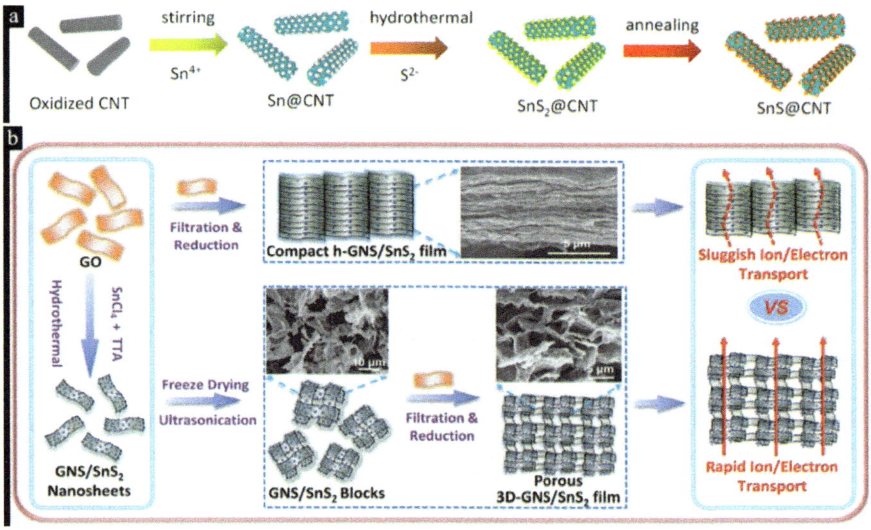

Fig. 6.20 **a** Schematic formation processes of the SnS@CNT film. Reproduced with permission Ref. [70]. Copyright 2020, The Royal Society of Chemistry. **b** Schematic formation processes of the 3D-GNS/SnS_2 film. Reproduced with permission Ref. [71]. Copyright 2020, Wiley–VCH

long-range conductivity. The SnS_2 nanocrystals bonded with GNS served as active sites for Na storage and ensured cycling stability (Fig. 6.20b).

Transition metal diselenides (TMDs), a promising class of transition metal chalcogenides, have attracted attention as potential anode materials for sodium-ion batteries (SIBs) due to their high theoretical capacity and cost-effectiveness [72]. In a novel electrode design, a dually protected anode was fabricated by constructing a CNT/$CoSe_2$/C composite [72]. The process involved decorating $CoSe_2$ nanoparticles onto individual CNTs, forming a 3D network, and externally protecting them with amorphous carbon layers through carbonization (Fig. 6.21a). The resulting hybrid sponges were compressed into a dense framework and directly used as freestanding anodes in coin-type cells (Fig. 6.21b). This unique composite harnessed the morphological, electronic, and surface properties to greatly enhance Na^+ storage. The highly conductive and compressible 3D CNT sponge provided multiple pathways for ionic and electronic conduction, ample space for active $CoSe_2$ loading and electrolyte infiltration, and accommodation for volume expansion during cycling. The external dual protection with carbon layers not only increased overall conductivity but also buffered volume expansion and controlled the thickness of the solid electrolyte interface (SEI) film by preventing direct contact between the active material and corrosive surroundings. Leveraging these features, the hybrid composite served as a freestanding anode for SIBs. Another transition metal diselenide, vanadium diselenide (VSe_2), was chosen as a high-capacity anode material for SIBs due to its theoretical capacity, abundance, and inherent safety. However, conventional powder electrodes of VSe_2 exhibited limited electrochemical performance due to blocked electron transport channels and material scattering during charge–discharge processes (Fig. 6.21c, d). To overcome these limitations, a binder-free film composed of ultrafine VSe_2 particles wrapped in N-doped carbon nanofibers (VSe_2/NCNFs) was prepared as the anode for quasi-solid-state SIBs [73]. The ultra-small VSe_2 nanoparticles facilitated rapid Na ion migration, while the highly graphitized carbon fibers provided fast electronic transmission and a stable host to alleviate volume expansion during cycling (Fig. 6.21e). The elastic coordination between VSe_2/NCNFs and the gel electrolyte further mitigated overall volume expansion. The freestanding VSe2/NCNFs anode, combined with a gel polymer, exhibited a high reversible capacity of 420.8 mAh/g at a current density of 0.05 A/g, excellent rate capability of 278.1 mAh/g at 5 A/g, and retained 206.8 mAh/g after 10,000 cycles at 5 A/g.

In recent years, there has been increasing interest in the development of flexible and wearable electronics that can seamlessly integrate with human tissues, offering a wide range of applications in human-interfaced devices. Flexible sodium-ion batteries (SIBs) have emerged as a promising solution, offering advantages such as ultrathin and lightweight form factors that can cater to the demands of wearable electronics. However, these flexible SIBs face challenges when it comes to frequent disassembling and recharging, hindering their ability to provide sustainable energy, especially in situations where there is no access to electricity or charging bases. To address this issue, Fan et al. have introduced a flexible self-charging sodium-ion full battery (SCSFB) that utilizes self-synthesized $Na_3V_2(PO_4)_3$@C as the cathode, commercial hard carbon as the anode, and a flexible $BaTiO_3$-P(VDF-HFP) film immobilized with liquid electrolyte of $NaClO_4$ as a built-in piezoelectric gel-electrolyte ($BaTiO_3$-P(VDF-HFP)-$NaClO_4$) [74]. This innovative design allows the flexible SCSFB to simultaneously harvest and store energy from body motion, making it a promising power source for self-powered wearable electronics. The formation process of the flexible piezoelectric gel-electrolyte is illustrated in Fig. 6.22b, where a mixed solution of $BaTiO_3$ piezoelectric particles and P(VDF-HFP) powders is deposited on a glass substrate, forming a $BaTiO_3$-P(VDF-HFP) film that is then polarized to generate the piezoelectric capability. The gel-electrolyte is prepared by immersing the film in liquid electrolyte. This unique design effectively prevents the leakage of liquid electrolyte, enhancing the safety of the flexible SCSFB. Figure 6.22c showcases the assembled SCSFB in a CR2032 coin-type cell as well as the flexible devices that can be easily bent or folded using Kapton tapes. In addition to their flexibility and electrochemical storage performance, these devices also exhibit remarkable self-charging capability through various stress patterns, including static compression, repeated bending, and continuous palm patting. Serially connected self-charged devices demonstrate the ability to power electronic devices effectively. The direct self-charging mechanism is deduced based on the introduction of electromagnetic field theory, where no rectifier is needed, and the battery is charged by the built-in piezoelectric component (Fig. 6.22d). The device operates in two stages: an initial electrochemical equilibrium state and an active state when external strains or deformations are applied, resulting in Na^+ migration and charging. The piezoelectric field, generated by the piezoelectric gel-electrolyte, persists after the external force

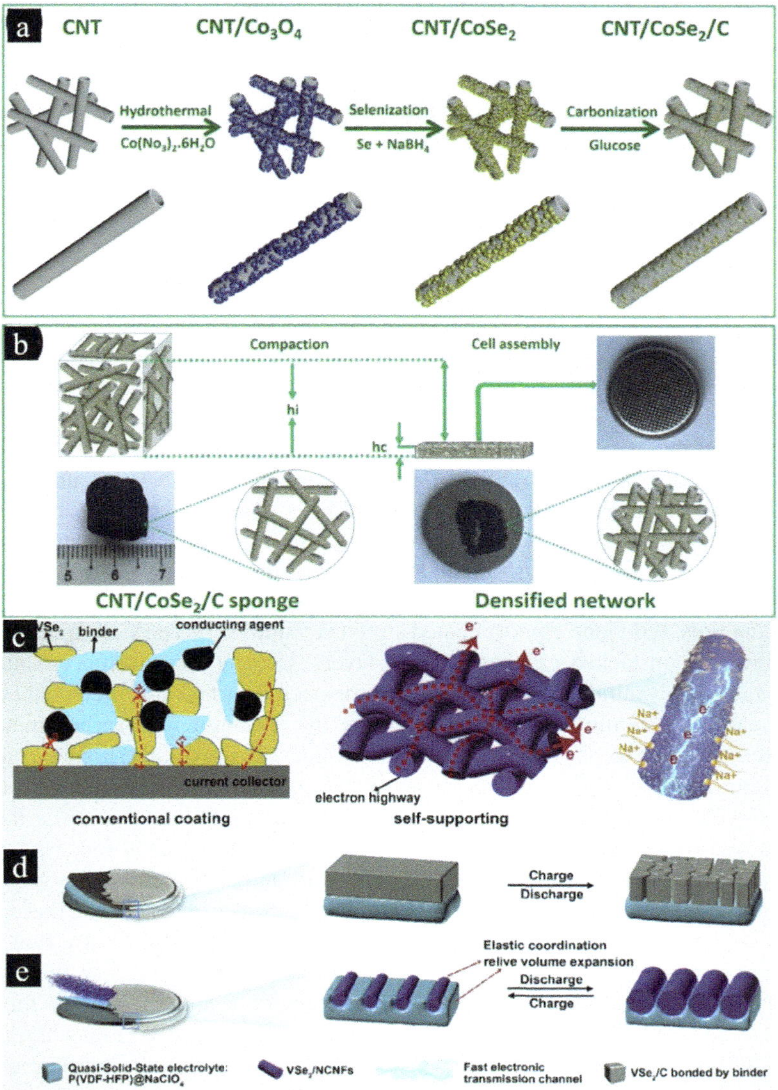

Fig. 6.21 **a** Schematic formation processes of the CNT/$CoSe_2$/C sponge. **b** Photographs of CNT/$CoSe_2$/C sponge before and after compaction during cell assembly with description of initial and final thicknesses. Reproduced with permission Ref. [72]. Copyright 2020, Wiley–VCH. **c** The electron transport path in a conventional coating electrode and VSe_2/NCNFs, respectively. **d** Changes in the charge/discharge process of powder materials bonded by binder. **e** Changes of binder free VSe_2/NCNFs and elastic quasi-solid electrolyte during charge/discharge process. Reproduced with permission Ref. [73]. Copyright 2020, Elsevier

is released, allowing for continued self-charging through internal residual strains. Upon reaching a new equilibrium, the piezoelectric field disappears, and a reverse migration of Na^+ occurs. The self-charging process can be repeated by reapplying an external force. This groundbreaking work presents an innovative approach to the development of sustainable and safe flexible SIBs for self-powered wearable electronics.

6.4 Li–S

Li–S batteries have gained attention as potential next-generation energy storage devices due to their high theoretical specific energy and cost-effectiveness. However, they face challenges such as high self-discharge, poor cycle stability, limited rate capability, and low sulfur utilization. To address these issues, various host materials including porous carbonaceous materials, CNT, graphene nanosheets, and conductive polymers have been utilized to prepare sulfur composite cathodes. Graphene, in particular, is a commonly used host material for sulfur confinement due to its high surface area and surface functionalization capabilities. Figure 6.23a–d illustrates different graphene-sulfur configurations, such as carbon-black-decorated graphene wrapping poly-(ethylene glycol)-coated structure, sandwich-type structure, gradient structure, and unstacked graphene double layers. Despite these advancements, the challenge of polysulfide dissolution remains, especially when sulfur content is high in the cathode. To mitigate this issue, coating the separator with carbon materials has been explored to physically absorb polysulfides. However, the nonpolar nature of carbon results in weak interaction with polar polysulfides, impacting the performance of the coating layer. To overcome this, metal oxides such as NiFe layered double hydroxide (LDH) functionalized on mesoporous nitrogen-doped graphene substrate (LDH/NG) and ZnO nanowires with carbon nanofiber mat (ZnO/C) have been investigated as polysulfide adsorbents in organic electrolytes. While these interlayers improve the chemical impact on confining polysulfides, they also face challenges of low density and high loading requirements, resulting in low sulfur loading. Yan et al. introduced a new strategy by utilizing a highly conductive lithium fluoride/GO (LiF/GO) solid electrolyte interface (SEI) coated separator (Fig. 6.23e). The LiF/GO SEI coated PP separator effectively mitigates polysulfide shuttling in Li–S batteries. Compared to batteries with GO coated PP separator and bare PP separator, the battery with LiF/GO coated PP separator exhibits an initial discharge capacity of 888 mAh/g and 82.3% capacity retention after 200 cycles, indicating significant improvement (Fig. 6.23f).

The low mass densities of nano sulfur and porous carbon hosts pose a challenge in achieving high volumetric density and maintaining high gravimetric energy density. Traditional slurry-casting processes for assembling porous carbon/sulfur cathodes often involve the use of metal current collectors, conductive agents, and polymer binders, which decrease volumetric-energy–density. To overcome this, a

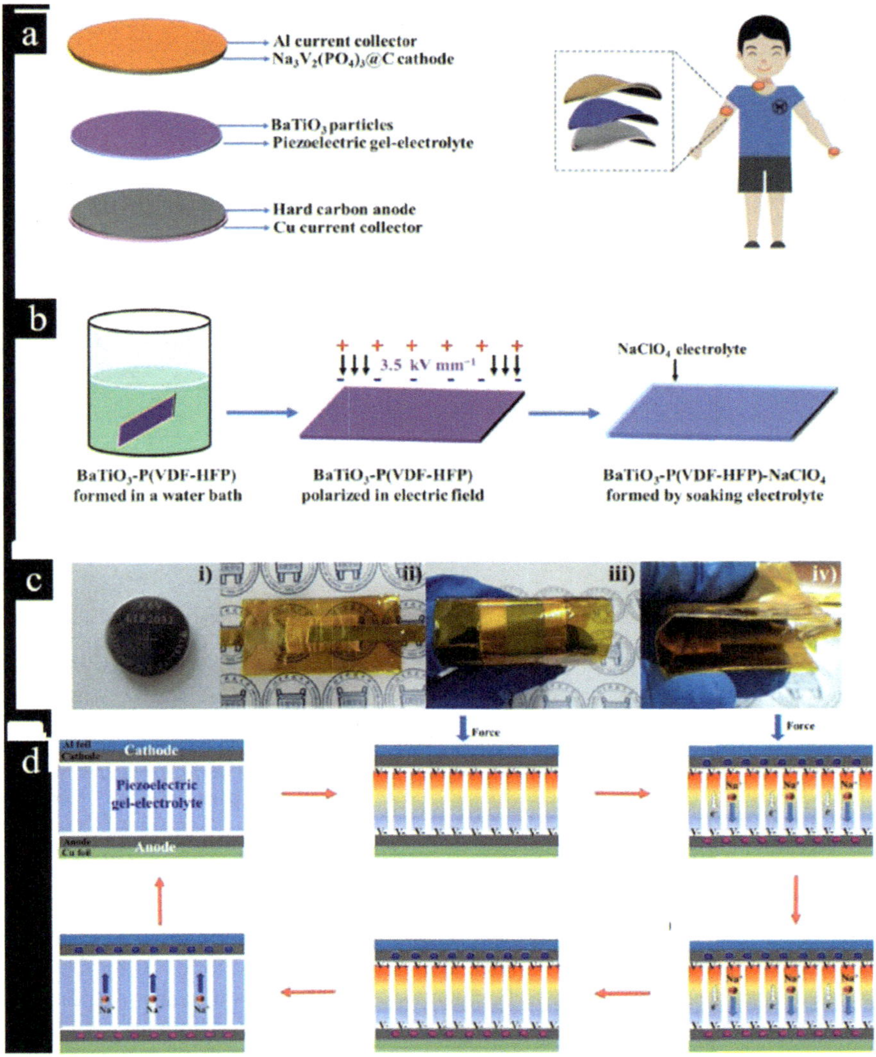

Fig. 6.22 **a** Schematic formation processes of the flexible SCSFB used to harvest/store external mechanical energy via various body-motions. **b** Schematic formation processes of $BaTiO_3$-P(VDF-HFP)-$NaClO_4$ piezoelectric gel-electrolyte. **c** Photograph of the as-assembled SCSFB. **d** Schematic self-charging process of the flexible SCSFB. Reproduced with permission Ref. [74]. Copyright 2020, The Royal Society of Chemistry

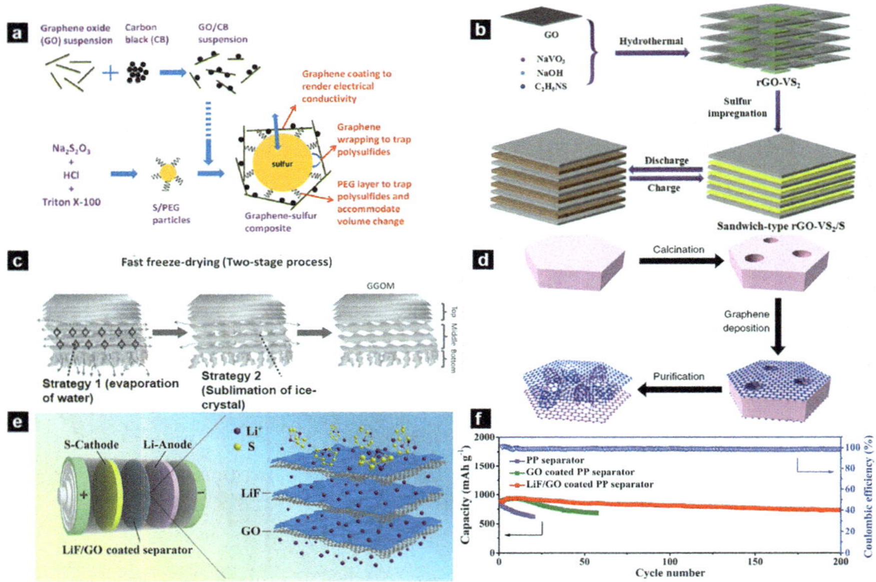

Fig. 6.23 Schematic of the synthesis steps for **a** a graphene-sulfur composite, with a proposed schematic structure of the composite. Reproduced with permission [75]. Copyright 2011, American Chemical Society; **b** the sandwich structured rGO-VS_2/S composite. Reproduced with permission [76]. Copyright 2017, Wiley–VCH; **c** the graded microstructure (GGOM). Reproduced with permission [77]. Copyright 2014, Wiley–VCH; **d** the unstacked double-layer templated graphene. Reproduced with permission [78]. Copyright 2013, Nature. **e** Schematic cell configuration of Li-batteries assembled with LiF/GO coated separator. **f** Cycling performance of the batteries with PP, GO coated PP, LiF/GO coated PP separators at 1 C. Reproduced with permission [79]. Copyright 2018, Wiley–VCH

reliable and efficient strategy involves constructing a binder-free, compact, conductive, free-standing, and integrated cathode by combining the sulfur host with an interlayer. In this regard, a flexible, binder-free cathode was developed using nitrogen-doped carbon foam (NCF) and carbon nanotubes (CNTs) as the scaffold, filled with sulfur nanoparticles encapsulated in poly(3,4-ethylenedioxythiophene) (PEDOT). Figure 6.24a illustrates the schematic of the flexible NCF/CNT/PEDOT@S hybrid cathode. Commercial melamine foam (MF) was used as a 3D template and carbon source, coated with polyaniline (PANI) to enhance nitrogen content and create smaller voids. The resulting MF-P/CNT was transformed into NCF/CNT after drying and carbonization. The wrinkled CNT layer improved conductivity and provided active sites, while the NCF/CNT film offered substantial space within the conductive framework. To achieve high electrochemical performance, the NCF/CNT was immersed in a PEDOT@S suspension, easily infused into the NCF/CNT foam due to its unordered framework and considerable void spaces. The flexible self-supporting electrode was successfully assembled for a flexible battery. Electrode flexibility and allocation were systematically investigated, demonstrating the retention of the monolithic structure of the NCF/CNT/PEDOT@S electrode even when twined around a glass rod (Fig. 6.25b). Pressing the electrode maintained its flexibility and absorbed more active materials (Fig. 6.24c). The soft-package Li–S battery, formed by the NCF/CNT/PEDOT@S electrode and lithium foil, was embedded in a bracelet, capable of lighting LEDs in bending and folding states (Fig. 6.24d–f). The battery's size in the bracelet is depicted in Fig. 6.24g. To further showcase practical applications, the performance of the flexible NCF/CNT/PEDOT@S battery was tested using a commercial smartwatch, demonstrating good functionality even when bent at different angles (Fig. 6.24h).

Bao et al. presented a compact, conductive, and integrated cathode (G/CNT-S//G/CNT) for high volumetric energy density Li–S batteries, achieved by compressing graphene/CNT (G/CNT) aerogels that served as sulfur host (G/CNT-S) and interlayer (G/CNT) simultaneously [81]. The fabrication process of the G/CNT-S//G/CNT integrated cathode is illustrated in Fig. 6.25a. Initially, uniform suspensions of GO and multi-walled CNT were freeze-dried to form 3D cross-linked GO/CNT aerogels. These aerogels were then rapidly reduced to 3D G/CNT aerogels via self-propagating combustion. The strongly cross-linked wrinkled graphene and CNT preserved the 3D interconnected porous networks even after reduction. The G/CNT-S compact electrode was prepared by immersing high-loading sulfur into the large surface area and macroporous G/CNT aerogels, followed by low-temperature annealing and mechanical compression. The mechanical compression was crucial to overcome the low mass density of G/CNT aerogels, enabling the formation of free-standing and compact films with increased total pore volume. Additionally, the 3D G/CNT aerogels were also pressed to create ultralight G/CNT compact films. Finally, the resulting G/CNT-S host and G/CNT interlayer were assembled to form the compact and integrated G/CNT-S//G/CNT cathode. The G/CNT-S film exhibited excellent flexibility without structure cracks when subjected to different bending angles (Fig. 6.25b, c), while the G-S film showed film fractures and limited flexibility, emphasizing the role of cross-linked CNT in enhancing structure stability. Figure 6.25d depicts a visualized permeation test that demonstrates the efficient inhibition of polysulfide diffusion by the G/CNT interlayer. The compressed ultralight G/CNT film acts as a flexible interlayer that synergistically suppresses polysulfide shuttling through chemical interaction and physical restriction.

In addition to hierarchically structured carbonaceous materials, metal oxides/sulfides have gained significant attention due to their ability to strongly adsorb polysulfide intermediates. However, the low conductivities of these materials can hinder the electrochemical kinetics, especially at high sulfur loadings. Therefore, current research focuses on two main aspects: (1) developing effective host materials with controlled morphologies to enhance interfacial interactions, and (2) increasing the conductivity of the Li–S system and catalytic conversion of polysulfides to improve electrochemical performance. Wang et al. proposed a novel flexible porous CNF film modified with graphene and ultrafine TiO_2 nanoparticles as a sulfur host [82]. The fabrication process is depicted in Fig. 6.26a, where a precursor film is prepared using electrospinning and subsequently carbonized and etched to form the flexible target

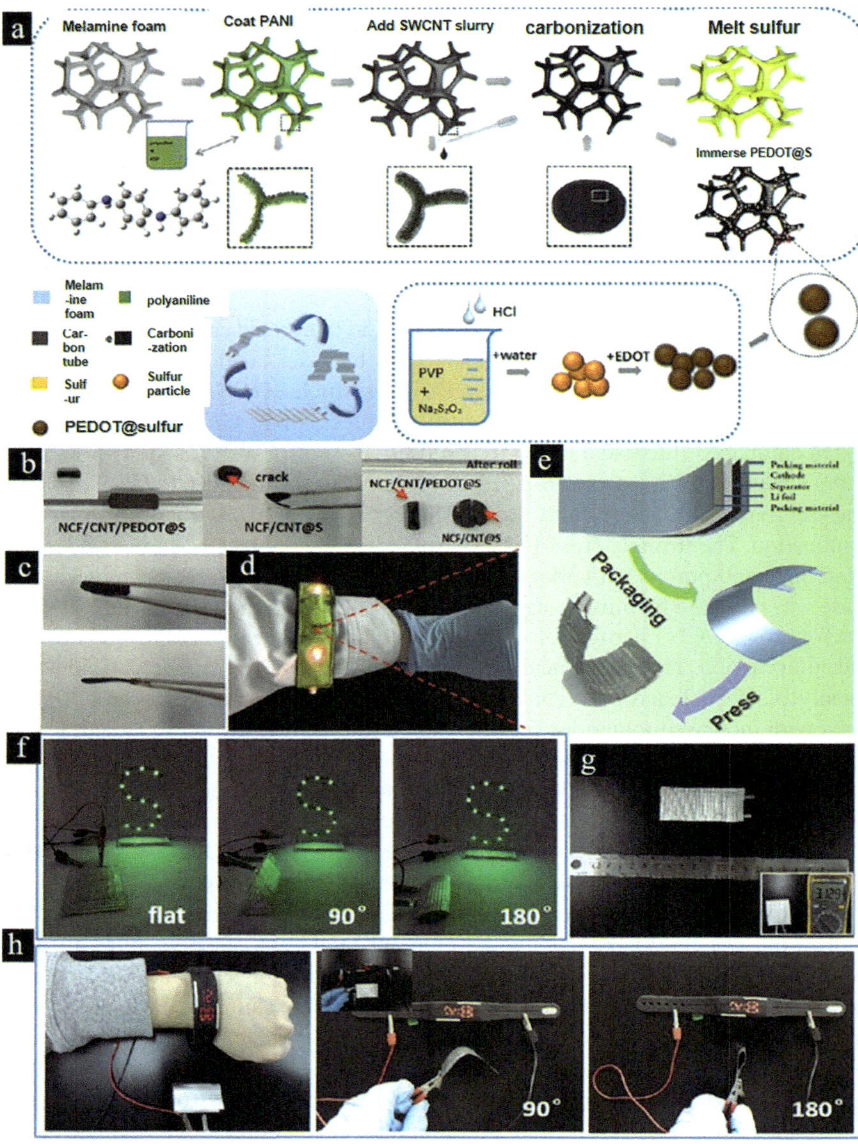

Fig. 6.24 **a** Schematic illustration of the fabrication of NCF/CNT/PEDOT@S cathode. **b** Photographs of the freestanding NCF/CNT/PEDOT@S and NCF/CNT@S electrodes curled around a glass rod. **c** Photographs of the NCF/CNT/PEDOT@S electrode before (top) and after (below) pressing. **d** Powering a flexible bracelet. **e** Schematic of the structure of the flexible battery. **f** Photographs of the soft-package Li–S battery lighting up 11 LEDs. **g** Size of the flexible battery for a bracelet. **h** Powering a smart bracelet with the flexible NCF/CNT/PEDOT@S battery at 90° and 180°. Reproduced with permission Ref. [80]. Copyright 2018, The Royal Society of Chemistry

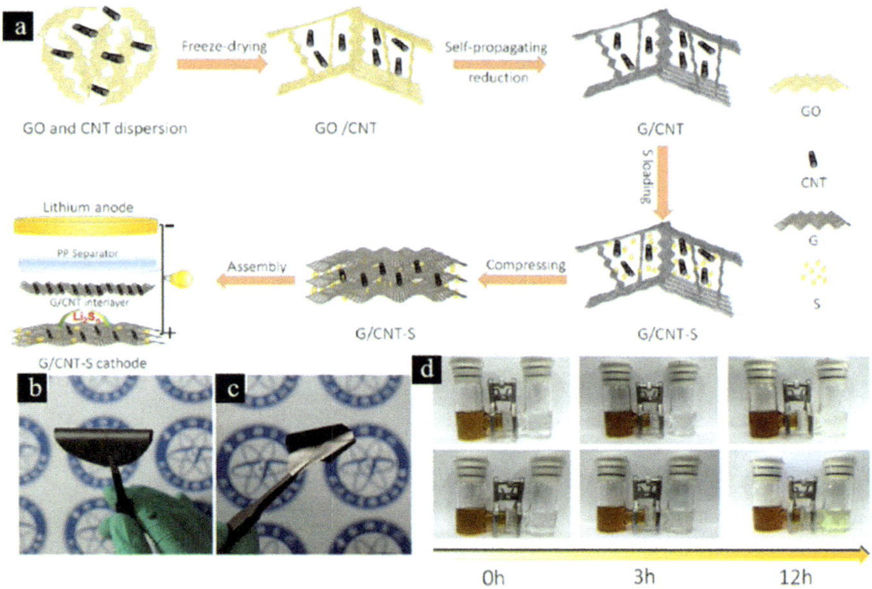

Fig. 6.25 **a** Schematic illustration of the fabrication of G/CNT-S//G/CNT cathode integrated G/CNT-S host with G/CNT interlayer. **b, c** Photographs of G/CNT-S films taken under different bending angles. **d** Polysulfide permeation measurements for G/CNT/PP (top) and individual PP (bottom) separators during the course of polysulfide diffusion from Li_2S_6/THF solution (left side) to pure THF (right side) of the U-shaped glass cell. Reproduced with permission Ref. [81]. Copyright 2019, Elsevier

film. The mesopores and ultrafine TiO_2 particles in the nanofiber structure play crucial roles during lithiation, as illustrated in Fig. 6.26b. Furthermore, a flexible pouch cell assembled with the S/TiO2/G/NPCFs film successfully powers a light-emitting diode device and withstands bending, demonstrating its potential for flexible Li–S batteries (Fig. 6.26c). This unique structure enables efficient sulfur dispersion, strong confinement of lithium polysulfides, and prolonged cycle life (Fig. 6.26d). HfO_2, a novel metal oxide with potential for enhancing interfacial redox reactions, is utilized as a polysulfide barrier in Li–S batteries by Wang et al. An ultrathin and cross-stacked CNT film modified with HfO_2 through atomic layer deposition effectively suppresses polysulfide shuttling (Fig. 6.26e). Electrodes incorporating a HfO_2/CNT interlayer exhibit improved electrochemical properties, including long-term cycling stability, high-rate performance, anti-self-discharge capabilities, and suppression of Li anode corrosion, even with high sulfur loadings. Another approach involves introducing VOx into the interlayer to anchor lithium polysulfides through strong chemical interaction [83]. Wang et al. obtained flexible VOx nanosphere@SWCNT hybrid films with dual-confinement functions for polysulfides via a facile hydrothermal method (Fig. 6.26f). Hollow VOx nanospheres, synthesized through a hydrothermal reaction, are infused with sulfur to form a core–shell structure (VOx@S), which is then mixed

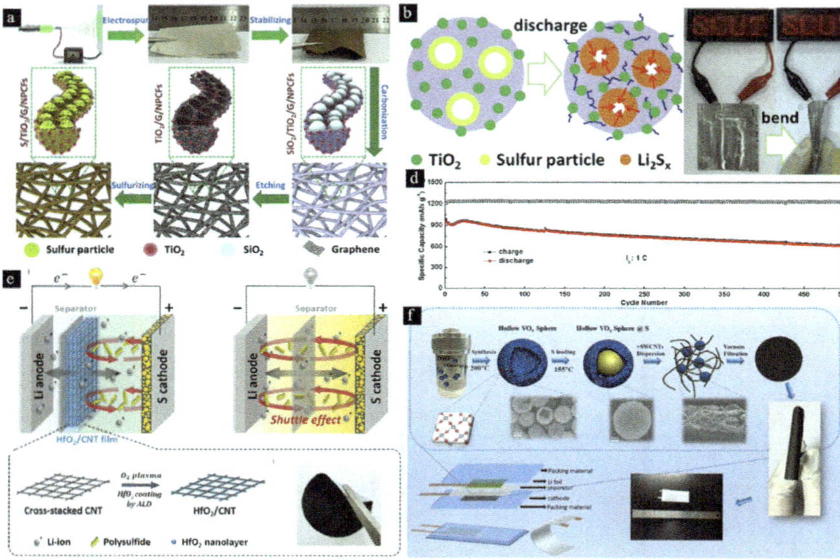

Fig. 6.26 **a** Schematic illustration of the fabrication of S/TiO_2/G/NPCFs cathode. **b** Schematic illustration of the functions of the mesopores and ultrafine TiO_2 particles with the nanofiber during lithiation. **c** Photographs of LED device lighten by the flexible battery during folding. **d** Long-term cycling performance of S/TiO_2/G/NPCFs electrode at 1 C. Reproduced with permission Ref. [82]. Copyright 2017, Elsevier. **e** Schematics of Li–S cells with a HfO_2/CNT interlayer (left) and a pristine separator (right). Reproduced with permission Ref. [84]. Copyright 2018, Elsevier. **f** Schematic illustration of the fabrication of VHS@S/SWCNT cathode. Reproduced with permission Ref. [83]. Copyright 2018, Wiley–VCH

with SWCNTs to create a freestanding and flexible film. The interweaving SWCNTs provide conductivity and structural integrity, while the small VOx nanospheres offer strong confinement and excellent adsorption abilities for polysulfides, resulting in improved cyclability.

6.5 Li-O_2 Battery

Li-air batteries have attracted significant attention for their high energy density, making them promising for energy storage and electric vehicles with long driving range. These batteries consist of a Li metal anode, a separator, an electrolyte, and an air cathode. During discharge and charge, Li_2O_2 is formed and decomposed, but the insulating nature of Li_2O_2 on the cathode surface can lead to limited capacity or battery failure. The high charge potential required for Li_2O_2 decomposition (44.2 V) can also cause electrolyte deterioration and reduce cycle life. To address these challenges, carbon-based materials have been widely investigated as optimal choices

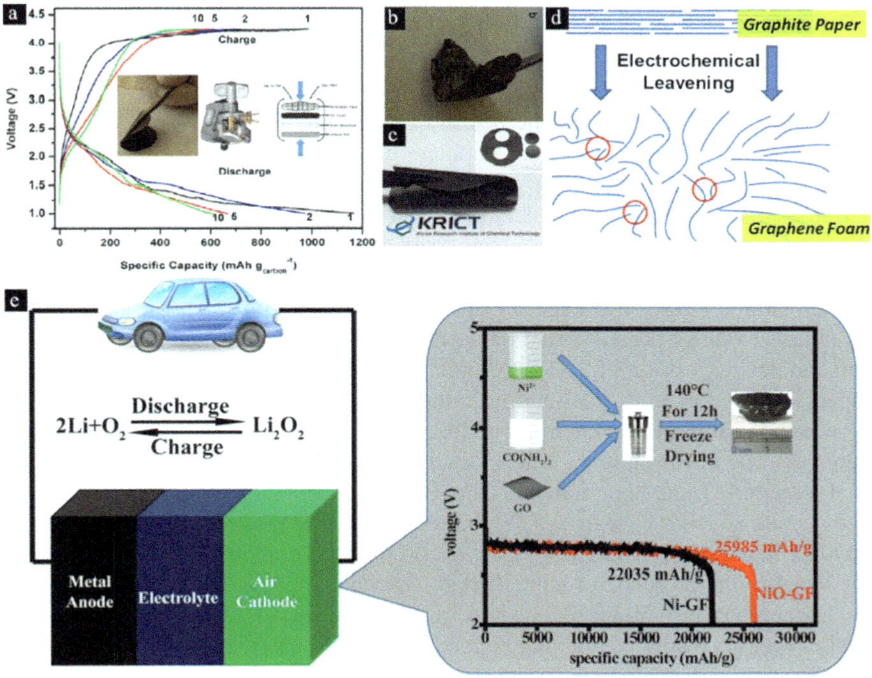

Fig. 6.27 **a** Charge–discharge profile of GO paper in ECC-air test cell. Inserts are ECC-air test cell and schematic illustration of the electrode assembles. Reproduced with permission [85]. Copyright 2014, Elsevier. **b**, **c** Photograph of the as-prepared graphene paper and GNP/GO paper, respectively. Reproduced with permission [86]. Copyright 2016, Elsevier. Reproduced with permission [88]. Copyright 2015, Elsevier. **d** schematic illustration for the electrochemical leavening synthesis of graphene foams. Reproduced with permission [87]. Copyright 2018, The Royal Society of Chemistry. **e** Charge–discharge profiles of NiO-GF and Ni-GF. Insert is the schematic illustration of NiO-GF. Reproduced with permission [89]. Copyright 2018, Elsevier

for air cathodes. Their excellent conductivity improves discharge–charge performance, while their low cost, high specific area, and lightweight properties ensure high capacity and energy density. However, their poor catalytic ability for the oxygen evolution reaction restricts achieving low charge potentials. Various approaches have been explored, including incorporating noble metals and oxides (Pt, Au, Pd, Ru, RuO_2) and transition metals and compounds (MnO_2, TiC, Ti_4O_7, Cu_2O, FeOOH, NiOOH, Ni_2CoO_4) into carbon-based composites. These strategies enable charge potentials around 3.5 V but face limitations. The use of catalysts that strongly bind O_2 or LiO_2 discharge intermediates can lead to undesired Li_2O formation, reducing reversibility. Recently, carbon-based films have gained interest for Li-air battery cathode materials, with two main groups being functional carbon films and N-doped carbon films, which will be discussed further in the following section.

6.5.1 Functional Carbon Film

Functional carbon films, including graphene film, CNTs, CNFs, and carbon microfibers, exhibit unique properties in non-aqueous Li-air batteries due to their structures and defects/vacancies. Graphene film, in particular, demonstrates high discharge capacity and round-trip efficiency attributed to its 3D 3-phase electrochemical area, diffusion channels for electrolyte and oxygen, and unique structure. Li et al. utilized a smooth and flexible graphene oxide (GO) paper electrode in non-aqueous Li-air batteries, prepared through vacuum filtration [85]. The GO paper electrode showed an initial discharge capacity of 1165 mAh/g and 612 mAh/g after 10 cycles, benefiting from its porous structure facilitating O_2 diffusion and efficient formation and decomposition of Li_2O_2. Akbulut et al. synthesized flexible pure graphene papers and used them as freestanding cathodes [86]. The graphene papers, prepared by vacuum filtration and chemical reduction, maintained flexibility and exhibited stable cycling performance up to 85 cycles with a capacity of 497 mAh/g. Kang et al. assembled graphene nanoplatelets (GNPs) with GO as a stabilizer to create free-standing GNP/GO papers [87]. These papers demonstrated higher discharge capacity than commercial carbon papers and reduced graphene oxide (rGO) paper, attributed to the presence of edge dangling s-bonds, defects in graphene, and high electric conductivity. Yan et al. synthesized graphene foam (3D-G electrode) by electrochemical leavening of graphite papers, controlling the level of structural defects [87]. The low-defect graphene foam showed a high round-trip efficiency of 80% and stable voltage profiles, while the high-defect foam exhibited lower efficiency and capacity. The presence of structural defects hindered charge transfer kinetics, affected chemical properties, and led to the growth of side products, ultimately degrading electrochemical performance.

Functional metal oxides offer attractive properties such as low cost, high average voltage, and environmental friendliness, making them desirable for catalysis applications. However, their low conductivity limits their use in lithium-air batteries. To overcome this limitation, a promising approach is to combine metal oxides with carbon-based materials. For instance, a 3D reticular NiO-graphene foam was synthesized using a simple hydrothermal method followed by high-temperature calcination (Fig. 6.28e). This composite material exhibited an impressive discharge capacity of 25,986 mAh/g at a current density of 100 mA/g. The exceptional specific capacity can be attributed to the synergistic effects of NiO, which lowers the reaction barrier, and the 3D network structure of graphene foam, which provides increased active sites and facilitates electron transfer.

CNTs, including SWCNTs and MWCNTs, have emerged as promising cathode materials for non-aqueous Li-air batteries due to their exceptional properties, such as high chemical and thermal stability, elasticity, tensile strength, and conductivity resulting from their unique structures. Wang et al. developed a freestanding CNT-based electrode using a facile impregnation method, achieving a discharge capacity of approximately 8300 mAh/g and a discharge plateau of around 2.75 V at a current density of 0.1 mA/cm^2 [90]. Moreover, at higher discharge rates of 0.2 and 0.5 mA/

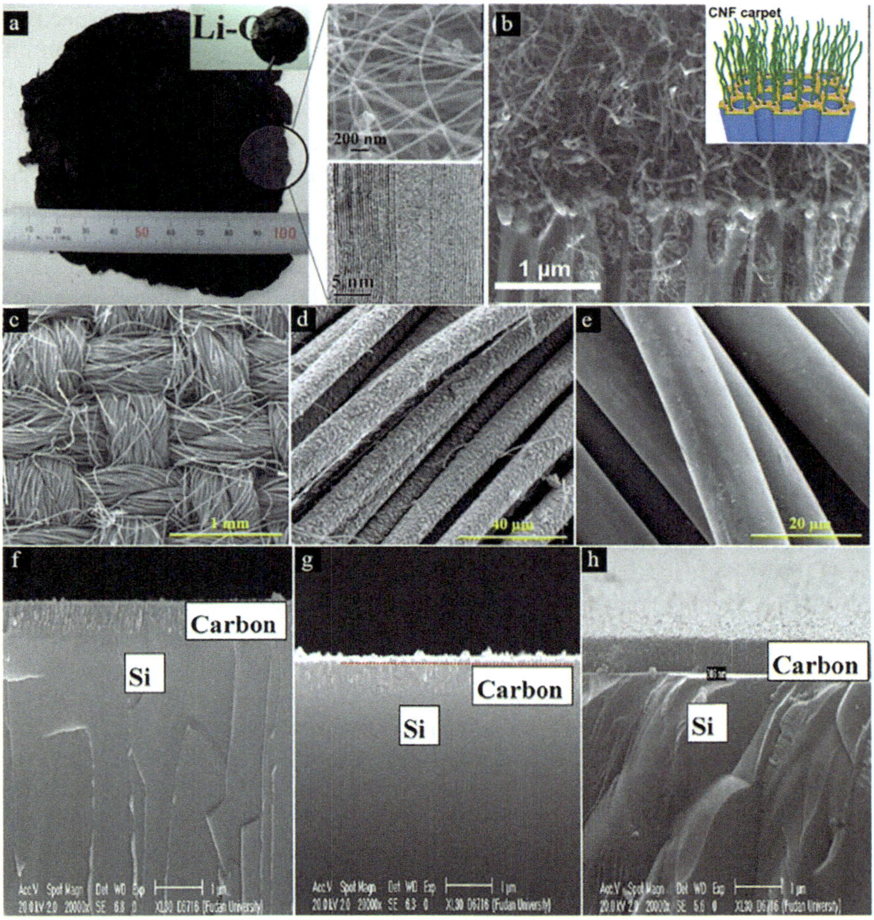

Fig. 6.28 **a** Photographs of the MWCNTP, SEM and HRTEM image of a multi-walled carbon nanotube. Reproduced with permission [90]. Copyright 2013, The Royal Society of Chemistry. **b** SEM image of the AAO filter after nanofiber growth. *Inset*: schematic representation of the electrode after catalyzed growth of CNFs. Reproduced with permission [93]. Copyright 2011, The Royal Society of Chemistry. SEM images of **c** pristine ACM electrodes, **d** discharging to 2 V, and **e** charging to 4.3 V. Reproduced with permission [91]. Copyright 2018, The Royal Society of Chemistry. Cross-sectional SEM images of **f** pristine DLC thin film electrodes, **g** discharging to 2.0 V, and **g** charging to 4.5 V. Reproduced with permission [92]. Copyright 2018, American Chemical Society

cm^2, the electrode demonstrated capacities of approximately 8000 mAh/g and 2000 mAh/g, respectively. In another study, Zhou et al. demonstrated a paper-based cathode composed of MWCNTs prepared through a floating catalyst method, exhibiting an impressive discharge capacity of approximately 34,600 mAh/g at a current density of 500 mA/g. Notably, the cathode maintained its capacity without loss during 50 cycles of continuous discharge and charge at 250 mA/g, with a cutoff capacity of 1000 mAh/g. The MWCNT paper consisted of interpenetrating MWCNTs, forming a thin and semi-transparent structure with void spaces that facilitated the storage of Li_2O_2 particles (Fig. 6.28a). Beyond graphene and CNTs, other functional carbon materials such as CNFs and nanostructured diamond-like carbon (DLC) have also been employed as cathode materials for Li-air batteries. Shao-Horn et al. developed a binder-free CNFs cathode by growing aligned hollow CNFs directly on ceramic porous substrates using atmospheric pressure CVD on porous anodized aluminum oxide (AAO) substrates coated with thin layers of Ta and Fe. The aligned CNFs, roughly perpendicular to the substrate surface, exhibited slight entanglement near the base of the fibers (Fig. 6.29b). The all-carbon-fiber electrode demonstrated high gravimetric energies of up to 2500 Wh/kg, representing a fourfold enhancement compared to state-of-the-art lithium intercalation compounds like $LiCoO_2$ (with 600 Wh/kg). The superior performance was attributed to the low carbon packing in the CNFs electrodes and efficient utilization of available carbon mass and void volume for Li_2O_2 formation. Aurbach et al. designed a hierarchical activated carbon microfiber (ACM) by calcining in CO_2 and utilized it as the cathode material in a non-aqueous Li-air battery [91]. The resulting cathode exhibited a maximum specific capacity of 4116 mAh/g and a charge voltage of 4.3 V at a current density of 0.025 mA/cm^2, outperforming conventional carbon composite air-electrodes with a discharge capacity of 2100 mAh/g and a charge voltage plateau of 4.5 V. SEM images of pristine, discharged, and charged ACM electrodes revealed unblocked millimeter and micron-sized channels even after the discharge process (Fig. 6.28c–e). Finally, Yang et al. prepared a DLC thin film electrode using radio frequency sputtering, demonstrating a discharge plateau around 2.7 V and a discharge capacity of 2318 mAh/g at a current density of 220 mA/g [92]. The excellent electrochemical performance was attributed to the presence of sp3-bonded carbon atoms in the DLC thin film, which played a crucial role in achieving a high discharge capacity and voltage plateau in the Li-air cell.

6.5.2 N-doped Carbon Film

Carbon materials can exhibit improved electrochemical properties when doped with non-metallic elements like nitrogen. The introduction of heteroatoms can alter the chemical and electronic characteristics of carbon-based materials, resulting in the formation of defects and functional groups. In 2012, Li et al. pioneered the use of nitrogen-doped graphene nanosheets (N-GNSs) in Li-air batteries. They achieved impressive discharge capacities of 11,660 mAh/g, 6,640 mAh/g, and 3,960 mAh/g at current densities of 75 mA/g, 150 mA/g, and 300 mA/g, respectively, surpassing the performance of pure GNSs. Chen et al. also demonstrated the significant enhancement of electrode reaction kinetics by incorporating chemical nitrogen and sulfur doping along with a 3D bicontinuous porous configuration in graphene [94]. Through

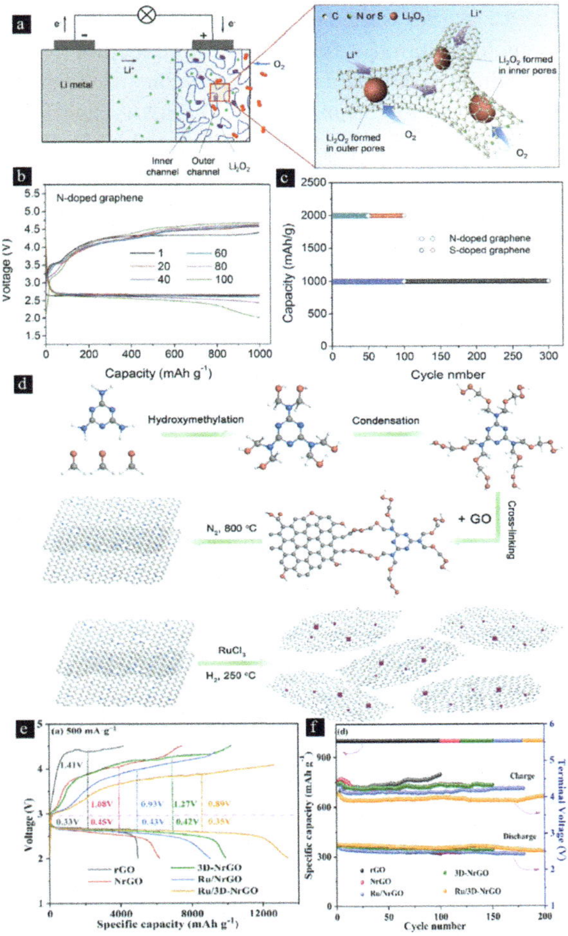

Fig. 6.29 **a** Schematic representation of the nanoporous graphene-based Li-air battery. **b** Charge/discharge profiles of nanoporous N-doped graphene recorded at the controlled capacity of 1000 mAh/g at 300 mA/g. **c** Cycling stability of the nanoporous N- and S-doped graphene. Reproduced with permission [94]. Copyright 2016, Wiley–VCH. **d** Schematic synthesis of N-doped GO film and Ru/N-doped rGO film. **e** Initial charge/discharge profiles of different cathodes at a rate of 500 mA/g. **f** The cycling performance depending on the cycle number of cathodes at a cutoff capacity of 1000 mAh/g and a charge/discharge rate of 200 mA/g. Reproduced with permission [95]. Copyright 2020, American Chemical Society

a nanoporous metal-based chemical vapor deposition (CVD) method, they synthesized free-standing and flexible nanoporous N-doped and S-doped graphene using benzene, pyridine, or thiophene as carbon, nitrogen, and sulfur sources. Cycling tests revealed that the nanoporous N-doped graphene exhibited a remarkable cycling lifetime of 300 cycles at a cut-off capacity of 1000 mAh/g (Fig. 6.29a–c).

A high-performance air cathode for lithium-oxygen batteries was designed, featuring a 3D architecture of N-doped graphene embedded with Ru nanoclusters [95]. The synthesis process of the Ru/N-doped rGO film is illustrated in Fig. 6.30c. Initially, graphene oxide (GO) was mixed with a melamine–formaldehyde resin dispersion, facilitating a cross-link reaction between oxygen-containing functional groups on GO and the resin's tertiary amine groups. Subsequent annealing and pyrolysis at high temperature resulted in the formation of covalently bonded N-doped GO.

The Ru species were then loaded onto the N-doped GO by impregnating it with a Ru precursor solution. After drying and reduction in a H_2 atmosphere, the Ru/N-doped rGO film was obtained. Remarkably, the discharge capacity of the Ru/N-doped rGO film reached 13,382 mAh/g at 500 mA/g, significantly surpassing other graphene electrodes without 3D linkages (Fig. 6.29e, f]. Furthermore, the Ru/N-doped rGO film exhibited a stable discharge voltage plateau at ~2.61 V and a low charge voltage plateau at ~3.85 V. The improved performance can be attributed to three factors: (1) the high content of doped N, which provides catalytically active sites; (2) the 3D linked graphene architecture with a high surface area and hierarchically porous structure, facilitating the accommodation of discharging products; (3) the incorporation of highly dispersed Ru nanoclusters, enhancing the oxygen evolution reaction (OER) activity and reducing charging overpotential.

In addition to N-doped graphene, N-doped carbon nanotubes (CNTs) and carbon nanofibers (CNFs) have emerged as promising candidates for cathode materials in Li-air batteries. Zhang et al. developed a self-supporting and flexible electrode by synthesizing hierarchical N-doped CNTs on a stainless-steel mesh (N-CNTs@SS) using a scalable one-step strategy. The N-CNTs@SS electrode exhibited remarkable physicochemical properties, including a discharge capacity of up to 9,299 mAh/g at 500 mA/g, excellent rate capability, and exceptional cycle stability for 232 cycles. Dai et al. employed a chemical vapor deposition (CVD) method to fabricate vertically aligned N-doped coral-like carbon fibers (VA-NCCF). The VA-NCCF electrode demonstrated a discharge voltage plateau of 40,000 mAh/g at a current density of 500 mA/g, accompanied by a low overpotential of 0.3 V at a current density of 100 mA/g. The authors attributed this performance to the special continuously coating layer of Li_2O_2 that effectively filled the interbranch space, enhancing the contact between the electrode and Li_2O_2 and resulting in a reduced overpotential.

In recent years, research efforts have focused on exploring composites consisting of heteroatom-doped carbons (such as O, N, P, S) and transition metal oxides, as well as metal-nitrogen-doped carbons (M-NC). Ramakrishna et al. developed a series of nitrogen/oxygen co-doped carbon nanotubes grown on carbon nanofiber films (MNO-CNT-CNFFs, where M = Fe, Co, Ni) using a facile free-surface electrospinning technique followed by in-situ growth carbonization [96]. The fabrication process of the free-standing MNO-CNT-CNFF is illustrated in Fig. 6.30a. A spinning precursor solution containing PAN, cellulose acetate (CA), metal acetylacetonate, and DMF was prepared to mass synthesize a metal-rich nanofiber film (M-NFF) (Fig. 6.30b). The obtained M-NFF was stabilized at 280 °C, followed by controlled carbonization with the presence of melamine, resulting in the formation of graphitic carbon nitride on the M-NFF (Fig. 6.30c). Concurrently, CA decomposed, leaving channels within the carbonized PAN fibers that exposed the transition metal compositions, facilitating the growth of CNTs and the formation of MNO-CNT active materials. The resulting MNO-CNT-CNFF exhibited flexibility, robustness, high voltage (~3.4 V), low overpotential (0.15 V), and long cycle life (120 h). The exceptional performance was attributed to the high catalytic activity of Fe-catalyzed CNTs and the efficient mass transport characteristics of the 3D carbon fiber films. Cobalt–nickel spinel oxide ($NiCO_2O_4$) demonstrated higher electrocatalytic activity, making it an

Fig. 6.30 **a** The schematic illustration of the fabrication process of MNO-CNT-CNFFs. Photographs of **b** M-NFFs, **c** the stabilized M-NFFs, and **d** the MNO-CNTCNFFs. Reproduced with permission [96]. Copyright 2017, American Chemical Society. **e** The schematic illustration of the fabrication process of NCO@NCF. **f** A photograph of hybrid films during different heat treatment processes and assembled cell structure. Reproduced with permission [97]. Copyright 2016, The Royal Society of Chemistry. **g** The schematic illustration of the fabrication process of Co_4N/CNF. **h** Proposed reaction mechanism with CNF and Co_4N/CNF electrode. **i** Schematic images of flexible Li-O_2 battery using Co_4N/CNF electrode (inset). **j** Voltage profile of flexible Li-O_2 batteries with different capacity restriction. Photographs of the LED turned on with **k** planar and **l** 180° bent conditions. Reproduced with permission [98]. Copyright 2018, American Chemical Society

excellent bifunctional catalyst for both oxygen reduction reactions (ORR) and oxygen evolution reactions (OER). Additionally, $NiCo_2O_4$ acted as a catalyst to reduce charge–discharge overpotentials in Li-air batteries. Yamauchi et al. utilized a single-nozzle co-electrospinning technique to prepare N-doped carbon nanofibers (NCF) covered with $NiCo_2O_4$ nanoparticles (NCO@NCF) (Fig. 6.30e) [97]. The core–shell polymer hybrid nanofiber mat-like film underwent a series of steps including electrospinning, solvent evaporation, pre-oxidation, carbonization, and re-oxidation, resulting in a self-supported film composed of N-doped carbon nanofibers covered with $NiCO_2O_4$ nanoparticles. The colors and shapes of the hybrid films with different calcination treatments are depicted in Fig. 6.30f. The NCO@NCF film exhibited good mechanical strength and could be directly used as a binder-free, self-supported cathode in Li-air batteries. The NCO@NCF film demonstrated a discharge capacity as high as 5304 mAh/g at 200 mA/g, surpassing the capacity of the binder-rich mashed NCO@NCF cathode (2920 mAh/g). Additionally, the NCO@NCF film displayed the lowest discharge and charge overpotentials (~0.22 V and ~0.95 V, respectively). The hierarchical porous structure facilitated O_2 diffusion and enhanced electrolyte infiltration, while the interconnected carbon nanofibers formed a conductive network that enabled high-speed electron transmission.

The utilization of N-doped metal oxides in energy conversion and storage devices offers advantages in electronic conductivity by lowering band gaps or achieving conduction and valence band overlap. Various nitride materials, such as titanium nitride, molybdenum nitride, cobalt nitride, nickel–iron nitride, cobalt-molybdenum nitride, and nickel–iron-molybdenum nitride, have been proposed for these applications. Particularly, metallic CO_4N catalysts have demonstrated remarkable stability and excellent oxygen evolution reaction (OER) activities in alkaline solutions due to the formation of a thin oxide layer that facilitates electrocatalysis and rapid charge transfer. Kim et al. designed a brush-like structure of metallic CO_4N nanorods anchored to an N-doped carbon nanofiber (CNF) membrane (CO_4N/CNF) by hydrothermal coating of cobalt fluoride hydroxide on CNF followed by nitridation annealing. The Li-O_2 cell based on the CO_4N/CNF electrode exhibited a high capacity of 11.9 mAh/cm^2 and enhanced cycling capability (>177 cycles). The reaction mechanisms of the CNF and CO_4N/CNF electrodes during testing are depicted in Fig. 6.30h. For the CNF electrode, initially reduced LiO_2 radicals weakly interacted with the CNF surface and were easily solvated in the electrolyte due to the relatively low LiO_2 adsorption energy. The recharge process led to the disappearance of particle-type Li_2O_2, while Li_2CO_3 and $LiRCO_3$ residues remained, hindering the dissociation and blocking active sites on CNF, resulting in sluggish oxygen reduction reaction (ORR) and OER. In contrast, the Co4N/CNF electrode tightly bound LiO_2 radicals, forming film-type Li_2O_2 layers on partially oxidized CO_4N surfaces via further reduction. A pouch-type Li-air cell consisting of the CO_4N/CNF membrane exhibited stable cycling performance in an air atmosphere for over 200 h without severe polarization (Fig. 6.30j). Moreover, the Li-air cell successfully powered a commercial green LED, maintaining functionality even when the battery was planar or bent (Fig. 6.30k, l). The superior performance was attributed to the suitable design and tailored properties of the CO_4N/CNF electrode, including improved charge transfer facilitated by metallic catalysts directly grown on N-doped CNF without the need for binders, numerous active sites provided by ultra-high density CO_4N nanorods for accommodating Li_2O_2, and the presence of an amorphous CoOx film on CO_4N that promotes ORR/OER with low charge overpotentials.

6.6 Li-Metal Battery

Lithium (Li) metal anodes have garnered significant attention for their potential applications in high-energy–density Li batteries, owing to their high theoretical capacity (3860 mAh/g) and lowest standard electrochemical potential (−3.04 V). However, several challenges hinder the practical implementation of Li metal anodes. The uncontrolled growth of dendrites, infinite volume expansion, and side reactions are among the key issues. Dendrite growth, in particular, poses a critical problem as it can penetrate the separator, leading to short circuits, thermal runaway, fires, and even explosions. Additionally, dendrites react with the electrolyte, irreversibly consuming active materials. The uneven dissolution of lithium dendrites results in dead Li, further reducing battery life.

Extensive efforts have been dedicated to suppressing dendrite growth, including modifications of liquid electrolyte additives to form a stable solid electrolyte interphase (SEI) layer on the lithium anode surface. However, the artificial SEI layer is often incapable of accommodating the morphological changes of the lithium anode and is prone to breakage during cycling. These broken sites are considered as nucleation and growth sites for lithium dendrites. Consequently, solid-state electrolytes and mechanical separators with high shear modulus have been developed to hinder dendrite formation. Unfortunately, most solid electrolytes exhibit low ionic conductivity and large interfacial impedance, limiting their effectiveness. While mechanical barriers prevent lithium dendrites from causing short circuits, they do not entirely suppress dendrite formation and the resulting dead lithium fractions during repeated Li plating/stripping.

In recent years, the design of functional three-dimensional (3D) porous host matrices has emerged as a viable approach to impede the growth of Li dendrites. The 3D structure provides a large specific surface area, facilitating low current density and uniform distribution of positive charges. Additionally, the porous architecture offers ample space for accommodating Li deposition, inhibiting dendritic growth and mitigating the volume change of Li metal during cycling. Among the various host materials, carbon and its derivatives have garnered significant attention as ideal candidates for Li deposition due to their lightweight nature, high electronic conductivity, mechanical strength, electrochemical stability, and low cost.

Notably, carbon-based materials have been explored as effective strategies for mitigating the challenges associated with Li metal anodes. In 2012, Stucky et al. reported the use of spatially heterogeneous carbon-fiber papers as dendrite-free current collectors for lithium deposition. These papers were synthesized by depositing a thin SiO_2 layer on a carbon-fiber paper, followed by a magnesiothermic reaction to convert SiO_2 into SiC. The resulting SiO_2/SiC-decorated 3D current collectors exhibited dendrite-free behavior even after deep lithium deposition at high current densities [99].

Furthermore, carbon spheres were employed as interlayers for protecting Li metal in 2014. Cui et al. pioneered this approach by coating the lithium metal anode with a monolayer of interconnected amorphous hollow carbon nanospheres. The stable

SEI formed on the top surface facilitated Li metal deposition beneath the carbon nanospheres. Moreover, the weak bond between the hollow carbon nanosphere layer and the copper substrate allowed for adjustments in the storage space for Li deposition. This study demonstrated that nanoscale interfacial engineering could be a promising strategy to address the intrinsic challenges of lithium metal anodes.

In 2016, Cui et al. introduced layered reduced graphene oxide (rGO) with nanoscale interlayer gaps as a stable host for Li metal anodes. Since then, research on carbon-based hosts for Li metal batteries has regained prominence. However, many carbon-based hosts lack sufficient Li affinity. To overcome this limitation, the coating of lithiophilic materials onto the surface of carbon-based hosts has been proposed as a solution. Figure 6.31 provides a timeline summarizing recent research efforts on carbon-based nanomaterials as Li metal anodes [100].

Functional mesoporous carbon-based films have emerged as promising host materials for Li metal anodes due to their unique physico-chemical and electrochemical properties. The porous nature of these films allows for tunable interfacial structures,

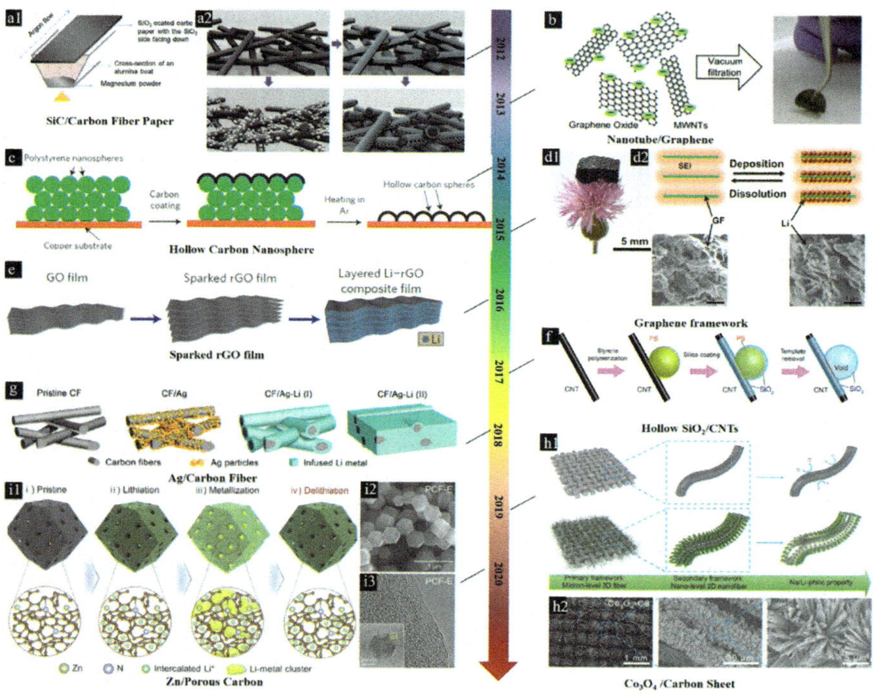

Fig. 6.31 Timeline showing recent research advances regarding carbon-based Li metal anodes. Reprinted with permission from Refs. [99, 100]. Copyright 2012, Elsevier BV. Copyright 2013, Wiley–VCH. Copyright 2014, Nature Publishing Group. Copyright 2015, American Chemical Society. Copyright 2016, Nature Publishing Group. Copyright 2017, American Chemical Society. Copyright 2018, Cell Publishing Group. Copyright 2019, Wiley–VCH. Copyright 2020, Wiley–VCH

providing a high specific surface area and conductivity. This facilitates a uniform charge distribution and regulated Li^{+} flux dispersion across the entire electrode, effectively preventing dendrite nucleation. The interface between the carbon-based film and Li metal plays a critical role in the plating and stripping process, allowing for fine regulation of lithium nucleation overpotential and interface impedance. This is achieved through modified electronic and ionic conductivity, interfacial energy, and other surface properties of the layered porous carbon-based anode.

Moreover, the interconnected micro/nanopores within the carbon-based film serve as effective hosts for Li storage at the micro/nanoscale. This enhances the electrochemical activity of Li while suppressing unfavorable side reactions between Li and the electrolyte. The porous structure of the film also helps inhibit drastic volume changes and provides a continuous pathway for charge transport during the plating and stripping processes. As a result, the stability of the electrode is improved, and the charge/discharge rate can be significantly increased.

Additionally, the layered structure of these devices promotes effective wetting and penetration of the electrolyte onto the anode surface, facilitating rapid charge transfer at the electrode–electrolyte interface. Furthermore, some layered devices exhibit specific physical or electrochemical properties, such as flexibility, elastic strength, shear modulus, and permeation selectivity. These structural features contribute to the suppression of dendrite formation and enhanced cyclability of the Li metal anode.

6.6.1 Bare Carbon Materials for Li Metal Anodes

Carbon nanomaterials, including graphene, carbon nanotubes/fibers (CNTs/CNFs), and carbon spheres, have been recognized as excellent hosts for Li metal anodes. Efforts have been devoted to exploring their potential in improving the performance of Li metal anodes. For instance, layered reduced graphene oxide (rGO) has shown promise in enhancing the deposition behavior and electrochemical performance of Li metal anodes. The synthesis of layered rGO involves Li-assisted reduction of a graphene oxide film obtained through vacuum filtration of colloidal dispersions using a membrane filter. This filtration method offers simplicity, reproducibility, and precise control over the film thickness.

When the rGO host comes into contact with molten Li, the lithiophilic nature of sparked rGO and the capillary force of the nanogaps allow for the infusion of Li into the matrix (Fig. 6.32a). This results in the division of dense Li into finer domains, reducing the thickness fluctuation to approximately 20%. Moreover, the integration of rGO leads to the formation of a high-surface-area Li anode, which effectively reduces the effective current density during cycling. Consequently, stable cycling with low hysteresis is achieved in both symmetric-cell and full-cell configurations, even when using a carbonate-based electrolyte.

The exceptional electrochemical performance of layered Li-rGO films can be attributed to their notable structure. Firstly, the stable scaffold provided by rGO significantly mitigates volume changes at the electrode level during cycling. Secondly, the

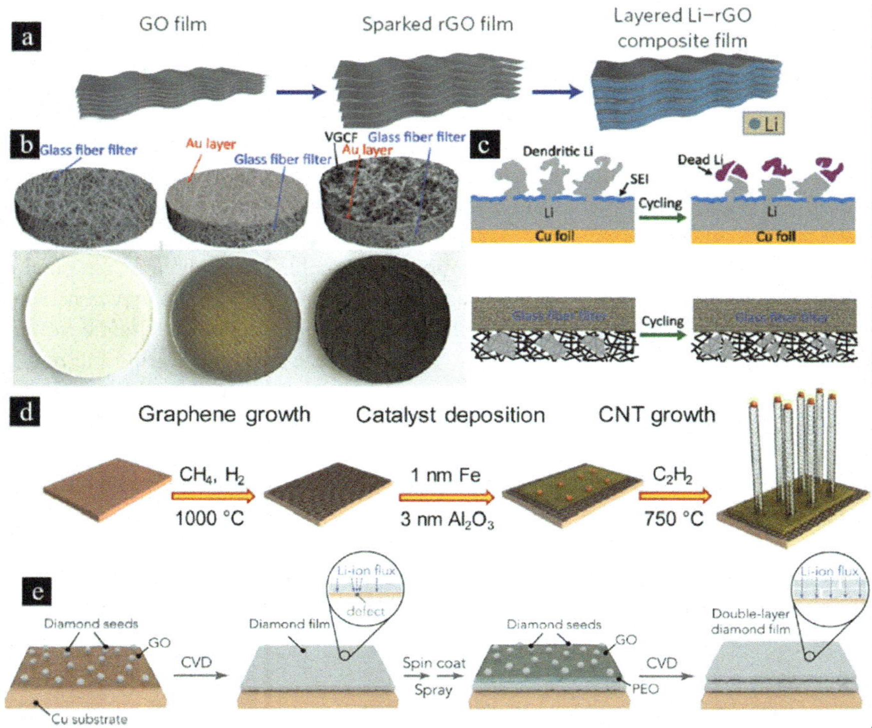

Fig. 6.32 **a** Schematic of the fabrication process for the layered Li-rGO composite electrode. Reproduced with permission Ref. [101]. Copyright 2016, Nature Publishing Group. **b** Schematic of the fabrication process for the VGCF@GF electrode. **c** Schematic illustration of Li deposition/stripping processes on the Cu foil and VGCF@GF electrode. Reproduced with permission Ref. [102]. Copyright 2018, Royal Society of Chemistry. **d** Schematic of the fabrication process for the GCNT electrode. Reproduced with permission Ref. [103]. Copyright 2017, American Chemical Society. **e** Schematic of the fabrication process for the ultrastrong double-layer nanodiamond electrode. Reproduced with permission Ref. [104]. Copyright 2018, Cell Publishing Group

lithiophilic nature of rGO, combined with its large surface area, ensures uniform Li infusion and deposition during synthesis and cycling. Additionally, the top rGO cap layer serves as an electrochemically and mechanically stable artificial interface, contributing to the stabilization of the solid electrolyte interphase (SEI) formed during cycling.

The superior kinetic properties of two-dimensional (2D) carbon nanoarchitectures, such as carbon nanotubes/fibers, have been demonstrated compared to carbon matrices with particle morphology. In a study by Zhao et al., they showcased the coating of a carbon fiber host on glass fiber filters (VGCF@GF) to achieve dendrite-free Li deposition at high area capacity (Fig. 6.32b). The preparation of the VGCF@GF electrode involved depositing a thin film of Au on the GF surface through magnetron sputtering, followed by the coating of the VGCF matrix to create the final electrode. By incorporating a functional protective layer (GF) and a well-established conducting host (VGCF), the VGCF@GF electrode with its various voids effectively mitigated the volume change during Li plating/stripping and maintained electron transport pathways among isolated lithium particles (Fig. 6.32c).

In another approach, a seamless graphene-carbon nanotube (GCNT) film electrode was synthesized using chemical vapor deposition (CVD). This process involved depositing iron nanoparticles and aluminum oxide, followed by CVD growth of CNTs using acetylene as the carbon source (Fig. 6.32d). The novel structure of the GCNT film allowed for the homogeneous distribution of a considerable amount of Li metal as a thin coating over CNT bundles. This design successfully suppressed dendrite formation during reversible plating and stripping operations. The resulting full battery based on GCNT-Li/sulfurized carbon (SC) exhibited exceptional performance in Li-metal batteries with high energy density (752 Wh/kg), significant areal capacity (2 mAh/cm^2), and excellent cyclic stability (80% retention after >500 cycles). This remarkable performance can be attributed to the synergistic combination of graphene and CNTs in the electrode design.

Diamond is a promising material for stabilizing Li metal, but not all diamond forms are suitable as an artificial interface. The use of diamond particles as electrolyte additives has limitations due to their large-scale physical integrity issues and high graphite content, which hinders the full realization of diamond's advantages in dendrite suppression and cycling efficiency enhancement. To address this, Cui et al. developed a double-layer nanodiamond film using microwave-plasma chemical vapor deposition (MPCVD) with colloidal nanodiamond seeds (Fig. 6.32e). In this film design, one layer with defects is shielded by an intact nanodiamond film in the other layer, promoting uniform Li-ion flux and ensuring mechanical homogeneity at the nanodiamond interface. The engineered nanoporous diamond thin film, with weak binding to the current collector, low electrical conductivity, and high electrochemical stability, allows for Li deposition beneath the interface layer, reducing parasitic reactions between Li metal and the electrolyte.

Apart from diamond, there are several other promising carbon hosts for accommodating Li metal. Table 6.5 provides a summary of different carbon host synthesis methods and their battery performance as anodes in Li-metal batteries.

While carbon-based materials are generally considered suitable hosts for Li plating due to their high electrical conductivity and chemical stability, many carbon substrates with designed nanostructures have shown unsatisfactory performance in Li plating. These bare carbon hosts exhibit lithiophobic behavior, resulting in isolated and randomly distributed nucleation sites for Li plating. Recent research has focused on heteroatom doping to introduce lithiophilic functional groups, such as pyridinic and pyrrolic nitrogen, into carbon-based hosts, thereby enhancing their Li affinity. Nitrogen (N) or sulfur (S) doping, as well as N/S-containing functional groups, have been employed to improve the performance of Li metal anodes. These dopants provide good electrical conductivity for rapid electron transport and offer defect structures for enhanced lithium storage.

Zhang et al. utilized N-doped graphene (NG) obtained by treating graphene in ammonia as a Li plating matrix to guide Li metal nucleation and inhibit dendrite

Table 6.5 Cycle performance based on carbon host anodes

Host materials	Fabrication method	Voltage hysteresis (mV)/current density (mA/cm^2)/capacity (mAh/cm^2)/	Cycling time in half cells (h)/current density (mA/cm^2)	Cycling stability of full batteries (cathode/cycle number/retention/rate)	References
Carbon cloth	Pressure	46/1/1, 150/5/1	–	NCM/300/83/1C	[105]
Graphite microtubes	CVD	14/1/10	3000/10	LFP/1000/52/0.5C	[106]
Flexible carbon microtube	Electrodeposited	50/5/5	1400/10	LFP/250/99.8/0.5C	[107]
Hollow carbon fibers	Carbonized cotton	30/2/1	1200/1	LFP/200/91.3/0.5C	[108]
Interconnected graphene	Vacuum filtration and CVD	–	300/1	–	[109]
N-doped carbon rod array	Carbonization	12/1/1, 40/2/1	620/2	LFP/100/100/1C	[110]
S-doped OMC nanospheres	Soft template method	20/0.5/1	1600/0.5	LFP/300/110/1C	[111]

growth. The presence of lithiophilic N-containing functional groups in NG promotes the nucleation process of Li, leading to dendrite-free morphology and impressive electrochemical performance. However, it is crucial for the lithiophilic sites to be uniformly distributed to ensure the uniformity of nucleation sites and prevent uncontrolled dendrite growth. Moreover, an appropriate amount of N doping is necessary to increase the lithiophilicity of the carbon matrix without compromising the structural integrity and electrical conductivity.

6.6.2 Carbon–Metal-Based Composite Materials for Li Metal Anodes

A promising approach to enhance the surface chemistry of nanocarbon materials and reduce nucleation overpotential is the incorporation of metal seeds. These metal nanoparticles serve as preferential sites for Li nucleation due to their low nucleation overpotential, thereby controlling the Li nucleation behavior and inhibiting dendrite growth. Cui et al. introduced selective Li deposition through heterogeneous seeded growth by designing a nanocapsule structure using hollow carbon spheres with embedded nanoparticle seeds [112]. In this approach, Au nanoparticles were chosen as nucleation seeds, and the lithium metal predominantly grew inside the hollow carbon spheres. The synthesis process involved immobilizing citrate-stabilized Au nanoparticles onto silica modified with 3-aminopropyltriethoxysilane, followed by coating with resorcinol formaldehyde resin and calcination to obtain amorphous carbon. Subsequently, the silica template was etched away using HF or KOH, resulting in hollow carbon shells with Au nanoparticles embedded inside. Compared to bare carbon nanocapsules, this selective deposition and stable encapsulation of Au nanoparticles effectively prevented dendrite formation and achieved a high Coulombic efficiency of up to 98% in alkyl carbonate electrolyte for over 300 cycles.

Researchers have drawn inspiration from Cui's groundbreaking research, which demonstrated that lithium nucleation over-potential was virtually eliminated when plating on Au foil. This discovery has motivated scientists to explore the potential of novel carbon/metal-based hosts. The performance of various carbon–metal-based host anodes is summarized in Table 6.6, showcasing their cycling capabilities and potential for application in lithium batteries.

Cui's group has also made significant contributions in the field of Li metal anodes by introducing 3D porous carbon matrices with Si lithiophilic coatings as hosts [113]. The Si coating interacts with molten Li, forming a lithium silicide alloy phase that facilitates the wetting of the carbon scaffold's surface and fills its porous structure (Fig. 6.33c). To assist this infusion process, a thin Si layer is deposited onto the scaffold surface using chemical vapor deposition (CVD) (Fig. 6.33c). Without a host scaffold, the shape and thickness of the Li electrode change with each Li-stripping/plating process, resulting in an unstable surface (Fig. 6.33d, e). In contrast, the Li/C

Table 6.6 Cycle performance of carbon–metal-based host anodes

Host materials	Fabrication method	Voltage hysteresis (mV)/current density (mA/cm^2)/capacity (mAh/cm^2)/	Cycling time in half cells (h)/current density (mA/cm^2)	Cycling stability of full batteries (cathode/cycle number/retention/rate)	References
Au-graphene aerogel	Hydrothermal method	62.4/2/2, 105.4/4/4	1800/2,1200/4	LFP/1100/50.2/2C	[115]
Ag-carbon nanofibers	Joule heating method	25/0.5/1	500/0.5	–	[116]
Ag–N-doped porous carbon	Polymer blowing-up	18.9/1/1	2000/1	LFP/70/100/1C	[117]
Ni–N-doped graphene oxide	Two-step pyrolysis	19/0.5/1, 30/2/1, 50/4/1	–	–	[118]
Co–N-doped carbon nanosheet	Room-temperature crystallization	6.5/0.5/1, 10/1/1, 20/2/1, 38/5/1	1000/5	LFP/300/100/5C	[119]

composite anode exhibits minimal volume change and maintains an intact and stable surface (Fig. 6.33e, g). The 3D carbon matrix offers a high surface area, reducing the actual current density and preventing the formation of dendritic Li. Its conducting interconnected network enables facile electron and ion transport. Additionally, the low density of the 3D matrix ensures a high specific capacity and energy density for the Li anode, while its flexibility and mechanical stability alleviate volume changes during cycling. Other coating modifications, such as Ag and Mg, have also been explored to improve the wetting of molten Li on the host scaffold by reacting with Li during the plating process. For instance, Zhang et al. developed a coralloid silver-coated carbon fiber-based composite Li anode (CF/Ag-Li) through Ag electroplating and molten Li infusion (Fig. 6.33h). The Ag particles not only confer lithiophilic properties to the carbon nanofibers, forming a Li metal composite anode (CF/Ag-Li), but also regulate Li nucleation and growth during plating. In-situ optical microscopy revealed the Li-stripping process of the fully infused CF/Ag-Li electrode, showing uniform dissolution of Li from the electrolyte side to the opposite side without any dead Li observed (Fig. 6.33i, j). Importantly, the thickness of CF/Ag-Li remains nearly constant after stripping, indicating minimal volume change. Scanning electron microscopy images after stripping and replating demonstrate uniform stripping morphology and dendrite-free replating morphology (Fig. 6.33k, l).

6.6.3 Carbon–Metal Oxides/Nitrides-Based Composite Materials for Li Metal Anodes

Simultaneously, the utilization of various metal oxides or nitrides nanomaterials with strong oxidizing properties, such as CO_3O_4, ZnO, Al_2O_3, CuO, SnO_2, TiN, etc., for decoration on the carbon framework is crucial to expand the applicability of the molten infusion strategy in carbon-based hosts. To facilitate a comprehensive comparison and trace the historical advancements, Table 6.7 presents the fabrication methods employed and battery parameters achieved for carbon-based Li anodes incorporating different metal oxide or nitride layers.

Cui and colleagues introduced a novel core–shell lithium host architecture composed of an outer shell of Al_2O_3 and an inner core of hollow carbon spheres (HCS). The HCS was synthesized using a hard-template method, and subsequently, Al_2O_3 thin-film coatings were deposited onto the HCS electrodes utilizing an ALD system (Fig. 6.34a). The schematic diagram in Fig. 6.34b illustrates the process of lithium plating and stripping on HCS and Al_2O_3-HCS electrodes. It is evident that the electrolyte could permeate the pores of the HCS, resulting in uncontrolled lithium deposition. Lithium plating occurred throughout the carbon sphere, leading to the formation of a thin SEI layer on both the lithium and lithium host surfaces. However, during the stripping process, the SEI on the lithium surface could break and accumulate within the electrode. With subsequent cycles, the SEI layer continued to grow and thicken, ultimately leading to battery failure due to side reactions that consumed

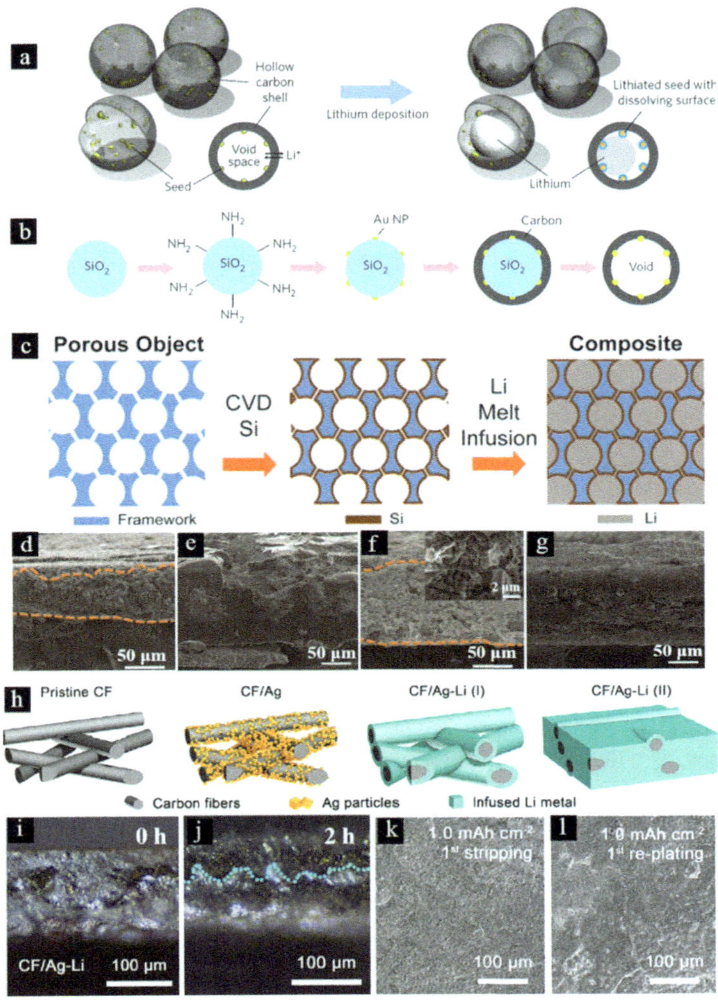

Fig. 6.33 **a** Schematic diagram of Li metal nanocapsules design. **b** Synthesis procedure of hollow carbon shells loaded with Au NPs inside. Reproduced with permission Ref. [112]. Copyright 2016, Nature Publishing Group. **c** Schematic illustration of composite Li/C anode by infusing Li into Si-coated carbon fiber. SEM image of bare Li electrode **d** after ~50% Li stripping and **e** after 30 cycles. SEM image of Li/C electrode **f** after ~50% Li stripping and **g** after 30 cycle. Reproduced with permission Ref. [113]. Copyright 2016, American Chemical Society. **h** Schematic diagram of the fabrication process for the CF/Ag-Li composite electrode. Cross-sectional *in-situ* optical microscopy photos of the **i** fresh CF/Ag-Li and **j** stripping process of CF/Ag-Li after 2 h. SEM image of CF/Ag-Li electrode after first Li **f** stripping and **g** replating. Reproduced with permission Ref. [114]. Copyright 2018, Cell Publishing Group

Table 6.7 Comparisons of electrochemical performance of carbon–metal oxides/nitrides-based hosts with different metal oxide/nitride layer as anode materials for LMBs

Host materials	Fabrication method	Voltage hysteresis (mV)/ current density (mA/cm^2)	Cycling time in half cells (h)/ current density (mA/ cm^2)	Cycling stability of full batteries (cathode/ cycle number/ retention/ rate)	References
Co_3O_4-carbon sheet	Hydrothermal method	75/2, 140/ 5, 200/10	800/1	LFP/200/ 88.4/2C	[120]
ZnO-carbon fiber cloth	Dipping	11/0.5, 18/1, 120/ 15	1800/1	LFP/200/ 77.9/1C	[121]
Al_2O_3-hollow carbon spheres	ALD	–	2160/ 0.5	–	[122]
Cu_xO-carbon felt	Electroplating	25/2	1000/2	NCA/ 100/73.9/ 1 C	[123]
MnO_2-graphene	CVD-hydrothermal method	96/1	1600/1	LCO/600/ 80/1C	[124]
$ZnCO_2O_4$/ ZnO-C@Ni foam	Calcination	88/2	2500/2	LFP/500/ 99.2/2C	[125]
TiN-carbon nanofiber	Electrospinning-dip-azotizing	30/1, 250/ 4	600/1	LFP/250/ 75.8/1 C	[126]
MO_2N-carbon nanofiber	Reduction-nitridation	24/3, 30/6	1500/6	NMC811/ 150/90/ 0.33 C	[127]

active lithium and electrolyte. In the case of ALD Al_2O_3-HCS, the introduction of the Al_2O_3 coating served as an artificial SEI layer. This coating facilitated lithium deposition inside the HCS while simultaneously preventing electrolyte infiltration. As a result, a thin and stable SEI layer was formed outside the lithium host during cycling.

In addition to Al_2O_3, ZnO has shown to enhance the interfacial wettability of lithium due to the reaction $Li + ZnO \rightarrow Zn + Li_2O_2$ and $Li + Zn \rightarrow LiZn$. Researchers have employed ZnO coatings on various substrates to improve lithiophilicity and enhance Li plating performance. Liu et al. utilized the ALD method to coat a ZnO thin layer on a polyimide membrane, resulting in a lithiophilic surface. Hu et al. improved the lithiophilicity of garnet-based solid-state electrolytes and carbonized wood by incorporating a ZnO layer. Deng et al. further enhanced the lithiophilicity of a cellular graphene scaffold by decorating it with ZnO nanoparticles.

Carbon cloth (CC) is considered an ideal host material for Li due to its flexibility, electrical conductivity, and mechanical and thermal stability. Li et al. demonstrated the synthesis of a 3D porous host (CC@ZnO) with a hierarchical architecture by coating ZnO nanowire arrays onto CC using a seed growth method (Fig. 6.34c). Electrochemical characterization confirmed the ability of the Li-CC@ZnO electrode to sustain stable Li plating and stripping while suppressing dendrite formation.

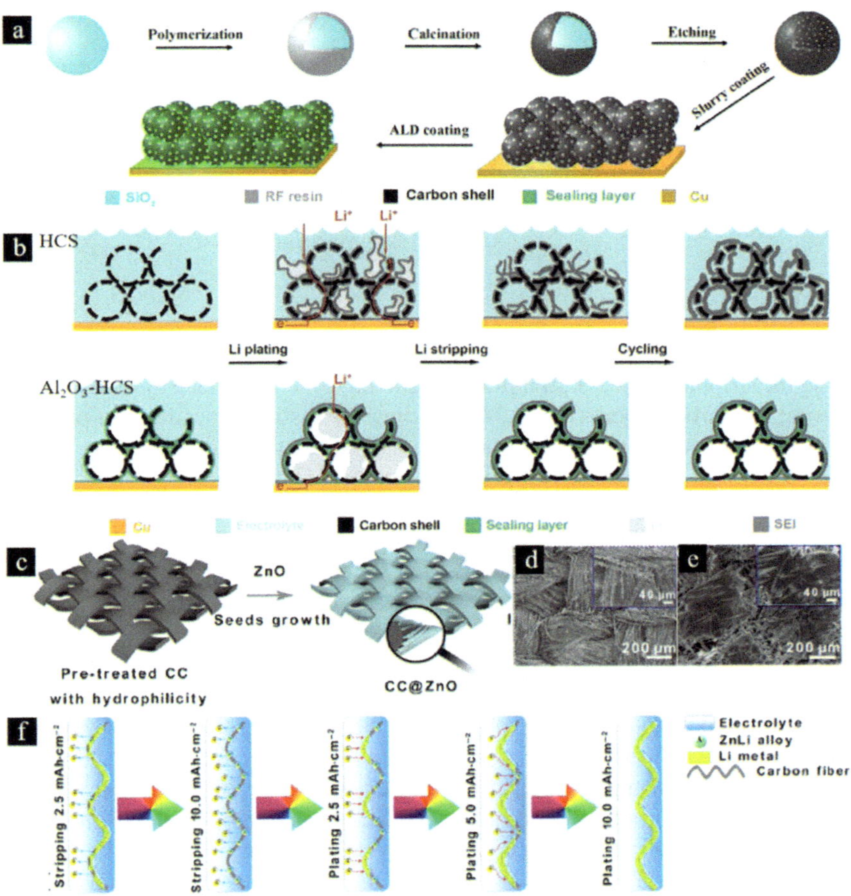

Fig. 6.34 **a** Schematic diagram of Al_2O_3-coated HCS electrode. **b** Schematic of lithium plating/stripping in HCS and Al_2O_3-coated HCS electrode. Reproduced with permission Ref. [128]. Copyright 2018, Nature Publishing Group. **c** Schematic illustration of Li-CC@ZnO electrode. SEM image of Li-CC@ZnO **d** after Li stripping and **e** plating at 2.5 mAh/cm^2. **f** Schematic demonstration of the Li plating/stripping behaviors for a Li-CC@ZnO electrode. Reproduced with permission Ref. [129]. Copyright 2018, Springer. **g** Schematic diagram of the fabrication process for ZnO@HPC electrode. **h** Schematic diagrams for the comparison of Li deposition in HPC and ZnO@HPC. Reproduced with permission Ref. [130]. Copyright 2017, Elsevier

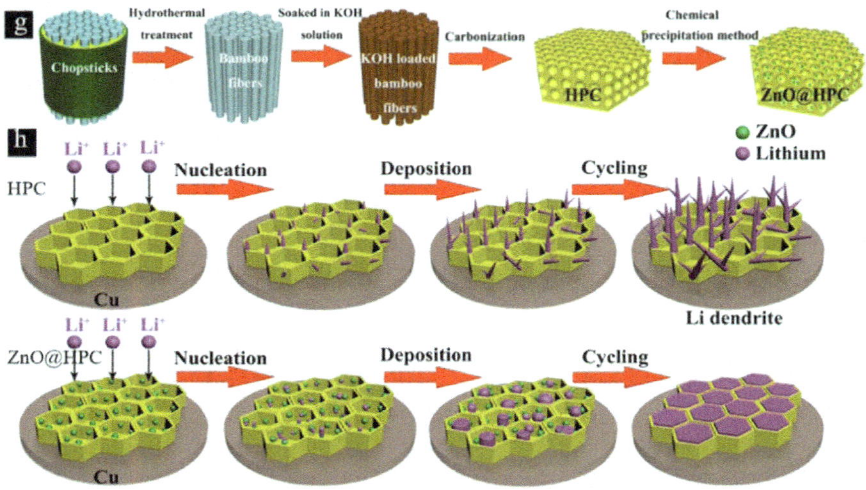

Fig. 6.34 (continued)

During the stripping process, Li initially dissolved from the surface of Li-CC@ZnO, exposing the carbon fibers (Fig. 6.34d). During the plating procedure, Li preferentially deposited on the exposed areas of the carbon fibers, gradually filling the free space between them at higher capacity (Fig. 6.34e). The surface appeared compact with no dendrites, except for some pores at the junction. Throughout the cycling, the Li-CC@ZnO electrode effectively regulated Li deposition within its pores, preventing dendritic nucleation and ensuring stable Li deposition and stripping, as illustrated in Fig. 6.34f.

Tao et al. demonstrated the use of a bamboo-derived 3D hierarchical porous carbon (HPC) decorated with ZnO quantum dots as a lithiophilic host for dendrite-free Li metal batteries. The fabrication process involved a KOH-assisted hydrothermal treatment of bamboo fibers, which dissolved the lignin and left behind the KOH-loaded bamboo fibers. After carbonization and precipitation reactions, the ZnO@HPC scaffold was obtained. Comparison of Li deposition within the HPC scaffold with and without ZnO nanoparticle decoration is depicted in Fig. 6.34h. Without ZnO, most Li plated on the outer layer of the carbon matrix, resulting in undesirable shapes and dendritic growth. The conductive nature of the carbon matrix promoted direct Li plating on its surface rather than filling the pores. In contrast, the ZnO@HPC scaffold effectively controlled the nucleation and growth of Li, resulting in flat and uniform deposition. Initially, Li ions preferred to nucleate on the ZnO seeds, and with further plating, Li alloyed with Zn to form LiZn. The highly conductive and lithiophilic LiZn induced subsequent Li growth within the 3D porous carbon, leading to controllable Li deposition and dendrite-free Li anodes. The 3D porous scaffold reduced local current density, and the lithiophilic ZnO quantum dots facilitated lithium deposition. Remarkably, the ZnO@HPC scaffold confined up to 131 mAh/cm^2 of lithium metal, exhibiting acceptable volume expansion, reduced overpotential, and effective dendrite suppression. In comparison to Li foil paired with $LiCOO_2$, the 3D Li within the ZnO@HPC scaffold displayed improved capacity and significantly lower voltage hysteresis. The ZnO-decorated 3D hierarchical porous carbon scaffold offers valuable insights into the design principles for metallic lithium anodes.

References

1. Mclean DA, Egerton L (1945) Paper capacitors containing chlorinated impregnants-stabilization by anthraquinone. Ind Eng Chem 37:73–79
2. Weeks JR (1950) Metallized paper capacitors. Proc IRE 38:1015–1018
3. Niu Z, Zhang L, Liu L et al (2013) All-solid-state flexible ultrathin micro-supercapacitors based on graphene. Adv Mater 25:4035–4042
4. Pech D, Brunet M, Taberna PL et al (2010) Elaboration of a microstructured inkjet-printed carbon electrochemical capacitor. J Power Sources 195:1266–1269
5. Hong K, ChoO M, Sang K (2015) Atomic layer deposition encapsulated activated carbon electrodes for high voltage stable supercapacitors. ACS Appl Mater Interfaces 7:1899
6. Pech D, Brunet M, Durou H et al (2010) Ultrahigh-power micrometre-sized supercapacitors based on onion-like carbon. Nat Nanotechnol 5:651–654
7. Zhang S, Zhu JY, Qing Y et al (2018) Ultramicroporous carbons puzzled by graphene quantum dots: integrated high gravimetric, volumetric, and areal capacitances for supercapacitors. Adv Funct Mater 28:1805898
8. Zhao Y, Liu M, Deng X et al (2015) Nitrogen-functionalized microporous carbon nanoparticles for high performance supercapacitor electrode. Electrochim Acta 153:448–455
9. Xiong Z, Liao C, Han W et al (2015) Mechanically tough large-area hierarchical porous graphene films for high-performance flexible supercapacitor applications. Adv Mater 27:4469–4475
10. Chang J, Adhikari S, Lee TH et al (2015) Leaf vein-inspired nanochanneled graphene film for highly efficient micro-supercapacitors. Adv Energy Mater 5:1500003
11. Strauss Marsh V et al. (2018) A simple route to porous graphene from carbon nanodots for supercapacitor applications. Adv Mater 30:1704449
12. Ma YJ, Zhang XG, Liang Z et al (2020) B/p/n/o co-doped hierarchical porous carbon nanofiber self-standing film with high volumetric and gravimetric capacitance performances for aqueous supercapacitors. Electrochim Acta 337:135800
13. Yao Y, Ma C, Wang J et al (2015) Rational design of high-surface-area carbon nanotube/microporous carbon core–shell nanocomposites for supercapacitor electrodes. ACS Appl Mater Interfaces 7:4817–4825
14. Jin H, Zhou L, Mak CL et al (2015) High-performance fiber-shaped supercapacitors using carbon fiber thread (cft)@polyanilne and functionalized cft electrodes for wearable/stretchable electronics. Nano Energy 11:662–670
15. Shang Y, Wang C, He X et al (2015) Self-stretchable, helical carbon nanotube yarn supercapacitors with stable performance under extreme deformation conditions. Nano Energy 12:401–409
16. Hu YS, Kong WH, Hong L et al (2004) Experimental and theoretical studies on reduction mechanism of vinyl ethylene carbonate on graphite anode for lithium ion batteries. Electrochem Commun 6:126–131
17. Wu SP, Han CP, Iocozzia J et al (2016) Germanium-based nanomaterials for rechargeable batteries. Angew Chem Int Ed 55:7898–7922
18. Liu L, Xie F, Lyu J et al (2016) Tin-based anode materials with well-designed architectures for next-generation lithium-ion batteries. J Power Sources 321:11–35

19. He M, Kraychyk K, Walter M et al (2014) Monodisperse antimony nanocrystals for high-rate li-ion and na-ion battery anodes: Nano versus bulk. Nano Lett 14:1255–1262
20. Dai R, Wang Y, Da P et al (2014) Indirect growth of mesoporous bi@c core-shell nanowires for enhanced lithium-ion storage. Nanoscale 6:13236–13241
21. Xu Y, Liu Q, Zhu Y et al (2013) Uniform nano-sn/c composite anodes for lithium ion batteries. Nano Lett 13:470–474
22. Luo B, Qiu T, Ye D et al (2016) Tin nanoparticles encapsulated in graphene backboned carbonaceous foams as high-performance anodes for lithium-ion and sodium-ion storage. Nano Energy 22:232–240
23. Chen P, Wu F et al (2014) Four-layer tin–carbon nanotube yolk–shell materials for high-performance lithium-ion batteries. Chem Sus Chem 7:1407–1414
24. Mao M, Yan F, Cui C et al (2017) Pipe-wire TiO_2-sn@carbon nanofibers paper anodes for lithium and sodium ion batteries. Nano Lett 17:3830
25. Chen SQ, Wang Y, Ahn H et al (2012) Microwave hydrothermal synthesis of high performance tin-graphene nanocomposites for lithium ion batteries. J Power Sources 216:22–27
26. Zou Y, Wang Y (2011) Sn@CNT nanostructures rooted in graphene with high and fast li-storage capacities. ACS Nano 5:8108–8114
27. Luo B, Wang B, Liang M et al (2012) Reduced graphene oxide-mediated growth of uniform tin-core/carbon-sheath coaxial nanocables with enhanced lithium ion storage properties. Adv Mater 24:1405–1409
28. Shen Z, Hu Y, Chen Y et al (2015) Tin nanoparticle-loaded porous carbon nanofiber composite anodes for high current lithium-ion batteries. J Power Sources 278:660–667
29. Chen P, WuY Wang F (2014) Four-layer tin-carbon nanotube yolk-shell materials for high-performance lithium-ion batteries. Chem Sus Chem 7:1407–1414
30. Qin J, He C, Zhao N et al (2014) Graphene networks anchored with sn@graphene as lithium ion battery anode. ACS Nano 8:1728–1738
31. Cheng F, Li W-C, Zhu J-N et al (2016) Designed synthesis of nitrogen-rich carbon wrapped Sn nanoparticles hybrid anode via in-situ growth of crystalline zif-8 on a binary metal oxide. Nano Energy 19:486–494
32. Zhu Z, Wang S, Du J et al (2014) Ultrasmall Sn nanoparticles embedded in nitrogen-doped porous carbon as high-performance anode for lithium-ion batteries. Nano Lett 14:153–157
33. Zhang G, Zhu J, Zeng W et al (2014) Tin quantum dots embedded in nitrogen-doped carbon nanofibers as excellent anode for lithium-ion batteries. Nano Energy 9:61–70
34. Liu YY, Lin DC, Liang Z et al (2016) Lithium-coated polymeric matrix as a minimum volume-change and dendrite-free lithium metal anode. Nat Commun 7:10992
35. Li W, Yang Z, Cheng J et al (2014) Germanium nanoparticles encapsulated in flexible carbon nanofibers as self-supported electrodes for high performance lithium-ion batteries. Nanoscale 6:4532–4537
36. Fang S, Shen L, Zheng H et al (2014) Ge–graphene–carbon nanotube composite anode for high performance lithium-ion batteries. J Mater Chem A 3:1498–1503
37. Liu X, Zhou L, Zhao Y et al (2013) Hollow, spherical nitrogen-rich porous carbon shells obtained from a porous organic framework for the supercapacitor. ACS Appl Mater Interfaces 5:10280–10287
38. Ma X, Zhou Y, Min C et al (2017) Synthesis of olive-like nitrogen–doped carbon with embedded Ge nanoparticles for ultrahigh stable lithium battery anodes. Small 13:1700403
39. Youn DH, Patterson NA, Park H et al (2016) Facile synthesis of Ge/N-doped carbon spheres with varying nitrogen content for lithium ion battery anodes. ACS Appl Mater Interfaces 8:27788–27794
40. Liu C, Li F, Ma LP et al (2010) Advanced materials for energy storage. Adv Mater 22:E28-62
41. Xiao X, Li X, Zheng S et al (2017) Nanostructured germanium anode materials for advanced rechargeable batteries. Adv Mater Interfaces 4:1600798
42. Mo R, Rooney D, Sun K et al (2017) 3D nitrogen-doped graphene foam with encapsulated germanium/nitrogen-doped graphene yolk-shell nanoarchitecture for high-performance flexible li-ion battery. Nat Commun 8:13949

43. Huo K, Wang L, Peng C et al (2016) Crumpled n-doped carbon nanotubes encapsulated with peapod-like Ge nanoparticles for high-rate and long-life li-ion battery anodes. J Mater Chem A 4:7585–7590
44. Wu L, Hu X, Qian J et al (2013) Sb–c nanofibers with long cycle life as an anode material for high-performance sodium-ion batteries. Energy Environ Sci 7:323–328
45. Yi Z, Han Q, Zan P et al (2016) Sb nanoparticles encapsulated into porous carbon matrixes for high-performance lithium-ion battery anodes. J Power Sources 331:16–21
46. Yi Z, Han Q, Cheng Y et al (2016) A novel strategy to prepare sb thin film sandwiched between the reduced graphene oxide and Ni foam as binder-free anode material for lithium-ion batteries. Electrochim Acta 190:804–810
47. Yushin NNG (2014) High-capacity anode materials for lithium-ion batteries: choice of elements and structures for active particles. Part Part Syst Charact 31:317–336
48. Yang F, Yu F, Zhang Z et al (2016) Bismuth nanoparticles embedded in carbon spheres as anode materials for sodium/lithium-ion batteries. Chemistry 22:2333–2338
49. Yin H, Li Q, Cao M et al (2017) Nanosized-bismuth-embedded 1d carbon nanofibers as high-performance anodes for lithium-ion and sodium-ion batteries. Nano Res 10:2156–2167
50. Su D, Dou S, Wang G (2015) Bismuth: a new anode for the Na-ion battery. Nano Energy 12:88–95
51. Wen Z, Cui S, Kim H et al (2012) Binding sn-based nanoparticles on graphene as the anode of rechargeable lithium-ion batteries. J Mater Chem 22:3300–3306
52. Park M-G, Song JH, Sohn J-S et al (2014) Co–sb intermetallic compounds and their disproportionated nanocomposites as high-performance anodes for rechargeable li-ion batteries. J Mater Chem A 2:11391–11399
53. Park AR, Park CM (2017) Cubic crystal-structured SnTe for superior Li- and Na-ion battery anodes. ACS Nano 11:6074–6084
54. Xin FX, Tian HJ, Wang XL et al (2015) Enhanced electrochemical performance of $Fe_{0.74}Sn_5$@reduced graphene oxide nanocomposite anodes for both Li-ion and Na-ion batteries. ACS Appl Mater Interfaces 7:7912–7919
55. Yu LT, Liu J, Xu XJ et al (2017) Metal-organic framework-derived Nisb alloy embedded in carbon hollow spheres as superior lithium-ion battery anodes. ACS Appl Mater Interfaces 9:2516–2525
56. Shi X, Song HH, Li A et al (2017) Sn-co nanoalloys embedded in porous n-doped carbon microboxes as a stable anode material for lithium-ion batteries. J Mater Chem A 5:5873–5879
57. Huang X, Cui S, Chang J et al (2015) A hierarchical tin/carbon composite as an anode for lithium-ion batteries with a long cycle life. Angew Chem 127:1510–1513
58. Wang B, Wen Z, Jin J et al (2017) A novel strategy to prepare Ge@C/Rgo hybrids as high-rate anode materials for lithium ion batteries. J Power Sources 342:521–528
59. Tian HZ, Seh ZW, Yan K et al (2017) Theoretical investigation of 2D layered materials as protective films for lithium and sodium metal anodes. Adv Energy Mater 7:1602
60. Deng J, Yan C, Yang L et al (2013) Sandwich-stacked SnO_2/cu hybrid nanosheets as multichannel anodes for lithium ion batteries. ACS Nano 7:6948–6954
61. Wang H, Min S, Wang Q et al (2017) Nitrogen-doped nanoporous carbon membranes with co/cop Janus-type nanocrystals as hydrogen evolution electrode in both acidic and alkaline environments. ACS Nano 11:4358–4364
62. Li D, Seng KH, Shi D et al (2013) A unique sandwich-structured c/ge/graphene nanocomposite as an anode material for high power lithium ion batteries. J Mater Chem A 1:14115–14121
63. Zhang H, Huang X, Noonan O et al (2017) Lithium-ion batteries: tailored yolk–shell sn@c nanoboxes for high-performance lithium storage. Adv Funct Mater 27:1606023
64. Song J, Yan P, Luo L et al (2017) Yolk-shell structured sb@c anodes for high energy Na-ion batteries. Nano Energy 40:504–511
65. Liu J, Yu L, Wu C et al (2017) New nanoconfined galvanic replacement synthesis of hollow Sb@C yolk–shell spheres constituting a stable anode for high-rate Li/Na-ion batteries. Nano Lett 17:2034–2042

66. Ngo DT, Le HTT, Kim C et al (2015) Mass-scalable synthesis of 3D porous germanium–carbon composite particles as an ultra-high rate anode for lithium ion batteries. Energy Environ Sci 8:3577–3588
67. Wu C, Maier J, Yu Y (2015) Sn-based nanoparticles encapsulated in a porous 3D graphene network: advanced anodes for high-rate and long life Li-ion batteries. Adv Funct Mater 25:3488–3496
68. Wang SQ, Xia L, Yu L et al (2016) Free-standing nitrogen-doped carbon nanofiber films: integrated electrodes for sodium-ion batteries with ultralong cycle life and superior rate capability. Adv Energy Mater 6:1502217
69. Kong B, Zu L, Peng C et al (2016) Direct superassemblies of freestanding metal-carbon frameworks featuring reversible crystalline-phase transformation for electrochemical sodium storage. J Am Chem Soc 138:16533–16541
70. Lee YH, Luu THT, Duong DL et al (2020) Monodispersed SnS nanoparticles anchored on carbon nanotubes for high-retention sodium-ion batteries. J Mater Chem A 8:7861–7869
71. Sang ZY, Yan X, Su D et al (2020) A flexible film with SnS2 nanoparticles chemically anchored on 3D-graphene framework for high areal density and high rate sodium storage. Small 16:2001265
72. Zhang K, Park M, Zhou L et al (2016) Urchin-like $CoSe_2$ as a high-performance anode material for sodium-ion batteries. Adv Funct Mater 26:6728–6735
73. Wu Y, Zhong W, Tang W et al (2020) Flexible electrode constructed by encapsulating ultrafine VSe_2 in carbon fiber for quasi-solid-state sodium ion batteries. J Power Sources 470:228438
74. Zhou D, Yang TT, Yang JQ et al (2020) A flexible self-charging sodium-ion full battery for self-powered wearable electronics. J Mater Chem A 8:13267–13276
75. Wang HL, Yang Y, Liang YY et al (2011) Graphene-wrapped sulfur particles as a rechargeable lithium-sulfur battery cathode material with high capacity and cycling stability. Nano Lett 11:2644–2647
76. Cheng ZB, Xiao ZB, Pan H et al (2018) Elastic sandwich-type rGO-VS2/S composites with High tap density: structural and chemical cooperativity enabling lithium-sulfur batteries with high energy density. Adv Energy Mater 8:1702337
77. Lv W, Li ZJ, Zhou GM et al (2014) Tailoring microstructure of graphene-based membrane by controlled removal of trapped water inspired by the phase diagram. Adv Funct Mater 24:3456–3463
78. Zhao MQ, Zhang Q, Huang JQ et al (2014) Unstacked double-layer templated graphene for high-rate lithium-sulphur batteries. Nat Commun 5:3410
79. Ni XY, Qian T, Liu XJ et al (2018) High lithium ion conductivity LiF/GO solid electrolyte interphase inhibiting the shuttle of lithium polysulfides in long-life Li-S batteries. Adv Funct Mater 28:1706513
80. Zhang M, Amin K, Cheng M et al (2018) A carbon foam-supported high sulfur loading composite as a self-supported cathode for flexible lithium-sulfur batteries. Nanoscale 10:21790–21797
81. Shi HD, Zhao XJ, Wu ZS et al (2019) Free-standing integrated cathode derived from 3D graphene/carbon nanotube aerogels serving as binder-free sulfur host and interlayer for ultrahigh volumetric-energy-density lithium-sulfur batteries. Nano Energy 60:743–751
82. Song X, Gao T, Wang S et al (2017) Free-standing sulfur host based on titanium-dioxide-modified porous-carbon nanofibers for lithium-sulfur batteries. J Power Sources 356:172–180
83. Zhang M, Yang Y, Zhang XH et al (2018) Flexible VOx nanosphere@SWCNT hybrid films with dual-confinement function of polysulfides for high-performance lithium-sulfur batteries. Adv Mater Interfaces 5:1800766
84. Kong WB, Wang DT, Yan LJ et al (2018) Ultrathin HfO_2-modified carbon nanotube films as efficient polysulfide barriers for Li-S batteries. Carbon 139:896–905
85. Cetinkaya T, Ozcan S, Uysal M et al (2014) Free-standing flexible graphene oxide paper electrode for rechargeable Li–O_2 batteries. J Power Sources 267:140–147
86. Seyma O, Tugrul C, Mahmud T et al (2016) Synthesis of flexible pure graphene papers and utilization as free standing cathodes for lithium-air batteries. Int J Hydrogen Energy 41:9796–9802

87. Zhang WY, Zhu JX, Ang HX et al (2013) Binder-free graphene foams for O_2 electrodes of Li-O_2 batteries. Nanoscale 5:9651–9658
88. Kim DY, Kim M, Kim DW et al (2015) Flexible binder-free graphene paper cathodes for high-performance Li-O_2 batteries. Carbon 93:625–635
89. Zhu TJ, Li XY, Zhang Y et al (2018) Three-dimensional reticular material NiO/Ni-graphene foam as cathode catalyst for high capacity lithium-oxygen battery. J Electroanal Chem 823:73–79
90. Chen Y, Li FJ, Tang DM et al (2013) Multi-walled carbon nanotube papers as binder-free cathodes for large capacity and reversible non-aqueous Li-O_2 batteries. J Mater Chem A 1:13076–13081
91. Etacheri V, Sharon D, Garsuch A et al (2013) Hierarchical activated carbon microfiber (ACM) electrodes for rechargeable Li–O_2 batteries. J Mater Chem A 1:5021–5030
92. Yang Y, Sun Q, Li YS et al (2011) Nanostructured diamond like carbon thin film electrodes for lithium air batteries. J Electrochem Soc 158:B1211–B1216
93. Mitchell RR, Gallant BM, Thompson CV et al (2011) All-carbon-nanofiber electrodes for high-energy rechargeable Li–O_2 batteries. Energy Environ Sci 4:2952–2958
94. Han J, Guo X, Ito Y et al (2016) Effect of chemical doping on cathodic performance of bicontinuous nanoporous graphene for Li-O_2 batteries. Adv Energy Mater 6:1501870
95. Liu M, Sun K, Zhang Q et al (2020) Rationally designed three-dimensional N-doped graphene architecture mounted with Ru nanoclusters as a high-performance air cathode for lithium–oxygen batteries. ACS Sustainable Chem Eng 8:6109–6117
96. Ji DX, Peng SJ, Safanama D et al (2017) Design of 3-dimensional hierarchical architectures of carbon and highly active transition metals (Fe Co, Ni) as bifunctional oxygen catalysts for hybrid lithium-air batteries. Chem Mater 29:1665–1675
97. Xue HR, Mu XW, Tang J et al (2016) A nickel cobaltate nanoparticle-decorated hierarchical porous N-doped carbon nanofiber film as a binder-free self-supported cathode for nonaqueous Li-O_2 batteries. J Mater Chem A 4:9106–9112
98. Yoon KR, Shin K, Park J et al (2018) Brush-like cobalt nitride anchored carbon nanofiber membrane: current collector catalyst integrated cathode for long cycle Li-O_2 batteries. ACS Nano 12:128–139
99. Ji X, Liu DY, Prendiville DG et al (2012) Spatially heterogeneous carbon-fiber papers as surface dendrite-free current collectors for lithium deposition. Nano Today 7:10–20
100. Kim J, Lee J, Yun J et al (2020) Functionality of dual-phase lithium storage in a porous carbon host for lithium-metal anode. Adv Funct Mater 30:1910538
101. Lin D, Liu Y, Liang Z et al (2016) Layered reduced graphene oxide with nanoscale interlayer gaps as a stable host for lithium metal anodes. Nat Nanotechnol 11:626–632
102. Yang Y, Xiong J, Zeng J et al (2018) VGCF 3D conducting host coating on glass fiber filters for lithium metal anodes. Chem Commun 54:1178–1181
103. Raji ARO, Salvatierra RV, Kim ND et al (2017) Lithium batteries with nearly maximum metal storage. ACS Nano 11:6362
104. Huang WH, Qing YJ, Wang NW et al (2018) Novel thermosensitive core-shell surface molecularly imprinted polymers based on SiO_2 for the selective adsorption of sulfamethazine. Materials 11:2067
105. Feng XT, Ashley J, Zhou TC et al (2018) Fluorometric determination of doxycycline based on the use of carbon quantum dots incorporated into a molecularly imprinted polymer. Microchim Acta 185:500
106. Jin S, Sun Z, Guo Y et al (2017) High areal capacity and lithium utilization in anodes made of covalently connected graphite microtubes. Adv Mater 29:1700783
107. Sun C, Wu T, Wang J et al (2018) Favorable lithium deposition behaviors on flexible carbon microtube skeleton enable a high-performance lithium metal anode. J Mater Chem A 6:19159–19166
108. Liu L, Yin YX, Li JY et al (2017) Free-standing hollow carbon fibers as high-capacity containers for stable lithium metal anodes. Joule 1:563–575

109. Nie X, Zhang AY, Liu YH et al (2019) Synthesis of interconnected graphene framework with two-dimensional protective layers for stable lithium metal anodes. Energy Storage Mater 17:341–348
110. Chen L, Chen H, Wang Z et al (2019) Self-supporting lithiophilic N-doped carbon rod array for dendrite-free lithium metal anode. Chem Eng J 363:270–277
111. Wu CJ, He J, Li YY et al (2018) Solid-phase extraction of aflatoxins using a nanosorbent consisting of a magnetized nanoporous carbon core coated with a molecularly imprinted polymer. Microchim Acta 185:515
112. Yan K, Lu Z, Lee H-W et al (2016) Selective deposition and stable encapsulation of lithium through heterogeneous seeded growth. Nat Energy 1:16010
113. Liang Z, Lin DC, Zhao J et al (2016) Composite lithium metal anode by melt infusion of lithium into a 3D conducting scaffold with lithiophilic coating. P Natl Acad Sci USA 113:2862–2867
114. Zhang R, Chen X, Shen X et al (2018) Coralloid carbon fiber-based composite lithium anode for robust lithium metal batteries. Joule 2:764–777
115. Yang T, Li L, Wu F et al (2020) A soft lithiophilic graphene aerogel for stable lithium metal anode. Adv Funct Mater 30:2002013
116. Yang CP, Yao YG, He SM et al (2017) Ultrafine silver nanoparticles for seeded lithium deposition toward stable lithium metal anode. Adv Mater 29:1702714
117. Sun Q, Zhai W, Hou G et al (2018) In situ synthesis of a lithiophilic Ag-nanoparticles-decorated 3D porous carbon framework toward dendrite-free lithium metal anodes. ACS Sustain Chem Eng 6:15219–15227
118. Zhai P, Wang T, Yang W et al (2019) Uniform lithium deposition assisted by single-atom doping toward high-performance lithium metal anodes. Adv Energy Mater 9:1804019
119. Zhou T, Shen J, Wang Z et al (2020) Regulating lithium nucleation and deposition via MOF-derived Co@C-modified carbon cloth for stable Li metal anode. Adv Funct Mater 30:1909159
120. Li S, Liu Q, Zhou J et al (2019) Hierarchical CO_3O_4 nanofiber–carbon sheet skeleton with superior Na/Li-Philic property enabling highly stable alkali metal batteries. Adv Funct Mater 29:1808847
121. Deng W, Zhu W, Zhou X et al (2018) Highly reversible Li plating confined in three-dimensional interconnected microchannels toward high-rate and stable metallic lithium anodes. ACS Appl Mater Interfaces 10:20387–20395
122. Xie J, Wang J, Ryoung LH et al (2018) Engineering stable interfaces for three-dimensional lithium metal anodes. Sci Adv 4:eaat5168
123. Yue X, Li X, Wang W et al (2019) Wettable carbon felt framework for high loading Li-metal composite anode. Nano Energy 60:257–266
124. Baozhi Y, Tao T, Srikanth M et al (2018) Nanoflake arrays of lithiophilic metal oxides for the ultra-stable anodes of lithium-metal batteries. Adv Funct Mater 28:1803023
125. Wu K, Zhao BL, Yang CK et al (2020) $ZnCO_2O_4$/ZnO induced lithium deposition in multi-scaled carbon/nickel frameworks for dendrite-free lithium metal anode. J Energy Chem 43:16–23
126. Lin K, Qin X, Liu M et al (2019) Ultrafine titanium nitride sheath decorated carbon nanofiber network enabling stable lithium metal anodes. Adv Funct Mater 29:1903229
127. Luo L, Li J, Asl HY et al (2019) A 3D lithiophilic MO_2 N-modified carbon nanofiber architecture for dendrite-free lithium-metal anodes in a full cell. Adv Mater 31:1904537
128. Xie J, Wang J, Lee HR et al (2018) Engineering stable interfaces for three-dimensional lithium metal anodes. Sci Adv 4:eaat5168
129. Wang X, Pan Z, Wu Y et al (2019) Infiltrating lithium into carbon cloth decorated with zinc oxide arrays for dendrite-free lithium metal anode. Nano Res 12:525–529
130. Jin C, Sheng O, Luo J et al (2017) 3D lithium metal embedded within lithiophilic porous matrix for stable lithium metal batteries. Nano Energy 37:177–186

Chapter 7
Application in Catalysis

A catalyst is a substance that can alter the speed of a chemical reaction (either increasing or decreasing it) without undergoing any chemical changes itself or affecting the chemical equilibrium. It plays a vital role in various industrial processes, including chemical, petrochemical, biochemical, and environmental protection industries. Catalysts are extensively utilized, accounting for approximately 90% of industrial processes, due to their ability to enhance reaction rates and efficiency while maintaining their own mass and chemical properties.

The continuous advancements in catalyst research and development have a profound impact not only within the field itself but also on society as a whole. This has made catalyst research, development, and application an essential and highly relevant topic for researchers worldwide. A notable example that exemplifies the significance of catalysts is the groundbreaking work of German chemist Haber in 1909. Haber's use of iron as a catalyst enabled the direct synthesis of ammonia from nitrogen and hydrogen, marking a significant milestone in artificial nitrogen fixation technology. The development of synthetic ammonia holds immense importance on multiple fronts. Firstly, it revolutionized nitrogen fertilizer production, providing a cost-effective alternative to relying solely on natural sources and contributing to the advancement of global agriculture. Secondly, synthetic ammonia served as a crucial raw material for the production of nitric acid and explosives, playing a pivotal role in industrial and military applications. Moreover, this breakthrough in chemical production stimulated the development of related theories and propelled advancements in high-temperature, high-pressure reactions, catalysis, and other associated technologies.

Catalysts play a crucial role in the field of energy by facilitating reactions and promoting efficient utilization and development of energy resources. In catalyst design, a key objective is to maximize the specific surface area to provide a greater number of active sites for enhanced catalytic performance. This is exemplified by the comparison between a solid sphere and a hollow spherical shell of the same diameter, where the latter offers a larger catalytic area and is naturally more favorable for

B. Kong et al., *Functional Mesoporous Carbon-Based Film Devices for Energy Systems*,
https://doi.org/10.1007/978-981-99-7498-6_7

reactions. Thus, designing catalysts with a large specific surface area becomes a critical factor influencing catalytic performance. Various methods, such as morphology control and core–shell structure design, can be employed to effectively increase the specific surface area, with mesoporous materials being particularly advantageous in this regard. Additionally, the cost of catalysts must be considered during the design process. For instance, platinum, a precious metal, is often used as a catalyst in hydrogen evolution reactions. Despite its excellent performance, the high price and limited availability of platinum pose challenges for large-scale applications. It becomes imperative to identify suitable alternatives that reduce costs while delivering superior hydrogen energy performance. Carbon-based materials have garnered significant attention due to their unique physical and chemical properties, accessibility, affordability, and other favorable attributes. Furthermore, powder catalysts are prone to detachment from reaction surfaces, exhibit poor mechanical stability, and pose challenges for industrial use. Hence, the direct design of thin film catalyst materials becomes desirable to enhance stability and facilitate practical applications. It is important to note that the development of efficient catalysts requires comprehensive consideration of various factors. Researchers must take into account all relevant aspects to design energy catalytic materials that exhibit excellent performance, cost-effectiveness, and environmental friendliness.

Apart from catalyst design, understanding the catalytic process mechanism is equally crucial. Investigating the catalytic mechanism enables us to gain deeper insights into the working principles behind catalysts' exceptional performance and provides valuable guidance for catalyst design. Currently, the study of catalytic mechanisms relies heavily on two key approaches: characterization methods and theoretical calculations. By effectively utilizing these tools, we can delve into the intricacies of the microscopic aspects and unravel the secrets hidden within. Characterization methods allow us to analyze the catalyst's structure, composition, and surface properties, providing valuable experimental data. On the other hand, theoretical calculations utilize computational models to simulate and predict the behavior and interactions of catalysts at the atomic and molecular levels. By combining these approaches, we can uncover the underlying mechanisms driving catalytic reactions and pave the way for further advancements in catalyst design and optimization.

In this chapter, our attention turns towards exploring the utilization of functionalized mesoporous carbon film materials for energy catalysis applications. Specifically, we will delve into various catalytic reactions, including hydrogen evolution, oxygen evolution, oxygen reduction, carbon dioxide reduction, and nitrogen reduction. Throughout this chapter, we aim to shed light on the underlying mechanisms of these reactions, highlight the challenges that must be overcome, and present the role of functionalized mesoporous carbon-based thin films in catalyzing these processes. By delving into the intricacies of these reactions and the associated catalysts, we hope to equip readers and researchers with a comprehensive understanding and spark inspiration for their future endeavors in this field.

7.1 Hydrogen Evolution Reaction

Hydrogen evolution reaction is one of the main reactions of water decomposition and is considered to be an efficient way to produce hydrogen on a large scale. It paves a new avenue for the preparation of clean energy.

7.1.1 Introduction

Energy crisis and environmental pollution pose significant challenges to countries worldwide, demanding the pursuit of green and sustainable solutions. As an alternative to conventional fossil fuels, clean and renewable energy sources are crucial for addressing these issues. Among the contenders, hydrogen energy emerges as a highly competitive energy carrier due to its exceptional attributes such as high specific energy, emission-free combustion, and recyclability [1]. However, a pressing concern remains: how can we achieve cost-effective and large-scale hydrogen production in a manner that is convenient, efficient, and environmentally friendly? Electrochemical water splitting has emerged as a popular method in recent decades to address this challenge. It enables the extraction of hydrogen from abundant water resources without polluting the environment. Moreover, it offers the advantage of seamless integration with other surplus or intermittent energy sources, such as wind or water-generated electricity. This promising approach holds the potential to revolutionize hydrogen production and meet the increasing global demand.

The hydrogen evolution reaction (HER) plays a crucial role in water splitting, yet it faces a significant challenge due to its thermodynamic uphill nature. It demands a substantial overpotential to drive the reaction, resulting in high energy consumption. To overcome this obstacle, catalytic materials based on precious metal platinum have been employed to lower the energy barrier and enhance the overall rate of the HER process. However, as previously mentioned, the high cost and limited availability of platinum hinder its widespread utilization. Therefore, it is essential to explore alternative HER catalysts that offer excellent performance while being environmentally friendly and cost-effective [2]. Finding such catalysts is of utmost importance to advance the field of water splitting and enable the large-scale production of hydrogen.

Significant progress has been made in the development of electrocatalysts for the hydrogen evolution reaction (HER). In this section, we will explore various aspects of HER, starting with an overview of the mechanism and fundamental concepts. We will then delve into the evaluation systems used for HER electrocatalysts, including a range of experimental measurement techniques. Furthermore, we will focus on recent advancements in HER catalysts, highlighting the utilization of mesoporous materials, carbon-based materials, thin films, and other innovative catalysts. We will examine the factors contributing to their remarkable catalytic performance, such as their unique properties and underlying theories. A comprehensive analysis of the relationships between catalytic activity, micromorphology, structure, composition, and synthetic methods will be provided. Finally, we will discuss the challenges, prospects, and future research directions for HER-related catalysts in the quest for sustainable and clean hydrogen energy.

7.1.2 Fundamentals of the HER

To gain a deeper understanding of the hydrogen evolution reaction (HER), it is crucial to explore its underlying reaction mechanism [3]. HER typically involves two distinct steps. The first step, known as the Volmer reaction, involves the addition of electrons to H_2O (in alkaline media) or H_3O^+ (in acidic media) to form a hydrogen atom on the active site of the electrode material surface (M). The second step comprises the actual formation of hydrogen, which can occur through two mechanisms. The first mechanism, known as the Heyrovsky reaction, involves the electrochemical desorption of hydrogen. The other mechanism, referred to as the Tafel reaction, involves the chemical desorption of hydrogen. Both reaction paths, Volmer-Heyrovsky and Volmer-Tafel, can take place in both acidic and alkaline solutions, with their occurrence dependent on the concentration difference of corresponding ions in the electrolyte, primarily influenced by hydrogen ions. To visualize these reaction paths, refer to Fig. 7.1

(1) electrochemical hydrogen adsorption (Volmer reaction) (7.1 and 7.2)

$$H_3O^+ + M + e^- \rightleftharpoons M{-}H^* + H_2O\,(\text{acidic medium}) \tag{7.1}$$

$$H_2O + M + e^- \rightleftharpoons M{-}H^* + OH^-(\text{alkaline medium}) \tag{7.2}$$

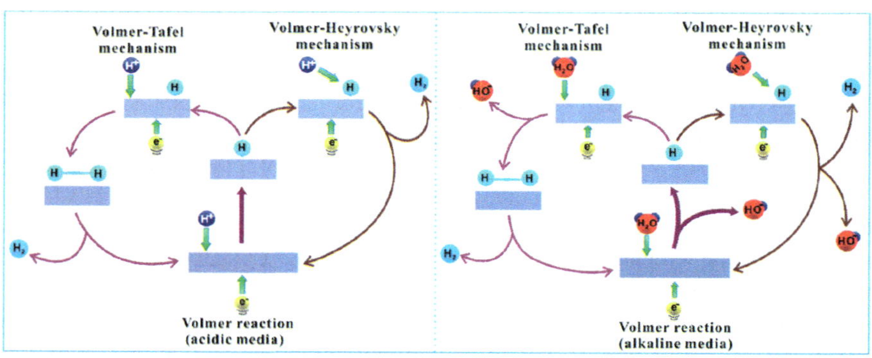

Fig. 7.1 Mechanism of hydrogen evolution on the surface of an electrode in acidic (left) and alkaline (right) solutions. Reproduced with permission. Copyright 2020 The Royal Society of Chemistry

(2) electrochemical desorption (Heyrovsky reaction) (7.3 and 7.4)

$$H_3O^+ + M{-}H^* + e^- \rightleftharpoons H_2 + M + H_2O \text{ (acidic medium)} \tag{7.3}$$

$$H_2O + M{-}H^* + e^- \rightleftharpoons H_2 + M + OH^- \text{ (alkaline medium)} \tag{7.4}$$

(3) chemical desorption (Tafel reaction) (Eq. 7.5)

$$2M{-}H^* \rightleftharpoons H_2 + 2M \text{(both acidic and alkaline mediums)} \tag{7.5}$$

The Tafel slope (b) parameter is used to measure the potential needed to increase or decrease the current density by 10 times, which can be used to reflect the specific mechanism in HER process. There are three specific cases here. Under the condition of the Volmer reaction is fast, if electrochemical desorption (Heyrovsky reaction) is the rate-determining step, then b should be 39 mV dec^{-1} at 25 °C, shown as following equation.

$$b = \frac{4.6\,RT}{3F} = 0.039\,V\,dec^{-1} \tag{7.6}$$

And when the chemical desorption (Tafel reaction) is the rate- determining step, then b should be 29 mV dec^{-1} (25 °C), which is deduced from the following equation.

$$b = \frac{2.3\,RT}{2F} = 0.029\,V\,dec^{-1} \tag{7.7}$$

If the Volmer reaction is a slow rate-determining step, then b should be 116 mV dec^{-1} (25 °C) derived from the following equation.

$$b = \frac{4.6\,RT}{F} = 0.116\,V\,dec^{-1} \tag{7.8}$$

Therefore, in the actual research work, the specific mechanism of HER process can be roughly judged by calculating the Tafel slope.

In catalytic reactions, there is a Sabatier principle, which means that the interaction between the catalyst and the reacting species is neither too strong nor too weak. If the interaction is too strong, the reaction product is difficult to desorption; and if the interaction is too weak, it is hard for the reaction species to bond to the catalyst. Under the guidance of this theory, we can make use of the density functional theory (DFT) to calculate the adsorption energy of H* and desorption energy of H_2 (ΔG_{H^*}), which can reflect the interaction on the surface of the catalyst with reaction species. Obviously, under ideal situation, ΔG_{H^*} should be close to zero, so that we can have the highest exchange current density (j_0). Based on this, the "volcano plot" becomes a very useful tool for choosing excellent HER catalysts [4]. In general, the trend of the "volcano plot" is identified based on the correlation between j_0 of practical

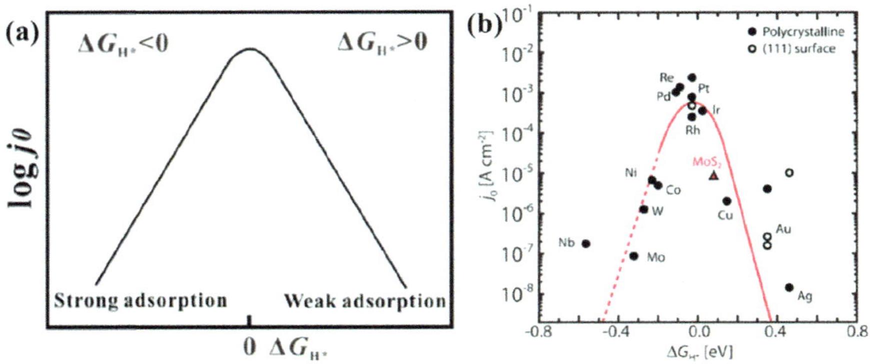

Fig. 7.2 **a** Relationship between j_0 and ΔG_{H^*} under the assumption of a Langmuir adsorption model. Reproduced with permission. Copyright 1958 Royal Society of Chemistry. **b** Dependence of j_0 on ΔG_{H^*} for HER on the surface of various metals, alloy compounds, and nonmetallic materials in acidic medium. Reproduced with permission. Copyright 2017 American Association for the Advancement of Science

experiments and ΔG_{H^*} from corresponding DFT calculations. As shown in Fig. 7.2, the peak of volcano plot is located at $\Delta G_{H^*} = 0$. $\Delta G_{H^*} > 0$ means adsorption of H* is relatively weak, resulting in exponential decrease of j_0 with the increasing value of ΔG_{H^*}. Conversely, at the region of $\Delta G_{H^*} < 0$, which means that the desorption of H_2 is relatively difficult, j_0 will decrease when the value of ΔG_{H^*} decrease [5].

7.1.3 Assessment of the HER Activity and Electrocatalyst

(1) Overpotential

In electrochemistry, an overpotential is the potential difference (voltage) of a half-reaction between the redox potential determined by thermodynamically and the redox reaction potential observed experimentally. In an electrolytic cell, the presence of overpotential means that the battery needs more energy to drive the reaction than thermodynamics expects. Overpotential can be divided into activation overpotential, concentration overpotential and resistance overpotential. According to the Nernst equation, under standard conditions, the equilibrium potential of the HER process (as compared to the reversible hydrogen electrode (RHE)) is zero. But in the actual experiment, we need a more negative potential (i.e., a large overpotential) to overcome the kinetic barrier and drive the HER. The difference between the equilibrium potential and the actual required potential is the overpotential corresponding to HER electrocatalyst. Therefore, the actual applied voltage can be expressed by the following formula:

$$E = E_{HER} + \eta. \tag{7.9}$$

At the same time, there are inevitable resistances in the whole system, which are the impedance from the electrolyte solution, the internal resistance of the electrocatalyst, and the contact resistance in the electrochemical system. The presence of these resistance will introduce ohmic potential drops and therefore the potential-current density needs to be corrected/compensated. Therefore, the expression of the actual potential should be as follows:

$$E = E_{HER} + iR + \eta. \tag{7.10}$$

In general, as shown in Fig. 7.3, we compare the performance of different electrocatalysts by overpotentials at three specific current densities, 1 (η_1), 10 (η_{10}) and 100 (η_{100}) mA cm^{-2} [6]. Among them, the η_{10} is most commonly used for performance comparisons because this current density corresponds to the 12.3% efficiency of solar water splitting device and has more practical comparative significance. Obviously, a smaller overpotential of η_{10} means better performance (at the same active material load and electrode area). The other two, η_1, is also called onset overpotential, which reflects the potential where HER starts with. The η_{100} is used to measure whether the catalyst performs well at high current density. If it performs well at high current, it means the catalyst is suitable for industrial application.

(2) Tafel plot

Tafel formula is the experience formula put forward by Tafel in 1905, its mathematical expression is:

$$\eta = a + b \lg I \tag{7.11}$$

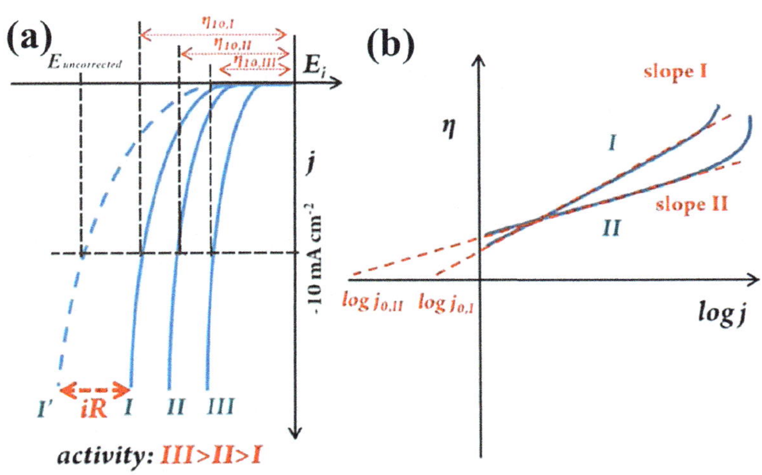

Fig. 7.3 **a** Schematic HER polarization curves on different electrocatalysts with iR correction and overpotentials indicated. **b** Schematic Tafel plots on different electrocatalysts with the Tafel slopes and exchange current densities indicated. Reproduced with permission. Copyright 2019 Wiley–VCH

where overpotential (η) and current density (I) are both the absolute value. There are two constants "a" and "b". The "a" indicates the overpotential value at current density for the unit (1 A/cm^{-2}). The value of "a" is related with the nature of the electrode material, electrode surface state and solution composition and temperature factors. According to the value of "a", we can compare whether it is easy for the electron transfer process of different electrode system. As for the "b", usually called Tafel slope, is used to measure the potential needed to increase or decrease the current density by 10 times. In other words, the smaller of "b" means the smaller overpotential required when the current density increases tenfold, and the better the performance. Therefore, Tafel slope is a key comparison parameter in electrochemical measurement.

(3) Electrochemical Impedance Spectroscopy

Electrochemical impedance spectroscopy (EIS) is to apply a different small amplitude frequency ac signal to the electrochemical system places, measure the ratio of ac signal voltage and current (the ratio is the impedance of the system) along with the change of sine wave frequency ω, or measure the impedance phase angle Φ along with the change of ω. Furthermore, the electrode process dynamics, double layer and diffusion are analyzed to study the mechanism of electrode materials and solid electrolyte and so on.

The electrochemical system is regarded as an equivalent circuit (Fig. 7.4), which is composed of the basic elements such as resistance (R), capacitance (C) and inductance (L) combined in series and parallel in different ways [7]. Through EIS, the equivalent circuit composition and the value of each element can be measured, and the electrochemical meaning of these elements can be used to analyze the structure of the electrochemical system and the nature of the electrode process, etc. The low frequency region depends on Warburg impedance, which is a characteristic of a diffused controlled electrode process. In the high frequency region, charge transfer resistance (R_{ct}) and dual layer capacitance will become important components. If the chemical system dynamics is slow, it will show a large R_{ct}, possibly showing a very limited frequency region where substance transfer is important. R_{ct} is related to the interface charge-transfer process of an electrode. Regarding HER, a smaller R_{ct} value suggests a faster reaction rate, leading to a smaller overpotential. Notably, EIS spectra should be recorded within the hydrogen evolution voltage (usually at η_{10}). In addition, vigorous turbulence such as drastically rotating or stirring should be avoided during the EIS measurements.

(4) Stability

Stability is another important indicator for the practical application of electrocatalysts. It determines whether the electrocatalysts can be used for a long time with high efficiency. There are generally three methods to measure the stability of a catalyst: linear sweep voltammetry (LSV), galvanostatic test, or potentiostatic test. LSV method is to compare the changes of LSV curve before and after a certain number of cycles (such as 5000 times). If the LSV curve is basically unchanged, then the

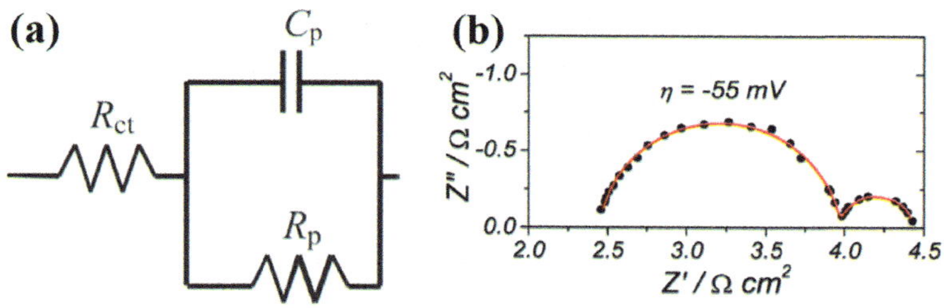

Fig. 7.4 **a** Electrical equivalent model of the Faradaic impedance of HER reaction. **b** Example of the complex plane plots obtained at the hanging meniscus Pd RDE in 0.1MH_2SO_4 saturated with H_2; points – experimental, lines – fits to the model of the HER. Reproduced with permission. Copyright 2017 Elsevier Inc.

material has excellent stability. As for potentiostatic test and galvanostatic test, in which a controlled voltage or current is applied to the system and performed over a period of time such as three days, we usually observe the corresponding current or voltage change. Taking potentiostatic as an example, we applied a voltage with current density of 10 mA cm^{-2}, and then observed the change of current density. If the current density was maintained at about 10 mA cm^{-2} after three days, it proved that the catalyst had excellent stability. On the contrary, if the current density drops rapidly, the catalyst is not stable.

(5) Faradic efficiency

The Faradaic efficiency represents the utilization efficiency of electrons related to the electrochemical reaction, and it can be calculated as the ratio of the experimental amount of H_2 produced to the theoretical value. Specifically, we can use a fluorescence sensor or volumetric method to obtain the experimental amount of H_2 produced under a constant oxidation current (I) within a certain time (t). On the other hand, the theoretical amount of H_2 production (n_{H2}) can be calculated on the basis of the following equation, where F is the Faraday constant.

$$n_{H2} = \frac{It}{2F} \tag{7.12}$$

It is clear that we aim for the highest Faraday efficiency (usually higher than 99%).

(6) Turnover frequency

In 1968, the turnover frequency (TOF) was first defined by Michel Boudart to evaluate the catalytic rate of catalysts, that is, the number of molecules reacting at each site per second. It is an indicator of the intrinsic catalytic activity of each catalytic site. The TOF value is usually used to evaluate the activity of HER, which can be determined by the following equation:

$$\mathrm{TOF} = \frac{J \times A}{2 \times F \times n} \tag{7.13}$$

where J is to measure the current density at A given overvoltage, A represents the area of working electrode, n is the number of moles of the active catalyst, and F is Faraday's constant.

(7) Electrochemical active surface area

The electrochemical active surface area (ECSA) is a key parameter of various electrochemical systems. Measuring the ECSA allows comparing the specific catalytic activity of different catalyst materials and investigating the catalyst utilization and degradation in different conditions and electrode architectures. Because the ECSA is proportional to the electrochemical double-layer capacitance (C_{dl}), a popular method to evaluate the ECSA of a catalyst is based on the following equation

$$\mathrm{ECSA} = \frac{C_{dl}}{C_S} \tag{7.14}$$

in which C_s represents the specific capacitance. The C_{dl} value can be extracted from CV curves at different scan rates. Specifically, current responses within a certain potential window without the faradaic processes are recorded with increasing scan rates usually from 4 to 100 mV s^{-1}. By plotting $\Delta j/2$ (Δj represents the difference between the anodic and the cathodic current density at a certain potential) as a function of the scan rate, we can obtain the slope, which corresponds to the C_{dl} value. Notably, because the nature of the real electrochemical surface area of the electrocatalysts is difficult to accurately determine, the ECSA is sometimes used only as an indirect estimation for comparisons.

(8) Density functional theory calculation for HER

As mentioned earlier, an ideal electrocatalyst should exhibit neither too strong nor too weak interaction with the material. Under this principle, then ΔG_{H*} should be 0 or very close to zero. Usually, we use the density functional theory (DFT) to calculate ΔG_{H*} (Fig. 7.5a) [8]. By constructing the appropriate catalyst model and the model of H* adsorption at the active site on the catalyst surface, the ΔG_{H*} value can be obtained by the equation:

$$\Delta G_{H^*} = \Delta E_{H^*} + \Delta E_{ZPE} - T\Delta S \tag{7.15}$$

in which ΔE_{H*}, ΔE_{ZPE} and ΔS represent the binding energy of catalyst surface with H*, the zero-point energy change, And entropy change during the adsorption of H*, respectively. Particularly, the value ΔE_{H*} can be achieved based on

$$\Delta E_{H^*} = E(\mathrm{surface} + H^*) - E(\mathrm{surface}) - 1/2\, E_{H2} \tag{7.16}$$

While both ΔE_{ZPE} and ΔS are correlated with the computed vibrational frequency.

In 1995, Norskov and his collaborators pioneered the D-band central theory to explain the interaction between transition metals and atoms (Fig. 7.5b) [9]. In the

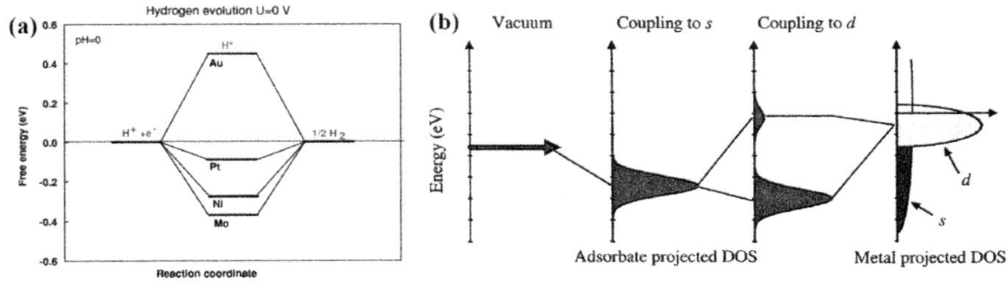

Fig. 7.5 **a** Free energy diagram for hydrogen evolution at equilibrium (U = 0). The data are for "standard" conditions corresponding to 1 bar of H_2 and pH = 0 at 300 K. Reproduced with permission. Copyright 2005 Electrochemical Society, Inc. **b** Schematic illustration of the formation of a chemical bond between an adsorbate valence level and the s and d states of a transitionmetal surface. Reproduced with permission. Copyright 2005, Springer

past few decades, this theory has been widely used in the field of catalysis, becoming one of the most influential theories. In HER process, the D-band center theory is sometimes used to explain the H* adsorption process, and sometimes used to explain the desorption process. When the H* orbital and the d orbital of the metal have interaction, they will split into bonding orbital and anti-bonding orbital, and the strength of the H* interaction of the metal (M-H) mainly depends on the anti-bonding orbital. As the center of the D-band gets higher, the antibonding orbital will be pushed up or even higher than the Fermi level, which makes the adsorption more stable. Therefore, in the case of ΔG_{H*} less than zero, considering desorption process of HER, simply put, if d-band center is lower, the desorption process is easier to take place, which is more favorable for HER process.

7.1.4 Mesoporous Material Electrocatalysts for HER

Researchers have long aimed to enhance the catalytic performance of the hydrogen evolution reaction (HER) by focusing on the design of catalysts with increased specific surface area. In this section, we will explore the utilization of mesoporous materials as catalysts to advance the performance of HER. Specifically, we will examine various types of mesoporous materials, including mesoporous carbon-based materials and mesoporous carbon-based film materials, among others. By delving into the application of mesoporous materials in the HER process, we aim to provide readers with valuable insights and guidance for the design and utilization of mesoporous materials in HER applications.

Transition metal compounds, including transition metal oxides, nitrides, phosphates, and sulfides, have garnered significant attention in research endeavors. Over the past decades, metal chalcogenides such as MoS_2, WS_2, FeS_2, and $CoSe_2$ have been extensively explored as electrocatalysts for the hydrogen evolution reaction (HER). However, unmodified iron sulfide materials have exhibited limited HER performance, primarily due to their low surface area and scarcity of active sites. Achieving high-surface-area metal sulfides poses a significant challenge, given the volume contraction and low electronic affinity between metal and sulfur. To overcome these limitations, the utilization of mesoporous materials has emerged as a promising approach in electrochemical water splitting catalysis. Mesoporous materials offer abundant accessible mesopores and exposure of active sites, facilitating efficient charge transfer and product diffusion. In this context, Suib and colleagues successfully synthesized mesoporous pyrite FeS_2 nanoparticles using a straightforward two-step synthetic protocol. This approach involved an inverse micelle sol–gel method followed by a low-temperature sulfurization treatment [10].

The X-ray diffraction (XRD) pattern confirmed the crystalline structure of pyrite FeS_2. Figure 7.6a displays the diffraction pattern of the mesoporous FeS_2, which matches the cubic phase of pyrite FeS_2 (JCPDS No. 421340). This indicates the successful conversion of Fe_2O_3 into FeS_2. The presence of the mesoporous structure in the FeS_2 material contributes to its lower crystallinity, as it disrupts the long-range structural order. Nitrogen sorption measurements were conducted to assess the pore size distribution and surface area. The isotherm of the mesoporous FeS_2 exhibits a type-IV adsorption isotherm (Fig. 7.6b), confirming the presence of a mesoporous structure according to the IUPAC classification. The calculated Brunauer–Emmett–Teller (BET) surface area and total pore volume of the mesoporous FeS_2 are 128 m^2 g^{-1} and 0.23 cm^3 g^{-1}, respectively. The pore size distribution, obtained using the Barrett–Joyner–Halenda (BJH) method (inset in Fig. 7.6b), shows a monomodal pore size of 3.4 nm for the synthesized FeS_2. In contrast, the isotherm of the commercial FeS_2 indicates a nonporous structure with a surface area less than 1 m^2 g^{-1}. Transmission electron microscopy (TEM) was employed to further illustrate the mesoporous structure and crystallinity. Figures 7.6c–e reveal the loose internal structure of FeS_2 spheres, confirming the presence of mesopores within the range of 4–7 nm in size. The crystallite sizes of the mesoporous FeS_2 spheres are estimated to be 15–20 nm, significantly smaller than those of commercial FeS_2 (103 nm). The corresponding element mapping images demonstrate the homogeneous distribution of Fe and S throughout the FeS_2 material (Fig. 7.6f–i).

The electrocatalytic performance of mesoporous FeS_2 in hydrogen evolution reaction (HER) was extensively studied using a three-electrode system in alkaline media (pH = 13). To provide a comprehensive comparison, the HER activities of commercial FeS_2, Pt/C (20%), and bare Ni foam were also evaluated. As shown in Fig. 7.7a, the mesoporous FeS_2 coated electrode demonstrated remarkable HER activity with a small overpotential of 96 mV to achieve a current density of 10 mA cm^{-2}, which is comparable to most reported metal chalcogenide HER catalysts in alkaline media. In contrast, bare Ni foam exhibited negligible HER activity, and the commercial FeS_2 required a large overpotential to attain a 10 mA cm^{-2} current density. Figure 7.7b indicates a calculated Tafel slope of approximately 78 mV dec^{-1} for mesoporous FeS_2, suggesting a fast Volmer–Heyrovsky mechanism. Electrochemical impedance spectroscopy (EIS) was utilized to investigate the electrode kinetics under HER operating conditions (Fig. 7.7c). The Nyquist plot showed a smaller charge transfer resistance (Rct) of mesoporous FeS_2 (7 Ω) compared to commercial FeS_2 (238 Ω).

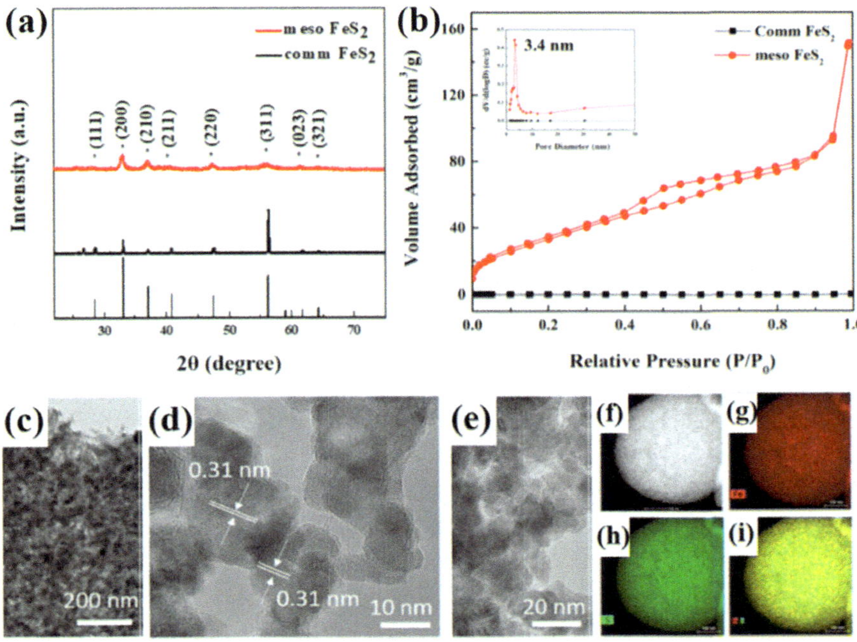

Fig. 7.6 **a** XRD patterns of meso FeS_2, comm FeS_2, and the JCPDS pattern for standard pyrite. **b** N_2 sorption isotherms and the corresponding pore size distribution of commercial FeS_2 and mesoporous FeS_2. **c–e** low- and high magnification TEM images, and **f** high-angle annular dark-field TEM image of meso FeS_2. **g–i** EDX elemental mapping images of Fe, S, and the merged image of Fe and S for the meso FeS_2. Reproduced with permission. Copyright 2017 American Chemical Society

The low Rct value signifies more favorable HER kinetics on the mesoporous FeS_2 electrode, highlighting the benefits of mesoporosity in terms of providing active sites and facilitating charge and mass transfer efficiency. The superior durability of the mesoporous FeS_2 material was confirmed by the chronoamperometric curve in Fig. 7.7d, which exhibited a sustained initial current density without significant decrease over 24 h of continuous operation at a constant overpotential of 100 mV. This indicates the excellent stability of the prepared mesoporous FeS_2. Given its exceptional electrocatalytic performance in HER, the mesoporous FeS_2 materials presented in this study, with their high surface area, hold promise as alternatives to Pt-based HER catalysts, thereby advancing the development of electrocatalytic water splitting.

Transition metal selenides (TMSs) have emerged as promising candidates for electrocatalysis due to their lower intrinsic electrical resistivity, enabling efficient charge transfer during the electrocatalytic process. Various nanostructures of TMSs, such as $CoSe_2$ nanoparticles, NiSe nanowires, and Ni_3Se_2 nanoforest, have demonstrated respectable hydrogen evolution reaction (HER) activities. However, there is still room

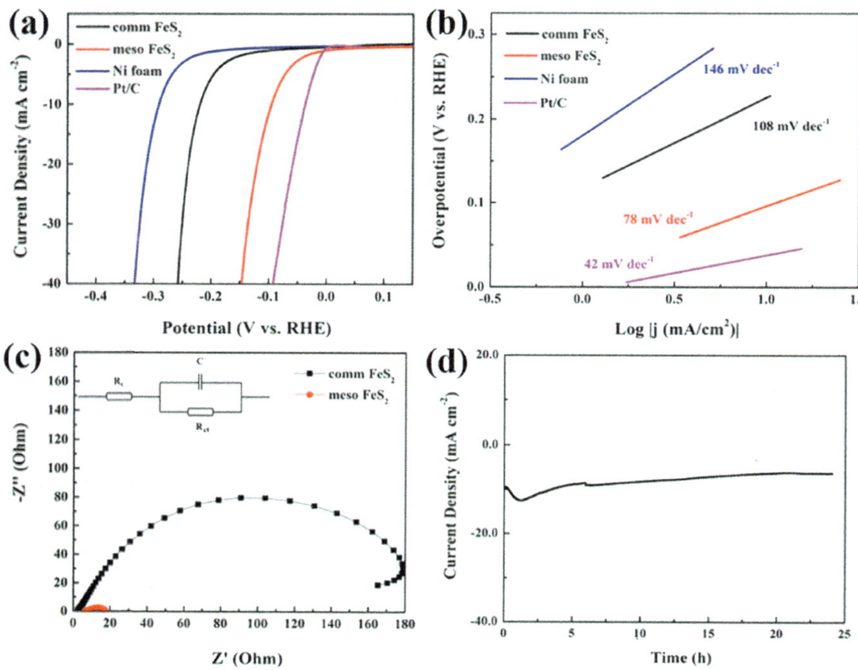

Fig. 7.7 HER performance of the mesoporous FeS_2 in alkaline media (pH = 13). **a** Polarization curves and **b** Tafel plots of the commercial FeS_2, mesoporous FeS_2, 20% Pt/C, and the bare Ni foam in 0.1 M KOH at a scan rate of 10 mV/s. **c** Nyquist plots of commercial FeS_2 and mesoporous FeS_2 at an overpotential of 200 mV. The inset shows the fitted equivalent circuit. **d** Chronoamperometry curve of the mesoporous FeS_2 at a constant overpotential of 100 mV. All the materials were loaded onto Ni foam at a mass loading of 0.53 mg cm^{-2}. Reproduced with permission. Copyright 2017 American Chemical Society

for improvement in their catalytic performance, including hydrogen production yield, overpotential for hydrogen evolution, and stability, in order to compete with noble metal catalysts. Additionally, the limited applicability of these electrocatalysts to specific pH ranges, either acidic or basic, hampers their utilization in different water electrolysis technologies with diverse electrolyte pH requirements. To address these challenges, Zhang et al. devised a novel approach to synthesize 3D mesoporous nanosheet networks of $Ni_{1-x}Co_xSe_2$ on Ni foam (denoted as $Ni_{1-x}Co_xSe_2$ MNSN/NF, where x ranges from 0 to 0.35). This was achieved through topological transformation of NiCo layered double hydroxide (LDH) nanosheet arrays followed by acid etching. By investigating the influence of $Ni_{1-x}Co_xSe_2$ stoichiometry on the HER performance, they discovered that $Ni_{0.89}Co_{0.11}Se_2$ MNSN/NF exhibited the highest catalytic activity and remarkable stability over a wide pH range.

To synthesize the desired nanosheet arrays, a hydrothermal method was employed, utilizing cobalt and nickel precursors at various concentrations in a methanol and water mixture. The resulting NiCo-LDH nanosheet arrays exhibited similar morphology regardless of the Co/Ni feeding ratios, as observed in the scanning electron microscopy (SEM) images shown in. The vertically aligned nanosheets, with a thickness of approximately 30 nm, were uniformly interconnected on the Ni foam skeleton, denoted as NiCo-LDH NSA/NF. Subsequent selenization led to a reduction in nanosheet thickness to around 24 nm while preserving the array structure. Acid treatment in a 0.1 M hydrochloric acid aqueous solution caused roughening and distortion of the nanosheets, resulting in a further reduction in thickness to approximately 12 nm, as shown in. Remarkably, the acid etching process resulted in the formation of numerous pores with sizes ranging from several to tens of nanometers. This formation of mesoporous $Ni_{0.89}Co_{0.11}Se_2$ nanosheets was attributed to the preferential dissolution of residual LDH ingredients in the acid aqueous solution. N_2 adsorption–desorption analysis confirmed the porous nature of the $Ni_{0.89}Co_{0.11}Se_2$ nanosheets, as indicated by the type IV isotherm and hysteresis loop observed. The specific surface area of the nanosheets was measured to be 37.32 $m^2\ g^{-1}$, with pore sizes ranging from 2 to 20 nm and an average pore size of 4.8 nm.

The catalytic performance of $Ni_{1-x}Co_xSe_2$ MNSN/NF integrated electrocatalysts with different Co/Ni ratios ($x = 0$, 0.07, 0.11, 0.20, and 0.35) was investigated using linear sweep voltammetry (LSV) curves. It was observed that the stoichiometry of $Ni_{1-x}Co_xSe_2$ played a crucial role in determining the catalytic performance. Among the different compositions, $Ni_{0.89}Co_{0.11}Se_2$ MNSN/NF exhibited the lowest overpotential (η_{10}) required to deliver a current density of -10 mA cm^{-2}. Based on this finding, the subsequent study focused on $Ni_{0.89}Co_{0.11}Se_2$ MNSN/NF. To compare its performance, bare Ni foam, NiCo-LDH NSA/NF, $NiSe_2$ MNSN/NF, $Ni_{0.89}Co_{0.11}Se_2$ nonporous film/NF ($Ni_{0.89}Co_{0.11}Se_2$ NPF/NF), and commercial Pt/C (20 wt% Pt/XC-72) were also evaluated. $Ni_{0.89}Co_{0.11}Se_2$ MNSN/NF exhibited a remarkably low η_{10} of 85 mV, significantly outperforming $NiSe_2$ MNSN/NF (143 mV), $Ni_{0.89}Co_{0.11}Se_2$ NPF/NF (206 mV), and NiCo-LDH NSA/NF (245 mV) with $Ni_{0.89}Co_{0.11}Se_2$ MNSN/NF displaying a Tafel slope of 52 mV dec^{-1}. This indicates that the HER mechanism follows the Heyrovsky mechanism ($H_2O + Hads + e^- \rightarrow H_2\uparrow + OH^-$), where Hads represents the adsorbed hydrogen atom on the catalyst surface. Another important aspect to evaluate an HER catalyst is its long-term stability. Electrolysis tests were conducted at two different overpotentials (100 and 150 mV) to assess the durability of $Ni_{0.89}Co_{0.11}Se_2$ MNSN/NF as an integrated catalyst. Notably, during the 30-h tests, the current density remained almost unchanged at both overpotentials, demonstrating the excellent catalytic stability of $Ni_{0.89}Co_{0.11}Se_2$ MNSN/NF in a basic medium.

$Ni_{0.89}Co_{0.11}Se_2$ MNSN/NF exhibits remarkable performance not only in alkaline conditions but also in acidic and neutral solutions. In 0.5 M H_2SO_4 solution (pH = 0), $Ni_{0.89}Co_{0.11}Se_2$ MNSN/NF displays a significantly lower η_{10} compared to $NiSe_2$ MNSN/NF (52 mV vs. 106 mV). The Tafel plots (inset of Fig. 7.4a) further reveal that Pt/C/NF follows the Tafel mechanism ($Hads + Hads \rightarrow H_2\uparrow$) with a Tafel slope of 31 mV dec^{-1}. On the other hand, $Ni_{0.89}Co_{0.11}Se_2$ MNSN/NF exhibits a Tafel slope of 39 mV dec^{-1}, indicating the dominance of the Heyrovsky mechanism. Notably, $Ni_{0.89}Co_{0.11}Se_2$ MNSN/NF also demonstrates high HER activity and durability in neutral conditions, making it the first selenide-based electrocatalyst capable of efficiently catalyzing HER in a neutral electrolyte. In 1.0 M phosphate buffered saline

(PBS) solution (pH = 7), the polarization curves reveal a remarkably low η_{10} of only 82 mV for $Ni_{0.89}Co_{0.11}Se_2$ MNSN/NF, which is much smaller than that of $NiSe_2$ MNSN/NF without Co incorporation (169 mV).

The activity of an HER catalyst in different electrolyte conditions depends on its ability to adsorb either H+ or H_2O molecules. To assess the impact of cobalt doping on the electronic structure, we calculated the adsorption energy of H+ and H_2O on the surfaces of $Ni_{1-x}Co_xSe_2$ and $NiSe_2$, respectively. The results showed that cobalt-doped $NiSe_2$ exhibited increased adsorption energy for H+ (0.81 eV vs. 0.56 eV) and significantly larger adsorption energy for H_2O (1.48 eV vs. 0.50 eV) compared to undoped $NiSe_2$. These calculations indicated that cobalt incorporation reduced the energy barriers for both H+ and H_2O adsorption. Additionally, the $|\Delta GH^*|$ values were calculated using DFT simulations, showing a much lower $|\Delta GH^*|$ value for $Ni_{21}Co_3Se_{48}$ compared to $NiSe_2$ (0.20 eV vs. 0.72 eV) and approaching that of Pt (0.09 eV). This suggests a favorable adsorption–desorption behavior of intermediate adsorbed hydrogen on $Ni_{21}Co_3Se_{48}$, in line with its improved HER performance.

The outstanding catalytic performance and stability of the 3D $Ni_{0.89}Co_{0.11}Se_2$ MNSN/NF can be attributed to several factors. Firstly, its unique 3D networked structure with mesoporous nanosheets increases the number of active sites and facilitates the access of HER-relevant species in the electrolyte to these sites. It also enables efficient release of formed gas bubbles from the catalyst surface. Secondly, the electronic structure engineering achieved through cobalt incorporation reduces the energy barriers for H+ and H_2O adsorption, resulting in high HER activity across all pH ranges. Furthermore, the Co doping enhances the electrical conductivity of $Ni_{0.89}Co_{0.11}Se_2$, ensuring fast charge transport during the catalytic process and leading to improved performance. These collective effects contribute to the exceptional catalytic performance and stability of the 3D $Ni_{0.89}Co_{0.11}Se_2$ MNSN/NF catalyst.

The aforementioned illustrations highlight the benefits and utilizations of mesoporous materials in the context of HER. The incorporation of mesoporous structures has emerged as a significant avenue for advancing HER catalysts. However, it is important to recognize that mesoporous materials alone may not provide comprehensive solutions to all challenges. Therefore, it is imperative to explore additional strategies and methodologies aimed at enhancing the overall performance of catalysts.

7.1.5 Carbon-Based Materials Electrocatalysts for HER

In recent years, the field of carbon materials has witnessed significant advancements, offering us a diverse range of materials to choose from. Notably, Andre Geim and Konstantin Novoselov's pioneering work at the University of Manchester led to the isolation of graphene from graphite through micromechanical stripping, earning them the 2010 Nobel Prize in Physics. Carbon nanotubes, discovered by Sumio Iijima in

1991, have also emerged as a crucial component in various materials, with single-walled and multi-walled carbon nanotubes playing prominent roles. Carbon-based materials possess numerous advantages, including excellent electrical conductivity, lightweight nature, high toughness, and low toxicity. Moreover, the carbon skeleton provides a large specific surface area, facilitating the exposure of more active sites and promoting efficient charge transfer and transformation of intermediate products. Furthermore, carbon-based materials can serve as carriers for various catalytically active substances such as nanoparticles and single atoms, broadening their applications as catalytic materials.

Over the past few decades, researchers have explored various non-precious 3d transition metals (TMs) as potential alternatives to precious-metal catalysts for hydrogen evolution reaction (HER). Among these alternatives, nitrogen-doped carbon-coated TMs nanoparticles (TMs@N–C) have garnered significant interest due to their unique structure and composition. The incorporation of carbon materials in this hybrid architecture serves a dual purpose: protecting the TMs from acid/alkaline corrosion and providing active sites for HER. The modulation of electron spin density and charge distribution by TMs and nitrogen species contributes to the catalytic activity of the carbon materials. Metal–organic frameworks (MOFs), composed of metal ions or clusters bridged by organic functional groups, have shown promise as templates or precursors for the synthesis of TMs@C composites. However, despite recent progress, MOFs-derived TM@C composites face several challenges. These include a low degree of carbon graphitization, limited surface area (often less than 100 $m^2\ g^{-1}$), metal/carbon aggregation at high temperatures, poor contact between carbon and active TM nanoparticles, and the presence of thick carbon layers (>5 layers), resulting in unsatisfactory catalytic performance compared to precious Pt catalysts. Furthermore, most TM@C composites reported to date demonstrate activity only in acidic media, limiting their applicability.

A novel approach proposed by Wu's group involves the utilization of graphene oxide (GO)-wrapped core–shell (Co, Zn)-bimetallic zeolitic imidazolate frameworks (ZIFs) as precursors [11]. The unique core–shell structure of ZIF-67@ZIF-8 facilitates the encapsulation of ultrafine Co nanoparticles within in situ formed nitrogen-doped carbon nanotubes (N-CNTs) with thin carbon layers. The presence of GO serves as a matrix to anchor the CNTs and create enough space for their growth, preventing particle and/or carbon aggregation at high temperatures. Furthermore, the coverage of GO by ZIF-67@ZIF-8 is believed to expose more catalytic sites and enhance mass transfer. The synthesis strategy for Co@N-CNTs@rGO hybrid composites is depicted. Initially, ZIF-8 nanotubes grow in situ on both sides of the GO sheet, resulting in a sandwich-like structure composed of truncated cubic ZIF-67@ZIF-8 crystals uniformly anchored on the GO, referred to as ZIF-67@ZIF-8@GO. Finally, the 3D hierarchical Co@N-CNTs@rGO composites are obtained through simultaneous thermal-induced reduction. Raman spectra of the thermally annealed products exhibit peaks at around 1352 cm^{-1} and 1589 cm^{-1}, corresponding to the disorder carbon (D-band) and graphitic carbon (G-band), respectively. The increased relative intensity ratio of D to G-band (ID/IG) in the order of A-ZIF-8@GO < A-ZIF-67@ZIF-8 < A-ZIF-67@ZIF-8@GO < A-ZIF-67@GO suggests that Co species are more effective than Zn species in promoting carbon graphitization and enhancing conductivity. The nitrogen adsorption–desorption isotherms (Fig. 7.11c) reveal that the A-ZIF-67@ZIF-8@GO composites have a larger BET surface area (384 m^2 g^{-1}) and pore volume (0.618 cm^3 g^{-1}) compared to A-ZIF-67@ZIF-8 and A-ZIF-67@GO composites. This indicates that the introduction of rGO and/or ZIF-8 promotes hierarchical porosity, providing abundant catalytic active sites and facilitating mass/charge transportation. The magnified field emission scanning electron microscopy (FESEM) images demonstrate the uniform extrusion of numerous carbon nanotubes with lengths ranging from 30 to 150 nm from all sides of the rGO sheets, resulting in a sandwich-like structure. This 3D hierarchical arrangement effectively prevents restacking and self-aggregation of rGO and CNTs caused by strong van der Waals interactions. High-resolution transmission electron microscopy (HRTEM) image further reveals the tight encapsulation of Co nanoparticles by "bamboo joint-like" CNTs with only 3–4 carbon layers.

Co@N-CNTs@rGO achieves a current density of 10 mA cm^{-2} with a significantly lower overpotential of 108 mV compared to A-ZIF-67@GO (173 mV), A-ZIF-67@ZIF-8 (258 mV), and A-ZIF-8@GO (410 mV). The Tafel slope of Co@N-CNTs@rGO is 55 mV dec^{-1}, indicating that the HER follows a Volmer–Heyrovsky mechanism. Analysis of the Nyquist plots reveals that Co@N-CNTs@rGO exhibits the smallest semi-circular diameter, indicating enhanced charge transfer ability and reduced resistance. The electrochemical surface area (ECSA) tests confirm that Co@N-CNTs@rGO possesses a larger catalytic surface. Long-term stability tests demonstrate negligible decay in current density (20 mA cm^{-2}) after 100 h. The ΔG values of HER on different catalysts are depicted. It is observed that bare Co and bare graphene exhibit either excessively strong or weak H* adsorption, resulting in low HER activity. However, encapsulating Co within graphene effectively tunes the ΔG values. Nitrogen doping on graphene also contributes to improve HER activity. Notably, the combination of N-doping, Co clusters, and graphene in Co@N-CNTs@rGO yields a ΔG value close to 0, indicating the highest HER activity among the five models. These findings and the proposed strategy offer new possibilities for designing and fabricating highly active transition metal-based electrocatalysts as alternatives to precious Pt for electrochemical water splitting.

Recently, single-atom catalysts have emerged as a promising research area alongside metal particles. Liu and colleagues employed atomic layer deposition (ALD) to deposit monodisperse Pt atoms on nitrogen-doped graphene nanosheets (NGNs) [12]. The density and dispersion of Pt single atoms (SAs) on NGNs could be easily adjusted by varying the number of ALD cycles. Two types of ALD treatments were carried out with 50 and 100 cycles, resulting in distinct morphologies and catalytic activities, denoted as ALD50Pt/NGNs and ALD100Pt/NGNs. Analyzing the ADF-STEM images (Fig. 7.8a), it is evident that ALD50Pt/NGNs exhibited Pt SAs (bright spots) and small Pt clusters distributed on the NGNs surface, while ALD100Pt/NGNs displayed a higher abundance of Pt clusters, including Pt nanoparticles (NPs) (Fig. 7.8b). Consequently, ALD50Pt/NGNs showed more pronounced improvement in HER performance in 0.5 M H_2SO_4 (Fig. 7.8c). X-ray absorption near-edge structure (XANES) analysis and density functional theory (DFT) calculations provided further insights into these samples. It was revealed that individual Pt atoms could strongly bind to N sites, forming entities with high binding energies. Bader charge analysis confirmed a charge transfer from Pt to N, resulting in an increased total vacant density of states with Pt 5d character in ALD50Pt/NGNs compared to ALD100Pt/NGNs and Pt/C. Importantly, the charge transfer is not solely dependent on the loading but also influenced by the size and distribution of deposited Pt, which can be controlled through ALD cycles. Additional calculations indicated that the unoccupied 5d orbitals of single Pt atoms could interact with the 1 s orbital of H atoms, leading to the formation of electron pairing and hydride, a key factor contributing to the superior HER activity. Furthermore, changes in work function, as evidenced by a Scanning-Kelvin-Probe experiment, may also impact the HER activity. Additionally, the strong interaction between Pt SAs and N sites enhances the stability of Pt SAs. The field of single-atom catalysts has seen rapid growth, encompassing Co, Ni, Fe, V, and other elements, with most of them being based on carbon materials. For instance, Li and co-workers achieved single Fe atom anchoring on graphdiyne (Fe/GD) through a two-step approach, demonstrating excellent HER activity (overpotential of 66 mV at 10 mA cm^{-2}) and stability (negligible losses after 60 h at 30 mA cm^{-2}) [13].

The utilization of carbon-based materials in HER catalysts has emerged as a widely adopted and successful approach. Various carbon-based materials, including metal nanoparticle@carbon composites and single-atom catalysts, have been developed to enhance HER performance. However, there are still areas that warrant further exploration and improvement. One such area is the optimization of metal particle or single-atom loading to maximize catalytic activity. Finding innovative methods to increase the loading capacity of these active species within carbon matrices could lead to further advancements in HER catalysis.

7.1.6 Mesoporous Carbon-Based Film Materials for HER

In the preceding sections, we have discussed mesoporous materials and carbon-based materials individually as potential HER catalysts. However, these materials often address only specific aspects of the problem at hand. Furthermore, when using these powder materials directly as catalysts, complex steps and processes are involved, and there is a risk of detachment from the electrode surface, making large-scale production inconvenient. To overcome these challenges, mesoporous carbon-based film materials have emerged as a promising alternative. These materials combine the advantages of mesoporous materials and carbon-based materials. Additionally, as film materials, they offer direct application, a larger macro catalytic area, and enhanced mechanical properties. In current research, significant progress has been made in the synthesis of such thin films using various methods discussed in Chap. 4. Although the design of mesoporous carbon-based film materials is still evolving, we will provide a few selected examples to illustrate their potential in the following sections.

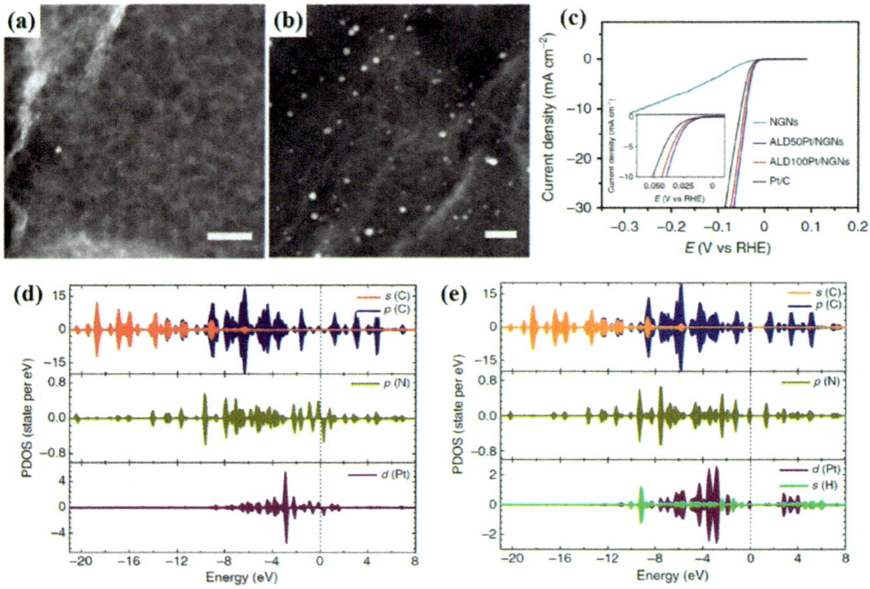

Fig. 7.8 ADF STEM images of ALDPt/NGNs samples with **a** 50 and **b** 100 ALD cycles. Scale bars, 10 nm. **c** The HER polarization curves for ALDPt/NGNs and Pt/C catalysts were acquired by linear sweep voltammetry with a scan rate of 2 mVs^{-1} in 0.5 M H_2SO_4 at room temperature. N_2 was purged before the measurements. The inset shows the enlarged curves at the onset potential region of the HER for the different catalysts. **d** Partial density of states (PDOS) of (**d**) non-H and **e** two H atoms adsorbed on a single Pt atom of ALDPt/NGNs. The Fermi level is shifted to zero. The upper part of the panel shows the PDOS of graphene, the middle part of the panel gives the PDOS of the N atom and the lower part of the panel exhibits the PDOS of the d orbital of Pt. Reproduced with permission. Copyright 2016, Nature Publishing Group

In recent years, two-dimensional materials such as MoS_2 and WS_2, which exhibit exposed edges, have emerged as highly promising electrocatalysts for the HER. To maintain the stability of the layered structure of MoS_2 during synthesis and enhance electrical contact with the active sites, carbon materials with excellent electrical conductivity have been utilized as supports. Examples include reduced graphene oxide (RGO), carbon paper, and active carbon. Increasing the surface area for catalyst loading is essential for improving the electrocatalytic efficiency of the HER. Binder-free films and three-dimensional (3D) electrode structures have proven effective in enhancing catalyst loading and utilization of catalytic sites, thereby boosting the HER catalytic activity. Carbon nanotubes (CNTs) have garnered significant attention as catalyst support materials due to their unique properties, nanoscale dimensions, and high surface area. Additionally, the utilization of 3D nanostructured architectures as catalyst support materials holds promise, as they provide a large specific surface area that facilitates rapid electrochemical reactions and efficient charge and mass transport.

Ramakrishna et al. have developed a straightforward wet-chemical method for synthesizing 3D CoS_2/RGO-CNT nanocomposite architectures with excellent HER activity. The synthesis process involves a hydrothermal approach to grow CoS_2 nanosheets on graphene substrates, followed by their combination with CNTs through vacuum filtration. The resulting CoS_2/RGO-CNT film, which is flexible and robust, is directly evaluated as a catalytic electrode for the HER. The uniform decoration of CoS_2 ultrathin nanosheets on RGO sheets, with RGO sizes in the range of several micrometers. BET and BJH tests confirm the high surface area of the hierarchical CoS_2/RGO composite, measuring 193.5 $m^2\ g^{-1}$, and reveal its mesoporous nature with a pore size of approximately 3 nm. Cross-sectional SEM images demonstrate the tight entrapment and uniform distribution of CoS_2/RGO composites within the porous CNT network. The strong electrical conductive pathways formed by the intimate contacts between composite particles and CNTs are crucial for catalytic applications. The resulting 3D hierarchical film exhibits excellent flexibility and mechanical strength, as demonstrated by the stress–strain curve, which shows an average tensile strength of 32 MPa and a Young's modulus of approximately 1.3 GPa at a strain of 2.0%. Moreover, the binder-free CoS_2/RGO-CNT composite exhibits significantly higher electrical conductivity (7.2×10^{-2} S m^{-1}) compared to a conventional electrode (2.5×10^{-3} S m^{-1}) utilizing Nafion as a binder. The enhanced strength and conductivity of the binder-free CoS_2/RGO-CNT composite contribute to improved catalytic stability and reduced electrode polarization. The polarization curve of the prepared CoS_2/RGO-CNT electrode displays low overpotentials of 142, 153, and 178 mV at current densities (j) of 10, 20, and 100 mA cm^{-2}, respectively. Additionally, the Tafel slope of the 3D CoS_2/RGO-CNT composite is approximately 51 mV dec^{-1}, suggesting that the Volmer–Heyrovsky mechanism governs the HER. A lower Tafel slope indicates that the catalyst requires a lower applied overpotential to generate the desired current. Furthermore, the CoS_2/RGO-CNT electrode exhibits the smallest charge transfer resistance (Rct) among all the composite catalysts, underscoring its superior HER activity.

Yan et al. have successfully developed a highly active electrocatalyst for the HER, consisting of composite nanowires comprising S-doped CoWP nanoparticles embedded in a (S, N)-doped carbon matrix. The catalyst was synthesized using Co- and W-based MOF (metal–organic framework) nanowires as the precursor $\{W(CN)_{83}Co_3(C_{13}H_{14}N_2)_6\}_n$. The MOF precursor, rich in N and S ligands, facilitated the doping of S in both the carbon shell and CoWP core, with additional N doping in the carbon shell. The synthesis process involved the fabrication of Hofmann-type Co and W-containing MOF nanowires (CoW-MOF) through a straightforward hydrothermal method, as depicted in Fig. 7.9a. The resulting CoW-MOF nanowires exhibited diameters of 200–600 nm and lengths of 1–3 μm. By pyrolyzing the CoW-MOF precursor, S-doped CoW nanoparticles were dispersed within S- and N-doped carbon nanowires (S-CoW@(S, N)-C). Subsequently, the S-CoW@(S, N)-C nanowires were phosphorized at 400 °C to transform the CoW nanoparticles into CoWP nanoparticles. The final product, S-CoWP@(S, N)-C, retained the nanowire morphology observed in the CoW-MOF nanowire precursor (Fig. 7.9b, c), with nanowire diameters ranging from 100 to 300 nm, which were thinner than those of the CoW-MOF nanowires. Transmission electron microscopy (TEM) characterization clearly revealed the presence of 5–15 nm nanoparticles embedded in a carbon matrix (Fig. 7.9d).

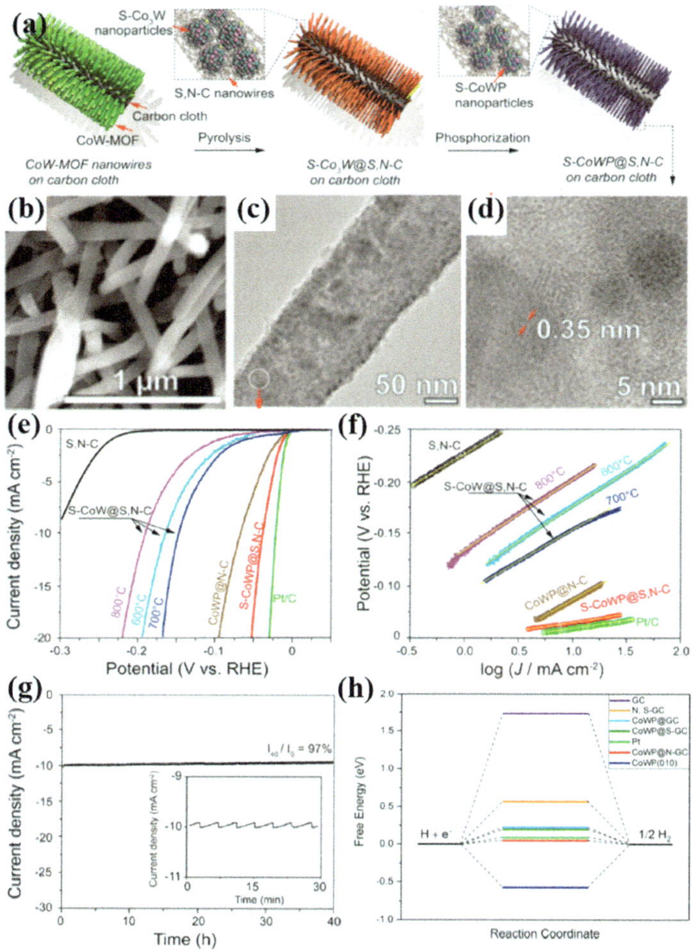

Fig. 7.9 **a** Schematic of the synthesizing route of S-CoWP@S, N-C. **b** SEM image of S-CoWP@S, N-C nanowires. **c** TEM image of a single S-CoWP@S, N-C nanowire. **d** HRTEM images of carbon wrapped S-CoWP nanoparticles within the nanowire. **e** HER polarization curves. **f** Corresponding Tafel plots of panel a. **g** Durability test for the S-CoWP@S, N-C nanowire sample at a potential of –35 mV (vs. RHE). The insert in panel g is the enlarged durability plot during 9.5–10 h. **h** Calculated free energies for hydrogen adsorption on the surface of graphitic carbon (GC), CoWP(010) facet, N- and S-doped GC (N, S-GC), CoWP(010) facet covered with pristine GC (CoWP@GC), Pt, and CoWP(010) facet covered with N- or S-doped GC (CoWP@NGC and CoWP@S-GC). Reproduced with permission. Copyright 2018 American Chemical Society

The S-CoWP@(S, N)-C nanowires were subjected to electrocatalytic performance evaluation for the HER in an acidic solution (Fig. 7.9e). Comparative analysis was conducted with Pt/C, S-CoW@(S, N)-C nanowires, and S- and N-doped carbon nanowires obtained by removing Co from the pyrolyzed Co-MOF (S, N-C). Notably, the S-CoWP@(S, N)-C nanowires achieved a stable current density of -10 mA cm^{-2} at an impressively low potential of -35 mV (vs. RHE), with a loading amount of 0.75 mg cm^{-2}, only 16 mV higher than that of Pt/C. The corresponding Tafel slope of 35 mV dec^{-1} was comparable to that of Pt/C (Fig. 7.9b). The significantly lower Tafel slope compared to its precursor (S-CoW@(S, N)-C) suggests that the rate-limiting process for the HER is the Heyrovsky-Tafel reaction, attributed to the enhanced H* absorption of metal phosphides. Furthermore, the S-CoWP@(S, N)-C nanowire catalyst exhibited exceptional stability, with over 98% current density retained during 40 h of potentiostatic testing at -35 mV (vs. RHE) (Fig. 7.9g). Notably, S-CoWP@(S, N)-C also demonstrated remarkable HER electrocatalytic activity in alkaline electrolyte, with an overpotential of 67 mV at $-$ 10 mA cm^{-2} using graphite as the counter electrode. The authors further investigated the ΔGH^* (Gibbs free energy change) of CoWP covered with S- or N-doped carbon. The carbon shell with a CoWP core exhibited a significantly reduced ΔGH^* value of 0.227 eV, much lower than that of graphitic carbon and pristine CoWP, indicating a synergistic effect between carbon and the CoWP nanoparticles. Similarly, S- or N-doped carbon-wrapped CoWP displayed even smaller ΔGH^* values of 0.196 and 0.050 eV, respectively, comparable to Pt ($\sim$0.09 eV), highlighting their potential for the HER. The integration of carbon's strong desorption capacity with CoWP's strong absorption capacity in the core–shell nanostructure formation led to a small ΔGH^*. Thus, the ΔGH^* of carbon composites can be effectively tuned by heteroatom doping and encapsulation of transition-metal compound nanoparticles. In conclusion, this method offers a promising approach for the development of robust and efficient electrocatalysts for solar-driven water splitting, with the potential for low-cost implementation.

In general, in view of the advantages of mesoporous carbon-based film materials, this material is bound to receive more and more attention and will have more and more applications in the future research.

7.2 Oxygen Evolution Reaction

Oxygen evolution is the process of producing oxygen after a reaction. In the following text, the detailed introduction, mechanism about oxygen evolution reaction and the application of mesoporous materials on this subject are introduced in the following text.

7.2.1 Introduction

Water splitting, which involves the hydrogen evolution reaction (HER) and the oxygen evolution reaction (OER), plays a vital role in producing hydrogen. In the previous section, we explored the HER, and now our focus will shift to the OER. Similar to the HER, the OER is thermodynamically uphill, necessitating a substantial overpotential to drive the reaction. The overall overpotential for water splitting comprises two components: the HER overpotential and the OER overpotential. Generally, the OER demands a higher overpotential compared to the HER. Hence, finding effective ways to minimize the overpotential in OER processes is crucial for enhancing overall water splitting efficiency.

Extensive research has focused on developing electrocatalysts for the oxygen evolution reaction (OER). Noble metal oxides like RuO_2 and IrO_2 have demonstrated exceptional catalytic performance. However, their high cost and limited availability, similar to platinum, hinder their widespread application on a large scale. Thus, there is a critical need to design and create OER catalysts that offer high performance, cost-effectiveness, and environmental friendliness.

In this section, we will delve into the mechanism of the OER and provide fundamental knowledge, including the measurement system used to evaluate catalytic performance. Moreover, we will showcase recent advancements in OER catalysts beyond noble metal catalysts. These include the utilization of mesoporous materials, carbon-based materials, thin films, and other cutting-edge materials as catalysts for the OER. We will explore why these materials exhibit excellent catalytic performance and discuss relevant theories. Additionally, we will explore the application of bifunctional catalysts in water splitting, illustrating their significance and potential.

7.2.2 Fundamentals of OER

(a) Mechanism of OER

It is widely accepted that OER can proceed through two different mechanisms: adsorbate evolution mechanism (AEM) and the lattice-oxygen-mediated mechanism (LOM). The AEM is typically assumed to involve four concerted proton–electron transfer reactions centered on the metal ion, as described in equation of 7.17–7.18 [14]. At each step, a proton is injected into the electrolyte, eventually combining with a transferred electron at the cathode. In particular, OH□ first adsorbs on the surface O vacancy site. The adsorbed OH (HO* species) then undergoes subsequent deprotonation to form O*. The following O–O bond formation step allows O* to react with another OH□ to form the HOO* intermediate. In the final step, O_2 is evolved through the deprotonation of HOO* with the regeneration of the active site.

$$OH^- + * \rightleftharpoons *OH + e^- \tag{7.17}$$

$$^*OH + OH^- \rightleftharpoons {}^*O + e^- + H_2O \tag{7.18}$$

$$^*O + OH^- \rightleftharpoons {}^*OOH + e^- \tag{7.19}$$

$$^*OOH + OH^- \rightleftharpoons O_2 + e^- + H_2O + {}^* \tag{7.20}$$

DFT calculation was often used to evaluate and compare the energy change of each step, which can reflect the catalytic activity of different electrocatalysts. The free energy differences for each step (ΔGi) during OER are theoretically calculated as following equations described:

$$\mathbf{\Delta G1} = G(^*OH) - G(^*) - \mu OH = E(^*OH) - E(^*) - E(H_2O) + 1/2E(H_2) - eU + \Delta G_{H^+}(pH) + \Delta(ZPE - T\Delta S) \tag{7.21}$$

$$\mathbf{\Delta G2} = G(^*O) - G(^*OH) + \mu OH = E(^*O) - E(^*OH) + 1/2E(H_2) - eU + \Delta G_{H+}(pH) + \Delta(ZPE - T\Delta S) \tag{7.22}$$

$$\mathbf{\Delta G3} = G(^*OOH) - G(^*O) - \mu OH = E(^*OOH) - E(^*O) - E(H_2O) + 1/2E(H_2) - eU + \Delta G_{H+}(pH) + \Delta(ZPE - T\Delta S) \tag{7.23}$$

$$\mathbf{\Delta G4} = G(^*O_2) - G(^*OOH) - \mu OH = E(^*O_2) - E(^*OOH) + 1/2E(H_2) - eU + \Delta G_{H+}(pH) + \Delta(ZPE - T\Delta S)$$

or

$$4^*(1.23eV - eU + \Delta G_{H^+}(pH)) - (\Delta G1 + \Delta G2 + \Delta G3) \tag{7.24}$$

where U represents the potential obtained against the reversible hydrogen electrode (RHE) under standard conditions. ΔEi is the reaction energy calculated directly using DFT. According to the $\Delta G_{H+}(pH) = -k \cdot BT \log(pH)$, the change of Gibbs free energy for a proton relative to the pH is obtained. ΔGi are calculated from zero-point energy (ZPE), entropy correction, and the DFT energy to $\Delta Gi = \Delta ZPEi + \Delta Ei - \Delta T\Delta Si$. To avoid the calculation including O_2 (gas), which is difficult to calculated within the GGA-DFT scheme, the sum of ΔG was fixed at the experimental ΔG value (4.92 eV) in $2H_2O = 2H_2 + O_2$. On the basis of above analysis, the theoretical η could be calculated from the ΔGi as the following equation presented:

$$\eta = \max(\Delta G1, \Delta G2, \Delta G3, \Delta G4, 5)/e - 1.23\,V \tag{7.25}$$

In principle, the OER on a given catalyst might be limited by any of these four reaction steps. The most positive value among ΔG1, ΔG2, ΔG3, and ΔG4 determines the overpotential of OER; then, the catalytic performance can be determined. An ideal OER catalyst requires all the four elemental steps with reaction free energies of the same magnitude at U = 0 (namely, 1.23 eV), but this ideal situation is almost impossible to achieve.

On the other hand, LOM proceeds on two neighboring metal sites, which is different from that in conventional AEM that proceeds on a single metal site (Fig. 7.10a, b). Firstly, two *OH on the metal sites undergo deprotonation, resulting in two metal–oxo species. Secondly, these two neighboring oxo species directly couple to form the O–O bond. Instead of combining with water or OH^- to form HOO*. Finally, O_2 is released with the bareness of two vacant metal centers that are subsequently occupied by OH^- [15]. In the past few decades, a heterogeneous water oxidation catalyst has not been used for conducting LOM to form O_2 by direct O–O coupling, because this process is considered to need a large activation barrier. In fact, this process has been widely accepted to operate in certain molecular catalysts involving transition metals such as perovskites.

(b) Assessment of the OER activity and electrocatalyst

Similar with the evaluation system of HER, the assessment of OER activity are as follows: (i) Overpotential is one of the most important descriptors to evaluate the performance of target electrocatalysts. (ii) Tafel slope (a smaller Tafel slope (b) indicates that current density can increase faster with smaller overpotential change). (iii) Electrochemical Impedance Spectroscopy (EIS). (iv) Stability (v) Faradic efficiency. (vi) Turnover frequency. (vii) Electrochemical active surface area (ECSA). (viii) DFT calculation for OER.

These test methods and evaluation indicators have been briefly introduced in the previous chapter, so there is no need to repeat them here. It should be noted that DFT calculation for OER is much more complex than that for HER, mainly due to the complex reaction mechanism of OER. We have described in the mechanism section how to calculate the Gibbs free energy variation for each step of OER. In addition to

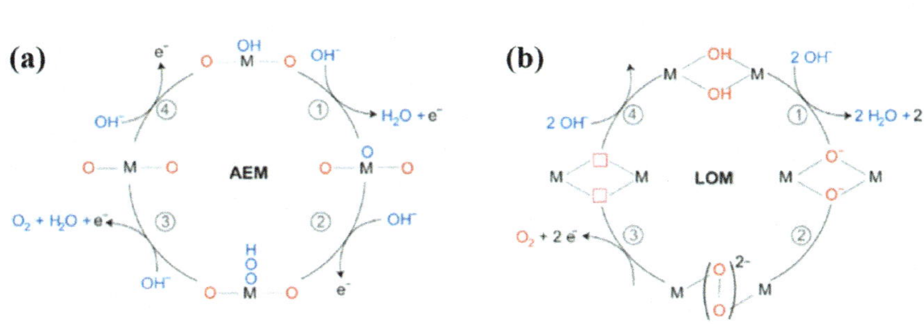

Fig. 7.10 **a** and **b** Schematic illustration of the OER mechanisms, namely, AEM and LOM, respectively. Reproduced with permission. Copyright 2020 American Chemical Society

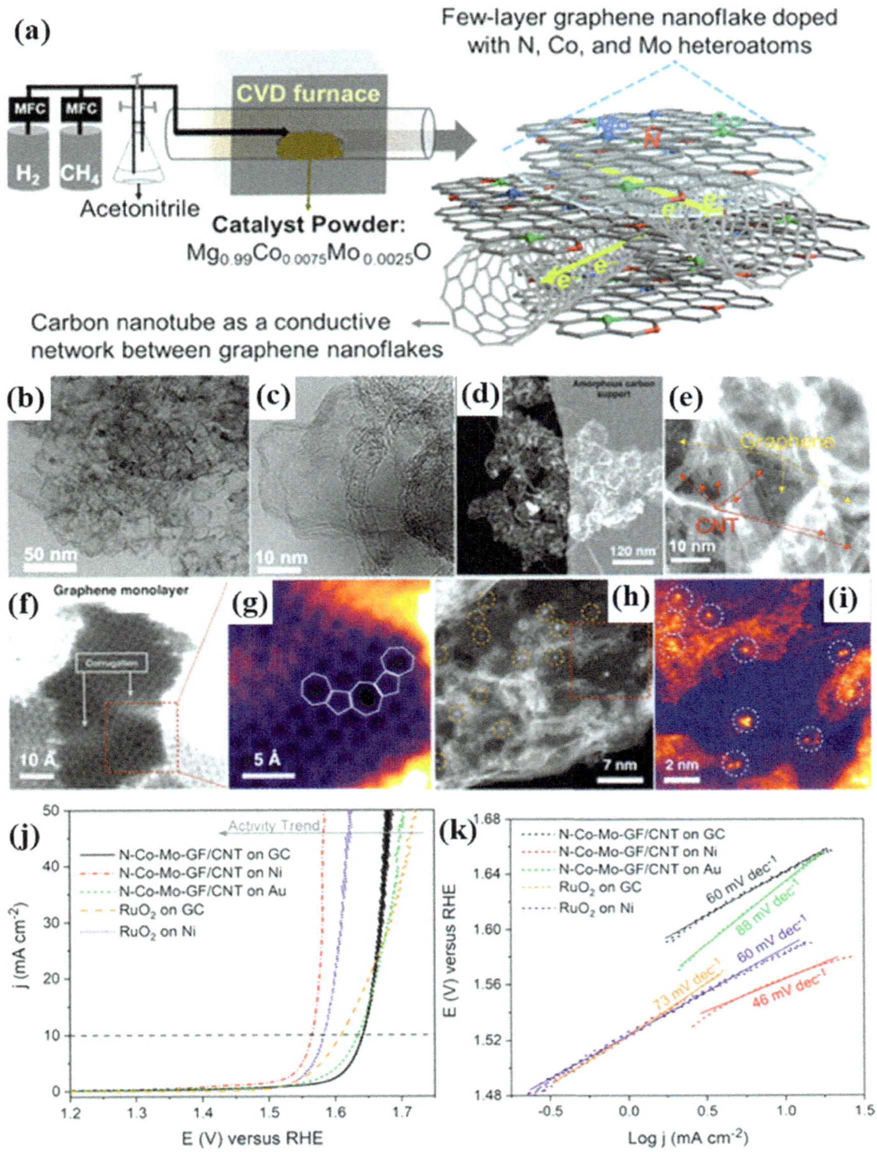

Fig. 7.11 **a** Scheme of the synthesis reactor used for the growth of the N–Co–Mo–GC/CNT and Co–Mo–GC/CNT materials (MFC is an abbreviation for mass flow controller). **b**, **c** HR-TEM and **d**–**i** STEM images of N–Co–Mo–GF/CNT. **b** TEM image of 30–50 nm GFs and **c** few-layer GF, **d** STEM image of the GFs interconnected by CNTs, with **e** showing CNTs bridging the GFs, **f**, **g** high-resolution STEM images taken from a graphene monolayer, showing the honeycomb structure, and **h**, **i** STEM images of double- or few-layer GFs. Individual metal atoms are observed as bright spots on the carbon support. **j** OER polarization curves and **k** corresponding OER Tafel plots of the N–Co–Mo–GF/CNT catalyst on GC, Ni, and Au compared to Pt/C on GC and Ni substrates in 0.1 M KOH. Reproduced with permission. Copyright 2020 American Chemical Society

energy change, we can also obtain catalyst D band center, charge distribution, and intermediate bond length by DFT calculation, which is helpful for us to compare the performance differences of catalysts. Of course, D-band central theory can still be applied to OER to explain the performance differences of catalysts.

When we apply a certain catalyst directly to water splitting as a bifunctional catalyst, the main performance indicators we are concerned with are still overpotential, Tafel slope, stability and so on. We can know from the Nernst equation that the theoretical driving potential for overall water splitting should be 1.23 V. In general, when commercial Pt/C and RuO_2/C are used as HER and OER catalysts, respectively, the potential is 1.56 V at a current density of 10 mA cm^{-2}, which can also be used as an indicator for comparison.

7.2.3 Mesoporous Materials Electrocatalysts for OER

Over the past years, numerous electrocatalysts for the oxygen evolution reaction (OER) have been developed, including transition metals, transition metal oxides, and transition metal phosphates, among others. Among these catalysts, mesoporous materials have garnered significant interest due to their potential for enhancing OER performance. In this section, we will explore the application, advantages, and underlying theories of mesoporous material catalysts in OER, highlighting their effectiveness through several illustrative examples.

Co_3O_4 has garnered considerable attention as a promising catalyst for the oxygen evolution reaction (OER) due to its reducible oxide nature, weak Co–O bond strength, and low barriers for oxygen vacancies. It is particularly effective when employed in a high-surface-area morphology, facilitating fast charge transport and making it an ideal candidate for direct reduction treatment to enhance its electrical and catalytic properties. To showcase this potential, Zheng et al. have conducted a study on mesoporous Co_3O_4 nanowires (NWs) grown hydrothermally, demonstrating the efficacy of $NaBH_4$ treatment in improving their catalytic and electrochemical performance. The mesoporous structure of the Co_3O_4 NWs offers a large surface area, enabling efficient reduction treatment and efficient mass and charge transfer. Additionally, the direct electrical contact with conductive substrates enhances charge carrier transport. Compared to pristine Co_3O_4 nanowires, the reduced Co_3O_4 nanowires exhibit significantly higher current density (13.1 mA cm^{-2} at 1.65 V vs. reversible hydrogen electrode) and a lower Tafel slope. Density-functional theory calculations further indicate that the presence of oxygen vacancies introduces new gap states, allowing for delocalization of electrons previously associated with Co–O bonds. This delocalization enhances electrical conductivity and electrocatalytic activity, contributing to the improved performance of the reduced Co_3O_4 nanowires.

Besides single transition metal compounds, bimetallic oxides, particularly spinel oxides, have garnered significant attention in OER research. Among them, mesoporous $MnCo_2O_4$, a spinel oxide, has been extensively investigated due to its structural flexibility and mixed valence states. Geng et al. have reported that mesoporous

$MnCo_2O_4$ exhibits predominately MnIV on the surface and MnIII in the bulk, while CoII is present in both the surface and bulk regions. This unique composition contributes to the high OER activity of mesoporous $MnCo_2O_4$. The overpotential at 10 mA cm^{-2} for $MnCo_2O_4$ is relatively small (0.40 V), comparable to that of Co_3O_4 (0.39 V). Although this specific spinel example may not exhibit the lowest overpotential compared to other reported OER catalysts, it holds significant research value as a bifunctional catalyst for both oxygen evolution reaction (OER) and oxygen reduction reaction (ORR) in oxygen electrocatalysis. Moreover, the synthesis method for this catalyst is simple and suitable for mass production, making it highly practical. Notably, the study of spinel catalysts in OER provides valuable insights into the design and understanding of corresponding materials for various electrochemical systems.

The microstructure and electrochemically active surface area of catalysts play a crucial role in their electrocatalytic activities. To enhance the catalytic performance, reducing the catalyst size is a promising strategy as it increases the surface area and exposes active sites. However, achieving ultrasmall nanoparticles is challenging due to aggregation or the presence of stabilizing ligands that can poison the catalyst. A solution to this challenge lies in the construction of 3D porous nanostructures, which provide high specific surface areas, enriched active sites, and facilitate the diffusion of electrolyte and ionic species. Zhang et al. propose a novel and facile method for preparing Ni–Fe–O-based catalysts with a unique mesoporous nanowire network morphology, utilizing the eutectic reaction in precursor design and the dealloying inheritance effect (DIE). The fabrication process of the Ni–Fe–O mesoporous nanowire network is depicted in their work. Precursors of $Al_{97}NixFe_{3-x}$ alloys (x = 2, 1.5, 1), rapidly solidified, were tailored for dealloying. In the subsequent 1 M NaOH dealloying process, selective removal of Al occurred, and the remaining Ni/Fe spontaneously oxidized, resulting in the growth of NiFe-based oxide ($NiFe_2O_4$-type) around the Ni core. Scanning electron microscopy (SEM) and transmission electron microscopy (TEM) were employed to investigate the nanostructures of $NixFe_{3-x}$–O (x = 2, 1.5, 1). The nanowire morphology observed in the as-dealloyed products inherited the intermetallic phases' structure from the precursors. Furthermore, all $NixFe_{3-x}$–O (x = 2, 1.5, 1) samples exhibited a network of bundled nanowires, resembling the appearance of coral. The ligaments were composed of parallel nanowires with diameters of approximately dozens of nanometers (Ni_2Fe_1–O: ≈ 30 nm; $Ni_{1.5}Fe_{1.5}$–O: ≈ 43 nm; Ni_1Fe_2–O: ≈ 21 nm). The samples also featured large-sized channels, hundreds of nanometers in width, facilitating electrolyte penetration and oxygen gas release. TEM analysis further confirmed the nanowire morphology in the $NixFe_{3-x}$–O (x = 2, 1.5, 1) samples, particularly evident, where well-defined mesoporous nanowire networks were clearly visible. These mesoporous nanowire networks provided large specific surface areas for the $NixFe_{3-x}$–O (x = 2, 1.5, 1) samples, which were assessed using N_2 adsorption/desorption measurements. The isotherms exhibited type IV characteristics with a hysteresis loop indicating that the samples possessed mesoporous structures. The calculated BET surface areas were 73.64, 83.41, and 114.84 $m^2\ g^{-1}$ for Ni_2Fe_1–O, $Ni_{1.5}Fe_{1.5}$–O, and Ni_1Fe_2–O, respectively. Furthermore, the pore size distribution analysis confirmed the mesoporous nature of the samples, with average pore diameters falling within the mesoporous range.

The iR-corrected polarization curves of the $NixFe_{3-x}$–O (x = 2, 1.5, 1) samples and reference materials for OER are shown. It is evident that XC-72 exhibits weak OER activity, while GCE shows almost no activity. In contrast, the $NixFe_{3-x}$–O (x = 2, 1.5, 1) catalysts demonstrate remarkable oxygen evolution on their surfaces. Notably, Ni_2Fe_1–O exhibits a more negative OER onset potential and significantly higher current density compared to $Ni_{1.5}Fe_{1.5}$–O and Ni_1Fe_2–O. Impressively, the Ni_2Fe_1–O catalyst achieves a current density of 10 mA cm^{-2} at an overpotential of only 244 mV, surpassing the performance of $Ni_{1.5}Fe_{1.5}$–O (290 mV), Ni_1Fe_2–O (316 mV), IrO_2 (339 mV). The Tafel slopes for Ni_2Fe_1–O, $Ni_{1.5}Fe_{1.5}$–O, and Ni_1Fe_2–O are 39, 39, and 47 mV dec^{-1}, respectively, which are much lower than those of IrO_2 (78 mV dec^{-1}) and XC-72 (138 mV dec^{-1}). Electrochemical impedance spectroscopy (EIS) was employed to investigate the underlying factors influencing the catalysts' activity. This work presents the Nyquist diagrams of the $NixFe_{3-x}$–O (x = 2, 1.5, 1) catalysts at 1.55 V versus RHE. Among the catalysts, Ni_2Fe_1–O exhibits the lowest Rct value, indicating a direct correlation with its high catalytic activity. A chronopotentiometric test was performed on the Ni foam-supported Ni_2Fe_1–O, with a constant current density of 10 mA cm^{-2}, revealing negligible change in the overpotential even after 60 h. Based on these results, it can be concluded that Ni_2Fe_1–O exhibits high OER activity and superior long-term durability, making it a promising electrocatalyst for OER. The exceptional OER performance of Ni_2Fe_1–O can be attributed to the synergistic effects of its 1D mesoporous nanowire morphology, unique 3D porous structure, and the presence of a metal core. Firstly, the unique mesoporous nanowire network structure enhances the specific surface area, facilitating efficient contact between the electrolyte and active catalysts. Additionally, the interconnected mesopores within this structure promote efficient charge/ion transfer and mass transport. Secondly, the continuous large pores among the nanowire arrays enable the penetration of the electrolyte and facile release of the produced oxygen gas. Lastly, the residual Ni metal core provides a direct pathway for electron transport, ensuring excellent catalyst conductivity.

7.2.4 Carbon-Based Materials Electrocatalysts for OER

Previously, we have discussed carbon-based materials as HER catalysts and some of their advantages. Here, we will focus on the use of carbon-based materials as OER catalysts. Through the synthesis, characterization, testing and other means of carbon-based materials to demonstrate its application prospects.

In recent years, nitrogen-doped carbon materials (NCs) have gained attention as a promising class of OER electrocatalysts due to the favorable electron affinity of nitrogen dopants. To enhance the catalytic performance of NCs, two main strategies are commonly employed. The first approach focuses on improving the intrinsic activity of each nitrogen-doped site, while the second approach aims to increase the number of effective active sites. However, the presence of high-density holes in holey graphene (HG) sheets can impede charge transfer and disrupt the sp2 conjugated structures. To overcome these limitations, Zhao et al. introduced a novel synthetic method to develop a superior metal-free electrocatalyst that exhibits bifunctional ORR and OER activities under both alkaline and acidic conditions. This approach involves a simple two-step process to directly grow ultrathin N-doped holey carbon (N-HC) layers on a graphene oxide (GO) substrate. The resulting hybrids, denoted as N-HC@G-x (x representing the treatment temperature), feature selectively doped pyridinic-N as the active sites for OER. Concurrently, the intercalated GO sheet is transformed into reduced graphene oxide (rGO), which provides mechanical support, stabilizes the N-HC layer, and facilitates efficient charge transfer. Among the synthesized hybrids, the optimized N-HC@G-900 exhibits remarkable OER activity in both alkaline and acidic media. Notably, this catalyst offers the additional advantages of controllable synthesis using affordable and non-toxic reagents.

The high-resolution transmission electron microscopy (HRTEM) images captured from N-C/Zn@G-hydro treated at 900 °C clearly exhibit the presence of distinct nanoholes (approximately 2 nm in size, marked by red circles), which arise due to the evaporation of zinc during the thermal treatment. The resulting N-HC@G-900 material, derived from the N-C/Zn@G-hydro precursor, demonstrates a significantly increased Brunauer–Emmett–Teller (BET) surface area of 256.44 $m^2\ g^{-1}$. Importantly, the BET surface area of N-C@G-900 (99.8 $m^2\ g^{-1}$) is much lower than that of N-HC@G-900, providing further evidence that the enhanced surface area primarily originates from the formation of holey carbon (HC) structures. In terms of electrocatalytic performance, the N-HC@G-900 catalyst outperforms N-C@G-900, as illustrated by the linear sweep voltammetry (LSV) curves. Notably, N-HC@G-900 exhibits an onset potential (Eonset) of 1.51 V and achieves a current density of 10 mA cm^{-2} (E10) at a potential of 1.58 V. These key OER performance metrics are comparable to those of the reference catalyst Ir/C (Eonset = 1.44 V, E10 = 1.54 V). Impressively, under high current densities, N-HC@G-900 demonstrates superior OER activity compared to Ir/C. Additionally, this work highlights the superior stability of N-HC@G-900 in alkaline OER conditions compared to Ir/C. Taken together, these significant performance indicators position N-HC@G-900 as a top-performing electrocatalyst for alkaline OER applications.

Furthermore, a study by Joo et al. introduces a novel approach to the design of highly integrated bifunctional oxygen electrocatalysts, utilizing graphitic nanoshell/mesoporous carbon (GNS/MC) nanohybrids that incorporate multiple structural motifs responsible for both oxygen evolution reaction (OER) and oxygen reduction reaction (ORR) activities [16]. The fabrication of these carbon-based bifunctional electrocatalysts involved the nanocasting method employing a mesoporous silica template. In this process, Ni-phthalocyanine (NiPc) and Fe-phthalocyanine (FePc) precursors were used. Specifically, the metal-Pc precursor was mixed with SBA-15 silica and subjected to pyrolysis at 1000 °C under a nitrogen (N_2) flow. After etching the silica template, a carbon-based material was obtained. Through systematic investigation, it was discovered that the GNS/MC sample, prepared using a NiPc:FePc mass ratio of 3:7 at a pyrolysis temperature of 1000 °C (referred to as Ni_3Fe_7–MC), exhibited a mesoporous carbon structure enriched with graphitic nanoshells, demonstrating optimal bifunctional activity for both OER and ORR. In terms of OER, the GNS/MC catalyst displayed an overpotential of 340 mV to achieve a current density of 10 mA cm^{-2}, significantly lower than that of monometal-doped Ni–MC (480 mV) and Fe–MC (500 mV) catalysts. This highlights the crucial role of both Ni and Fe metals in enhancing OER activity. Moreover, the GNS/MC catalyst outperformed the precious metal-based Pt/C (660 mV) and Ir/C (370 mV) catalysts.

As mentioned previously, carbon-based materials have demonstrated their significance in facilitating various electrochemical reactions through the incorporation of transition metal compounds, single atoms, and other elements. Notably, carbon nanotubes (CNTs) and graphene exhibit exceptional conductivity and can be effectively doped with heteroatoms, which possess distinct electronegativities compared to carbon. Heteroatom doping in graphitic carbon nanomaterials, regardless of whether the dopant has a lower electronegativity (e.g., B, P, S, and transition metals) or higher electronegativity (e.g., N) than carbon, induces polarization in neighboring carbon atoms, alters the electronic properties of carbon frameworks, and enhances the catalytic activity of carbon-based electrocatalysts in various electrochemical reactions. To further explore this concept, Kauppinen et al. present a straightforward and scalable one-step synthesis method for producing a novel electrocatalyst composed of few-layer graphene nanoflakes (GFs) and CNTs doped with N, Co, and Mo heteroatoms (referred to as N–Co–Mo–GF/CNT) [17]. The synthesis of the mesoporous N–Co–Mo–GF/CNT catalyst involved a chemical vapor deposition (CCVD) synthesis reactor, as depicted in Fig. 7.10a. In this process, $Mg_{0.99}(Co_{1-x}Mo_x)_{0.01}O$ (with x $\sim$ 0.25) was employed as the catalyst for synthesizing graphitic carbon nanomaterials. Interestingly, the CCVD synthesis also led to the growth of crumpled graphene structures, likely resulting from the direct catalytic decomposition of CH_4 on the MgO oxide. Notably, the growth conditions were optimized to obtain a hybrid material of graphene and CNTs, rather than standalone CNTs or graphene layers.

The high-resolution transmission electron microscopy (HRTEM) images captured in Fig. 7.11b, c showcase the N–Co–Mo–GF/CNT hybrid material, highlighting the presence of graphene nanoflakes (GFs) with sizes ranging from 50 to 70 nm. Interestingly, Fig. 7.11d, e visually demonstrate how the carbon nanotubes (CNTs) act as bridges, connecting and enhancing the conductivity of the graphene nanoflakes within the material. Analysis of the monolayer graphene through Fig. 7.11f, g did not reveal the presence of metallic Co and Mo atoms, while nitrogen (N) was directly identified in the STEM image. Additionally, in the multilayer structures displayed in Fig. 7.11h, i, individual metal atoms were observed. By employing the Barrett–Joyner–Halenda (BJH) method, the cumulative pore volume and pore-size distributions of the N–Co–Mo–GF/CNT were calculated, unveiling an average pore width of 11.5 nm. Furthermore, the BET method determined a high specific surface area of 911 m^2 g^{-1} for the material.

Figure 7.11j compares the OER polarization curves of the prepared N–Co–Mo–GF/CNT catalyst on different substrates (GC, Ni, and Au) with the well-established RuO_2 catalyst on GC and Ni substrates. When tested on the GC substrate, RuO_2 exhibited slightly higher activity than N–Co–Mo–GF/CNT at low current densities

(<20 mA cm^{-2}). However, at high current densities (>20 mA cm^{-2}), the synthesized catalyst outperformed RuO_2, requiring a lower potential (1.68 V) compared to RuO_2 (1.72 V) to achieve a current of 50 mA cm^{-2}. Notably, when coated on a Ni substrate, the OER activity of the N–Co–Mo–GF/CNT catalyst was significantly enhanced. The onset overpotential and the overpotential required to reach 10 mA cm^{-2} (η_{10}) decreased by approximately 50 and 77 mV, respectively, on Ni compared to the GC substrate. This enhancement can be attributed to the high-surface-area mesoporous structure of the N–Co–Mo–GF/CNT catalyst, allowing the electrolyte to permeate the underlying layers and the Ni/catalyst interface. This creates a synergistic effect, leading to the formation of new active sites with higher reactivity and lower onset overpotential. The Tafel slope values indicate that the N–Co–Mo–GF/CNT catalyst on Ni exhibited the most active electrode, followed by GC and Au substrates. The N–Co–Mo–GF/CNT catalyst on Ni required overpotentials of only 330 and 350 mV to reach 10 and 50 mA cm^{-2}, respectively. Recent reports have highlighted substrate effects enhancing the OER activity of RuO_2. However, this study demonstrates that for a mesoporous electrolyte-permeable catalyst like N–Co–Mo–GF/CNT, the role of the substrate can be even more significant. It suggests that further advancements in the activity of mesoporous carbon-based catalysts can be achieved by incorporating suitable active metals either onto the substrate or the catalyst surface, facilitating the formation of new active sites at the metal/catalyst interface.

7.2.5 Mesoporous Carbon-Based Film Materials for OER

Mesoporous carbon-based thin film materials offer numerous advantages as OER catalysts. One key advantage is their ability to provide a high specific surface area, which allows for the exposure of a larger number of active sites and promotes efficient electrolyte penetration. Moreover, carbon-based materials possess desirable traits such as lightweight nature and excellent electrical conductivity. These materials can be easily loaded with various active components, enhancing their catalytic performance. Additionally, the mechanical properties of film materials are highly advantageous, enabling their direct utilization as electrodes in electrochemical reactions. In this section, we will showcase several examples that demonstrate the effectiveness of mesoporous carbon-based thin film materials as OER catalysts.

Carbon cloth, a metal-free 3D electrode, holds great promise as a substrate for supporting various catalysts. Its exceptional mechanical strength, lightweight nature, flexibility, and high electrical conductivity set it apart from other commonly used substrates. Furthermore, being metal-free, it eliminates concerns regarding the release of metal ions, making it environmentally friendly. Tong et al. introduced a novel approach to obtain a 3D self-supporting metal-free monolithic porous carbon cloth doped with N-heteroatoms, known as 3D NiD-PCC, through a series of thermal reduction and etching processes. The process involved hydrothermally growing Ni-based precursor nanosheets on the 3D carbon cloth, followed by annealing in a nitrogen environment to obtain a Ni nanoparticle (NP) embedded carbon cloth (referred to as NiC). Subsequently, the NiD-PCC was produced by overnight etching with conc. HCl, which removed the Ni NPs, resulting in a metal-free electrocatalyst with high BET surface area and pore volumes. The unique aspect of this material lies in its monolithic structure, where the entire current collector serves as both the catalytic active surface and the electrode, offering enhanced stability and resistance to pulverization issues over prolonged usage.

SEM analysis of the NiD-PCC sample revealed a roughened surface characterized by uniformly distributed pores, as depicted in Fig. 7.12b. These pores exhibited diameters ranging from 70 to 150 nm. TEM images in Fig. 7.12c, d further confirmed the presence of these pores, along with an exfoliated surface structure of the NiD-PCC. The formation of such pores and exfoliated surfaces was expected to contribute to a significantly higher BET surface area and increased pore volumes. Analysis of nitrogen adsorption–desorption isotherms showed that the specific BET surface area of the NiD-PCC reached 135 $m^2\ g^{-1}$, which is sixteen times greater than that of BCC (8 $m^2\ g^{-1}$). Moreover, the NiD-PCC exhibited a calculated pore volume of 0.15 $m^3\ g^{-1}$, with prominent peaks observed around 5 nm, while BCC recorded a pore volume of 0.01 $m^3\ g^{-1}$.

The OER activities of the electrocatalysts were assessed in a 1.0 M KOH solution, utilizing the self-supporting metal-free NiD-PCC as the working electrode. The obtained LSV curve of the NiD-PCC catalyst exhibited an onset overpotential of 250 mV, and it achieved a current density of 10 mA cm^{-2} at a potential of 1.59 V (corresponding to an overpotential of 360 mV), as illustrated in Fig. 7.12e. These values were notably higher than those observed for BCC and NiC (1.65 V, 420 mV), as well as surpassing the performance of many other reported carbon-based electrocatalysts. To further investigate the OER mechanism of the catalyst, Tafel plots were generated by plotting log(j) against the corresponding potentials, as depicted in Fig. 7.12f. Analysis of the Tafel plots revealed that the NiD-PCC exhibited a Tafel slope of 98 mV dec^{-1}, which was significantly lower than that of BCC (186 mV dec^{-1}) and NiC (155 mV dec^{-1}), indicating excellent catalytic kinetics for the NiD-PCC. Moreover, the durability of the NiD-PCC catalyst was assessed using the chronopotentiometry method at a constant current density of 10 mA cm^{-2} over a prolonged period of 28 h, as shown in Fig. 7.11g. The electrocatalyst demonstrated remarkable durability, maintaining a consistent overpotential of 360 mV even after 28 h of continuous operation.

In the pursuit of catalyst design, there is a growing interest in developing multifunctional electrocatalysts that can facilitate multiple electrochemical reactions simultaneously, including OER, HER, and even ORR. In this context, we specifically explore the role of these multifunctional catalysts in OER.

To address this, Wu et al. introduced a facile and efficient pseudomorphic replication approach to synthesize a unique hybrid material called Co@N-CS/N-HCP@CC. The synthesis process involves the formation of oriented 2D ZIF-67 nanosheets grafted with 3D ZIF-67 polyhedra architecture on the surface of carbon cloth (CC), followed by a low-temperature pyrolysis step. The resulting hybrid material consists of ultrafine cobalt nanoparticles embedded within both 2D N-doped carbon nanosheets and 3D N-doped hollow carbon polyhedra.

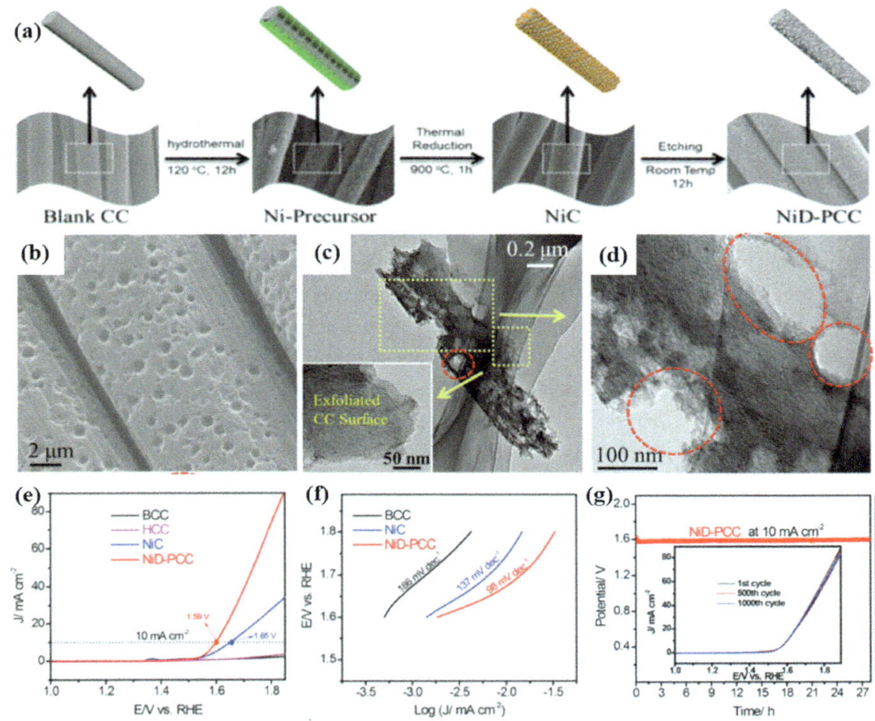

Fig. 7.12 **a** Schematic illustration of the synthesis of the monolith 3D NiDPCC. **b** SEM images of the NiD-PCC. **c** TEM image of the NiD-PCC. The inset is the HRTEM image of the NiD-PCC displaying the exfoliated section of the right-yellow daze section of the sample. **d** Magnified TEM image of the left yellow daze section. **e** LSV plots in 1 M KOH at a scan rate of 10 mV s^{-1}, **f** Tafel plots of the catalysts. **g** Stability test of NiD-PCC at a current density of 10 mA cm^{-2} for 28 h. The inset is the CV stability test up to 600 cycles. Reproduced with permission. Copyright 2016 The Royal Society of Chemistry

Initially, ultrathin cobalt layered double hydroxides (Co-LDH) nanosheets are vertically grown on the CC surface through a fast electrodeposition process using a $Co(NO_3)_2$ aqueous solution. Subsequently, the Co-LDH@CC serves as a template, which undergoes a phase structure reconstruction and nanocrystal growth by coordination with 2-MeIM at room temperature. This leads to the formation of CC-supported 2D ZIF-67 nanosheets grafted with 3D ZIF-67 dodecahedra (2D/3D ZIF-67@CC) precursors. Finally, the hierarchical Co@N-CS/N-HCP@CC composites are obtained by subjecting the 2D/3D ZIF-67@CC precursors to a one-pot pyrolysis at 450 °C for 2 h under flowing N_2 gas.

The FESEM images of the A-S-50 sample (where A-S-50 represents Co-LDH@CC coordinated with 2-MeIM for 50 min, followed by pyrolysis) are presented. It can be observed that A-S-50 successfully retains the morphology of the precursor, wherein the 2D nanosheets grafted with 3D polyhedra are uniformly distributed on the surface of the 1D carbon fiber, with the polyhedra exhibiting a rough surface. The ultrafine nanoparticles are confined within the hollow polyhedral matrix, as depicted, with an average size of approximately 5 nm. HRTEM analysis reveals a lattice fringe interspacing of 0.206 nm in the nanoparticles, corresponding to the (111) lattice plane of metallic cobalt.

Impressively, the Co@N-CS/N-HCP@CC catalyst exhibits exceptional electrochemical performance in water splitting. The LSV curves demonstrate the lowest overpotential of 248 mV required to achieve a current density of 10 mA cm^{-2}, surpassing the other tested composites. Furthermore, the Tafel slope of the Co@N-CS/N-HCP@CC catalyst is the smallest at 68 mV dec^{-1}, indicating excellent reaction kinetics. Additionally, the catalyst displays a remarkably low charge transfer resistance (Rct) value of 16.7 Ω, highlighting its superior charge transfer ability.

Encouraged by these outstanding results, we performed overall water splitting using the Co@N-CS/N-HCP@CC catalyst as both the anode and cathode in a two-electrode configuration electrolyzer. Notably, the catalyst achieved a current density of 10 mA cm^{-2} at an operating potential as low as 1.545 V in 1.0 M KOH, which is even lower than that of the state-of-the-art IrO_2@CC||Pt/C@CC couple (1.592 V). Moreover, the Co@N-CS/N-HCP@CC catalyst exhibits excellent durability, enabling sustained water splitting even with a single battery.

Li et al. present an innovative and straightforward approach for constructing cobalt nanoparticle-encapsulated 3D conductive films (Co/CNFs) through in situ thermal emission from monolithic cobalt metal foil [18]. This unique Co/CNFs material serves as a flexible and binder-free electrode with highly efficient activities for oxygen reduction reaction (ORR), oxygen evolution reaction (OER), and hydrogen evolution reaction (HER). The fabrication process of the Co/CNFs catalyst is depicted in their work. Initially, the melamine precursor is placed on the surface of the Co foil. Upon heating, melamine undergoes conversion into nitrogen carbon, forming a nitrogen-doped carbon layer on the Co foil at lower pyrolysis temperatures. As the temperature increases, volatile Co species are released from the Co foil surface and are trapped by the vacancies and unsaturated sites of the nitrogen-doped carbon layer. Subsequently, these Co species aggregate to form Co nanoparticles (NPs) on the carbon surface, acting as seeds for the growth of nitrogen-doped carbon nanotubes (CNTs). This results in the formation of a 3D assembly of Co NP-embedded carbon nanotube films (Co/CNFs). The carbon film can be easily detached from the Co foil, demonstrating its flexibility and strength. The Co/CNFs (1000) film exhibits a hierarchical structure, with randomly arranged nanotubes on the surface.

The superior OER activity of the Co/CNFs (1000) catalyst is evident, where it shows a significantly lower overpotential of 0.32 V at a current density of 10 mA cm^{-2}. This performance is comparable to that of recently reported catalysts with similar characteristics. Furthermore, the Tafel slope calculated from the linear sweep voltammetry (LSV) curves indicates that the Co/CNFs (1000) catalyst has the smallest value (79 mV dec^{-1}) among all the prepared samples, even outperforming RuO_2. Additionally, the Co/CNFs (1000) catalyst exhibits the lowest charge transfer resistance and the largest double-layer capacitance, further highlighting its outstanding electrochemical properties. While the focus of this discussion has been on its performance in OER, the Co/CNFs (1000) catalyst also demonstrates promising potential in other electrochemical reactions such as oxygen reduction reaction (ORR), which will be briefly explored in the following section.

7.3 Oxygen Reduction Reaction

Oxygen reduction (ORR) is an important reaction that helps clean pollutants from the air and can also be used to generate electricity. Oxygen reduction is a dissipative reaction that uses oxygen as a reducing agent, returning oxygen atoms to the environment in their most commonly used form (i.e. oxygen in aqueous solutions).

7.3.1 Introduction

Oxygen reduction reaction mainly refers to a process in which oxygen molecules gain electron to form a certain product. This reaction is the cornerstone of many important energy-conversion devices, such as fuel cells and metal–air batteries, as well as some industrial processes like chlor-alkali electrolysis. Because of people's demand for energy and environmental pollution, fuel cell has become an ideal device for energy storage and conversion. Fuel cells are efficient because they convert the Gibbs free energy in the chemical energy of fuel into electric energy through electrochemical reaction and are not limited by the Carnot cycle effect. In addition, fuel cells use fuel and oxygen as raw materials; At the same time, there is no mechanical transmission parts, so there is no noise pollution, the emission of harmful gas is very little. Therefore, from the perspective of energy conservation and ecological protection, fuel cell is the most promising power generation technology. In fuel cells, such as proton exchange membrane fuel cells (PEMFC), one of the important reactions is the oxygen reduction reaction. However, the ORR in cathode, however, is usually very sluggish, which greatly affected the further development of fuel cell and their practical application.

For example, in the typical H_2–O_2 fuel cells, the hydrogen oxidation reaction (HOR) at anode and ORR at cathode can be simplified as two half-cell reactions:

$$\text{HOR: } 2H_2 \rightarrow 4H^+ + 4e^-, E_0 = 0\,\text{V(vs. RHE)} \tag{7.25}$$

$$\text{ORR: } O_2 + 4H^+ + 4e^- \rightarrow 2H_2O, E_0 = 1.23\,\text{V(vs. RHE)} \tag{7.26}$$

As we know, the dynamics of these two half reactions are critical to the output of the overall fuel cell. In practice, however, both reactions require large activation energies, especially the ORR. In general, the precious metal platinum is considered to be a good catalyst for these two reactions. But as has been said before, the Pt faces scarcity and high prices. In one fuel-cell reactor, catalyst Pt accounts for about 50% of the cost. In addition, Pt as ORR catalyst has poor stability, and is vulnerable to methanol poisoning, which will form CO and thus seal off the catalytic active site. Therefore, the rational design of ORR catalyst with excellent performance, low price and environmental friendliness has become a research hotspot.

Over the past few decades, considerable efforts have been made to find more suitable catalytic materials for ORR [19]. For the Pt-based catalyst, the size and structure were optimized to reduce the dosage while maintaining the performance. In addition, it is more important to utilize earth-abundant elements to get efficient catalysts through the design of compounds based on these elements such as C, N, Fe, Co, Ni and so on.

Similarly, in this section, we will mainly introduce the reaction mechanism of ORR as well as some basic testing methods and measurement criteria. Then, we will introduce the applications of mesoporous materials, carbon-based materials and mesoporous carbon-based film materials as catalysts in ORR, and discuss the reasons for their excellent catalytic performance. As an important reaction in fuel cells and metal air cells, we will also briefly describe its role and performance in these devices. We hope that this energy saving will give readers a deeper understanding of ORR and inspire them to research and design the corresponding catalyst.

7.3.2 Fundamentals for ORR

Generally, the ORR involves either four proton-electron transfers to reduce oxygen to water, desirable for fuel cells, or a two–proton-electron pathway, attractive for the production of hydrogen peroxide [20]. The four-electron pathway can proceed via several mechanisms. A direct four-electron mechanism can either be dissociative or associative in nature, depending on the oxygen dissociation barrier on the catalyst surface. An indirect four-electron mechanism involves first the two-electron pathway to hydrogen peroxide, followed by further reduction to water:

Four-electron pathway:

Dissociative:

$$O_2 + 2^* \rightarrow 2O^* \tag{7.27}$$

$$2O^* + 2H^+ + 2e^- \rightarrow 2OH^* \tag{7.28}$$

$$2OH^* + 2H^+ + 2e^- \rightarrow 2H_2O + 2^* \tag{7.29}$$

Associative:

$$O_2 + {}^* \rightarrow O_2^* \tag{7.30}$$

$$O_2^* + H^+ + e^- \rightarrow OOH^* \tag{7.31}$$

$$OOH^* + H^+ + e^- \rightarrow O^* + H_2O \tag{7.32}$$

$$O^* + H^+ + e^- \rightarrow OH^* \tag{7.33}$$

$$OH^* + H^+ + e^- \rightarrow H_2O + \ ^* \tag{7.34}$$

Here, we will first introduce some basic contents of the four-electron pathway, because both fuel cells and metal air cells are involved in the four-electron process. As for the two-electron pathway that is beneficial to hydrogen peroxide production, we will elaborate on them in the following sections.

The thermodynamic equilibrium potential of ORR is 1.23 V, which means a zero-net current at that point. However, due to the energy barriers during each step and the sluggish kinetics, the experimentally applied potential is different with theoretical potential, which is often known as overpotential. The overpotential is directly linked with the fuel-cell efficiency, where a high overpotential would lead to the low efficiency.

Typically, the thin-film rotating disk electrode (RDE) and rotating ring-disk electrode (RRDE) are used for testing ORR [21]. The curve can be divided into three parts in which ORR kinetics are controlled in different ways. In the kinetic-controlled area, the rate of ORR is quite slow and the current density increases slightly as the potential decreases. In mixed kinetic- and diffusion-controlled zone, the reaction accelerates as the potential drops, reflecting in a remarkable increase of the current density. In the diffusion- controlled zone, the current density is determined by the diffusion rate of reactants and reaches a platform at a certain rotating speed. There are usually two key parameters that are compared to different catalysts, that is, onset potential (E_{onset}) and half-wave potential ($E_{1/2}$). In a nutshell, the more positive these two potentials are, the more active this catalyst is. However, E_{onset} is often defined in a variety of ways. One definition is that 5% of the limit diffusion current should be E_{onset}. Also, a potential at which the current density exceeds a threshold value of 0.1 mA cm^{-2} was also suggested. The definition of half-wave potential is relatively clear, which generally refers to the electrode potential corresponding to the current of half of the limiting diffusion current.

The electrons transfer numbers (n) in the ORR process was calculated by Koutecky-Levich equation:

$$\frac{1}{J} = \frac{1}{J_K} + \frac{1}{J_L} = \frac{1}{B\omega^{0.5}} + \frac{1}{J_K} \tag{7.35}$$

$$B = 0.2nFC_0D_0^{2/3}\upsilon^{-1/6} \tag{7.36}$$

In which J is the measured current density, J_K is the kinetic current density, J_L is the diffusion-limiting current density, ω is electrode rotation rate, n is transferred electron number, F is the Faraday constant, C_0 is the bulk concentration of O_2, D_0 is the diffusion coefficient of O_2 and ν is the kinematic viscosity of the electrolyte.

In experiment, the RRDE was used, thus the apparent electron transfer numbers and the H_2O_2 yield could be calculated by the following equation,

$$H_2O_2\,(\%) = 200 \times \frac{I_R/N}{I_R/N + I_D} \tag{7.37}$$

$$n = \frac{4I_D}{I_R/N + I_D} \tag{7.38}$$

In which I_D is the current density of the disk, I_R is the current density of the ring, and N is the H_2O_2 collection coefficient at the ring (0.37).

The kinetic current was calculated using the following equation:

$$J_k = \frac{J \times J_L}{J_L - J} \tag{7.39}$$

In which J is the measured current density, J_K is the kinetic current density, J_L is the diffusion-limiting current density. The kinetic current densities of different catalysts are often compared at relatively high potential (e.g. 0.85 V vs. reversible hydrogen electrode (RHE)).

For an ORR of a four-electron process, we can think of it as a reverse process of OER, which will provide a better understanding when we consider its DFT calculation. We can still compare the performance of different catalysts by calculating the energy difference of each intermediate based on the OER method described earlier to obtain the energy change of the reaction process. In addition, we can compare factors such as bond lengths when oxygen molecules adsorb. Of course, the D-band center theory is still a very important tool to explain the difference in ORR catalytic performance. By calculating the density of electronic states (DOS) and the D band center, it is helpful for us to understand the catalytic principle of the catalyst in a deeper level and to design a better catalyst.

7.3.3 Mesoporous Materials Electrocatalysts for ORR

In recent years, there has been extensive research on porous carbon materials co-doped with transition metals and heteroatoms for the purpose of oxygen reduction reaction (ORR). Both theoretical and experimental approaches have been employed to study these materials. One key focus has been on the development of hierarchically ordered macro/meso/microporous structures that can facilitate rapid mass transfer and efficient utilization of active centers. Wang et al. introduced a novel synthesis method for the fabrication of Fe-doped ordered macro/microporous single-crystalline ZIF-8, referred to as OMS-Fe-ZIF-8. Subsequently, the OMS-Fe-ZIF-8 material was subjected to carbonization, resulting in the formation of atomically dispersed FeN_4-doped hierarchical ordered porous carbon, denoted as FeN_4/HOPC. The FeN_4/HOPC material exhibited a three-dimensional interconnected hierarchical ordered porous carbon framework, with Fe atoms uniformly distributed at the atomic level.

The synthesis process of FeN_4/HOPC-c-1000 is depicted. Initially, PS templates were immersed in a low concentration Fe^{3+} methanol solution and dried under vacuum, resulting in the deposition of iron on the PS surface. Subsequently, the iron-doped PS template was immersed in another methanol solution containing zinc nitrate and 2-methylimidazole, followed by soaking in a mixed solution of methanol and ammonia to induce the formation of the Fe-ZIF-8 crystalline phase. After removing the PS template with THF, the Fe-doped ordered macro/microporous single-crystalline ZIF-8 (OMS-Fe-ZIF-8) precursor was obtained. Carbonization of OMS-Fe-ZIF-8 at 1000 °C led to the synthesis of FeN_4-doped hierarchical ordered porous carbon (FeN_4/HOPC) as the catalyst. The OMS-Fe-ZIF-8 exhibited a porous tetrakaidecahedron single-particle morphology, similar to previous reports. Upon pyrolysis at 1000 °C, the morphology and pore structure of FeN_4/HOPC-c-1000 were well preserved, with a certain degree of collapse. The channel diameter shrank from 118 to 96 nm, and the honeycomb crystal transformed into an interconnected carbon skeleton. Additionally, the thin walls between macropores fractured, leading to the formation of numerous mesopores. The ordered interconnected macro/mesopores promote efficient mass transfer on the electrode surface. This work present the BET surface area, pore volume, and pore size distributions of FeN_4/HOPC-c-1000 and FeN_4/C. Notably, FeN_4/HOPC-c-1000 exhibited a higher specific surface area (1483.2 $m^2\ g^{-1}$) and pore volume (1.59 $cm^3\ g^{-1}$) compared to FeN_4/C (757.4 $m^2\ g^{-1}$ and 0.54 $cm^3\ g^{-1}$). The type-IV N_2 adsorption–desorption isotherms of FeN_4/HOPC-c-1000 indicated the presence of mesopores, which formed during pyrolysis through the fracturing of thin walls between macropores.

The presence of atomically dispersed Fe sites in the carbon skeleton is evident from the atomic-scale bright spots observed. These sites play a crucial role as active sites during the ORR process. To analyze the coordination structure of Fe in FeN_4/HOPC-c-1000, the authors compared the Fe K-edge EXAFS spectra of FeN_4/HOPC-c-1000 with that of Fe_2O_3 and Fe foil. In the case of Fe_2O_3, a single peak at 2.75 Å corresponding to Fe–Fe backscattering is observed, while in the Fe foil, two strong peaks at 2.18 Å and 4.28 Å represent Fe–Fe interactions. However, in FeN_4/HOPC-c-1000, no distinct Fe–Fe peak is observed, indicating the absence of Fe clusters formed during the pyrolysis process. This supports the presence of atomically dispersed Fe atoms, which are likely coordinated with either oxygen (Fe–O) or nitrogen (Fe–N). To further elucidate the coordination structure of Fe, EXAFS fitting was performed based on the first shell. The well-resolved single peak at 1.65 Å can be attributed to Fe–N coordination rather than Fe–O coordination.

The ORR activity of samples synthesized under various conditions was evaluated using LSV in O_2-saturated 0.5 M H_2SO_4, with 20 wt% Pt/C as the reference catalyst. The optimized FeN_4/HOPC-c-1000 catalyst displayed a respectable half-wave potential of 0.80 V, which is only 20 mV more negative than that of commercial Pt/C. The electron transfer number of FeN_4/HOPC-c-1000, determined from the slopes of Koutecky-Levich (K-L) plots, approached the ideal value of 4.0 within the potential range of 0.65–0.25 V. Additionally, the average yield of H_2O_2 and the average number of transferred electrons for FeN_4/HOPC-c-1000 were calculated as 1.75% and 3.97, respectively. To assess methanol crossover resistance and catalyst stability, chronoamperometric measurements were conducted at the respective half-wave potentials of FeN_4/HOPC-c-1000 and Pt/C in O_2-saturated 0.5 M H_2SO_4 with the RDE at a rotation rate of 1600 rpm. After 24 h, the current of the Pt/C catalyst significantly decreased to 53% of the initial value due to Pt nanoparticle dissolution/agglomeration on the support. In contrast, the current of FeN_4/HOPC-c-1000 remained at 83% of the original value, top), demonstrating the excellent stability of the Fe-N-C structure in acidic solution. Moreover, upon injecting 10 mL of methanol into 100 mL of 0.5 M H_2SO_4 electrolyte, the ORR current of the Pt/C catalyst experienced a significant decline due to methanol poisoning, while only a temporary oscillation occurred for FeN_4/HOPC-c-1000, indicating its favorable methanol tolerance.

When considering wider applications in mobile devices, it is important to conduct tests under 1 bar H_2-air conditions. The FeN_4/HOPC-c-1000 catalyst exhibited a remarkable power density limit of 0.42 W cm^{-2} at a cell voltage of 0.57 V, surpassing the performance of FeN_4/C which achieved 0.35 W cm^{-2} at 0.47 V. This indicates the superior output performance of FeN_4/HOPC-c-1000. Additionally, FeN_4/HOPC-c-1000 demonstrated good durability in fuel cell operation. These improved performance characteristics can be attributed to the high utilization rate of catalytic centers and the extended mass transfer pathway in FeN_4/HOPC-c-1000, facilitated by its exceptionally large specific surface area and the ordered interconnected 3D hierarchical porous structure, respectively.

In an innovative approach, Zhang et al. have developed a novel method to create a series of 2D layered meso-M/N-C/N-G nanocomposites. This process involves utilizing KIT-6/N-graphene as the template, 4,4-bipyridine as the nitrogen and carbon source, and a metal nitrate as the metal source [22]. The resulting meso-M-N-C/N-G nanocomposites exhibit several desirable characteristics, including a high specific surface area, a uniform distribution of ultra-small M-N-C nanoparticles measuring less than 5 nm, and the presence of mesopores with a size of 3.6 nm. These mesopores are formed through the assembly of nanoparticles on the surface of 2D N-graphene nanosheets. Notably, among the various versions of the meso-M/C-N/N-G nanocomposites, the Fe- and Co-based variants, particularly the meso-Fe-N-C/N-G nanocomposites, display the highest activity for the oxygen reduction reaction (ORR). These catalysts demonstrate superior performance and durability compared to the Pt/C catalyst in both acidic and alkaline environments, suggesting their potential as viable ORR catalysts for high-performance proton exchange membrane fuel cells (PEMFCs) (Fig. 7.13b–e).

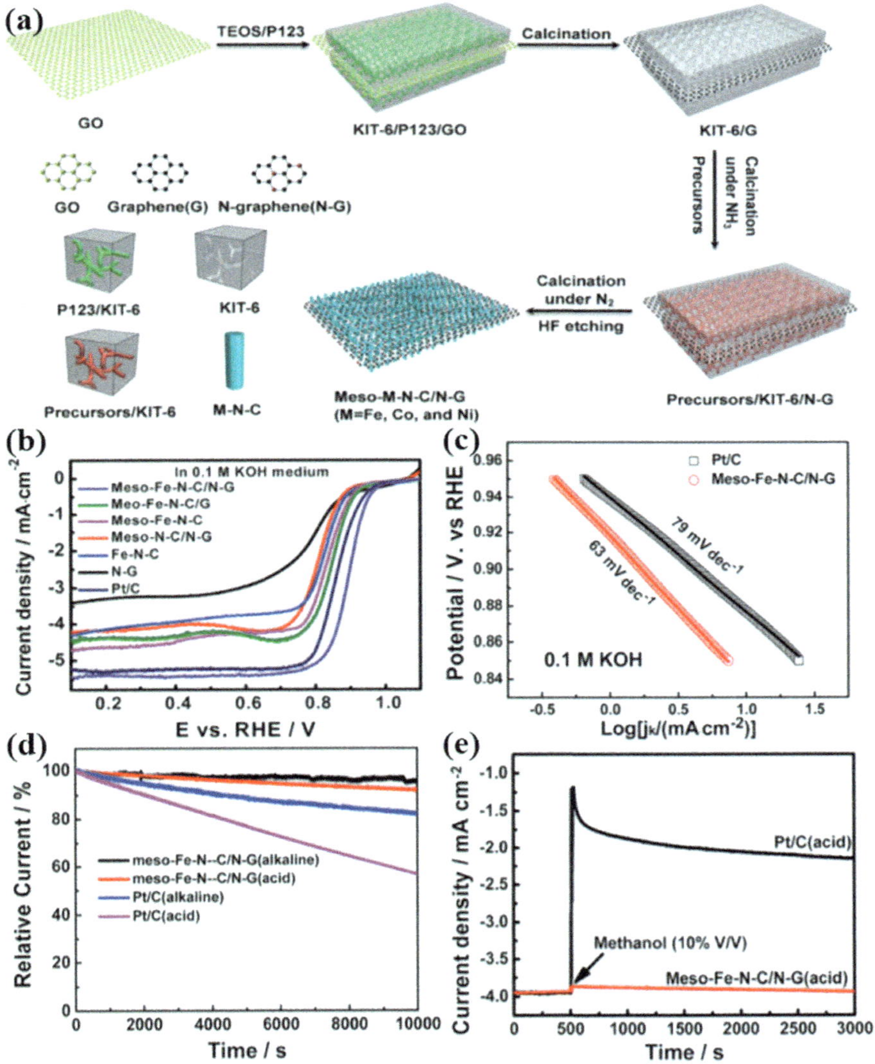

Fig. 7.13 **a** Schematic illustration showing the synthesis process of meso-M/N-C/N-G nanocomposites. **b** LSV curves of different catalysts at a rotation rate of 1600 rpm in O_2-saturated 0.1 M KOH media with a scan speed of 10 mV s^{-1}. **c** Tafel plots of meso-Fe-N-C/N-G nanocomposites and Pt/C (20 wt%) catalyst in O_2 saturated 0.1 M KOH solution. **d** Current–time chronoamperometric response of meso-Fe-N-C/N-G and Pt/C at 0.5 V versus RHE in O_2-saturated 0.1 M KOH and $HClO_4$ media. **e** Chronoamperometric response of meso-Fe-N-C/N-G and Pt/C at 0.5 V versus RHE in O_2-saturated 0.1 M $HClO_4$ media. The i–t curves are measured at a rotating speed of 1600 rpm. Reproduced with permission. Copyright 2017 The Royal Society of Chemistry

7.3.4 Carbon-Based Materials Catalysts for ORR

As previously mentioned, carbon-based materials can serve as a class of host for single-atom catalysts, which is more obvious in ORR catalysts. FeN_4, as mentioned in mesoporous materials, is also an atomic catalyst supported on a carbon base. We already know that noble metals have excellent catalytic performance for ORR, so when designed as a single atom catalyst, can they maintain excellent performance while effectively reducing the cost?

In a groundbreaking study, Chen et al. successfully synthesized a novel heterogeneous catalyst consisting of single iridium atoms (Ir-SAC) for highly efficient oxygen reduction reaction (ORR) catalysis. The synthesis of Ir-SAC involved utilizing zeolite imidazolate frameworks-8 (ZIF-8) as the host material, which possesses a cavity diameter of 11.6 Å. The active site host was combined with in-situ encapsulated Ir acetylacetonate, with a molecular diameter of approximately 9.8 Å, as a guest species. The schematic diagram provides a visual representation of this process.

Importantly, the porous structure of the host material ensures accessibility of the active metal centers, maximizing the utilization of the catalyst during ORR reactions. To verify the atomic isolation of Ir in the obtained Ir-SAC, aberration-corrected HAADF-STEM (high-angle annular dark-field scanning transmission electron microscopy) imaging was employed. The high-resolution HAADF-STEM image clearly exhibits numerous speckled bright dots, confirming the atomic dispersion of Ir.

To further corroborate the atomic dispersion and identify the atomic coordination environment, extended X-ray absorption fine structure (EXAFS) and X-ray absorption energy near edge structure (XANES) spectroscopy were employed. Commercial IrO_2 and Ir/C-Com catalysts were used as benchmarks for comparison. The Fourier transforms (FTs) of the EXAFS spectra reveal that both Ir-SAC and IrO_2 exhibit a dominant peak at approximately 1.5 Å, corresponding to the Ir-N/O scattering path. In contrast, Ir/C-Com shows a prominent peak at 2.3 Å, indicative of the Ir-Ir scattering path. The absence of Ir-Ir scattering in the Ir-SAC spectrum provides strong evidence for the atomic isolation of Ir.

Further analysis was conducted using EXAFS wavelet transform (WT), which not only offers radial distance resolution but also k-space resolution, enabling discrimination of backscattering atoms. Consistent with the FT analysis, the WT analysis of Ir-SAC reveals a single intensity maximum at around 3.8 $Å^{-1}$, attributed to the M-N contributions, further confirming the absence of crystalline metal structures. The best-fitting analysis demonstrates that the main peak at 1.5 Å corresponds to Ir-N and Ir-O first-shell coordination, with coordination numbers of 4.2 and 0.8, respectively. Additionally, a minor satellite peak at 2.01 Å can be interpreted as the Ir-C contribution, suggesting a square-pyramidal configuration for the Ir-N/O bonding.

To assess its electrochemical performance, the Ir-SAC catalyst was subjected to rigorous evaluation using both rotating disk electrode (RDE) and rotating ring-disk electrode (RRDE) techniques in a 0.1 M $HClO_4$ solution, alongside other catalysts including Ir/N-C, Ir/C, Ir/C-Com, and commercial Pt/C (Pt/C-TKK). This work show cases the ORR polarization curves obtained for these catalysts.

Among the Ir-based catalysts, Ir/C-Com displayed the lowest ORR activity, characterized by onset potential (Eonset) and half-wave potential ($E_{1/2}$) values of 0.81 V and 0.59 V, respectively. In contrast, the newly developed Ir-SAC exhibited remarkable catalytic performance, surpassing expectations. It demonstrated an ultra-high Eonset of 0.97 V and an $E_{1/2}$ of 0.864 V, outperforming many previously reported non-platinum alternatives in acidic electrolytes. Notably, the Ir-SAC even exhibited superior ORR activity compared to Pt/C, as evident from its more positive $E_{1/2}$. This exceptional performance is achieved with significantly lower metal loading on the electrode (0.8 μg Ir cm^{-2} compared to 6 μg Pt cm^{-2}), making the Ir-SAC highly promising for practical applications. It exhibits a substantially higher mass activity of 12.2 A mg^{-1} Ir at 0.85 V, representing a remarkable 27-fold improvement over Pt/C (0.423 A mg^{-1} Pt).

Moreover, the turnover frequency (TOF) was calculated to assess the intrinsic activity per active site and compare it with other notable catalysts. Impressively, the Ir-SAC achieved a record-high TOF of 24.3 e^{-} site -1 s^{-1} at 0.85 V versus RHE, which is 5.6 times higher than that of commercial Pt/C (3.63 e^{-} site -1 s^{-1}). These findings highlight the exceptional electrochemical performance of the Ir-SAC catalyst and its potential as a cost-effective alternative to platinum-based catalysts.

To unravel the underlying factors behind the exceptional electrocatalytic performance of Ir-SAC and understand the fundamental distinction between single atoms and particles, density functional theory (DFT) calculations were conducted. These calculations shed light on the origins of the catalytic behavior observed.

The adsorption energy plays a crucial role in determining the ORR activity. Stronger adsorption energies result in difficulties in the desorption process, leading to reduced ORR activity. This is evident in the ORR free energy diagrams of the Ir_{19} cluster, where uphill energy profiles with OH* desorption as the rate-determining step were observed. In contrast, the free energy diagram of ORR on Ir-SAC exhibited a consistent downhill energy profile via an associative pathway, with a limiting potential (U) of 0.43 V. This aligns with the higher ORR activity demonstrated by Ir-SAC. Comparatively lower overpotentials were obtained on Ir-SAC (0.80 V) compared to Ir-C-SAC (2.20 V) and Ir_{19} cluster (1.64 V), indicating faster ORR kinetics on Ir-SAC.

The charge density difference diagram of IrN_4 clearly illustrates the electron transfer from Ir atoms to the neighboring N atoms. This electron transfer renders the originally electroneutral Ir positively charged, which significantly differs from the metallic state observed in the Ir_{19} cluster. The position of the d-band center, which correlates strongly with the metal-adsorbate interaction, was analyzed through density of states (DOS) calculations. The d-band centers were estimated to be $-$2.89 eV and -2.65 eV for Ir-SAC and Ir_{19} cluster, respectively. The downshifted d-band center results in decreased adsorption energy, consistent with the reduced ΔGOH^*. The greater electron delocalization observed in the Ir_{19} cluster contributes to its stronger adsorption capability towards OH, as confirmed by the DOS overlap of Ir 5d and O 2p states. In the case of Ir-SAC, the coordinated N atoms localize the d-electrons of Ir, thereby reducing its adsorption energy towards OH.

By combining experimental findings with theoretical calculations, this study provides valuable insights into the application of single-atom catalysts in ORR and elucidates their underlying mechanisms. These findings serve as a valuable reference for future research in this field.

Biomass-derived carbon materials offer a diverse range of sources, with wood being a prominent example. In this study, raw wood, rich in cellulose/hemicellulose microfibers and lignin, is utilized to produce metal-free mesoporous N-doped carbons (N/HPC) with a high surface area and hierarchical pores. The preparation involves a straightforward pyrolysis process that takes advantage of enzyme-catalyzed hydrolysis of cellulose microfibers, generating abundant preformed micropores. The calculated BET surface area highlights N/E-HPC-900 as the highest (1039 $m^2\ g^{-1}$).

N/E-HPC-900 exhibits remarkable ORR/OER activity, with a cathodic peak and equivalent onset potential comparable to that of the benchmark Pt/C catalyst. Moreover, They demonstrate that the kinetic current density (jK) at 0.7 V (vs. RHE) surpasses that of the reference Pt/C catalyst (18.1 vs. 8.0 mA cm^{-2}). These findings position N/E-HPC-900 as a promising cathode material for Zn–air batteries. A reversible Zn–air battery based on this material demonstrates a current capacity of 801 mA h g^{-1} and sustains continuous operation for 110 h without any discernible performance degradation.

This non-templating approach proves highly effective, scalable, and well-suited for converting low-cost biomass into high-value products suitable for various energy-related applications.

7.3.5 Mesoporous Carbon-Based Film Materials for ORR

In order to enhance mass and electron transport in practical metal-air batteries, active electrocatalysts are typically loaded onto 3D conductive substrates such as carbon cloth or carbon paper (CP). However, this process requires the use of additional binding additives, resulting in a complicated procedure and increased electrode resistance. An alternative approach would be the in situ generation of graphene nanosheets (GNSs) from the conductive substrate, followed by nitrogen (N) doping to create active sites. This would allow the GNSs to directly serve as a self-standing electrode, simplifying the electrode manufacturing process. Furthermore, the integrated structure and natural active sites of GNSs would significantly enhance the electrode's catalytic stability and electron conductivity. In a successful development by Huang et al. a defect-enriched and PN-dominated carbon-based nanomaterial is achieved through the in situ exfoliation of graphene from carbon paper, followed by high-temperature ammonia treatment [23].

The carbon fiber surface exhibits a smooth and flattened appearance. Through chemical exfoliation and heat treatment, the adhesives are partially destroyed and etched away, resulting in a rough and porous surface. This indicates the creation of defective sites in D-CP@G. The morphology of the CP-based material remains largely unchanged after ammonia treatment, as seen in DN-CP@G. N_2 adsorption and desorption measurements at 77 K were conducted to characterize the specific surface area and pore size of the CP-based materials. Pristine CP shows a low adsorption capacity and an overlap of adsorption-sorption isotherms, indicating poor porosity and low surface area (11.6 $m^2\ g^{-1}$). In contrast, D-CP@G and DN-CP@G exhibit high adsorption capacity and characteristic H_1 and H_3 isotherms, suggesting a higher surface area (60.5 $m^2\ g^{-1}$ for D-CP@G, 67.3 $m^2\ g^{-1}$ for DN-CP@G) and a hierarchically porous structure spanning from micro- to mesopores, respectively.

The ORR activity of the materials follows the order of pristine CP, D-CP@G, and DN-CP@G, as evidenced by their respective half-wave potential ($E_{1/2}$) and limiting current density. The n values obtained from the K-L plots are 2.4, 3.2, and 3.7 for pristine CP, D-CP@G, and DN-CP@G, respectively. The high n value of DN-CP@G indicates its efficient ORR process with minimal by-products. DN-CP@G exhibits not only good ORR activity but also high stability in 0.1 M KOH solution. Taking advantage of its excellent ORR/OER performance and unique integrated 3D porous structure, DN-CP@G is utilized as a self-standing air electrode in the cathode of a Zn-air battery. As depicted, the assembled battery maintains a high open-circuit voltage (OCV) of 1.43 V over an extended period. For comparison, a mixture of Pt/C and Ir/C catalysts (1:1 mass ratio) is tested under the same conditions. This work also shows that during the initial cycling test, the mixed Pt/C + Ir/C electrode exhibits higher discharge voltage and lower charge voltage compared to the DN-CP@G electrode. However, the performance of the Pt/C + Ir/C electrode gradually deteriorates with continuous cycling, resulting in a voltage gap of 1.41 V after 40 cycles, indicating low voltaic efficiency and poor durability. In contrast, the DN-CP@G electrode maintains highly stable discharge and charge voltage plateaus throughout the long-term cycling test. Even at the 250th cycle, the voltage gap of the DN-CP@G electrode is only 0.98 V, nearly identical to the initial cycle (0.95 V), demonstrating its long cycle life and high voltaic efficiency. Additionally, an all-solid-state (ASS) Zn-air battery is fabricated using the DN-CP@G electrode directly as the air cathode. Remarkably, the ASS Zn-air battery exhibits stable discharge and charge cycles at a current density of 1 mA cm^{-2}, further confirming the outstanding electrocatalytic activity and stability of the DN-CP@G electrode, as discussed above.

In a coincidental discovery, Yu et al. present a highly effective approach to enhance the ORR and OER activities of commercial carbon cloth (CC) through combined in situ texturizing and surface engineering techniques. They demonstrate the direct utilization of CC as a cost-effective, efficient, robust, self-supporting, and bifunctional air electrode in practical rechargeable Zn-air batteries. To further expand its application, the authors construct a flexible all-solid-state Zn-air battery, where N, S-CC serves as an additive-free air electrode positioned alongside a piece of zinc foil coated with alkaline gel electrolyte. Remarkably, the assembled solid-state battery exhibits excellent flexibility and achieves a high open-circuit voltage of up to 1.25 V. The charge/discharge cycles for a single solid-state battery can be maintained for 120 cycles at 5 mA cm^{-3} under various bending conditions. As a demonstration, two solid-state batteries connected in series successfully power a blue LED watch. These results highlight the promising potential of N, S-CC-based solid-state batteries in the field of smart and flexible electronics.

Due to limited testing technology, most ORR catalysts are powder catalysts. Self-supporting or thin film catalysts have been reported but not much. In the case of film catalysts, the film/self-supporting material is often cut into the appropriate shape and affixed to the ring disk electrode. Nevertheless, we believe that self-supporting/thin film catalysts will receive considerable attention. As the above examples, this kind of catalyst has good mechanical properties, which makes it more widely used in various practical applications, such as all solid flexible metal air battery. With the improvement of testing technology, the development of mesoporous carbon-based film materials as ORR catalysts will become a worthy direction of research.

7.3.6 Hydrogen Peroxide Production

Hydrogen peroxide is one of the 100 most important chemicals in the world. But current industrial production of hydrogen peroxide need a large amount of energy consumption, and involves an energy-intensive anthraquinone oxidation/reduction, which requires complex and large-scale infrastructure and generates a substantial volume of waste chemicals. Another option that is currently getting a lot of attention and may be used as an alternative to industrial production is to obtain hydrogen peroxide directly from oxygen using an electrochemical method. If we can selectively control the oxygen reduction process, that is, put the oxygen reduction through a two-electron process, then we can get the hydrogen peroxide directly without any contamination.

The mechanism of H_2O_2 production (two-electron pathway) can be expressed by following equations:

In alkaline medium:

$$O_2 + H_2O + e^- \rightarrow OOH^* + OH^- \quad (7.40)$$

$$OOH^* + e^- \rightarrow HO^- \quad (7.41)$$

In acidic medium:

$$O_2 + H^+ + e^- \rightarrow HOO^* \quad (7.42)$$

$$HOO^* + H^+ + e^- \rightarrow H_2O_2 + * \quad (7.43)$$

The electrochemical performance of oxygen reduction to H_2O_2 was usually investigated by the rotating ring-disk electrode (RRDE) technique in a three-electrode cell system. Typically, a glassy carbon disk Pt ring electrode is employed as the working electrode. To quantify the amount of H_2O_2 generated, the ring is set at ≈1.2 V versus RHE, where the oxygen reduction current is negligible and the H_2O_2 oxidation on the ring electrode is mass transport limited. This gives rise to a positive current at the ring electrode, as H_2O_2 is produced at the disk. Hydrogen peroxide yield and electron transfer number are two important indicators. The formula for calculating these two parameters is the same as that for the four-electron process. And the Faradaic efficiency (λ%) can be defined as the ratio of charge converted to H_2O_2 to the total charge transferred:

$$\lambda\% = (i_R/N)/i_D \times 100 \tag{7.44}$$

Generally, the as-synthesized H_2O_2 can decompose through either electrochemical peroxide reduction reaction(PRR) or chemical peroxide disproportionation reaction (PDR):

$$HO^- + H_2O_2 + 2e^- \rightarrow 3OH^- \tag{7.45}$$

$$H_2O_2 + 2H^+ + 2e^- \rightarrow 2H_2O \tag{7.46}$$

$$2HO^- \rightarrow 2OH^- + O_2 \tag{7.47}$$

$$2H_2O_2 \rightarrow 2H_2O + O_2 \tag{7.48}$$

which represents a serious problem for H_2O_2 production via the two-electron ORR in a fuel cell. The chemical disproportionation experiment can be carried out by dispersing catalysts in a H_2O_2 solution and measuring the H_2O_2 concentration by titration method at specified time intervals.

Achieving high activity and selectivity in electrocatalysts is crucial for the two-electron route. While noble metals and their alloys have shown excellent performance, their limited availability poses challenges for large-scale applications. Carbon-based materials offer a promising alternative as they are abundant, versatile, and electrochemically stable. In this regard, Cui et al. present a novel strategy to enhance the activity and selectivity of carbon catalysts for H_2O_2 production through surface oxidation [24]. By oxidizing commercially available carbon nanotubes (CNTs), they observe a significant reduction in overpotential and an enhanced selectivity of up to 90% in both basic and neutral media. Electrochemical results in a basic electrolyte (0.1 M KOH, pH ~ 13) are shown in Fig. 7.14a–c. The oxygen reduction currents measured on a disk electrode demonstrate that the O-CNTs exhibit higher current and remarkably lower overpotential (~130 mV lower at 0.2 mA) compared to commercial CNTs. The rapid increase in current for O-CNTs indicates fast ORR kinetics. Furthermore, the O-CNTs demonstrate significantly higher H_2O_2 selectivity (~90% in the potential range of 0.4–0.65 V) compared to CNTs (~60%). Long-term testing (~10 h) confirms the stability of O-CNTs with negligible changes in activity or selectivity, as measured on both the ring and disc electrodes. These findings highlight the potential of surface-oxidized carbon catalysts for efficient and selective H_2O_2 synthesis.

In their study, Cui et al. utilize density functional theory (DFT) calculations to investigate the activities of various oxygen functional groups for oxygen reduction reaction (ORR) towards H_2O_2 production [24]. They introduce nine different configurations of oxygen functional groups, including carboxyl (–COOM, M=H and Na), carbonyl (C=O), etheric (O–C–O), and hydroxyl (–OH), at different positions on the graphene surface (Fig. 7.14d). The descriptor $\Delta GOOH^*$ is employed, and an activity volcano is plotted to illustrate the activities of the different oxygen functional groups. The limiting potential (UL), which serves as a metric of activity, is defined as the lowest potential at which all the reaction steps have a downhill free energy. The theoretical overpotential is determined as the maximum difference between the limiting potential and equilibrium potential. The calculated UL as a function of $\Delta GOOH^*$ for the two-electron ORR to H_2O_2 is depicted in Fig. 7.14e. The structures located on the right side of the volcano exhibit weak binding of OOH*, making the formation of OOH* the limiting step. Conversely, the structures located on the left side bind OOH* strongly, resulting in the formation of HO^{2-} as the limiting step. Hence, the binding strength of OOH* to the surface is a crucial factor in determining the ORR activity. The maximum limiting potential is found to be 0.70 V, representing zero overpotential at the top of the volcano. The computed values in Fig. 7.14e indicate that the –OH functional group has minimal contribution to the ORR. However, the C–O–C groups on the basal plane and at the graphene edge (O basal 1 and O edge) exhibit high activity for the two-electron reduction of oxygen to H_2O_2, with overpotentials of 0.02 V and 0.06 V, respectively, comparable to previously reported noble metal catalysts. These experimental and theoretical findings offer valuable insights for catalyst development in the production of industrial chemicals using clean and renewable electrical energy.

Additionally, Bao et al. have developed a simple technique to synthesize carbon materials containing isolated patches of hexagonal boron nitride (h-BN) within the carbon structure, and these materials have shown great potential as catalysts for H_2O_2 production [25]. In their study, they performed density functional theory (DFT) calculations and discovered that the interface between the h-BN domains and the graphene lattice of the host material exhibits unique catalytic properties that promote the selective generation of H_2O_2 through the two-electron oxygen reduction reaction ($2e^-$ ORR). This finding provides valuable insights into the high catalytic activity observed in the carbon materials incorporating h-BN domains.

In the field of oxygen reduction reaction (ORR), understanding the underlying elementary reaction steps is crucial for designing catalysts that can selectively drive different pathways, such as the two-electron reduction to H_2O_2 or the four-electron reduction to H_2O. Wang et al. have made significant progress in this area by investigating transition metal (TM) single atom coordination motifs as a means to control the ORR pathways [26]. They anchored TM single atoms (Fe, Pd, Co, and Mn) into vacancies of carbon nanotubes (TM-CNT) and coordinated them with neighboring C, O, or N atoms to fine-tune the reaction pathways. Among the TM-CNT catalysts,

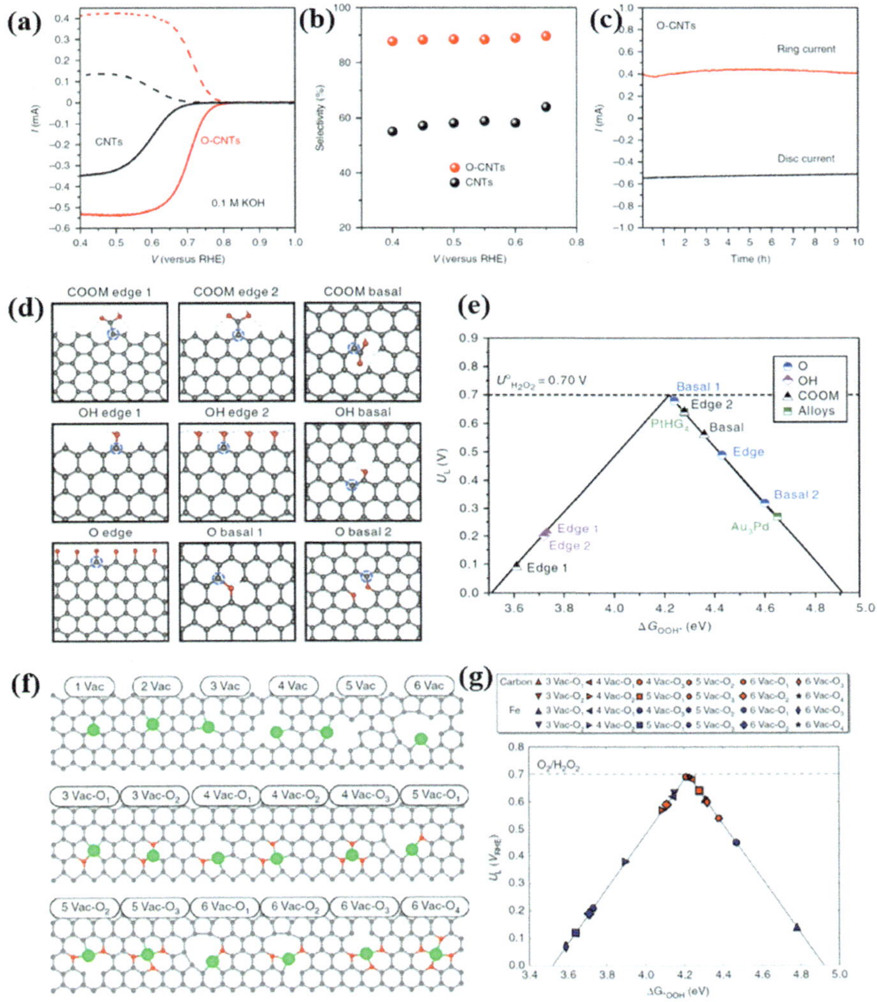

Fig. 7.14 **a** Polarization curves at 1,600 r.p.m. (solid lines) and simultaneous H_2O_2 detection currents at the ring electrode (dashed lines) in 0.1 M KOH. **b** Calculated selectivity of these two catalysts at various potentials at 0.1 M KOH. **c** Stability measurements of O-CNTs at 0.1 M KOH. **d** Different oxygen functional group type configurations examined in this study. The carbon atoms denoted by a blue circle are the active sites under investigation (M=H and Na). **e** Calculated two-electron (solid black) ORR-related volcano plot for the electro-reduction of oxygen to H_2O_2 displayed with the limiting potential plotted as a function of ΔG_{OOH*}. The equilibrium potential for the two-electron ORR is shown as the dashed black line. Reproduced with permission. Copyright 2018, Nature Publishing Group. **f** All examined configurations for single Fe atom coordinated in two-dimensional carbon material with and without O species. Green, red, and gray colors denote Fe, O (or N), and C atoms, respectively. **g** The calculated ORR activity volcano plot for $2e^-$ pathway to H_2O_2. Red and bluesymbols indicate *OOH adsorption at C and Fe, respectively. The equilibrium potential of O_2/H_2O_2 is shown as a black dashed line. Some of the points are not shown in the volcano plot since their binding energies are out of the range. Reproduced with permission. Copyright 2019, Nature Publishing Group

Fe-CNT demonstrated remarkable performance in terms of activity and selectivity for H_2O_2 generation. It achieved an unprecedented onset potential of only 0.822 V versus the reversible hydrogen electrode (vs. RHE) to achieve a current density of 0.1 mA cm^{-2} for H_2O_2 generation in 0.1 M KOH on a rotating ring-disc electrode (RRDE), while maintaining a H_2O_2 selectivity of over 95% in both alkaline and neutral pH conditions. By altering the neighboring coordination from O to N, the TM-CNT catalyst successfully shifted the ORR pathway from the two-electron to the four-electron reduction, leading to the production of H_2O. This demonstrates the broad tunability of reaction pathways in this materials platform. Density functional theory (DFT) calculations revealed that the catalytically active C and Fe sites in Fe-C-C-O and Fe-C-C-N motifs, respectively, are responsible for the H_2O_2 and H_2O pathways (Fig. 7.14f, g). Interestingly, the incorporation of Fe atoms significantly enhanced the catalytic activity for H_2O_2 generation compared to systems with only O dopants in various Fe-C-C-O motifs, as revealed by the DFT calculations.

7.4 Carbon Dioxide Reduction Reaction

Carbon dioxide reduction is an effective means to reduce carbon emissions by means of resource utilization. After the reduction of carbon dioxide to carbon monoxide, other energy can be stored in the chemical bonds, allowing the carbon dioxide to be converted into a high value-added fuel or chemical. In the following text, the carbon dioxide reduction reaction was introduced.

7.4.1 Introduction

Due to the current environmental and energy problems, the utilization of electrochemical technology for carbon dioxide reduction reaction (CO_2RR) has been developing rapidly in recent years. The use of fossil fuels, industrial production, automobile exhaust and so on will bring carbon dioxide emissions. Unlike nature's carbon metabolism, it's a closed carbon cycle. Organisms have the ability to transfer carbon between different oxidation states to store and release energy. More importantly, there are natural photosynthesis that can reduce the released carbon dioxide. However, in industry production or human life, the link of CO_2 reduction is missing. In other words, we just release more CO_2, but only the photosynthesis in nature reduces CO_2, which greatly destroys the original carbon cycle. As a result, the amount of carbon dioxide in the atmosphere will continue to rise, causing many serious environmental problems and affecting human survival. CO_2, as a greenhouse gas, can cause the greenhouse effect, which is the main factor of climate change in the world today. The greenhouse effect will cause the global average temperature to rise, and the most direct result will be the melting of the Antarctic and Arctic glaciers, leading to the extinction of many species. In addition, the greenhouse effect will cause the rise of sea level. Many coastal cities such as Shanghai and New York will eventually be flooded if left unchecked. In this case, CO_2 capture and conversion or recycling is particularly important.

In recent years, significant progress has been made in the development of various materials for CO_2 capture, including ionic liquids (ILs), zeolites, porous carbons, porous organic polymers, covalent organic frameworks (COFs), and metal-organic frameworks (MOFs). Notably, Professor Omar Yaghi has made significant contributions in the field of CO_2 capture using MOFs and COFs [27]. While CO_2 capture is an important aspect, our focus here is on CO_2 conversion, specifically, CO_2 reduction. Direct electrochemical reduction of CO_2 offers several advantages over hydrogenation of CO_2 or CO using electrochemically produced H_2. Firstly, it combines the electrochemical water splitting and subsequent thermal hydrogenation into a single electrochemical process. Secondly, it enables the synthesis of products that are not easily achievable through thermally driven methods. Lastly, electrochemical CO_2 reduction processes can often be operated at or near room temperature and ambient pressure. These characteristics make electrochemical CO_2 reduction processes more suitable for decentralization compared to thermal processes. Additionally, they can be designed to accommodate the fluctuating electricity production from intermittent renewable sources, allowing for quick adjustment of production rates.

Unlike the previously mentioned HER, OER, and ORR, CO_2 reduction does not show that particular precious metals have extremely outstanding performance. Relatively speaking, Cu-based materials are studied more as catalysts. Therefore, the research and development of high performance catalysts for CO_2 reduction is also the focus of this field.

7.4.2 *Fundamentals of CO_2RR*

(1) Mechanism of CO_2RR

In order to better understand the process of CO_2RR and design the corresponding catalyst, the research on the mechanism of CO_2RR provides us with a good channel. In the electrochemical CO_2RR, the cathodic reaction can usually express as the following general equation [28]:

$$\mathrm{xCO + nH^+ + ne^- \rightarrow product + yH_2O} \tag{7.49}$$

Generally, the anodic reaction in electrochemical CO_2RR must be the oxygen Evolution reaction. Such an overall reaction can be expressed as follows

$$\mathrm{xCO_2 + yH_2O \rightarrow product + zO_2} \tag{7.50}$$

And since x and n are all variable in the equation, obviously, unlike OER and HER, which only has a specific product, CO_2RR can have a variety of useful products.

Here we list some common reaction products that also have industrial value and their corresponding equations:

$$CO_2 + 2H^+ + 2e^- \rightarrow HCOOH(aq) \tag{7.51}$$

$$CO_2 + 2H^+ + 2e^- \rightarrow CO(g) + H_2O \tag{7.52}$$

$$CO_2 + 6H^+ + 6e^- \rightarrow CH_3OH(aq) + H_2O \tag{7.53}$$

$$CO_2 + 4H^+ + 4e^- \rightarrow C(s) + 2H_2O \tag{7.54}$$

$$CO_2 + 8H^+ + 8e^- \rightarrow CH_4(g) + 2H_2O \tag{7.55}$$

In addition to these products that only contain one carbon atom, the production of more reduced products such as multicarbon oxygenates and hydrocarbons (C_{2+}) is highly desirable because of their higher energy density and wider applicability 4, 5.

$$2CO_2 + 2H^+ + 2e^- \rightarrow (COOH)_2(s) \tag{7.56}$$

$$2CO_2 + 8H^+ + 8e^- \rightarrow CH_3COOH(aq) + 2H_2O \tag{7.57}$$

$$2CO_2 + 10H^+ + 10e^- \rightarrow CH_3CHO(aq) + 3H_2O \tag{7.58}$$

$$2CO_2 + 12H^+ + 12e^- \rightarrow C_2H_5OH(aq) + 3H_2O \tag{7.59}$$

$$2CO_2 + 12H^+ + 12e^- \rightarrow C_2H_4(g) + 4H_2O \tag{7.60}$$

Such products are also of great significance from a point of view of fundamental research, since they involve the formation of the C–C bond, a key challenge in heterogeneous catalysis. For the formation of C_{2+} products, two pathways have been identified with regard to the C–C coupling step: (1) *CO dimerizes to form *OCCO species at low overpotentials, and (2) *CO is hydrogenated to form *CHO species, a common intermediate for the formation of methane, which takes place at high overpotentials. The first pathway was recently confirmed by an in situ spectroscopic observation of the hydrogenated CO dimer intermediate (*OCCOH) on the Cu (100) surface [29].

(2) Some chemical challenges for CO_2RR

CO_2 requires input energy to reduce it to more desirable products. The thermodynamic cost of reducing CO_2 is comparable to that for the hydrogen evolution reaction

(HER): for example, products such as carbon monoxide and ethylene are thermodynamically formed at −0.11 and +0.07 V (versus RHE, reversible hydrogen electrode), respectively (Fig. 7.15b) [30]. However, in practice the energy input required for electrochemical CO_2 reduction is greater than the thermodynamic ideal. Higher energy input, that is, electrochemical overpotential beyond the thermodynamic requirement, is required to drive the CO_2RR at appreciable rates.

On the other hand, since CO_2RR involves many different and multi-step reaction paths, the overall overpotential is greater and the reaction kinetics slower. Because of the multi-step and multi-path, it is difficult to control one step to improve the reaction rate, which is more challenging for the design of catalysts. In addition, multiple pathways lead to a variety of reaction products and even side reactions such as HER, so selectivity in the CO_2RR process is also an urgent problem to be solved.

(3) Important data for CO_2RR activity

When testing the performance of CO_2RR catalysts, several important parameters are often used to compare the performance differences of CO_2RR catalysts. First of all, as we mentioned in the previous content, overpotential is a very important parameter to measure the activity of a catalyst. In order to compare the selectivity of catalysts, Faraday efficiency is usually another important indicator in CO_2RR tests. Faraday efficiency is the portion of the electrical current going to a specific product in steady-state electrolysis. Besides, the partial current density at a given potential, which is the total current times the Faradaic efficiency. A more active catalyst is one that achieves a given partial current density at a lower overpotential, and/or provides a larger partial current density at a given overpotential. Other parameters, such as stability, ECSA, etc., are also used as performance indicators for CO_2RR catalyst. DFT calculation can help us better explore and understand the reaction path for a desired product. A major goal of applied electrocatalysis research is the development of electrode materials and conditions that are selective and active (as well as cheap and stable) for production of desired products.

7.4.3 *Mesoporous Carbon-Based Film Materials for CO_2RR*

In the CO_2RR section, we will not introduce the applications of some mesoporous material catalysts and carbon-based material catalysts alone, although these materials have been widely reported. We will directly introduce the application of mesoporous carbon based thin films as CO_2RR catalysts.

An ideal catalyst for CO_2 electroreduction should possess a large accessible surface area, highly reactive sites, and excellent stability to meet the requirements of industrial applications, such as high working overpotential, high current density for target product formation, and scalability. However, most of the previously reported catalysts for CO_2 electroreduction require fixation on a conductive substrate using insulating polymer binders like Nafion or PVDF, resulting in increased complexity

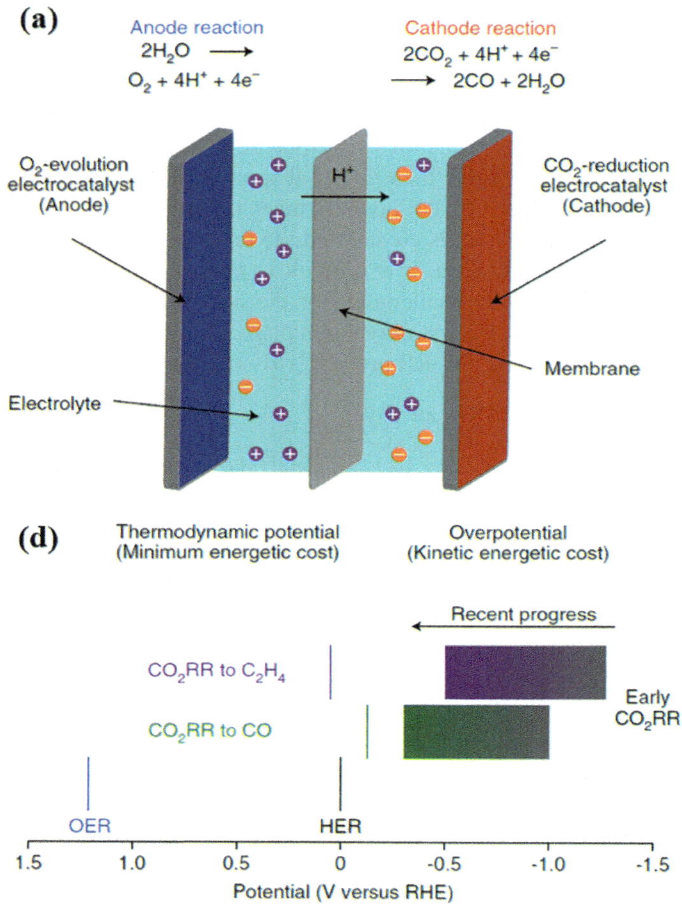

Fig. 7.15 **a** A typical electrochemical CO_2 recycling system. The anode and cathode electrocatalysts perform oxidation and reduction, respectively. The membrane separates these compartments, allowing proton transport to the cathode. The electrolyte contains positive and negative species that facilitate charge and CO_2 transport. **b** Energetic considerations for CO_2 reduction. On the reversible hydrogen electrode (RHE) scale, the thermodynamic potentials for OER (+1.23 V), HER (0 V), CO_2 to CO (−0.11 V), and CO_2 to C_2H_4 (+0.07 V) are compared, alongside the typical kinetic overpotentials needed for CO_2RR. Progress in lowering these overpotentials (at current densities of 5 mA cm^{-2}) are shown with the shaded bars. Reproduced with permission. Copyright 2019, Nature Publishing Group

and cost. Therefore, there is a need to develop a scalable approach that can transform inexpensive raw materials into efficient and stable catalysts. Li et al. present an effective method for constructing a hierarchical and atomistic catalyst through solid-state diffusion between bulk Ni and an N-doped carbon (N-C) layer. Notably, the resulting carbon paper is self-supported and can be directly used as a binder-free electrode for CO_2 reduction [31].

Initially, a thin film of melamine is deposited on the surface of a Ni foil using spray painting equipment. By precisely controlling the heating rate up to 1000 °C, the melamine film gradually transforms into a C_3N_4 structure that covers the Ni foil. At high temperatures, the C_3N_4 is catalytically converted to N-C by the Ni source. The strong Lewis acid-base interaction between Ni-N coordination facilitates the formation of abundant unsaturated carbon vacancies on the high-temperature N-C layer, allowing the surface Ni atoms to diffuse into the N-C soil and occupy these vacancies. The Ni 'seeds' derived from the bulk Ni act as catalysts, promoting the growth of N-CNTs on the N-C layer to achieve a hierarchical structure, wherein the one-dimensional nanotubes grow perpendicularly to the two-dimensional N-C layer. These flexible and self-standing carbon papers can be easily peeled off from the Ni foil surface, resulting in freshly prepared carbon papers (F-CPs). Subsequently, an acid-leaching step is employed to dissolve the majority of Ni seeds within the N-C soil, yielding hierarchical carbon papers (H-CPs) with abundant N-CNTs on both sides. Aberration-corrected high-angle annular dark-field scanning transmission electron microscopy (HAADF-STEM) images of H-CPs reveal the atomic dispersion of Ni atoms predominantly over the nanotube surface, demonstrating their resistance to acid leaching. Mechanical tests conducted on H-CPs demonstrate their excellent mechanical properties. Fourier transform k3-weighted c(k) function analysis of the extended X-ray absorption fine structure (EXAFS) spectra for H-CPs shows a significant decrease in the dominating Ni-Ni bonding (at 2.15 Å) and the appearance of a new Ni-N coordination at 1.42 Å. BET analysis confirms that H-CPs exhibit a higher surface area of 143.97 $m^2\ g^{-1}$ compared to F-CPs (113.21 $m^2\ g^{-1}$) due to the dissolution of Ni nanoparticles. This increase in specific surface area leads to an enhanced CO_2 adsorption capacity, as evidenced by the CO_2 adsorption isotherms, which show an increase from 3.85 to 7.64 $cm^3\ g^{-1}$ at 100 kPa.

The CO_2 electroreduction experiments were conducted using a two-compartment gas-tight cell, which utilized a Nafion-115 proton exchange membrane to prevent the oxidation of the generated products. To exclude the influence of HCO_{3-}, linear sweep voltammetry measurements were performed on F-CPs and H-CPs in N_2-saturated 0.5 M $NaClO_4$ solution and CO_2-saturated 0.5 M $KHCO_3$ solution (Fig. 7.15a). In the CO_2-saturated $KHCO_3$ solution, the H-CPs exhibited significantly higher activity with a low onset potential of 0.32 V versus RHE. Furthermore, the H-CPs maintained excellent faradic efficiency (FE) across a wide range of working potentials (0.7–1.2 V), indicating effective suppression of the competitive hydrogen evolution reaction (HER) over the single Ni sites (Fig. 7.15b). The main products catalyzed by the H-CPs and F-CPs were CO and H_2, and no liquid product was detected by 1H NMR analysis. Notably, the CO_2-to-CO catalytic behavior of the H-CPs outperformed most of the previously reported catalysts, especially at high current densities (Fig. 7.15c). Impressively, the H-CPs exhibited long-term stability over 40 h of continuous operation at 1.0 V, without any significant decay in both FE and current density (Fig. 7.15d).

The activity differences between single-atom (SA) and non-single-atom states of Ni in CO_2 electroreduction (CO_2RR) were further elucidated through spin-polarized density functional theory (DFT) calculations. The active sites of Ni SAs were modeled as various NiNx moieties (x = 1–4) embedded in a graphene nanosheet, while the Ni nanoparticles (NPs) were represented by a Ni_{309} cluster (Fig. 7.16e). The corresponding free energy profiles of all five systems, as well as the proposed reaction paths of CO_2RR on NiNx, are summarized in Fig. 7.16f. The calculations revealed that the reduction of CO_2 to COOH* on the surfaces of all five systems is energetically uphill, indicating that the initial electronic step (CO_2 activation) is the rate-limiting step for both Ni SAs and Ni NPs. Notably, the Ni_{309} cluster exhibited the most favorable ΔG (0.1 eV) for this step, followed by NiN (0.59 eV), NiN_3 (1.10 eV), NiN_2 (1.11 eV), and NiN_4 (1.64 eV). The formed COOH* species can be readily further reduced to CO* species on both NiNx and Ni_{309} clusters through proton-coupled electron transfer. For the four types of NiNx, the desorption steps of CO* all have negative ΔG values, indicating efficient removal of the formed CO* species from the catalyst surface to complete the reduction process. In contrast, the desorption of CO* from the Ni_{309} cluster is challenging ($\Delta G = 1.13$ eV) due to the strong binding strength of CO*, which is unfavorable for efficient catalysis. Therefore, the calculations reveal that Ni NPs would exhibit poorer catalytic performance compared to Ni SAs in CO_2 reduction to CO. Notably, according to the calculations, NiN_4 displayed a remarkably positive ΔGH^* value of 1.52 eV, while the chemical adsorption of H* was not stable on the single-atom Ni sites of the other NiNx. This indicates significant suppression of competitive H_2 production over all four single-atom Ni sites, supporting the experimental observation of preferential CO_2RR over the hydrogen evolution reaction (HER) (Fig. 7.17).

He et al. present a highly efficient Ni single-atom/porous carbon fiber membrane catalyst (NiSA/PCFM) for CO_2 electroreduction to CO using an electrospinning method [32]. The fabrication process of the NiSA/PCFM membrane is outlined in Fig. 7.18a, involving electrospinning and heat treatment. Notably, the NiSA/PCFM membranes produced in their laboratory apparatus have a minimum size of ~280 cm^{-2} (Fig. 7.18b). The NiSA/PCFM membrane exhibits robustness and flexibility, as demonstrated in Fig. 7.18b, retaining its initial structure even under bending stress. To evaluate the CO_2 electroreduction activity, NiSA/PCFM, NiSA/CFM, and PCFM powders were drop-casted onto a carbon paper using conventional powdering and binding methods for linear sweep voltammetry (LSV) tests and electrolysis in an H-type cell. These powdered samples were labeled as P-NiSA/PCFM, P-NiSA/CFM, and P-PCFM. Significantly, both P-NiSA/PCFM and P-NiSA/CFM exhibit excellent faradic efficiency (FE) for CO generation within the working potential range of -0.5 to -1.0 V (versus RHE), with P-NiSA/PCFM achieving a maximum FE of 95% at -0.7 V. It should be noted that the NiSA/PCFM membrane possesses remarkable mechanical strength, allowing it to serve directly as a working electrode in an H-type cell system. The highest CO faradaic efficiency obtained by NiSA/PCFM is 96% at -0.7 V, similar to that of P-NiSA/PCFM (95%). However, NiSA/PCFM exhibits a significantly higher CO partial current density of 56.1 mA cm^{-2} at -1.0 V (Fig. 7.18e), surpassing those of P-NiSA/PCFM, P-NiSA/CFM, and P-PCFM.

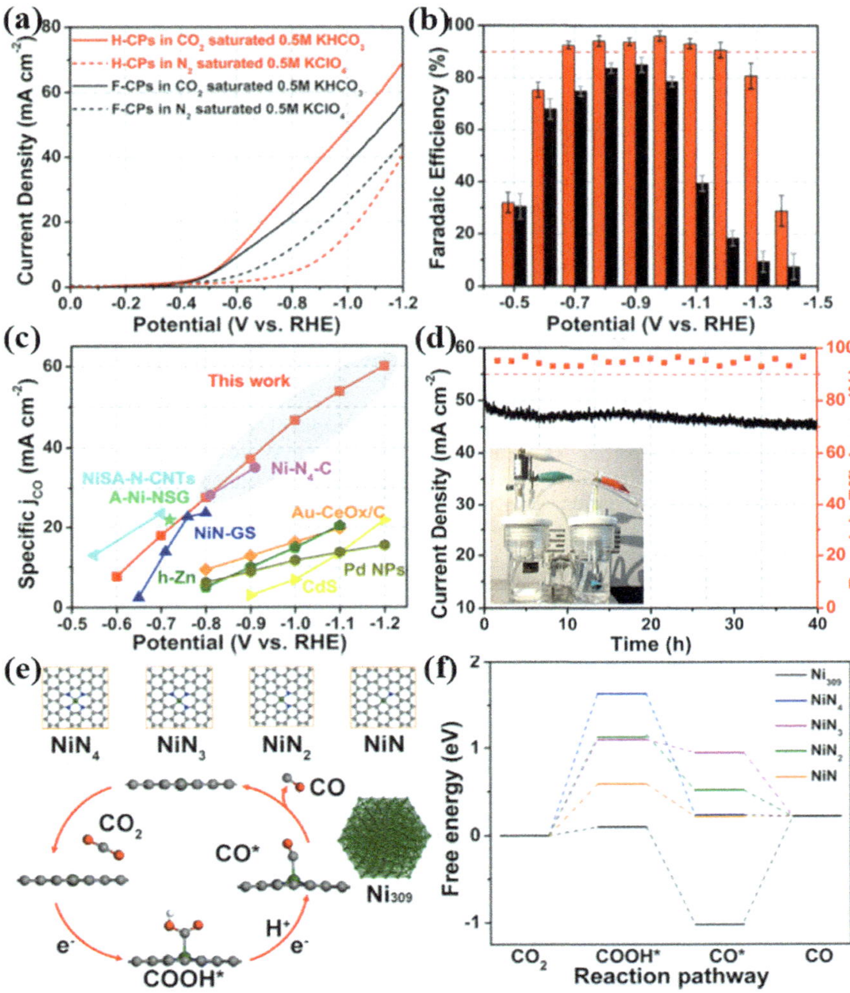

Fig. 7.16 **a** Linear sweep voltammetry curves in N_2-saturated (dotted line) 0.5 M $KClO_4$ or CO_2-saturated (solid line) 0.5 M $KHCO_3$ electrolyte at a scan rate of 10 mV s^{-1}. **b** FEs of CO and for H-CPs (red) and F-CPs (black) at different applied potentials. **c** Specific current density of H-CPs compared with those of other state-of-the-art catalysts for CO_2 to CO reduction (highlighted in the high potential region). **d** Stability test of H-CPs at a potential of -1.0 V versus RHE during 40 h. **e** Optimized atomic structures of Ni_{309} cluster, different NiN_x structures, and proposed reaction paths for CO_2 electroreduction by H-CPs. C, O, Ni, N, and H atoms are represented by gray, red, green, blue, and white spheres, respectively. **f** Free energy diagram for the conversion of CO_2 to CO on the Ni SAs and Ni_{309} cluster at $U = 0$ V versus RHE. Reproduced with permission. Copyright 2019 Elsevier Inc.

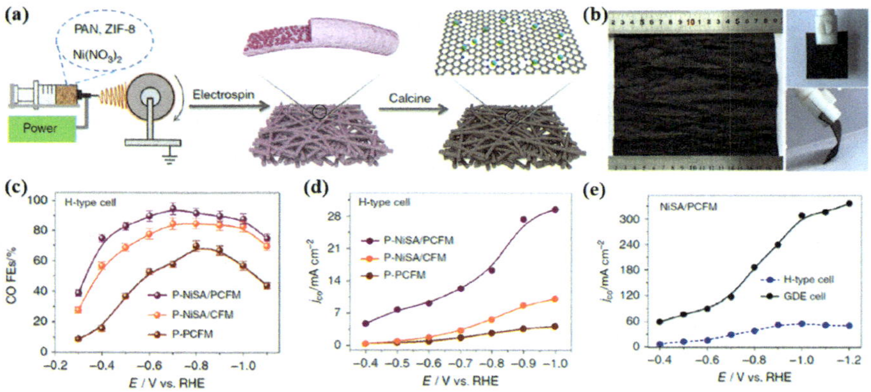

Fig. 7.17 **a** Synthesis strategy of NiSA/PCFM. **b** Digital images of NiSA/PCFM membrane. **c** CO faradaic efficiency and **d** partial current densities for three catalysts at various cathode potentials in H-type cell. **e** Partial current densities of NiSA/PCFM at various cathode potentials in different cells. Reproduced with permission. Copyright 2020, Nature Publishing Group

Moreover, using a gas diffusion electrode (GDE) device further enhances the partial CO current densities. For instance, with an 88% CO FE, NiSA/PCFM achieves a CO partial current density of 308.4 mA cm^{-2} in a GDE cell at −1.0 V, which is more than 5 times higher than that observed in the H-type cell. Additionally, the partial CO current density in the GDE cell can reach an even higher value of 336.5 mA cm^{-2} with 83% CO FE at −1.2 V. Remarkably, the NiSA/PCFM membrane exhibits excellent durability, with negligible drops in both CO FE and partial current densities during 120-h tests, retaining over 95% of the initial values. To gain insights into the observed activity differences among the three catalysts, spin-polarized density functional theory (DFT) calculations were performed on well-defined N-C and Ni-N_4-C structures. The resulting free energy profiles (ΔG) are summarized. The simulations reveal that the conversion of CO_2 to adsorbed COOH* intermediates on both N-C and Ni-N_4-C structures is energetically uphill, indicating that the initial electronic step is the rate-determining step for both systems. Notably, the Ni-N_4-C structure exhibits a more favorable ΔG (0.7 eV) than N-C (1.3 eV) for this step, underscoring the superior performance of NiSA/PCFM.

He et al. present a facile method for the synthesis of through-hole carbon nanofibers (TCNFs) incorporated with abundant and uniformly distributed Cu single atoms (CuSAs) as a highly stable catalyst for efficient electrochemical CO_2 reduction to produce further products such as methanol and acetic acid [33]. The fabrication process of flexible and self-supported CuSAs/TCNFs is depicted in Fig. 7.18a, resulting in a surface area of approximately 310 cm^2 under laboratory conditions. Remarkably, the CuSAs/TCNFs exhibit structural integrity even under bending stress, demonstrating comparable strength and flexibility to carbon cloth. Moreover, these catalysts demonstrate excellent tensile strength, tolerating approximately 1 MPa in stress–strain tests.

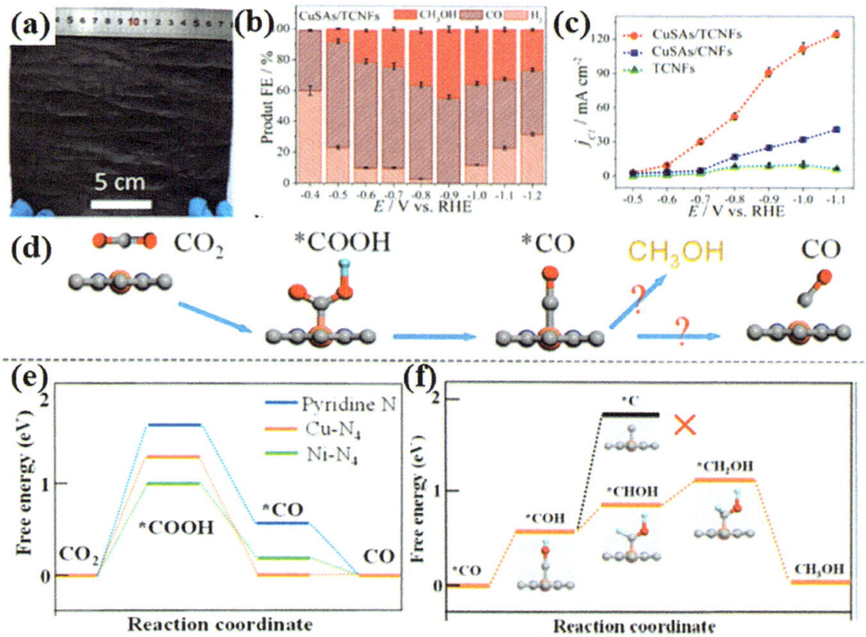

Fig. 7.18 **a** Digital image of CuSAs/TCNFs membranes. **b** Faradaic efficiencies of all products at CuSAs/TCNFs. **c** Partial current density of three samples. **d** Optimized atomic structures of CuSAs/TCNFs and proposed reaction paths for CO_2 electroreduction. **e** Free energy diagram of CO_2 to CO on pyridine N, Ni-N_4, and Cu-N_4 structure. Free energies for conversion of *CO to CH_3OH on Cu-N_4 structure. Orange, gray, dark blue, red, and light blue spheres stand for Cu, C, N, O, and H atoms, respectively. Reproduced with permission. Copyright 2019 American Chemical Society

The electrocatalytic activity of CuSAs/TCNFs, CuSAs/CNFs, and TCNFs catalysts for CO_2 reduction was investigated in a CO_2-saturated 0.1 M $KHCO_3$ solution using an undivided cell configuration. The onset potential of CuSAs/TCNFs was observed to be −0.41 V vs reversible hydrogen electrode (RHE), which was significantly more positive compared to TCNFs. Gas-phase products were periodically quantified using gas chromatography (GC) by directly venting the CO_2-purged cathode into the instrument, while liquid-phase products were analyzed using 1H NMR spectroscopy. GC analysis revealed that CO and H_2 were the predominant gas-phase products catalyzed by CuSAs/TCNFs, CuSAs/CNFs, and TCNFs. In addition, methanol was detected as a liquid-phase product in the presence of CuSAs/TCNFs, as confirmed by 1H NMR and 13C NMR spectra. The Faradaic efficiencies (FEs) of the products showed a strong dependence on the working potential, with CuSAs/TCNFs achieving the highest FE for methanol (44%) at −0.9 V. Intriguingly, nearly pure methanol (44%) and CO (56%) were obtained as liquid- and gas-phase products, respectively, at this potential. Furthermore, the through-hole structure of CuSAs/TCNFs was found to play a crucial role in enhancing the CO_2RR performance, as evidenced by the higher CO partial current density compared to CuSAs/CNFs. The stability of CuSAs/TCNFs was tested through long-term electrolysis at a stationary cathode potential of −0.9 V for 50 h, demonstrating a steady current density of approximately 90 mA cm^{-2} with negligible decay throughout the stability test.

Figure 7.18d–f illustrates the proposed reaction paths for CO_2 reduction and the corresponding free energy profiles for the different systems. Based on DFT calculations, it was found that the rate-determining step (RDS) for all three samples is the conversion of CO_2 to adsorbed COOH*. The Ni-N_4 structure exhibits the most favorable free energy (0.98 eV) for this step, followed by Cu-N_4 (1.17 eV) and pyridine N (1.62 eV). Subsequently, the COOH* intermediate can easily be transformed into a CO* species with the assistance of a proton and an electron. Both Ni-N_4 (−0.21 eV) and pyridine N (−0.54 eV) exhibit negative free energies for CO* desorption, indicating efficient removal of the CO* intermediate from the catalyst surface and the production of CO. In contrast, Cu-N_4 shows a slightly positive free energy (0.12 eV) for CO* desorption, suggesting an endergonic step. Consequently, the CO* intermediate on Cu-N_4 is more likely to be further reduced to methanol rather than being released as CO. Additionally, the DFT calculations were extended to investigate the reaction mechanism of CH_3OH and CH_4 on the CuSAs/TCNFs catalyst. It was found that the reduction of *COH to *CHOH, which is a key step for CH_4 formation, has a moderate free energy barrier (~0.86 eV) on the Cu-N_4 structure. On the other hand, the step of *COH to *C exhibits a much higher free energy barrier (~1.88 eV), exceeding the barriers for any steps in CH_3OH production. Therefore, the single-atom Cu sites on the CuSAs/TCNFs catalyst tend to favor the production of CH_3OH rather than CH_4.

The above examples demonstrate the application of mesoporous carbon-based films as CO_2RR catalysts. At present, most researches are still focused on C_1 products. In the future development, the formation of C_2 products and their corresponding mechanisms can be further studied. By the way, in the field of carbon dioxide reduction, we can also consider artificial photosynthesis, such as combining nanomaterials with photosynthetic bacteria. Professor Peidong Yang at the University of California, Berkeley has done a lot of outstanding work in this field.

7.5 Nitrogen Reduction Reaction

7.5.1 Introduction

At the beginning of this chapter, we describe the application of catalyst in ammonia synthesis industry, which has provided great help for human survival and life over the past 100 years. Ammonia, as a precursor of chemical fertilizer and fuel, has been widely used in agriculture, chemical industry, pharmacy, military and other fields. But at present, the Haber–Bosch process is still the main way to synthesize ammonia around the world. Although this technology has been a great success and has benefited many people around the world, there are still some problems with this technology. The process consumes a lot of energy because the reaction conditions require high temperatures and pressures. At the same time, the technology requires a lot of infrastructure and emits a lot of polluting gases such as carbon dioxide. It would be a huge boost to the synthetic ammonia industry if there were a way to produce ammonia with low energy consumption, clean and high yield.

Nitrogen reduction reaction (NRR), which usually means the direct photocatalytic or electrocatalytic conversion of N_2 and water molecules into NH_3, is regarded as a potential alternative to the Haber–Bosch process [34]. The electrocatalytic NRR was discovered by Davy et al. in 1807, and has received more and more attention and research in recent years. Nevertheless, it must be admitted that eNRR cannot achieve high conversion efficiency and ammonia yield at present. Therefore, proper catalyst design and further study of the corresponding mechanism can help us to improve the performance of eNRR faster. Of course, this is both a challenge and an opportunity. Because there are few studies on NRR catalyst, there is a large space for the follow-up research and promotion.

At present, there are three main problems in water phase NRR under mild conditions. First of all, N≡N key large (about 941 kJ/mol), is difficult to fracture. Secondly, water molecules in aqueous solution will compete with electrons and produce hydrogen evolution reaction (HER), which will reduce their electrochemical Faraday efficiency. Thirdly, under the condition of normal temperature and pressure, the solubility of nitrogen in water is very low, and its electrocatalytic reaction is significantly affected by the mass transfer process. Therefore, in order to improve the conversion and selectivity of nitrogen reduction, need to design reasonable catalyst system, in order to improve the mass transfer of N_2 and the chemical adsorption on the surface of the catalyst, accelerate the activation of N≡N, aquaculture and inhibit decomposition hydrogen side effects.

7.5.2 *Fundamentals of NRR*

(1) Mechanisms of NRR

The overall reaction of NRR can be expressed as follows:

$$N_2 + 6H^+ + 6e^- \rightarrow 2NH_3 \tag{7.61}$$

Similar with CO_2RR, the reaction of the cathode during electrolysis is OER.

Normally, like ORR, NRR is done with two types of reaction, dissociative and associative. In dissociative mechanism, N and N are broken up before dissociative reaction, separate nitrogen atoms are absorbed on the catalyst surface, and each single atom is converted to NH_3 during hydrogenation.

Dissociative pathway:

$$2^* + N_2 \rightarrow 2^*N \tag{7.62}$$

$$N^* + H^+ + e^- \rightarrow^* NH \tag{7.63}$$

$$*NH + H^+ + e^- \rightarrow *NH_2 \tag{7.64}$$

$$*NH_2 + H^+ + e^- \rightarrow \; * + NH_3 \tag{7.65}$$

The association mechanism, when hydrogenated nitrogen molecules, two nitrogen atoms remain combined with the status of each other, and there are two possible hydrogenated ways: one is the hydrogenation priority in the furthest away from the surface of a nitrogen atom (assuming the N_2 molecule is end coordination model), to generate an NH_3 and leaving a nitrogen atom adsorption on the catalyst surface (or metal nitrides M≡N units), is then hydrogenated second NH_3 is given. Another way of hydrogenation is that the nitrogen atoms in the two nitrogen centers are hydrogenated individually in turn until one of the nitrification is converted to NH_3 and N–N bond is broken.

Associative pathway:

$$* + N_2 \rightarrow *N_2 \tag{7.66}$$

$$*N_2 + H^+ + e^- \rightarrow *N = NH \tag{7.67}$$

$$\text{(I) } *N = NH + H^+ + e^- \rightarrow *NNH_2 \tag{7.68}$$

$$*NNH_2 + H^+ + e^- \rightarrow *N + NH_3 \tag{7.69}$$

$$*N + H^+ + e^- \rightarrow *NH \tag{7.70}$$

$$*NH + H^+ + e^- \rightarrow *NH_2 \tag{7.71}$$

$$*NH_2 + H^+ + e^- \rightarrow \; * + NH_3 \tag{7.72}$$

$$\text{(II) } *N = NH + H^+ + e^- \rightarrow *HN = NH \tag{7.73}$$

$$*HN = NH + H^+ + e^- \rightarrow *HN - NH_2 \tag{7.74}$$

$$*HN - NH_2 + H^+ + e^- \rightarrow *H_2N - NH_2 \tag{7.75}$$

$$*H_2N - NH_2 + H^+ + e^- \rightarrow *NH_2 + NH_3 \tag{7.76}$$

$$*NH_2 + H^+ + e^- \rightarrow * + NH_3 \tag{7.77}$$

In the associative process, if we're in case (II), another possible product is hydrazine. If the barrier changes in favor of the direct desorption of hydrazine from the catalyst surface, hydrazine will be present in the product.

(2) Assessment of NRR activity

In the NRR process, the testing method of LSV curve is basically the same as HER or OER. However, in addition to the overpotential comparison, LSV curves under N_2 and Ar atmospheres are often compared in NRR, and the potential range and NRR activity of nitrogen reduction reaction can be judged by the difference of LSV curves under two atmospheres. LSV curve under Ar atmosphere represents only HER occurs, the LSV curves under N_2 atmosphere represents its HER and NRR occurred at the same time, the current difference between the two curves can be seen as contribution by NRR reaction. Thus, to some extent, current differential polarization curve can represent the catalyst on the catalytic activity of nitrogen reduction reaction size, two curves by superposition gradually separate the overlapping of secondary voltage range is big probability is the voltage range of nitrogen reduction reaction occurs. Combined with the measured ammonia yield at different voltages, the catalytic activity of the catalyst was explained.

In NRR, two important indicators are ammonia production and Faraday efficiency. Commonly used in ammonia production detection method with ion chromatography and spectrophotometric method, ion chromatography method has many advantages, such as high sensitivity and good reproducibility), but is expensive and complex instruments, spectrophotometry was more widely used for lower cost, the method of Nessler's reagent and indophenol blue method is the most common. The formula of ammonia production is as follows:

$$R = \frac{[NH_4^+] \times V}{t \times S} \tag{7.78}$$

where R (mol cm^{-2} s^{-1}) is the ammonia formation rate, [NH_4^+] (mol L^{-1}) is the ammonia concentration in the electrolyte, V (L) is the volume of the electrolyte, t (s) is the reaction time and S (cm^{-2}) is the effective area of the electrode.

The calculation formula of Faraday's efficiency:

$$FE = \frac{3 \times F \times n(NH_4^+)}{Q} \tag{7.79}$$

where F (C mol^{-1}) is Faraday's constant, nNH_4^+ (mol) is the amount of ammonia in the electrolyte, and Q (C) is the applied charge.

7.5.3 Catalysts for NRR

Since there are not many NRR catalysts at present, we will not introduce a classification. In this section, we mainly introduce several NRR catalysts to briefly explore this field, hoping to enlighten readers.

To enhance the nitrogen reduction reaction (NRR) activity and achieve a higher Faradaic efficiency and NH_3 yield rate, Chen et al. introduced a novel strategy involving the modulation of electron density in copper (Cu) nanoparticles over a wide potential range, while simultaneously suppressing the hydrogen evolution reaction (HER) activity. The researchers employed a modified solvothermal method to synthesize polyimide (PI) nanoflowers, which were subsequently subjected to different levels of condensation at 300 °C (PI-300), 400 °C (PI-400), and 600 °C (PI-600) to control the degree of conjugation. These PI materials served as supports for depositing Cu nanoparticles through a wet impregnation technique.

Initial observations indicated that the Cu/PI electrode, exemplified by Cu/PI-300, exhibited higher current density under N_2 flow compared to the reference measurement in Ar (Fig. 7.18a), suggesting a potential selectivity towards N_2 reduction. Electrochemical testing of the Cu/PI-300 electrode with an optimized Cu content of 5% revealed the highest NRR Faradaic efficiency of 6.56% at a potential of −0.3 V versus RHE (Fig. 7.18b). Furthermore, the NH_3 yield rate on the Cu/PI-300 electrode, with a catalyst loading of 5 mg cm^{-2}, was further increased to 17.2 $\mu g\ h^{-1}\ cm^{-2}$ at an optimized potential of −0.4 V versus RHE.

Cu/PI-300 exhibited a lower electronic density on the Cu surface, resulting in an enhanced polarization of adsorbed N_2 molecules. This was evident from the significant differences in electron density observed. The strengthened interaction between N_2 and the Cu surface with lower electronic density was also reflected in the increased adsorption energy. Another crucial aspect in achieving selectivity for the catalyst is the suppression of the hydrogen evolution reaction (HER) during the nitrogen reduction reaction (NRR). In basic electrolyte conditions, the adsorption of water molecules and the desorption of OH^- ions typically dominate the overall mass transfer process in the HER. However, a Cu surface with an electron-deficient nature leads to a strong electrostatic interaction with OH- ions, which is unfavorable for HER in alkaline solutions. Consequently, Cu/PI-300 exhibited the poorest HER performance among all Cu/PI samples in this study.

Furthermore, the addition of the first hydrogen atom to the pre-adsorbed N_2 ($*N_2$) and subsequent steps in the NRR were facilitated by Cu clusters with even lower electronic density, as indicated by the significantly reduced Gibbs free energy change (ΔG) (from 2.30 eV on pristine Cu to 1.76 eV on Cu-0.04e^- and 1.60 eV on Cu-0.06e^-) (Fig. 7.19e). The dissociation step from $*NNH_4$ to $*NH_2$ was facilitated due to the electron-deficient Cu surface, leading to a lower free energy. This approach of inducing electron deficiency offers valuable insights into the rational design of cost-effective catalysts for the N_2 reduction reaction, achieving both high selectivity and activity.

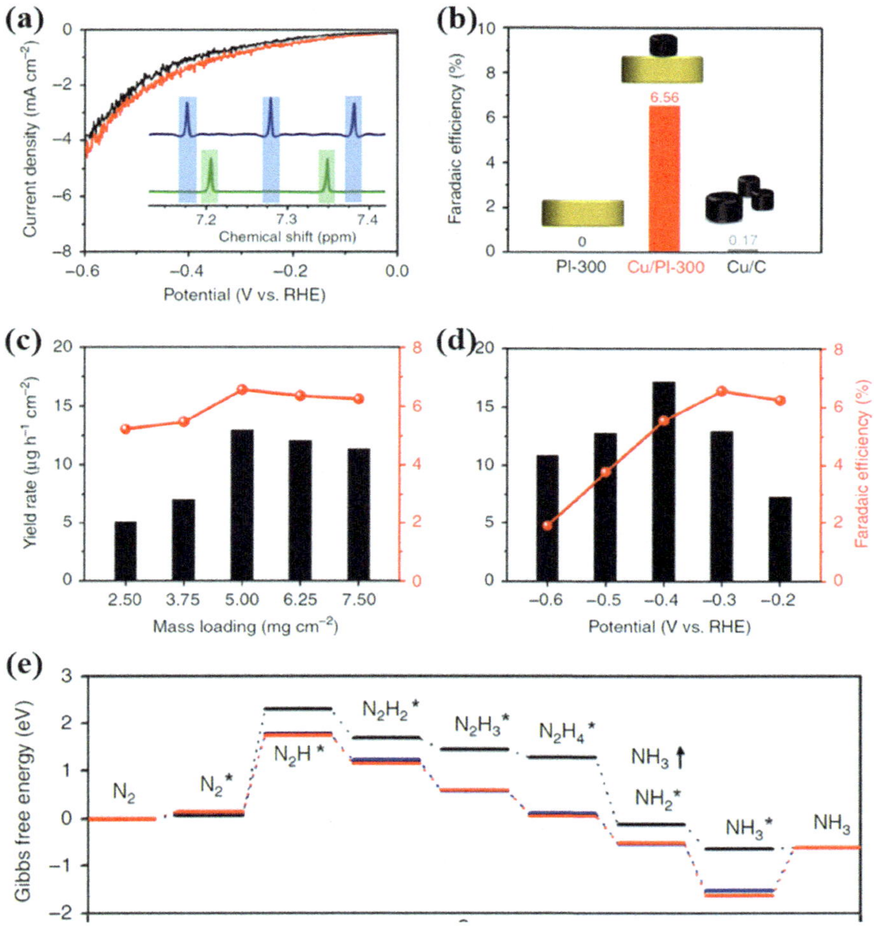

Fig. 7.19 **a** The linear sweep voltammogram (LSV) curves of Cu/PI-300 (catalyst loading: 5 mg cm^{-2}) measured at a scan rate of 10 mV s^{-1} under the Ar and N_2 atmosphere. **b** The Faradaic efficiencies of bare PI-300, Cu/PI-300, and Cu/C for NH_3 generation at −0.3 V versus RHE within 6 h. The Faradaic efficiencies (spheres) and NH_3 yield rates (bars) of Cu/PI-300 with varied catalyst loadings at −0.3 V versus RHE (**c**) or with a fixed catalyst loading (5 mg cm^{-2}) at different work potentials (**d**) within 6 h. Insets: (a) 1H NMR spectra of both $^{14}NH_4^+$ and $^{15}NH_4^+$ produced from the NRR reaction using $^{14}N_2$ and $^{15}N_2$ gas respectively. **e** Free energy diagrams of each step of NRR process on pristine Cu (black), Cu-0.04e$^-$ (blue) and Cu-0.06e$^-$(red) models. Reproduced with permission. Copyright 2019, Nature Publishing Group

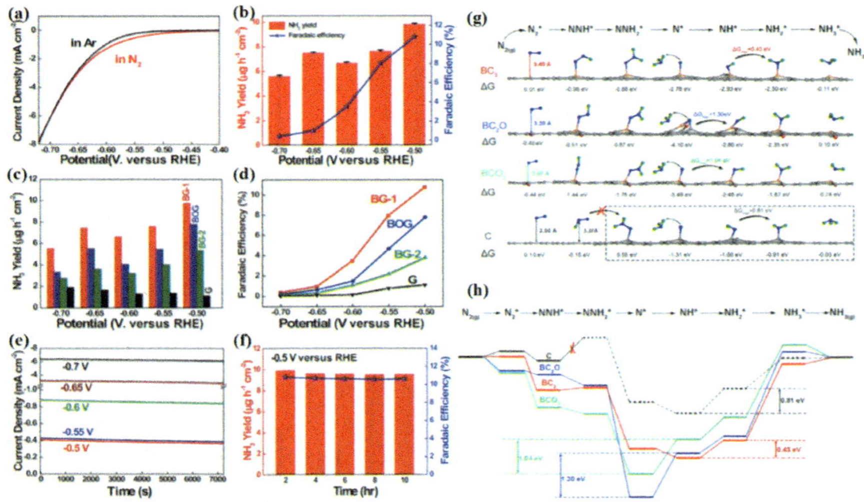

Fig. 7.20 **a** LSV curves of BG-1 in Ar- and N_2-saturated solutions at the scan rate of 1 mV s^{-1}. **b** The NH_3 production rates (left y axis) and FE_{NH3} (right y axis) of BG-1. The error bars represent the average of three independent measurements. **c** The NH_3 production rates of BG-1, BOG, BG-2, and G at different potentials. **d** The FE_{NH3} values of BG-1, BOG, BG-2, and G at different applied potentials. **e** Chronoamperometric curves of BG-1 at different applied potentials. **f** NRR stability test of BG-1 at −0.5 V versus RHE. **f** Reaction pathways and the corresponding energy changes of NRR on BC_3, BC_2O, BCO_2, and C, respectively. The dotted rectangular box indicates the steps that cannot take place. (C) Free energy diagrams of NRR on BC_3, BC_2O, BCO_2, and C, respectively. Reproduced with permission. Copyright 2018 Elsevier Inc.

To mitigate the intensified side reaction of HER during the nitrogen reduction reaction (NRR), it is crucial for the catalyst surface to facilitate strong adsorption of N_2. Zheng et al. present an innovative approach using two-dimensional boron-doped graphene (BG) as an excellent electrocatalyst for NRR, which promotes the adsorption of N_2 [35]. By incorporating boron atoms into the graphene (G) framework, the original sp2 hybridization and conjugated planar structure of graphene are preserved. The introduction of boron as a dopant induces electron deficiency in the graphene lattice, significantly enhancing the electrocatalytic activity. The synthesis of BG involves the thermal reduction of graphene oxide and H_3BO_3 in a mixed H_2/Ar gas environment. Various mass ratios of H_3BO_3 to graphene oxide, namely 5:1, 1:10, and 0:1, were employed, resulting in the formation of BG-1, BG-2, and undoped G (referred to as G), respectively. For comparison, a mixture of H_3BO_3 and graphene oxide (5:1) was annealed in an Ar atmosphere, leading to higher oxygen content and designated as BOG.

Distinct current densities were observed in the linear sweep voltammetry (LSV) curves of BG-1 in Ar-saturated (black curve) and N_2-saturated (red curve) electrolytes, indicating the reduction of N_2 within a potential range of 0.45 to 0.70 V

(Fig. 7.20a). To determine the potential reduction products, NH_3 and N_2H_4, quantification was carried out using various methods such as the sodium salicylate-sodium hypochlorite method, Nessler's reagent spectrophotometry, and the 4-(dimethylamino) benzaldehyde spectrophotometric method. Interestingly, NH_3 was the sole product observed, with no presence of N_2H_4. Further analysis of NH_3 yields and the Faradaic efficiency of NH_3 ($FENH_3$) at different applied potentials was performed for the BG-1, BOG, BG-2, and G electrocatalysts (Fig. 7.20b–d). Remarkably, the BG-1 sample exhibited superior performance with a NH_3 production rate of 9.8 $\mu g\ hr^{-1}\ cm^{-2}$ and an $FENH_3$ of 10.8% at 0.5 V, which were five and ten times higher, respectively, than those achieved by the undoped G catalyst. Similarly, the other two BG samples (BOG and BG-2) also demonstrated significant enhancements in both NH_3 production rates and $FENH_3$ values.

To explore the effect of different boron structures on the nitrogen reduction reaction (NRR), the researchers conducted first-principle calculations. Among various NRR pathways, the distal pathway in the associative mechanism (associative I) was found to be more favorable in metal-free catalysts as it allows for the stabilization of intermediates. The binding energies and equilibrium distances between an adsorbed N_2 molecule and a boron (or carbon) atom were calculated as follows: 0.01 eV, 3.40 Å for BC_3; 0.40 eV, 3.20 Å for BC_2O; 0.44 eV, 3.00 Å for BCO_2; and 0.1 eV, 3.50 Å for G (undoped graphene). The reduced distance of adsorbed N_2 on boron-doped graphene (BG) structures and the more negative nitrogen binding energies suggest that the incorporation of boron dopants enhances the adsorption of N_2 and stabilizes the B-N_2 interaction, thereby promoting the NRR. Additionally, the formation of adsorbed intermediate NH* was identified as the rate-limiting step for both BCO_2 (with an energy barrier of 1.04 eV, N*/NH*) and BC_2O (with an energy barrier of 1.30 eV, N*/NH*), indicating a significant uphill energy. On the other hand, the formation of adsorbed intermediate NH_2* was found to be the limiting step for BC_3 (with an energy barrier of 0.43 eV, NH*/NH_2*). Notably, BC_3 exhibited the lowest reaction energy barrier (0.43 eV), suggesting its superior catalytic performance for the NRR among the considered structures. This study opens up new possibilities for the development of efficient metal-free catalysts for the NRR and other electrocatalytic reactions.

Mesoporous carbon-based membrane films have shown potential as catalysts for various electrochemical reactions, including the nitrogen reduction reaction (NRR). While there are limited reports on their application in the NRR, some studies have explored their use as catalysts in this context. Ding et al. introduced a flexible TiO_2 nanofibrous membrane as a self-supported active matrix for loading transition metal electrocatalysts. By employing coordination-driven epitaxial growth, CoS nanosheets were uniformly confined onto individual nanofibers, forming a double-active heterojunction. This interfacial engineering strategy resulted in a close coupling between CoS and TiO_2, facilitating efficient charge transfer and fast reaction kinetics at the heterointerface. Carbon nanoplating was further employed to integrate the CoS@TiO_2 heterojunction, enhancing the conductivity and structural integrity of the electrocatalyst. This interfacial design led to the development of a mechanically robust and electrically conductive self-supported electrocatalyst, denoted as

C@CoS@TiO_2. The C@CoS@TiO_2 electrocatalyst exhibited remarkable performance in terms of ammonia yield and Faradaic efficiency, as well as superior long-term durability. The adsorption of a nitrogen molecule on the CoS surface enables the transfer of electrons between the 3σg orbital of nitrogen and the unoccupied d orbital of CoS. This electron transfer facilitates the breakage of the inert N–N bonds. CoS, as a narrow-band gap sulfide with a NiAs-type structure, is considered a promising candidate for both associative and dissociative mechanisms in the NRR. The highest ammonia yield and Faradaic efficiency were achieved at 0.55 V versus RHE, reaching values of 8.09×10^{-10} mol s^{-1} cm^{-2} and 28.6%, respectively. Furthermore, the C@CoS@TiO_2 nanofibrous membrane exhibited excellent repeatability over five successive cycles, highlighting its potential for practical applications.

Based on the examples above, we can have a preliminary understanding of the NRR catalyst. In addition, perhaps we can imitate artificial photosynthesis by learning from nature to fix nitrogen artificially.

References

1. Pivovar B (2019) Catalysts for fuel cell transportation and hydrogen related uses. Nat Catal 2:562–565
2. Chen Z, Qing H, Zhou K et al (2020) Metal-organic framework-derived nanocomposites for electrocatalytic hydrogen evolution reaction. Prog Mater Sci 108:100618
3. Zhu J, Hu L, Zhao P et al (2019) Recent advances in electrocatalytic hydrogen evolution using nanoparticles. Chem Rev 120:851–918
4. Parsons R (1958) The rate of electrolytic hydrogen evolution and the heat of adsorption of hydrogen. Trans Faraday Soc 54:1053
5. Seh ZW, Kibsgaard J, Dickens CF et al (2017) Combining theory and experiment in electrocatalysis: Insights into materials design. Science 355(6321):eaad4998
6. Gao Q, Zhang W, Shi Z et al (2018) Structural design and electronic modulation of transition-metal-carbide electrocatalysts toward efficient hydrogen evolution. Adv Mater 31:1802880
7. Lin D, Lasia A (2017) Electrochemical impedance study of the kinetics of hydrogen evolution at a rough palladium electrode in acidic solution. J Electroanal Chem 785:190–195
8. Noerskov JK, Bligaard T, Logadottir A et al (2005) Trends in the exchange current for hydrogen evolution. ChemInform 152(3):J23
9. Nilsson A, Pettersson LGM, Hammer B et al (2005) The electronic structure effect in heterogeneous catalysis. Catal Lett 100:111–114
10. Miao R, Dutta B, Sahoo S et al (2017) Mesoporous iron sulfide for highly efficient electrocatalytic hydrogen evolution. J Am Chem Soc 139:13604–13607
11. Chen Z, Wu R, Liu Y et al (2018) Ultrafine Co nanoparticles encapsulated in carbon-nanotubes-grafted graphene sheets as advanced electrocatalysts for the hydrogen evolution reaction. Adv Mater 30:1802011
12. Cheng N, Stambula S, Wang D et al (2016) Platinum single-atom and cluster catalysis of the hydrogen evolution reaction. Nat Commun 7(1):13638
13. Xue Y, Huang B, Yi Y et al (2018) Anchoring zero valence single atoms of nickel and iron on graphdiyne for hydrogen evolution. Nat Commun 9(1):1460
14. Song J, Wei C, Huang Z-F et al (2020) A review on fundamentals for designing oxygen evolution electrocatalysts. Chem Soc Rev 49:2196–2214
15. Grimaud A, Hong WT, Shao-Horn Y, Tarascon J-M (2016) Anionic redox processes for electrochemical devices. Nat Mater 15:121–126

16. Cheon JY, Kim K, Sa YJ et al (2016) Graphitic nanoshell/mesoporous carbon nanohybrids as highly efficient and stable bifunctional oxygen electrocatalysts for rechargeable aqueous Na-Air batteries. Adv Energy Mater 6:1501794
17. Tavakkoli M, Flahaut E, Peljo P et al (2020) Mesoporous single-atom-doped graphene-carbon nanotube hybrid: synthesis and tunable electrocatalytic activity for oxygen evolution and reduction reactions. ACS Catal 10:4647–4658
18. Yang Z, Zhao C, Qu Y et al (2019) Trifunctional self-supporting cobalt-embedded carbon nanotube films for ORR, OER, and HER triggered by solid diffusion from bulk metal. Adv Mater 31:1808043
19. Wan C, Duan X, Huang Y (2020) Molecular design of single-atom catalysts for oxygen reduction reaction. Adv Energy Mater 10:1903815
20. Jiang Y, Ni P, Chen C et al (2018) Selective electrochemical H_2O_2 production through two-electron oxygen electrochemistry. Adv Energy Mater 8:1801909
21. Xia W, Mahmood A, Liang Z et al (2015) Earth-abundant nanomaterials for oxygen reduction. Angew Chem Int Ed 55:2650–2676
22. Huo L, Liu B, Zhang G et al (2017) 2D layered non-precious metal mesoporous electrocatalysts for enhanced oxygen reduction reaction. J Mater Chem A 5:4868–4878
23. Hang C, Zhang J, Zhu J et al (2018) In situ exfoliating and generating active sites on graphene nanosheets strongly coupled with carbon fiber toward self-standing bifunctional cathode for rechargeable Zn-Air batteries. Adv Energy Mater 8:1703539
24. Lu Z, Chen G, Siahrostami S et al (2018) High-efficiency oxygen reduction to hydrogen peroxide catalysed by oxidized carbon materials. Nat Catal 1:156–162
25. Chen S, Chen Z, Siahrostami S et al (2018) Designing boron nitride islands in carbon materials for efficient electrochemical synthesis of hydrogen peroxide. J Am Chem Soc 140:7851–7859
26. Jiang K, Back S, Akey AJ et al (2019) Highly selective oxygen reduction to hydrogen peroxide on transition metal single atom coordination. Nat Commun 10(1):3997
27. Ding M, Flaig RW, Jiang H-L, Yaghi OM (2019) Carbon capture and conversion using metal–organic frameworks and MOF-based materials. Chem Soc Rev 48:2783–2828
28. Nitopi S, Bertheussen E, Scott SB et al (2019) Progress and perspectives of electrochemical CO_2 reduction on copper in aqueous electrolyte. Chem Rev 119:7610–7672
29. Gao D, Arán-Ais RM, Jeon HS, Roldan Cuenya B (2019) Rational catalyst and electrolyte design for CO_2 electroreduction towards multicarbon products. Nat Catal 2:198–210
30. Ross MB, De Luna P, Li Y et al (2019) Designing materials for electrochemical carbon dioxide recycling. Nat Catal 2:648–658
31. Zhao C, Wang Y, Li Z et al (2019) Solid-diffusion synthesis of single-atom catalysts directly from bulk metal for efficient CO_2 reduction. Joule 3:584–594
32. Yang H, Lin Q, Zhang C et al (2020) Carbon dioxide electroreduction on single-atom nickel decorated carbon membranes with industry compatible current densities. Nat Commun 11(1):593
33. Yang H, Wu Y, Li G et al (2019) Scalable production of efficient single-atom copper decorated carbon membranes for CO_2 electroreduction to methanol. J Am Chem Soc 141:12717–12723
34. Zheng G, Yan J, Yu G (2019) Nitrogen reduction reaction. Small Methods 3:1900070
35. Yu X, Han P, Wei Z et al (2018) Boron-doped graphene for electrocatalytic N_2 reduction. Joule 2:1610–1622

Indian Statistical Institute Series

The **Indian Statistical Institute Series**, a Scopus-indexed series, publishes high-quality content in the domain of mathematical sciences, bio-mathematics, financial mathematics, pure and applied mathematics, operations research, applied statistics and computer science and applications with primary focus on mathematics and statistics. Editorial board comprises of active researchers from major centres of the Indian Statistical Institute.

Launched at the 125th birth Anniversary of P.C. Mahalanobis, the series will publish high-quality content in the form of textbooks, monographs, lecture notes, and contributed volumes. Literature in this series are peer-reviewed by global experts in their respective fields, and will appeal to a wide audience of students, researchers, educators, and professionals across mathematics, statistics and computer science disciplines.

Indranil Mukhopadhyay · Partha Pratim Majumder

Statistical Methods in Human Genetics

STATISTICAL INDIAN INSTITUTE
UNITY IN DIVERSITY

Springer

Indranil Mukhopadhyay
Human Genetics Unit
Indian Statistical Institute
Kolkata, West Bengal, India

Partha Pratim Majumder
Human Genetics Unit
Indian Statistical Institute
Kolkata, West Bengal, India

ISSN 2523-3114 ISSN 2523-3122 (electronic)
Indian Statistical Institute Series
ISBN 978-981-99-3222-1 ISBN 978-981-99-3220-7 (eBook)
https://doi.org/10.1007/978-981-99-3220-7

This Springer imprint is published by the registered company Springer Nature Singapore Pte Ltd.
The registered company address is: 152 Beach Road, #21-01/04 Gateway East, Singapore 189721, Singapore

We dedicate this book to our mothers.

Preface

This book is primarily meant for the practicing biologist, particularly, the human geneticist. Human geneticists and other biologists use a variety of statistical methods to draw inferences from the results of their experiments. However, many scientists use statistical packages without really understanding the statistical methods. They are often unable to carry out customized statistical analyses demanded by the data, because the statistical package will not allow customisation.

In this book, we have tried to address these limitations in two ways. First, throughout the book, we have used data gleaned from a few types of experiments. For example, an experiment was performed to measure the expression of some genes. Such experiments are now very popular. Hundreds of papers are published every year to measure gene expression in various tissues and to compare differences in gene expression levels between tissues or in the same tissue under different environmental exposures or conditions. Expression differences are of paramount importance in the study of human disease. Study of gene expression differences can inform the genetic basis of pathophysiology.

By using data of a single type of experiment, we believe that the focus of the reader will be glued to the statistical methodologies presented, instead of having to grapple with multiple data types and experimental designs. However, we have also used other data sets to explain a few other statistical methods required beyond the realm of gene expression analysis. When our reader can focus on the statistical methodology more sharply, we believe that the learning will also be better, and applications will be more innovative. Throughout the book, we hand-hold readers to understand statistical methodologies which the readers also will find useful in analysing their data. We start with statistical techniques of data exploration, data summarisation, and visualisation, and slowly step into more intricate statistical methods, always focusing on applications to genetics. We have provided descriptions of statistical methods for the analysis of absolute and differential levels of gene expression when the properties of the population distribution of the level of gene expression are known. We also describe methods when the properties are unknown or do not conform to a 'standard' distribution. Additionally, in modern genetic data analysis, some specialised concepts and methods are used, that do not normally arise outside of the field of genetics. For

example, the concept of ‘population stratification’ arises in view of heterogeneity of ethnicities of humans and consequent differences in frequencies of some relevant genetic variables. In association analysis, ignoring these frequency differences can play havoc on inferences; handling these required some special methodologies. We explain some of these specialised statistical methods, as well; in particular, some that are important and contemporaneous.

Although human genomic data are often multivariate, it has not been possible for us to do full justice to multivariate statistical methods. However, we did not wish to ignore such methods altogether. We felt it important for us to introduce the reader to the problems of analysis when multiple variables are simultaneously considered for analysis and to describe some preliminary methods of multivariate statistics.

Concurrently with describing modern statistical methods, we also develop R programs, starting from the scratch (that is, introducing R), so that the reader can perform analysis of her/his own data as required, without having to mould the data to fit the requirements of a statistical package. R is free, and as a community-developed product is available in the public domain.

We hope that this book will become the standard text for biologists, those who carry out quantitative biological research, especially in human genetics.

Kolkata, India

Indranil Mukhopadhyay
Partha Pratim Majumder

Contents

About the Authors

Indranil Mukhopadhyay is a Professor at the Human Genetics Unit, Indian Statistical Institute, Kolkata, India. He has earned his Ph.D. from the University of Calcutta, Kolkata, India. With more than 50 research papers published in several national and international journals of repute, his research interests are in multi-loci genetic association study, data integration of several genetic data sets, single-cell data analytics, and mathematical statistics. He has considerable experience in teaching undergraduate- and graduate-level courses in statistical methods in human genetics and supervising the work of doctoral students working in this domain. He serves as a Member of the Research Advisory Committee of ICAR-IASRI and Member of the International Biometric Society, the Indian Society of Human Genetics, and the Indian Society for Medical Statistics, among others.

Partha Pratim Majumder is the founder of the National Institute of Biomedical Genomics, Kalyani, West Bengal, India. He is currently a National Science Chair of the Government of India. He concurrently holds academic positions in many national institutes. He serves on the governing boards and executive committees of the Human Genome Organisation, Human Cell Atlas, and International Common Disease Alliance. Earlier, he has served as Member of governance of the International Genetic Epidemiology Society, the Indian Society of Human Genetics, the Indian Society for Medical Statistics, and the American Society of Human Genetics, among others. He is an elected Fellow of all science academies of India, the International Statistical Institute, and The World Academy of Sciences. He has served as the President of the Indian Academy of Sciences. His research interests are in genetic epidemiology, human biomedical genomics, human genome diversity and evolution, and statistical genetics.

He has been awarded the G. N. Ramachandran Gold Medal (2021) by the CSIR; Barclay Memorial Medal (2020) by The Asiatic Society; Sir P. C. Ray Memorial Medal (2020) by the University of Calcutta; Golden Jubilee Commemoration Medal (2018) by the Indian National Science Academy; Centenary Medal of Excellence (2014) by the School of Tropical Medicine; TWAS Prize in Biology (2009) of The World Academy of Sciences, Trieste; G. D. Birla Award for Scientific Research

(2002) by the K. K. Birla Foundation, New Delhi and Kolkata; the Om Prakash Bhasin Award in Biotechnology (2001) by the Om Prakash Bhasin Foundation, New Delhi; the Ranbaxy Research Award in Applied Medical Sciences (2000) by the Ranbaxy Science Foundation, New Delhi; and the New Millennium Science Medal (2000) by the Indian Science Congress Association and the Council for Scientific and Industrial Research (CSIR), the Government of India.

Chapter 1
Introduction to Analysis of Human Genetic Data

1.1 Need for Genetic Data Analysis

Biological data present a multitude of challenges in terms of presentation and analysis. The most interesting feature is that data may come from diverse but related different domains including botany, zoology, fishery, genetics, genomics, neuroscience, anthropology, etc. Types of data differ; often a data set comes along with some unique characteristics of its own. However, each data set has a common goal: how to analyse the data properly and scientifically so that we are able to extract important information that is not apparent by a cursory examination of the data. It seems imperative to extract and exploit the full information contained in the data to come out with a nice story that explains and throws insight into the actual problem for which the data have been generated. Naturally, careful study of the data should lead to further more important questions about the problem concerned. Sometimes, it may lead to more interesting questions that are not intended to be studied initially, but that would shed light on important features unknown at least at the beginning of the experiment or the data analysis exercise. Whenever we have data, usually we have a tendency to jump on it and calculate some statistical measures as analysis of that data. However, it should start not with calculations but by asking questions, as many as you can. We need to understand the background of the experiment and the hypothesis and/or objective that triggers this experiment. Understanding the whole scenario is essential to appreciate the problem at hand. Thus, first, we should ask questions to dissect it, to get a holistic view as well as specific features of data and then only can start the appropriate analysis.

So, it is essential to identify the unique features of the data and frame the scientific problems in terms of several questions that would lead to detailed analysis. But the analysis completely depends on the relevance of the questions to the problem attempted at being solved, availability of the data, reliability, and quality of the data generated and stored, and knowledge of statistical analytic techniques. Data management and storage should also claim an important place in data analytics, especially in today's world of data science. Moreover, during the process of generation of data

I. Mukhopadhyay and P. P. Majumder, *Statistical Methods in Human Genetics*,
Indian Statistical Institute Series, https://doi.org/10.1007/978-981-99-3220-7_1

some data might be missing, fully or partially; however, these data should not be discarded as we can get some information rather than no information from this type of partially missing data. However, statistical analysis of data with missingness is more complicated but should be undertaken, although advanced statistical techniques are required for this purpose.

On the other hand, in order to carry out an appropriate protocol for data analysis, one needs to know the basic structure and nature of the data, other important features that might be a distinguishable character for this particular data set, the way it has been generated and stored, etc. Thus, a complete understanding of the whole experiment and data generation process is essential to proceed further with the data analysis and eventual conclusion relating to the scientific questions asked.

Thus, the most important issue is to learn how to understand the problem clearly so that we can generate relevant questions relating to the data. Note that these questions might be of general nature, but care should be taken to identify the special questions that are only relevant to the data set considered for the analysis. This problem formulation stage should be done with utmost care and caution. Entire downstream analysis depends on this very crucial stage. Inadequate understanding of the data and the objective of its analysis may lead to erroneous results, which when interpreted, may not show any parity to the actual problem being investigated; the entire exercise would become futile. Hence, after the analysis, we need to interpret the results in relation to the actual problem and questions asked at the beginning of this exercise. If we see that interpretation of results really seems meaningful in relation to the problem, we can say that statistical analysis has been done appropriately. If we are not satisfied, we have to revisit the entire exercise from the beginning. At the end, we need to tell a story based on our data analysis.

Human genetic data analysis, the main focus of this book, is no more different than analysis of other types of data. However, because of advances in technologies, the diversity of human genetic data is large. Some popular data sets are voluminous and present major challenges in data handling and manipulation, in preparation for statistical analysis. It is interesting to note that Mendel's laws on heredity were formulated on the basis of careful statistical analysis of data collected from carefully designed experiments. During that period, statistical analysis had not flourished, in fact, statistics was not really a recognised discipline. However, it was the genius of Mendel that triggered the first statistical data analysis in genetics leading to fundamental laws of genetics being derived.

Statistics, as many people think, is not really a bunch of complicated mathematical methods with a fearsome look full of complicated symbols and notations. Rather it is a beautiful garden where flowers and green leaves play with gentle breeze full of ideas and thoughts. We should have an open mind to embrace new ideas, try to be patient and understand the scientific problem, and spend some time only for thinking about the intricate and interesting features of our data and experiment along with scientific question, and the solution will provide more light than we have ever thought.

In this book, we explain admixture of general purpose statistical methods and some specific methods applicable to human genetic data. We have kept the use of notations and symbols to a minimal level and almost omitted any detailed mathemat-

ical derivations. However, we have included derivations of a few important results in the appendix. We start with a few typical data sets and explain the types of questions that can be generated from the data. Later, we extend the data set and naturally also the question set, in order to get clearer answers to the questions. We consider mainly data sets that consist of gene expression data and genetic data with genotypes and other variables. Each chapter starts with a set of questions and the entire chapter is then devoted to a discussion of statistical methodology that can be used to address those questions and will eventually lead to the appropriate analysis. As a chapter progresses, more intricate sets of questions are considered along with their solutions through statistical analysis.

1.2 Nature of Analysis

Choice of techniques of statistical analysis of genetic data depends on the nature of data. We must have a clear understanding about the nature of the data and the scientific questions that underlie the generation of the data set; otherwise, the entire exercise of data analysis may lead to misleading results. Note that some typical features of the data may pose considerable challenges that may not even be solved through standard or common statistical methods. A common tendency in such situation is to bypass or ignore this special feature and do a quick statistical analysis using common methods. However, this would end up with very weak or even misleading conclusions that may not be replicable in the future. Although a set of data may have potential and be sufficiently informative to results producing valid inferences, we may lose that only due to our ignorance or reluctance to understand the data carefully and address the challenges posed by the data. Let us discuss this with one example.

To check whether a gene has any effect on a disease, we may collect data from tissues harvested from affected organs (for example, from the liver of a liver cancer patient and the patient's blood) from the same patient and measure their gene expression values. It is not unnatural that while processing the biological samples, some data might not be available due to degradation of sample quality or technological artefacts. Suppose that finally gene expression data for both normal and tumour samples are available for 20 individuals whereas for 4 individuals data for normal tissues are lost or unavailable. One may opt for two methods of analysis of such data.

Option 1 Analyse the data only based on 20 paired samples for which full data is available and ignore the remaining 4 samples.

Option 2 Consider the data of normal tissues for 20 individuals and tumour samples of 24 individuals. Then do a statistical test that should be applied to two independent samples.

However, both the above options for analysis are wrong, or at least inappropriate. Why? Note that since normal as well as tumour tissues are collected from the same individual, the data on expression levels of genes are expected to be biologically related (and also statistically). The second option is untenable since it does not use the

information on the possible relationship between levels of gene expression between tumour and normal samples obtained from the same patients. In fact, this is a wrong method of statistical testing. On the other hand, in the first option, we are discarding information that are in four tumour samples although we have spent money and time to generate the data. Thus, neither of the above tests should be applied in this case. If one does so, this is only due to ignorance of statistical knowledge or reluctance to exploit the full information that is available in the entire sample. However, it is not very difficult to analyse such data, although it would not be a standard one. So we should be honest, open-minded, and enthusiastic to learn how to use statistics with its full power to come up with the best possible analysis given the intricate nature and challenges posed by the data.

It is crucial to understand the data. The very first step in the central dogma of genetics is related to DNA sequence being transcribed to mRNA. The most common variant at one nucleotide position that we use is single nucleotide polymorphism (SNP). Note that a SNP has exactly two alleles giving rise to three genotypes, one of which is observed in an individual. Let two alleles at a bi-allelic locus be A and T and hence the possible genotypes be AA, AT, or TT. Note that here we consider AT and TA genotypes as a single class AT. So genotype data have no numerical value, but only three categories. For the sake of analytical convenience, we may assign numbers 0, 1, and 2 respectively, but care should be taken while interpreting the result. The main reason being one can assign any numbers to these three categories and the final result should not be dependent on what numbers we assign to genotypes.

This categorial variable can be converted to a meaningful variable that takes only a few discrete values. If we consider the number of minor alleles in a genotype at bi-allelic locus, it can take values '0', '1', and '2' depending on the number of minor allele in the genotype. In the above example, if T is the minor allele, the genotypes AA, AT, and TT would produce 0, 1, and 2 values respectively. Thus, it becomes a discrete variable. However, whether we work with discrete variable or the original categorical variable depends on the problem that we are trying to address.

So genotype data are categorical, having three categories. However, sometimes during the data collection, data for a few other variables may be collected naturally. For example, when we collect blood or tumour tissue from a person, we may collect data on her/his age, gender, smoking or food habit, etc. These auxiliary variables, also known as 'covariates', contain important information that should be included in the analysis to get a better insight into the problem that is attempted to be solved using the biospecimens.

In another scenario, to study whether a gene is somehow affecting a disease, we generate gene expression data. These data are continuous in the sense that these can take any value, up to a certain place of decimal, within a given interval. Suppose a particular set of gene expression data reveals that the expression level of a gene varies between 0 and 10.95 (for instance). The analysis will differ completely from that used for genotype data. Moreover, it also depends on whether normal and tumour samples are collected from the same individual or different individuals.

The above examples are only a few in the bigger set of data types generated through experiments. Thus, data might be discrete, categorical, continuous as well

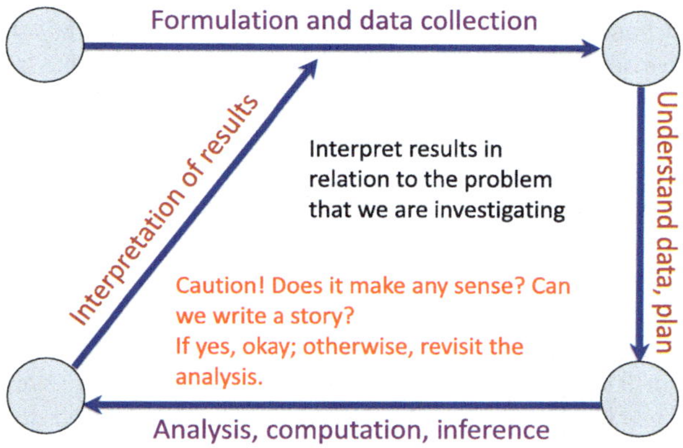

Fig. 1.1 A schematic of data analysis protocol

as a mixture of all these. We should take utmost care to understand the nature of data, spend sometime to find statistical methods that should be most appropriate, try to use the maximum possible information contained in the sample, at the end interpret the results based on domain knowledge and see whether it sounds meaningful with respect to the actual problem that initiated this analysis. (Fig. 1.1) We shall take our decisions on this experiment to the next step, may be based on information available from other sources; then only we will be able to advance ourselves towards deciphering the genetic architecture of the disease or the characteristic.

1.3 R: A Versatile Tool for Genetic Data Analysis

Most current data sets are too voluminous to be amenable to manual analysis; using a computer has become inevitable. Broadly, we may use

(1) programming language and write programs of several statistical measures and mathematical functions, run our own program to carry out the analysis, or
(2) a software package, pull required functions from the package, and use them to perform the analysis.

Both approaches have their own drawbacks as well as advantages. Software can be used when we do routine calculations. It is usually menu-driven, user-friendly and one has to remember the way a function is to be used. This part is not very difficult. But when we want to do novel analysis based on some novel scientific questions and ideas, software usage is limited because we have to find appropriate functions and use them one by one, each time going to the menu, pulling that function, exercising it, and finally combining the result in the same order. In most cases, menus or functions are used by clicking the mouse on the appropriate tab.

On the other hand, writing ones own code has a tremendous advantage. We can always customise our code according to our needs and questions asked, whether novel or standard or a combination of both. But in order to do that, we need to learn how to write programs using a particular computing language like C, C++, FORTRAN, etc. This requires extensive training.

To bridge the gap between the two, the computing world offers something which can be treated as a combination of the two approaches. Among many efforts, Roy Ihaca and Robert Gentleman of Auckland University, New Zealand, developed a programming language, named it as **R**, and gave it to the world absolutely free. Usually, this is known as "R programming language and environment".

R is very user-friendly; a little bit of training in programming can make anyone comfortable to use it. Data entry or reading data from files can be done very easily using a command consisting of only a few words. Basic operations like addition, multiplication, etc., and many simple mathematical functions like logarithm, exponential, etc., can be easily done with commands similar to the operation name itself. Multitude of simple and complex statistical measures, statistical testing procedures, fitting regression lines, etc., are done using standard R commands. Moreover, it is convenient to merge a few R commands to do more sophisticated data analysis. We can also save the results in an easy-to-understand format and produce nice graphs leading to a better understanding of data and its analysis. The recent introduction of R studio and tidyverse has made our lives more simple and provides us with tools for extensive analysis of data.

This book is written with a focus to do our own calculation using R. A brief exposure to R is given in Appendix A. Although it is given in Appendix, spending some time towards learning R might be very effective and comforting while doing data analysis.

1.4 Some Remarks

The remainder of the book deals with statistical methods to analyse genetic data with the help of R only. In each analysis, we provide the R command and the output generated. All graphs used in this book are produced only using R. This will give an idea about the beauty and power of R language in generating useful graphs and diagrams and to do any kind of statistical analysis. The discussion is motivated by the data itself through generation of adequate questions or queries, analysing them with the logic of the method chosen, and its execution using R along with the interpretation of the results. This would provide a holistic approach to data analysis using statistical intuition, subject knowledge, and computational methodology.

Exercise

1.1 Write three advantages of using R in statistical analysis of data.

1.2 Is there any issue in using standard software for statistical analysis? If so what are they?

1.3 The main steps in statistical data analysis are (1) actual numerical computation, (2) understanding the problem, (3) our expectation about solving a few relevant questions, (4) inference through statistical analysis, and (5) interpretation of results. Order these steps from the starting step of data analysis to the end. If any step is missing, what is that step and where does it come in the sequence of data analysis steps?

1.4 Is it wise to always convert one type of data to another? Justify your answer with examples.

1.5 Explain with an example how you can convert a continuous data into a categorical data. Is it wise to do that? Justify your answer.

1.6 As described in Sect. 1.2, give another concrete example where we may do common mistake during data analysis. Try to suggest what should be the appropriate method that you can think of? You need not explain or develop any new statistical method, but an idea about the solution would be fine.

1.7 Suppose we have collected data on 50 individuals on four different variables. Missing observations are artefacts for many mechanisms, some known, and some might even be unknown. During the analysis, it is observed that some observations are missing. What do you do if

   (a) information of two individuals are missing completely for all four variables.
   (b) information for three individuals are missing for only two variables.
   (c) information for three individuals are missing for only one variable, however, for one individual, although all information is available but we have enough reason to be suspicious of the fact that data on one variable might be wrong.

1.8 Suppose a person gives some data to a statistician. The statistician immediately calculates a few measures and do a quick analysis without asking any question. Do you support the act of that statistician? Justify your answer.

1.9 "Statistical methods are always generic, not problem specific"—A scientist makes an interpretation of this statement as if one knows some statistical methods, s/he can do data analysis for any data set whatsoever. What are your views about this? Explain clearly, may be with relevant examples.

1.10 "In most cases we can categorise or discretise continuous data into categorical or discrete variable. Hence it is enough to know statistical methods that deal with categorical and discrete variables." Do you agree with this? Justify your answer with some examples.

Chapter 2
Basic Understanding of Single Gene Expression Data

Various physiological processes of a living organism are controlled by proteins. Information on production of proteins is encoded in the DNA of each cell of an organism. The information in the DNA is transcribed into RNA and then translated into proteins. This is known as the central dogma of biology.

In a cell, not all genes are transcribed. Subsets of genes that are transcribed provide a cell with its identity, cell-type. Cells are specialised and activities carried out within a cell depends on RNA molecules produced within cell. Transcriptome is the collection of RNA molecules transcribed within a cell. Technology tried to address the difficulty of characterising the transcriptome of a single cell in the following way. Usually a large set of cells from a tissue is collected and the transcriptome of this bulk of cells is characterised. Even in a single tissue, characteristics of the bulk transcriptome - that is, which genes have been transcribed and at what levels—are highly variable across individuals, which may be beyond differences in age, gender, life-exposures and ethnicity. Characteristics of the transcriptome can also indicate the state of health or disease of the tissue in an individual.

The transcriptome is complex, because there are different types of RNA, called RNA species, not all of are messenger RNA (mRNA) which are directly involved in the production of proteins. Messenger RNA (mRNA) is called so because it encodes proteins via the genetic code. Genetic code is actually DNA information used to produce amino acids, which are the building blocks of a protein. Therefore, mRNA is the most widely studied RNA molecule.

These studies which were conducted using methods that were low-throughput and involved considerable human intervention evolved over time. Initially studies were limited to measuring single transcripts using methods known as northern blotting or quantitative polymerase chain reaction. Rapidly, technologies and methods were developed to quantify genome-wide gene expression levels called transcriptomics. Carried out by hybridization-based methods using microarray technologies producing high-throughput data at a modest cost (Schena et al., 1995), these experiments and the resulting data were fraught with many errors and artefacts (Casneuf et al., 2007; Shendure and Ji 2008). To resolve these problems, sequence-based methods

I. Mukhopadhyay and P. P. Majumder, *Statistical Methods in Human Genetics*,
Indian Statistical Institute Series, https://doi.org/10.1007/978-981-99-3220-7_2

were developed to directly determine transcriptome sequences. However, many limitations remained.

Introduction of high-throughput massively-parallel sequencing technology revolutionised the field of transcriptomics. RNA analysis now rapidly done by sequencing of cDNA (Wang et al., 2009) called RNA-sequencing (RNA-seq) provided a deep and quantitative view of gene expression, alternative splicing, and allele-specific expression. Typically, an RNA-seq experiment involves three steps:

- isolating RNA from cells collected from a tissue.
- converting RNA to cDNA sequencing library. This is a critical step in which the desired RNA molecules (mRNA but not ribosomal RNA) are selected.
- mRNAs are then converted to cDNA by using reverse transcriptase, fragmented, and adaptors required for sequencing are ligated to the fragments; this creates the sequencing library using next-generation sequencing (NGS) platform.

A whole new world of accurate and massive data ushered in for comparing changes that result from different physiological conditions or in various disease states. The RNA-seq data provides an accurate, high-resolution view of the global transcriptional landscape in platforms like NGS and formats like FASTQ.

Thus, gene expression data provide an indication of the activity of genes that can be used for many genetic and genomic studies. Analysis of gene expression data has become extremely important during the last few decades.

But before analysing the data we need to understand the data as clearly as possible. Here Table 2.1 represents a set of observations corresponding to the expression of a single gene for a set of individuals. Although data are collected for a number of cases and controls, to start with we only consider the control data. Later on we shall discuss analysis of gene expression data for both cases and controls and also for multiple genes.

In a specific study, after the execution of various primary steps of data analysis (alignment, assembly, estimation of expression level), RNA-seq data of a study involving n individuals can be summarised in the form of a data matrix $((x_{ij}))$, where x_{ij} denotes the expression level of gene i $(i = 1, 2, \ldots, p)$ where p is the total number of genes in a human genome in the j-th individual $(j = 1, 2, \ldots, n)$. In the context of a disease study, instead of the absolute expression level of the i-th gene in the j-th patient, the differential expression level of the gene compared to

Table 2.1 Gene expression values for a single gene

2.67	0.74	4.60	1.95	1.64	2.57	1.69	0.23	2.94	2.30	3.32	3.42	4.00	2.66	9.21
1.40	4.91	1.45	1.99	6.96	2.41	4.57	1.98	4.06	4.64	2.90	1.66	4.97	4.03	9.01
3.58	2.81	0.02	4.29	2.60	6.67	4.00	1.24	3.72	3.32	5.75	3.99	4.07	2.81	0.01
2.16	3.38	4.21	0.79	1.08	3.53	4.35	3.45	2.95	1.61	3.36	0.15	2.04	2.78	3.46
1.56	0.57	3.46	2.37	5.36	1.80	3.15	1.86	4.04	2.77	2.51	3.47	3.25	2.42	9.98
1.16	0.71	1.70	4.49	6.10	2.39	4.27	0.88	2.86	3.62	4.22	2.52	2.75	1.44	2.31

either a matched unaffected individual or a tissue that is unaffected by the disease collected from the same patient and this is presented as data matrix $((x_{ij}))$.

However, for the sake of simplicity, in Table 2.1, we consider only a set of observations $\{x_1, x_2, \ldots, x_n\}$ as the gene expression values of a particular gene corresponding to n different individuals. Note that we assume that these n individuals are selected randomly, i.e. without any preference to anybody, from a larger group of individuals. Gene expression values for this larger group, is commonly known as 'population' under study and the study is conducted through the sample of n expression values corresponding to n randomly chosen individuals.

Note that a population is a large group of individuals or items and we are interested in studying this population and deciphering hidden patterns. However, usually it is too large to collect information on all items. So we draw a random sample of smaller size, which would be manageable for analysis, and study these sample values to infer about the population. It is important that we must ensure that sample represents the salient features of the population. Usually a random sample is a good one, which is a collection of items or individual values selected without attaching any preference to any member of the population. This forms the basis of data analysis.

However, in some situations we may not have a random sample in the truest sense. Certain ascertainment criteria are considered while collecting such sample. In such a case, we should be very careful to consider the effect of such criteria and adequate correction is necessary while analysing the data. Without this correction, analysis might be biased and there is a high chance that we miss some important features. Sometimes sample generation mechanism has some subtleties that might get ignored during routine analysis leading to wrong or vague conclusions. Thus deep understanding of data and the mechanism of its generation is essential to exploit the full information contained in the sample.

2.1 Generating Questions

Just looking at the data would not reveal its inherent pattern or information. However, based on domain knowledge and common intuition we can think of several questions that would be helpful in understanding the data as well as extracting salient features of the data. Research begins with questions. And the solutions or attempts to answer these questions are essential to understand the data, its nature and inherent pattern, the truth it may reveal, and for further investigation. So we begin with a few questions about the data. Although here we refer to the gene expression data, this should be a common theme before actually starting any analysis of any kind of data.

Here data means statistical data; the observations corresponding to different units that are subject to some random factors, some known, some may be unknown, but will always contain some degree of variation among them. Variation is the most important characteristic for any statistical data; without it we cannot do any statistical analysis.

With respect to the data presented in Table 2.1, a few questions that can be deemed important are:

(1) What is the general nature of the data, if we look at it very naively? Can we represent the data using a diagram to reveal some general patterns?
(2) Are all observations in a reasonable range or some observations show unusual behaviour compared to the rest of the data?
(3) Can we have an idea about the gene expression on an average, that can be considered as a representative value? Can we devise any measure that depicts this idea?
(4) Can we have any idea about the spread of the observations? Can we devise any measure to present this idea?
(5) Can we compare this so called average value with any previously known knowledge, may be based on literature survey or other studies?
(6) Can we infer about the same for any measure that represents the overall scatteredness of the data?
(7) A very important technical question might be: Can we think that observations come from a normal distribution? Or does there exist a random variable that follows normal distribution, from which these samples are drawn? This question is for a technical reason, but has great impact on the methods of data analysis. In fact the answer to this question will guide the entire downstream analysis of data.
(8) When we have two samples, for cases and controls, can we compare the general features between these two groups? How valid is that comparison?
(9) If we have data on multiple genes for a group of individuals, can we think of a subgroup of genes that might act in a similar fashion?

Many other questions can be asked, but for the time being we shall restrict ourselves only to these and explore the methods of addressing them in the best possible way.

2.2 Visualising Data

Experiments generate data i.e. raw data, as it is termed. They are usually, but not always, numerical observations. Looking at hundreds (or less) of such numerical figures only add more complexity in understanding. Hence first we have to present the raw data so that it might reveal some information, patterns, insight, even if at a crude level. This presentation should give some confidence that at least we should be able to get something out of it that relates to our primary question(s). Note that these primary questions or objectives need to be sorted out at the very beginning, during the planning of the experiment that would generate the data.

So, presentation of data is very important. It would guide our analysis plan. And visualisation of data, if possible, should be the most effective and attractive way of presentation. However, visual presentation should be as simple as possible, but at the same time should reveal salient features of the data. Hence, it is expected that every part of the diagram presenting the data should have some meaning of its own; there should not be any part left redundant. This visualisation provides a feeling of

the data, which is the first and perhaps the most important step in data analysis. A number of visual representation techniques are available; we discuss here two most popular and informative way of presenting this kind of data (Table 2.1).

Note that gene expression data are continuous in nature, meaning that it can take any value within a particular interval. This interval may be large or small, but the variable i.e. gene expression in this case, can take any value in that interval. This kind of variable that can take any value within a certain range, is known as "continuous variable". A random sample for this variable will constitute frequency data. It is so called because within a small interval we may have more than one, in fact a large or moderate number of observations. Thus this type of data is known as "frequency data".

Naturally larger the data size, more will be the information contained in the data set, better insight can be obtained about the data and the domain knowledge related to it. However, due to several resource constraints, like time of collecting samples and associated cost, it may not always be possible to go beyond a reasonable size of data, sometimes restricting even to a small sample size. Thus appropriate statistical analysis can only throw light to draw inferences about the question asked and it should start with some exploratory analysis, may be in terms of naive diagrammatic representation of the data or some elementary analysis.

Two most common diagrams used frequently are **histogram** and **boxplot**. Before discussing how to draw these diagrams, we need to understand the general feature of frequency data; then only we can develop diagrams for its representation.

2.2.1 Frequency Data

A value in a data set may occur more than once or a number of observations in a data set may lie within an interval. Such types of data, where an observation may occur more than once, are called "frequency data". A continuous variable is a quantity or entity that can take any possible values within a certain interval. Some examples of continuous variables are: level of mRNA expression, viral load in 1 mL of blood, length of a leaf, weight of fish caught, etc. Naturally this type of variable can take any real number within a range or interval thus retaining the notion of continuity in a common sense and hence is called continuous variable. In some situations, especially when the underlying variable is continuous, each observation may occur only once, but a few observations are very close to each other. This type of data is also an example of frequency data. In Table 2.1 data on level of gene expression can take any value within a certain range, 0.01 to 9.98. This indicates that the data set has minimum value of 0.01 and maximum value of 9.98. All other values in this data set lie within these two values. There is no real number within this interval 0.01 to 9.98 that this variable cannot potentially take, although in reality, it takes a few numbers because the sample size is small, the measuring instruments can measure only up to second place of decimal; there are other experimental artefacts also.

But how do we find the minimum and maximum values of the data set? That's easy with R. This is the first time in this book where we are going to use R. A quick look at the Appendix would help in working with the actual data. First we would need to input the data so that we can read it within R. For instance, we can store the entire data in a variable named 'x', (say). *min(x)* and *max(x)* functions calculate minimum and maximum of the data stored in x respectively.

```
> x<- c(2.67, 0.74,4.60,1.95,1.64,2.57,1.69,0.23,2.94,2.30,3.32,9.21,1.40,4.91,
1.45,1.99,6.96,2.41,4.57,1.98,4.06,4.64,2.90,9.01,3.58,2.81,0.02,4.29,2.60,6.67,
4.00,1.24, 3.72,3.32,5.75,0.01,2.16,3.38,4.21,0.79,1.08,3.53,4.35,3.45,2.95,1.61,
3.36,3.46,1.56,0.57, 3.46,2.37,5.36,1.8,3.15,1.86,4.04,2.77,2.51,9.98,1.16,0.71,
1.70,4.49,6.10,2.39,4.27,0.88, 2.86,3.62,4.22,2.31)     ## input data and store in 'x'
> min(x)                    ## finding minimum of values stored in 'x'
[1] 0.01
> max(x)                    ## finding maximum of values stored in 'x'
[1] 9.98
```

Once we have an idea about the possible values of the variable in the sample or the range of variables in which the data are expected to lie, we should proceed with the analysis part. A visual representation would be the first step that might guide the analysis because it gives an impression (idea) of the distribution of observations in the given sample (minimum to maximum).

2.2.2 *Histogram*

The most commonly used diagram to represent data corresponding to a continuous variable is known as "histogram". Histogram provides information and an idea about the 'distribution' of observations, its shape and many other features. Now, 'distribution' means how the data (sample observations) are distributed over different small intervals (non-overlapping) within the range of data. To do this, we first divide the range $(maximum - minimum)$ into a number of intervals, called class intervals. The smallest and the highest values of a class are known as lower class limit and upper class limit respectively for that class. We calculate the frequency of each class i.e. number of observations belonging to each class and hence a frequency distribution is obtained. Here we distribute the entire data over several classes thus constructed. However, these classes may be disjoint, meaning that although for the given data set we can accommodate all the values into classes, but theoretically some values, being real numbers, may fall between the upper limit of a class-interval and the lower limit of the next class-interval. To get rid of this problem and make it theoretically stronger in covering the whole range of values, we change the lower limit and upper limit of a class by lower class boundary and upper class boundary respectively.

Let k be the number of classes, f_i be the the frequency of i-th class, and L_i and U_i be respectively the lower and upper class limits of the i-th class ($i = 1, \ldots, k$). Then

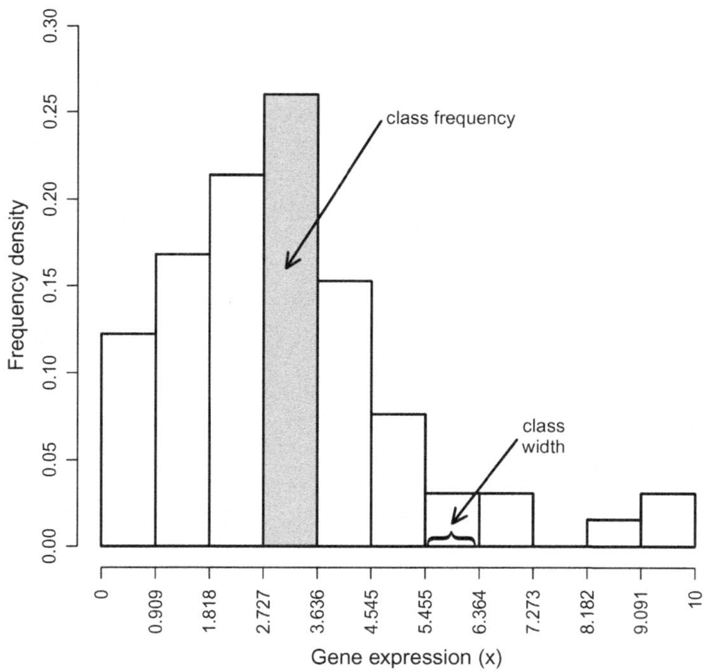

Fig. 2.1 Histogram of single gene expression data for 90 samples

the lower (LCB) and upper (UCB) class boundaries of the i-th class are given by respectively, $LCB_i = L_i - (L_i - U_{i-1})/2$ and $UCB_i = U_i + (L_{i+1} - U_i)/2$. Let w be the class-width of each class defined as $UCB - LCB$, assuming that the class widths are same for all classes. However, this assumption can be relaxed. We count the number of observations in each class, known as the class frequency. We also calculate frequency density as the ratio of class frequency to class width (f_i/w), for each class. This gives an idea about the density of values in a class. We then plot the class boundaries on the horizontal axis and draw a rectangle with height being equal to the frequency density and breadth as class width. This diagram is known as histogram (Fig. 2.1). Clearly, the area of each rectangle is equal to the class frequency and the total area under the histogram is the total frequency, because,

$$\text{Area of } i\text{-th class} = \text{frequency density} \times \text{class width} = \frac{f_i}{w} \times w = f_i = \text{ class frequency of } i\text{-th class}$$

$$\text{and total area under histogram } = f_1 + f_2 + \cdots + f_k = n = \text{ total frequency.}$$

Now let's try to understand this histogram in a very simple way. No complicated mathematics, only a careful look at the histogram while keeping in mind about the nature and context of our data set.

It is intuitively clear from the Fig. 2.1 that the class [2.727, 3.636] has the highest frequency and the frequencies in the classes to the left of this class decreases gradually whereas that to right decreases relatively sharply. Thus there is a concentration of observations around [2.727, 3.636] and also a long tail exists to the right spreading up to 10. Also some values to the right end seems little strange compared to the general nature of the data set. Thus it creates a suspicion whether observations in classes lying to the extreme right end are natural to obtain from the experiment that generates the data.

Another interesting feature can be observed if we study the histogram in depth. Suppose we draw a histogram using relative frequency density in the vertical axis. Relative frequency density of a class is defined as the ratio of relative frequency (=class frequency/total frequency) to the class width of the corresponding class. Since without any loss of generality we can assume that class widths for all classes are same, only one class width value would suffice for the entire calculation. Naturally the area of each rectangle represents the relative frequency of that class. Area under the histogram is total relative frequency, which is always 1, irrespective of the total frequency, because

$$\text{Area of } i\text{-th class} = \text{relative frequency density} \times \text{class width} = \frac{f_i}{nw} \times w = \frac{f_i}{n} = \text{relative class frequency of } i\text{-th class}$$

$$\text{and total area under histogram} = \frac{f_1}{n} + \frac{f_2}{n} + \cdots + \frac{f_k}{n} = 1 = \text{total relative frequency (since } f_1 + \cdots + f_k = n).$$

Drawing a histogram using relative frequency has immense advantage in downstream analysis (as we shall see later) mainly due to the fact that the total area of the histogram is always 1 whatever be the value of n. The value 1 is interpreted as the total relative frequency (or total normalised frequency). This gives us a nice opportunity to compare two frequency distributions, even when the total frequencies are not same for the two data sets.

Now, if we were to increase the number of observations i.e. total frequency and at the same time reduce the class width, we will get more and more dense rectangles but width being smaller and smaller, area of each rectangle would be smaller; the total area would always remain equal to 1 in each case, irrespective of the number of classes as well as total number of observations. Hence in the limiting case, when the total frequency is extremely large and the class width is extremely small, we get a smooth curve as the histogram. This is known as **frequency curve**. This frequency curve is very important in statistics and forms the basis of the underlying probability structure for the frequency distribution of a continuous variable. This curve can be modelled in a probabilistic framework and hence easy to study extensively. So histogram, which is based on real data, can be viewed as a basic tool to advance towards a theoretical probability structure to which we can assign sample observations as realisations of random experiments. This transition from real data to theoretical sophistication is a beautiful journey in statistics and may bring flexibility, comfort, and mathematical justification while doing data analysis.

Note that by constructing a frequency distribution of a continuous variable we lose individual values taken by the variable and this information cannot be retrieved unless we again go back to the original raw data, which is a humongous task. However, this sacrifice in losing some information is intended for analysis of data and answering the general questions laid down in Sect. 2.1. Moreover, as in most cases, for a judiciously chosen class width, since the observations lie densely within a class, losing individual values will not affect much to get an overall idea about the distribution of sample observations.

R code for drawing histogram is simply 'hist()'. But this histogram is drawn using frequency as the height of each rectangle rendering difficulty in interpretation of area of each rectangle and the entire area under the histogram. Moreover, it uses a default choice of the number of classes.

Now, the question is: what is the ideal number of class intervals? Ideally we should organise data so that the number of class intervals is neither too large nor too small. Otherwise the distribution would either be too smooth or too jagged, i.e. it would lose its latency. An empirical formula for the number of class intervals is an integer value closest to square root of the number of observations.

```
> br <- round(seq(0,10,length=12,3)     ## creating user-defined class boundaries
> hist(x,breaks=br, freq=F)       ## draws histogram with relative frequency density
```

R code for drawing a histogram is given in the box above. First we split the range of observations into 11 classes. 'br' gives the boundary points of the consecutive classes. Note that 'length=12' would be the number of boundaries which is one more than the number of classes. 'hist' gives the R-code for histogram; 'freq=F' is required when we draw histogram using relative frequency density (always recommended). It gives us Fig. 2.1 except the labels inside the graph and grey shaded region (for this we need to write little more sophisticated R code). Figure 2.2 presents three histograms with varying number of classes; 10 is closest value to the square-root of n in this case. So the choice of k as an approximation to $\sqrt{n}$ works! However, the user is free to choose their own number of classes while drawing a histogram.

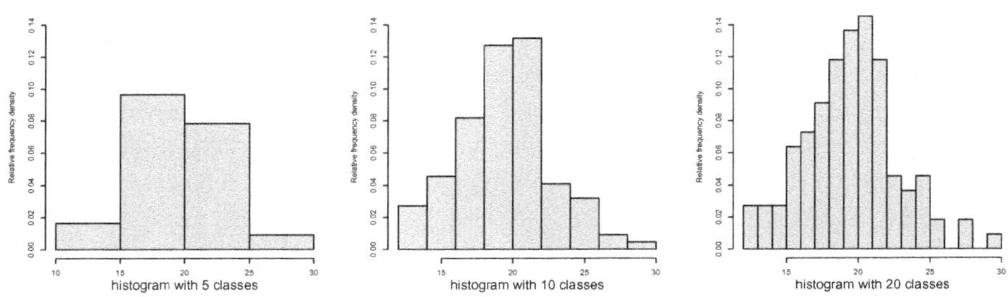

Fig. 2.2 Histogram with different number of classes

Clearly, a little bit of difference in the number of classes in histogram for the same data will not matter much because this is a random sample and it represents general features of the population only approximately.

Note that on R console, each line starts with a '>' sign. This sign will appear automatically when you press enter. R Code for drawing histogram is given in Appendix B.

2.2.3 Ogive

The frequency distribution of a continuous variable like the gene expression values can also be represented through another very useful diagram, known as '**ogive**'. This diagram represents the 'cumulative frequency', less-than type and/or greater-than type for the variable concerned. Cumulative frequency of less-than (greater-than) type of a class is defined as the number of observations that are less (greater) than or equal to the upper (lower) class boundary of that class.

We plot less-than type cumulative frequency for a given class corresponding to its upper class boundary and join the consecutive points by straight lines. The graph thus obtained is called an **ogive**. Ogive also emphasizes the fact that in constructing or representing a frequency distribution of a continuous variable we assume that the class frequency is distributed uniformly throughout the class, which in fact, is the fundamental assumption behind a continuous frequency distribution. In a similar fashion, we can plot the cumulative frequency greater-than type corresponding to the lower class boundary of that class and join the consecutive points by straight lines.

The first ogive looks increasing in nature while the second one is decreasing as we go along higher values of class boundaries. So these two ogives must intersect at a point. An interesting interpretation can be given to the point of intersection of these two ogives thus obtained. Let this point be denoted by O. Drop a perpendicular from O which cuts horizontal axis at M and the corresponding value is F_M at the vertical axis. Clearly the number of observations more than M along the less-than type ogive is same as the number of observations below M along the greater-than type ogive. Naturally there are 50% observations that are less than M as well as greater than M and hence $F_M = 45$ as total frequency is 90. This point M is called median of the given data set. We shall discuss some interesting properties of median later.

At this point, we expect that readers have some preliminary knowledge of R programming or they should read Appendix A at the end of this book, which contains a brief tutorial on R. The R code for obtaining ogives as in Fig. 2.3 based on the data in Table 2.1, is given in Appendix B.

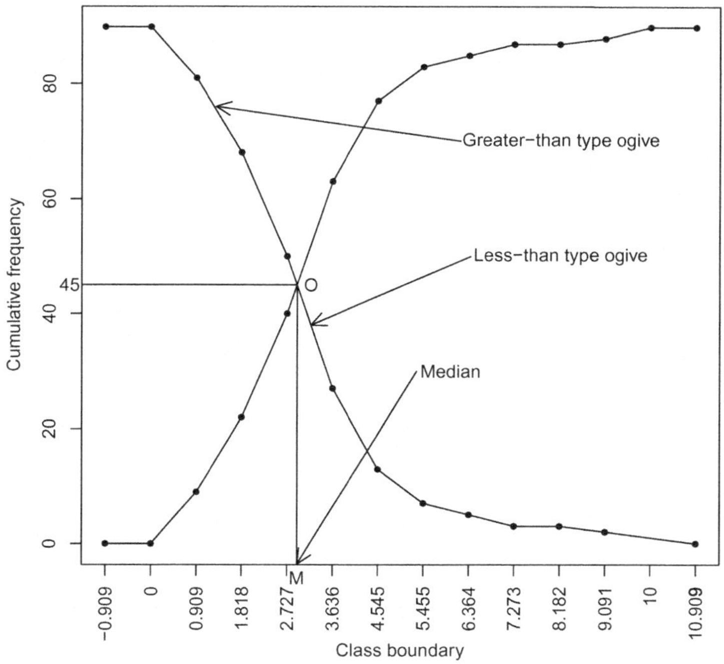

Fig. 2.3 Ogive of single gene expression data for 90 samples

2.2.4 *Boxplot*

A natural tendency of data is to cluster around some particular value(s) as well as to cover a range of values. These two features are probably most important in statistics to understand the general characteristics of the data. Before discussing some measures to represent these features, we would like to discuss a nice visual representation provided by a plot, known as '**boxplot**'. This plot provides summary measures of the data in one single diagram and can also be used to compare two or more distributions with respect to these characteristics.

A boxplot consists of a rectangle with a thick horizontal bar representing the median of the frequency distribution. The lower and upper horizontal lines of the middle rectangle represent first and third quartiles respectively. The two tails stretching towards the top and bottom are known as whiskers. These whiskers indicate how far the data extends on either side of the first quartile and third quartile. One can draw custom-made boxplots by controlling the whiskers to cover a substantial number of observations. There are some points that are scattered along the lines of whiskers but far beyond it. These points are known as 'outliers'. However, these outliers can be controlled by the user so that one can identify some important outliers that should not be deleted while doing the analysis. Boxplot is a nice tool to detect outliers present in

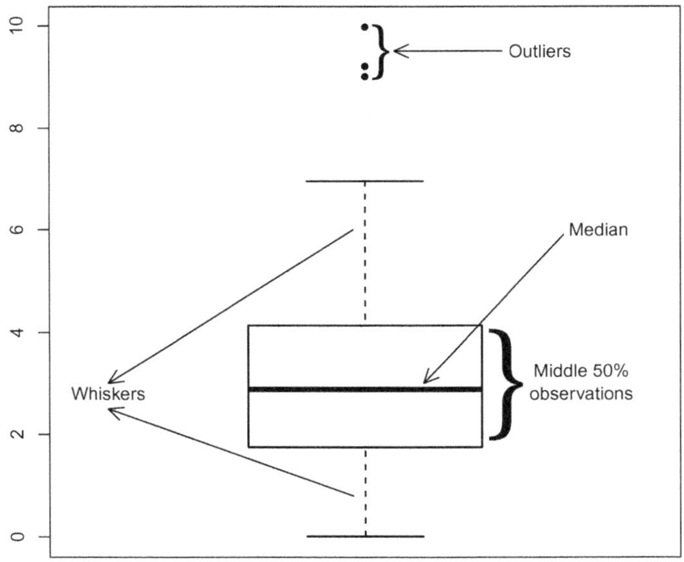

Fig. 2.4 Boxplot of single gene expression data for 90 samples

the data. Figure 2.4 represents the boxplot of a single gene expression data as given in Table 2.1.

The R command for drawing a boxplot is simply 'boxplot(b)', where 'b' contains the data list for Table 2.1.

```
> x<- c(2.67, 0.74,4.60,1.95,1.64,2.57,1.69,0.23,2.94,2.30,3.32,9.21,1.40,
4.91,1.45,1.99,6.96,2.41,4.57,1.98,4.06,4.64,2.90,9.01,3.58,2.81,0.02,4.29,
2.60,6.67,4.00,1.24,3.72,3.32,5.75,0.01,2.16,3.38,4.21,0.79,1.08,3.53,4.35,
3.45,2.95,1.61,3.36,3.46,1.56,0.57,3.46,2.37,5.36,1.8,3.15,1.86,4.04,2.77,2.51,
9.98,1.16,0.71,1.70,4.49,6.10,2.39,4.27,0.88,2.86,3.62,4.22,2.31)    ## input
                                                              data and store in 'x'
> boxplot(x)                ## draws boxplot using values stored in 'x'
```

However, the diagram for boxplot in Fig. 2.4 is drawn using a series of little more sophisticated commands as given in Appendix B.

The boxplot in Fig. 2.4 shows that the distribution is symmetric about 2.8 (approximately) and has almost equal spread with respect to this value. The upper whisker is longer than the lower whisker indicating that the data are more scattered towards the right side of median than the left. It also indicates the presence of outliers which are marked as solid points along the line of whiskers on the upper side. This means that there are a few (in this case 3) values of gene expressions that are much deviated than the general data cloud and need special attention.

The experimenter who generated these data should check how these *unusual* data points or outliers, as they are commonly called, are generated in order to identify

possible errors associated with these points. In most situations we would just delete these outliers and work with rest of the data as the number of outliers is less and hence we assume that it should not affect the general nature of the data. Moreover, these data points may be due to artefacts of the experiment. However, special attention should be given to identify the actual reason for the presence of such data. If it is justified that the presence of these outliers is a natural phenomenon, then we should not omit them; rather special care should be taken to extract inference from such data.

2.3 Summary Measures

While visual representation of data is very useful in understanding the nature of the data, these do not provide succinct quantitative descriptions of the data. Therefore, a small number of quantitative and objective measures that capture the nature of the data are required. These can explain the nature and trend of data, decipher the hidden pattern that is not evident immediately and provide future guidelines about the scientific experiments as a whole. However, ideas behind these measures would come from the diagrammatic representations of the data. Diagrams provide an overall presentation of data, some hints about salient features, and an indication of some hidden information that needs to be unearthed and explored. So, developing a few quantitative measures should start with adequate visual representation(s) of data.

2.3.1 Measures of Central Tendency

A careful look at a histogram reveals that area of some rectangles in the middle portion is much more than those in the tail regions. This clearly indicates the abundance of observations around a value in the middle region of the entire range of observations. Interestingly this is a common scenario in almost every data set. Thus we can say that for any data set, there is a tendency of the observations to cluster around some central value(s). Now our task is to develop some appropriate measures that can capture this feature of the data.

Consider again two more histograms (Fig. 2.5) or frequency curves (limiting forms of histograms). These histograms are drawn by increasing total frequency and decreasing class widths to have a feel about the structure of the data. All histograms and frequency curves reveal that the data have a high concentration around some central value, which in these case (Fig. 2.1) is about 2.9 (approximately). To apprehend this tendency, we define some measures, known as measures of central tendency, because they are expected to provide a quantification of the tendency of data to cluster around some central value(s). It is clear that both histograms reveal more or

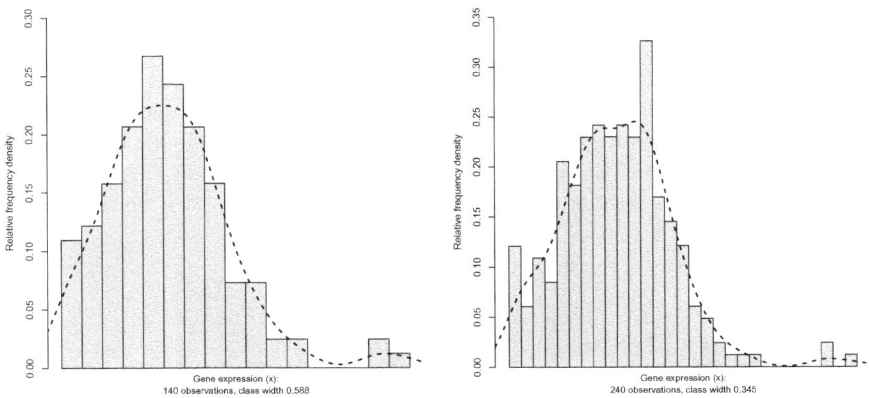

Fig. 2.5 Histograms of expression data with increasing sample size and decreasing class widths. Frequency curves are drawn using dotted lines

less the same overall picture about the data and hence population. Thus a moderate number of appropriately chosen sample, more specifically a random sample, would be enough to reveal the main features.

Arithmetic mean

10 children in my class were nice throughout the week. I gave 4.8, 4.5, 5.6, 5.7, 4.4, 4, 4.9, 5, 5.8, 5.3 dollars respectively (5.3$ means 5$ and 3 cents) to 10 children so that they could buy candies during the weekend. Later, next Monday the children were although happy but a little upset because not all of them got the same amount from me. I realised my mistake. Actually the amounts I gave to the ten children were not same. I had met each child at a different period and had randomly handed them some amount. So the question is: if I were to give the same amount, how much money would each child get?

The answer is easy. Total amount is 50$ that needs to be distributed among 10 children. So each child is supposed to get 5$. This is what is called arithmetic mean. In other words, if we distribute the total amount equally among a number of children (or individuals/units), each one would receive an amount that is equal to the arithmetic mean.

Thus a simple and most commonly used measure of central tendency is known as 'arithmetic mean' or simply 'mean', defined as $\bar{x} = \frac{1}{n}\sum_{i=1}^{n} x_i$ where $x_1, x_2, \ldots, x_n$ are n observations corresponding to n units sampled from the population. By applying this formula we get the mean gene expression value to be 3.1005 for the data given in Table 2.1. For a frequency distribution in class intervals, mainly for a continuous distribution, the arithmetic mean is calculated assuming that the class frequency is distributed uniformly around the mid-point of each class in that interval. Hence the

formula is given by: $\bar{x} = \frac{1}{N} \sum_{i=1}^{n} x_i f_i$ where N is the total frequency and f_i is the frequency of the i-th class for which the mid-point is x_i, for $i = 1, \ldots, n$.

Since nowadays, we calculate everything using computer programming or by a software, there is no need to calculate mean using the second formula. We can easily calculate the mean for a set of observations, however large or small. Another practical tip: in a real data set sometimes a few observations are missing and are recorded as 'NA'. But R can tackle it directly; calculating 'mean' while eliminating those 'NA' values.

```
> mean(x)                ## calculates mean of values stored in 'x'
> mean(x,na.omit=T)      ## if data set contains 'NA'
```

Although mean is intuitively an appealing measure of central tendency, it suffers from some drawbacks. In a set of experimental data, some values are observed to be very different from most other values. These few (might be one or two) observations can pull the arithmetic mean towards them thus distorting the position of mean around which majority of the values lie. The mean, therefore, is highly affected by such extreme observations, known as 'outliers'. In this data set, it is clear from the histogram and especially from the boxplot that there are few outliers that may affect the arithmetic mean. So in this situation mean might give misleading value as a measure of central tendency and should not be used in such situations.

Median

To avoid the impact of outliers in measuring central tendency, another measure known as 'median' is preferred in many situations. The median is the middle-most observation when the data are arranged in increasing or decreasing order. Suppose n is odd, i.e. $n = 2m + 1$ for some integer m. If $x_1, x_2, \ldots, x_n$ are observations already arranged in increasing order, the median is defined as $\tilde{x} = x_{m+1}$. If n is even, i.e. $n = 2m$ for some integer m, then any value between the two middle-most observations can be taken as median. However, for the sake of simplicity, we can define median as the mean of the two middle-most observations, i.e. $\tilde{x} = (x_m + x_{m+1})/2$. Clearly, median is not much affected by the presence of extreme values. The median gene expression value is 2.835. The R command for calculating median is '$median(x)$' (note how simple the commands are to calculate mean or median!).

Suppose that we have been given a frequency distribution for a continuous variable instead of raw data points. In such a situation, we need to use a formula to calculate median. This formula is derived using the basic assumption that the class frequency is distributed uniformly around the mid-value (class mark) of each class interval. This is given by: $\tilde{x} = x_l + (N/2 - F_l)w/f_0$ where x_l is the lower class boundary of the class for which cumulative frequency (less than type) is less than or equal to $N/2$, N is the total frequency, w is the class width, F_l is the cumulative frequency up to x_l, and f_0 is the frequency of the median class. Here we assume that the class

widths are the same for all classes in the frequency distribution constructed based on the given data.

However, like mean, we usually calculate median using computer programming or software and hence the formula for calculating median using grouped frequency data is not required at all.

```
> median(x)          ## calculates median of values stored in 'x'
> median(x,na.omit=T)                   ## if data contains 'NA'
```

Clearly mean is much more than median for the given data set. It indicates presence of outliers that is also confirmed by boxplot of the data. Hence, depending on our objective, we have to decide the appropriate measure of central tendency for further downstream analysis.

2.3.2 *Measures of Dispersion*

From Figs. 2.1 and 2.4 it is clear that although data are clustered around central values like 3.1005 or 2.835, observations are spanned over a range of values as seen from the span of the histogram or boxplot. Note that outliers may belong towards either or both ends of histogram or boxplot; one end shows more stretched compared to other end. Thus we also need to study the nature of dispersion or scatteredness present in the data along with the mean. Note that if a data set has no dispersion, all values would be same and hence there is no need to do any statistical analysis whatsoever. In fact the study of dispersion is most important in Statistics, but taking careful attention to central tendency at the same time. This simple philosophy triggers many branches in statistics and rigorous research is still continuing to explore several aspects and features that germinate from it. To develop some measures of the degree of dispersion in a data set, we need to understand its meaning and interpretation along with the inherent nature of data.

Central tendency is always observed for any kind of data. But there is another inherent characteristic without which no statistical analysis is possible or meaningful. I realised it with a real life experience, a scary incident indeed! It happened to me long back. While trekking in the mountains we came across a beautiful small river, the width of the river was not so wide. Our leader said "Cross the river. It's not very deep." Being very excited to cross a river bare feet, I walked into the river boldly (and of course carelessly)! I had the confidence that the river was not too deep; there was no question of drowning, or falling into the cold water. But I did tumble, fall, get soaked and wet in that freezing water. Why? I ignored what is known as '**dispersion**'. Our leader intended to mean that the average depth is not too much. That was indeed true, but at some places it was deeper than rest of the places while in some places it was very shallow. Mistakenly I fell into the deeper region, tumbled and got

drenched in the freezing water! Because there was dispersion. So dispersion or how data are scattered is extremely important and is the main reason of statistical analysis.

Mean deviation

If there is no dispersion in the data, all observations would have been same and this value is a representative of the entire data sets, a measure of central tendency. Because there is dispersion, values are different. The concept is so simple! Hence dispersion is always observed and should be studied around a measure of central tendency. We can take the difference of each observation from any chosen measure of central tendency, say A. Thus, we consider $x_i - A$ for all $i = 1, \ldots, n$ and take their mean. However some of these differences are positive and some are negative, hence the sum might be approximately or exactly equal to 0, which is counter intuitive. Moreover, we are only interested in measuring the extent of dispersion without considering whether an observed value is lower or higher than the measure of central tendency i.e., ignoring the sign of differences. We need to remember that we are interested in knowing how far an observation is from a central tendency measure, not the direction (positive or negative). So, instead of just the difference, we can take absolute value of each difference or deviation from an appropriate measure of central tendency. An average or mean of such deviations can be taken as a nice measure of dispersion present in the data. This is known as 'mean deviation about A' and is given as:

$$MD_A = \frac{1}{n}\sum_{i=1}^{n}|x_i - A|$$

where A is an appropriately chosen measure of central tendency.

It can be shown that MD_A is minimum if we take $A = \tilde{x}$, i.e. median, which is also a measure least affected by outliers. Thus we define another good measure of deviation, somehow less affected by outliers, as mean deviation about median,

$$MD_{\tilde{x}} = \frac{1}{n}\sum_{i=1}^{n}|x_i - \tilde{x}|.$$

Similarly we can define mean deviation about mean as

$$MD_{\bar{x}} = \frac{1}{n}\sum_{i=1}^{n}|x_i - \bar{x}|.$$

Note that it always true that $MD_{\tilde{x}} \leq MD_{\bar{x}}$ for any data set. For the gene expression data (Table 2.1), $MD_{\bar{x}} = 1.358$ and $MD_{\tilde{x}} = 1.341$. R code to calculate mean deviation about mean and median is given below.

```
> md.median=mean(abs(x - median(x)))          ## mean deviation about median
> md.median                                   ## showing output on the screen
> round(md.median,3)            ## if you want the output up to three decimal places
> md.mean=round(mean(abs(x - mean(x))),3)  ## mean deviation about mean up
                                              to three decimal places
```

Standard deviation

Although $MD_{\tilde{x}}$ has an advantage that it is somewhat less affected by outliers, its major disadvantage is that it is not really easy to handle mathematically for further analysis. Working with mean deviation sometimes may be difficult. To make our lives simpler, we consider squaring the values of these differences (instead of absolute deviations) and take their mean. However, there is a problem. While squaring the differences, we have also squared the unit in which observations are measured. An easy solution to get rid of this problem is to just take the square-root of the mean thus obtained. Hence we get a measure of dispersion with respect to the chosen average A, as given by,

$$SD_A = \sqrt{\frac{1}{n}\sum_{i=1}^{n}(x_i - A)^2}.$$

This is known as root mean square deviation about A. Since for different choices of A, we get different values and it can be very large for some choices of A, it is preferable to take the value of A for which this will be minimum. It can be shown mathematically that SD_A would be minimum if we take $A = \bar{x}$. Hence, we define standard deviation as a measure of dispersion as

$$s_x = \sqrt{\frac{1}{n}\sum_{i=1}^{n}(x_i - \bar{x})^2}.$$

The standard deviation of gene expression values for the data in Table 2.1 is found to be 1.865 using the R command for standard deviation as '$sd(x)$'. Remember that this command calculates the standard deviation taking $n - 1$ in the denominator instead of n. So a simple multiplication factor would give s_x as we defined just now.

```
> sd(x)                    ## denominator being n-1
> sd(x)*sqrt((n-1)/n)      ## denominator being n, as defined
```

Note that if the sample size is moderately large, it does not matter whether we take n or $n - 1$ in the denominator; both results would be approximately same. Standard

deviation as obtained in R using 'sd()' is 1.8649 and that using divisor $n = 90$ as 1.8545. Note that $MD_{\tilde{x}} = 1.341$, which is naturally less than s_x as it is less affected by outliers. It can also be proved mathematically that for any data set, $MD_{\tilde{x}} \leq s_x$. However, standard deviation is the most popular and widely used measure of dispersion.

Quartile deviation

Based on the above discussion, we need to device another measure of dispersion that is least affected or not affected at all by outliers. It is clear that this measure should be constructed using median or measures like median that is not affected by outliers. To define such a measure, that is least affected by outliers, we first define quartile for a distribution. We define Q_1, and Q_3 as 1st, and 3rd quartile respectively, as the value of the variable such that the number of observations less than that value is 25% and 75% respectively. Similarly we can define Q_2 as the value below which there are 50% observations. Clearly Q_2 is the median and the interval (Q_1, Q_3) contains middle 50% observations. Naturally any measure based on Q_1 and Q_3 would be free from the effect of outliers. Longer this interval, more disperse the data would be. Thus we define a new measure of dispersion based on quartiles only, known as **quartile deviation**, as:

$$QD = \frac{Q_3 - Q_1}{2}$$

For the data set in Table 2.1, QD is found to be 1.07.

Note that although this measure of dispersion is completely robust to the presence of outliers, it depends only on the middle 50% observations and ignores 50% observations that lie towards the ends or 'tails' of the distribution. Naturally this incorporates serious loss of information from the data as a price to remove the effect of outliers completely. This might be one reason for not using quartile deviation very frequently as a measure of dispersion. Sometimes $Q_3 - Q_1$ is known as **interquartile range**.

```
> QD=(quantile(x,0.75)-quantile(x,0.25))/2     ## calculates and stores
                                                  quartile deviation
> QD                              ## showing quartile deviation on the screen
```

Till now we have discussed the intuitive approach to address the question (1)–(4) as mentioned in Sect. 2.1 and the corresponding mathematical formula based on statistical intuitive argument. Recalling these questions, a careful look at histograms and/or boxplots reveals the general features of data, apply our thought process and the development of measures of central tendency and dispersion is followed naturally.

2.4 Points to Remember

We have discussed a few visual tools for representation of data for a single variable and some basic summary measures to understand the characteristics of data. The data we consider here is gene expression data for a single gene for a number of individuals randomly selected from a population. We must emphasise that these visual tools as well as summary measures are equally important and applicable to understand the salient features of any data set corresponding to a continuous variable. This should be the starting point of statistical analysis of the data set arising out of any random experiment that gives continuous frequency data. We should always give serious attention to the presence and effect of outliers as in the later sections we see how important and sometimes distorting effects it can cast in the data analysis.

Now one very simple but trivial question is whether one should or should not analyse the data. Suppose an experiment generates values all equal to 2.57; we cannot statistically analyse that data. The only reason is absence of any variation and we doubt whether the data has been generated out of a random experiment. Such data are not amenable to statistical data analysis. If values in the data are different (may not be all values but mostly) and there is a prior knowledge about data generation process that ensures introduction of randomness, we must do statistical analysis in the best possible way. Anything between these two extreme scenarios should be guided by the jurisprudence of the experimenter, nature of data, knowledge about the data generation process, and most importantly the hypothesis and objective of the experiment.

Exercise

2.1 Based on some ideas and queries, a scientist performed an experiment that gives data on expression values of 12 genes from 102 individuals. Can you guess what should be the questions that triggered this experiment? Try to come up with as many questions as possible.

2.2 In Question 2.1, after generating the data it is observed that data from all genes from all individuals were not obtained. Can you give some idea about the data structure and relevant questions that might be absent previously?

2.3 Consider the following data set.

data set 1:

3.5, 3.7, 5.7, 4.2, 5.4, 2.9, 2.3, 8.6, 3.4, 3.1, 4.5, 7.8, 6.9, 6.2, 2.6, 6.0, 5.0, 10.3, 9.9, 11.9, 7.3, 3.0, 3.9, 2.0, 6.7, 4.0, 5.7, 1.4, 6.0, 2.4, 3.7, 9.2, 3.8, 4.5, 9.9, 6.3, 2.0, 5.2, 4.0, 1.4, 2.0, 4.4, 7.9, 1.6, 12.6, 3.9, 4.0, 3.3, 4.7, 2.9, 1.3, 4.5, 3.7, 4.9, 2.2, 2.4, 2.5, 1.7, 3.7, 3.3

(a) What is the type of the data?

(b) If it is continuous, draw a histogram with 6 classes and also draw a boxplot. If not, don't go for Questions 2.3 (c) and justify your answer.

(c) Draw the above histogram and boxplot in one graph. [hint: use the R command 'par(mfrow=c(1,2))']
(d) What is the percentage of observations that is less than 5.2?
(e) What is the percentage of observations that is more 3.7?
(f) What is the percentage of observations that are greater than or equal to 4.5?
(g) What percentage of observations lies between 3.84 and 6.12?
(h) Calculate appropriate measures of central tendency and dispersion.
(i) Comment on your findings.
(j) Calculate mean eliminating upper 10% and lower 10% observations.

2.4 Summary measures are defined mathematically so that we can calculate them using a set of observations. However, interpretation is necessary. Interpret arithmetic mean, median and standard deviation in simple language. Use some examples to explain your interpretation.

2.5 The hemoglobin level (g/dl) of a sample of 50 men are given below.
17.0, 17.7, 15.9, 15.2, 16.2, 17.1, 15.7, 17.3, 13.5, 16.3, 14.6, 15.8, 15.3, 16.4, 13.7, 16.2, 16.4, 16.1, 17.0, 15.9, 14.0, 16.2, 16.4, 14.9, 17.8, 16.1, 15.5, 18.3, 15.8, 16.7, 15.9, 15.3, 13.9, 16.8, 15.9, 16.3, 17.4, 15.0, 17.5, 16.1, 14.2, 16.1, 15.7 15.1, 17.4, 16.5, 14.4, 16.3, 17.3, 15.8
Calculate Mean, Median, Range, Variance, Standard deviation and Quartile deviation.

2.6 In reference to Question 2.5, let x_i denote the hemoglobin level (g/dl) of i-th individual; $i = 1, 2, \ldots, 50$. Define a new variable as $y_i = x_i - \bar{x}$, $i = 1, 2, \ldots, 50$. Calculate Mean and Variance of y.

2.7 Here is a frequency distribution of the Apgar scores for 100 low birthweight infants. Find the Mean and SD of Apgar scores.

Argar Score	Frequency
0	6
1	1
2	3
3	4
4	5
5	10
6	11
7	23
8	24
9	23
Total	100

2.8 In data set 1 (Question 2.3), suppose we remove the observation 12.6 from the data set and calculate new mean and new standard deviation based on remaining observations. It is seen that both new mean and new standard deviation are smaller than the original mean and original standard deviation. However, if

we remove the observation 4.7 from data set 1, we see that both new mean and new standard deviation are larger than the original mean and original standard deviation. Can you explain these situations?

2.9 Suppose there are 100 observations in a data set. We can calculate arithmetic mean $\bar{x}$, all three quartiles i.e., Q_1, Q_2(=median), Q_3 and standard deviation s. Now replace top 25% observations by Q_3 and the bottom 25% observations by Q_1. What would be the effect of such operation on the summary measures mentioned above. Justify your answer clearly.

2.10 Consider the following two data sets.

data set 2:
10.01, 14.18, 9.95, 11.71, 12.26, 9.86, 13.33, 10.70, 7.46, 11.19, 8.07, 6.23, 7.17, 7.77, 9.95, 12.16, 10.55, 10.03, 14.11, 9.84, 9.36, 11.18, 9.61, 9.99, 10.98, 8.90, 11.09, 10.42, 11.28, 12.85, 10.10, 10.39, 8.99, 10.41, 9.31, 9.11, 10.62, 11.18, 10.35, 6.02, 4.06, 10.54, 8.95, 5.58, 10.40, 9.65, 12.45, 12.15, 9.66, 13.30, 8.12, 11.73, 11.01, 13.53, 10.80, 11.14, 11.60, 10.39, 9.39, 8.38, 11.16, 13.00, 9.72, 6.38, 8.83, 8.33, 9.51, 5.35, 9.04, 13.35, 9.26, 11.56, 10.09, 8.62, 9.75, 10.11, 10.09, 10.37, 8.84, 6.32

data set 3
10.68, 10.92, 12.27, 17.62, 15.55, 12.04, 12.91, 11.20, 13.55, 11.72, 16.27, 16.88, 15.52, 17.14, 14.16, 15.28, 18.92, 13.32, 19.80, 15.24, 12.57, 17.13, 13.27, 17.37, 11.51, 14.10, 15.29, 11.56, 13.43, 12.27, 13.90, 13.67, 12.49, 13.46, 15.40, 13.30, 15.86, 14.10, 12.28, 12.54, 16.97, 14.06, 14.75, 17.78, 15.34, 9.52, 15.91, 19.21, 12.72, 16.85, 18.64, 15.90, 12.35, 13.76, 15.78, 18.18, 9.63, 18.68, 17.11, 12.69, 11.15, 16.57, 11.81, 15.71, 17.29, 13.69, 16.46, 13.82, 9.18, 11.24, 20.09, 14.52, 13.93, 11.85, 15.34, 17.12, 12.92, 12.46, 13.46, 14.15, 14.97, 9.57, 16.12, 21.06, 15.37, 15.02, 16.44, 14.43, 19.24, 19.51

(a) Present these two data sets diagrammatically, preferably in a single graph, so that we can compare them. Write down your observations.
(b) Using the commands for Fig. 2.2 or other commands, can you add colours to different parts of the histogram drawn for data set 2, change the labels for x-axis and y-axis, change the legends and shade area of each rectangle?
(c) Draw two boxplots for data set 2 and data set 3 side by side on the same graph, and distinguish them by colours as well as legends.
(d) Analyse the data based on appropriate measures of central tendency and dispersion. Analysis means not only to look at the features of each data set, but a comparison between two data sets is essential.
(e) Comment on your findings.
(f) Calculate third quartile for both data sets and comment on the nature of distribution.
(g) Calculate first, second and third quartiles for both data sets and comment on the nature of distribution considering all these calculated measures.
(h) Identify at which values (at least approximately) the peak for frequency curve would occur for both data sets? Comment on your findings.

2.11 Can you explain intuitively why mean deviation about median is less than the mean deviation about mean?

2.12 Define a measure of central tendency ignoring top 10% and bottom 10% of the observations. Write an R code to calculate such a measure.

2.13 Draw a histogram for data set 3 in Question 2.10 using only 9 classes of equal width. Change the colour of inside and border of the histogram.

2.14 Draw Fig. 2.3 with less-than type ogive in red colour and greater-than type ogive in blue colour. Also try to make the code little shorter.

2.15 In drawing ogive, why do we join two consecutive points by a straight line? "In an ogive, the first and last line segments are always parallel to the horizontal axis" - justify this statement.

2.16 Draw two ogives using data set 3 as in Question 2.10. Calculate median from the ogive.

2.17 From the ogive drawn based data set 3 in Question 2.10, calculate percentage of observations that are

(a) less than 15,
(b) greater than 17, and
(c) between 12 and 18.

Do the same from the histogram also and check whether your answers match.

2.18 For the above three data sets, we want to study the dispersion.

(a) Calculate quartile deviation for each data set and comment on your findings.
(b) Do you think your comments remain more or less same if you compare their dispersions using standard deviations? Justify your answer.

2.19 Suppose we calculate means $\bar{x}_1$ and $\bar{x}_2$ based on two data sets of sizes n_1 and n_2 respectively. If we merge the two data sets, what would be the arithmetic mean ($\bar{x}$, say) of the combined group in terms of known quantities? Show that $\bar{x}$ lies between $\bar{x}_1$ and $\bar{x}_2$. Extending this problem to k groups with means $\bar{x}_1, \ldots, \bar{x}_k$ with sample sizes for the groups as $n_1, \ldots, n_k$ respectively, find the expression of the arithmetic mean when all groups are merged together. Show that,

$$\min_{i=1,\ldots,k} \{\bar{x}_1, \ldots, \bar{x}_k\} \leq \bar{x} \leq \max_{i=1,\ldots,k} \{\bar{x}_1, \ldots, \bar{x}_k\}$$

2.20 Based on a set of 20 observations, an experimenter calculates arithmetic mean as 5.68. However, after a few days, while checking the entire process, it becomes clear that by mistake the experimenter has taken a value 12.4 while calculating the mean. Since this value is wrongly taken, so a correction is necessary. Moreover, at this stage the raw data are not available any more. The experimenter wants to recalculate the mean just by removing this wrong observation. How can he calculate the mean removing 12.4 and using only the available measures (not the data)? What would be the new corrected mean?

2.21 In Question 2.20, suppose it becomes known that the experimenter uses 12 instead of 2.4. How does he modify his calculation? What would be the corrected value of mean?

2.22 The weight (Kg.) of 10 individuals are recorded as: 52, 62, 54, 82, 49, 68, 62, 34, 97, 72. After scrutiny, it is found the lowest and the highest weights are wrongly reported. The true highest is 82 Kgs and the second highest weight is 79 Kgs. If the average weight is 67 Kgs, what is the true lowest weight?

2.23 Let $z_i = x_i + y_i, i = 1, 2, \ldots, 7$. Show that the mean of z is the sum of mean of x and mean of y using the data. Also calculate the variance of z. Is it equal to the sum of $var(x)$ and $var(y)$?

x	21	71	32	78	56	73	74
y	82	172	102	167	134	160	155

References

Casneuf, T., Van de Peer, Y., & Huber, W. (2007). In situ analysis of cross-hybridisation on microarrays and the inference of expression correlation. *BMC Biioinformatics, 8.*

Schena, M., Shalon, D., Davis, R. W., & Brown, P. O. (1995). Quantitative monitoring of gene expression patterns with a complementary DNA microarray. *Science, 270*(5235), 467–70.

Shendure, J., & Ji, H. (2008). Next-generation DNA sequencing. *Nature Biotechnology, 10,* 1135–45.

Wang, Z., Gerstein, M., & Snyder, M. (2009). RNA-Seq: A revolutionary tool for transcriptomics. *Nature Review Genetics, 10*(1), 57–63.

Chapter 3
Basic Probability Theory and Inference

Most people have an apprehension that probability theory is a very difficult area of mathematics or statistics. However, if we don't go into the details and intricacies of probability theory, it is very intuitive and easy to understand. In fact, probability theory forms the core of real data analysis. It is indeed vast. But to begin with, a simple understanding of probability theory is not at all difficult and can be framed or constructed from real-life phenomena. People always wonder: (1) why do we need probability theory? (2) does it have anything to do with a real-life situation or experiment?

Well, the concept of probability is probably at the core of any experiment that generates data. To understand its concept clearly, we need to know a few important ideas, which are common in daily life.

3.1 Random Experiment

Probability is there in our everyday life. We make probabilistic statements every now and then, knowingly or unknowingly. When you go out, your mother tells you to carry an umbrella because it might rain after a little while. Looking at the sky, she predicts not the rainfall but the possibility of raining. A cloudy sky might lead her to say that there is a high chance of rain today; so you should carry an umbrella.

From these statements, two facts are obvious: (1) she predicts based on the condition of the sky, how it was yesterday and/or how it looks today and (2) her prediction is not with certainty, rather there is some amount of uncertainty associated with her statement. This uncertainty is nothing but the backbone of probability theory that provides a measure of the quantification of uncertainty about an event whose outcome may not be always predicted accurately, but can be guessed.

Usually, whatever be the nature of data, they are subject to some random causes that give rise to fluctuations present in the data. We call such an experiment 'random

I. Mukhopadhyay and P. P. Majumder, *Statistical Methods in Human Genetics*, Indian Statistical Institute Series, https://doi.org/10.1007/978-981-99-3220-7_3

experiment'. Precisely, a random experiment is an experiment whose outcome cannot be predicted beforehand with certainty, but we might know all possible outcomes that this experiment would generate.

The most classical example of a random experiment is the 'coin tossing experiment'. If we toss a coin, we don't know its outcome; however, we know that it must be either 'head' or 'tail'. If we throw a die, we know that the upper face will show one of the numbers from the set $\{1, 2, 3, 4, 5, 6\}$; but we cannot say with certainty which one will appear. This uncertainty is due to the inherent randomness of the experiment, in this case observing the number on the upper face when we throw a die. Similarly choosing a person at random (i.e. without attaching any preference to anybody) from a group of persons having three categories of blood pressure, viz., low, normal, and high blood pressure, is another example of a random experiment. We know that a randomly selected person will have either low or normal or high blood pressure (all possible cases), but we would be able to know the actual status of blood pressure only when a person is selected and her/his blood pressure level is measured. These are some examples of a random experiment.

3.2 Event

An 'event' may be defined as an outcome or a set of outcomes or a combination of outcomes of a random experiment. Consider a random experiment where we toss a coin two times and note the outcome. Let H and T denote 'head' and 'tail' respectively. So, $\{HH\}$ denotes two heads, $\{HT\}$ denotes head in the first toss and tail in the second toss, etc. These events are called 'elementary events'. In this case, all possible elementary events are:

$$\{HH\}, \{HT\}, \{TH\}, \{TT\}$$

However, we can think of another event A as '*getting at least one head*'. So, here $A = \{HH, HT, TH\}$. This event, which is a combination of elementary events is sometimes known as a 'composite event'. Whatever be the name, its treatment in probability theory is basically same; the only advantage of an elementary event is that probability of elementary events are easy to calculate and probability of composite events can be derived using probabilities of some elementary events that constitute the composite event.

Consider a die throwing experiment. When we throw a die, '*getting number 4*' is an elementary event. But '*getting odd number*' means observing either 1 or 3 or 5, i.e. it consists of three elementary events and is a composite event. Similarly, '*getting an odd number which is less than 5*' is an example of a composite event. In the case of each of these events, we don't know what we get if we throw a die. Uncertainty again! But we understand that the degree of uncertainty varies depending on events.

Now that we have an idea about random experiments and associated events, we should try to define probability of an event, be it elementary or composite.

3.3 Idea and Definition of Probability

It is now clear that there is some amount of uncertainty in almost all real-life phenomena or experiments. We need quantification of this uncertainty as the degree of uncertainty is not same in all cases. Probability is basically how we can quantify the uncertainty in an experiment, or rather the uncertainty to observe an event. Before starting a game, the referee tosses a coin in the air and the captain of one team shouts 'head'. The choice of 'head or 'tail' may depend on her/his prior belief, but it is expected that she/he has an equal chance of winning or losing the toss; otherwise, it would be unfair. Assume that the coin is a fair coin. So, we say that chance of getting a 'head' in a single toss is 50%, i.e. probability of winning (or losing) is $\frac{1}{2}$. 'head' can occur in only one way out of two possible ways as the coin has only two faces marked as 'head' and 'tail'. We can interpret this in another way. If we toss 10 times, we expect to get an equal number of heads and equal number of tails approximately (we can never be sure!). Considering this, the classical definition of probability of an event A (getting 'head' in a toss is an event) is given by,

$$P(A) = \frac{\text{number of outcomes favourable to the event } A}{\text{total number of events}}$$

Note that here we assume all outcomes are equally likely to occur.

However, this classical definition has a few drawbacks. Without going into the details of it, we can discuss this with an example. Suppose that in my wrist watch the big hand strikes at 12. What is the probability that it will stop functioning before it strikes 6?

Looking carefully it reveals that the big hand may stop at any moment between two 12s; that means the number of total cases is really infinite. But the number of favourable cases is also infinity because there are enumerable points on the wristwatch between 12 and 6 where it can stop. So, probability of the desired event cannot be calculated, if we stick to the classical definition.

But anyone can give the correct answer as $\frac{1}{2}$! It is easy to see that the totality of cases is the entire circumference whereas favourable situation is the half circumference (Fig. 3.1), because the grey curve represents the favourable space, whereas the total circumference is the total possible space. Clearly, the length of the grey curve is exactly half of the total circumference; hence, the probability would be $\frac{1}{2}$.

Thus, classical definition is not a general one; one can give a more precise definition of probability. The formal definition of probability requires some mathematical concepts, which we state without details.

Let $(\Omega, \mathscr{A}, P)$ be a probability space where Ω is the sample space, a collection of all outcomes, $\mathscr{A}$ is the σ-field generated by the class of subsets of Ω, and $P(.)$ is a set function satisfying the following conditions:

(1) $P(A) \geq 0$ for any $A \in \mathscr{A}$,
(2) $P(\Omega) = 1$, and

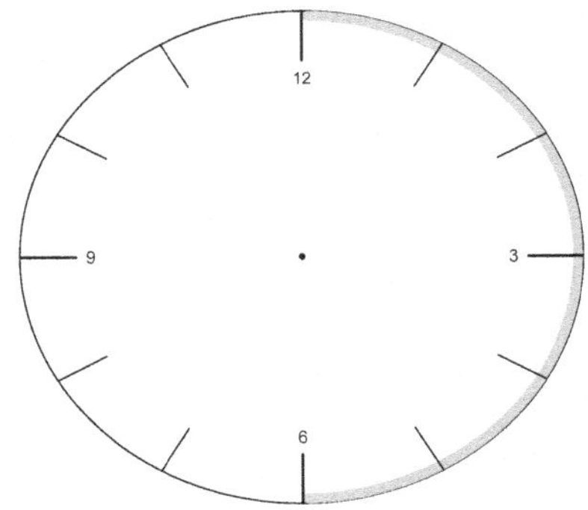

Fig. 3.1 Clock showing how the probability that the big hand might stop between 12 and 6. Grey part is the favourable region whereas the entire circumference is the total possible region where the big hand might stop

(3) If $A_1, \ldots, A_k$ be mutually exclusive events such that $A_i \in \mathscr{A}$, $\bigcup_{i=1}^{k} A_i \in \mathscr{A}$, then $P\left(\cup_{i=1}^{k} A_i\right) = \sum_{i=1}^{k} P(A_i)$.

A few examples

Ex-3.1 A coin is tossed two times. The possible elementary events are: $\{HH\}$, $\{HT\}$, $\{TH\}$, $\{TT\}$. Assuming the coin is unbiased, i.e. getting a head is as likely as getting a tail, we might expect to observe $\{HH\}$ only once if we repeat the experiment four times. So,

$$P(\{HH\}) = \frac{1}{4}.$$

Let A be the event of getting at least one H. Here, occurrence of any one of $\{HH\}$, $\{HT\}$, and $\{TH\}$ would indicate the occurrence of the event A and there are altogether four such elementary events when we toss the coin two times. Hence,

$$P(A) = \frac{3}{4}.$$

Ex-3.2 A group consists of 26 unaffected and 3 affected people. Select one person at random. What is the probability that the selected person is affected? Number of events favourable to our desired event is 3 because selection of any one of three affected people will satisfy our criterion. On the other hand, the total number of selections is 29 and any one of 29 people may be selected at random. Hence,

$$P(\text{selecting one affected people}) = \frac{3}{29}.$$

3.4 Some Facts

Now we state a few facts, or results (without proof) related to the probability theory. These are very useful to proceed further. Although we are not giving any proof of these results, however, we try to explain them through diagrams so that a clear idea would emerge. A few of them come directly from the definition of probability. But, first, we define a few operations using sets (or events) as follows:

Complement of a set: A^c, which indicates non-occurrence of A. Let A be an event that an odd number appears when a die is thrown. So, $A^c = \{$ numbers that are not odd $\} = \{2, 4, 6\}$.

Union of sets: $\{A \cup B\}$, i.e. $\{A \text{ or } B\}$ indicates that either A occurs or B occurs or both occur. In die throwing experiment with one die, if $A = \{$ number appears is odd $\} = \{1, 3, 5\}$, $B = \{$ multiple of 3 appears $\} = \{3, 6\}$, then $A \cup B$ means any one of the numbers in the set $\{1, 3, 5, 6\}$ would appear.

Intersection of sets: $\{A \cap B\}$, i.e. $\{A \text{ and } B\}$ indicates that both A and B occur. In the previous example, $A \cap B = \{3\}$, because only 3 satisfies both A and B.

Mutually exclusive sets: Two sets A and B are mutually exclusive or disjoint sets if $A \cap B = \phi$, where ϕ denotes a null set. In throwing a die, let $A = \{$ an odd number appears $\} = \{1, 3, 5\}$ and $B = \{$ an even number appears $\} = \{2, 4, 6\}$. Therefore, $A \cap B = \phi$, i.e. null set; hence A and B are mutually exclusive sets or disjoint sets.

Figure 3.2 presents a diagrammatic representation of these four types of set operations. Note that any set operation when applied on a set or a number of sets, will give another set.

Now we state a few facts.

(1) $P(A \cup B) = P(A) + P(B) - P(A \cap B)$.
(2) For any A and B such that $A \subset B$, $P(A) \leq P(B)$.
(3) $0 \leq P(A) \leq 1$.
(4) $P(A^c) = 1 - P(A)$.
(5) $P(\phi) = 0$ where ϕ is a null set.
(6) If A and B are mutually exclusive events, then $P(A \cap B) = 0$.

Mathematical proofs along with intuitive 'proofs' are given in Appendix C.

A Few Examples

We now give a few examples that use these facts. If we can define an event as a set, it is very easy to calculate probabilities of several events (sets). First, we try to find out probabilities of simple or elementary events. Then using them and the above-stated facts, we deduce probability of a little more complicated event that can be a combination of elementary events.

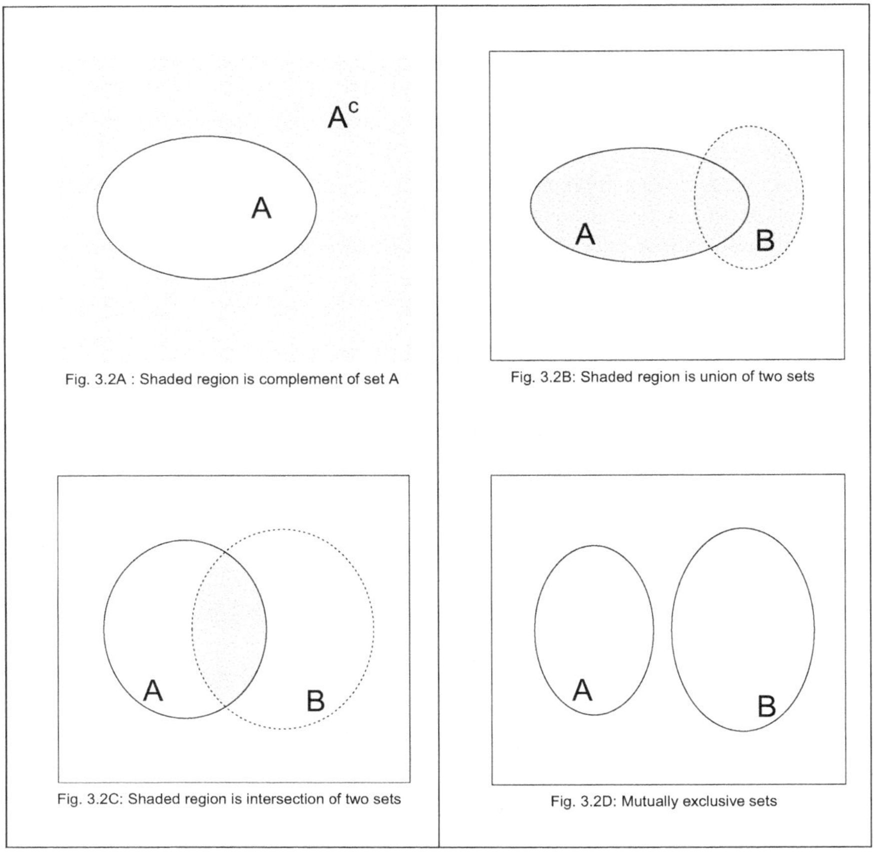

Fig. 3.2 A few set operations

Ex-3.3 A die is thrown. Let $A = \{$ number appears is odd $\}$, $B = \{$ number appears is a multiple of 3$\}$. Clearly, $A = \{1, 3, 5\}$, $B = \{3, 6\}$ so that $(A \cap B) = \{3\}$. Now,

$$P(A) = \frac{\text{number of cases favourable to } \{1, 3, 6\}}{\text{total number of cases}} = \frac{3}{6}; \text{ similarly, } P(B) = \frac{2}{6},$$

$$P(\text{number appears is odd and a multiple of 3}) = P(A \cap B) = \frac{1}{6},$$

$$\therefore\ P(\text{number appears is a multiple of 3 or an odd number})$$

$$= P(A \cup B) = P(A) + P(B) - P(A \cap B) = \frac{3}{6} + \frac{2}{6} - \frac{1}{6} = \frac{4}{6} = \frac{2}{3}.$$

Ex-3.4 Let X = genotype of a randomly selected person at a bi-allelic locus with alleles D and d. Define $A = \{DD\}$, $B = \{Dd\}$, and $C = \{dd\}$. Also it is

known that $P(A) = 0.3$, $P(B) = 0.2$, and $P(C) = 0.5$. Then,

$$P(X \text{ is homozygous}) = P(A \text{ or } C) = P(A \cup C) = P(A) + P(C) - P(A \cap C) = 0.3 + 0.5 - 0 = 0.8$$

Note that A and C cannot occur simultaneously because a person cannot have two genotypes DD and dd at the same locus; hence, $P(A \cap C) = 0$.

Ex-3.5 A coin is tossed three times. Let $A = \{\text{at least one head appears}\}$. Noting all possible outcomes when a coin is tossed three times, we have $A = \Big\{\{HHH\}, \{HHT\}, \{THH\}, \{HTH\}, \{HTT\}, \{THT\}, \{TTH\}\Big\}$.
Clearly $P(A) = \frac{7}{8}$. However, it is sometimes difficult to count all such cases favourable to A, especially when the number of tosses is large. So we use complementation rule to evaluate $P(A)$ very easily.
Note that $A^c = \{\text{ no head appears}\} = \{TTT\}$ and naturally $P(A^c) = \frac{1}{8}$.
Therefore, we have,

$$P(A) = 1 - P(A^c) = 1 - \frac{1}{8} = \frac{7}{8} \qquad \text{[matches with our direct calculation!!!]}$$

Ex-3.6 In Example 3.4, let $F = \{\text{ genotype contains at least one } D \text{ allele }\}$.
Clearly, $F = C^c$, where, as given in the previous Example 3.4, $C = \{\text{ genotype contains no } D \text{ allele }\} = \{dd\}$.
So, $P(F) = P(C^c) = 1 - P(C) = 1 - 0.5 = 0.5$.

Before proceeding further, we introduce another concept named 'random variable' that is related to probability theory. The idea of a random variable and subsequent treatment help in calculation of probability of many composite events very easily.

3.5 Random Variable

Let us consider the classical coin tossing experiment. Suppose we toss a fair coin three times and we are interested to evaluate probabilities of several events related to this experiment. But, first let's write down all possible outcomes and the corresponding probabilities (Table 3.1).

Now, we define a quantity X as

$$X : \text{ number of heads in three tosses of a fair coin}$$

Clearly, X takes the values 0 or 1 or 2 depending on the outcome. For example, if $\omega = \{HHH\}$, X takes the value 3, i.e. $X(\omega) = 3$; similarly, if $\omega = \{HTT\}$, X

Table 3.1 Outcomes and corresponding probabilities when a fair (unbiased) coin is tossed three times

Event (ω)	HHH	HHT	HTH	HTT	THH	THT	TTH	TTT
Probability	$\frac{1}{8}$	$\frac{1}{8}$	$\frac{1}{8}$	$\frac{1}{8}$	$\frac{1}{8}$	$\frac{1}{8}$	$\frac{1}{8}$	$\frac{1}{8}$

Table 3.2 Outcomes, random variable realisations and corresponding probabilities when a coin is tossed three times

ω	HHH	HHT	HTH	HTT	THH	THT	TTH	TTT
Probability	$\frac{1}{8}$	$\frac{1}{8}$	$\frac{1}{8}$	$\frac{1}{8}$	$\frac{1}{8}$	$\frac{1}{8}$	$\frac{1}{8}$	$\frac{1}{8}$
X	3	2	2	1	2	1	1	0

becomes 1, i.e. $X(\omega) = 1$ and so on. We can present the possible values that X takes depending on ω in Table 3.2.

By introducing X in this way, we may have some advantages. We can describe any event, elementary or composite in terms of value(s) of X and calculate the probability corresponding to it. For example,

$$P(\text{exactly 2 heads}) = P(HHT, HTH, THH) = \frac{3}{8} = \boldsymbol{P(X = 2)}$$

$$P(\text{at least 2 heads}) = P(HHH, HHT, HTH, THH) = \frac{4}{8} = \boldsymbol{P(X \geq 2)}$$

This 'X' is known as 'random variable'; it varies, i.e. takes different values depending on different outcomes, and they are random in the sense that outcomes are random, i.e. cannot be predicted beforehand.

But the major advantage is that, sometimes, we can evaluate these probabilities without even looking at (or listing) all possible outcomes of an experiment! Looks surprising, but, yes, it's true. We shall discuss it later. Rather, now we shall consider a few examples that will use the rule of probability theory (discussed in Sect. 3.4).

Ex-3.7 A die is thrown. Let us define a few events and evaluate their probability explaining each event in terms of a random variable X (say), as given below. Define a random variable X as the number appearing on the upper face of the die when it is thrown once.

$$A = \{\text{number appearing is more than } 1\} = (X > 1) = \{2, 3, 4, 5, 6\}$$
$$B = \{\text{number appearing is an odd number}\} = (X \text{ is odd}) = \{1, 3, 5\},$$
$$C = \{\text{number appearing is more than } 4\} = (X > 4) = \{5, 6\},$$
$$D = \{\text{number appearing is more than } 5\} = (X > 5) = \{6\}$$
$$P(A) = P(X > 1) = 1 - P(X = 1) = 1 - \frac{1}{6} = \frac{5}{6}$$
$$P(B \cap D) = P(\{X = \{1, 3, 5\}\} \; and \; \{X > 5\}) = P(\phi) = 0,$$
[because no number satisfies both B and D simultaneously]
$$P(B \cup D) = P(B) + P(D) - P(B \cap D) = P(\{X = \{1, 3, 5\}\}) + P(\{X > 5\}) = \frac{3}{6} + \frac{1}{6} = \frac{2}{3}$$
$$P(B \cup C) = P(B) + P(C) - P(B \cap C) = P(X = \; odd) + P(X > 4) - P(X = 5) = \frac{1}{2} + \frac{2}{6} - \frac{1}{6} = \frac{2}{3}$$

We come back to probability theory again and introduce another idea, ‘conditional probability’.

3.6 Conditional Probability

The name suggests that there should be ‘something’, may be an event which presents a condition. Probability of an event that we plan to calculate is related somehow to this condition. Since we always evaluate probability of an event, this conditioning should be with respect to another event. Let us discuss this with the help of an example.

Two boys Harry and Biltu are playing with a die with six faces numbered $\{1, 2, \ldots, 6\}$. Harry has already thrown the die and knows that the outcome is an odd number, i.e. either of 1 or 3 or 5. He asks Biltu “Hey, the number appeared is odd. Now can you tell me the probability that it may be a number greater than 1?”

Well, the answer is simple! Given that the number appeared is odd and there are three such numbers, only possible numbers that are greater than 1 are 3 or 5. Hence, Biltu immediately replies, it’s easy! It is $\frac{2}{3}$.

Using notation of events, define two events A and B associated with the same random experiment. We are interested in evaluating $P(A$ given that B has occurred). We denote this by $P(A|B)$ and is defined as:

$$P(A \cap B) = P(B)P(A|B) \text{ i.e. } P(A|B) = \frac{P(A \cap B)}{P(B)} \text{ provided } P(B) > 0.$$

Figure 3.3 presents the idea behind conditional probability. A portion of A lies within B. While calculating conditional probability of A given that B has occurred

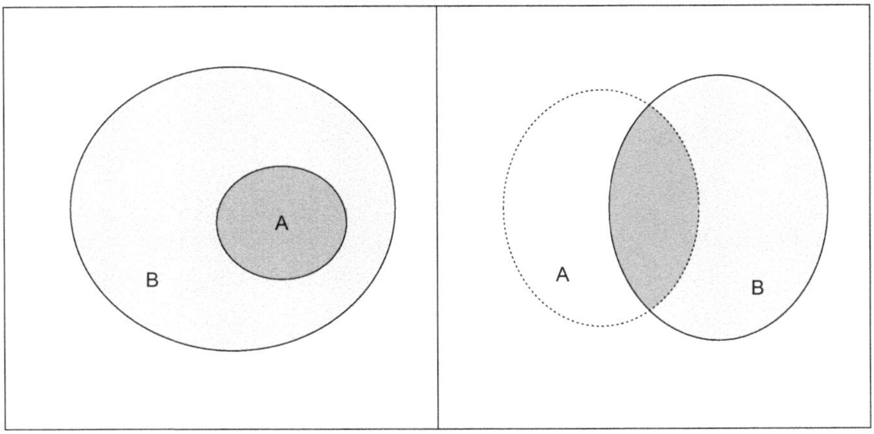

Fig. 3.3 Conditional probability diagram: dark-shaded region indicates the required event, whereas lighter-shaded region indicates the conditioning event

already, we focus only on the dark-shaded region, i.e. portion of A within B and it is compared with respect to totality of B.

In the context of the above example, denote $A = (X > 1)$ and $B = (X \text{ is odd})$ and we want to evaluate $P(A|B)$. Using the definition conditional probability and noting that $P(A \cap B) = \{3, 5\}$, the probability that Harry asks for can be deduced as

$$P(A|B) = \frac{P(A \cap B)}{P(B)} = \frac{2/6}{3/6} = \frac{2}{3}.$$

It is exactly the same answer that Biltu got by direct calculation. So depending on the situation, we can evaluate conditional probability either directly or using the formal definition. Both must give the same result.

Ex-3.8 It is seen that the number of individuals having genotypes AA, Aa, and aa at a bi-allelic locus are 30, 123, and 365 respectively. However, the number of affected individuals in these three groups respectively are 1, 4, and 29. Given that a person's genotype is aa, what is probability the person is affected?

The number of persons with genotype aa is 365, out of which 29 are affected. So, the required probability is $\frac{29}{365}$, by direct calculation. Now, we calculate this using the formal definition. First define $B = \{$ person with genotype $aa\}$ and $C = \{$ person is affected$\}$. Hence,

$$P(B \cap C) = P(\text{ a person with genotype } aa \text{ is affected}) = \frac{29}{518}$$

$$\text{and } P(C) = P(\text{ a person has genotype } aa) = \frac{365}{518}.$$

$$\therefore P(B|C) = \frac{P(B \cap C)}{P(C)} = \frac{29/518}{365/518} = \frac{29}{365}.$$

Ex-3.9 Probability that a person will suffer from a disease if the person also carries a particular mutation is 0.7. The disease prevalence is 0.08. What is the probability that an affected individual carries the mutation?

Let $M = \{$ person carries mutation $\}, D = \{$ person suffers from the disease $\}$.
Here, it is given that $P(D|M) = 0.7$ and $P(D) = 0.08$.
Hence, $P($ an affected person carries the mutation $) = P(D \cap M) = P(D|M)P(M) = 0.7 \times 0.08 = 0.056$.

Although from the above example, it seems that direct evaluation is easy, in many cases, the use of formal definition helps us a lot in understanding and evaluating probabilities in many complicated problems (to be seen later)!

3.6.1 *Breaking a Probability Down by Conditioning*

In real life, an event may occur with many other events. In Example 3.8, selection of affected people is related with three groups with different genotypes. We cannot select an affected person directly; we have to select an affected person from any of three groups. In other words, the probability of selecting an affected person varies between genotype groups. Sometimes, it would be very useful if we can express an event (or set) in terms of other sets. Naturally, that motivates us to write the probability of the desired set in terms of probabilities of other sets, if possible. However, the probabilities of other sets should be evaluated easily; otherwise, this exercise has no value.

Look at Fig. 3.4. Part of A occurs with B and the rest of A occurs with B^c. It is clear that $P(A)$ should be in terms of other probabilities, more specifically in terms of several conditional probabilities. We try to evaluate $P(A)$ using other probabilities.

$$\begin{aligned} P(A) &= P(\{A \text{ and } B\} \text{ or } \{A \text{ and } B^c\}) \\ &= P(\{A \cap B\} \cup \{A \cap B^c\}) \\ &= P(A \cap B) + P(A \cap B^c) \\ &\qquad [\text{since } A \cap B \text{ and } A \cap B^c \text{ are mutually exclusive events}] \\ &= P(B)P(A|B) + P(B^c)P(A|B^c) \\ &\qquad [\text{using definition of conditional probability}] \end{aligned}$$

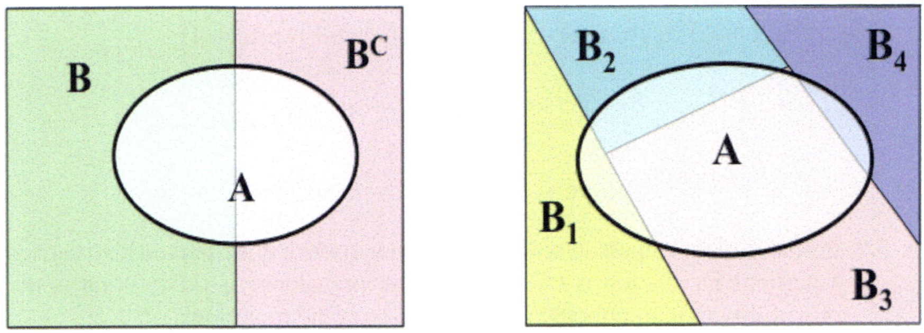

Fig. 3.4 An event A can be written in terms of two events and more than two events

Now suppose that we try to extend this idea when A occurs with more than two events, say $B_1, B_2, \ldots, B_k$ with $k \geq 1$. Again looking at Fig. 3.4 and proceeding similarly, we have,

$$\begin{aligned} P(A) &= P\big[(A \cap B_1) \cup (A \cap B_2) \cup \cdots \cup (A \cap B_k)\big] \\ &= P(A \cap B_1) + P(A \cap B_2) + \cdots + P(A \cap B_k) \\ &= P(B_1)P(A|B_1) + P(B_2)P(A|B_2) + \cdots + P(B_k)P(A|B_k) = \sum_{i=1}^{k} P(B_i)P(A|B_i) \end{aligned}$$

Ex-3.10 Suppose in a population at a bi-allelic locus with two alleles A and a, the genotype probabilities (commonly known as genotype frequencies) for genotypes AA, Aa, and aa are p_{AA}, p_{Aa}, and p_{aa} respectively. Also, suppose the conditional probability of being affected for a given genotype g is f_g, where $g = AA, Aa, aa$. What is the disease prevalence?

Here, $D = \{\text{ disease}\}$, $f_{AA} = P(D|AA)$, $f_{Aa} = P(D|Aa)$, $f_{aa} = P(D|aa)$, $p_{AA} = P(AA)$, $p_{Aa} = P(Aa)$, and $p_{aa} = P(aa)$. Hence, the disease prevalence or the probability of disease becomes

$$\begin{aligned} P(D) &= P\big[(D \cap AA) \cup (D \cap Aa) \cup (D \cap aa)\big] \\ &= P(AA)P(D|AA) + P(Aa)P(D|Aa) + P(aa)P(D|aa) = \\ &\quad p_{AA}f_{AA} + p_{Aa}f_{Aa} + p_{aa}f_{aa} \end{aligned}$$

Ex-3.11 There are two urns. Urn 1 contains 4 white and 3 black balls, while Urn 2 contains 5 white and 6 black balls. One ball is drawn at random and we don't know from which urn it is drawn. What is the probability that the ball is white?

First, define a few events that will facilitate the understanding and hence computation of the required event. Let $A = \{\text{ball drawn is white}\}$, $B_1 = \{\text{Urn 1 is selected}\}$, and $B_2 = \{\text{Urn 2 is selected}\}$.
Since we don't know from which urn the ball is finally drawn, we have to consider all possibilities, i.e. simultaneous occurrence of A with both B_1 and B_2. Since we can first select any urn randomly, we have

$$P(B_1) = P(B_2) = \frac{1}{2}.$$

Again, by the given conditions, we have

$$P(A|B_1) = \frac{4}{7},\ P(A|B_2) = \frac{5}{11},$$

and hence, $P(A) = P\big[(A \cap B_1) \cup (A \cap B_2)\big] = P(B_1)P(A|B_1)+$

$$P(B_2)P(A|B_2) = \frac{1}{2}.\frac{4}{7} + \frac{1}{2}.\frac{5}{11} = \frac{79}{154}$$

3.6.2 *Bayes' Theorem*

Continuing with Example 3.8, one can ask a very pertinent question: if we know the disease prevalence and genotype frequencies in a population, can we find the probability that a person has genotype AA if a person is known to be affected?

This type of question has a deep interpretation. Here, we try to find the probability of a *cause* when *an event has occurred* by looking at *prior probabilities* of several causes and occurrence of the current event under each such cause. This question gave rise to one of the most celebrated theorems in probability theory, the 'Bayes' Theorem', postulated and proved by Reverend Bayes.

We use the same set-up as described in Sect. 3.6.1 for both two events as well as k events.

Theorem 1 $P(B|A) = \frac{P(B)P(A|B)}{P(B)P(A|B)+P(B^c)P(A|B^c)}$

Proof

$$P(B|A) = \frac{P(B \cap A)}{P(A)} = \frac{P(B \cap A)}{P\big[(A \cap B) \cup (A \cap B^c)\big]} = \frac{P(B \cap A)}{P(A \cap B) + P(A \cap B^c)}$$
$$= \frac{P(B)P(A|B)}{P(B)P(A|B) + P(B^c)P(A|B^c)}$$

For more than two conditioning events $B_1, B_2, \ldots, B_k$, we have Bayes' theorem as:

Theorem 2 $P(B_1|A) = \frac{P(B_1)P(A|B_1)}{\sum_{i=1}^{k} P(B_i)P(A|B_i)}$

Proof follows exactly in the same way as in Theorem 1. Thus, through Bayes' theorem, we get the *posterior probability* of B_i in terms of *prior probabilities* $P(B_i),\ i = 1, 2, \ldots, k$ and the conditional probabilities $P(A|B_i),\ i = 1, 2, \ldots, k$.

Ex-3.12 In Example 3.10, what is the probability that genotype of an affected person is AA?

In the same set-up on Example 3.10, we have

$$P(AA|D) = \frac{P(AA)P(D|AA)}{P(AA)P(D|AA) + P(Aa)P(D|Aa) + P(aa)P(D|aa)} = \frac{p_{AA} f_{AA}}{p_{AA} f_{AA} + p_{Aa} f_{Aa} + p_{aa} f_{aa}}$$

Ex-3.13 Two different coins, coin 1 is unbiased and coin 2 is biased with $P(Head|\text{ coin } 2) = 0.7$. David selects one coin at random and toss it once. The coin shows '$Tail$'. What is the probability that the unbiased coin, i.e. coin 1 is used?

This is a direct application of Bayes' theorem. However, we don't need to guess whether we apply Bayes' theorem or not. It will reveal itself if we start writing down several events associated with this problem and evaluate the probability of the required event.
Let $A_1 = \{\text{ Coin 1 is used}\}$, $A_2 = \{\text{ Coin 2 is used}\}$, and $B = \{\text{ Tail appears}\}$. So, we have $P(A_1) = P(A_2) = 0.5$, since a coin is selected at random. Also $P(B|A_1) = 0.5$ and $P(B|A_2) = 1 - 0.7 = 0.3$. Hence,

$$P(A_1|B) = \frac{P(A_1 \cap B)}{P(B)} = \frac{P(A_1)P(B|A_1)}{P\big[(A_1 \cap B) \cup (A_2 \cap B)\big]} = \frac{P(A_1)P(B|A_1)}{P(A_1)P(B|A_1) + P(A_2)P(B|A_2)} = \frac{0.5 \times 0.5}{0.5 \times 0.5 + 0.5 \times 0.3} = \frac{0.5}{0.5 + 0.3} = \frac{5}{8}$$

Ex-3.14 In connection with Example 3.11, suppose we know that the ball drawn is white. We want to know the probability that it comes from Urn 2.

Again direct application of Bayes' theorem yields the result. We have

$$P(B_2|A) = \frac{P(B_2)P(A|B_2)}{P(B_1)P(A|B_1) + P(B_2)P(A|B_2)} = \frac{\frac{1}{2}\cdot\frac{5}{11}}{\frac{1}{2}\cdot\frac{4}{7} + \frac{1}{2}\cdot\frac{5}{11}} = \frac{35}{79}$$

3.7 Independence

Sometimes, it may so happen that the conditioning event has nothing to do with the event under consideration. That means the conditional probability of A given the event B is same as just $P(A)$, i.e. $P(A|B) = P(A)$. In such a situation, we can say that B has no influence on A and hence the events A and B are 'independent'. Formally, we can say that A and B are independent if (assuming $P(A) > 0$ or $P(B) > 0$),

$$P(A \cap B) = P(A).P(B)$$

This is equivalent to $P(A|B) = P(A)$ because, if A and B are independent,

$$P(A \cap B) = P(B).P(A|B) = P(A).P(B) \text{ if } P(A|B) = P(A)$$

$$\text{Similarly, } P(A \cap B) = P(A).P(B) \implies P(B)P(A|B) = P(A).P(B) \text{ i.e.}$$

$$P(A|B) = P(A)$$

Ex-3.15 X = outcome of a die roll, Y = outcome of another die roll. Any event involving X and any event involving Y are independent.

Ex-3.16 E_1 = genotype of an individual is AA, E_2 = genotype of that individual's father is AA. Clearly, E_1 and E_2 are not independent because father always transmits one allele to his offspring.

Ex-3.17 F_1 = an individual gets 'a' allele from his mother, F_2 = the individual gets an 'A' allele from his father. Here, F_1 and F_2 are independent events because mother transmits her allele to offspring independent of whatever allele is transmitted by father. Implicitly we assume that random mating occurred.

Ex-3.18 Suppose a coin is tossed 10 times. Probability of getting a head in a single trial (i.e. single toss) is p, where $0 < p < 1$. What is the probability that each of first 3 tosses will give head and each of the last seven tosses will give tail?

Clearly, $P($ first toss gives head $) = p$. $P($ second toss gives head $)$ is again p, because the outcome of second toss is not in any way dependent on the outcome of the first toss.
So, Probability of getting head in first two tosses $= p \times p = p^2$, since the two events are independent. Hence, proceeding this way, we have the required probability as $p^3(1-p)^7$.

3.8 Discrete Probability Distributions

Let's go back to random variable. Referring to Table 3.2, we have further questions: although we understand that random variable might ease our calculation and description of probabilities of several events, but can it offer any real advantage? Table 3.2 shows that we have to look at the probabilities of each outcome in the table and evaluate probability of an event in terms of the random variable X! This cannot be a real advantage because we are depending on all outcomes to evaluate the probability of an event. This becomes a nearly impossible (or impossible) task if the number of tosses becomes large. For 10 tosses of a coin, the total number of outcomes is 1024; very difficult to list them! Fortunately, there is a much simpler way to handle these kinds of situation. We can write a function, a mathematical function that will immediately provide the probability that X takes a particular value. Let us explain this using an example.

Consider the coin tossing experiment where we toss an unbiased coin three times. Clearly, each trial, i.e. throwing a coin once, has two outcomes, 'Head' or 'Tail', may be denoted in general by 'Success' or 'Failure' respectively. You may also designate a 'Tail' by 'Success'. It depends on the problem and how one decides to frame it. Since we are using the same coin and tossing is done under identical conditions, we may assume that the probability of getting 'Success' remains same from trial to trial. Moreover, since the outcome of a toss does not depend on any previous tosses, we assume that trials are independent. These kinds of trials are known as 'Bernoulli trials'. Under these assumptions, we define a random variable X as the number of 'heads' in three trials (i.e. tosses). Then the probability that X takes a value x, i.e. $P(X = x)$ is given by a function

$$P(X = x) = f(x) = \binom{3}{x}\left(\frac{1}{2}\right)^{x}\left(\frac{1}{2}\right)^{3-x}, \quad x = 0, 1, \ldots, 3$$

Using different values of $x = 0, 1, 2, 3$, we have the respective probabilities as

$$P(X = 0) = \left(\frac{1}{2}\right)^{3} = \frac{1}{8}, \quad P(X = 1) = 3\left(\frac{1}{2}\right)^{1}\left(\frac{1}{2}\right)^{2} = \frac{3}{8},$$
$$P(X = 2) = 3\left(\frac{1}{2}\right)^{2}\left(\frac{1}{2}\right)^{1} = \frac{3}{8}, \quad P(X = 3) = \left(\frac{1}{2}\right)^{3} = \frac{1}{8}$$

which matches with that given in Table 3.2.

In general, if we throw the coin n times, i.e. n Bernoulli trials, and probability of 'Success' in a single trial is $p (0 < p < 1)$, we have

$$P(X = x) = f(x) = \binom{n}{x} p^{x} q^{n-x}, \quad x = 0, 1, \ldots, n; \quad q = 1 - p, 0 < p < 1 \tag{3.1}$$

The evaluation of the expression in (3.1) is easy. Following Example 3.18, the probability that the first x tosses yield heads (successes) and next $n - x$ trials yield

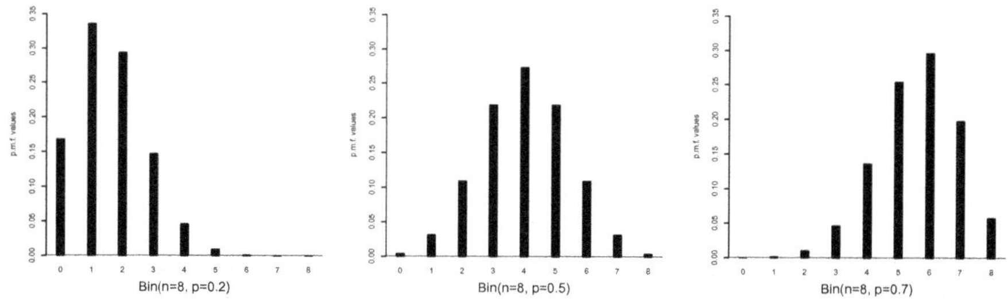

Fig. 3.5 p.m.f. of Binomial distribution with $n = 8$ and different values of p

tails (failures) is $p^x(1-p)^{n-x}$. Now we relax our idea. The x successes can occur in any x positions out of n trials in $\binom{n}{x}$ ways. Hence, the required probability of observing x successes in n tosses becomes same as given in (3.1), with $q = 1 - p$.

Note that the above function $f(x)$ has two important properties:

(1) $f(x) \geq 0$ for all x.
(2) $\sum_x f(x) = 1$.

This kind of function $f(x)$ corresponding to a random variable X (say) satisfying the above two properties is known as a 'probability mass function (p.m.f.)' and gives the probability of different events (described in terms of X) very easily directly through $f(x)$.

The particular function in (3.1) is known as p.m.f. of a Binomial random variable; we say that X follows Binomial distribution with parameters n and p and denote this by $X \sim Bin(n, p)$.

To get an idea how the probabilities vary depending on the values of X, we present p.m.f. by means of a diagram, known as 'Column diagram'. This simply plots a vertical bar whose height represents the probability $P(X = x)$ for all possible values of x (Fig. 3.5).

From Fig. 3.5 it seems that binomial probabilities are symmetrical when $p = 0.5$. However, it has a tail towards right and left respectively for $p = 0.2$ and $p = 0.7$.

3.9 Expectation and Variance

We had introduced a term 'parameter' in Sect. 3.8. In Binomial distribution, there are two parameters, n and p. A parameter characterises a distribution and we need to know the value of a parameter in order to know about the probability distribution. In fact, if we know the value of all parameters in a probability distribution, we would be able to know everything about that distribution. Moreover, some important features of a probability distribution are always in terms of parameters. Analogous to arithmetic mean, a measure of central tendency, here also we can conceptualise similar idea with respect to a probability distribution. This is called 'Expectation'.

3.9.1 Expectation

The expectation of a random variable is the long-term average (mean) outcome. With reference to the example of tossing a fair coin three times, i.e. for a $Bin(3, \frac{1}{2})$ distribution we can think that the number of 'heads' is 0, 1, 2, and 3 with probabilities $\frac{1}{8}, \frac{3}{8}, \frac{3}{8}$, and $\frac{1}{8}$ respectively. Although we are not sure which outcome we are going to observe while conducting the experiment, however, we can expect to see the value taken by X. Also, note that chance of observing 1 is three times that for 0 and so on. Hence, expected value of X, i.e. expectation of X, denoted by $E(X)$ becomes

$$E(X) = 0.\frac{1}{8} + 1.\frac{3}{8} + 2.\frac{3}{8} + 3.\frac{1}{8} = \frac{3}{2} = 1.5$$

This can be looked as a weighted arithmetic mean of possible values that a discrete random variable can take with weights as the corresponding probabilities.

Note that $E(X) = \frac{3}{2}$, i.e. a fraction, although in real life we never see fractional value taken by X when $X \sim Bin(3, \frac{1}{2})$. However, this is what we can expect; it may not necessarily be an integer even when the random variable under consideration takes only integer values. Expectation, as the term indicates, gives the value that we would think or expect to get even when we just know the distribution of the random variable under consideration. This value, may or may not be realised when the experiment is actually performed; however, we can expect to get a realisation close to the 'expectation of the variable', i.e. $E(X)$.

It can be shown that when $X \sim Bin(n, p)$, we have $E(X) = np$. This expression can be obtained from an intuitive point of view. The probability of observing a 'success' in a single trial is p, i.e. p is the number of success in one trial. Hence, in n trials, the number of successes should be np (just use unitary method!).

Ex-3.19 A die is thrown and X denotes the number appearing on the upper face of the die. What is the expectation of X?
Here X takes the values $1, 2, \ldots, 6$, each with probability $\frac{1}{6}$. So the p.m.f. of X is:

$$f(x) = P(X = x) = \frac{1}{6}, \quad x = 1, 2, \ldots, 6$$

Hence $E(X) = 1.\frac{1}{6} + 2.\frac{1}{6} + \cdots + 6.\frac{1}{6} = \frac{7}{2} = 3.5$.

Ex-3.20 A person is selected from a group of 28 individuals out of which 5 are affected by a disease. What is the expected number of affected persons in the sample?

Here X is the number of affected persons in a sample of size 1 (only one person is selected). So, X takes values either 0 or 1. Clearly, $X \sim Bin(1, \frac{5}{28})$ and so $E(X) = 1 \times \frac{5}{28} = \frac{5}{28}$.

3.9.2 *Variance*

Similar to the idea of expectation, we can think of variance of a random variable as the long-term spread (variation) of its outcomes. With reference to the previous example, i.e. for a $Bin(3, \frac{1}{2})$ distribution, we can think that the number of 'heads' is different if we repeat the experiment and there is an inherent probability associated with each observation. Since the outcomes vary randomly but within the set of values $\{0, 1, 2, 3\}$, we can think of a measure of variability analogous to variance as a measure of dispersion. Hence, variance of X, denoted by $V(X)$ becomes

$$V(X) = E\big(X - E(X)\big)^2 = (0-1.5)^2.\frac{1}{8} + (1-1.5)^2.\frac{3}{8} + (2-1.5)^2.\frac{3}{8} + (3-1.5)^2.\frac{1}{8} = \frac{3}{4} = 0.75$$

Since expectation for a $Bin(3, \frac{1}{2})$ random variable is 1.5, we expect to see 1.5 but there is a variation of 0.75 or a standard deviation $\sqrt{0.75} = 0.866$ associated with this.

It can be shown that when $X \sim Bin(n, p)$, we have $V(X) = npq$.

Ex-3.21 In Example 3.19, what is $V(X)$?

Here, X takes the values $1, 2, \ldots, 6$, each with probability $\frac{1}{6}$ and $E(X) = 3.5$. Hence,

$$E(X) = (1-3.5)^2.\frac{1}{6} + (2-3.5)^2.\frac{1}{6} + \cdots + (6-3.5)^2.\frac{1}{6} = 2.917.$$

3.9.3 *A Few More Discrete Distributions*

There are some frequently occurring discrete random variables with corresponding probability distributions that might be deemed appropriate in explaining real-life situations or phenomena. Among them, the most popular ones are Poisson and negative binomial random variables. If X follows a negative binomial distribution, it indicates the number of failures before the rth success. So in this case, there is no bound to the number of trials; it continues until we get r successes. We present the p.m.f., $E(X)$ and $V(X)$ for these distributions (along with Binomial distribution) in Table 3.3.

Table 3.3 Probability mass functions of three discrete distributions: Binomial, Poisson, and negative binomial

Distribution	$f(x) = P(X = x)$	x	Parameter(s)	$E(X)$	$V(X)$
Binomial distribution	$\binom{n}{x} p^x q^{n-x}$ where $q = 1 - p, 0 < p < 1$	$x = 0, 1, \ldots, n$	n, p	np	npq
Poisson distribution	$\frac{e^{-\lambda}\lambda^x}{x!}, \lambda > 0$	$x = 0, 1, 2, \ldots$	λ	λ	λ
Negative binomial distribution	$\binom{x+r-1}{r-1} p^r q^x$ where $q = 1 - p, 0 < p < 1$	$x = 0, 1, 2, \ldots$	p	rq/p	rq/p^2

3.10 Continuous Probability Distributions

We have discussed a few variables like number of heads in three tosses of a coin, number of A alleles in the genotype of an individual at a particular locus, etc. However, as stated earlier, all these variables can take only isolated or specified values depending on the random experiment. Consider another variable, say, height of a student. Now the value of height varies from one student to another (in a particular population) but it can take any value within a certain range. Suppose if the population is kindergarten students in a city, we can think that the values of height for different students might vary between 2 feet and 4 feet. But it can be any value (in decimals or fraction) within this range. This kind of variable is called 'continuous random variable'. Another example related to genetics might be gene expression values. It is a continuous variable and can take any value within its own range. Note that although we may get a set of distinct values up to a few places after decimal, however, it is possible to get a more accurate value if we use more sophisticated instrument for measuring. Thus, in this sense, it is continuous as there is always a possibility of observing any value, which is not the case for a discrete variable.

But how do we attach probability to such a continuous random variable? Recall the discussion in Sect. 2.2.2 and Fig. 2.1. We know that a histogram provides a graphical presentation of frequency distribution of a continuous variable X (say). Careful understanding of histogram and Fig. 2.1 reveal some interesting facts. If we draw histogram using relative frequency density on the vertical axis, the total area under the histogram is one. The total area always remains same, i.e. its value is 1 always, whatever be the number of observations or the number of classes. Now letting class width (ω) of each class very small and at the same time increase the number of observations (n), histogram would approach to a smooth curve having this property (Fig. 2.5). This limiting form of histogram, i.e. 'frequency curve', is obtained (theoretically) when $\omega \to 0$ and $n \to \infty$. Let's denote this frequency curve by $f(x)$ where x represents the value of the variable of interest; in our case, it may be gene expression value. Clearly, $f(x)$ has two important properties that are quite apparent: (1) $f(x) \geq 0$ for all x, and (2) total area under $f(x)$ over its entire range is 1.

Now we define a continuous random variable X that can take any value within a feasible range (i.e. a set of values) depending on the random experiment. Examples of such random variables may be height, weight, HDL level, total cholesterol, gene expression values, etc. Any realisation of X based on a random experiment is denoted by x and the limiting form of the histogram is denoted by $f(x)$. We may say that $f(x)$ is a probability density function (p.d.f.) of X and it always satisfies the two properties as given below.

1. $f(x) \geq 0$ for all x, and
2. $\int_a^b f(x)dx = 1$, where a and b are minimum and maximum possible values that x can take.

This function, $f(x)$ is called a 'probability density function' (p.d.f.) of x because at a point c, $f(c)$ represents the limiting relative frequency density (in a limiting histogram), and hence can be considered as 'probability density' at c. The probability that X lies between any two values c and d (say) with $c < d$ is given by

$$P(c \leq X \leq d) = \int_c^d f(x)dx.$$

So probability is interpreted as area under the curve $f(x)$ between two values just as area under histogram represents relative frequency between two values.

Clearly, probability at a particular point is zero as area on a point is always zero. So, we have

$$P(X = x) = 0 \text{ for all } x$$

$$\Longrightarrow P(c \leq X \leq d) = P(c < X \leq d) = P(c \leq X < d) = P(c < X < d) = \int_c^d f(x)dx \text{ for } c < d,$$

because this is the area under the curve $f(x)$ between c and d. Proceeding similarly as in discrete case, we can define expectation and variance, denoted by μ and σ^2 respectively as

$$\mu = E(X) = \int_a^b xf(x)dx, \quad \sigma^2 = V(X) = \int_a^b (x-\mu)^2 f(x)dx$$

3.10.1 *Normal Distribution*

In statistical analysis, the most common and widely used distribution for a continuous variable is known as 'normal' distribution. Suppose that a random variable X has a normal distribution. Then the p.d.f. of X is

$$f(x) = \frac{1}{\sqrt{2\pi}} e^{-\frac{1}{2}\left(\frac{x-\mu}{\sigma}\right)^2}, \quad -\infty < x < \infty, -\infty < \mu < \infty, \sigma^2 > 0.$$

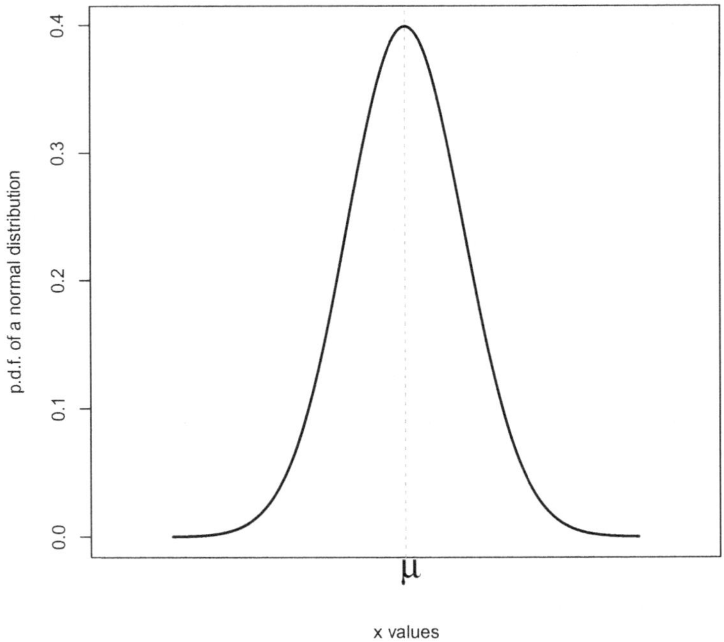

Fig. 3.6 p.d.f. of a normal distribution with mean at μ

If a continuous random variable X follows a normal distribution with mean μ and variance σ^2, we denote this by $X \sim N(\mu, \sigma^2)$. If we look at the graph of normal distribution (Fig. 3.6), immediately some nice features emerge. Its p.d.f. is symmetric about μ, two tails are falling down on both sides smoothly and, at a regular manner, it has a peak at mean value indicating that mean and median are same, most of the values, i.e. around 99.73% values will lie between $\mu - 3\sigma$ and $\mu + 3\sigma$ whatever be the values of μ and σ, etc. It can be shown that $E(X) = \mu$ and $V(X) = \sigma^2$.

Calculation of normal probability

Sometimes, it is essential to calculate probability of a certain event based on a normal random variable. Let us explain this using an example.

Ex-3.22 Suppose gene expression values for a gene are known to follow a normal distribution with mean 1 and variance 0.2. We might be interested in evaluating probabilities of different events like:

22.1 What is the probability that the gene expression value might be less than or equal to 0.82?
22.2 What is the probability that the gene expression value might be greater than 3.2?
22.3 What is the probability that for a randomly chosen person, the gene expression value will lie between 0.6 to 1.3?

Well, to answer these questions, we would need to know how to calculate $P(a \leq X \leq b)$ for any $a < b$ when $X \sim N(\mu, \sigma^2)$. This requires evaluation of the integral $\int_a^b f(x)dx$ where $f(x)$ is the p.d.f. of a normal distribution. Unfortunately, there is no closed form by which we can immediately calculate this probability for given values of a and b. But fortunately, we know R language and we can use R to calculate probability of any event when $X \sim N(\mu, \sigma^2)$. The standard R code is available for evaluating one-sided probability, i.e. for $P(X \leq x)$ or $P(X > x)$. Note that it is immaterial to consider $<$ or $\leq$ as explained earlier. Let's try to calculate the above probabilities.

The R code for calculating probability of $X \leq x$ is given by

```
> pnorm(x,mean=μ, sd=σ)
```

Here, 'norm' stands for 'normal' distribution, mean, and sd are numerical values of mean and standard deviation of X respectively, to be given as input inside the function 'pnorm( )'. Using this function, our command for **22.1** would be

```
> pnorm(0.82,mean=1, sd=sqrt(0.2))          ## P(X ≤ 0.82|X ~ N(1, 0.2))
```

which gives the value 0.3436609; 'sqrt(a)' indicates the 'square root of a'. If we want the value up to four places after decimal, we can modify the command as

```
> round(pnorm(0.82,mean=1, sd=sqrt(0.2)),4)
```

This gives the value $P(X \leq 0.82|X \sim N(1, 0.2)) = 0.3437$. Note that 'round(a, 4)' function rounds the value of 'a' up to four places after decimal.

(22.2) R code for calculating probability that gene expression value might be greater than 3.2 is given by

```
> pnorm(3.2,mean=1, sd=sqrt(0.2),lower.tail=F)
```

which gives the value $4.341614e - 07$ means the value is 0.0000004341614. To get the right-sided probability, we have to write 'lower.tail=F' inside the 'pnorm( )' function; default is for calculation of left-sided probability. Hence, this gives the value $P(X > 3.2|X \sim N(1, 0.2)) = 4.3416 \times 10^{-07}$.

(22.3) To calculate $P(0.6 \leq X \leq 1.3)$, we first calculate $P(X \leq 1.3)$ and subtract $P(X \leq 0.6)$ from it. Since $P(0.6 \leq X \leq 1.3) = P(X \leq 1.3) - P(X < 0.6)$, we have the R code as

```
> pnorm(1.3, mean=1, sd=sqrt(0.2),lower.tail=T) - pnorm(0.6, mean=1,
sd=sqrt(0.2),lower.tail=T)
```

and the value is 0.5632858. Note that here we have written "lower.tail=T" inside the 'pnorm' function; but it is not necessary as long as we calculate the left-sided probability. However, to avoid confusion, it's better to explicitly write "lower.tail=T" or "lower.tail=F" depending on whether we want left-sided or right-sided probability respectively.

Again, if we want the value up to three places of decimal, the R code should be

```
> round(pnorm(1.3, mean=1, sd=sqrt(0.2),lower.tail=T) - pnorm(0.6,
mean=1, sd=sqrt(0.2),lower.tail=T), 3)
```

which gives the value 0.563.

3.10.2 *Few Other Distributions*

We now present a few more distributions, mainly pictorially. Although 'Normal distribution' is the most popular and widely applicable probability distribution of a continuous random variable, some other distributions may occur sometimes in many real-life situations. However, there is no hard and fast rule to identify the appropriate distribution, which the underlying variable would follow. We can draw a histogram or boxplot and looking at the shape and features of the diagram, we might be able to guess a probability distribution. After this step, we usually estimate the parameters of the distribution and fit that distribution to the data. Next step is to check or justify whether the fit is good. There are some statistical techniques to evaluate the goodness of fit of the probability distribution to the dataset. Sometimes, based on prior knowledge we may assume the probability distribution. It is important to note that larger the sample size, better will be the justification about the goodness of fit. Although there are many such probability distributions, we consider three more probability distributions, which are common and have inetresting features.

$$Normal\ distribution: \quad X \sim N(\mu, \sigma^2): \quad f(x) = \frac{1}{\sigma\sqrt{2\pi}} e^{-\frac{1}{2}\left(\frac{x-\mu}{\sigma}\right)^2}, \quad -\infty < x < \infty$$
$$Laplace\ distribution: X \sim Laplace(\mu, \sigma): \quad f(x) = \frac{1}{2\sigma} e^{-\frac{|x-\mu|}{\sigma}}, \quad -\infty < x < \infty$$
$$Gamma\ distribution: X \sim Gamma(\alpha, \beta): \quad f(x) = \frac{\alpha^\beta}{\Gamma(\beta)} e^{-\alpha x} x^{\beta-1}, \quad x > 0$$
$$Cauchy\ distribution: X \sim Cauchy(\mu, \sigma): \quad f(x) = \frac{\sigma^2}{\pi} \frac{1}{\sigma^2+(x-\mu)^2}, \quad -\infty < x < \infty$$

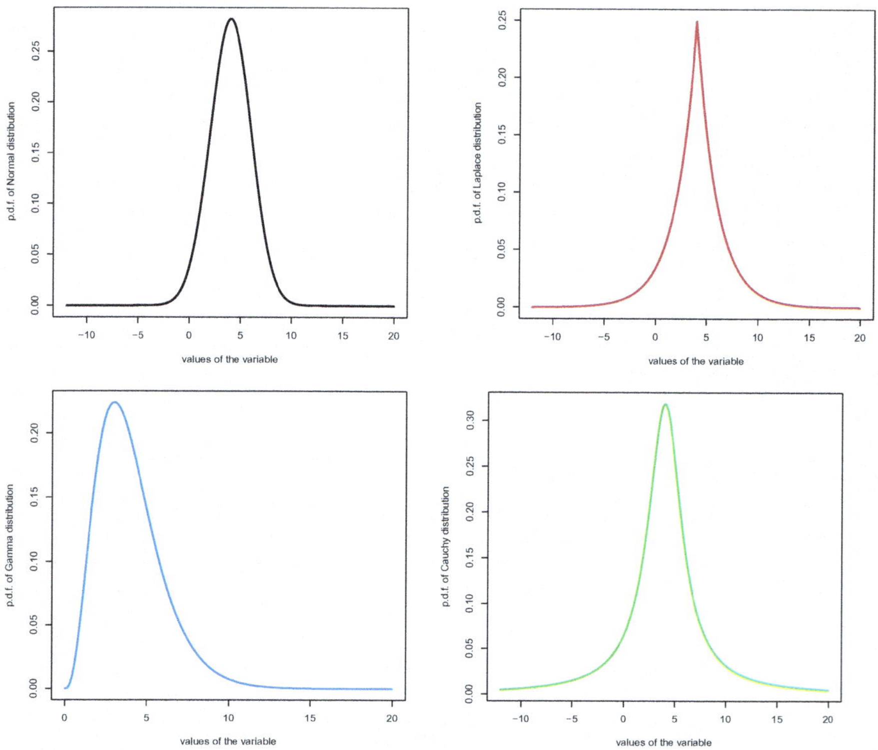

Fig. 3.7 p.d.f.s for four distributions

There are specific reasons behind choosing these distributions. Figure 3.7 presents the p.d.f.s of these four probability distributions for some specific values of respective parameters. We use the same location and dispersion parameters for all four distributions so as to have a clear idea when all four p.d.f.s are drawn on same plot, i.e. superimposing all four graphs. Clearly, p.d.f. of a normal distribution is symmetric about μ (here $\mu = 5$) with the peak gradually and smoothly falling on both sides in a symmetric manner. Thus, the curve is symmetric about μ. We choose the Laplace distribution because like normal distribution it is also symmetric about mean but it has a high kurtosis, i.e. a fatter tail on both sides compared to normal distribution. We select the gamma distribution because it is positively skewed having a longer tail towards the right side of the mean and realisations start from 0. Cauchy distribution is chosen because it has much fatter tails and although symmetric about median, its mean and variance do not exist. This distribution has a special role in statistics and is used to demonstrate some interesting facts which would need a detailed explanation.

A few interesting observations are immediately visible from Fig. 3.7. Normal, Laplace and Cauchy distributions are symmetric about 5, i.e. if we fold around the value 5, the right side and left side will fall on top of each other. But gamma distri-

bution is not symmetric; it is tilted to the left having a peak at a value less than 5. But peak occurs for other distributions exactly at 5. Laplace distribution has a very sharp peak at 5. Careful attention also reveals that beyond -10 to the left and beyond 20 to the right, a few values may still be observed. Normal distribution has almost no value, whereas there is substantial contribution at these regions for Laplace and Cauchy distributions. Hence, these two distributions are characteristically different from normal distribution. Gamma distribution has a long tail towards the right indicating the existence of some large values. These facts, although very simple to observe, play a very important role in any data analysis. Sometimes, another distribution is used in some situations, known as 'Exponential distribution'. This distribution can be viewed as a special case of gamma distribution, with $\beta = 1$. Thus, the p.d.f. of an exponential distribution is with parameter α :

$$f(x) = \frac{1}{\alpha} e^{-x/\alpha}, \quad x > 0, \alpha > 0.$$

It can be shown that for an exponential distribution with parameter α, mean is α, and variance is α^2.

3.10.3 Important Results Related to Probability Distributions

On many occasions, we need to standardise the data, i.e. subtract mean from every observation and divide it by its standard deviation. For normal distribution, even after standardisation, the variable follows a normal distribution with mean 0 and variance 1; but this is not true for all distributions. Moreover, if X and Y are two independent random variables both following normal distributions, then any linear function of X and Y also follows a normal distribution. Some very interesting results are available for normal distribution. These results have immense implications and use in data analysis. We state a few results below, without proof.

Theorem 3 *If* $X \sim N(\mu, \sigma^2)$, *then* $Z = \frac{X-\mu}{\sigma} \sim N(0, 1)$.

Theorem 4 *Suppose* $X \sim N(\mu_1, \sigma_1^2)$, $Y \sim N(\mu_2, \sigma_2^2)$, *and* X *and* Y *are independently distributed. Then for any constants (non-random, real numbers)* a, b, *and* c, $a + bX + cY \sim N(a + b\mu_1 + c\mu_2, b^2\sigma_1^2 + c^2\sigma_2^2)$.

Now, let's draw a random sample from X. Let it be $X_1, X_2, \ldots, X_n$, where n is the sample size. During developing statistical testing procedures for some problems, we shall see that a few quantities are of utmost important. These quantities are functions of $X_1, X_2, \ldots, X_n$. Note that $X_1, X_2, \ldots, X_n$ constitute a random sample from X. Random sample means, first, we draw an item from the population and measure the value of the variable X; then we draw another one but it has nothing to do with the first observation, and so on. Thus, observations correspond to random variables that are copies of the original variable X and hence they are *identically* distributed. Moreover, since these identically distributed random variables $X_1, X_2, \ldots, X_n$ constitute

a random sample, they are independent of each other. These two facts lead to some nice results. We present these results as theorems (without proof).

Theorem 5 *Let $X_1, X_2, \ldots, X_n$ be independently distributed random variables with $E(X_i) = \mu_i$ and $V(X_i) = \sigma_i^2$ for all $i = 1, 2, \ldots, n$. Then, we have*

$$E\Big(\sum_{i=1}^{n} X_i\Big) = \sum_{i=1}^{n} \mu_i, \text{ and } V\Big(\sum_{i=1}^{n} X_i\Big) = \sum_{i=1}^{n} \sigma_i^2.$$

Corollary 1 *If $\mu_i = \mu$ for all $i = 1, \ldots, n$, we have, $E\big(\sum_{i=1}^{n} X_i\big) = n\mu$. Similarly, if $\sigma_i^2 = \sigma^2$ for all $i = 1, \ldots, n$, we have $V\big(\sum_{i=1}^{n} X_i\big) = n\sigma^2$.*

Theorem 6 *Define $\bar{X} = \frac{1}{n}\sum_{i=1}^{n} X_i$, as the sample mean. Clearly, $\bar{X}$ is also a random variable. Then $\bar{X}_i \sim N(\mu, \sigma^2/n)$ if $X_i \sim N(\mu, \sigma^2)$ i.i.d. for $i = 1, 2, \ldots, n$.*

Remarks: Theorem 5 is true for discrete as well as continuous variables as long as expectation and variance exist for all variables.

Theorem 5 can be envisaged intuitively also. To see this easily, assume without any loss of generality that all variables are positive. Now if we add two variables, the values corresponding to the sum of variables become larger than the individual variables. Hence it is expected that expectation of the sum of variables would be large compared to that for individual variables. Similarly, variance becomes large as larger values would show more variability in the data obtained from sum of variables. Hence, these results become intuitively more natural.

Theorem 7 *Let $X_1, \ldots, X_n$ are independently and identically distributed (i.i.d.) random variables each following a normal distribution with mean 0 and variance 1, i.e. $X_i \sim N(0, 1)$, $i = 1, \ldots, n$ and $X_1, \ldots, X_n$ are independent. Then $\sum_{i=1}^{n} X_i^2 \sim \chi_n^2$.*

So, we get a new distribution, called χ_n^2 distribution, pronounced as 'chi-square distribution'. As per definition, if Y follows a chi-square distribution with n **degrees of freedom**, or notationally $Y \sim \chi_n^2$, the p.d.f. of Y is given by

$$f(y) = \frac{\left(\frac{1}{2}\right)^{n/2}}{\Gamma(n/2)} e^{-y/2} y^{\frac{n}{2}-1}, \quad y > 0,$$

where $\Gamma(m)$ is a standard 'Gamma' function defined as $\Gamma(m) = \int_0^\infty e^{-x} x^{m-1} dx$. Although it is not essential to memorise the p.d.f. of Y, but we should have a clear idea about what is called 'degrees of freedom'.

Degrees of freedom: Drosophila, a fruit fly is flying freely in a room. A room is identified by length, width, and height, along x-axis, y-axis together xy-plane (floor) and z-axis (in the direction of floor to ceiling vertically). The position of the fly at any point inside the room is determined by these three axes. If we drop any axis, we cannot identify its position. So the degrees of freedom of the fly is three. Now assume that for some reason (restriction), the fly was to fall into a cup of water, soaking its wings, rendering it unable to fly, but capable of crawling on the floor. Now, its position can be determined only by x- and y-axes; we don't need the z-axis, which is no more active due to its wet wings, i.e. one restriction. Hence, in this case, the degrees of freedom reduce to 2.

Thus, we can think the degrees of freedom as the number of independent variables involved in the expression. Without any restriction, the position of the fly is a function of three independent axes. But with one restriction degrees of freedom is reduced to 2, i.e. one less than three. For a set of n independent variables, degree of freedom is n. If we put 1 restriction, it becomes $n - 1$. In general, if we put k $(1 < k < n)$ independent linear restrictions, the degrees of freedom come down to $n - k$.

Now that, we have normal and χ^2_n distribution at our hands, we have a very important result, as stated below.

Theorem 8 *Define* $s^2 = \frac{1}{n-1}\sum_{i=1}^{n}(X_i - \bar{X})^2$, *as the sample variance. Clearly,* s^2 *is also a random variable. Then*

(1) $\frac{(n-1)s^2}{\sigma^2} = \sum_{i=1}^{n}(X_i - \bar{X})^2/\sigma^2 \sim \chi^2_{n-1}$, *and*

(2) $\bar{X}$ *and* s^2 *are independently distributed.*

Note that although s^2 involves n independent variables $X_1, \ldots, X_n$, but they are subject to one linear restriction, i.e. $\sum_{i=1}^{n}(X_i - \bar{X}) = 0$. Hence, the degrees of freedom associated with $\frac{(n-1)s^2}{\sigma^2}$ reduces to $n - 1$. The second part of Theorem 8 is very important and it can be exploited to develop some statistical tools in data analytics. Independence of sample mean and sample variance is a typical characteristic for normal distribution that makes our life simple in dealing with many data analysis problems in genetics and other areas.

Two more distributions that we would need to know are t and F distribution. We mention them without the explicit expression for their probability density functions. These distributions involve two independent variables.

Theorem 9 *Suppose* $X \sim N(0, 1)$, $Y \sim \chi^2_n$, *and* X *and* Y *are independently distributed. Then,* $\frac{X}{\sqrt{Y/n}} \sim t_n$, *where* t_n *is known as* t *distribution with* n *degrees of freedom.*

Theorem 10 *Suppose* $X \sim \chi^2_m$, $Y \sim \chi^2_n$, *and* X *and* Y *are independently distributed. Then,* $\frac{X/m}{Y/n} \sim F_{m,n}$, *where* $F_{m,n}$ *is known as* F *distribution with degrees of freedom* m *and* n.

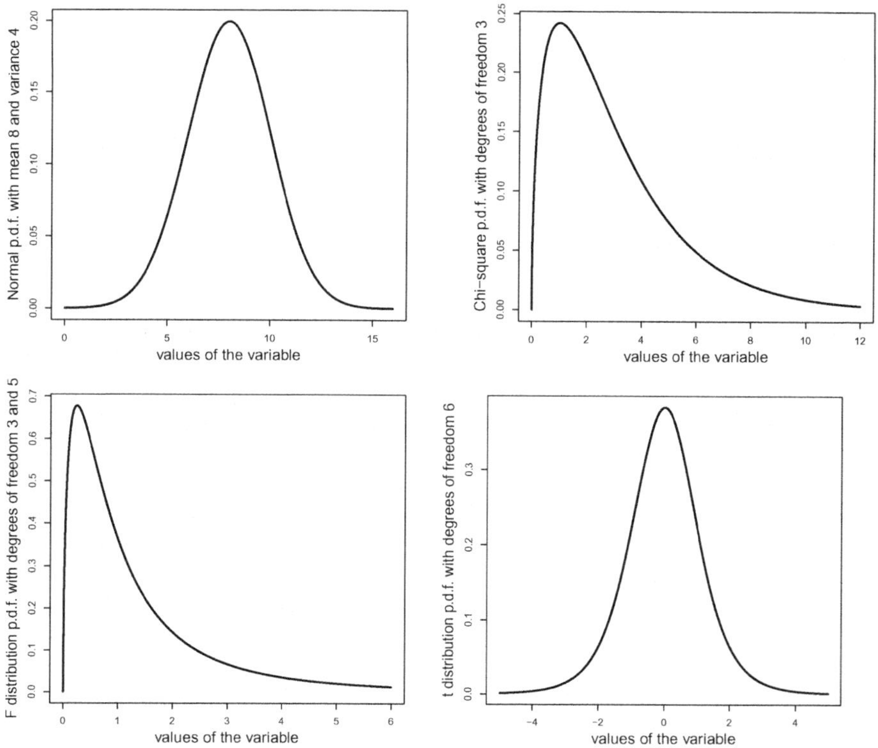

Fig. 3.8 p.d.f.s for four probability distributions: Normal Chi-square, F, and t

The following theorem deals with two independent χ^2 variables. It states that if the two chi-square variables are independent, their sum also follows chi-square distribution with degrees of freedom being the sum of degrees of freedom of the respective variables.

Theorem 11 *Suppose* $X \sim \chi^2_m$, $Y \sim \chi^2_n$ *and* X *and* Y *are independently distributed. Then,* $X + Y \sim \chi^2_{m+n}$

Although we did not give the exact form of the p.d.f of t or F distribution, Fig. 3.8 presents graphs of their density functions. It is clear from the graphs, that t-distribution is symmetric about 0 but have much fatter tails on either side compared to normal distribution whereas F distribution and χ^2 distribution are not symmetric. In fact, F distribution has a longer tail towards the right side and its peak lies on the left side of its range. Naturally, mean, variance, and other properties of χ^2_n, $F_{m,n}$, and t_n distributions depend on the degrees of freedom. Height of the peak also depends on degrees of freedom.

With this knowledge, we can now proceed to develop a few testing methods required to address several problems relating to gene expression data and other continuous variables that occur in practice in any discipline.

Exercise

3.1 Two dice are thrown. Naturally, we observe two numbers appearing on the upper faces of two dice. What is the probability that

(a) sum of these two numbers is 7,
(b) product of these two numbers is 12, and
(c) maximum of these two numbers is 5?

3.2 Two dice are thrown. What is the probability that sum of the numbers appearing on the upper faces of two dice is 7 when it is known that

(a) one die shows 4,
(b) at least one die shows 4,
(c) the difference of two numbers is 3, and
(d) minimum of two numbers is 1?

3.3 A group consists of 56 individuals of which 17 are affected by a disease. One person is selected at random. What is the probability that the person is affected?

3.4 A group consists of 72 individuals of which 18 are affected by a disease. Five persons are selected at random. What is the probability that

(a) two are affected,
(b) at least one is affected, and
(c) two are affected when it is known that at least one is affected?

3.5 A fair coin is tossed 4 times. What is the probability that we can get

(a) three consecutive heads once,
(b) two consecutive heads at least once,
(c) three consecutive tails once, and
(d) all tosses yield tails?

3.6 Explain why there is a mistake in each of the following statements:

(a) The probability that David will pass the yearly examination is 0.72 and the probability that he will not pass is −0.28.
(b) The probability that it will be raining tomorrow is 0.82 and the probability that it will not be raining tomorrow is 0.32.

(c) The probability that last year's champion in a football tournament will win its first match in this year's tournament is 0.76, the probability that it will tie the game is 0.09, and the probability that it will win or tie is 0.92.
(d) The probabilities that a person will make 0, 1, 2, 3, or 4 spelling mistakes while typing a page are 0.13, 0.24, 0.38, 0.16, and 0.19 respectively.
(e) The probabilities that a student passes in English literature and mathematics are 0.46 and 0.39 respectively, whereas probability passing in either one of these two subjects is 0.87.

3.7 The probability that the annual maintenance of a new PCR machine will be rated as very easy, easy, average, difficult, or very difficult are, respectively 0.07, 0.3, 0.36, 0.16, and 0.11. Find the probabilitues that annual maintenance of the machine will be rated as:

(a) average or better;
(b) average or worse;
(c) difficult or very difficult;
(d) neither very difficult nor very easy.

3.8 From a pack of cards, suppose three cards are drawn at random. What is the probability that

(a) all three are aces?
(b) none of these three is an ace?
(c) two are black and one is red?
(d) each of the three cards belongs to different suits?
(e) each of the three cards belongs to different denominations?
(f) all three cards are neither black nor aces?
(g) all of them are spades when it is known that they belong to any black suit?
(h) If the cards are drawn randomly one by one and first two cards are red, what is the probability that the third card drawn will also be red?

3.9 A box contains 5 red and 5 white balls while another box contains 7 red and 4 white balls. A box is selected at random and one ball is drawn from that box. What is the probability that selected ball is white?

3.10 In problem 3.9, suppose it is seen that the ball drawn is white, but we don't know from which box it was drawn. Find the probability that it was selected from the second box.

3.11 A population consists of three different ethnic groups A, B and C (say) in 2 : 3 : 5 proportions. Proportion of persons suffering from a rare disease in these three groups is 0.02, 0.01, and 0.005 respectively. What is disease prevalence in that population? Suppose one person is selected and is found to affected. What is the probability that the person belong to ethnic group C?

3.12 Mental stress may be related to high blood pressure. Probability that a person gets stressed is 0.3. Probability of observing high blood pressure when a person is stressed is estimated as 0.7 whereas the same probability is only

0.1 when the person is not stressed. Suppose probability that the indicator of blood pressure shows high is 0.1. What is the probability that the person was under stress when blood pressure was measured as high?

3.13 Proportion of individuals suffering from cardiovascular disease is 0.12. A random sample of 24 individuals is taken from that population. What is the probability that

(a) exactly 4 of have cardiovascular disease?
(b) at least 2 of them have cardiovascular disease?
(c) 3 individuals are suffering from the disease when it is known that 2 have cardiovascular disease?
(d) 3 individuals are suffering from the disease when it is known that at least 2 have cardiovascular disease?
(e) the number of individuals suffering from the disease exceeds the average number affected by the disease?

3.14 Suppose HDL level is assumed to be normal with mean 50 and variance 6. Suppose you measure HDL of a person chosen at random from the population. What is the probability that the HDL for that person

(a) is more than 55?
(b) is less than 40?
(c) lies between 35 and 45?

3.15 Suppose we select 20 individuals at random from the population whose LDL follows a normal distribution with mean 120 and variance 10. What is the probability that

(a) exact 4 individuals are having LDL more than 150?
(b) at least 4 individuals have LDL more than 150?
(c) at most 4 individuals LDL more than 150?

How many of them are expected to have LDL level (a) more than 140, and (b) lies between 130 to 140?

3.16 Suppose I change the numbers on the faces of a die. Instead of $\{1, 2, 3, 4, 5, 6\}$, I printed the numbers as $\{1, \frac{1}{2}, \frac{1}{3}, \frac{1}{4}, \frac{1}{5}, \frac{1}{6}\}$. Find the expectation and variance of the number of upper face(s) if

(a) the die is thrown only once,
(b) the die is thrown twice, and
(c) the die is thrown 10 times.

Can you get the answers of 3.16(b) and 3.16(c) without evaluating the probabilities, just by using 3.16(a)? If so, explain how and show that you arrive at the same answers. If not, explain why you cannot do?

3.17 Out of a group of 50 persons, 12 have some cardiovascular problems, 8 has hypertension and 3 have both. A person is selected at random. What is the probability that the person has

(a) both the problems?
(b) either hypertension or cardiovascular disease?
(c) neither hypertension nor cardiovascular problems?
(d) has cardiovascular problem but not hypertension?

3.18 In a family father has genotype Aa and mother has genotype Aa at a particular bi-allelic locus. Probabilities that a person with genotypes AA, Aa, aa has a disease that is somehow caused by the genotype at this locus are 0.9, 0.4 and 0.001 respectively. What is the probability that a newborn baby might have the disease? What is the expected number of children having the disease if the couple has three children? You can assume random mating and allele frequency of allele A is 0.1.

3.19 Twenty-four people had a blood test and the results are shown below.
A, B, B, AB, AB, B, O, O, AB, O, B, A, AB, A, O, O, AB, B, O, A, AB, O, B, A

(a) Construct a frequency distribution for the data.
(b) If a person is selected randomly from the group of twenty-four people, what is the probability that his/her blood type is not O?

3.20 Suppose chance of surviving from a disease is 0.95 if the person has no co-morbidity. However, the chance reduces to 0.62 and 0.34 if the person has cardiovascular problem and both cardiovascular issue and hypertension. In the population, it is known that prevalence of either one of these two is 0.12 whereas that for only cardiovascular issue is 0.08 and that for hypertension only is 0.06. What is the chance of survival if a person is affected by the disease? Suppose a person is selected from the population and is found to be affected by the disease. What is the chance that he is also suffering from hypertension?

3.21 A company has 12 statisticians, 9 mathematicians, and 23 computer scientists. The manager, who is a management person, wants to build a team consisting of 6 people for a project on data science. What is the probability that the team consists of

(a) 3 statisticians, 2 mathematicians and 9 computer scientists?
(b) no mathematician at all?

After selecting a team of 3 statisticians, 2 mathematicians and 9 computer scientists for the first project, the manager wants to gather another team for a second project. For this second project what is the probability that team consists of 5 statisticians, 3 mathematicians, and 8 computer scientists? Two persons are selected at random from the entire group of scientists. What is the probability that both of them are computer scientists and are idle at that moment?

3.22 For any two events A and B, if A and B are independent events, show that A^c and B^c are also independent. Give an example.

3.23 Sometimes, to get answers to a sensitive question, a typical method, known as **randomised response technique** is used. Suppose we want to know probabilities of some events relating to smokers among school students. We construct 40 cards and write 'I smoke at least once a week' on 15 cards and 'I do not smoke at least once a week' on rest of the cards. Note that the number 15 is arbitrary. A sample of 294 students is selected randomly. Now we ask each student in the sample to select randomly one of the 40 cards and respond 'yes' or 'no' without disclosing the question. Here, there is no way that the experimenter knows what is written on the card chosen by each student. Selection of card is made with replacement. Let A be the event that a student gives 'yes' and S denote that a randomly selected student smokes at least once a week.

(a) Establish a relationship between $P(A)$ and $P(S)$.
(b) If 90 students answered 'yes' out of 294 students, an estimate of $P(A)$ can be given as $\frac{90}{294}$. Find an estimate of $P(S)$ using the result in part (a).

3.24 Suppose lengths of sides that are perpendicular to each other in a right angled triangle follow two independent normal distributions, each with mean 0 and variance 1. Length of each side is measured from its midpoint; hence it can be negative when measured to the left of the midpoint. What is the distribution of the square of the hypotenuse? If we consider 100 such triangles, how many you can expect to have hypotenuse with a length of more than 1.25?

3.25 Suppose height (X) of a group of college students follows a normal distribution with mean 167 cm and variance 15 cm^2. What is the probability that a student's height is (a) 172 cm, (b) 165 cm, and (c) either 170 cm or 171 cm or 172 cm? Given reasons in favour of your answers. Also calculate using R function $P(X > 172)$ and $P(X \leq 162)$. You will see that these two probabilities are same! Justify with reasons.

3.26 Suppose a sampling scheme suggests to continue sampling unless we get 4 affected individuals from a population. Assuming that the probability of an individual to be affected is 0.1, what is the probability that

(a) at least 4 selections are required?
(b) more than 8 selections are required?
(c) 7th selection will produce 4th affected person?
(d) 5 more selections are required when it is known that the first 5 selections did not give any affected individual?
(e) 5 more selections are required when it is known that the first 5 selections give only 2 affected individuals?

3.27 A special case of negative binomial distribution can be thought of when a random variable indicates the number of failures before the first success. What is the p.m.f. of X? Also find $E(X)$ and $V(X)$.

3.28 Suppose I toss a coin to select one out of two individuals. Probability of getting head in a single trial by this coin is known to be $\frac{3}{4}$. Since the coin is biased, I am unable to select one person unbiasedly, i.e. with probability $\frac{1}{2}$. So I devised

a scheme. The coin would be tossed two times and if $\{HH\}$ or $\{TT\}$ appears, no selection is made and tossing is continued. Person 1 is selected if $\{HT\}$ appears, and person 2 is selected if $\{TH\}$ appears.

(a) Show that under this scheme, the probability of selecting person 1 is $\frac{1}{2}$.
(b) What is the probability that person 1 will be selected on 8th toss?
(c) Since there is a possibility that the tossing might continue without any limit, what is the probability that person 2 will not at all be selected?
(d) What is the probability that a person will be selected on 10th toss of the coin?

3.29 There is a screening test for prostate cancer that looks at the level of PSA (prostate-specific antigen) in the blood. There are a number of reasons besides prostate cancer that a man can have elevated PSA levels. In addition, many types of prostate cancer develop so slowly that they are never a problem. Unfortunately, there is currently no test to distinguish the different types and using the test is controversial because it is hard to quantify the accuracy rates and the harm done by false positives. For this problem, we'll call a positive test a true positive if it catches a dangerous type of prostate cancer. We'll assume the following numbers:
Rate of prostate cancer among men over 50 is 0.0005,
True positive rate for the test is 0.9,
False positive rate for the test is 0.01. Let T be the event a man has a positive test and let D be the event a man has a dangerous type of the disease. Find $P(D|T)$.

3.30 Plot the p.d.f. of exponential distributions with parameters $\alpha = 1$ and $\alpha = 3$ on the same plot using R functions. Comment on your findings.

3.31 Plot the p.d.f. of a χ^2 distribution with varying degrees of freedom, on the same graph. Comment on your findings.

Chapter 4
Analysis of Single Gene Expression Data

By now, we have some idea on how we should look at the data sets at the very outset. In order to get some feel about the data we need to explore it through some visual representation tools (histogram, boxplot, etc.) and a few suitable summary measures (mean, standard deviation, etc.) that may be guided by relevant graphs. These reveal some characteristics about the data set as given in Table 2.1. Now we disclose one important feature about the data given in Table 2.1.

The values in Table 2.1 are not the actual values; but they are log-transformed. We take the logarithm with base 2 for the raw data to get values in Table 2.1. This triggers a natural question: why have we taken transformed values instead of original values? Moreover, out of millions of transformations (theoretically), why do we suddenly prefer to use logarithm transformation? So we need to explain the choice of transformation and why this is regarded as the 'best' for this data set.

Knowledge on data generation in the case of microarray data already indicates that a log transformation might help managing the gene expression data. But is there any way to develop or choose a transformation objectively looking at the data?

Before discussing the transformation we demonstrate an exploratory tool in order to determine whether a set of observed values follows a particular distribution. Here by distribution, we mean, strictly speaking, a probability distribution of a random variable from which the data are supposed to be drawn (randomly). We have already discussed the notion of random variable and its corresponding probability distribution. This will help us to develop an exploratory visual tool that might guide us in identifying the distribution.

4.1 Q-Q Plot

'Quantile-Quantile plot' or commonly known as 'Q-Q plot' is a nice visual tool that explores the degree to which a particular theoretical distribution fits to a set of

I. Mukhopadhyay and P. P. Majumder, *Statistical Methods in Human Genetics*, Indian Statistical Institute Series, https://doi.org/10.1007/978-981-99-3220-7_4

numerical observations. The name suggests that this technique heavily depends on quantiles of the distribution. We define p-th sample quantile (z_p) as an observed value such the proportion of the number of observations that are less than or equal to z_p is p for $0 < p < 1$. Analogously ζ_p is said to be the p-th population quantile if $P(X \leq \zeta_p) = p$. Note that for $p = 0.5$, ζ_p is the population median while its sample analogue is the median obtained from the actual sample data. First, we explain the idea of Q-Q plot using a normal distribution and then its extension towards identifying any distribution will be immediate.

Suppose a random variable X follows a normal distribution with mean 5 and variance 4. However, Theorem 3 establishes that we can easily standardise $N(\mu, \sigma^2)$ variable to a standard normal variable. Hence, without any loss of generality, we can work assuming that $X \sim N(0, 1)$. We then calculate population quantiles (ζ_p) for several values of p. On the other hand, based on sample observations drawn from the same population, we also calculate several z_p's, the sample quantiles for different values of p. Naturally, if the observations come from the same normal distribution, the sample quantiles should be very close to their corresponding population quantiles. If we plot population quantiles in the horizontal axis and the corresponding sample quantiles in the vertical axis, we can expect that all points should ideally lie on a straight line passing through the origin making an angle of 45° with the horizontal axis. Any significant deviation of the points from the line would indicate that the distribution differs from the normal distribution. Larger the deviation, higher would be the distance of the sample observation from the normal distribution. To calculate sample quantiles, we must standardise each sample value, i.e. instead of considering $\{x_i, \ i = 1, \ldots, n\}$, we should consider $\{y_i = \frac{x_i - \bar{x}}{s}, \ i = 1, \ldots, n\}$. Plot these sample quantiles corresponding to population quantiles calculated using a standard normal distribution. Let's demonstrate this idea first using a sample drawn from a normal distribution.

We draw 50 observations, drawn from $N(0, 1)$ distribution and calculate sample quantiles and population quantiles for the same values of p (Table 4.1). Using the information as in Table 4.1, we can plot corresponding to population quantiles. Also, we draw a straight line that passes through the point (0, 0) having an angle 45° with the horizontal axis. This is what is known as 'Q-Q plot' (Fig. 4.1).

Clearly, the points are very close to the dotted line, which guides us to declare that data come from a normal distribution. Note that points may not necessarily lie on the dotted line, but they are close enough, which is a strong indication in favour of our decision. Although we can work with actual values, for normality checking, we can work with standardised values and compare to a standard normal distribution. The R code for from drawing Fig. 4.1 is given below.

Table 4.1 Data and calculation of sample and population quantiles from N(0,1) distribution

100p%	ζ_p	z_p
10%	−1.28	−1.29
20%	−0.84	−0.67
30%	−0.52	−0.45
40%	−0.25	−0.26
50%	0.00	−0.04
60%	0.25	0.21
70%	0.52	0.52
80%	0.84	0.86
90%	1.28	1.34

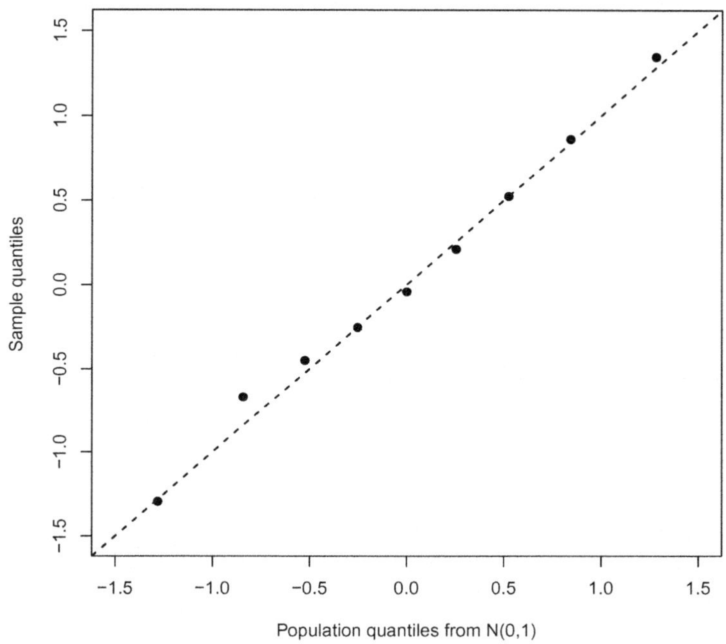

Fig. 4.1 Q-Q plot to check whether standardised data of Table 4.1 come from a standard normal distribution

```
> x<- c(9.31, 6.25, 9.04, 8.72, 9.75, 9.02, 9.71, 10.24, 11.39, 12.49, 11.03, 7.38,
9.67, 11.43, 12.11, 5.38, 9.69, 7.98, 9.43, 11.27, 10.85, 9.28, 9.18, 11.07, 12.21, 12.36,
9.62, 9.73, 7.60, 6.57, 10.10, 8.70, 9.98, 10.59, 11.12, 10.42, 9.42, 12.91, 6.35, 10.17,
8.43, 8.86, 10.70, 10.72, 12.68, 10.61, 7.14, 8.66, 11.63, 11.53, 8.95, 7.48, 8.92, 9.63,
13.04, 8.44, 10.15, 9.23, 7.75, 8.96, 11.67, 10.66, 9.56, 8.60, 11.28, 9.28, 7.60, 9.68,
14.09, 7.03)                                        ## data input
> y<- (x-mean(x))/sd(x)                             ## standardises data
> zp<- quantile(y,probs = seq(0, 1, 0.1))           ## sample quantiles for
                                                       p = 0.1, 0.2, ..., 0.9
> zetap<- qnorm(seq(0, 1, 0.1))                     ## population quantiles from N(0,1) for
                                                       p = 0.1, 0.2, ..., 0.9
> plot(zetap,zp,xlim=c(-1.5,1.5),ylim=c(-1.5,1.5),pch=19,xlab="Population quan-
tiles from N(0,1)",
   ylab="Sample quantiles")                         ## plot z_p versus ζ_p
> abline(0,1,lwd=1.5,lty=2)                         ## draws line through (0, 0) having 45° angle
                                                       with horizontal axis
```

Table 4.1 and Fig. 4.1 demonstrate the closeness of data points from normal distribution. However, it is worth exploring the pattern of sample quantiles based on data from distribution other than normal through Q-Q plot. First, we generate 100 observations from each of four distributions viz. Normal, Laplace, Gamma, and Cauchy distributions. Calculate sample quantiles for different values of $p (0 < p < 1)$ after standardising the observations, i.e. if $\{x_i, \ i = 1, \ldots, n\}$ are the observations, we calculate sample quantiles for $y_i = (x_i - \bar{x})/s_x, \ i = 1, \ldots, n$. Plot these sample quantiles corresponding to population quantiles calculated on the basis of a standard normal distribution (Fig. 4.2).

The purple dotted line corresponds to the ideal fit to a normal distribution. We see that black points are closest to the purple line, whereas others are deviating from it, with red points showing the most deviation. Red points corresponds to Cauchy distribution which is known to have a very fat tail and thus far from a normal distribution. Thus, Q-Q plot gives us a nice visual tool to check whether sample observations in a data set came from a normal distribution. If the points fall very close to the line drawn through the origin and subtends an angle 45° to the right of horizontal axis (i.e. x-axis), we can say that the data came from a normal distribution.

Note that using Q-Q plot, we can check whether the data come from any particular distribution. Assume a probability distribution, calculates the population quantiles based on this probability distribution, and plot them against the corresponding sample quantiles calculated on the basis of the given data. Here also if we see that points are very close to the line drawn through origin with an angle 45° with the horizontal axis, we can say that the given sample observations have come from the chosen probability distribution. Closer the curve (for a large number of sample quantiles, it looks like a curve) to the line, stronger would be the evidence that data came from the assumed distribution.

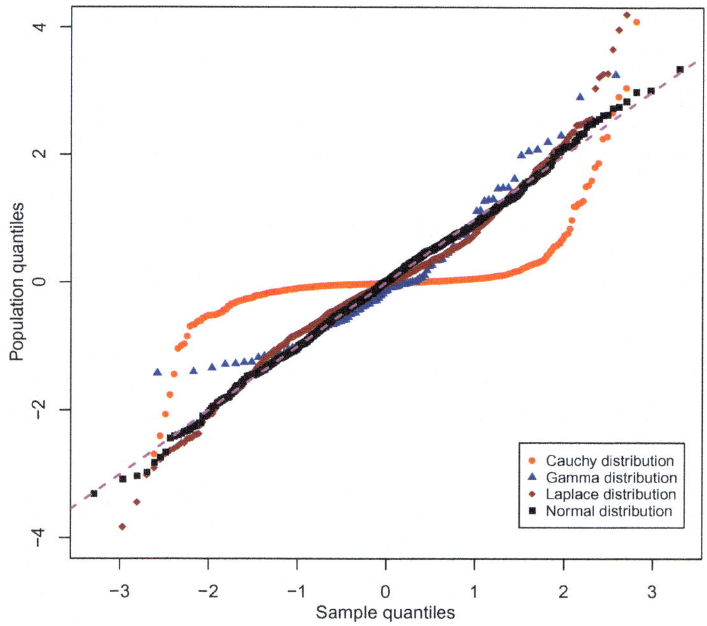

Fig. 4.2 Q-Q plots for some distributions

```
> qqplot(x,y)          # 'x' contains random sample from the distribution, 'y'
                         contains data
```

The R code for drawing a Q-Q plot is very simple: '*qqplot*()'. Use the command '?*qqplot*()' in R terminal and a full help page will open. Use this function, play around it to explore how fascinating plots can be drawn, try to get the Fig. 4.1 as given here, and customise it using your own data.

4.2 Transformation of Variables

Usually, in the paradigm of gene expression data analysis, if x is a value or realisation of the random variable under consideration, sometimes we start working with $log_2(x)$ instead of x. However, this is not a random transformation choice.

If we look at histogram or boxplot (Fig. 4.3) of the raw observations, we see an abundant presence of so-called 'outliers'; but it is not wise to think that actual experiment results in so many outliers. Moreover, the actual outliers are lost in the crowd and a histogram or boxplot fails to reveal any features of the data. Actually, the

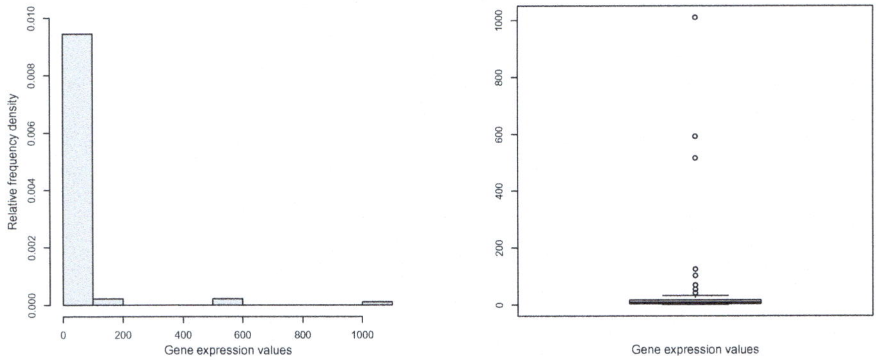

Fig. 4.3 Histogram and boxplot of gene expression data for original observations

tremendous variation present in the data is due to artefacts of the actual experiment that generates the gene expression data.

Clearly, we cannot get adequate information from the graphs. It is also clear that except a few, all observations are clustered in a small interval (with respect to the scale of measurement), whereas there are two or three clusters with very large values. Boxplot also fails to reveal some idea as the so-called outliers are distributed over a long range towards the right of the mean and other values that constitute majority of the data are concentrated in a very narrow region. The coverage of most values is too narrow to reveal any information about the data. Moreover, it is unrealistic to think that most of the data in reality show very small dispersion as we know that microarray experiment or other gene expression data generating experiments always induce lots of artefacts and hence scatteredness to the data. So there is a need to filter them out or stabilise the observations, at least to some extent, and to dampen the affect of the large outliers and make them compatible with the data. The most effective way to do this is to transform the data. Thus, transformation of data before doing any analysis attracts serious attention among scientists.

There is also another reason in favour of transforming raw data. We know that if we are sure that the data arising out of a continuous variable follows a normal distribution, doing statistical analysis would be much easy. There are lots of statistical tests and methods applicable to data only when it follows a normal distribution. Thus, normal distribution plays a crucial role in analysis. If from a simple Q-Q plot, we see that the data are nowhere near a normal distribution and we can validate this idea using a statistical test (to be discussed later), we are not in a position to use the myriads of available methods at our disposal assuming that the underlying distribution as normal. However, it is exciting to know that in many situations, appropriately transformed values follow a normal distribution while the original values do not. So there is a sound technical reason to consider the possibility of undertaking such transformation before starting any analysis of data.

However, the type of transformation should be determined very carefully, preferably in an objective manner. Box and Cox (1964) proposed a rule of transformation. Suppose y is a variable that takes values $y_1, y_2, \ldots, y_n$. We want to transform these values so that the transformed values should very closely follow a normal distribution. Let $y^{(\lambda)}$ be the transformed variable from the original variable y through a parameter λ. We have to find a λ such that the transformed variables would be:

$$y^{(\lambda)} = \begin{cases} \frac{y^{\lambda}-1}{\lambda} \text{ if } \lambda \neq 0 \\ \log y \text{ if } \lambda = 0 \end{cases}$$

Estimation of λ is difficult and demands lot of calculation and assumption. Rather we choose different values of λ and draw a 'Q-Q plot' to check whether it would be a good fit to a normal distribution. There might be enumerable choices of λ that we can try to find the appropriate transformation, but a little bit of brain storming based on the general nature of the data would reduce this burden to a great extent. Clearly, the values of raw gene expression data may be very large with high variation. So the appropriate value of λ cannot be very large. We draw Q-Q plots for $\lambda = 2, 0.9, 0, -0.9, -2$ and see which is closest to the normal distribution. Note that we calculate population quantiles assuming normal distribution while drawing such a Q-Q plot. Naturally, a set of values coming from a normal distribution would always be very close to a straight line passing through the points (0, 0) and (1, 1), i.e. a line passing through origin making an angle 45° with the horizontal axis. In Fig. 4.4, purple line is passing through the black points (stars) in the most representative way. Black points are drawn for $\lambda = 0$, i.e. they are log-transformed values. Thus, it establishes that $\lambda = 0$ is the most appropriate choice and hence the transformation that makes the given raw data closest to a normal distribution is logarithm transformed values.

Application of transformation of variables is extremely important in statistical analysis, especially when the distribution of the variable considered is not a normal distribution and sometimes it is very difficult to know the distribution even approximately. In such situations transforming a variable would make the data more manageable for analysis. Moreover, it is seen that in many cases it is possible to convert original distribution to a normal distribution, at least approximately for the given data set. Once we establish that the transformed data follows normal distribution to a certain degree of confidence, a whole new world of statistical analysis methods would open in front of us and many important information can be extracted from the given data, which at the beginning looks very complicated to analyse.

However, it is important to note that any information obtained after the analysis of data is based on the transformed data and not on the raw data. Once we get some information or inference from the transformed data, we must interpret it in terms of the original variable, although sometimes it is difficult to do so. So from now on the analysis of the data will be done using log-transfomed observations as given in Table 2.1. For the rest of the book, we always mean log-transformed values as gene expression data, unless otherwise mentioned.

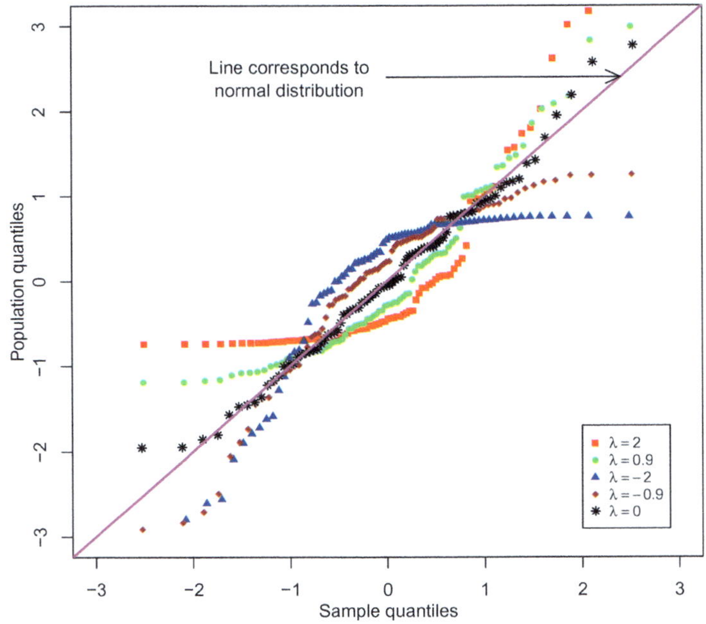

Fig. 4.4 Q-Q plots showing that logarithm transformation is the best one

4.3 A Few Testing Problems

Questions 5–8 in Sect. 2.1 are related to testing some facts about the variable considered, which in our case is gene expression data. We address different types of testing problems, how they can be solved mainly from intuitive consideration, and how best we can interpret the results. But first, we quickly discuss some basic concepts and criteria that a testing procedure should satisfy in order to get meaningful and reliable results. Note that testing of hypothesis covers a major part of data analysis. It is seen sometimes that for the statistical problem under consideration, existing testing methods might fail to give confidence to a statistician because of the intricate nature of the data and/or the hypothesis or objective of the problem. In that case, we might have to develop new method of testing only specific to that problem. So it is imperative to understand the basic ideas behind testing of such statistical hypothesis.

4.3.1 Basics of Testing of Hypothesis

We used to play cricket matches, everyday, during the afternoon after school hours. We made two teams and these two teams would play a series of games. Madhu, the captain of Team A observed that the other captain Shaun always called 'head'

during the toss to determine which team would bat first. Batting first seemed to be more exciting. Moreover, Shaun used the same coin for the toss every day. Madhu observed that Shaun won the toss every time but once during the last six occasions. So he suspected that the coin Shaun uses was in all probability a biased one, i.e. the chance of getting 'head' was more than getting a 'tail'. Now the question is, how does one check that the coin is fair (unbiased), i.e. the chance (probability) of getting a 'head' is same as getting a 'tail' in a single toss.

Well, a little bit of thinking gives us a very simple strategy. Just toss the coin a fixed number of times, say 20. We expect to see around 10 heads (may be slightly more or slightly less). However, if the number of heads is very large, say more than 17, we may say that the coin is not fair. It shows more chance of getting a head than a tail. This is nothing but statistical testing of a hypothesis.

In the above example, we test a hypothesis, i.e. the coin is fair or unbiased against a suspicion that the coin is not fair. The first one, the hypothesis that we want to check is called 'null hypothesis' (denoted by H_0) and the other one, which we consider in case of a false null hypothesis, is called 'alternative hypothesis' (denoted by H_1). The strategy that determines whether null hypothesis is true or false is called the testing methodology or strategy.

A scientific hypothesis is usually generated based on available prior information (in most cases partial information). Often a pilot study is conducted to generate specific hypotheses. The hypothesis that is to be tested is called the null hypothesis. A decision on the null hypothesis is taken on the basis of observations collected as a result of a random experiment from the population under study. Usually, one has an alternative hypothesis in mind as well, which is automatically accepted (for most practical situations or further downstream analysis), if the null hypothesis is rejected. So based on a random sample of a fixed size, we can either reject the null hypothesis or accept it or can say that we need some more evidence to come to a firm decision. In this section, we shall discuss, in brief, the principles and methodology of testing of hypothesis.

For any null hypothesis, usually denoted by H_0, any one of the four possible situations can arise:

(1) Reject H_0 when H_0 is false: a correct decision.
(2) Reject H_0 when H_0 is true: an incorrect decision.
(3) Do not reject H_0 when H_0 is false: an incorrect decision.
(4) Do not reject H_0 when H_0 is true: a correct decision.

The second and third decisions are called Type I error and Type II error, respectively. In ancient Indian stories, there are a few instances of these errors. In 'Abhigyan Shakuntalam', King Dushyanta married Shakuntala when he met her in hermitage while hunting in a jungle. He gave her his royal ring to serve as her identity. However, while going to meet him, she lost the ring in the river. When she reached the palace and met him, the king failed to recognise her. Thus, Dushayanta committed a Type I error as he rejected Shakuntala as his wife when, in fact, she was his true wife.

In Mahabharta, Dronacharya was fighting for the Kauravs. During the war, an elephant named 'Aswathama', his son's namesake was killed. Yudhishthira on advice

of Krishna, went to Dronacharya and pronounced that 'Aswathama' was dead and added that it was an elephant in a low murmur. Dronacharya, on listening the first part of ever truthful Yudhishthira's sentence, presumed that his son was dead, and stopped fighting. Thus, Droncharya could be said to have committed a Type II error, i.e. accepting a statement when it was not true.

However, we should devise our testing method to avoid frequently committing these two types of errors. In other words, in good decision-making, the probabilities of these types of errors should be very low. However, for a fixed sample size, we cannot minimise the probabilities of both of these types of errors. Heuristically, this fact can be argued in a very simple way. Note that Type I error is related to rejection of H_0, whereas Type II error is concerned with not rejecting H_0. These two types of decisions are complementary to each other. Hence, if we decrease (or increase) the probability of one type of errors, it would automatically increase (or decrease) the probability of other type of error. So we cannot minimise the probabilities of both types of errors at the same time using a fixed sample.

On the other hand, we cannot afford to develop a testing procedure that would ensure a high probability of one type of error, be it Type I or Type II. One simple and effective way to get rid of this problem is to fix the probability of Type I error at a very low level, say 0.05 or 0.01, known as 'level of significance' and minimise the probability of type II error. We can also fix the probability of Type II error and minimise Type I error. Whether we choose the first option or the second one, depends on the problem. Given a statistical testing problem, first identify which error is more serious. Keep the probability of that error at a very low level and minimise the probability of the other error. This will ensure that by adopting this testing method, we are not committing a serious mistake in any way. Usually, it is seen that in most cases type I error is more serious than type II error.

We define the statistical Power of a test as:

$$Power = 1 - P(Type\ II\ error)$$

and develop our testing method by maximising the power and keeping probability of Type I error fixed at a prespecified level of significance.

Now we use a statistic (defined as a function of sample observations) that would be used for testing the null hypothesis and also for estimating the population parameter. Note that H_0 carries some information about the parameter. So based on data on a set of samples, we first calculate the value of the test statistic. Then, we calculate the probability of obtaining the observed value of statistic or higher (lower) assuming that H_0 is true. This probability value is known as the p-value. The p-value has a very important role in getting inference through testing of hypothesis; however, it should be interpreted carefully.

Another very important but often not focused upon is the design by which we draw the sample observations. In almost all testing problems we draw the samples randomly, i.e. following a random mechanism or design, when we collect sample observations from the population of interest. As an example, suppose we want to test whether systolic blood pressure of students of class XI in a particular school

is 120. For this, we may go to that class and select a number of students, say 12, just by looking at their roll numbers or sitting arrangement. This looks like a random sample because we do not know the names of any students. However, this is not really a random sample, because the mechanism by which the students are drawn is not random. Instead, a meaningful design would be as follows. We write the roll number of each student on a piece of pear, fold it and put it in a jar. After shuffling the jar we pick up one paper, unfold it, note down the roll number, measure the blood pressure of the student corresponding to that roll number, and return the paper in the jar. Repeat this procedure until we have picked 12 papers following the same procedure. Finally, we get 12 observations of blood pressure that constitute our sample. This, in fact, is a random sample, because there is a clear random mechanism or experiment at the back of the selection process. Note that we may not return the paper after each draw; even in that case it remains a random sample, although the design is slightly different.

Thus, in a testing paradigm, we always work with a sample whose observations are drawn randomly. This clearly indicates that drawing an observation or its value has nothing to do with the values of other observations. Thus, the selection mechanism induces randomness or *independence* among the observations. In other words, sample observations (or the identical random copies of original variable) are independent of each other. This independence plays a very important role in the development of different testing procedures based on different hypotheses. Henceforth, unless otherwise mentioned, we assume that samples are independently distributed or simply, we work with random or independent samples in all hypothesis testing problems.

4.3.2 Interpretation of p-value

Associated with any testing problem, a widely known and used quantity is p-value. Basically, p-value gives the strength of our decision regarding H_0. If the p-value is very small, we may be highly confident in rejecting H_0. On the other hand, large p-value indicates that the evidence is too weak to reject H_0. Usually, but not always, it is compared to the level of significance, which is taken to be very small. Sometimes people involving in such analysis consider a rule of thumb to compare p-value to a small value, usually 0.05 or 0.01. But in such cases, serious confusion may arise if not interpreted correctly.

Suppose the p-value corresponding to a test statistic for a testing problem is 0.00003. We can immediately reject H_0 as the p-value is extremely small, compared to 0.05. However, a p-value of 0.000000023 still provides stronger evidence in favour of rejecting H_0 than the previous one. On the other hand, if it is 0.1543, we can say that it is too large compared to 0.05 and hence we do not reject H_0.

Now consider two situations; one produces a p-value 0.048 and the other gives 0.052. Considering the level of significance as 0.05, thumb rule suggests that we reject H_0 in the first case and do not reject H_0 in the second case. Strictly speaking, these conclusions are not very meaningful. In fact, a p-value of 0.048 shows weak

evidence in favour of rejection of H_0. On the other hand, if we do not reject H_0 in the second case only because it is slightly more than 0.05, we will make serious mistake. Note that 0.048 and 0.052 fall on either side of 0.05 with an absolute difference 0.002. This difference may occur due to a change in a single observation or a random fluctuation only in one observation. In other words, very minor changes in a single observation changes the decision, which should not be appropriate if we just use our common sense. Thus, in such situations, when p-value is reasonably close to 0.05, we can say that there exists a marginal significance of the effect of some factors that distorts the value specified by H_0. Instead of taking a very firm decision about H_0 at this stage, we need to investigate it more carefully either using more samples or doing an in-depth study of the entire phenomenon or understanding the study design that generates the data or by some other process. We should also check the assumptions under which the test is carried out and look for other powerful tests from the statistics literature, if the situation so demands.

Looking at the definition as well as the cited example it is clear that lower the p-value, stronger is the evidence against H_0.

We will now discuss some examples of testing methods that are commonly used. Careful attention should be given to the assumptions under which the tests are derived, the actual methodology of the test, and how the resulting statistic is used to make the final inference.

4.3.3 *Test for Mean: Single Sample Problem*

In the context of the gene expression data as given in Table 2.1, suppose the experimenter who generated this data set is interested to see whether the mean of expression values is around 4. Let X be the random variable denoting gene expression of a gene under consideration. The observed data set is assumed to be a random sample from X. Then the null and alternative hypotheses are:

$$H_0 : \mu = \mu_0 \text{ against } H_1 : \mu \neq \mu_0, \text{ where } \mu_0 = 4, \text{ in this case.}$$

Since we have no prior information about what might be the true value of mean if H_0 is rejected, i.e whether it is more or less than μ_0, we use the two-sided alternative hypothesis.

First, we assume that X follows a normal distribution but unfortunately the standard deviation is unknown. This assumption of normal distribution must be validated by either a statistical test or by a strong belief based on prior knowledge about the nature of gene expression values for the gene. Q-Q plot might give an idea whether the data follow a normal distribution. However, a statistical test for testing whether a normal distribution is applicable will be discussed later.

Here, our hypothesis is related to the population mean μ. So we should start by calculating the sample mean $\bar{x}$ based on the observed values. Then we check whether the sample mean is either too small or too large compared to population mean μ_0,

under null hypothesis. Alternatively, we can look at the difference of μ_0 from $\bar{x}$, i.e. the quantity $\bar{x} - \mu_0$. Any difference should be judged with respect to the scale of measurement. And we know that standard deviation is a good scaling measure. Hence, since this difference is in regard to its standard deviation ($\sigma/\sqrt{n}$), we divide the difference by $\sigma/\sqrt{n}$ and thus get our test statistic as:

$$\tau = \frac{\bar{x} - \mu_0}{\sigma/\sqrt{n}}.$$

It can be easily shown (Theorem 3) that τ follows $N(0, 1)$, a normal distribution with mean 0 and variance 1, under H_0. So we reject H_0 if observed τ is too large or the p-value corresponding to τ is very small compared to $\alpha (0 < \alpha < 1)$, the chosen level of significance. Throughout this book, we take α to be equal to 0.05, unless otherwise mentioned. However, the above testing procedure is based on the assumption that the standard deviation is known. But in reality, this assumption is rarely valid as it is difficult to know the value of σ apriori. Thus, it is safe to assume that the standard deviation is unknown and in that case we cannot calculate the value of the test statistic, even under H_0, because we don't know the value of σ.

Since σ is unknown, we replace σ^2 by its unbiased estimator $s^2 = \frac{1}{n-1}\sum_{i=1}^{n}(x_i - \bar{x})^2$ and get the test statistic as:

$$t = \frac{\bar{x} - \mu_0}{s/\sqrt{n}},$$

which follows a t distribution with $n - 1$ degrees of freedom, under H_0. Thus, we reject H_0 if the observed value of t is either too small or too large or the p-value corresponding to the value of t is very small compared to α. This statistical procedure to test whether mean is equal to a specific value when standard deviation is unknown and the parent distribution is normal, is known as the famous **one-sample t-test**. This is based on t-distribution, which was first derived by a great statistician Student (his real name was William Sealy Gosset) in 1908. Hence, this test is sometimes known as Student's t-test.

If we look at the histogram of the observations, in Fig. 2.1, it may not be very clear that 4 is widely different from the calculated mean 3.1005 as there is a relatively long tail towards the right side. That is the reason we need to do statistical tests to ensure any strong evidence in favour of or against our belief.

Based on the observations in Table 2.1, we see that $t = -4.5755$ and degrees of freedom is 89. Hence, the corresponding p-value is 0.000015. Since the p-value is extremely small compared to 0.05, we can straightway reject H_0 concluding that, on the basis of the given data, there is significant evidence to believe that the mean gene expression value is different from 4.

```
> t.test(x, mu=4, alternative="two.sided")   ## single sample t-test with two sided
                                                alternative hypothesis
```

The output on the screen contains the value of the test statistic and the p-value. This p-value can be compared to the chosen level of significance leading to the decision based on the given data. Note that if we have prior information that the mean might be greater than 4, we should take alternative hypothesis as $H_1 : \mu > 4$ and in the R code, replace 'two.sided' by 'greater' to get the result.

Advanced readers may ask why the test statistic follows a t-distribution. We explain this as a theorem below.

Theorem 12 *Let* $X_i \sim N(\mu, \sigma^2), \quad i = 1, \ldots, n$ *i.i.d. Then* $\frac{\bar{X}-\mu}{s/\sqrt{n}} \sim t_{n-1}$.

Proof By Theorem 6 and Theorem 8, $\bar{X} \sim N(\mu, \sigma^2/n)$, $\frac{(n-1)s^2}{\sigma^2} \sim \chi^2_{n-1}$ and $\bar{X}$ and s^2 are independently distributed. It follows immediately using Theorems 6–8 that $\frac{\bar{X}-\mu}{\sigma/\sqrt{n}} \sim N(0, 1)$, $\frac{(n-1)s^2}{\sigma^2} \sim \chi^2_{n-1}$ and $\bar{X}$ and s^2 are independently distributed.

Take, $X = \frac{\bar{X}-\mu}{\sigma/\sqrt{n}}$, and $Y = \frac{(n-1)s^2}{\sigma^2}$. Using Theorem 9 we have,

$$\frac{\frac{\bar{X}-\mu}{\sigma/\sqrt{n}}}{\sqrt{\frac{(n-1)s^2}{\sigma^2}}} = \frac{\bar{X} - \mu}{s/\sqrt{n}} \sim t_{n-1}$$

Corollary 2 *When* H_0 *is true, i.e. under* H_0, $\mu = \mu_0$, *putting* $\mu = \mu_0$, *we have,*

$$\frac{\bar{X} - \mu_0}{s/\sqrt{n}} \sim t_{n-1}.$$

4.3.4 Wilcoxon Single Sample Test

Single sample t-test provides a method for testing whether the population mean is equal to a specified value when observations are randomly collected from a normal distribution. However, sometimes data are collected randomly but from a distribution which is not a normal distribution. In such situations, we cannot do the above t-test as the test statistic does not follow a t-distribution, even under H_0. In this connection, it is worthwhile to discuss a statistical test for mean or more generally for an appropriate test for location parameter when the random variable does not follow a normal distribution.

Since we assume here that the random variable describing the population does not follow a normal distribution, we cannot guarantee the existence of population mean. For example, if the variable follows Cauchy distribution, it's mean does not exist. Since we don't know the population distribution, it would be safe to assume that mean may or may nor exist. We should work with the most general situation. Hence, in such cases, we do not assume the existence of mean. However, it is obvious that median always exists and it is a very good measure of central tendency. Hence, in a paradigm where we don't have the liberty to assume normal distribution, we

consider median as our parameter of interest as a measure of central tendency for the population.

Suppose there is a strong evidence or prior knowledge to believe that the variable does not follow a normal distribution. In such a case we can do another test for location parameter, known as **Wilcoxon rank sum test** for a single sample. This test is based on the sum of ranks of absolute deviation of the values from the value of location parameter as specified by H_0. This kind of test does not use any assumption about the distribution of the random variable considered, except that it has a continuous distribution, i.e. the variable is an absolutely continuous random variable.

Let $\tilde{\mu}$ be the population median for X. So our hypotheses of interest are,

$$H_0 : \tilde{\mu} = \tilde{\mu}_0 \text{ against } H_1 : \tilde{\mu} \neq \tilde{\mu}_0,$$

where $\tilde{\mu}_0$ is value of median specified by H_0. Let R_i be the rank of d_i where $d_i = |x_i - \tilde{\mu}_0|$, $i = 1, \ldots, n$. We consider a test statistic as the sum of ranks for which $x_i - \tilde{\mu}_0$ is positive and calculate the corresponding p-value. Thus, our test statistic becomes,

$$W = \sum_{i=1}^{n} rank\{d_i : x_i > \tilde{\mu}_0\}.$$

For the data in Table 2.1, suppose we take $\tilde{\mu}_0$ as 3.5 and the Wilcoxon test statistic is found to be 1273 with p-value 0.00184. We naturally reject H_0 concluding that, on the basis of given data, there is a strong evidence to believe against the assumed median value 3.5 at 5% level of significance, even when we do not make or validate any assumption about the nature of the distribution of the variable. R code for Wilcoxon rank sum test would provide all information about this test including the p-value.

```
> wilcox.test(x, mu=4, alternative="two.sided")    ## single sample Wilcoxon
                                                      rank sum test with
                                                    ## two sided alternative
                                                      hypothesis
```

Note that nowhere in this test we are using any fact or property of a normal distribution or any other distribution. These kinds of tests are not based on any assumption of parent distribution and hence do not depend on any parameter of the actual distribution. Hence, these are called 'non-parametric' tests contrary to other 'parametric' tests like t-test, etc. that are dependent on parameters of the assumed distribution.

4.3.5 *Test for Variance: Single Sample Problem*

Sometimes we might be interested to see whether the variance of the distribution from which the samples are drawn is equal to a specific value. Here we assume that the parent distribution of the random variable is normal. Moreover, if the value of the variance is known based on prior knowledge, this should be validated based on given observations. Suppose here we want to test

$$H_0 : \sigma^2 = \sigma_0^2 \text{ against } H_1 : \sigma^2 > \sigma_0^2.$$

where σ_0^2 is the value of the variance specified by H_0. We assume that based on prior information, we are only interested to see whether true variance exceeds a certain value. It is believed that if the true variance is less than or equal to a certain value, the data seems to be stable to a certain permissible degree. Based on prior information, suppose we can assume that $\sigma_0^2 = 3$.

We start with comparing the sample variance $s^2 = \frac{1}{n-1}\sum_{i=1}^{n}(x_i - \bar{x})^2$ to σ_0^2 and consider the ratio s^2/σ_0^2 as our test statistic. We reject H_0 if observed value of s^2/σ_0^2 is too large compared to 1. It is known that when H_0 is true, by Theorem 8, $(n-1)s^2/\sigma_0^2$, a multiple of the ratio of sample variance to population variance (under H_0), follows a χ^2 distribution with $n-1$ degrees of freedom. Since the sample size is 90, the degrees of freedom here is 89. The p-value corresponding to this statistic is obtained as 0.1445 and hence we have no reason to reject H_0 concluding that the true variance of gene expression values can be taken as 3 (at level of significance 0.05).

Suppose we don't know the function which does this test in R, we can easily write our own code.

```
> n<- length(x)                    ## number of observations
> test.statistic<- (n − 1) ∗ var(x)/(sigma ∗ sigma)        ## sigma=σ0
> pvalue <- pchisq(test.statistic,n-1,lower.tail=F)        ## calculates p-value
> pvalue                           ## prints p-value on R terminal

> pchisq((length(x)-1)*var(x)/3,length(x)-1,lower.tail=F)  ## all previous
                                   ##commands in one line to calculate p-value
```

The function '*pchisq*()' calculates the probability that a χ^2 variable is greater than the observed value of the test statistic, which is obtained by '$(length(x) - 1) * var(x)/3$' with degrees of freedom '$length(x) - 1$' for a right-sided test

('$lower.tail = F$'), where '$length(x)$' is the sample size. The output is only the p-value, which is 0.1445.

4.3.6 *Test for Equality of Two Means*

Consider the data given in Table 4.2. Gene expression values are given for two different groups of people. Two groups may be cases or controls, i.e. affected and unaffected individuals with respect to a particular disease, or two ethnic groups, etc. The two groups are considered independent in the sense that observations in one group are not at all related or affected by observations in the other group. Moreover, observations in each group are random samples from the respective populations.

We want to check whether the mean expression levels are same in the two distributions from which the two independent samples of sizes $n_1(=30)$ and $n_2(=24)$ are drawn. So our hypotheses of interest are:

$$H_0 : \mu_1 = \mu_2 \text{ against } H_1 : \mu_1 \neq \mu_2.$$

We assume that both distributions are normal, i.e. if X and Y are the two variables, we assume that $X \sim N(\mu_1, \sigma_1^2)$ and $Y \sim N(\mu_2, \sigma_2^2)$ and X and Y are independently distributed. To do an exact test, we have to assume that $\sigma_1^2 = \sigma_2^2$ although the common value is not known. There is a statistical test that can be used to validate the assumption about the variances. Under these assumptions, we consider the difference $\bar{X} - \bar{Y}$ and compare it to $\mu_1 - \mu_2$, which is 0 under H_0. However, since this difference is in regard to its standard deviation, which is unknown, we consider a test statistic under H_0 as:

$$t = \frac{\bar{x} - \bar{y}}{s\sqrt{(\frac{1}{n_1} + \frac{1}{n_2})}}, \quad \text{where} \quad s^2 = \frac{(n_1 - 1)s_1^2 + (n_2 - 1)s_2^2}{n_1 + n_2 - 2}.$$

where $s_1^2 = \frac{1}{n_1-1}\sum_{i=1}^{n_1}(x_i - \bar{x})^2$ and $s_2^2 = \frac{1}{n_2-1}\sum_{i=1}^{n_2}(y_i - \bar{y})^2$.

Table 4.2 Gene expression values for two independent samples

First sample					Second sample			
2.67	0.74	4.60	1.95	1.64	3.39	2.63	3.63	3.94
1.40	4.91	1.45	1.99	6.96	6.28	3.70	5.54	6.54
3.58	2.81	0.02	4.29	2.60	5.24	2.19	5.96	6.24
2.16	3.38	4.21	0.79	1.08	6.04	5.15	4.26	3.50
1.56	0.57	3.46	2.37	5.36	4.41	3.10	4.11	4.97
1.16	0.71	1.70	4.49	6.10	6.01	2.65	5.45	7.00

It can be shown that under H_0, the test statistic will follow a t-distribution with $n_1 + n_2 - 2$ degrees of freedom. Hence, we can calculate the p-value corresponding to it, which is 0.000044. So we reject H_0 concluding that based on the data in Table 4.2, there exists strong evidence to believe that the average gene expression values are different in the two populations.

```
> x<- c(2.67, 0.74, 4.60, 1.95, 1.64, 1.40, 4.91, 1.45, 1.99, 6.96, 3.58, 2.81, 0.02,
4.29, 2.60, 2.16, 3.38, 4.21, 0.79, 1.08, 1.56, 0.57, 3.46, 2.37, 5.36, 1.16, 0.71, 1.70,
4.49, 6.10)          ## first sample data input
> y<- c(3.39, 2.63, 3.63, 3.94, 6.28, 3.70, 5.54, 6.54, 5.24, 2.19, 5.96, 6.24, 6.04,
5.15, 4.26, 3.50, 4.41, 3.10, 4.11, 4.97, 6.01, 2.65, 5.45, 7.00)
                                        ## second sample data input
> t.test(x, y, alternative = "two.sided", var.equal=T)          ## two sample t-test
```

Note that we must mention that '$var.equal = T$' within the R function; otherwise it will do a test that will be valid only when the variances are unequal and unknown.

Output of $t.test$ gives the test statistic value as -4.4602 with corresponding p-value 4.406×10^{-5}. Since the p-value is very small compared to the chosen level of significance 0.05, we reject the null hypothesis at level of significance 0.05 concluding that, in the light of the given data, it seems that the population means are not equal to each other.

For advanced readers, we show how the test statistic follows t-distribution in the following theorem.

Theorem 13 *Let $X_1, \ldots, X_{n_1}$ be i.i.d. random samples from $X \sim N(\mu_1, \sigma^2)$, $Y_1, \ldots, Y_{n_2}$ be another i.i.d. random samples from $Y \sim N(\mu_2, \sigma^2)$. Assume that the two samples are also independent of each other. Then,*

$$\frac{\bar{X} - \bar{Y} - (\mu_1 - \mu_2)}{s\sqrt{\frac{1}{n_1} + \frac{1}{n_2}}} \sim t_{n_1+n_2-2}.$$

Proof Clearly, $\bar{X} \sim N(\mu_1, \sigma^2/n_1)$, $\bar{Y} \sim N(\mu_2, \sigma^2/n_2)$, and $\bar{X}_1$ and $\bar{X}_2$ are independently distributed. So by Theorem 4, $\bar{X} - \bar{Y} \sim N(\mu_1 - \mu_2, \sigma^2(\frac{1}{n_1} + \frac{1}{n_2}))$.

Again, by Theorem 8 $\frac{(n_1-1)s_1^2}{\sigma^2} \sim \chi^2_{n_1-1}$, $\frac{(n_2-1)s_2^2}{\sigma^2} \sim \chi^2_{n_2-1}$ and they are independent. Hence, by Theorem 11,

$$\frac{(n_1 - 1)s_1^2}{\sigma^2} + \frac{(n_2 - 1)s_2^2}{\sigma^2} \sim \chi^2_{n_1+n_2-2}$$

$$\text{i.e. } \frac{(n_1 + n_2 - 2)s^2}{\sigma^2} \sim \chi^2_{n_1+n_2-2} \qquad \text{where } s^2 = \frac{(n_1 - 1)s_1^2 + (n_2 - 1)s_2^2}{n_1 + n_2 - 2}$$

Note that sample means are independent as they come from two independent distributions. Moreover, sample mean and sample variance are also independently distributed

for normal distributions. Hence, it is clear that $\bar{X} - \bar{Y}$ and s^2 are independently distributed random variables. Hence, by Theorem 9,

$$\frac{\frac{\bar{X}-\bar{Y}-(\mu_1-\mu_2)}{\sqrt{\sigma^2\left(\frac{1}{n_1}+\frac{1}{n_2}\right)}}}{\sqrt{\frac{(n_1+n_1-2)s^2}{\sigma^2}/(n_1+n_{2-2})}} \sim t_{n_1+n_2-2} \quad \text{i.e.} \quad \frac{\bar{X}-\bar{Y}-(\mu_1-\mu_2)}{s\sqrt{\frac{1}{n_1}+\frac{1}{n_2}}} \sim t_{n_1+n_2-2}$$

Corollary 3 *Suppose H_0 is true, i.e. under H_0, $\mu_1 = \mu_2$. Hence, putting $\mu_1 = \mu_2$, we have, under H_0,*

$$\frac{\bar{X}-\bar{Y}}{s\sqrt{\frac{1}{n_1}+\frac{1}{n_2}}} \sim t_{n_1+n_2-2}.$$

4.3.7 *Test for Equality of Two Variances*

In the two-sample t-test, along with normality assumption, we have also assumed that population variances are equal. This may or may not be true. Before doing a two-sample t-test, we have to check whether this assumption is valid. To validate our assumption of whether the two variances are equal, we have to develop a statistical test for $H_0 : \sigma_1^2 = \sigma_2^2$ against $H_0 : \sigma_1^2 \neq \sigma_2^2$.

Under the assumptions of normality, we take the ratio of sample variances, i.e. s_1^2/s_2^2 as our test statistic. If H_0 is true, the ratio of population variances is one. Hence, under H_0, we can expect that corresponding ratio of sample variances should be very close to 1. So, we reject H_0 if the observed value of s_1^2/s_2^2 is either too large or too small compared to 1. It can be shown that when H_0 is true, this statistic will follow an F-distribution with degrees of freedom $(n_1 - 1, n_2 - 1)$, i.e.

$$\frac{s_1^2}{s_2^2} \sim F_{n-1-1,n_2-1} \text{ under } H_0.$$

Hence, we can easily calculate p-value corresponding to the value of the test statistic using the F distribution with degrees of freedom $n_1 - 1$ and $n_2 - 1$. Here the value of the test statistic is 1.637 with the corresponding p-value as 0.1144. Thus, based on the given data we can believe that our assumption about the equality of variances is correct. Hence, the population variances are equal for two distributions and we can do a standard two-sample t-test to see whether the average gene expressions for the two populations are same, based on the data given in Table 4.2.

```
> pf(var(x)/var(y), length(x)-1, length(y)-1, lower.tail=F)     ## calculates
                                                p-value using F distribution
```

For advanced readers, we derive the distribution of the above test statistic under H_0.

Theorem 14 *Under the above set-up,* $\frac{s_1^2}{s_2^2} \sim F_{n_1-1,n_2-1}$ *under* H_0.

Proof Note that $\frac{(n_1-1)s_1^2}{\sigma_1^2} \sim \chi^2_{n_1-1}$ and $\frac{(n_2-1)s_2^2}{\sigma_2^2} \sim \chi^2_{n_2-1}$. Moreover, s_1^2 and s_2^2 are independently distributed. Then by Theorem 10,

$$\frac{\frac{(n_1-1)s_1^2}{\sigma_1^2}/(n_1-1)}{\frac{(n_2-1)s_2^2}{\sigma_2^2}/(n_2-1)} \sim F_{n_1-1,n_2-1} \quad \text{i.e.} \quad \frac{s_1^2}{s_2^2}.\frac{\sigma_2^2}{\sigma_1^2} \sim F_{n_1-1,n_2-1}.$$

Hence, under $H_0 : \sigma_1^2 = \sigma_2^2$, we have $\frac{s_1^2}{s_2^2} \sim F_{n_1-1,n_2-1}$.

4.3.8 *Wilcoxon Two-Sample Test for Locations*

The two-sample t-test can be done only when the distributions are normal and variances are equal but unknown. But in many real-life situations data from either one or both populations do not follow normal distribution. Naturally, t-test in such a situation is not recommended and may mislead the entire conclusion. In such cases, there is a nice but simple test based on comparing the ranks of the observations from two groups, known as Wilcoxon two-sample test for equality of location parameters. Following similar logic as in Wilcoxon single sample test, we are going to develop a non-parametric test where no assumption is made about the distributions except that they are absolutely continuous. We consider median instead of mean as the location parameter.

Like Wilcoxon single sample test, here we consider the null hypothesis as $H_0 : \tilde{\mu}_1 = \tilde{\mu}_2$, i.e. whether the medians of two distributions are equal. Since we are not allowed to assume any distribution of the random variables considered, we have to construct a test using ranks of the observations. Thus, our hypotheses of interest become,

$$H_0 : \tilde{\mu}_1 = \tilde{\mu}_2 \text{ against } H_0 : \tilde{\mu}_1 \neq \tilde{\mu}_2.$$

Note that the decision about H_0 is taken using a test statistic and its distribution under H_0. First assume that H_0 is true. This indicates that two samples are drawn from the same distribution and so we have a random sample of size $n_1 + n_2$. Rank all these observations. Consider the sum of ranks corresponding to the observations that come from the second distribution. If R_i is rank of the i-th observation from the second sample, we can construct the test statistic as:

$$S = \sum_{i=1}^{n_2} R_i.$$

Naturally, if $\tilde{\mu}_2 > \tilde{\mu}_1$, the distribution for the second sample is shifted to the right of that for the first sample, although a little overlap might be there. Consequently, S would be very large. On the other hand, if $\tilde{\mu}_2 < \tilde{\mu}_1$, S would be very small. This if H_0 is true, S will neither be small nor large. So we reject H_0 if S is either too large or too small. We can calculate p-value directly using the R function.

If we ignore that observations in Table 4.2 follow normal distributions, we can do Wilcoxon two-sample test to test $H_0 : \tilde{\mu}_1 = \tilde{\mu}_2$ against $H_1 : \tilde{\mu}_1 \neq \tilde{\mu}_2$; the p-value is obtained as 0.00006 which indicates that medians of the two groups are significantly different. The most important point to remember is that this test can be done when there is no assumption about the distribution concerned.

Sometimes, many people also do Mann-Whitney test for testing the same hypothesis. Note that although the test statistic is somewhat different than Wilcoxon test, basically one is obtained from the other through a linear relation. Hence, the p-value obtained on the basis of Mann-Whitney test is exactly same as that obtained through Wilcoxon test. Here nowhere have we assumed any specific probability distribution of X and Y.

```
> wilcox.test(x, y, alternative = "two.sided")
```

First assume that our chosen level of significance is 0.05. Based on the data set (Table 4.2), the value of test statistic is 138, whereas the corresponding p-value is 0.0000601. The very small value of p-value compared to the level of significance indicates that on the basis of the given data, there is enough evidence to conclude that the medians and hence the central tendencies of two data sets are significantly different.

4.3.9 *Test for Normality*

The above discussion establishes one important fact that basic testing methods depend very much on the assumption whether the data conforms to a normal distribution, especially when the data are of continuous type. We have seen that test for location (mean or median) can be done using t-test if the data are from normal distribution; otherwise, Wilcoxon test is appropriate. Hence, the appropriate test for a given problem depends on our knowledge of the parent population, i.e. distribution from which the data are supposed to come. To this end, although Q-Q plot can provide a nice visual tool to understand whether the variable follows a normal distribution, this is not a statistical test, in the true sense. However, there are some statistical tests available for this purpose. Among them, two are the most popular and we discus them in detail.

4.3.9.1 Kolmogorov-Smirnov Test

Before discussing any test for normality, we have to understand one important fact. Normal distribution is entirely specified by its mean and variance, denoted respectively by μ and σ^2. These are called parameters, that characterise a probability distribution describing the population under consideration. For an exponential distribution, whose p.d.f. is given by $f(x) = \frac{1}{\theta}e^{-x/\theta}, \ x > 0$, the only parameter is θ and it is enough to know everything about this distribution if we know the value of θ.

However, if we want to know from which distribution the data are coming, we cannot think of any particular distribution at the outset. That means we neither have any knowledge about the possible distribution nor any idea about the parameters nor even the number of parameters that it should have. Sometimes even population mean or variance may not exist for a typical distribution; however, any quantile-based measures of central tendency or dispersion always exist. For example, mean does not exist for a Cauchy distribution, but median does. On the other hand, although we know whether the variable under consideration is discrete or continuous, it would be too ambitious to think of the existence of p.d.f (p.m.f.) in case the variable is continuous (discrete). There are some instances where neither p.d.f. nor p.m.f. exists although the distribution function always exists. Thus, a distribution function seems to be a promising quantity that could have been used to develop a test to identify whether the data corresponding to random variable come from a normal distribution or from a specific distribution. This is also a non-parametric test as we are not assuming any p.d.f. (or p.m.f.) of the random variable from which data are generated.

Thus, to meet our objective, we can generalise the question as how to develop a test, more specifically a non-parametric test, to see whether the two distributions are same for two random variables. One is the specific distribution that we think to be true, whereas the other distribution is the actual distribution from which we have the actual data. Since we don't know the actual distribution, we have no idea about its p.d.f. But the fact that probability distribution function always exists enables us to develop a statistical test based on an analogous concept of distribution function that can be deduced only from the data. This quantity is known as the 'empirical distribution function'. We compare it to the known or our objective distribution. This idea is the basis of the famous Kolmogorov-Smirnov (KS) test.

Suppose that $\{X_1, \ldots, X_n\}$ is a random sample from the population whose distribution is given by $F(x)$. The empirical distribution function $\hat{F}_n$ is based on the given observations. We define $\hat{F}_n(x)$ for each x as:

$$\hat{F}_n(x) = \frac{\text{number of observations } \leq x}{n}$$

If we want to test that the random sample is from a normal distribution, we can frame this problem as a testing of hypothesis problem. Here,

$$H_0 : X \sim \text{Normal distribution against } X \nsim \text{Normal distribution.}$$

Let $\Phi(x)$ denote the distribution function, i.e. probability distribution of a standard normal distribution (i.e. a normal distribution with mean 0 and variance 1). To test the above H_0, Kolmogorov and Smirnov proposed the test statistic, based on empirical distribution function, as:

$$D_n = \sup_x |\hat{F}_n(x) - \Phi(x)|.$$

The test statistic D_n measures the maximum difference between the empirical distribution and standard normal distribution. Larger the value of D_n, less likely would H_0 be true.

Note that in most situations the sample mean is not very close to 0, neither the sample variance is close to 1. So first we standardise the sample values by subtracting sample mean from each observation and divide them by the sample standard deviation. We then apply Kolmogorov-Smirnov (KS) test to the standardised data. The p-value corresponding to this test is obtained either by using asymptotic distribution of D_n under H_0 or by exact calculation for small samples.

```
> ks.test((x-mean(x))/sd(x), "pnorm")      ## Kolmogorov-Smirnov test
                                              for checking normality
```

Here, it is observed that the value of KS statistic for the data considered in Table 2.1 is 0.0959 with corresponding p-value 0.3796. Since the p-value is very large compared to the chosen level of significance 0.05, we can safely conclude that the data have been drawn from a normal distribution. Here we have taken a two-sided alternative as there is no prior information about stochastic ordering of the distribution with respect to a standard normal distribution. 'pnorm' in R code indicates that we are trying to see whether the data conforms to a normal distribution.

4.3.9.2 Shapiro-Wilk Test

Another interesting test for normality was proposed by Shapiro and Wilk (1965) based on the results and principles of linear model theory in statistics. Before discussing this test, we first introduce the concept of order statistics. If we arrange the observations in increasing order, we can assume that there exists a random variable $Y_1 = min\{X_1, \ldots, X_n\}$; i.e. the realisation of Y_1 is the minimum of the set of observations. Similarly, we can define Y_n as the maximum 'order statistic' as $Y_n = \max\{X_1, \ldots, X_n\}$. In general, we have $Y_1, Y_2, \ldots, Y_n$ as the set of order statistics corresponding to the random sample $\{X_1, X_2, \ldots, X_n\}$. Shapiro-Wilk (SW) test is described based on these order statistics.

It compares the sample variance for the ordered observations to its corresponding population-level variance. Let $m' = \{m_1, \ldots, m_n\}$ be the vector of expected values of a standard normal order statistics and $V = ((v_{ij}))$ be the corresponding variance-covariance matrix. Then the test statistic is proposed as:

$$W = \frac{(\sum_{i=1}^{n} a_i y_i)^2}{\sum_{i=1}^{n} (y_i - \bar{y})^2},$$

where $a' = (a_1, \ldots, a_n) = \frac{m'V^{-1}}{(m'V^{-1}V^{-1}m)^{\frac{1}{2}}}$.

As we know that for normal distribution, measures of skewness ($\sqrt{\beta_1}$) and kurtosis (β_2) are 0 and 3 respectively indicating that it is a symmetric distribution and has moderate kurtosis (i.e. 'mesokurtic'). So one can have a fairly good idea about the distribution just by looking at the value of $\sqrt{\beta_1}$ and β_2. Extensive empirical sampling studies have shown that the value of W for non-null distributions usually tends to shift to the left of that for the null case. Note that here our null hypothesis of interest is that the data conforms to a normal distribution. It is also interesting to note that W lies between $na_1^2/(n-1)$ and 1.

Moreover, the probability density function of W for $n = 3$ is given by $f(w) = \frac{3}{\pi}(1-w)^{-\frac{1}{2}}W^{-\frac{1}{2}}$, $\frac{3}{4} \leq w \leq 1$. Thus, we can directly use $f(w)$ to calculate the p-value for $n = 3$. However, for $n > 3$, no compact form of the distribution is available even under H_0. Thus, we have to calculate the p-value approximately. Based on extensive simulations, however, it has been observed that this p-value, although approximate, is fairly accurate and can be used for all practical purposes.

Applying this test to the data, we see sometimes that the p-value is very small and we have no option but to conclude that the data do not follow a normal distribution. However, when we apply Kolmogorov-Smirnov test, we get a larger p-value that directs us to accept that the data follow a normal distribution. This apparent contradiction is actually due to the presence of outliers. If we remove the outliers and apply Shapiro-Wilk test to the data given in Table 2.1, we would get the p-value as 0.5074, thus matching our inference from KS test. So the KS test is more robust to the presence of outliers, whereas SW-test is more sensitive to it. The reason is that the KS-test is based on empirical distance, whereas SW-test is based on moments of higher order. Naturally, moments are highly affected by the outliers, thus making SW-test more sensitive to outliers compared to KS-test. Hence, we recommend applying both these tests for checking normality and accordingly taking a decision. This also shows the importance of studying outliers rather than ignoring them.

```
> shapiro.test((x-mean(x))/sd(x))     ## Shapiro-Wilk test for checking normality
```

4.4 Points to Remember

Not only in genetics or genomics, but for any statistical testing problem, we need to remember a few facts. Knowing what problem we are going to solve, we should try to understand the nature of the scientific problem and the objective of the experimenter. This could be done by asking many questions. A rigorous discussion might reveal a

few questions that demand statistical analysis. We must ensure at this point whether we are required to do any statistical testing of hypothesis. If so, first frame and define the null hypothesis and alternative hypothesis appropriately. Considerable time and effort should be devoted to do this; otherwise, the entire downstream analysis might not be correct and lead to misleading conclusion. Note that we must decide on both null and alternative hypotheses, keeping in mind the problem and its objective. Only then do we proceed to do sample collection and eventual testing of hypothesis. Framing or changing any hypothesis after data collection is not permitted; that might amount to manipulating the data and the results.

Once we are sure about the hypothesis to be tested, we must check whether the data show enough evidence that it comes from a normal distribution. If it follows a normal distribution, we may proceed to carry out our hypothesis testing assuming normal distribution. We must be careful about the assumptions on variance when testing for mean(s). Especially, in case of two-sample problems, we must check whether the variances are equal. If they are same, then only we can apply the two-sample t-test. On the other hand, if the assumption of normal distribution is not valid based on the given data, we cannot do a t-test; rather we should proceed to do non-parametric tests like Wilcoxon test. Thus, knowledge about the parent probability distribution is extremely important; it triggers and controls the entire analysis protocol.

For checking assumption of normal distribution, we should not forget to standardise the data before applying Kolmogorov-Smirnov test or Shapiro-Wilk test. However, for Shapiro-Wilk test, it is not mandatory to standardise the values because this test is specific to checking normal distribution only.

Finally, interpretation of p-value should be done cautiously and carefully. A simple comparison of p-value with respect to the chosen level of significance is not recommended. We need to understand the problem, how serious the type I error rate is, etc., to come to a conclusion based on p-value. We should fix the maximum permissible type I error rate (i.e. level of significance) before performing the statistical test, even before the data collection process or experiment. It should not be changed or adjusted after the test or looking at the p-value. Such a thing would amount to ignorance of the problem or unethical manipulation of results or both.

Exercise

4.1 Consider the data set 1 as given in Exercise-2.3.

(a) Draw a Q-Q plot to check whether the data come from a normal distribution. Ensure your observation by a statistical test.

(b) Perform an appropriate test to check whether we can say that the population mean is around 3.9. Write your conclusion properly.

4.2 Consider the data set:
16.40, 13.14, 11.34, 18.80, 13.07, 8.12, 11.72, 10.53, 11.06, 6.06, 8.27, 12.05, 13.16, 8.87, 12.38, 13.44, 15.72, 16.22, 13.64, 10.44, 11.95, 7.69, 12.57, 9.13, 16.70, 13.35, 11.50, 10.09, 14.93, 15.47, 8.19, 7.24, 10.52, 12.14, 14.69, 8.84, 13.95

(a) Check whether the data come from a normal distribution.
(b) Draw a Q-Q plot and comment.
(c) Discuss in detail the statistical test that you want to use for testing $H_0 : \mu = 14$ against $H_1 : \mu > 14$. Write your conclusion.
(d) Calculate power of the test assuming values of μ as 12, 12.2, 12.4, 12.6, ..., 14.8, 15. Draw a graph showing powers against these μ values and comment. This graph is known as a 'power curve'.
(e) Do a statistical test for $H_0 : \sigma^2 = 4$ against $\sigma^2 \neq 4$.

4.3 Two independent samples from populations that are normally distributed produced the following statistics: for sample 1 the sample size was 25, the sample mean was 34.2 and the sample standard deviation was 12.6. For sample 2, the sample size was 27, the sample mean was 49.1 and the sample standard deviation was 19.4. Assume that population variances are equal. Carry out a statistical test for the null hypothesis that the two population means are equal, in favour of the two-sided alternative?

4.4 Consider data set 2 and data set 3 as in Exercise-2.10.

(a) Test whether the variances in the two data sets are equal.
(b) Do a statistical test to see whether each data set comes from normal distribution.
(c) Perform a test to check equality of means after doing appropriate preliminary analysis.
(d) Write your comments based on your analysis.

4.5 For the data set:
4.46, 3.54, 4.02, 4.60, 4.09, 4.32, 3.98, 4.48, 4.44, 4.43, 4.28, 3.75, 4.58, 4.49, 3.61, 4.59, 3.99, 4.27, 3.98, 3.94, 4.83, 3.08, 4.56, 4.43, 4.00, 4.59, 4.76, 4.51, 3.69, 4.72, 4.12, 4.50, 3.80, 4.30, 4.82, 4.18, 4.54, 4.39, 4.02, 4.70, 3.89, 3.79, 4.45, 3.18, 4.44, 4.45, 4.26, 4.02, 4.25, 3.82
Check whether the above data follows a normal distribution. If not, find a suitable λ in the Box-Cox transformation so that the transformed variable follows a normal distribution. Compare mean and variance of both original observations and transformed observations. Comment on your findings.

4.6 For the data set given in problem 4.5, suppose we want to do an appropriate test for location.

(a) What statistical test you would adopt? Justify.
(b) Suppose the experimenter wants to do the same type of hypothesis based on original observations. What should be the most appropriate test in this case and why?

(c) Write a short report on the conclusions obtained in (a) and (b).

4.7 Consider the set of observations:
10.56, 8.58, 10.91, 9.72, 8.21, 8.58, 10.98, 10.36, 9.17, 12.11, 7.73, 14.01, 12.50, 9.21, 10.29, 9.45, 8.98, 8.54, 9.62, 10.60, 12.87, 6.34, 13.12, 8.26, 15.09, 7.66, 10.65, 14.72, 10.30, 7.42

(a) What are the values of first- and fourth-order statistics?
(b) What are the values of $F_n(x)$ at $x = 8.58, 12.50, 9.62$, and 14.72? Draw a graph of $F_n(x)$ against different given values of x. While drawing this graph, write R code of your own to first calculate values of $F_n(x)$ for all possible values of x and then draw an appropriate graph.
(c) Calculate $\Phi(x)$ values assuming $\mu = \bar{x}$ and $\sigma^2 = s^2$ for all given values of x. On the same graph obtained in 4.7(b), draw the graph of $\Phi(x)$ using a different colour and line type. Comment on your findings based on the two graphs thus obtained.
(d) To check whether the data follow a normal distribution, you apply both Kolmogorov-Smirnov test and Shapiro-Wilk test using R codes 'ks.test(x)' and 'shapiro.test(x)'. Note down the p-values for KS-test and SW-test. Do you see anything contradictory? If so, where is the fallacy? Explain and solve the fallacy. What is the right conclusion?
(e) Now test whether the location parameter may be taken as 10.43.

4.8 Choose a level of significance as 0.05. You are testing whether the data conform to a normal distribution. Based on a given data set, you get very small p-value for SW-test and relatively large p-value for KS-test. Naturally, you reject H_0 based on SW-test while you are unable to reject H_0 on the basis of KS-test. This seems an apparent contradiction. What might be the reasons and how do you solve it?

4.9 Interpretation of p-value is extremely important in testing statistical hypothesis. Assume that the level of significance for the test is 0.01. How do you interpret if the p-value is (a) 0.2415, (b) 0.03, (c) 0.008, (d) 0.0012, and (e) 0.0000023?

4.10 A data set contains the following observations:
3.52, 5.10, 4.34, 5.22, 3.55, 4.22, 4.89, 3.89, 4.29, 5.77, 5.74, 4.15, 4.78, 4.39, 4.80, 2.87, 1.86, 4.09, 5.04, 2.21, 2.22, 3.20, 4.63, 4.40, 4.20, 3.57, 4.82, 5.67, 6.22, 4.45, 5.95, 5.52, 4.10, 3.39, 4.45, 6.27, 4.46, 5.15, 4.16, 4.14, 2.51, 2.04, 2.74, 4.51, 2.81, 4.98, 4.53, 3.35, 3.99, 3.73, 4.82, 3.53, 3.90, 5.21, 3.42, 4.03, 4.00, 4.68, 4.11, 3.07
Answer the following questions, preferably using R.

(a) Draw a histogram and write down the main features that are immediately clear from the histogram.
(b) Arrange the observations in increasing order and subtract sample mean from each of them. Let us denote the transformed value as y. Thus, we get $y_1, y_2, \ldots, y_n$, where n is the number of observations in the given data set.

(c) Calculate empirical distribution function at each point, i.e. at each value of y, calculate

$$\hat{F}(y) = \frac{\text{number of observations } \leq y}{n}.$$

(d) Calculate $F(y) = P(Y \leq y)$ for all such y values when $Y \sim N(0, 1)$, i.e. using 'pnorm()' function of R.
(e) Plot $\hat{F}(y)$ and $F(y)$ against y on the same graph and interpret your result.
(f) Again subtract sample median from ordered observation of x values when they are arranged in increasing order. Let these values be $z_1, z_2, \ldots, z_n$.
(g) Calculate empirical distribuion function $\hat{F}(z)$ based on $z_1, z_2, \ldots, z_n$ as done in (c).
(h) Calculate $F(z) = P(Z \leq z)$ for all such z values if $Z \sim Cauchy(0, 1)$, i.e. using 'pcauchy()' function of R.
(i) Plot $\hat{F}(z)$ and $F(z)$ against z on the same graph and interpret your result.
(j) What is your overall conclusion?

4.11 Suppose a single sample t-test based on n_1 and n_2 observations produces p-values P_1 and P_2 respectively, In both cases, sample means are the same and so are the sample variances. Alternative hypothesis is greater than the type in both tests. What is the relation between P_1 and P_2 if $n_1 < n_2$?

4.12 Let type I error rate for testing $H_0 : \theta = 4$ against $H_1 : \theta < 4$ be 0.05. Power of the test at $\theta = 2$ is 0.69. How does the power get affected if the type I error rate is reduced to 0.01? If the test statistic (T) follows an exponential distribution with parameter θ, what is the type I error rate if we reject H_0 when $T < 2$?

4.13 An experiment was conducted by students at a university in the fall of 1994 to explore the nature of the relationship between a person's heart rate and stepping up and down on steps of various heights. One of the objectives of this study was to see whether the height of steps matters. Students were randomly assigned to two groups: Low-steps, where the height of steps was 5.75 inches (this group was coded as '0'), and High-steps, where the height of steps was 11.5 inches (this group was coded as '1'). Student performed the exercise (stepping up and down) for three minutes, after which their heart rates were measured. The investigators had hypothesised that the average heart rate would be different between the two groups. A two-sample t-test was used to analyze the data. Answer the following questions:

(a) Explain why we use a two-sample t-test and not a z-test or a paired t-test.
(b) Is this a one-sided or two-sided test?
(c) The sample mean, SD of heart rates for low steps and high steps are 85, 7 and 97, 22 respectively. 20 and 25 students are assigned to the low and high steps group respectively. Test the investigator's hypothesis at 5% level of significance.

References

Box, G. E. P., & Cox, D. R. (1964). An analysis of transformation. *Journal of the Royal Statistical Society. Series B (Methodological)*, *26*(2), 211–252.

Shapiro, S. S., & Wilk, R. S. (1965). An analysis of variance test for normality (complete samples). *Biometrika, 52*, 591–611.

Chapter 5
Analysis of Gene Expression Data in a Dependent Set-up

Gene expression data provides us with a big picture of the activity of the genes under study. However, it is of great interest to see whether the same gene behaves in the same way in tumour tissue as well as normal tissue. Or two genes in the same or different pathways behave similarly. Especially in studying cancer, it is expected that at least for some genes, expression patterns vary in different types of tissues or cells from different regions. In other words, a few genes might not show any significant differential signature in different tissues, while others might. One gene might act in synergy with another gene and show some significant activity or change in expression level. Naturally, this demands a thorough understanding and detailed study to get a clear picture of the gene expression pattern.

Multitude of genes exist in any living organism. Human genome alone consists of around 20000–30000 genes. They usually act in a dependent manner. It is expected that there are different types of interplay among genes, some act together in a pathway, some act based on other types of interactions, sometimes based on epigenetic triggers and other processes, etc. Various biological mechanisms and modalities play an important role at almost every stage, some of them being known, but probably mostly unknown. Thus, it is not a wise idea to study the expression of a gene individually; rather in order to get a more vivid picture, we need to study the expression data of two or more genes together.

This chapter is devoted to studying two genes together or the same gene in two different tissues. In this context, the first task should be to identify what are the possible queries that can be addressed through appropriate statistical analysis of the above phenomena. Needless to mention that as we move from simple to complex data, i.e. single gene to two or multiple genes, the statistical analysis might also move from simple to a more complicated type. This is necessary in deciphering the true picture of the phenotype and gene expression scenario. It is clear that data structure becomes more complex than what has been discussed already and demands careful attention, rigorous study, and new concepts and methods that are mandatory for undertaking appropriate analysis. However, the development of such statistical analysis methods emerges from simple intuition, the natural flow of domain knowledge and the nature of the problem.

I. Mukhopadhyay and P. P. Majumder, *Statistical Methods in Human Genetics*,
Indian Statistical Institute Series, https://doi.org/10.1007/978-981-99-3220-7_5

5.1 Understanding the Data

Modern technology provides us an easy way to generate gene expression data. The data may come from two different tissues, viz, normal and tumour from the same individual. Since we get observations corresponding to the same gene but from two different regions of the same individual, it is natural to think that observations are correlated. Similarly, the expression data for two genes from a group of individuals may also be correlated as both genes belong to the same individual. However, two genes may belong to the same pathway or different pathways, same chromosome or different chromosomes. In other words, two or more genes may be co-regulated, i.e. action of one gene is triggered or influenced by another gene, not necessarily in the same direction. In fact, we can generalise this by considering the entire gene expression profile of thousands of genes from each individual thus giving rise to a huge amount of data. We will discuss its analysis later. For the moment, we concentrate on the first two types of scenarios based on two genes.

Whatever be the situation, it is imperative to understand that the nature of the data before starting the analysis. Table 5.1 presents data for a single gene expression from two different regions or tissues for a group of individuals whereas Table 5.2 gives gene expression data for two genes in a number of individuals. We can also think of a more general data set where data are generated for two groups of individuals, viz., cases and controls. For the moment, we assume that the data are given only for controls.

First and the most important step in the analysis of data should be to critically understand the data. This will reveal important features of the data. While exploring data we might be aware of the information that is not available at this moment. It is also important to know what is known and what is not known. Lack of a few particular features is not the weakness of the data. But it demands an appropriate data analysis protocol. The best way to proceed is to ask as many questions about the data as possible; then try to address them, objectively and through statistical tools. At the end, we expect to get a broader picture with an emphasis on special features of the data that may lead to further scientific questions.

5.2 Generating Questions

It is always advisable to generate important questions before starting any analysis when data are presented. These questions should be generated using common sense coupled with domain knowledge and expertise. Moreover, we need to understand the problem and the hypothesis or the objective of the data generation experiment. For the data presented in Tables 5.1 and 5.2, some questions might be as follows.

(1) For normal-tumour pair of samples, can we compare the activity of the gene on the basis of their expression data using any suitable diagram(s)?

Table 5.1 Gene expression data for normal and tumour samples from same individuals

Normal	Tumour	Normal	Tumour	Normal	Tumour	Normal	Tumour	Normal	Tumour	Normal	Tumour
3.26	4.53	2.85	5.13	4.81	3.71	4.3	3.45	1.62	4.66	3.75	1.31
2.58	1.28	0.95	3.92	3.8	4.9	3.77	1.94	1.16	3.19	3.94	6.56
0.85	3.56	0.56	5.79	2.11	3.04	2.63	1.43	4.15	4.57	2.31	4.2
1.96	2.08	2.94	0.93	3.86	7.91	7.2	6.02	2.54	4.79	3.64	6.18
5.12	1.71	2.29	0.8	0.35	2.66	2.85	2.59	2.95	3.41	6.85	7.87
3.58	2.08	3.44	6.21	2.13	2.01	2.39	4.39	3.94	2.98	4.84	5.05
3.61	3.58	0.25	2.06	4.74	4.8	4.16	4.81	4.7	4.39	1.41	2.71
2.84	2.29	1.74	3.66	6.71	7.15	2.22	2.12	4.61	4.64	5.49	9.15
4.29	0.17	2.27	8.24	4.73	4.26	0.53	4.13	3.24	2.42	1.89	9.34
3.68	1.72	3.24	6.76	0.41	4.33	5	3.45	1.34	4.41	3.42	1.35

Table 5.2 Gene expression data for two genes from same individuals

Gene1	Gene2	Gene1	Gene2	Gene1	Gene2	Gene1	Gene2	Gene1	Gene2	Gene1	Gene2
0.54	2.27	1.29	1.52	2.78	0.89	2.94	4.12	4.83	5.41	2.75	1.41
4.67	6.61	4.23	4.25	1.63	0.31	3.12	3.91	1.42	5.34	0.1	5.44
5.86	2.91	0.15	2.38	−0.06	5.97	5.17	4.83	2.49	2.71	1.81	4.95
1.98	4.83	0.77	0.27	2.55	4	8.93	2.94	1.78	3.12	1.84	6.99
0.31	2.32	2.59	1.57	2.25	3.41	2	4.57	4.15	3.98	1.98	1.99
4.59	4.53	5.22	7.1	0.69	4.64	3.81	4.32	4.08	3.14	4.15	10.12
2.84	4.33	4.93	1.12	0.01	4.75	3.29	3.41	0.86	1.78	2.96	3.32
1.96	4.09	1.73	5.83	2.52	2.34	4.36	3.61	4.6	7.47	0.01	1.65
2.69	2.93	3.31	7.14	0.85	5.49	4.95	4	2.39	3.44	3.39	7.04
3.74	7.48	4.55	3.99	6.08	4.49	0.69	4.25	−0.28	1.39	1.94	6.92

(2) Is there any observations that can be considered unreasonable or special with respect to the general nature of the data?
(3) Is there any special reason to study data for normal-tumour pair together instead of studying them individually?
(4) Can we have any idea of the general nature of the data with respect to central tendency, dispersion, or any other feature?
(5) Can we compare the expression level averages or their measures of dispersion?
(6) Can we think of the same set of questions (1)–(4) for expression of two genes as given in Table 5.2?
(7) Is there any reason to believe that the activity of Gene 1 has no relation with that of Gene 2? Is it possible to validate the belief using some statistical procedures or more objectively?
(8) Can we think that the observations come from a normal distribution? Or do there exist normal random variables from which these samples are drawn? This

question is for a technical reason, but has a great impact on the methods of data analysis.

(9) Can we somehow guess or predict the expression of Gene 2 when we have information on Gene 1? If so, how reliable would be our prediction?

Other questions can be asked for, but we concentrate to address these queries. Once this is done, the reader will get an idea how to address other relevant questions based on the given data set.

5.3 Visually Inspecting the Data

Only visualisation itself sometimes gives a lot of information about the general nature of the data that can be used as a base knowledge to do further analysis. Here also we consider only log-transformed values, i.e. the data given in Tables 5.1 and 5.2 are already log-transformed. This transformation dampens the effect of outliers to some extent at least and provides a rough idea about the underlying probability distribution of the variable. Choice of such a transformation is decided objectively; usually, we consider Box-Cox transformation.

Each of these data sets is a set of frequency data for a continuous variable. First, we can study histograms and boxplots for gene expression in normal and tumour tissues separately. Similarly for Table 5.2, we can think of same diagrammatic representation for the two genes separately. We draw histograms using relative frequency density to make an adequate comparison because the area under the histogram would be equal to one. This histogram would provide a sense of probability distribution. Also note that here we first study histograms for each gene although observations are given in pairs for each individual. Later on, we move on to discussing presentation of two genes together. We start with the simplest possible diagrams.

5.3.1 Histogram and Boxplot

First, we draw two histograms using relative frequency density for direct comparison of two frequency distributions. Here, we should have a pair of observations for each unit or individual, i.e. we collect data for normal and tumour samples from the same individual or data for expression of two genes from each individual. However, due to many reasons, a few observations might sometimes be missing for one variable, while a few for other variables. This unwanted but unavoidable situation leads to unequal sample sizes for the two variables (normal and tumour or corresponding to two genes). So, it is recommended that we draw histograms using relative frequency density so that we can compare two samples visually through histograms.

If we look at the histograms (Fig. 5.1) carefully, we observe that the gene expression values for normal samples fall in the range bewteen 0 and 8, whereas the range

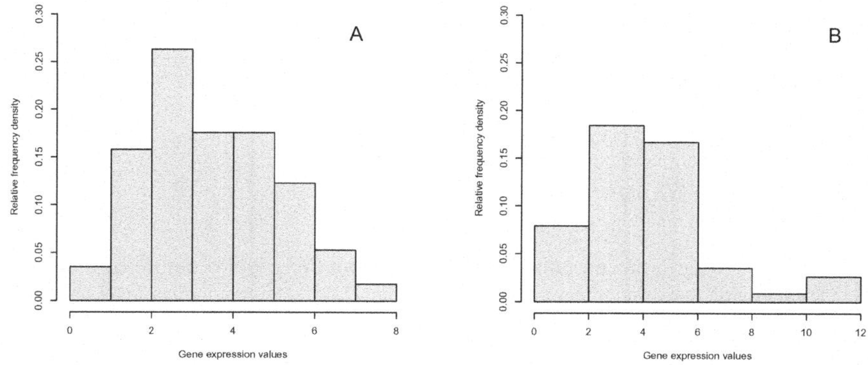

Fig. 5.1 Histograms of gene expression data for **A** normal samples, and **B** tumour samples

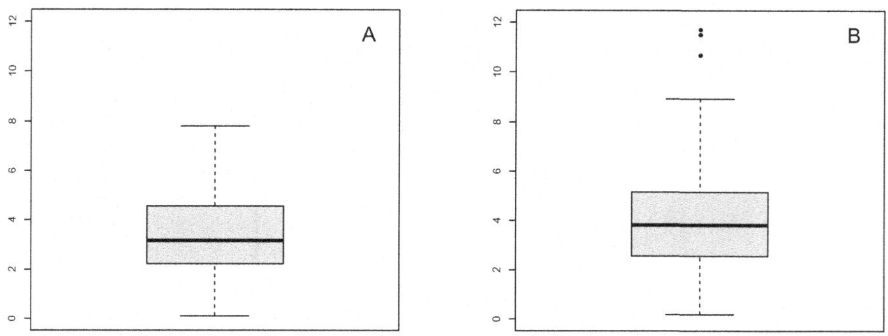

Fig. 5.2 Boxplots of gene expression data for **A** normal samples, and **B** tumour samples

for tumour samples is 0–12. However, whether the overall gene expression level is significantly large in tumour compared to normal tissue is an important question. This question should be addressed objectively through an appropriate statistical method of testing, which we shall discuss later in this chapter. The histogram reveals that relative frequency spectrum has a peak in the middle (slightly left) and it falls gradually on either side of the peak for both samples. However, it is not clear only from histograms whether the distributions are very close to normal distributions. Moreover, although gene expression values for tumour samples extend more towards the right, it fails to provide any information about whether any or some observations are unusually large. That means histogram sometimes cannot give a clear idea whether there are few outliers in the data. For this, we have to draw boxplots.

A comparison of two boxplots (Fig. 5.2) reveals that the gene expression values for tumour samples have more dispersion compared to normal samples and cover a wider range of values. They are more variable. More variability means more movement. We need to interpret that in the context of the given problem. Median for tumour samples is slightly larger than that for normal samples, but whether this difference is

significant needs more sophisticated statistical analysis. Moreover, a few outliers are also present in tumour sample data. Thus, along with histogram, sometimes boxplots are necessary to get some more information, especially for the detection of outliers.

5.3.2 *Finding Relationship Between Genes*

However, these diagrams can only give an idea about the general nature of the data, about its location, extent of spread, presence of outliers, shape of the distribution, etc., separately for each gene or under each condition. But it fails to throw any light on the need of studying them together since we have considered one data set at a time without considering what happens to the other. It completely ignores the movement of one data with respect to another or any kind of relation between the two. Note that, for each individual (say i), we can always have data (x_i, y_i) for expressions of two genes or for expression of the same gene in two types of tissues.

Let's have a look at a very simple plot that requires only common sense, not any complicated statistical or mathematical idea. From Table 5.1, just plot the points (x_i, y_i), i.e. gene expression value in tumour (y_i) against that in normal tissue (x_i) in a graph as given Fig. 5.3.

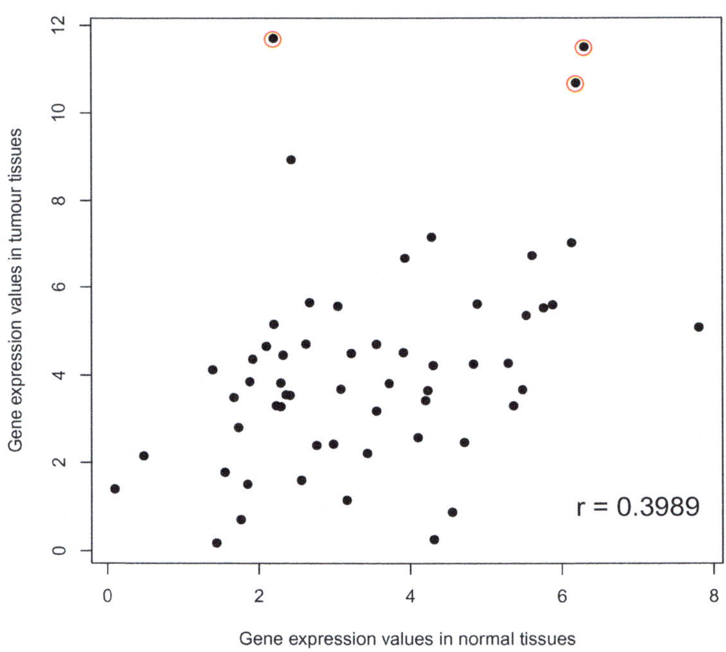

Fig. 5.3 Scatter diagram of gene expression in normal-tumour samples

The general pattern of the data cloud is to move from left the bottom corner to the top right corner with a substantial amount of spread. This clearly shows that expression in tumour tissue increases with the expression in normal tissue. Thus, we are left with no option but to study these two genes together; otherwise, we risk losing a lot of information on how one acts with respect to the other or the influence of one variable on the other. This diagram, obtained just by plotting the points in a two-dimensional rectangular coordinate system, is known as '**scatter diagram**'.

Let us now study this scatter diagram. This is a very simple but extremely important diagram involving two variables, having immense significance in statistical analysis. Many interesting and informative features can be identified throughout this diagram. Clearly, the scatter diagram shows that the expression of gene in tumour tissues has a tendency to increase with that in normal tissues. This pattern will be more evident if we ignore the so-called 'outliers' (red encircled points in Fig. 5.3).

A scatter diagram represents the movement of a variable with respect to another variable. This type of tendency may be more or less prominent depending on data for different pairs of variables. Naturally, one may encounter several other patterns as described through a few other scatter diagrams in Fig. 5.4. We may have an increasing tendency of one variable with the other (Fig. 5.4A) meaning that as x value increases, corresponding y value also increases. Similarly, Fig. 5.4B presents a decreasing tendency. On the other hand, there might exist a stagnant relation meaning that one variable behaves completely independently or indifferently of whether another increases or decreases (Fig. 5.4C). Another interesting type of relation may also be revealed depending on the data. Figure 5.4D presents a typical scenario where, as the value of one variable increases, the value of another variable first decreases (increases) and after a certain value starts increasing (decreasing). Figure 5.4D presents such a scenario. Clearly, these types of relations indicate the presence of what is called 'linear relation' or 'near linear relation'. If one variable behaves in a non-linear fashion with the movement of the other variable, it also indicates no or very weak linear relation; however, a relationship is there!

After studying these diagrams, we are sure that there is a very important concept that needs to be studied thoroughly. If we look at the scatter diagrams (Fig. 5.4) carefully, we see that Fig. 5.4C represents no relation, whereas Fig. 5.4D indicates the existence of a non-linear, more specifically, a quadratic relationship between two variables x and y. Moreover, patterns in Fig. 5.4A and that in Fig. 5.4B are similar as per as the trend is concerned (one increasing and another decreasing), but it is clear that points are more dispersed in Fig. 5.4B. So only the pattern or existence of a relationship is not enough; strength or extent of this relationship is also important and we should not ignore that.

So, we need to devise a quantity that will measure this relationship between two variables; rather the degree of linear relationship between two variables. Hence, we define a quantity, known as '**correlation coefficient**', denoted by r_{xy} or r_{yx} or simply r, when there is no confusion about the variables under study. The formula for correlation coefficient is given as

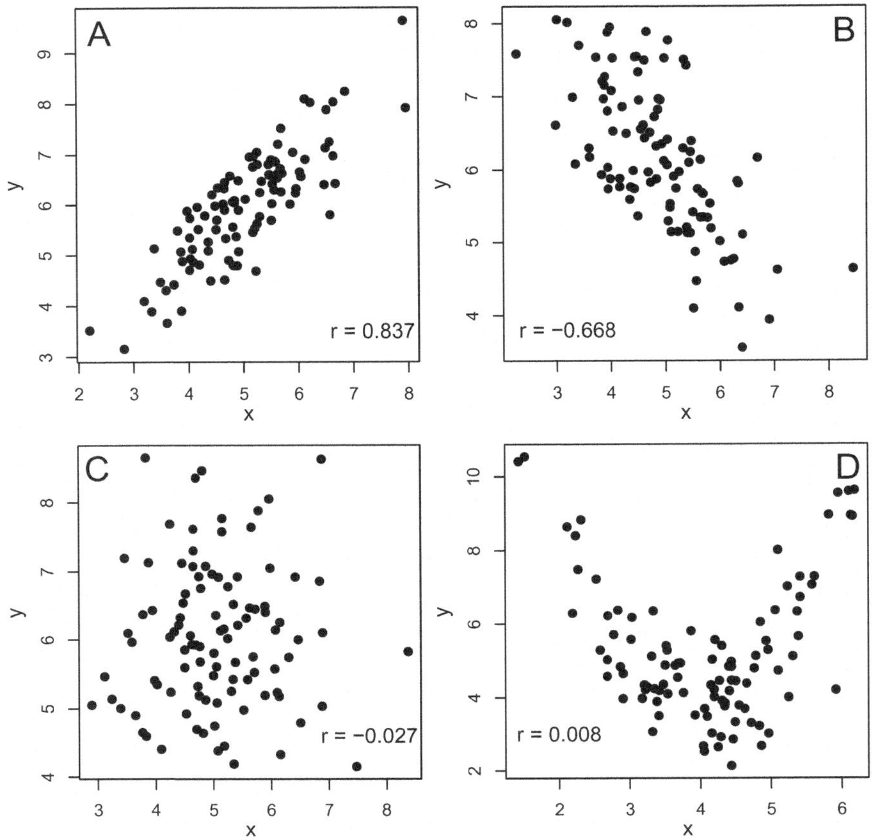

Fig. 5.4 Scatter diagrams: **A** positive linear relation, **B** negative linear relation, **C** no relation, **D** non-linear relation

$$r = \frac{\frac{1}{n}\sum_{i=1}^{n}(x_i - \bar{x})(y_i - \bar{y})}{\sqrt{\frac{1}{n}\sum_{i=1}^{n}(x_i - \bar{x})^2}.\sqrt{\frac{1}{n}\sum_{i=1}^{n}(y_i - \bar{y})^2}} = \frac{cov(x, y)}{s_x s_y}$$

where $cov(x, y) = \frac{1}{n}\sum_{i=1}^{n}(x_i - \bar{x})(y_i - \bar{y})$ is known as 'covariance', an important measure to be used later in statistical analysis. Note that correlation coefficient is just the scaled covariance, the standard deviations of x and y are being used as scales.

Because of this scaling and using deviation from mean of the raw values instead of considering the raw values directly, r enjoys a nice property that it always lies between -1 to $+1$, i.e. $-1 \leq r \leq +1$. This result has a huge significance; whatever

be the sets of values of two variables, it doesn't matter whether they are large or small or a mix of two, correlation coefficient cannot lie beyond these limits. If we go back to the paired data for which scatter diagrams were drawn, we see that the value of correlation coefficients are 0.837 and −0.668 for Figs. 5.4A, B respectively. These values clearly indicate the nature and degree of linear relationship. Figure 5.4A revealed more linearity that Fig. 5.4B in terms of explaining linear relationships. On the other hand, correlation coefficients are very close to 0 for both Figs. 5.4C, D, although the first one shows no relationship, whereas the second one indicates a strong presence of non-linear relationship. This justifies the notion that correlation coefficient represents the '*degree of linear relationship*' and *not any relationship* between two variables. Also note that the value of r is 0.3989 for data in Fig. 5.3, whereas it is 0.837 for Fig. 5.4A. This is a clear indication that higher the value of r, the two variables are more strongly linearly related. This in turn explains the more apparent scatteredness in Fig. 5.3.

This discussion triggers another possibility that there might exist some non-linear relations for which r would be approximately or exactly 0. In such a situation, we can use other measures like '**correlation index**' or '**correlation ratio**' to determine the strength of the non-linear relationship. However, we are not doing a detail discussion about these measures as we rarely need them in practice.

5.3.3 Regression

When two variables are linearly related, it is natural to think whether the knowledge of the value of one variable can be used to guess the value of the other variable. The stronger the relationship, better would be the guess. It is also natural to have an idea about the extent of linear relationship between two variables in the sense that how well we can predict one variable using the other or whether we can predict at all. A scatter diagram supported with evidence through the value of correlation coefficient value would guide us in deciding whether we can predict one variable using the other. This prediction should be done through a mathematical or statistical model. Here, by model, we mean that two variables are related by a mathematical equation (at least approximately). Any pattern other than completely scattered points might be indicative of doing such an exercise of prediction or explaining the relationship between two variables through a model.

For this, we can try to fit a very simple equation, i.e. a straight line through the scattered points in an optimum way so that this line would represent the phenomenon depicted through (x, y)-points, in a 'best' possible way. All points on the (x, y) plane need not lie on a straight line, because there are usually intrinsic errors in measurement or other factors that influence the observations, even if the two variables are highly linearly related. However, we may expect that points should be close to the fitted straight line. Closer the points, stronger would be the relationship revealed through the fitted equation.

If y_i be the observed value and Y_i is the so-called 'true value' corresponding to the i-th unit, then $y_i - Y_i$ is the difference that is due to other factors, including errors in measurement. This error component is an essential part of any model. To make it more clear, we can say that a mathematical model has two parts: (1) a part that can be explained through evidence (in this case data) and (2) another part that remains unexplained however rigorous the model might be. Since we don't know the actual model representing the relationship between two variables, we will never know it, however large data sets we might collect. So, we are only interested in reducing the error part of the model, that too based on our data (available evidence); it can never be reduced to zero or even if so, we will never know.

Clearly, larger the difference between the true (unknown) value and the observed value, larger would be the contribution of error part. Thus we are interested in the distance between observed and true values and hence we can take $|y_i - Y_i|$ or $(y_i - Y_i)^2$ as a measure of such distance. Now an average or sum of these distances for all observations can give us an idea of the overall deviation from the true values. The principle in finding the line of best fit, the linear regression equation, passing through a scatter of points is to minimise this 'average distance' and estimate the unknown quantities that are involved in the linear equation. Figure 5.5 provides a detailed pictorial representation of this scenario.

Let $y = \alpha + \beta x$ be the true equation between x (independent variable or explanatory variable) and y (response or dependent variable or study variable) and so for the i-th unit, we have the model:

$$y_i = \alpha + \beta x_i + e_i; \;\; i = 1, \ldots, n$$

where e_i is the error or the amount of deviation from the true value associated with y_i. Naturally, e_i is unknown and so we have to minimise some summary measures of e_is to get the estimates of α and β. So, we minimise $\sum_{i=1}^{n} e_i^2 = \sum_{i=1}^{n} (y_i - Y_i)^2$ and get the estimates of α and β, denoted by $\hat{\alpha}$ and $\hat{\beta}$ respectively, as follows:

$$\hat{\alpha} = \bar{y} - \hat{\beta}\bar{x} \;\; \text{and} \;\; \hat{\beta} = \frac{\sum_{i=1}^{n}(x_i - \bar{x})(y_i - \bar{y})}{\sum_{i=1}^{n}(x_i - \bar{x})^2} = \frac{cov(x, y)}{var(x)}$$

The fitted regression equation using the estimates of α and β becomes:

$$Y = \hat{\alpha} + \hat{\beta}x.$$

Figure 5.5 shows the best possible linear regression of y on x in the sense that the purple line minimises the error sum of squares for all values. However, a careful look reveals some more interesting facts. Although we claim that this would be the best fit (in least square sense), there are more points below the purple line than above

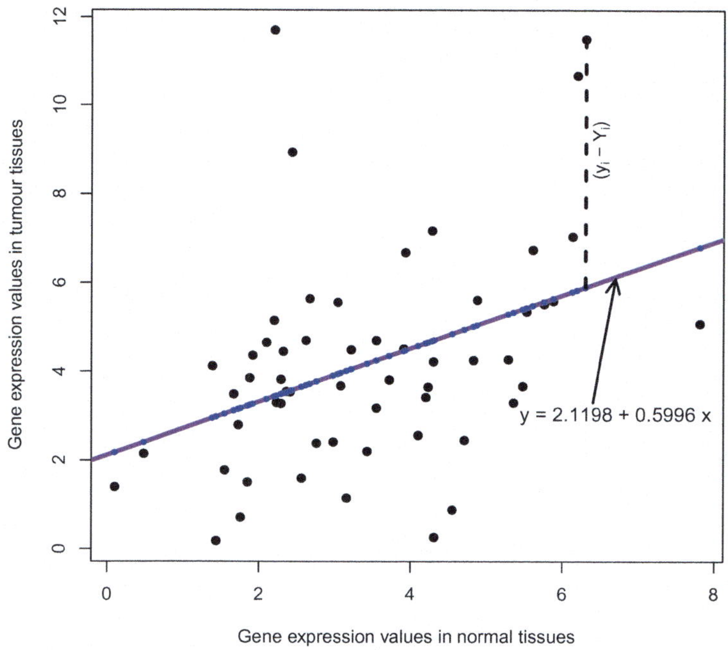

Fig. 5.5 Least squares regression line to predict gene expression tumour using normal

it (Fig. 5.5). The reason is the presence of outliers (seen from boxplot Fig. 5.2B). So, we do the same exercise of removing the outliers and obtain Fig. 5.6. Here, the purple line is the least square regression that is obtained using all observations, whereas the red dotted line represents the same without outliers.

Thus Fig. 5.6 reveals the impact of outliers. It is clear from this analysis that only a few outliers can shift the regression line. In many biological data, the impact of outliers is sometimes very significant; mere presence of only a few such values can make a substantial change in the regression lines. It may give a completely misleading picture if we routinely or blindly do regression analysis using a data set without any exploratory analysis of data. Thus, we should always be cautious; first try to see and understand the data points very carefully, may be using some diagrammatic representation, and then take actions accordingly. There are other types of effects of outliers; but in each situation, the decision can be altered by its presence. It is to be noted that the number as well as the values of outliers play a very important role in statistical analysis.

If we want to fit a linear regression of y on x, we can use R code.

```
> lm(y ~ x)          ## fitting a linear regression model of y on x
```

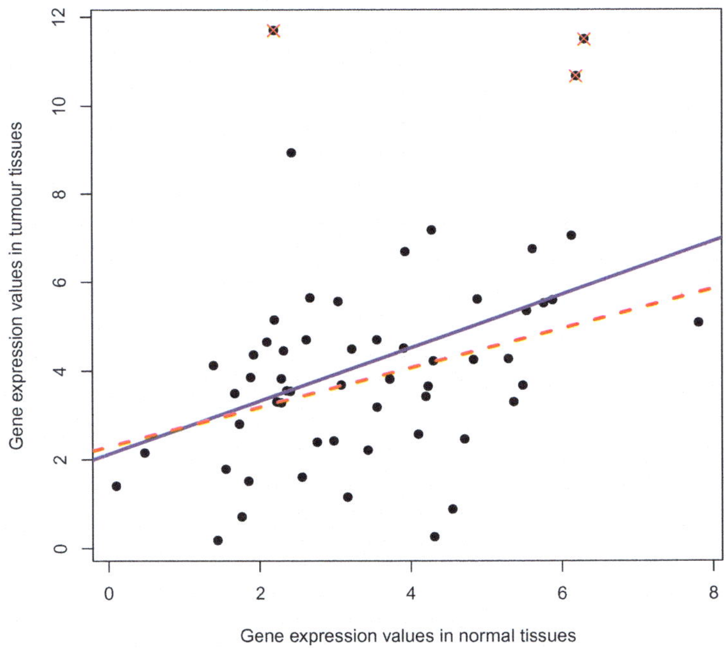

Fig. 5.6 Least squares regression line to predict gene expression tumour using normal: purple line is for with outliers, and red line is for without outliers

The R code is very simple. However, it is better to store the entire output and extract the relevant information as and when required. A summary of output is available using R code. It is also important to see the scattered points in the diagram along with the fitted regression equation.

```
> a<- lm(y~x)          ## stores output of lm() function in 'a'
> summary(a)           ## displays summary of lm() function
> plot(x,y)            ## displays scatter diagram
> abline(a)            ## superimposes fitted linear regression equation on
                          scatter diagram
```

We shall discuss a few more issues, especially regarding testing in the context of regression analysis later in this chapter. A curious reader might ask a very relevant question at this point. If we recall the pattern given in Fig. 5.4D, it clearly indicates that y (variable in vertical axis) is related as a quadratic equation in x (variable in horizontal axis). A linear regression cannot be a good fit to this data. But we can fit a model that is a quadratic equation in x, as

$$y_i = \alpha x + \beta x_i + \gamma x_i^2 + e_i, \quad i = 1, 2, \ldots, n$$

Although y is modelled as a quadratic function of x, this model still belongs to the class of 'linear models' because this is a linear function of parameters present in the equation that explains the relationship of y using x.

The same 'lm()' function of R code will work. Unfortunately, R function 'abline()' will not work here as the fitted equation is not a straight line. However, using R commands, we can easily calculate the predicted value based on fitted regression equation and plot on the same scatter diagram. Note that $a\$coefficients$ gives the estimated values of the parameters α, β, and γ in the same order from the output of 'lm()' function.

```
> z<- x^2                 ## stores x^2 values in z
> a<- lm(y~ x+z)                ## stores output of lm() function in 'a'
> summary(a)             ## displays summary of lm() function
> plot(x,y)             ## displays scatter diagram
> y.pred<- a$coefficients [1] + a$coefficients [2]*x + a$coefficients [3]*z
                                          ## calculates predicted
                                              ## y values
> lines(x,y.pred)            ## superimposes fitted equation on scatter diagram
```

5.4 Some Diagnostic Testing Problems for Paired Data

To decipher the complex interplay of two genes together or study the nature of one gene in two different tissues in a dependent set-up, we first need to determine whether, at all, we should consider the two together or separately. Whatever be the situation, we can also look for whether the nature of two genes is same in terms of their expression values. Variations in observations ensure that there are two underlying random variables that occur together although each one has its own characteristics at the same time. Thus, one important technical question arises in this situation. What is the probability distribution of the two variables taken together? In other words, we want to have an idea about the joint probability distribution of two variables representing expression values of both the genes in the same tissue or same gene in two different tissues. The joint probability distribution of two variables forms the basis of data analysis. To answer the questions, more specifically, Questions 2–7 in Sect. 5.2, we need to study them in the light of hypothesis testing in statistics. Note that the gene expressions for two genes constitute two populations that might depend on each other. Similarly, we can think that two underlying distributions exist for gene expression of the same gene in two different tissues along with a correlation between them, since pairs of observations come from the same individual. There are many such examples in biological experiments where this type of data is very natural.

5.4.1 Test for Normal Distribution

Before undertaking a study on any interesting feature of the two populations, we need to understand the underlying probability distributions. Based on our knowledge so far, it seems that if we can ensure that they follow normal distribution, our job will be simpler. In such situations, we can easily apply many standard statistical tests while addressing many relevant questions.

However, assuming that the two populations might be related, we have to study the joint probability distribution of the expression of two genes rather than looking at the distribution of individual genes. We know that a normal distribution is the most common probability distribution for a continuous random variable. In a bivariate situation, i.e. when we have two continuous variables that seem to be related to each other, we can think of an extension of a univariate normal distribution. This extension or generalisation in the case of a bivariate set-up is known as 'bivariate normal distribution'. Let us assume that (X, Y) jointly follow a bivariate normal distribution having a p.d.f. as

$$f(x, y) = \frac{1}{2\pi\sigma_1\sigma_2\sqrt{1-\rho^2}} e^{-\frac{1}{2(1-\rho^2)}\left\{\left(\frac{x-\mu_1}{\sigma_1}\right)^2 - 2\rho\left(\frac{x-\mu_1}{\sigma_1}\right)\left(\frac{y-\mu_2}{\sigma_2}\right) + \left(\frac{x-\mu_2}{\sigma_2}\right)^2\right\}}, \quad -\infty < x, y < \infty,$$

where $\mu_1 = E(X), \mu_2 = E(Y), \sigma_1^2 = V(X), \sigma_2^2 = V(Y)$, and ρ is the population correlation coefficient between X and Y. We denote this by $(X, Y) \sim N_2(\mu_1, \mu_2, \sigma_1^2, \sigma_2^2, \rho)$. A few interesting and useful properties of a bivariate normal distribution are given in the following theorem.

Theorem 15 *If* $(X, Y) \sim N_2(\mu_1, \mu_2, \sigma_1^2, \sigma_2^2, \rho)$, *then*

(1) $X \sim N(\mu_1, \sigma_1^2)$, *and* $Y \sim N(\mu_2, \sigma_2^2)$,
(2) $a + bX + cY \sim N(a + b\mu_1 + c\mu_2, b^2\sigma_1^2 + 2bc\rho\sigma_1\sigma_2 + c^2\sigma_2^2)$ *for any real constants* b *and* c, *and*
(3) X *and* Y *are independent if and only if* $\rho = 0$.

The first part of Theorem 15 implies that for a bivariate normal distribution, each of the individual variables marginally follows a univariate normal distribution. However, the converse may not be true, i.e. if each of X and Y follows a normal distribution with some mean and some variance, their joint distribution may not always follow a bivariate normal distribution. The second part of Theorem 15 implies that any linear function of two univariate or bivariate normals follows a univariate normal distribution. This is a very useful result and has extensive use in statistical inference.

Equipped with the above information, when we have two variables that might be related, we should check whether expressions for each of the variables follow a normal distribution. This should be the first step in the data analysis.

Let X and Y denote the random variables corresponding to the expression values in two populations. Although they have a joint (bivariate) distribution, we first test $H_0 : X \sim N(\mu_1, \sigma_1^2)$ for some μ_1 and σ_1^2, against $H_1 : X \nsim N(\mu_1, \sigma_1^2)$. We can use Kolmogorov-Smirnov test or Shapiro-Wilk test for normality as discussed in

Sect. 4.3.9. Similarly, we can also test $H_0 : Y \sim N(\mu_2, \sigma_2^2)$ for some μ_2 and σ_2^2 against $H_1 : Y \nsim N(\mu_2, \sigma_2^2)$. If either of these tests are rejected, we can conclude that the variable (corresponding to test that is rejected) does not follow a normal distribution. Hence, we can definitely conclude that the joint probability distribution of (X, Y) is not normal, or rather not 'bivariate normal'.

However, acceptance of both the hypotheses does not confirm that the joint distribution is bivariate normal, because there are a few situations where the joint distribution is not normal although the marginal distribution of each of the variables is normal. Checking normality through the Kolmogorov-Smirnov test or Shapiro-Wilk test would not suffice; we have to do another test which would directly check whether X and Y jointly follow a bivariate normal distribution, i.e. $H_0 : (X, Y) \sim N_2(\mu_1, \mu_2, \sigma_1^2, \sigma_2^2, \rho)$ for some $\mu_1, \mu_2, \sigma_1^2, \sigma_2^2$ and ρ.

Literature survey reveals that there are quite a few statistical tests that can check whether a bivariate data conforms to bivariate normal distribution. We briefly discuss three tests for checking bivariate normality. Let $\{(X_i, Y_i),\ i = 1, \ldots, n\}$ be n pairs of random observations from (X, Y). Our test of hypotheses are: $H_0 : (X, Y) \sim N_2(\mu_1, \mu_2, \sigma_1^2, \sigma_2^2, \rho)$ against $H_1 : (X, Y)$ do not follow bivariate normal distribution.

5.4.1.1 Doornik-Hansen Test

Data are given for two variables as paired data, i.e. for each unit, we have observations for both variables. We always calculate sample means and sample variances for both variables. Since each datapoint is a paired data, correlation coefficient is another feature which we must consider in order to understand the salient features of bivariate data.

We know that normal distribution is also symmetric and has moderate kurtosis; these measures should also be taken care of while constructing appropriate test statistic for checking bivariate normality. Doornik and Hansen 2008 considered all these measures and proposed a test statistic. Another interesting feature of this test is that it follows a χ^2 distribution with 4 degrees of freedom asymptotically, i.e. for a large sample size, the probability distribution of the test statistic can be approximated by that for a χ^2 variable with 4 degrees of freedom. The p-value is calculated using this asymptotic distribution. Based on data in Table 5.1, we see that the value of Doornik-Hansen test statistic is 7.6083 with corresponding p-value as 0.107. It is important to note that this test is very sensitive to outliers and we have removed outliers from the data before implementing the test. To apply this test, one must install a special package 'normwhn.test' that contains the R-function for Doornik-Hansen test. R function 'normality.test1()' is applied on x, which contains the data matrix, with columns as values for respective variables.

```
> library("normwhn.test")
> normality.test1(x)
```

5.4.1.2 Generalised Shapiro-Wilk Test

Based on sample means, variances and covariance of paired observations, Alva and Estrada (2009) generalised Shapiro-Wilk test to check for multivariate normality. Since here we are interested in bivariate normality, we describe briefly the test for bivariate observations.

Let $\{(X_1, Y_1), (X_2, Y_2), \cdots, (X_n, Y_n)\}$ be the n pairs of observations corresponding to (X, Y). Define sample variance-covariance matrix, consisting of sample variances and covariance as

$$S = \begin{pmatrix} s_{xx} & s_{xy} \\ s_{xy} & s_{yy} \end{pmatrix}$$

where $s_{xx} = \frac{1}{n}\sum_{i=1}^{n}(X_i - \bar{X})^2$, $s_{yy} = \frac{1}{n}\sum_{i=1}^{n}(Y_i - \bar{Y})^2$, $s_{xy} = \frac{1}{n}\sum_{i=1}^{n}(X_i - \bar{X})(Y_i - \bar{Y})$. Consider another symmetric positive definite matrix C such that $CSC' = I_2$, where I_2 is an identity matrix of order 2.

Now we define new variables Z_1 and Z_2 such that

$$\begin{pmatrix} Z_1 \\ Z_2 \end{pmatrix} = C \begin{pmatrix} X - \bar{X} \\ Y - \bar{Y} \end{pmatrix}.$$

Generalised Shapiro-Wilk statistic is defined as

$$W^* = \frac{1}{2}(W_1 + W_2),$$

where W_i is the univariate Shapiro-Wilk statistic based on Z_i for $i = 1, 2$.

This test is a generalised one that tests whether bivariate normal distribution fits to a data set. It is not possible to get a compact analytical expression for the distribution of the test statistic even when H_0 is true. However, there is an in-built R code available for computation of p-value using W^*. Based on the data in Table 5.1, we obtain the value of the test statistic as 0.9783 with corresponding p-value 0.3836. This test is also very sensitive to outliers and we have applied it after removing outliers. Here also we must first install the package 'mvShapiroTest' before applying the R code and then use the command 'mvShapiro.test()' for testing bivariate normality of the data.

```
> library("mvShapiroTest")
> mvShapirto.Test(x)
```

5.4.1.3 Royston Test

Royston proposed another test for bivariate normality using a modification of Shapiro-Wilk test (Royston 1982; Royston 1992). If the kurtosis of the data is greater than 3, it uses Shapiro-Francia test for (Shapiro & Francia 1972) leptokurtic distributions. On the other hand, if the kurtosis is less than 3, it uses Shapiro-Wilk test for platykurtic distributions. We apply this test to our data in Table 5.1 and obtain the value of the test statistic as 1.2722 with corresponding p-value 0.5316. This test is also sensitive to outliers. However, one must install the package 'royston' before using the R code.

```
> library("royston")
> royston.test(x)
```

There are a few other tests available in the literature. The performance of the above three test statistics is almost same. More importantly, they are sensitive to outliers present in the data. So, it is always advisable to remove the outliers from the data set before applying these tests for normality.

Sometimes, we collect data where each data point has more than two components. These are known as multivariate data. All the above tests can be used to test for multivariate normality, the analogue version of normal distribution in case of higher dimension. Thus statistics literature is very rich in providing some useful tests that should be used during any data analysis, be univariate or bivariate or multivariate. We shall discuss the appropriate statistical analyses for multivariate data later.

We have discussed three such tests using the data given in Table 5.1. But in order to demonstrate the performance of each test, we need to do extensive simulations. We calculate the empirical power and type I error rate for each of these tests assuming that data are coming from different distributions. The distributions chosen cover symmetric as well as positively skewed distribution. It would be an interesting exercise to do some more simulations covering a wider range of distributions.

We first simulate n pairs of bivariate data taking $n = 50, 75, 100, 125$ from a bivariate normal distribution with means 0, 0, variances 1, 1, and covariance 0.6. Apply each test to calculate the test statistic and its corresponding p-value. Repeat this experiment for 1000 times and calculate the proportion of times the p-value is less than 0.05, the chosen type I error rate. It is expected that the type I error rate should be close to 0.05 to ensure that each test really preserves the level of significance as an upper bound.

To calculate power, we generate data from the same bivariate normal distribution and then consider the absolute values of observations generated. If y is an observation, we take $|y|$. This would introduce positive skewness in the new data set. Naturally, now the data cannot be claimed as a random sample from a bivariate normal distribution. We calculate test statistic and its corresponding p-value for each test. Based on 1000 repetitions, we calculate the proportion of times p-value is smaller than 0.05, which would give the empirical power. If the data come from a null distribution, i.e. bivariate normal distribution itself, this proportion would give the type I error rate of

Table 5.3 Simulation study to compare type I error rate and power for Shapiro-Wilk, Doornik-Hansen, and Royston tests

	Shapiro-Wilk				Doornik-Hansen				Royston			
n	50	75	100	125	50	75	100	125	50	75	100	125
Type I error	0.049	0.044	0.053	0.04	0.046	0.046	0.054	0.051	0.07	0.067	0.066	0.062
Power	0.762	0.962	0.994	0.999	0.668	0.914	0.981	0.997	0.845	0.981	1	1

the test. Table 5.3 gives a comparative study of these three tests in respect of power and type I error rate.

It is observed that all three tests are very powerful even for a moderate sample size. However, Royston test has slightly inflated type I error rate, but at the same time, it is more powerful. So any of these three tests can be applied in practice. The main reason for the more or less same performance of all three tests is probably that here we are dealing with bivariate data, i.e. each data point has two dimensions only. This might be more prominent if we consider multivariate data with higher dimensions. In any case, we can use any of these tests to check normality in data having more than one dimension.

5.4.2 Are Genes Correlated?

In order to start our investigation about the behaviour of genes, we have to first understand whether the two genes, in some sense, are acting together or one gene might influence the other, or the same gene in two different tissues forms bivariate data with some correlation between them. As we see that correlation coefficient is a nice measure that can determine the degree or extent of linear relationship between two variables, we start with this measure. Note that use of correlation coefficient only ensures the existence of linear relationship between two variables.

To see whether the two variables are related, we can do a statistical test to check any absence of correlation. So, we can propose a null hypothesis $H_0 : \rho = 0$ against $H_1 : \rho \neq 0$. Here, ρ is the population correlation coefficient between two variables of interest. Naturally, we can think that correlation coefficient value calculated on the basis of sample observations would provide information about ρ, since the samples are drawn at random from the population. Thus, we can consider sample correlation coefficient r as a test statistic and reject H_0 if the observed r is too large or too small compared to 0, the value of ρ specified by H_0. Although assuming bivariate normal distribution, we can find the probability distribution of r when H_0 is true, it is slightly difficult to calculate the p-value. However, a small trick makes our life simple!

First, let's assume that (X, Y) jointly follow a bivariate normal distribution. Then, we can take another function

$$\psi(r) = r\sqrt{n-2}/\sqrt{1-r^2},$$

which is a non-decreasing function of r. Naturally, the fact that "r is too large or too small" is equivalent to "$\psi(r)$ is too large or too small". Moreover, it can be shown that $\psi(r)$ follows a t-distribution with $n-2$ degrees of freedom, when H_0 is true.

Thus, the p-value can be easily calculated using t-distribution with $n-2$ degrees of freedom. In our data set of normal-tumour gene expression, $r = 0.4227$ and hence $\psi(r) = 3.6126$ and the corresponding p-value is 0.00062. Since p-value is extremely small compared to the chosen level of significance 0.05, we can immediately say that on the basis of the given data set (Table 5.1), it seems that there is a strong correlation between the expression values of the two genes.

Based on our discussion in the previous two subsections, we can safely assume the expression values (X, Y) follow a bivariate normal distribution with means μ_1, μ_2, variances σ_1^2, σ_2^2, and correlation coefficient ρ. Note that all calculations are done without removing outliers present in the data, as we see from boxplots previously. However, even if remove outliers, the p-value for testing $H_0 : \rho = 0$, is 0.00143, which is still very small and hence there is no change in the conclusion. It is always recommended to understand the presence and importance of outliers before doing any test.

R code for the above testing problem is given below, where x and y are two series of observations for which we want to check $H_0 : \rho = 0$. Note that here the alternative hypothesis is $H_1 : \rho \neq 0$ and hence we have to do two-sided test. Since t-distribution is symmetric about 0, we can calculate the p-value based on the right tail probability and then multiply by 2.

```
> r<-cor(x,y)          ## calculates sample correlation coefficient
> test.statistic<-r*sqrt(length(x)-2)/sqrt(1-r*r)          ## calculates test
                                                             statistic ψ(r)
> p.value<- 2*pt(abs(test.statistic),length(x)-2,lower.tail=F)       ## calculates
                                                                        p-value

> cor.test(x,y)          ## test for zero correlation coefficient using
                            in-built R function
```

This test is developed under the assumption that the two variables jointly follow a bivariate normal distribution. However, when this assumption is invalid, we should not apply this test. In fact, to check whether the two variables are related, we cannot even do a test for the absence of correlation because that may not ensure the independence of the variables. Only in the case of bivariate normal distributions, it is necessary and sufficient that zero value of correlation coefficient indicates that there is no relation between the two variables. However, if the test for bivariate normality fails, we cannot use the above t-test to ensure independence between two variables. We have to look for another statistical test suitable for a general set-up.

5.4.3 Test of Independence

Based on the discussion in the previous section, it is clear that we have to consider a general test of independence in order to see whether two variables are related or independent. But before that, we first explain the notion of independence mathematically and then propose an adequate test.

5.4.3.1 What is Independence?

The joint probability distribution of two variables X and Y is denoted by $F_{X,Y}(x, y)$ where

$$F_{X,Y}(x, y) = P(X \leq x, Y \leq y) \text{ for any } x \text{ and } y.$$

This is known as "joint probability distribution function" of (X, Y). On the other hand, we can always think of the marginal probability distribution of the variables X and Y separately as

$$F_X(x) = P(X \leq x) \text{ for any } x \text{ and } F_Y(y) = P(Y \leq y) \text{ for any } y.$$

These are called "probability distribution function" of respective variables. Naturally, there should be a relation between the joint distribution function and the marginal distribution functions.

From the basic probability theory, we know that two events A and B are said to be *independent* if $P(A \cap B) = P(A).P(B)$. Now take $A = \{X \leq x\}$ and $B = \{Y \leq y\}$. Then from the notion of independence of events, we can immediately say that two variables X and Y are *independent* or *independently distributed* if

$$F_{X,Y}(x, y) = P(X \leq x, Y \leq y) = P(A \cap B) = P(A)P(B) = F_X(x)F_Y(y) \quad \text{for all } (x, y) \in \mathscr{R}^2.$$

Here, $\mathscr{R}^2$ denotes the two-dimensional Euclidean space or simply a two-dimensional plane that we know from high school level.

As a consequence of the above definition of independence, we can simplify it when both X and Y are discrete random variables with p.m.f.s $f_X(x) = P(X = x)$ and $f_Y(y) = P(Y = y)$ respectively or both are continuous variables with p.d.f.s $f_X(x)$ and $f_Y(y)$ respectively. Let the joint p.m.f. or p.d.f of (X, Y) is $f(x, y)$. Then the equivalent definition of independence is

$$X \text{ and } Y \text{ are independent } if\ and\ only\ if\ f(x, y) = f_X(x).f_Y(y) \text{ for all } x, y.$$

So given a joint distribution it is important to know whether the two variables are independently distributed. If not, we have no way but to consider them jointly in all subsequent analyses for a better understanding of phenomenon. In most analyses related to genetic or genomic data, two variables occurring from the same experiment

or sometimes otherwise, are related because of several reasons. One such reason might be the synergistic behaviour of joint actions of two variables that is intrinsic to biological mechanism.

Ex-5.1: Let X and Y denote the numbers appearing respectively on die 1 and die 2 when both are rolled once. Clearly, X and Y are independent because outcome on die 2 has no way dependent on the outcome of die 1. Here, $P(X = 2, Y = 3) = P(X = 2).P(Y = 3) = \frac{1}{6}.\frac{1}{6} = \frac{1}{36}$. In fact this is true for any x and y, i.e.

$$P(X = x, Y = y) = P(X = x).P(Y = y) = \frac{1}{36} \text{ for any } x, y = 1, 2, \ldots, 6.$$

Ex-5.2: Suppose $(X, Y) \sim N_2(\mu_1, \mu_2, \sigma_1^2, \sigma_2^2, \rho)$, where $\rho = 0$. Then, for any $(x, y) \in \mathscr{R}^2$, we have putting $\rho = 0$ in the p.d.f. of bivariate normal distribution,

$$f(x, y) = \frac{1}{2\pi\sigma_1\sigma_2} e^{-\frac{1}{2}\{(\frac{x-\mu_1}{\sigma_1})^2 + (\frac{x-\mu_2}{\sigma_2})^2\}} = \frac{1}{\sqrt{2\pi}\sigma_1} e^{-\frac{1}{2}(\frac{x-\mu_1}{\sigma_1})^2} . \frac{1}{\sqrt{2\pi}\sigma_2} e^{-\frac{1}{2}(\frac{x-\mu_2}{\sigma_2})^2} = f_X(x).f_Y(y)$$

where $f_X(x)$ and $f_Y(y)$ are p.d.f.s of $N(\mu_1, \sigma_1^2)$ and $N(\mu_2, \sigma_2^2)$ variables respectively.

Hence, if $\rho = 0$, X and Y are independently distributed in the case of bivariate normal distribution.

Ex-5.3: When we throw a die, a number appears on the upper face of the die. Let A be an event denoting that the number is greater than 4 and B denote the number 6. Here, $P(A) = \frac{2}{6} = \frac{1}{3}$ and $P(B) = \frac{1}{6}$. Now, we need to see whether these two events are independent,

$$P(A \cap B) = P(\text{number appearing is greater than 4 and it is 6}) = P(\text{number is 6}) = \frac{1}{6} \neq P(A)P(B).$$

Hence, A and B are not independent.

5.4.3.2 Nonparametric Test of Independence

If two variables X and Y are independently distributed, correlation coefficient (ρ) between them must be equal to zero. But $\rho = 0$ does not necessarily imply that the two variables are independent. However, it can be easily shown that if the correlation coefficient between two variables is zero, the variables are independent provided they follow a bivariate normal distribution (Ex-5.2). If they have other joint distribution, they may not be independent even though the correlation coefficient is zero.

By now it is clear that if the two variables do not jointly follow a bivariate normal distribution, we cannot do a test for the absence of correlation as this may not necessarily indicate the independence between them. So, now we think in terms of a more general framework. Instead of considering a test for correlation coefficient, we can think a test of independence if accepted, would imply that there is no relation between the two variables.

But why do we do any test for independence? If somehow we can prove that there is enough evidence that underlying variables X and Y (say) have no relation or dependence among them, we can treat them as completely unrelated variables. Hence our problem boils down to considering them individually. All relevant univariate studies would be enough to throw some light on the given scenario. Thus, knowing whether variables are independent, will control the entire downstream analysis and hence the conclusion.

Note that the test for independence is slightly less powerful since we do not assume any particular form of distribution. Naturally, we lose some information that is conveyed by this assumption, but it is better to sacrifice this information rather than wrongly assuming a distribution which would give more adverse effect on the entire statistical analysis and subsequent decision-making process. So, here we concentrate on describing some nonparametric tests that do not depend on normality assumption.

Puri and Sen (1971) developed a few tests to see whether two variables are independent. This test is based on ranks of the observations of X and Y, i.e. based on ranks of $(x_i, y_i),\ i = 1, \ldots, n$. Two test statistics are available in R code. Thus, in our example, if X and Y are two variables indicating gene expression values for normal and tumour samples from the same group of individuals, we want to test whether X and Y are independent, assuming no particular distribution. Thus, our hypotheses of interest are

$$H_0 : F_{X,Y}(x, y) = F_X(x)F_Y(y) \text{ for all } x, y \text{ against } H_0 : F_{X,Y}(x, y) \neq F_X(x)F_Y(y) \text{ for at least some } x \text{ and } y.$$

The p-value corresponding to the test based on normal score is 0.00036, whereas that for rank test is 0.00098. Since in each case the p-value is very small, we can conclude that there is enough reason to believe that the two variables are not independent.

If x is the data matrix for which the first two columns contain the values for x and y, we can use in-built function in R; but, first, we must install the package ‘ICSNP’.

```
> b<-ind.ctest(x,1,2,scores="normal")
> b$p.value
> b<-ind.ctest(x,1,2,scores="rank")
> b$p.value
```

Although this type of nonparametric test for independence is not widely used in practice, but this is extremely important. Usually, people are reluctant in checking the distributional assumption behind every test that they employ in analysis of data; however, it should be remembered that standard tests are valid only under some particular assumptions about the parent probability distribution. The evidence against or in favour of this must be explored very carefully, as the rest of the analysis will completely depend on this study. We emphasise that the Puri-Sen test should be employed to test for independence in case tests for bivariate normality ensures that the distribution is other than normal.

One reason for the reluctance of using these tests frequently by the scientific community is probably that it involves lots of calculation and understanding of the statistical tests. Needless to mention that these tests are complicated compared to standard tests. So one must write computer programs in order to use them in practice, which in turn requires substantial knowledge of programming languages. However, this obstacle has now been overcome completely because of the availability of in-built functions in R.

5.5 Some Standard Paired Sample Testing Problems

5.5.1 Are Two Means Equal?

Once we determine that two variables are correlated and they follow bivariate normal distribution, we can proceed one step further to see whether they are the same (at least approximately) with respect to their central tendencies. Here we assume that two variables $(X, Y) \sim N_2(\mu_1, \mu_2, \sigma_1^2, \sigma_2^2, \rho)$ and we have a random sample $\{(x_1, y_1,), \ldots, (x_n, y_n)\}$. Our hypotheses of interest for testing equality of two means are

$$H_0 : \mu_1 = \mu_2 \text{ against } H_1 : \mu_1 \neq \mu_2.$$

The reason for taking both-sided alternative in H_1 is the complete lack of prior knowledge about the nature of means if they were not equal. We naturally assume the stringent and natural situation where the population variances and correlation coefficient are all unknown.

Note that a simple idea can make development of a test very easy. If we define a variable as $U = X - Y$, then, by the second part of Theorem 15, we can easily show that $U \sim N(\mu, \sigma^2)$, where $\mu = \mu_1 - \mu_2$ and $\sigma^2 = \sigma_1^2 - 2\rho\sigma_1\sigma_2 + \sigma_2^2$. Now since $\mu = 0$ under H_0, we can look at the testing problem as testing $H_0 : \mu = 0$ against $H_1 : \mu \neq 0$ when σ^2 is unknown and $U \sim N(\mu, \sigma^2)$. Naturally, for the i-th individual, we have an observation $u_i = x_i - y_i$ for the random variable U. So, the sample observations are now modified as $\{u_1, u_2, \ldots, u_n\}$. This test boils down to a standard test for mean in case of univariate normal distribution, which is nothing but one sample t-test (discussed in Sect. 4.3.3). Hence, the test statistic would be:

$$t = \frac{\sqrt{n}\bar{U}}{s}$$

and $t \sim t_{n-1}$ under H_0. We reject H_0 if its observed value is too large or too small or the calculated p-value is extremely small. This test is known as 'paired t-test'. Note that although we can express $\bar{U}$ and s^2 in terms of measures based on X and Y

as $\bar{U} = \bar{X} - \bar{Y}$ and $s^2 = \frac{1}{n-1} \sum_{i=1}^{n} (U_i - \bar{U})^2 = s_1^2 - 2rs_1s_2 + s_2^2$, there is no need to evaluate them; we can directly use $\bar{U}$ and s^2 calculated on the basis of $\{u_1, u_2, \ldots, u_n\}$ or use the in-built function in R.

In reference to data in Table 5.1, we denote that x and y represent gene expression values for normal and tumour tissues respectively. Based on our data set, we see that the p-value is 0.0139 for paired t-test. Sometimes, prior knowledge indicates that tumour gene expression may be greater than that for normal tissue. This knowldge, if available with reasonable confidence, may be used to modify the alternative hypothesis. In this case, we may consider alternative hypothesis as $H_1 : \mu_1 < \mu_2$. P-value is calculated based on left tail probability of the null distribution of the paired t statistic. Since p-value is small (although not extremely small) compared to the chosen level of significance 0.05, we conclude there is some evidence in the data set that tumour gene expression is slightly larger than that in normal tissue. Further study is required to see more functionality of this gene in the context of the disease. R code for this paired sample t-test is given below. If we use $H_1 : \mu_1 \neq \mu_2$, we should write "two.sided" instead of "less" inside 't.test' function.

```
> t.test(x, y, paired = T, alternative = "less")        ## summary output of t-test
> t.test(x, y, paired = T, alternative = "less")$p.value      ## p-value of t-test
```

5.5.2 *Test for Locations for Non-Normal Distribution*

The above-discussed paired t-test is a nice way to check whether the two location parameters, as described by means, for a bivariate normal distribution, are equal. However, like the two-sample problem (Sect. 4.3.8), this test may not be valid or may produce an unrealistic result when the random variables jointly do not follow a bivariate normal distribution. We now describe a nonparametric test for location parameters in such a scenario. This does not require knowledge of the bivariate distribution.

As mentioned earlier, it is natural to test whether two medians are equal, as mean may not exist always. Thus, if $\tilde{\mu}_X$ and $\tilde{\mu}_Y$ are two medians for X and Y respectively, our null hypothesis of interest is: $H_0 : \tilde{\mu}_X = \tilde{\mu}_Y$. A very simple distribution-free test can be obtained if we apply a little tweaking to the null hypothesis and instead consider $H_0 : \tilde{\mu}_{X-Y} = 0$ against $H_1 : \tilde{\mu}_{X-Y} \neq 0$, where $\tilde{\mu}_{X-Y}$ is the population median of $X - Y$. Here, we implicitly assume that median of differences is approximately equal to the difference of medians, which is not always exactly true.

For this set of hypotheses, we define $D_i = X_i - Y_i$ for $i = 1, \ldots, n$, i.e. we define the difference of each paired observation as a new variable. Count the number of positive d_i's and consider it as our test statistic; denote it by T. Naturally, if H_0 is

true, there would be more or less an equal number of positive and negative values, whereas, if H_1 is true, the observed value of T is either too large or too small. So, we calculate the p-value as $P(T > t|H_0) + P(T < n - t|H_0)$, where under H_0, $T \sim Binomial(n, 0.5)$. This is known as the sign test for paired sample.

Assuming that our data set does not conform to a bivariate normal distribution, we do this sign test for paired data and obtain the p-value as 0.2619. So, we can conclude that based on the given data there is no reason to believe that their locations are significantly different. Note that this test is less powerful than an exact test based on the actual distribution of the random variables.

It is important to remember that this test is basically a test for proportion for a binomial random experiment and we can use the in-built R code for binomial test for single sample proportion.

```
> d<-x-y
> binom.test(sum(d>0), n=length(d), p=0.5, alternative = “two.sided")
```

This test is carried out without removing outliers. Since we are using medians, this is not seriously affected by their presence. If we do the same test removing the outliers, the conclusion remains same as the p-value in such a case is 0.2976.

If there is reason to believe that the difference of medians is significantly different from the median of differences, we cannot apply this test. However, we can use bootstrap technique for testing equality of two medians in such a scenario.

5.5.3 *Regression-Based Testing*

When two variables are related, more specifically when we see that there exists a significant correlation coefficient between them, we can always think of fitting an appropriate regression of one variable (response variable) on the other (independent or explanatory variable). Domain knowledge should be used to identify the response and explanatory variables. Interpretation of the results would depend on this choice.

Suppose Y and X are response and explanatory variables respectively in a regression framework. Using least square technique as discussed in Sect. 5.3.3, we can easily find an optimum regression equation that explains the effect of X on Y. But the major question is whether it is worthwhile to find such a regression line. To answer this, we would need to test whether the regression coefficient is significantly different from zero. Naturally, zero regression coefficient or a value very close to 0 would indicate a very weak (or no) linear relation making linear regression study not worthwhile.

Therefore, our testing problem would be

$$H_0 : \beta = \beta_0 \text{ against } H_1 : \beta \neq \beta_0,$$

where β is the regression coefficient and β_0 is its value as specified by null hypothesis. $\beta_0 = 0$ indicates the non existence of linear regression of Y on X. In this context, it is worth mentioning that β and ρ are related through the relation $\beta = \rho\sigma_y/\sigma_x$. Thus testing $H_0 : \beta = 0$ is equivalent to testing $H_0 : \rho = 0$, indicating the absence of linear relationship.

To develop an appropriate test, a natural test statistic would be $\hat{\beta}$, the least square estimate based on paired observations $\{(x_i, y_i),\ i = 1, \ldots, n\}$, which is given by:

$$\hat{\beta} = \frac{\sum_{i=1}^{n}(x_i - \bar{x})(y_i - \bar{y})}{\sum_{i=1}^{n}(x_i - \bar{x})^2}.$$

Since $\hat{\beta}$ is a linear function of $y_i,\ i = 1, \ldots, n$, it can be shown that $\hat{\beta}$ follows a normal distribution, under the assumption of normality of random error components, i.e. $\hat{\beta} \sim N(\beta, \frac{\sigma^2}{S_{xx}})$, where $S_{xx} = \sum_{i=1}^{n}(x_i - \bar{x})^2$. Following the logic explained in Sect. 4.3.3, we should take the standardised version of $\hat{\beta}$. But the presence of error variance σ^2 in the denominator makes the calculation impossible as we don't know the exact value of σ^2. So, we replace it with a good (unbiased) estimator $\hat{\sigma}^2$, where is $\hat{\sigma}^2 = \frac{1}{n-2}\sum_{i=1}^{n}(y_i - \hat{\alpha} - \hat{\beta}x_i)^2$ and n is the sample size. Thus, our final test statistic would be:

$$T = \frac{\hat{\beta} - \beta_0}{\sqrt{\hat{\sigma}^2/S_{xx}}}.$$

It can be shown that the probability distributions of $\hat{\sigma}^2$ and $\hat{\beta}$ are independent and hence under H_0, T follows a t-distribution with $n - 2$ degrees of freedom.

Suppose we want to predict the gene expression in tumour tissue based on that in normal tissue. We can use the regression equation of Y on x and test whether this regression equation can predict the value of Y reasonably well. For the data given in Table 5.1, after removing outliers, the p-value associated with this test is seen to be 0.00166. So, we can conclude that the regression coefficient is significantly different from 0 meaning that the prediction is possible. The outcome of R code of fitting regression equation contains lot of information. In fact, it always provides p-value for testing $H_0 : \beta = 0$.

```
> lin.reg<- lm(y~x)      ## fits linear regression and stores output in 'lin.reg'
> summary(lin.reg)       ## provides summary of output including p-values
```

This would give a lot of information about the fitting of regression equation of y on x. The p-value corresponding to the variable x (or explanatory variable) will be indicative of the existence of linear regression and we can take decision accordingly.

5.6 Points to Remember

Bivariate data occur frequently in the context of genetics. Study of individual variables would be good but it does not take into account the correlation between them. Thus, we might lose information if we avoid analysis considering the two variables at the same time. Statistics is very rich in this area. There are a myriad of statistical methods and tools that can be used in many scenarios of bivariate data. We have seen in Table 5.1 that gene expression of same gene in normal as well as tumour tissue is expected to be correlated as both observations come from the same individual. Similarly, data in Table 5.2 also reveal the same feature because gene expression of two genes comes from the same individual. In this case, correlation might be more pronounced and meaningful if the two genes belong to the same pathway.

This chapter is devoted to analysing gene expression data or data on two variables when they may be dependent. Thus, given any such bivariate data set, we should first check whether the two variables follow bivariate normal distribution. Life would be relatively easy if we get an affirmative answer based on test for normality in bivariate samples. In such a case, a simple t-test for the absence of correlation would ensure the dependence status between two variables. Otherwise, we have to do nonparametric tests like Puri-Sen tests to check whether they are independent.

It is most likely that data generated for two seemingly related variables from the same set-up or experiment, follow a bivariate distribution. However, after doing a test for independence, we have to proceed accordingly. If assumption for bivariate normal distribution is valid based on the given data, standard available tests based on bivariate normal distribution can be performed to have more knowledge about central tendency or dispersion of two random variables. Otherwise, we have to do appropriate nonparametric tests. As mentioned earlier, it is better to do nonparametric test when bivariate normality does not hold rather than wrongly applying tests that are only suitable for bivariate normal distribution.

In this chapter, we have discussed a few statistical tests that are not very commonly used. The reason behind this is probably our reluctance to go for a distribution other than bivariate normal distribution, mainly for computational issues and looking at some slightly complicated mathematical formula. Computational challenges, that one faced earlier, have largely been addressed by the use of R, or other software/programming language. Now suitable methods for analysis of bivariate data are available with computational ease. The only thing is that we have to understand the problem and we should have a clear idea on what we really want to analyse. Hence, it is advisable to use statistical treatment of the data that is deemed most appropriate.

Exercise

5.1 For data in Table 5.2, draw histograms and boxplots for each gene and on a single plot. Write a short report on your observations.

5.2 Consider the following data: (12.97, 35.5), (6.99, 52.36), (7.47, 37.93), (7.93, 5.13), (10.33, 7.96), (12.76, 14.87), (15, 48.35), (13.14, 54.91), (6.18, 47.5), (9.65, 0.96), (9.51, 10.05), (13.33, 36.25), (11.14, 5.4), (9.85, 13.17), (12.04, 16.96), (6.8, 56.8), (12.51, 33.51), (13.49, 30.2), (7.54, 29), (8.42, 19.76), (8.6, 15.29), (11.34, 8.98), (7.28, 46.34), (7.3, 42.98), (11.67, 23.7).

(a) Calculate the correlation coefficient between two variables.
(b) Draw a scatter diagram.
(c) Fit a linear regression equation to this data set and plot it on the scatter diagram.
(d) Do you think fitting this regression equation is appropriate? If so, why? If not, modify your answer with justification.
(e) After fitting the above linear regression equation of y on x, plot the residuals against x variable. What can you say from the diagram? Assuming these residuals are independent, perform a test to see whether they follow normal distribution with mean 0.
(f) The data set contains 25 pairs of observations, corresponding to 25 randomly chosen individuals. Split the data set randomly into two data sets: data set 1 contains 12 pairs whereas data set 2 contains remaining 13 pairs. Fit separate linear regression lines of y on x for each of two data sets. Now plot these regression lines and also the one that has been obtained using all observations as in (c) on the same graph. Comment on your findings with justification. Do you understand why it is recommended to draw random samples?
(g) Based on this bivariate data set, check whether the underlying variables follow a bivariate normal distribution?

5.3 For the data in 5.2, assume that they come from bivariate normal distribution. Apply a statistical test to check whether the two variables are independent. If you use Puri-Sen test to the same data set, what would be your conclusion? Comment on the basis of two conclusions based on the two tests.

5.4 If the correlation coefficient between two variable is $+1$, what would the scatter diagram look like?

5.5 Draw a scattered diagram using R where correlation coefficient is close to zero, but there exists a relationship between the two variables.

5.6 For Table 5.1, calculate r, the correlation coefficient, fit an appropriate linear regression line, and test for bivariate normality. Interpret your results.

5.7 In Table 5.2, suppose we want to know that gene 1 can be explained through gene 2 (at least partially) through a linear regression equation. Write this clearly as a problem of testing of hypothesis. Do the testing and calculate the p-value for this test and interpret the result. Develop a test to see whether, on an average, gene expression for gene 2 is more the that for gene 1 by 2.4 and interpret your result.

5.8 Generate a bivariate non-normal data using R and check for normality.

5.9 Consider two data sets on gene expression values (may be transformed):
Data set-1: 2.802, 5.324, 6.363, 6.176, 4.976, 6.261, 5.068, 4.483, 5.324, 6.576, 5.599, 7.377, 5.217, 6.072, 5.490, 6.118, 7.523, 7.702, 6.977, 5.293, 6.309, 6.047, 6.653, 6.527, 6.372, 7.019, 4.337, 3.995, 7.669, 3.773, 6.350, 7.188, 6.752, 7.085, 6.479, 7.423, 4.264, 5.362, 8.094, 8.410, 8.555, 9.605
Data set-2: 4.078, 5.287, 2.818, 5.755, 5.376, 4.116, 5.009, 4.107, 3.888, 6.835, 4.625, 6.448, 2.803, 5.185, 4.483, 4.838, 3.489, 4.463, 4.803, 3.645, 3.726, 5.938, 6.507, 4.771, 5.706, 4.936, 6.650, 4.328, 6.096, 4.095, 1.931, 4.091, 3.789, 4.774, 5.493, 5.429, 6.285, 5.810, 5.409, 6.533, 1.661, 4.776

(a) Test whether the variances are same in the two sets.
(b) Perform an appropriate test to see whether location parameters for the two distributions are same?
(c) Can you use a Kolmogorov-Smirnov test to check whether the probability distributions are same in two data sets?
(d) After you perform the above two testing procedures, suppose the experimenter tells you that the values in the two data sets actually form a single data set from the same samples for tumour and normal tissues (order of samples collected being same for tumour and normal). How would you modify your test in (b)?

5.10 Take logarithm transformation of the data given in 5.2 above. Fit a linear regression line of the second component on first component for pairs of data points. Do a statistical test with $H_0 : \beta = 2.3$ against $H_1 : \beta \neq 2.3$ and comment on your findings.

5.11 For a bivariate data set on two variables x and y, $\{(x_i, y_i),\ i = 1, 2, \ldots, n\}$ values are not available. However, we have data on sum of the variables and their differences, i.e. we have data as $\{(u_i, v_i),\ i = 1, 2, \ldots, n\}$ where $u_i = x_i + y_i$ and $v_i = x_i - y_i$ for each $i = 1, 2, \ldots, n$. We can easily calculate the correlation coefficient between u and v, say r_{uv}. Can we retrieve the correlation coefficient between x and y from r_{uv}? If so, how? Justify your answer. After fitting linear regression equation of v on u,we get the regression coefficient of β as a function of data on $(u_i, v_i), i = 1, 2, \ldots, n$. What would be the regression coefficient of y on x in terms of β already obtained?

5.12 A study of 100 patients is performed to determine if cholesterol levels are lowered after 3 months of taking a new drug. Cholesterol levels are measured on each individual at the beginning of the study and 3 months later. The cholesterol change is calculated which is the value at 3 months minus the value at the beginning of the study. On an average the cholesterol levels among these 100 patients decreased by 15.0 and the standard deviation of the changes in cholesterol was 40. What can be said about the two-sided p-value for testing the null hypothesis of no change in cholesterol levels? Provide justifications for your answer.

References

Doornik, J. A., & Hansen, H. (2008). An Omnibus test for univariate and multivariate normality. *Oxford Bulletin of Economics and Statistics, 70*, 927–939.
Puri, M. L., & Sen, P. K. (1971). *Nonparametric methods in multivariate analysis*. John Wiley & Sons Inc.
Royston, J. P. (1982). An extension of Shapiro and Wilk W test for normality to large samples. *Applied Statistics, 31*(2), 115124.
Royston, J. P. (1992). Approximating the Shapiro-Wilk W-Test for non-normality. *Statistics and Computing, 2*(117–119), 121133.
Shapiro, S. S., & Francia, R. S. (1972). An approximate analysis of variance test for normality. *Journal of the American Statistical Association, 67*, 215–216.
Villasenor-Alva, J. A., & Gonzalez-Estrada, E. (2009). A generalization of Shapiro-Wilk's test for multivariate normality. *Communications in Statistics: Theory and Methods, 38*(11), 1870–1883.

Chapter 6
Tying Genomes with Disease

Genes, genome, etc. are common words that feature in any genetic analysis. It is now believed that a disease, be it monogenic or complex, has manifested due to the genetic architecture of the organism. Humans are no different. There are around 15000–20000 genes in the human body. These genes are scattered over the human genome which consists of 22 homologous pairs of chromosomes (autosomes) and one pair of sex chromosomes. Quite naturally, we assume that one set of chromosomes that is transmitted from the father and the other set that is transmitted from the mother are completely independent. The holy grail of any genetic study is to first identify the gene(s) responsible to cause a disease or has some role in at least increasing the disease risk. It was with this question in mind, that a genetic association study was launched with the primary focus of identifying a gene or a set of genes that are associated with the disease or phenotype in general. Note that we have no idea where the disease gene is located on the human genome and this is what we try to find out. Eventually, this identification of genes that are responsible for disease might lead to specific drugs targeting to cure the disease. In this quest, we need data. Our hope is that appropriate data can answer, or at least throw some light, in deciphering the genetic architecture. Data consist not only of genetic data, but also phenotype data along with information on covariates, environment, etc., that are deemed relevant to our study of interest.

6.1 Characteristics of Genomic Data

The basic and very naive idea about the genetic structure is that genes are located in different regions of the chromosomes. However, one gene might present itself in different forms. These forms are known as 'alleles', which may be looked at as competing genes. Depending on the allele type and/or the number of a particular allele, disease status or risk might change. Change in alleles is the main source that adds variation in genetic data, making it amenable to statistical analysis. Thus we may be interested to see whether any particular allele is more (or less) observed in affected

I. Mukhopadhyay and P. P. Majumder, *Statistical Methods in Human Genetics*,
Indian Statistical Institute Series, https://doi.org/10.1007/978-981-99-3220-7_6

individuals compared with unaffected individuals. There are other types of data in this context, like gene expression data, methylation data, etc. We already have had some exposure to gene expression data and its analysis methods. On the other hand, data collection mechanism and study design are also important to fully understand the data. In this paradigm, sometimes we collect families and collect genetic data from all available individuals in a family. Sometimes we collect independent or unrelated individuals, among them a group affected by the disease while others are normal or free from the disease. Thus several characteristics of data prevail and we need to understand them. In order to proceed further with the study of genetic data, it is essential to represent them mathematically. This mathematical representation emerges very naturally.

6.2 Representing Mathematically

The human genome map tells us the precise position of a gene with other information like number of its alleles. In general this known information at a position or locus on genome helps us in identifying a gene that may be associated with the disease. This polymorphic entity with known locus is known as marker. A marker is the building block of genetic analysis. Usually it remains unchanged during transmission from parents to offsprings, i.e. from one generation to another.

Nowadays we usually consider Single Nucleotide Polymorphisms (SNP). SNP is defined as a marker having exactly two alleles at a particular nucleotide position on a chromosome. We know that DNA consists of four nucleotides, Adenine (A), Guanine (G), Cytosine (C), and Thymine (T), organised like beads on a string; two strings are twisted to form a double helix structure. Suppose a SNP has two alleles A or C. It indicates that in the population we can observe only A and C as alleles for this SNP marker and no other allele is possible. This information is known to us beforehand, from the human genome map or from other sources. Although at a nucleotide position we might expect to see any one of four nucleotides, i.e. A, T, G, or C, we can only observe two bases (or nucleotides) for a SNP. Thus a SNP always has two alleles. In all subsequent discussions, we use SNP and a gene interchangeably, i.e. our treatment for these two entities would be the same, unless otherwise mentioned. In fact, a gene is a more general term, having deep biological interpretation. And we can always think of different forms in which a gene might exist in the population. So 'a SNP has two alleles' and 'a gene has two alleles' have the same meaning and will be treated similarly in all subsequent discussions, be it mathematical or intuitive.

Suppose that a gene (or a marker) has two alleles, 'A' and 'a' (say). This 'A' is not necessarily adenine of DNA; it is just a generic notation for an allele. Since we observe two alleles on two chromosomes of the same pair, there are in all three possible combinations of two alleles. This combination is known as 'genotype'. Thus with two alleles 'A' and 'a', three genotypes are possible, viz, AA, Aa, and aa. The probability of observing an allele in the population is called *allele frequency* and similarly we can define *genotype frequency* as the probability of a particular

genotype. The allele with smaller allele frequency is termed as *minor allele* and its corresponding frequency is called *minor allele frequency (MAF)*.

Given that a marker or a gene has two alleles, we can define a random variable X as the number of minor alleles of the genotype of an individual at a particular locus. Clearly X is a random variable as the individual is selected randomly from the population without any prior knowledge of her/his genetic profile.

Let the minor allele be 'a' and the other allele be 'A', at a bi-allelic locus. So we can write the genotype frequency as

$$P(AA) = P(X = 0), \quad P(Aa) = P(X = 1) \text{ and } P(aa) = P(X = 2).$$

Thus if a marker has two alleles A and a with allele frequencies respectively as 0.8 and 0.2, the minor allele frequency is 0.2 ($P(a) = 0.2$).

But how does the data set look like? A small data set (Table 6.1) may provide an idea about the data. It contains case-control status (Col. 1), genotype data at four loci, each being a SNP (Col. 2–5), a covariate 'age' (Col. 6), and phenotype (Col. 7) which is cholesterol level for each individual. There are 50 cases and 50 controls in this data set. All individuals are selected randomly or independently of each other.

6.3 Generating Questions

Now that we have a data set with genotypes at four loci and age for two groups, cases and control, we need to ask questions on what we can or should do while analysing the data.

(1) How many people are there in the data set? How are they divided into two groups?
(2) What is the genotype distribution in controls? Do they mimic the population features?
(3) Are genotype distributions same in these two groups and for all four loci?
(4) Since there are four loci, are they related in any sense?
(5) Is there any association between disease status and genotype at any locus?
(6) Is there any effect of age on the disease status?
(7) Can we say that cholesterol level is somehow influenced by genotypes at one locus or all loci, or is it just influenced by age only?

There might be other questions; let's focus on these for the time being. The reader may proceed with other questions and find the solution once we resolve these issues.

Table 6.1 Data for genetic association

CC status	$G1$	$G2$	$G3$	$G4$	Age	Cholesterol	CC status	$G1$	$G2$	$G3$	$G4$	Age	Cholesterol
0	AA	Aa	AA	AA	41	158.68	0	Aa	AA	aa	Aa	34	161.82
0	AA	AA	AA	Aa	44	160.11	0	AA	AA	AA	Aa	38	163.98
0	AA	AA	Aa	AA	38	159.93	0	AA	AA	aa	aa	33	160.49
0	AA	AA	aa	Aa	33	160.24	0	Aa	AA	aa	AA	38	159.56
0	AA	AA	Aa	AA	32	162.48	0	AA	AA	aa	Aa	47	161.11
0	AA	AA	aa	AA	41	163.04	0	Aa	AA	AA	Aa	39	164.20
0	AA	AA	Aa	Aa	39	158.65	0	Aa	AA	Aa	AA	36	159.91
0	Aa	AA	Aa	Aa	47	158.55	0	AA	AA	Aa	AA	35	159.99
0	AA	AA	aa	Aa	45	157.78	0	AA	Aa	Aa	Aa	44	159.53
0	AA	AA	aa	AA	36	158.47	0	Aa	AA	Aa	Aa	45	161.68
0	AA	AA	aa	aa	41	160.81	0	AA	aa	Aa	Aa	37	162.87
0	AA	Aa	AA	AA	38	160.56	0	aa	AA	aa	Aa	36	161.35
0	AA	AA	aa	AA	37	161.02	0	Aa	AA	Aa	Aa	39	164.08
0	AA	AA	aa	AA	34	160.15	0	AA	Aa	Aa	AA	30	155.44
0	Aa	Aa	Aa	Aa	41	162.33	0	AA	Aa	Aa	aa	39	159.62
0	AA	Aa	aa	Aa	41	159.93	0	AA	AA	AA	Aa	47	165.57
0	Aa	AA	Aa	AA	32	160.24	0	AA	AA	Aa	Aa	39	161.68
0	AA	AA	Aa	Aa	43	162.48	0	AA	Aa	aa	AA	37	163.42
0	AA	aa	AA	AA	38	163.04	0	AA	Aa	aa	AA	44	163.44
0	AA	AA	Aa	aa	41	159.53	0	AA	Aa	aa	Aa	44	159.24
0	Aa	Aa	AA	aa	36	159.24	0	Aa	AA	AA	Aa	46	158.94
0	AA	AA	Aa	AA	38	158.78	0	Aa	AA	AA	Aa	42	161.57
0	AA	AA	aa	AA	39	162.67	0	AA	AA	Aa	AA	41	162.45
0	AA	AA	Aa	AA	40	159.32	0	AA	AA	aa	AA	43	160.83
0	Aa	AA	AA	AA	44	161.81	0	AA	Aa	aa	Aa	43	157.58
1	Aa	AA	Aa	aa	58	178.94	1	Aa	Aa	Aa	aa	49	179.83
1	AA	AA	Aa	aa	53	182.33	1	AA	AA	aa	aa	48	184.18
1	aa	AA	AA	aa	48	178.40	1	Aa	AA	Aa	aa	61	180.31
1	AA	AA	aa	aa	55	184.32	1	AA	aa	Aa	aa	52	175.58
1	AA	AA	Aa	aa	43	175.68	1	aa	AA	aa	aa	64	181.13
1	aa	AA	Aa	aa	55	183.95	1	Aa	AA	AA	aa	50	180.66
1	aa	AA	AA	aa	61	186.60	1	AA	AA	aa	aa	50	180.10
1	AA	Aa	aa	aa	42	183.24	1	AA	Aa	AA	aa	58	184.03
1	AA	AA	aa	Aa	47	183.56	1	AA	Aa	AA	Aa	43	180.73
1	AA	Aa	Aa	Aa	58	179.33	1	aa	Aa	aa	Aa	57	175.91
1	Aa	AA	Aa	Aa	45	177.09	1	AA	AA	Aa	Aa	58	182.69
1	AA	AA	Aa	Aa	56	183.69	1	aa	aa	Aa	Aa	53	177.30
1	AA	AA	aa	Aa	51	175.67	1	AA	Aa	Aa	Aa	41	184.00
1	AA	Aa	Aa	Aa	60	175.95	1	AA	AA	aa	Aa	45	178.69
1	aa	AA	AA	Aa	48	182.57	1	Aa	AA	AA	Aa	60	182.73
1	Aa	Aa	AA	Aa	46	177.30	1	Aa	AA	aa	Aa	72	175.89
1	Aa	Aa	Aa	Aa	43	177.51	1	Aa	Aa	Aa	Aa	48	180.56
1	AA	AA	AA	Aa	59	179.89	1	Aa	AA	aa	Aa	49	181.70
1	AA	Aa	aa	AA	59	178.61	1	aa	AA	aa	AA	53	183.24

(continued)

Table 6.1 (continued)

CC status	$G1$	$G2$	$G3$	$G4$	Age	Cholesterol	CC status	$G1$	$G2$	$G3$	$G4$	Age	Cholesterol
1	AA	Aa	aa	AA	59	173.33	1	aa	AA	AA	AA	58	179.03
1	Aa	AA	AA	AA	53	180.20	1	Aa	AA	aa	AA	71	184.00
1	aa	AA	aa	AA	67	180.11	1	aa	Aa	Aa	AA	62	178.83
1	AA	Aa	aa	AA	55	181.07	1	aa	AA	AA	AA	52	177.23
1	AA	Aa	Aa	AA	52	179.48	1	Aa	Aa	Aa	AA	52	174.78
1	Aa	Aa	aa	AA	39	179.70	1	AA	AA	Aa	AA	49	173.97

6.4 Relation Between Allele Frequency and Genotype Frequency

It is clear that genotype is simply a pair of alleles occurring in a pair of chromosomes for an individual. So we can expect that genotype and allele frequencies are somehow related. In this connection there is a basic theorem in population genetics which forms the basis of genetic analysis. This is known as **Hardy-Weinberg equilibrium (HWE)**. Before discussing HWE, we have to first understand the transmission probability, i.e. how to calculate the probability of a genotype of an offspring when the genotypes of her/his parents are known in a family or pedigree. To explain transmission probabilities, for the sake of simplicity, we first consider only a nuclear family. A nuclear pedigree or family consists of parents and their offspring only.

Here we assume random mating among the parents. So transmission of an allele from father to its offspring is completely independent of transmission of an allele from mother to the same offspring. In probability theory jargon, we say that the two events, transmission of allele from father and that from mother to offspring are independent of each other. Nothing surprising! This is quite simple. This simple fact is the key to evaluating the probability of observing offspring genotypes. In Fig. 6.1A,

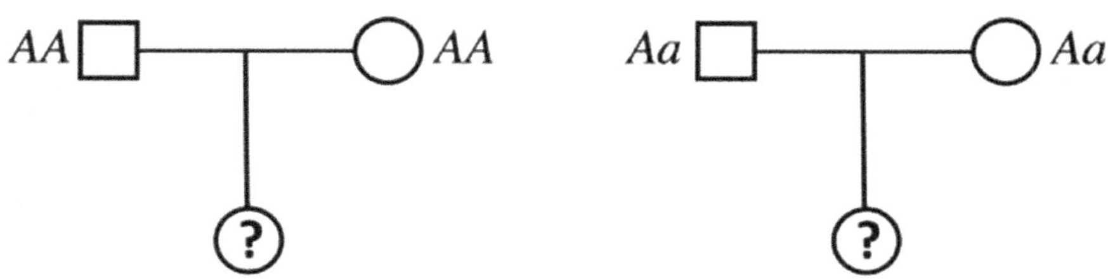

Fig. 6.1(A) Fig. 6.1(B)

Fig. 6.1 Two nuclear families with parental genotypes

genotypes of father is AA and of mother is AA. Let us discuss and calculate some probability of offspring genotype for some combinations of parental genotypes.

$$\begin{aligned} P(\text{Offspring has genotype } AA) &= P(\text{Father transmits } A, \text{Mother transmits } A) \\ &= P(\text{Father transmits } A) \times P(\text{Mother transmits } A) \\ &= 1 \times 1 = 1 \end{aligned}$$

Hence we have for Fig. 6.1A,

$$P(\text{Offspring's genotype is } Aa) = P(\text{Offspring's genotype is } aa) = 0.$$

If we look at Fig. 6.1B, however, for offspring, any genotype among $\{AA, Aa, aa\}$ is possible.

$$\begin{aligned} P(\text{Offspring has genotype } AA) &= P(\{\text{Father transmits } A, \text{Mother transmits } A) \\ &= P(\{\text{Father transmits } A) \times P(\text{Mother transmits } A) = \frac{1}{2} \times \frac{1}{2} = \frac{1}{4} \end{aligned}$$

If we want calculate that offspring has genotype Aa, it receives A allele from father (or mother) and then definitely receives a allele from mother (or father). Clearly parents transmit different alleles to the offspring. Hence,

$$\begin{aligned} & P(\text{Offspring has genotype } Aa) \\ &= P(\{\text{Father transmits } A, \text{Mother transmits } a\} \textbf{or} \{\text{Father transmits } a, \text{Mother transmits } A\}) \\ &= P(\text{Father transmits } A, \text{Mother transmits } a) + P(\text{Father transmits } a, \text{Mother transmits } A) \\ &= \frac{1}{2} \times \frac{1}{2} + \frac{1}{2} \times \frac{1}{2} = \frac{1}{2} \end{aligned}$$

Proceeding similarly, it can be shown that $P(\text{Offspring has genotype } aa) = \frac{1}{4}$.

6.4.1 *Hardy-Weinberg Equilibrium for an Autosomal Locus*

Genes are transmitted from parents to offspring generation after generation. It is expected that there should be a pattern or rule that governs this natural process. However, we know that many population genetics phenomena may trigger certain and subtle changes in the genetic profiles of individuals in a population. Among them mutation, migration, selection, etc. are a few genetic forces that can change the allele and/or genotype frequency in any generation. However, it may so happen that these changes are observed only in a particular generation, i.e. it was absent in previous generations. To keep it simple, first we need to explore whether there is really any rule that can be associated with genotype frequencies at a locus when forces other

than random mating are absent. Fortunately the answer is a positive one; this leads to one of the most celebrated and simplest laws in population genetics. This was discovered by G. H. Hardy and W. Weinberg independently in 1908 and is named after them.

Statement of Hardy-Weinberg equilibrium (HWE): In the absence of migration, mutation and selection, and if random mating is practiced, genotype frequencies at an autosomal locus remain unchanged from one generation to another.

Implication of HWE: This simple law can be proved mathematically and has far reaching influence on a genetic study. It says that if random mating occurs in the present generation, genotype frequencies become stable in the next generation and will remain constant in the subsequent generations provided in every generation there will be random mating only. Thus if at any generation random mating assumption is violated or any phenomenon like mutation or migration, etc. occurs, we will have to wait for only one generation of random mating and also have to wait for mutation, migration, etc. be become absent, for genotype frequencies to become stable again. Thus at the very beginning of a genetic study involving genotypes and alleles, we have to check whether HWE exists in that population for the markers under study. We can proceed for further study with a marker only if HWE is satisfied for that locus.

Suppose in a generation, for a bi-allelic locus, the genotype frequencies are $P(AA) = Q$, $P(Aa) = H$, and $P(aa) = R$ so that $Q + R + H = 1$. Then, it can be shown that after one generation of random mating, genotype frequencies change but become stable. The new genotype frequencies are:

$$P(AA) = \left(Q + \frac{H}{2}\right)^2, \quad P(Aa) = 2\left(Q + \frac{H}{2}\right)\left(R + \frac{H}{2}\right), \text{ and } P(aa) = \left(R + \frac{H}{2}\right)^2.$$

If random mating continues in subsequent generations and there are no other external factors affecting allele frequency or genotype frequency, these frequencies will remain unchanged.

The relation between genotype frequency and allele frequency can also be established as a consequence of HWE. If we take $p = \left(Q + \frac{H}{2}\right)$ and hence $P(AA) = p^2$. Genotype frequency of a particular genotype can be thought of as a function of allele frequencies. Hence, we can interpret it in another way. We have already noted that transmission of an allele from father to offspring is completely independent of transmission from mother to the same offspring, if we assume random mating among parents. That means that an offspring genotype AA may be viewed as having formed by combining two A alleles chosen at random from the population of alleles. Hence,

$$P(AA) = P(Father \to A, Mother \to A) = P(Father \to A)P(Mother \to A)$$
$$= p \times p = p^2$$

Note that $1 - p = 1 - \left(Q + \frac{H}{2}\right) = R + \frac{H}{2} = q$ (say). To get the probability of Aa genotype, the same argument continues. However, we have to be careful of the fact that if the father transmits A allele, the mother must transmit a allele whereas if the father transmits a allele, the mother must transmit A allele to the offspring; only in that case the genotype becomes Aa. Since both alleles are drawn randomly from the population with allele frequencies $P(A) = p$ and $P(a) = q$, we have

$$P(Aa) = P(Father \to A, Mother \to a) + P(Father \to a, Mother \to A)$$
$$= pq + qp = 2pq$$

Then, proceeding as above, we have the relation between genotype frequency and allele frequency under HWE, as

$$P(AA) = p^2, \quad P(Aa) = 2pq, \quad P(aa) = q^2 \text{ where } q = 1 - p, \ 0 < p < 1.$$

In the above discussion of obtaining genotype frequency from allele frequencies, we have implicitly assumed that allele frequency for males and females in the population are same.

A rigorous proof of HWE is given in the Appendix D. In the entire discussion, we have considered only bi-allelic locus on autosomal chromosomes. The same rule applies to a locus with multiple alleles.

Suppose a locus has k alleles $A_1, A_2, \ldots, A_k$ where $k > 2$. Then, if the population is in HWE, the genotype frequencies would be

$$P(A_i A_i) = p_i^2, \ i = 1, \ldots, k, \text{ where } P(A_i) = p_i, i = 1, \ldots, k, \text{ and}$$
$$P(A_i A_j) = 2p_i p_j, \quad \text{for } i \neq j = 1, \ldots, k$$

Moreover, our consideration of HWE in this form is only true for locus at autosomal chromosomes, which are 22 pairs in number in humans. The same rule does not apply to locus on X chromosome. So it demands a separate discussion about HWE for X-linked locus.

6.4.2 *HWE for X-linked Locus*

Suppose a bi-allelic locus on X chromosome has two alleles, say 'A' and 'a'. Clearly a female person has two copies of alleles thus giving rise to three genotypes AA, Aa, and aa, as in the case of autosomal chromosome. But in males, only one copy

of allele is observed because a male person has only one copy of X chromosome; his other chromosome is Y. It is thus natural to consider different allele frequencies of A for males and females. Let $p_f(t)$ and $p_m(t)$ be the allele frequencies for the allele A in females and males, respectively, at generation t. We are interested to know what would happen to these allele frequencies in the next generation under the assumption of random mating only (no other genetic force is present).

Note that a female with genotype AA gets one A allele from her father and another A allele from her mother, because she has two X chromosomes. Since we assume random mating, the probability that a female has AA combination in the $t+1$-th generation is the product of the probabilities of randomly selecting one A allele from the female population at generation t and that for the male population in generation t. Following the same arguments, we have probabilities of all genotypes in generation $t+1$ as

$$P(AA \text{ genotype among females in generation } t+1) = p_f(t)p_m(t),$$
$$P(Aa \text{ genotype among females in generation in generation } t+1) = p_f(t)\big(1-p_m(t)\big) + \big(1-p_f(t)\big)p_m(t), \text{ and}$$
$$P(aa \text{ genotype among females in generation in generation } t+1) = \big(1-p_f(t)\big)\big(1-p_m(t)\big).$$

On the other hand, a male person can receive his only copy of X chromosome from his mother and his other chromosome is Y which comes from his father. So he has only one allele at this locus on the X chromosome. Hence, for males, we have, following the previous arguments,

$$P(A \text{ among males in generation } t+1) = p_f(t), \text{ and}$$
$$P(a \text{ among males in generation } t+1) = \big(1-p_f(t)\big)$$

Genotype frequencies are not of the same pattern in females and males. This motivates us to look at allele frequency in generation $t+1$ directly.

Now, in generation $t+1$, frequency of A depends on both females and males; hence we consider $p_f(t+1)$ as the average allele frequency among females and males from generation t. On the other hand, for males in generation $t+1$, the frequency of A allele depends only on that among females in generation t. Thus we have

$$p_f(t+1) = \frac{1}{2}\big(p_f(t)+p_m(t)\big), \text{ and}$$
$$p_m(t+1) = p_f(t) \tag{6.1}$$

Since allele frequency for A seems different in females and males in generation $t+1$, even after random mating, in order to hold a law, we expect that some kind of average allele frequency might have been stable. We consider $\frac{1}{3}p_m(t+1)+$

$\frac{2}{3}p_f(t+1)$, which a weighted average of $p_f(t+1)$ and $p_m(t+1)$ and investigate this quantity.

$$\begin{aligned}\frac{1}{3}p_m(t+1)+\frac{2}{3}p_f(t+1) &= \frac{1}{3}p_f(t)+\frac{2}{3}\left\{\frac{1}{2}p_f(t)+\frac{1}{2}p_m(t)\right\} \qquad \text{[from (6.1)]}\\ &= \frac{1}{3}p_m(t)+\frac{2}{3}p_f(t)=\cdots\cdots=\frac{1}{3}p_m(0)+\frac{2}{3}p_f(0)\end{aligned}$$

Thus it is clear that although allele frequencies are different among females and males in every generation, however, their average (weighted average) remains stable over generations after random mating occurred in t-th generation.

We might be interested to know the nature of difference of allele frequencies among females and males and how it varies over generations. A little bit of calculation would reveal a startling feature! From (6.1), we have,

$$\begin{aligned}p_m(t+1)-p_f(t+1) &= p_f(t)-\frac{1}{2}\big(p_f(t)+p_m(t)\big)=\frac{1}{2}\big(p_f(t)-p_m(t)\big)\\ &= \left(-\frac{1}{2}\right)\big(p_m(t)-p_f(t)\big)=\left(-\frac{1}{2}\right)^2\big(p_m(t-1)-p_f(t-1)\big)\\ \cdots\cdots &= \left(-\frac{1}{2}\right)^n\big(p_m(t-n+1)-p_f(t-n+1)\big)\to 0 \text{ as } n\to\infty\end{aligned}$$

Thus in the long run, allele frequency of A at a locus on X chromosome will be the same for females and males. Note that although we have worked with a particular weighted average of female and male allele frequencies, these weights can be interpreted easily. Each male carries one X chromosome whereas each female carries two X chromosomes. Hence, we have chosen the weights according to the contribution of A alleles by males and females. These weights can also be obtained by mathematically assuming a general weighted average with unknown weights w_1 and w_2 (say), subject to $w_1 + w_2 = 1, 0 < w_1, w_2 < 1$; we omit this detailed mathematical derivation for simplicity and empasise on the intuitive justification.

We discuss a few problems on HWE below.

Ex-6.1. Four alleles in a randomly mating population are 0.10, 0.25, 0.35 and 0.30. In a population of size 20000, what is the expected genotype distribution in that population?

Let the four alleles be A_1, A_2, A_3 and A_4 with frequencies 0.10, 0.25, 0.35 and 0.30 respectively. To get the expected genotype frequency for a particular genotype in the population, we find the respective genotype frequency assuming HWE and then multiply it by 20000. Let n_{ij} be the expeceted genotype frequency of genotype A_iA_j for $i, j = 1, \ldots, 4$. Hence, we have,

$n_{11} = 20000 \times (0.1)^2 = 200,\ n_{22} = 20000 \times (0.25)^2 = 1250,\ n_{33} = 20000 \times (0.35)^2 = 2450,$

$n_{44} = 20000 \times (0.3)^2 = 1800,\ n_{12} = 20000 \times 2(0.1)(0.25) = 1000,\ n_{13} = 20000 \times 2(0.1)(0.35) = 1400,$

$n_{14} = 20000 \times 2(0.1)(0.3) = 1200,\ n_{23} = 20000 \times 2(0.25)(0.35) = 3500,\ n_{24} = 20000 \times 2(0.25)(0.3) = 3000,$

$n_{34} = 20000 \times 2(0.35)(0.3) = 4200.$

Ex-6.2. At a bi-allelic autosomal locus, the allele frequency for two alleles A and a are respectively 0.22 and 0.78. Assuming Hardey-Weinberg equilibrium,

(a) what is the percentage of heterozygous people in the population?
(b) what is the percentage of homozygous recessive people in the population?

It is given that $q = P(A) = 0.22$ and $p = P(a) = 0.78$. Assuming HWE, $P(AA) = (0.22)^2 = 0.0482$, $P(Aa) = 2 \times 0.22 \times 0.78 = 0.3432$, and $P(aa) = (0.78)^2 = 0.6084$.

(a) Percentage of heterozygous people in the population $= 100 \times P(Aa) = 34.32\%$.

(b) Since A is the minor allele in this case, we can assume A is the disease allele. Hence, percentage of homozygous recessive people in the population $= 100 \times P(AA) = 4.82\%$.

Ex-6.3. Suppose a population in Hardy-Weinberg equilibrium contains 11% recessive homozygote individuals for a certain trait. In a population of 30000,

(a) what is the percentage of homozygous dominant individuals in the population?
(b) what is the percentage of heterozygous individuals in the population?

Let q be the recessive allele frequency. Since HWE is maintained in the population, we have $q^2 = 0.11$ and hence $q = 0.3317$ and hence the frequency of the dominant allele is $p = 1 - q = 0.6683$.

(a) Percentage of homozygous dominant individuals in the population $= 100 \times p^2 = 100 \times (0.6683)^2 = 44.66\%$.

(b) Percentage of heterozygous individuals in the population $= 100 \times 2pq = 44.33\%$.

6.4.3 *Estimation of Allele Frequency*

It is essential to see whether HWE exists for a marker in a given population under study. Since we deal with genotype data, a statistical test is welcome. HWE explains relationship between genotype frequency and allele frequency. We have to exploit this information in order to check HWE. Technology provides us only genotype data

and not allele level data. Genotype proportions in the sample are available and hence we need to devise a method to estimate allele frequency from genotype data.

Suppose we select n individuals at random from a population and genotype them at a number of markers. Let us consider one marker having two alleles A and a and also let n_{AA}, n_{Aa} and n_{aa} be the number of individuals having genotypes AA, Aa, and aa respectively. Clearly $n_{AA} + n_{Aa} + n_{aa} = n$. However, since we have no idea about the allele frequency beforehand, which is usually the case, we have to estimate allele frequency from the given sample on genotypes.

Gene Counting Method

Allele frequency of an allele in the population indicates the probability of observing a particular allele when one allele is selected at random. It is simply the population proportion. Thus allele frequency can be estimated easily by just counting the number of alleles present in the sample. Note that each individual having a homozygous genotype, say AA, contributes two A alleles whereas a person with Aa genotype contributes only one A allele. Thus the total number of A alleles in the sample is $2n_{AA} + n_{Aa}$. For a bi-allelic locus, a sample of n individuals contributes a total of $2n$ alleles. Since we know a sample proportion is usually a good estimate of the corresponding population proportion, we have the estimates of allele frequencies as

$$\hat{p} = \hat{P}(A) = \frac{2n_{AA} + n_{Aa}}{2n} \text{ and } \hat{q} = \hat{P}(a) = 1 - \hat{p} \tag{6.2}$$

Thus genotype frequencies under HWE may be obtained as

$$\hat{P}(AA) = \hat{P}^2, \quad \hat{P}(Aa) = 2\hat{p}\hat{q} \text{ and } \hat{P}(aa) = \hat{q}^2 \tag{6.3}$$

There is no restriction on sample size. Sometimes for a rare allele, recessive genotype with two copies of rare allele may not be observed if the sample size is small. However, as long as at least two types of genotypes are observed in the sample, we can estimate the allele frequencies. Of course, a larger sample would increase the efficacy of the estimate.

Maximum Likelihood Method

A more statistically sound and intuitively appealing way to estimate allele frequency from genotype frequency is to use likelihood function for the data and then maximise it to get the maximum likelihood estimator (MLE) for allele frequency. Given the data as described in the previous section and noting the genotype probabilities under HWE, the likelihood function is given by

$$L(p|n_{AA}, n_{Aa}, n_{aa}) = \frac{n!}{n_{AA}!n_{Aa}!n_{aa}!}(p^2)^{n_{AA}}(2pq)^{n_{Aa}}(q^2)^{n_{aa}} = \frac{2^{n_{Aa}}n!}{n_{AA}!n_{Aa}!n_{aa}!}p^{2n_{AA}+n_{Aa}}q^{n_{Aa}+2n_{aa}} \tag{6.4}$$

$$\text{or } \log(L(p|n_{AA}, n_{Aa}, n_{aa})) = \log\left(\frac{n!}{n_{AA}!n_{Aa}!n_{aa}!}\right) + (2n_{AA}+n_{Aa})\log(p) + (2n_{aa}+n_{Aa})\log(q) + n_{Aa}\log 2 \tag{6.5}$$

Maximising likelihood (or log-likelihood) function with respect to p, we get the estimate of allele frequency exactly same as that obtained by gene counting method!

6.4.4 Mean and Variance of Allele Frequency Estimator

In statistics, we are always curious to know whether an estimator is good (in some sense). Naturally an estimator performs well if its expectation is same as the population parameter which we want to estimate. This indicates, ideally, that if we draw many samples and calculate this estimate based on each sample, we can expect these values to cluster around the parameter value, although we don't know the value of the parameter. This property, known as 'unbiasedness' is a very popular measure to judge whether an estimator is good. On the other hand, we also hope that the variance of the estimator should be small, or become smaller and smaller as the sample size gets larger. Naturally this leads to evaluating the expectation and variance of any estimator that might take an important role in further downstream analysis. If we somehow find that an estimator is not so good, we should explore for a better estimator.

In the above context, we now try to evaluate mean, i.e. expectation and variance of $\hat{p}$, the estimator obtained by gene counting or MLE method. Since this requires some rigorous calculation using a few results, we provide the detail proof in the Appendix D. It can be shown that

$$E(\hat{p}) = p \text{ for all } p \text{ and } V(\hat{p}) = \frac{pq}{n} \text{ for all } p.$$

Ex-6.4. Genotype frequencies at a bi-allelic locus for 100 individuals are as follows:

$$DD: 50, \quad Dd: 40, \quad dd: 10$$

Estimate the frequency of allele D and calculate its standard error.

Let p be the allele frequency for allele D. Then using the maximum likelihood method, as in Sect. 6.4.3, we have

$$\hat{p} = \frac{2 \times 50 + 40}{2 \times 100} = \frac{140}{200} = 0.7$$

$$\hat{V}(\hat{p}) = \frac{\hat{p}\hat{q}}{n} = \frac{0.7 \times 0.3}{200} = 0.00105 \text{ implies standard error}(\hat{p}) = \sqrt{0.00105} = 0.032$$

6.4.5 Test for HWE

Now we are in a position to check whether HWE exists in a population of interest. This can be framed as a statistical testing problem. The null and alternative hypotheses of interest are

$$H_0 : \text{HWE is satisfied against } H_1 : \text{HWE is not satisfied.}$$

Since we know genotype frequencies in terms of allele frequencies when HWE is satisfied, we can reformulate the hypotheses as

$$H_0 : P(AA) = p^2, P(Aa) = 2pq, P(aa) = q^2 \text{ against } H_1 : \text{at least one inequality in } H_0.$$

As we have seen earlier, in any testing problem the main component is the data at hand; it contains information about the phenomenon on which our testing problem depends. In this situation, our data consists of genotypes of n randomly chosen individuals at a particular locus. Hence using this data we can easily count the number of individuals having a particular genotype (Table 6.2).

If HWE holds for the marker under study, we expect that the expected frequencies should be very close to the observed frequencies. The expected genotype proportions under HWE, which are p^2, $2pq$, and q^2 for AA, Aa, and aa genotypes respectively. Hence, we can easily get the expected genotype frequencies as np^2, $2npq$ and nq^2 respectively. However, there is only one problem: we don't know the value of p, i.e. allele frequency of allele A. Again we can do a simple trick; just replace the allele frequency by its corresponding estimate based on the sample that we have. The whole scenario can be presented nicely as in Table 6.2.

Now we are interested to see how close the observed and expected frequencies are for each genotype. To check that we should consider the squared difference between the two. However, since this is in regard to the expected (or observed) frequency, we should divide the squared difference by the expected frequency and define a measure of closeness as

$$\frac{(n_{AA} - n\hat{p}^2)^2}{n\hat{p}^2} + \frac{(n_{Aa} - n2\hat{p}\hat{q})^2}{n2\hat{p}\hat{q}} + \frac{(n_{aa} - n\hat{q}^2)^2}{n\hat{q}^2} = T \text{ (say)} \tag{6.6}$$

Table 6.2 Observed and expected frequency table for testing HWE

	AA	Aa	aa
Observed	n_{AA}	n_{Aa}	n_{aa}
Expected	$n\hat{p}^2$	$2n\hat{p}\hat{q}$	$n\hat{q}^2$

It is clear that if HWE is satisfied, i.e. if H_0 is true, the value of T should be small; otherwise it would be large. Thus we reject H_0 if the observed value of T is large. Using statistical theory it can be shown easily that T follows a χ^2 distribution with 1 (one) degree of freedom approximately (asymptotically) when the sample size is large. So we calculate p-value as $P(\chi_1^2 > t)$ where t is the observed value of T and reject H_0 if the p-value is very small compared to the chosen level of significance. On the other hand if the p-value is large, we do not reject H_0 concluding that HWE is satisfied.

For the genotype data for $G1$ in Table 6.1, estimate of MAF is $\hat{p} = 0.27$. Now to test whether HWE is satisfied, we consider only control individuals. Thus within controls, the value of the χ^2-statistic that is used for testing purpose is 0.0192, using (6.6). Hence the p-value for this test is 0.8897. Since this value is quite large compared to the level of significance 0.05, we can conclude that based on the given data, it seems that HWE is satisfied for the locus $G1$. We have to check at all loci whether HWE exist. Repeating the same procedure, we see that the p-values for $G2$, $G3$, and $G4$ are 0.447, 0.0001, and 0.7756 respectively. So we can say that HWE fails to exist for the locus $G3$ whereas for other loci HWE is satisfied. So we should look for possible issues on why at locus $G3$ HWE is violated. However, for further study we can proceed without the genotype data on $G3$ or resolve this issue by relooking the entire process.

Note that testing for HWE is important and we should work with those markers for which HWE is satisfied. It is not advisable to consider markers where HWE is not satisfied based on the above testing method. In such situations, we should look for other reasons like problems during data generation, technical issues, etc. and start further downstream analysis only after resolving these issues.

There are a few important points to note. In testing for HWE, we should work only with genotypes of control individuals although case data are available. If a locus is associated with the disease or phenotype of interest, frequency of a particular allele and hence a particular genotype may be more frequent in cases than that in controls. This may show that occurrence of this allele in cases is different than that in the population and hence case individuals may not constitute a random sample from the population. The basic requirement of HWE testing is that the sample should be random which may be violated if case individuals are included. Hence sacrificing a few samples is recommended rather than inducing non-randomness in the data. This gives a nice but important information that increasing sample size is not always welcome; it depends on the nature of sample and the related problem. However, we may assume that case individuals in our sample constitute a random sample from the population of affected people. This subtlety of randomness property needs to be understood correctly; otherwise some corrections are necessary during the analysis if this assumption is violated.

Another important issue is the level of significance. Note that we only know the distribution of the test statistic T asymptotically, i.e. only when the sample size is large, T follows a χ^2 distribution. So the entire test is carried out at level of significance α *approximately*.

6.4.6 Study Design

Study designs are normally conceptualised to estimate effect sizes of various exposure factors on disease outcomes. In the context of genetic factors, the exposure is to the genes of the parents for a simple Mendelian disease, such as sickle-cell disease. However, for a complex disease, e.g. diabetes or cardiovascular disease, in addition to genetic factors, exposures to various environmental factors modulate disease outcome. Thus, some cardinal features of study design are to unravel effects of factors on disease outcome that are common between a simple Mendelian disease and a complex disease. However, for a complex disease study designs may need to be altered to gain statistical efficiency for estimating effect sizes.

The study designs that are commonly used to identify factors that are associated with diseases are observational. This contrasts with experimental study designs that underlie estimation of the impact of a drug for treating a disease. For a simple Mendelian disorder, we wish to identify a genomic alteration that is associated with a disease outcome. A study design that was popular for this purpose was a family study design. Observations on genotypes at several loci along with disease status were obtained on members of multi-generational families. Co-segregation of alleles at a locus with the disease was estimated using 'linkage analysis'. A co-segregating locus would be declared to be associated or 'close' to the disease locus because suppression of recombination resulting from physical contiguity of the loci would result in co-segregation. Even though this is a powerful study design, unless data are available on large, multi-generation families, the efficiency of detection of co-segregation is severely compromised. Thus, 'linkage analysis' and family study design have made way for the case-control study design. It may be noted that for many diseases, notably cystic fibrosis, disease genes have been mapped using family study designs and linkage analysis.

The case-control study design is also observational, in which association of an exposing factor (genotype or allele) is inferred by contrasting two groups of individuals—those with disease (cases) and those without (controls). In this study design, therefore, a set of individuals with disease are identified. If the disease is common, then individuals with disease can be recruited by surveying members of a population. However, if the disease is rare, then recruitment through a population survey becomes inefficient; in such a situation, since those afflicted with the disease are likely to seek advice from doctors, recruitment of cases can be made from hospitals and clinics. Generally, controls can be easily recruited from the general population. If we were to try to identify a gene for sickle-cell disease today, using the case-control study design we would first recruit a set of individuals with sickle-cell disease and then a set of normal individuals without any haemoglobin disorder. In these two sets of individuals, we would collect data on genotypes at a large number of loci and test whether there is significantly higher preponderance of a genotype or an allele among those with sickle-cell disease compared to those who are normal.

Even if a disease is a single-gene defect, there may be many factors that need to be understood and taken into account in formulating a study design. For example,

there are some mental-health disorders that are single gene, but do not manifest at birth. They are late-onset diseases and the age at onset is variable. Great care needs to be taken in the recruitment of controls. An individual may be much younger than the mean or the median age at onset, and hence will not exhibit characteristics of the disease even if the person carries the disease-predisposing allele. Such an unaffected person may be recruited as a control in a case-control study. This will adversely impact on the ability to identify the genetic factor associated with the disease. For a late-onset disease, it is better to recruit controls who are much above the mean or median onset-age of the disease. Huntington disease is a rare genetic disorder in which parts of the brain degenerate as the person carrying the gene mutation gains age. The disease causes rapid, jerky body movements and the loss of mental skills (dementia). Symptoms of Huntington's disease usually develop between ages 30 and 50 years. An individual who is 18 years old may be free of the disease even if the person is a carrier of the disease gene. In a case-control study design, such a person will be misclassified as a control, when in theory the person should be classified as a case. This misclassification will result in many controls carrying the disease gene, thereby reducing the power of detecting the disease gene using the case-control study design that relies on detecting the significance of the difference in the proportion of the disease gene among cases and controls.

For a complex disease for which there are multiple genetic predisposing factors and possibly some known environmental exposures also modulating the disease outcome, a case-control study has to be designed with great care. For example, for heart disease, it is known that smoking significantly enhances disease risk. For such a disease, matching is the key to identify genetic risk factors over and above the risk imposed by smoking. A patient needs to be carefully assessed for exposures to known environmental factors. An unaffected control needs to be chosen in such a way that the environmental exposures of this person match those of the patient. Thus, if we have recruited a patient who has suffered a heart attack and has been a smoker, we need to recruit another smoker of the same age and gender as the patient who has not suffered from heart disease. By matching, we are effectively ruling out the possibility that environmental factors (here, smoking) could have resulted in the disease outcome in the case. By effectively eliminating this possibility, we can now explore for genomic factors that are significantly associated with the disease outcome using statistically valid analytical procedures that are described elsewhere in this book.

'Matching' may be practical in designing a study if the number of environmental factors that are known to modulate disease outcome is small. If this number is beyond four or five, then it may become impossible to find controls that match with all the exposure characteristics of cases. Under such circumstances, elimination of the effects of differences in environmental exposures between cases and controls has to be done statistically by using regression analysis, either prior to or simultaneously while conducting association analysis of disease outcome with genomic factors.

The case-control study design may be severely compromised if there are many unknown environmental factors, possibly in addition to known factors, exposures to which can modulate disease outcome. An example is liver disease. It is known that infection with hepatitis viruses causes liver disease. Also, exposure to alcohol.

However, there are possibly many dietary factors or water pollutants that can also cause liver disease. If this is the reality, then the study design of choice is a longitudinal, cohort study design. In this design, a large number of individuals are selected at an early age and are longitudinally followed up. At each time-point of follow-up, assessments of exposures to relevant environmental factors (hepatitis virus infection, water pollutants, dietary chemicals, etc.) and of outcome of the disease under study are made. When adequate numbers of individuals with and without disease are noted, then the collected data are analysed to identify significant environmental factors that impact on disease outcome and also to identify significant genetic factors after elimination of the effects of identified environmental factors that are significant.

6.5 Genetic Association

The phrase 'genetic association' itself is self explanatory. In simple language, we need to explore the phenomenon of a gene being somehow associated or related or connected to a disease or a phenotype of interest. In other words, genetic association means whether a gene is associated with a disease or any phenotype of interest. The phenotype may be quantitative or qualitative in nature. For example, we may be interested to know whether a gene is associated with cholesterol level. Here cholesterol is the phenotype which is quantitative in nature and can take any value within a certain interval. Thus the phenotype cholesterol level is a continuous variable. On the hand, we may want to know whether any gene is associated with a disease, e.g. cardiovascular disease. Here the disease status, i.e. whether a person is affected or unaffected is the phenotype of interest. This is an example of qualitative phenotype. The phenotype disease status has two categories, viz 'affected' or 'unaffected'. We can write '0' for unaffected and '1' for affected individual. Note that '0' and '1' represent the two categories and not their exact numerical values.

A detailed study of genetic association helps us in identifying, or at least having some idea about possible gene(s) that may somehow be related with the phenotype or disease. It basically narrows down search for disease associated gene(s) so that a more focused downstream study may be undertaken to understand and decipher the genetic architecture in more detail.

Since phenotype data may be categorical as well as continuous, study of genetic association requires different statistical tools depending on the nature of data, and of course, the problem concerned. Thus we need to discuss various analytical methods.

6.5.1 *Genetic Association for Qualitative Phenotype*

We first discuss genetic association for qualitative phenotype. To study genetic association, we first select a random sample from the population. We usually that the disease is rare and not a very common phenomenon. Hence the population mostly

Table 6.3 Data structure for genetic association at a marker

	AA	Aa	aa	Total
Control	n_{11}	n_{12}	n_{13}	n_{10}
Case	n_{21}	n_{22}	n_{23}	n_{20}
Total	n_{01}	n_{02}	n_{03}	n

consists of normal or unaffected individuals; however, it contains a few affected individuals as well. We have to select random samples from two specific groups, affecteds (cases) and unaffecteds (controls). A random sample from the population can be regarded as controls. However, we need to select a group of cases, supposedly as a random sample among the case individuals in the population. Thus individuals belonging to the same group or different groups are independent of each other. Once we select an individual, we can find the genotypes of multiple markers across her/his genome. For the sake of simplicity, we now concentrate only on one marker. Our main research question is whether a marker is associated with the disease.

After genotyping at a marker for individuals, we have information (data) about their genotypes. If the marker under study has nothing to do with the affection status, genotypes in cases and those in controls are distributed randomly or exhibit the same pattern. We assume that genotypes are in HWE among controls. If there is no association between the marker and the disease, we would expect the same among case individuals in that particular marker. This means that when the marker is not associated with the disease, we observe same genotype distribution in cases and controls. However, if this marker has some influence in increasing or decreasing the disease risk, genotype distributions in the two groups should be markedly different. For example, for a recessive disease, suppose a is the minor allele that increases the disease risk. So we expect to find more aa genotypes in cases than in controls.

Thus our main objective is to see whether a marker is associated with the disease, i.e. whether genotype distributions are different between cases and controls. This basic problem in genetics can be thought or reformulated as a statistical testing problem. Here the hypotheses of interest are

$$H_0 : \text{ marker is not associated with the disease} \quad \text{against } H_1 : \text{ marker is associated with the disease}$$

In Table 6.3, $n_{i0} = n_{i1} + n_{i2} + n_{i3}$ for $i = 1, 2$ and similarly $n_{0j} = n_{1j} + n_{2j}$ for $j = 1, 2, 3$. So the number of cases is n_{20} whereas number of controls is n_{10} and the total number of individuals is n. Now the arguments follow exactly the same way as in case of testing HWE. Here it is natural to say that when H_0 is true, frequencies of respective genotypes are same in the two groups. The test statistic that would be used for testing the above hypothesis, should reflect different scenarios under H_0 and H_1 appropriately. Keeping the idea of testing HWE, we can consider a test statistic, which is based of the idea of comparing observed and expected frequency in each

cell in Table 6.3, as

$$T = \sum_{i=1}^{2} \sum_{j=1}^{3} \frac{(n_{ij} - E_{ij})^2}{E_{ij}} \tag{6.7}$$

where n_{ij} and E_{ij} respectively denote observed and expected frequency for the (i, j)th cell in Table 6.3.

To calculate E_{ij}, the expected frequency of a cell, first assume that H_0 is true, i.e. there is no genetic association. So we can expect that the proportion of genotype AA in controls is the same as that in the entire sample. Hence we have, under H_0,

$$\frac{n_{11}}{n_{10}} = \frac{n_{01}}{n} \implies n_{11} = \frac{n_{10} n_{01}}{n} \tag{6.8}$$

Similarly, for each (i, j)th cell, assuming H_0 to be true, we have $n_{ij} = \frac{n_{i0} n_{0j}}{n}$, $i = 1, 2, j = 1, 2, 3$. Substituting (6.8) in (6.7), we get after elementary algebraic simplification,

$$T = \sum_{i=1}^{2} \sum_{j=1}^{3} \frac{(n_{ij} - \frac{n_{i0} n_{0j}}{n})^2}{\frac{n_{i0} n_{0j}}{n}} = n\left(\sum_{i=1}^{2} \sum_{j=1}^{3} \frac{n_{ij}^2}{n_{i0} n_{0j}} - 1\right) \tag{6.9}$$

It can be shown that $T \sim \chi^2_2$ under H_0, approximately (asymptotically). Based on this null distribution of the statistic T we can calculate the p-value approximately and take a decision. Note that this method provides a very simple test for genetic association of a marker with the disease (qualitative phenotype).

Based on the data given in Table 6.1, we can easily test whether a locus is associated with the disease. Here the first column indicates the status of individual '0' and '1' indicating control and case respectively. R code is available for this χ^2 test. First we read the data, store it and then apply the test between two columns; one being case-control (CC) status and the other the genotype column (locus) of interest. Suppose the data are stored in a file named 'data.csv' with column headings indicating the contents of each column.

```
> a<-read.csv("data.csv",header=T)        ## reading the datafile
> names(a)          ## heading of columns
> head(a)          ## first six rows
> test.1<-chisq.test(a$CC.status,a$G1)         ## store results of the test
> test.1$p.value           ##displays p-value on screen
```

The p-value obtained is 0.00211 which is very small compared to the level of significance 0.05. So we reject H_0 of no association at level 0.05 approximately and conclude that on the basis of the given data, it seems that the locus G1 is associated with the disease.

We should note a few important things here. The level of significance for the test is *approximately* $\alpha = 0.05$, because the test statistic follows a χ^2 approximately (asymptotically). Hence, the p-value calculated on the basis of χ^2 distribution is only approximate. However, if the sample size is large, the approximation is good and can be used for all practical purposes.

The data set contains other loci and we can repeat the same procedure to other loci also. But a word of caution! We should not test for association at G3 because Hardy-Weinberg equilibrium is not satisfied for this locus (Sect. 6.4.5) and hence we should remove this locus from our analysis.

Now that we have a nice test for basic genetic association using qualitative phenotype (disease status), this triggers plenty of questions such as:

- This is a test of genetic association at only one marker. What to do when we have more than one marker? Can we just repeat this test for each marker and take decision at a fixed level, say $\alpha = 0.05$?
- If the markers are close enough, can we incorporate the closeness of markers in the test?
- When we collect data, in this case usually blood (or other tissues) for generating genotype data, we may also collect data on other covariates like age, gender, other environmental variables (smoking habit, food habit, etc.). How do we incorporate the information on these covariates while developing a genetic association test using a qualitative phenotype?
- However, there may exist other phenotypes that are not qualitative in nature. Can we develop a test for association for such trait? If so, can we ask the same questions as above, especially while dealing with a quantitative phenotype?

6.5.2 *Genetic Association in Presence of Covariates*

As discussed earlier, data for genetic association may contain variables other than genotypes and disease status. These variables might be important. Suppose a marker is associated with the disease but depends on the age of onset. For example, although blood sugar issues may be seen at a young age, it is, in general, more prevalent among the elderly population. A random sample includes individuals across different age groups. If the disease of interest is seen more frequently among persons of older ages, it might so happen that a genetic locus is actually not associated but our statistical test shows strong association. This is definitely a false positive; this occurs because we fail to incorporate or eliminate the effect of age while testing for the association. So in order to identify the presence of any genetic association with the disease, we should eliminate the effect of the covariate(s). A detailed and dedicated discussion and methods are also given in Chap. 7.

Naturally we cannot extend our previous discussion of using χ^2 test as it depends on cell frequency alone. On the other hand, regression technique is a nice approach to eliminate the effect of other variables. Thus keeping all these in mind, we have

to explore an adequate method of testing whether a marker locus is associated with the qualitative phenotype or disease status. However, we have studied regression analysis only when the response variable is continuous in nature. We need to develop an analogous idea and subsequent statistical methods in case of categorical data. But first let's introduce a new concept based on probability theory.

6.5.3 Odds Ratio

The new concept that we introduce is called 'odds' in favour or against an event. For instance, the odds that 'I can work out an algebra problem is 1 to 3' means out of 4 problems, I can do 1 and I cannot do 3. If we were to say "There is a 75% chance that it will rain tomorrow", then 3 out of 4 times we can say that it will rain. This means that it will rain three times out of 4 probable chances of raining. This can also be restated as "the odds of raining tomorrow is 3 to 1". This can also be understood in terms of probabilities of events associated with the problem. In the above example,

$$\text{Odds of raining tomorrow} = \frac{P(\text{raining})}{P(\text{not raining})} = \frac{3/4}{1/4} = \frac{3}{1} = 3.$$

Similarly the odds against raining tomorrow $= \frac{1/4}{3/4} = \frac{1}{3}$. We also interpret this in another way. The chances of raining tomorrow is three times more than the chance of not raining. Thus unlike probability, odds of any event may take any positive real number as its value, since it is the ratio of two probabilities. It is clear that 'odds' is more welcome to common people than 'probability' as it carries a very easy interpretation in connection with the real world.

Now in the context of disease prevalence, we can define the odds of a person being affected as the ratio of the probability that a person is affected to that of the person who is not affected. However, we can now easily extend this concept to incorporate another covariate that is available during the data collection and might occur naturally.

Suppose the probability that a young person is affected by a disease is 0.3 whereas the same probability is 0.8 for an elderly person. Thus we have two odds, one for the young and another for the old person, which are respectively $\frac{0.3}{0.7}$ and $\frac{0.8}{0.2}$. Now we define another measure, 'odds ratio', as

$$\text{odds ratio} = \frac{\text{odds for elderly person having disease}}{\text{odds for young having disease}} = \frac{0.8/0.2}{0.3/0.7} = 9.33 \quad (6.10)$$

This indicates that the chance of having the disease in an elderly person are 9.33 times more than that for a younger person. Note that each odds is defined in terms of conditional probability, i.e.

Table 6.4 Smoking status and affectedness

	$Y = 0$	$Y = 1$	$P(X = x)$
$X = 0$	p_{11}	p_{12}	p_{10}
$X = 1$	p_{21}	p_{22}	p_{20}
$P(Y = y)$	p_{01}	p_{02}	1

$$\text{odds ratio} = \frac{P(\text{person is affected}|\text{person is old})/P(\text{person is unaffected}|\text{person is old})}{P(\text{person is affected}|\text{person is young})/P(\text{person is unaffected}|\text{person is young})}$$

In order to define odds ratio in a general manner, we define the probability distribution of two jointly distributed random variables. Let X and Y be two random variables taking only two values 0 and 1 (say). As an example we can think that Y denotes the disease status (0 means healthy and 1 means affected) whereas X denotes the smoking status (0 for non-smoking and 1 for smoking). To study any relationship between disease and smoking habit, we have to consider both variables jointly (Table 6.4).
where $p_{01} = p_{11} + p_{21}$, $p_{02} = p_{12} + p_{22}$, $p_{10} = p_{11} + p_{12}$, $p_{20} = p_{21} + p_{22}$ and $p_{01} + p_{02} = p_{10} + p_{20} = p_{11} + p_{12} + p_{21} + p_{22} = 1$. As explained earlier, odds ratio can be deduced in terms of probability, this eventually leads to a definition of odds ratio. We need to recall the conditional probabilities with reference to Table 6.4 to express odds ratio in terms of p_{ij}s. First note that

$$P(Y = 1|X = 1) = \frac{P(Y = 1, X = 1)}{P(X = 1)} = \frac{p_{22}}{p_{20}}.$$

Similarly, we have

$$P(Y = 0|X = 1) = \frac{p_{21}}{p_{20}},\ P(Y = 1|X = 0) = \frac{p_{12}}{p_{10}},\ P(Y = 0|X = 0) = \frac{p_{11}}{p_{10}}$$

$$\text{Thus, } OR = \text{odds ratio} = \frac{\text{odds}|X = 1}{\text{odds}|X = 0} = \frac{P(Y = 1|X = 1)/P(Y = 0|X = 1)}{P(Y = 1|X = 0)/P(Y = 0|X = 0)} = \frac{\frac{p_{22}}{p_{20}}/\frac{p_{21}}{p_{20}}}{\frac{p_{12}}{p_{10}}/\frac{p_{11}}{p_{10}}} = \frac{p_{11}p_{22}}{p_{12}p_{21}} \quad (6.11)$$

Naturally based on the data structure, we can easily get an estimate of odds ratio just by replacing the population parameters by their corresponding estimates. Since the estimate of p_{ij} is $\hat{p}_{ij} = n_{ij}/n$ for $i, j = 0, 1$, an estimate of odds ratio would be

$$\widehat{OR} = \frac{n_{11}n_{22}}{n_{12}n_{21}} \quad (6.12)$$

'Odds' and 'odds ratio' play important roles in dealing with categorical variables along with other continuous and/or categorical variables. Especially, it is of utmost

importance, when the response is a categorical variable in a regression set up. Let us look at this afresh.

Note that $0 < \text{odds}(Y|x) < \infty$ and hence $-\infty < \log(\text{odds}(Y|x)) < \infty$. Sometimes it is advantageous to work with $\log(OR)$ instead of simple OR. It can be shown that for large sample size,

$$\log(\widehat{OR}) \sim N(\log OR, \sigma^2) \text{ approximately where } \sigma^2 \approx \frac{1}{n_{11}} + \frac{1}{n_{12}} + \frac{1}{n_{21}} + \frac{1}{n_{22}} \tag{6.13}$$

Although our study variable Y is categorical, in many real situations it may depend on other variables or covariates. Since we are expecting that the response variable depends on covariate (X) and hence the odds, we can think of a regression model of y on x. However, Y is a categorical variable whereas log(odds) can take any real value; it would be easy to consider a regression equation of log(odds) on x. Naturally the easiest way is to consider a linear regression model of log(odds) on x as

$$\log(\text{odds}) = \alpha + \beta x \tag{6.14}$$

A little mathematical simplification gives more insight into this model. Noting that $P(Y = 1|x) + P(Y = 0|x) = 1$, after simplification we get

$$\log\left(\frac{P(Y=1|x)}{P(Y=0|x)}\right) = \alpha + \beta x \implies \frac{P(Y=1|x)}{P(Y=0|x)} = e^{\alpha+\beta x}$$

$$\implies P(Y=1|x) = \frac{e^{\alpha+\beta x}}{1+e^{\alpha+\beta x}} \text{ and } P(Y=0|x) = \frac{1}{1+e^{\alpha+\beta x}} \tag{6.15}$$

Thus we can say that given x, $Y \sim Binomial(1, p(x))$ where $p(x) = \frac{e^{\alpha+\beta x}}{1+e^{\alpha+\beta x}}$

Here, X takes only two values, 0 and 1. So putting values of x as 0 and 1 in (6.15), we have

$$\text{Odds ratio} = \frac{P(Y=1|X=1)/P(Y=0|X=1)}{P(Y=0|X=0)/P(Y=0|X=0)} = \frac{e^{\alpha+\beta}}{e^{\alpha}} = e^{\beta}.$$

Hence, e^{β} is nothing but the odds ratio (OR). We also have

$$\beta >=< 0 \implies e^{\beta} >=< 1. \tag{6.16}$$

Thus expression in (6.16) may generate a testing method related to the odds ratio. Note that (6.14) looks similar to a linear regression equation except that here the response variable is the log(odds) that is linearly dependent on the explanatory variable (covariate) x. This is known as 'logistic regression', which should be used when the response variable is categorical instead of a continuous variable.

Ex-6.5 In a group of 120 smokers (exposed) and 280 non-smokers (not exposed), we look at lung cancer (event) status. If 39 smokers have lung cancer and 6 non-smokers have lung cancer, what is the odds ratio for lung cancer?

$$\text{odds for lung cancer among smokers} = \frac{\hat{P}(\text{lung cancer}|\text{smokers})}{\hat{P}(\text{no lung cancer}|\text{smokers})} = \frac{39/120}{81/120} = \frac{13}{27},$$

$$\text{odds for lung cancer among non-smokers} = \frac{\hat{P}(\text{lung cancer}|\text{non-smokers})}{\hat{P}(\text{no lung cancer}|\text{non-smokers})} = \frac{6/280}{274/280} = \frac{3}{137},$$

$$\text{Hence, odds ratio} = \frac{\text{odds for lung cancer among smokers}}{\text{odds for lung cancer among non-smokers}} = \frac{13}{27} \cdot \frac{137}{3} = 21.988.$$

6.5.4 *Statistical Test Relating to Odds Ratio*

To test whether a qualitative phenotype (e.g. disease status) is associated with genotypes in the presence of other covariates like age, gender, etc., we can calculate odds ratio and perform a statistical test for odds ratio. The asymptotic distribution of sample OR as given in (6.13) provides a method for such testing. Thus the hypotheses of interest are

$$H_0 : OR = 1 \text{ against } H_1 : OR \neq 1$$

We can use a test statistic

$$T = \frac{\log(\widehat{OR}) - 0}{\sigma} \text{ which follows } N(0, 1) \text{ under } H_0, \text{ approximately (asymptotically),} \tag{6.17}$$

where the expression for σ^2 is given in (6.13). Based on a data set, we can estimate the OR, calculate p-value, and take a decision accordingly. Note that here the level of significance is approximately α as the entire method depends on the asymptotic distribution of $\log(\widehat{OR})$, i.e. assuming that the sample size is large.

We can extend this idea of testing for association in the presence of other covariates. Let us discuss this with an example. Based on the data in Table 6.1, let Y denote the disease status and X be a random variable taking values 0, 1 or 2 for genotypes AA, Aa or aa respectively. Also, let Z be another variable denoting age. Note that although we can think of age as a continuous variable, we can include more variables, be it categorical or continuous or both. In this situation we should be interested in the association of phenotype with genotype after adjusting the effects of other variables. Hence a regression type model is essential in this case; we can easily consider a logistic regression of Y on X and Z. Thus our model would be

$$\log(odds(y|x, z)) = \beta_0 + \beta_1 x + \beta_2 z \tag{6.18}$$

Note that log($odds$) is also known as 'logit'. It looks like a linear regression of log ($odds$) on other variables and we can use the least square method. However, given the values of explanatory variables, especially when the auxiliary variable is continuous, we may get only one value of the explanatory variable. Thus it is always better to estimate the parameters using the likelihood method. Noting that Y_i can take values 0 or 1 for $i = 1, \ldots, n$, the likelihood function for (6.18) is given as

$$P(y_1, \ldots, y_n | x_1, \ldots, x_n, z_1, \ldots, z_n) = \prod_{i=1}^{n} P(Y_i = 1 | x_i, z_i) = \prod_{i=1}^{n} \left(\frac{e^{\beta_0 + \beta_1 x_i + \beta_2 z_i}}{1 + e^{\beta_0 + \beta_1 x_i + \beta_2 z_i}} \right)^{y_i} \times \left(\frac{1}{1 + e^{\beta_0 + \beta_1 x_i + \beta_2 z_i}} \right)^{1 - y_i}$$

Maximising with respect to the parameters β_0, β_1, and β_2 we get their estimates. However, it involves solving simultaneous equations that are not even linear. Hence iterative or other procedures are required to get the final result. However, a simple R command fits the logistic regression directly and provides the estimates of all parameters along with their p-values.

Clearly β_1 is the logarithm of odds ratio. Hence p-value associated with $\hat{\beta}_1$ can be used directly to test $H_0 : \beta_1 = 0$ against $H_1 : \beta_1 \neq 0$. This is a test for genotypic association eliminating the effect of other variables.

In the data set as in Table 6.1, we see that age is another covariate available for individuals. So it is of interest to see whether there is any effect of this covariate, i.e. age while investigating the genetic association of any genetic locus. Naturally we cannot directly apply χ^2 test as it does not have any provision to include the effect of age. However, we can use logistic regression incorporating age as another covariate and test whether the odds ratio is one for a particular genetic locus. Hence, to see genetic association of G1 locus with the disease, our model becomes

$$\log(odds(y|G1, \text{age}) = \beta_0 + \beta_1\, G1 + \beta_2\, \text{age}$$

Using in-built R code we can easily test $H_0 : \beta_1 = 0$ against $H_1 : \beta_1 \neq 0$.

```
> assoc<-glm(a$CC.status ~ a$G1+ a$Age, family="binomial")     ## fitting
                                                   logistic regression
> summary(assoc)                    ## displays summary of association
                                     testing using logistic model
```

It shows that the p-value corresponding to age is very small whereas that for genotypes p-value is very large compared to 0.05, the chosen level of significance. Hence, although we previously found an indication of association of G1 with disease, it is probably not the actual association; the signal is mainly due to the effect of age.

6.5.5 *Genetic Association for Quantitative Phenotype*

Quantitative phenotype (or variable) usually carries more information than qualitative variable. Instead of labelling a person as having 'high', 'low', or 'normal' blood pressure, it is always better to have the exact value of her/his blood pressure as measured by a sphygmomanometer. Categorising a continuous variable into a few categories would lead to loss of information. Thus, if we have genotype data at a locus along with a quantitative phenotype of interest, we may explore genetic association of the quantitative phenotype with the genetic locus.

Here no concept of case or control exists; we only have numerical, mostly continuous values of the quantitative phenotype. Genotype data are distributed over three categories viz, with genotypes 'AA', 'Aa', or 'aa' for a bi-allelic locus. All phenotype values are distributed in these three categories.

6.5.5.1 Analysis of Variance Method

Assuming that genotype at a marker locus has three categories (e.g. AA, Aa, and aa), our interest would be to see whether the quantitative phenotype values are distributed homogeneously in these three categories for a sample of individuals. That means the phenotype values show strong homogeneity within each class corresponding to genotypes. We can very easily represent this scenario using some mathematical notations.

Suppose individuals having a particular genotype form a class or group. Let $i = 1, 2, 3$ respectively denote genotypes AA, Aa and aa. Also let y_{ij} be the quantitative phenotype value for j-th person having genotype i. Then we write a model as

$$y_{ij} = \mu_i + e_{ij}, \qquad j = 1, \ldots, n_i, \ i = 1, 2, 3 \tag{6.19}$$

where μ_i is the effect genotype class i. e_{ij}s are assumed to be random errors associated with (i, j)-th observation. For statistical analysis we assume that $e_{ij} \sim N(0, \sigma^2)$.

A few important points to note:

- Number of individuals belonging to different genotype classes may not the be same. So n_i values are not same for $i = 1, 2, 3$.
- Population genotype frequencies must depend on allele frequencies. This fact should also be reflected in the sample.
- If the effect of genotype is present, μ_is would be different leading to different means for individuals in different genotype classes.
- It is clear that if the phenotype is not associated with the genotype at the locus of interest, different genotypes have no special effects on phenotype. Thus $\mu_1 = \mu_2 = \mu_3$ in this situation, where no effect of genotype is present on the quantitative phenotype.
- However, each y_{ij} has the same (constant) variance, i.e. $V(y_{ij}) = \sigma^2$. This is a very important assumption under which we develop the analysis method.

Based on the points discussed just now, we can frame the hypotheses of testing genetic association as

$$H_0 : \mu_1 = \mu_2 = \mu_3 \text{ against } H_1 : \text{ at least one inequality in } H_0. \tag{6.20}$$

In other words here we are testing equality of means in different genotype classes assuming variances are constant. This is standard Analysis of Variance (ANOVA) method.

A little mathematical deduction would give the test statistic as

$$\mathscr{F} = \frac{MSB}{MSE} = \frac{SSB/2}{SSE/(n-3)} \tag{6.21}$$

where $SSB = \sum_{i=1}^{3} \sum_{j=1}^{n_i} \left(\bar{y}_{i0} - \bar{y}\right)^2$, $SSE = \sum_{i=1}^{3} \sum_{j=1}^{n_i} \left(y_{ij} - \bar{y}_{i0}\right)^2$ with $\bar{y}_{i0} = \frac{1}{n_i} \sum_{j=1}^{n_i} y_{ij}$, $\bar{y} = \frac{1}{n} \sum_{i=1}^{3} \sum_{j=1}^{n_i} y_{ij}$, and $n = \sum_{i=1}^{3} n_i$.

This statistic although looks complicated at the first sight, may be understood from an intuitive point of view. If H_0 is true, there is no effect of genotype on phenotype and hence we can expect that all observations come from one class, i.e. they are homogeneous. So the class means, i.e. $\bar{y}_{10}$, $\bar{y}_{20}$ and $\bar{y}_{30}$ should be very close to each other and hence they should not differ much from the grand mean, i.e. $\bar{y}$. On the other hand, if H_0 is not true, class means vary and might be different from the grand mean. A measure of such variation may be given by (assuming there are k classes)

$$SSB = \sum_{i=1}^{k} \sum_{=1}^{n_{ij}} (\bar{y}_{i0} - \bar{y})^2.$$

Clearly SSB would be small if H_0 is true and large if H_0 is not true. However, as in testing of hypothesis for other problems, the crucial question is: how large is really large? To understand this, we need to look at the scenario from another perspective.

Note that whatever be the differences, it is due to random sample only, provided H_0 is true. For each class, amount of total variation in that class is measured by $\sum_{j=1}^{n_i} (y_{ij} - \bar{y}_{i0})^2$ and combining all classes, we have $SSE = \sum_{i=1}^{k} \sum_{j=1}^{n_i} (y_{ij} - \bar{y}_{i0})^2$ as a measure of variation. For the i-th class, $\bar{y}_{i0}$ may be taken as a representative value for that class and so $\sum_{j=1}^{n_{ij}} (\bar{y}_{i0} - \bar{y})^2$ may be taken as a measure of total variation present in the i-th class. Hence, we define a measure of such variation based on the entire data, as

$$SSE = \sum_{i=1}^{k} \sum_{j=1}^{n_{ij}} (\bar{y}_{i0} - \bar{y})^2.$$

Clearly we can get an estimate of variation σ^2 from SSE. In fact, it can be shown that:

$$E\left(\frac{SSE}{n-k}\right) = \sigma^2.$$

Note that if H_0 is not true, this SSB should be much larger compared to the variation due to random sampling and they should be close to each other if H_0 is true. Thus the ratio SSB/SSE or its scaled variation may be taken as a test statistic. Hence our final test statistic becomes $\mathscr{F} = \frac{MSB}{MSE}$ as defined in (6.21).

Now it is clear that if there is no association, i.e. H_0 is true, we can expect that class means ($\bar{y}_{i0}$) would be close to each other and hence MSB would be small with respect to MSE. On the other hand, if there is any association, class means would differ substantially from each other and hence their squared deviation from grand mean, i.e. MSB would be much larger than MSE. Thus if the ratio MSB/MSE is very large, we reject H_0 claiming that there is enough evidence of significant genetic association with the phenotype.

It can be shown that when H_0 is true, the test statistic $\mathscr{F}$ as given in (6.21) follows an F distribution with degrees of freedom 2 and $n-3$. Thus in practice we calculate the p-value using $F_{2,n-3}$ distribution and take the decision accordingly.

For the data set given in Table 6.1, we want to examine any possible existence of associate of the quantitative phenotype, cholesterol level in this case, with genotype locus G1 (say). First we group the observations with respect to the genotypes and then apply ANOVA technique as discussed.

The very small p-value 0.000639 is indicative of the presence of association of G1 with cholesterol level. We can repeat the same for all other loci. Note that we completely ignore the case-control status here because when we have the quantitative phenotype, it is advisable not to search for any qualitative characteristic like affectedness of an individual. We use R code for the calculation using the ANOVA method.

```
> y1<-a$Cholesterol[which(a$G1=="AA")]      ## extracting data for
                                              genotype AA
> y2<-a$Cholesterol[which(a$G1=="Aa")]            ## extracting data for
                                                   genotype Aa
> y3<-a$Cholesterol[which(a$G1=="aa")]            ## extracting data for
                                                   genotype aa
> y<-c(y1,y2,y3)                 ## creating phenotype data for analysis
> label<-c(rep(0,length(y1)),rep(1,length(y2)),rep(2,length(y3)))      ## creating
                                                   labels for classes
> anova<-aov(y~label)                ## applying R function to do ANOVA
> summary(anova)                              ## displays the output
```

6.5.5.2 Regression Based Test

Sometimes we use a slightly different test to check whether there is any association at a locus. If a locus is associated with the disease, usually it is the minor allele that seems to be the disease allele or responsible for increasing the disease risk. It is natural to think that increased number of minor alleles in a genotype induces more adverse effects. If we use the number of minor alleles as indicators of genotypes, i.e. if we consider a variable X that takes values 0, 1, or 2 for genotypes AA, Aa, or aa respectively where 'a' is the minor allele, increase in the value of X increases the disease risk, if associated. Thus, we can use regression approach to test for genetic association. Thus we consider a linear regression model as

$$y_i = \alpha + \beta x_i + e_i, \qquad i = 1, \ldots, n \tag{6.22}$$

where $e_i \sim N(0, \sigma^2)$ for all $i = 1 \ldots, n$.

Using this model we can test for $H_0 : \beta = 0$ against $H_1 : \beta \neq 0$. We can calculate the p-value based on t distribution and take decision accordingly. However, this implicitly assumes that minor alleles, if associated, have a linear effect of the phenotype which, may not always be the case.

6.5.5.3 Kruskal-Wallis Test

There is one issue about which we must be very cautious. This method is applicable only when the underlying assumptions are valid. Unless the variances are same for all classes, we cannot apply this ANOVA method nor when the variables do not follow a normal distribution. However, in such situations we can perform a nonparametric test, called Kruskal-Wallis test. This is nonparametric analogue of ANOVA. Although a little less powerful than ANOVA, we have no option but to use this test when assumptions of ANOVA are not valid.

A test for equality of variances of three groups and a test for normal distribution using Kolmogorov-Smirnov test or Shapiro-Wilk test is advised before applying ANOVA. Violation of any one goes against the use of ANOVA in this case. In our given data set, normality test for each group of genotypes fails and hence we should not use ANOVA; rather we use Kruskal-Wallis test, the nonparametric version of ANOVA. Here we just mention how to do Kruskal-Wallis test in R; we shall discuss this test in detail later.

```
> kruskal.test(y,label)$pvalue      ## provides p-value for Kruskal-Wallis test
```

The p-value 0.006773 is very small compared to 0.05. So we reject H_0 of equality of medians in three genotype groups concluding that G1 might be associated with cholesterol level.

6.6 Problems in Association Testing

The above discussion guides us into the nice world of genetic association study. Especially it converts some very simple idea to testing the existence of genetic association of a locus with either qualitative or quantitative phenotype. But that's not all. Let's consider one real situation. Suppose there are one million markers across the genome. We have genotype data for all loci and also phenotype data. Now apply an appropriate test that we learned. It is expected that p-values for a few tests are small compared to the level of significance (fixed before the test) and for others, p-values are large enough. This operation, in its simplest form can be described in four steps, as follows.

- Collect a group of cases and controls.
- Genotype at each marker.
- Perform case-control association test at each marker.
- Get the p-values and take decision at chosen level of significance, say 0.05 for each locus. If we see that the value is very small compared to 0.05, we feel excited because we find a locus associated with the phenotype or disease!

But, unfortunately, we are missing a few things here. There are several issues related to the inherent population structure from where the sample is taken, choice of level of significance, power of the association test, sample size required to achieve a high power, etc. Unless we focus on these issues, we might end up with too many false positives or fail to identify some true signals. Thus, we need to understand the problems and their eventual solutions so that our results may be replicated in future and hold promise for identification of disease associated loci.

6.6.1 Multiple Testing

Recall the meaning of type I error rate (or level of significance). If the type I error rate is 0.05, it indicates that the probability that our test identifies a marker to be significant when, in fact, it has no effect at all. This is known as 'false positive'. Then we may think that about 5% significant results may be false positives. Let's imagine a scenario. Assuming that type I error rate is 0.05, out of 1 million markers, our testing method can identify around $1000000 \times 0.05 = 50000$ markers to be false positives and we don't know which are the false positives really! That's too much. We can't afford to waste our time and money and other resources by searching too many false positives. Don't worry. There are some nice and easy statistical ways that have been developed specifically to deal with this problem.

The problem, just discussed, is known as multiple testing problem. Because of a very large number of tests, our initial choice of 'small' level of significance does not remain small enough to control the number of false positives. Note that this entire problem occurred because of our choice of α, known as level of significance for the

test. Type I error rate cannot not exceed the level of significance. Playing with this α might provide a solution. We discuss two most popular and effective approaches on how to control the number of false positives.

Bonferroni correction: Bonferroni conceived this problem about a century ago and suggested a very simple way to tackle this. Let's assume for each test, the level of significance is denoted by α. He suggested that instead of taking α for each testing, just take α/k where k is the number of tests. So if there is 1 million markers giving 1 million tests, take $\alpha = 0.05/1000000 = 5 \times 10^{-8}$ as the level of significance for each test. Naturally this value of α ensures that the approximate number of false positives among one million markers would be only 0.05. This is a simple and good solution.

However, there are some drawbacks to this method. In fact, this problem persists for all multiple testing procedures, to a varying degree. We know that if type I error rate decreases, type II error rate increases and consequently the power of the test decreases. Thus, Bonferroni correction might lead to serious compromise of the ability of the test to capture the associated loci, especially those having low or moderate effects. So, although we control the number of false positives, while doing so, we miss a few truly associated signals (markers). Although Bonferroni correction is a very simple and easy solution to the problem of multiple testing, it is too conservative giving too much penalty in finding true associations. Moreover, in order to ensure a reasonably high power, we require an extremely large sample, which is sometimes very difficult to collect. Addressing this issue, a few methods have been developed that are less conservative compared to Bonferroni correction.

Benjamini-Hochberg (BH) correction: This is probably the most popular method that is supposed to control the number of false positives in a much less conservative way than Bonferroni's method. While Bonferroni correction keeps the level of significance for each test at a fixed level, BH correction deviates from this idea. The type I error rate should change with the nature of p-value keeping the global type I error rate fixed at level α. The procedure of BH correction is as follows.

- Calculate the p-values for k markers, say $P_1, P_2, \ldots, P_k$.
- Arrange them in increasing order of magnitude, i.e. $P_{(1)}, P_{(2)}, \ldots, P_{(k)}$, where $P_{(1)}$ is the minimum of $P_1, P_2, \ldots, P_k$, $P_{(2)}$ is the next higher minimum, and so on, $P_{(k)}$ being highest one.
- Reject the i-th test if $P_{(i)} < \frac{i}{k}\alpha$.

It is seen that this method is less conservative in the sense that power is not too low compared to the Bonferroni method. However, we need a much larger sample to achieve the power that can be obtained in single marker analysis. Although there are some more methods that are improvements over the BH method, this is very popular mainly due to its simplicity and wide applicability in many scenarios.

We now apply these multiple testing corrections to see whether any of the genetic loci G1, G3, and G4 survives the correction. R code is very much useful as it has an in-built function to calculate corrected p-values.

```
> test.1<-chisq.test(a$CC.status,a$G1)
> test.2<-chisq.test(a$CC.status,a$G2)
> test.3<-chisq.test(a$CC.status,a$G3)
> pvalue<-c(test.1$p.value,test.2$p.value,test.4$p.value)
> p.adjust(pvalue,"bonferroni")
[1] 0.006352761 1.000000000 0.062294624
> p.adjust(pvalue,"BH")
[1] 0.006352761 0.417810511 0.031147312
```

Adjusted p-value by BH method seems less conservative as the corrected p-value for G3 is less than that obtained by Bonferroni method. If we fix level of significance as $\alpha = 0.05$, based on corrected p-values, it seems that both G1 and G3 are associated with the disease according to BH method. Note that the decision should be taken based on only one method of multiple correction chosen by the experimenter and we recommend the use of the BH correction especially when the number of tests is large.

6.6.2 Population Stratification

Solution to controlling Type I error rate in order to avoid too many false positives leads to the discovery of a wide range of multiple testing correction methods. However, there is no denying that we require a larger sample to achieve a reasonably high power when we have multiple markers to be tested. However collecting larger samples and generating genotype data at each marker for all individuals, is not so easy. The reason is not only cost and limited resources, but there are also some intrinsic characteristics of the data that need to be taken care of while collecting samples. Nowadays with multi-centric research facilities sometimes resources are available. But we need to understand and consider the population characteristics while collecting data. One such problem is due to 'population stratification'. Let's consider an example.

Suppose two different ethnic groups are mixed in a population. We don't have any knowledge of that. Consider case-control genotype data at a bi-allelic locus, where the number of control individuals with genotypes AA, Aa, and aa are 25, 68, and 58 respectively, whereas that among case individuals are 28, 70, and 50 respectively. The p-value corresponding to χ^2 using (6.9) is 0.6834. It means that at the level of significance 0.05 approximately, we have no evidence to believe that the genotype distribution among cases and controls are different, i.e. this genotype locus is not associated with the phenotype.

However, unknowingly, this is not the real situation that we see in the data. A closer look at the background information of sampled individuals reveals that there are actually two ethnic groups mixed together in the population. Assuming that the

Table 6.5 Genotype frequencies for two ethnic groups at a bi-allelic locus with alleles A and a

	Group I			Group II			Combined group		
	AA	Aa	aa	AA	Aa	aa	AA	Aa	aa
Control	8	38	44	17	30	14	25	68	58
Case	21	34	29	7	36	21	28	70	50
Total	29	72	73	24	66	35	53	138	108

ethnicity of an individual is known, Table 6.5 presents the genotype frequencies of cases and controls among the two ethnic groups.

Doing a simple association testing for each group by the χ^2 statistic, we see that the p-values for Group I and Group II are 0.0115 and 0.0487 respectively indicating the presence of association (may be marginal) in each group. But there is no association signal when these groups were merged together. So sub-grouping or heterogeneity may play an important role in masking the real situation with respect to genotype association. Hence, it becomes necessary to reconsider the entire problem.

Note that this population substructure is hidden in the population and we had no idea about that during data collection. It is also difficult sometimes to get any such idea. An example may be that mixing occurred in a previous generation and the genes were inherited within each subpopulation over subsequent generations. So this hidden or unknown population structure may give a completely erroneous result. This problem is known as 'Simpson's paradox'. And we need to be very careful while doing any association studies. The same is true for quantitative phenotype data also.

Let's look at the problem a little more carefully. It is the stratification or existence of multiple subgroups within the same population that creates the fallacy. This stratification is hidden, i.e. not known. So the best way to deal with this situation is to devise a method that tries to adjust the effect of stratification or do an association testing to eliminate the effect. The most popular method encapsulated in a software that is used to address this issue is 'eigenstrat'.

Keeping in mind all these issues that put few question marks about the simple association method using χ^2, standard ANOVA or regression methods, a few sophisticated methods have been developed and research is still continuing.

6.6.3 Polygenic Risk Score

Genome wide association studies have paved the path in identifying a large number of genetic variants, mostly single nucleotide polymorphisms that are significantly associated with many complex diseases. These variants usually have a small effect and hence correspond to a small fraction of truly associated variants. It is expected that the genotype profile of an individual should provide an indication of the predictive power of the phenotype. To this end, polygenic risk score (PRS) has been developed

for providing an estimate of genetic liability for an individual based on his/her genetic profile (Choi et al. 2020). By genetic profile, here we mean the set of genotypes at different loci of the individual.

According to the classic PRS method (Dudbridge 2013; Purcell et al. 2009) PRS is calculated as a linear combination of genotype scores at different loci weighted by the effect sizes of those loci. Genotype score at a locus is usually taken as the number of risk alleles of that individual. The set of loci includes common, intermediate, and rare variants that are associated with the disease. Naturally the effect sizes vary. The most challenging part of PRS is the proper estimation of the effect sizes that would be included in the linear combination.

Let there be K loci with effect sizes $\hat{\beta}_1, \hat{\beta}_2, \ldots, \hat{\beta}_K$ obtained through genetic association study. For continuous trait, we can use a standard regression model $y = \beta_0 + \beta_i g_i + \gamma' \boldsymbol{x} + \epsilon$ where g_i is the genotype at locus i, $\boldsymbol{x}$ contains other covariates, y is the phenotype and ϵ is the error in the model. For qualitative trait, we use logistic regression, $logit\big(P(Y = 1)\big) = \beta_0 + \beta_i g_i + \gamma' \boldsymbol{x} + \epsilon$, where symbols have the same meanings as before.

Based on genome-wide association study on a training set of samples, we can easily evaluate the effect sizes, i.e. $\hat{\beta}_i$ for $i = 1, 2, \ldots, K$. Now given the genotype score g_i at i-th locus for an individual, we have the PRS score of that individual as

$$PRS = \sum_{i=1}^{K} \hat{\beta}_i g_i$$

Usually the target individual, i.e. the individual for whom we are interested to calculate the PRS, should be outside the training sample; otherwise the calculated PRS might be inflated and may not show an unbiased picture for genetic susceptibility. However, it is desirable that the target individual should match other features as that observed in the data set that is used for estimation of β_is. PRS provides a predictive power for genetic liability. This is a score assigned to each individual based on genotype profile at a large number of loci. However, it is clear that the estimates of effect sizes play the most important role. Sometimes use of estimates from public databases is advocated. These estimates are calculated on large data sets. However, ethnicity is also an important factor. Moreover, presence of other covariates, existing LD structure (Baker et al. 2018) among the markers, etc. are some influencing factors that need to be taken care of while estimating β_is. Several other advanced methods are also developed to fine tune these estimates. It is also seen that instead of considering only the markers that reach the genome-wide significance, we can achieve more predictive power if we use a large number of SNPs in the calculation of PRS.

6.7 Some Advanced Association Testing Methods

Apart from multiple testing or population stratification problems, there are some other issues. Solutions to them, even partially, might lead to a better or more powerful method of genetic association. Needless to say, these methods would also consider some other features inherent to the data and the genetic architecture. These additional features, if known, should be explored and exploited properly to come up with more sophisticated and powerful methods.

It is easy to conceive that if two loci are situated at a close proximity on the same chromosome, they might be correlated. In other words, one might affect the other and this degree of influence varies. If there is extremely high correlation between them, we may ignore one and proceed to association testing with one marker. But if the correlation is moderate, ignoring one locus might incur loss of information. Single marker test does not consider this correlation at all. It is conjectured that if we consider two or more markers together, this holistic approach might be more realistic and informative. Another issue, which is more conceptual, is that it is the gene in which we are ultimately interested. Gene is not only the unit of inheritance, but it triggers also all downstream activities that is manifested in a phenotype. So instead of single marker at a time, a gene (of course through multiple markers) is more informative in deciphering disease gene association.

Honouring these ideas we discuss two methods that are well accepted and widely used in the context of studying the possible association of a disease with a gene or a user-defined small genomic region. Information about gene or genomic region of interest comes through a set of markers, mostly SNPs, that lie in its close proximity.

6.7.1 *Kernel-Based Association Test (KBAT)*

The goal of kernel-based association test (Mukhopadhyay et al. 2010) is to test association between a qualitative phenotype and a set of SNP markers that belong to a small genomic region or gene. In this method, no assumption is made about the direction of SNP effects. Since it considers multiple SNPs at a time, the correlation structure among the SNPs has already been taken into account implicitly. However, if two SNPs are extremely highly correlated, it is recommended to take only one marker from the two. This test, however, is not a standard test and depends on what is called 'genotype similarity score'.

The data consists of n_1 cases and n_2 controls genotyped at K markers. Naturally, individuals belonging to the case group are assumed to be homogeneous with respect to their genotype distribution at each locus. The same is true for the control group. But there is heterogeneity between the two groups if there is any association; otherwise they should be homogeneous. The test is based on 'kernel' or genotype similarity score, that is defined based on genotypes at a locus for two individuals.

The most commonly used kernel is 'allele match (AM) kernel'. One advantage of AM kernel is that it does not require the knowledge of risk allele. We can also define other kernels. If g_i and g_j are the genotypes of i-th and j-th individuals at a locus, AM kernel takes the value 4 if g_i and g_j are same; 2 if one is homozygote and the other heterozygote, and 0 if they don't share any common alleles. These genotype scores are the backbone of the test. We can evaluate such scores between pairs of individuals in controls and also in cases. Moreover, we can combine single marker score to generate multi-loci score between a pair of individuals. If the overall score in the control group is close to that for the case group, we can say that there is no evidence of association for the set of loci considered together; otherwise there may exist an association. Note that through this test we try to investigate association not at a single marker level, but whether the gene or genomic region of interest as a whole is associated with the qualitative phenotype (may be disease).

Let $h_l^k(g_i, g_j)$ be a measure of genotype similarity between i-th and j-th individuals with genotypes g_i^k and g_j^k at k-th locus ($k = 1, 2, \ldots, K$) respectively in group l ($l = 1, 2$). Here $l = 1$ denotes cases and $l = 2$ denotes controls. Within each group of size n_l, there are altogether $m_l = n_l(n_l - 1)/2$ distinct similarity scores for m_l possible pairs. Keeping the idea of ANOVA in mind, we can write the above discussion in terms of a linear model as

$$y_{l(ij)} = \mu + \alpha_l + e_{l(ij)}; \quad i < j = 1, 2, \ldots, n_l; \quad l = 1, 2 \tag{6.23}$$

where $y_{l(ij)}$ is the genotype similarity score for (i, j) pair of individuals in the l-th group, μ denotes the overall grand mean or the general effect for pairs of individuals, α_l is the group specific treatment effect for similarity scores over the general effect, and $e_{l(ij)}$ are the error components.

We assume

(i) $\alpha_1 + \alpha_2 = 0$,
(ii) $V(e_{l(ij)}) = \sigma^2$ for $l = 1, 2$,
(iii) $Cov(e_{l(ij)}, e_{l'(i'j')}) = \begin{cases} \rho\sigma^2 & \text{for } i \neq i';\ j = j';\ l = l' \text{ or } i = i';\ j \neq j';\ l = l' \\ 0 & \text{otherwise} \end{cases}$

Although it looks like an ANOVA model, it violates the basic assumption because the error variables are not always uncorrelated, and $y_{l(ij)}$s are not normally distributed. This prohibits the direct application of ANOVA. However, we use the basic philosophy behind ANOVA that the variation between groups might be more compared to that within groups if there is a significant effect of groups, on an average. Note that $\bar{U}_l^k = \sum_{i<j} h_l^k(g_i^k, g_j^k)/m_l$ is the average genotype similarity scores of all pairs of individuals in l-th group at k-th marker. Corresponding to k-th marker, we define measures of within and between groups variations as

$$SSW_k = \sum_{l=1}^{2} \sum_{i<j} \left(h_l^k(g_i^k, g_j^k) - \bar{U}_l^k\right)^2 \text{ and } SSB_k = \sum_{l=1}^{2} m_l\left(\bar{U}_l^k - \bar{U}_k\right)^2.$$

In ANOVA we compare between sum of squares to within sum of squares or a scaled version of them. For testing the association of K markers, the proposed KBAT statistic is

$$\mathscr{F} = \frac{\sum_{k=1}^{K} SSB_k}{\sum_{k=1}^{K} SSW_k} \tag{6.24}$$

This test statistic resembles with the standard ANOVA statistic except that in ANOVA we use MS (mean squares) instead of SS (sum of squares). So a few variations of KBAT can be thought of. For example, we can consider the sum of ratios of individual statistics instead of the ratio of the sum of squares. Hence, an alternative form of KBAT may be $\mathscr{F}_A = \sum_{k=1}^{K}(SSB_k/SSW_k)$. Another important issue is involved in KBAT statistic. In standard ANOVA under normality assumption, the test statistic follows an F distribution under null hypothesis of no association. However, the distribution of KBAT even under null hypothesis $H_0 : \alpha_1 = \alpha_2 = 0$ is not known although it can be shown that it is right-sided test provided $\rho > 0$. Usually for all the kernels considered in KBAT statistic, $\rho > 0$, i.e. we reject the null hypothesis of no association if observed $\mathscr{F} > F_\alpha$ where the threshold value F_α should be calculated either using the null distribution of the test statistic or empirically. Since the null distribution is unknown, we can obtain the value using permutation technique or using the asymptotic distribution of the test statistic under H_0. It is always less time consuming to use the asymptotic distribution while calculating the p-value, especially when there is an issue of multiple testing. Note that multiple testing burden also occurs here as there are many genes or gnomic regions to be tested; however, the burden is much less than that for single locus tests. For example, with 1 million markers distributed over 1000 genes, the number of individual single marker tests is 1 million whereas it is only 1000 for KBAT statistic. It is seen that the asymptotic distribution of another slightly modified version of KBAT statistic is χ^2 with K degrees of freedom, as given in the theorem below.

Theorem 16 *Let* $x = \hat{p}_k^2$, $y = 2\hat{p}_k\hat{q}_k$, *and* $z = \hat{q}_k^2$. *Define* $\hat{\mu}_{12,k} = 16(x^3 + y^3 + z^3) + 8(1 - (x^3 + y^3 + z^3) - (4xz - x^2z - xz^2)) - 4(xy^2 + zy^2 + y(1 - y)^2)$. *Take* $\hat{\sigma}_{1k}^2 = \hat{\mu}_{2k} - \hat{\mu}_{1k}^2$ *and* $\hat{\sigma}_k^2 = \hat{\mu}_{12,k} - \hat{\mu}_{1k}^2$. *Then* $\hat{\sigma}_{1k}^2$ *and* $\hat{\sigma}_k^2$ *are consistent estimators of* σ_{1k}^2 *and* $\hat{\sigma}_k^2$ *respectively,* $k = 1, \ldots, K$. *If* $\frac{n_2}{n_1} \to \nu$ *as* $n_1 \to \infty$, *then as* $min(n_1, n_2) \to \infty$,

$$T = \beta(n_1, n_2).\frac{K(1+\nu^2)}{2\nu(1+\nu)}.\frac{\sum_{k=1}^{K} SSB_k/\hat{\sigma}_{1k}^2}{\sum_{k=1}^{K} SSW_k/\hat{\sigma}_k^2} \xrightarrow{L} \chi_K^2, \text{ where } \beta(n_1, n_2) = \frac{\binom{n_1}{2}\binom{n_2}{2}}{n_1}$$

Theorem 16 indicates that if t is the observed value of the modified KBAT statistic T, we can easily calculate the p-value based on the asymptotic distribution of T. Hence the p-value (approximately) is given by value $= P(T > t|H_0) \approx P(\chi^2_K > t)$. The notation '$\xrightarrow{L}$' indicates 'convergence in distribution,' i.e. T is asymptotically distributed as a χ^2 variable with K degrees of freedom.

Since using χ^2 distribution, the p-value can be calculated very fast, it gets rid of doing the computationally intensive permutation technique and also provides the p-value up to a very high degree of accuracy.

6.7.2 *Sequence Kernel Association Test (SKAT)*

Proposal of KBAT (Mukhopadhyay et al. 2010) is a paradigm shift in genetic association study in the sense that it advocates the use of information from multiple markers at a time and hence is more powerful compared to single marker association tests. However, it has a disadvantage. We know that in many studies, along with genotype and phenotype data, information on other covariates like age, sex, etc. are also available. It is very difficult to adjust this KBAT statistic for other covariates. A different multi-loci association test is proposed (Wu et al. 2009) that takes care of the presence of other covariates, known as sequence kernel association test (SKAT).

Like KBAT, for a small genomic region or a gene consisting of multiple SNP markers, we can test the joint effect of these markers through SKAT adjusting the effects of other covariates like age, sex, etc. To describe this method, we first consider a linear model for quantitative trait as

$$y_i = \alpha_0 + \boldsymbol{\alpha}'\boldsymbol{X}_i + \boldsymbol{\beta}'\boldsymbol{G}_i + \epsilon_i, \quad i = 1, \ldots, n \tag{6.25}$$

where X_i is the covariate and G_i is the genotype vector for the i-th individual and ϵ_i is the error associated with the response variable y_i, $i = 1, \ldots, n$. For qualitative trait like case-control status, etc., we consider a logistic model as,

$$logit\ P(y_i = 1|\boldsymbol{X}_i, \boldsymbol{G}_i) = \alpha_0 + \boldsymbol{\alpha}'\boldsymbol{X}_i + \boldsymbol{\beta}'\boldsymbol{G}_i, \quad i = 1, \ldots, n \tag{6.26}$$

Here $\boldsymbol{\beta} = (\beta_1, \ldots, \beta_K)'$ is the effect of genotype vector $\boldsymbol{G}$ on the phenotype. Therefore the vector G_i contains the genotypes at K loci for individual i, $i = 1, \ldots, n$.

Our main purpose is to test

$$H_0 : \beta_1 = \cdots = \beta_K = 0 \text{ against } H_1 : \text{ at least one } \beta_i \neq 0.$$

However to construct a more powerful test, we assume that each β_j $(j = 1, \ldots, K)$ follows an arbitrary distribution with mean 0 and variance $w_j\tau$. This in turn implies that we can modify our hypotheses as

$$H_0 : \tau = 0 \text{ against } H_1 : \tau \neq 0.$$

Here we assume a random effect for β_is. We can now easily construct a commonly used score statistic for testing the null hypothesis. The score statistic for testing H_0 is

$$Q = (\boldsymbol{y} - \hat{\boldsymbol{\mu}})' H (\boldsymbol{y} - \hat{\boldsymbol{\mu}}) \tag{6.27}$$

where $H = GWG'$, $\hat{\boldsymbol{\mu}}$ is the predicted mean of $\boldsymbol{y}$ under H_0, i.e. $\hat{\boldsymbol{\mu}} = \hat{\alpha}_0 + X\hat{\boldsymbol{\alpha}}$ for continuous trait and $\hat{\boldsymbol{\mu}} = logit^{-1}(\hat{\alpha}_0 + X\hat{\boldsymbol{\alpha}})$ for dichotomous trait, G is the matrix of K genotypes for all n individuals, and H is the kernel matrix. We can define this H in many ways. However, it is convenient to use genotype similarity score between individual i and j as $H_{ij} = \sum_{k=1}^{K} w_k G_{ik} G_{jk}$ and H_{ij} is the (i, j)-th element of the matrix H and $W = diag(w_1, \ldots, w_k)$ with w_i being the weight of the ith variant. Based on prior information or intuitively we can define H in many ways suitable for the specific purpose of the study.

In order to calculate the p-value for a study, we can adopt a computation intensive method like permutation of bootstrap method. However, we can calculate the p-value much faster using the asymptotic distribution of the SKAT statistic, as given below.

Theorem 17 *Let $B = V - V\tilde{X}(\tilde{X}'V\tilde{X})^{-1}\tilde{X}'V$ where $\tilde{X} = (\mathbf{1}\ \ X)$ is $a\, n \times (K+1)$ matrix. Also note that for continuous phenotype, $V = \hat{\sigma}_0^2 I$ where $\hat{\sigma}_0^2$ is a consistent estimator under H_0, and for dichotomous trait, $V = diag\Big(\hat{\mu}_{01}(1-\hat{\mu}_{01}), \ldots, \hat{\mu}_{0n}(1-\hat{\mu}_{0n})\Big)$ with $\hat{\mu}_{0i} = logit^{-1}(\hat{\alpha}_0 + \hat{\alpha}' X_i)$. Then under H_0,*

$$Q \sim \sum_{i=1}^{n} \lambda_i \chi_{1,i}^2 \ \textit{asymptotically as } n \to \infty,$$

where $(\lambda_1, \lambda_2, \ldots, \lambda_n)$ are the eigenvalues of $P^{1/2} H P^{1/2}$ and $\chi_{1,i}^2$ are independent χ_1^2 variables for $i = 1, \ldots, n$.

However there is no direct way to calculate the p-value based on this asymptotic distribution even under H_0. SKAT uses an approximate method as proposed by Davies (1980). So the p-value calculated in SKAT uses two levels of approximations; one for the asymptotic distribution of Q and another for p-value calculation by Davies' method. However, it can be used in practice as the sample size is usually very large nowadays.

6.8 Points to Remember

We have learnt how to analyse genetic data especially genotype data along with some covariates to identify possible association with the disease or the phenotype of interest. However, some important issues emerged during our discussion.

We first identify whether the phenotype is qualitative or quantitative and appropriate tests should be chosen based on the nature of phenotype. Methods for qualitative phenotype are usually different from that for quantitative phenotype and we need to be careful about that. After the test at multiple loci, it is necessary to do multiple testing correction; otherwise our analysis might produce too many false positives, thus wasting resources. However, before doing any analysis, we must check for other factors like population stratification, HWE, etc. so that the ultimate decision would not be masked due to unwanted source(s) of influence.

We should have a global view of the problem, data, and expected outcome. This would lead to tackling the problem at hand in the best possible way. Although single marker test is a first step for this kind of genetic analysis, after checking HWE and/or satisfying other quality control criteria, it is worthwhile to try some other sophisticated methods that might throw more light on the entire scenario. Research is still continuing in this area. Hence, search for recently developed methods of association and other related problems is welcome, especially when it has potential to strengthen our conclusion or discovering new potential biomarkers.

New methods may sometimes look complicated because it considers several issues simultaneously. However, in most cases, a new and more appropriate statistical method comes with user-friendly programming code or as an R function that can be easily used. So we should not have any problem in implementing the newly developed method if deemed most appropriate for a problem.

Exercise

6.1 For the following pedigree,

(a) What are the possible genotypes of person 3? Find out the probability of each genotype of person 3.
(b) What is the probability that person 5 has genotype AA when it is known that person 3 has genotype Aa?
(c) What is the probability that person 5 has genotype AA?
(d) Given that person 5 has genotype Aa, what is the probability that person 3 has genotype Aa?
(e) Further it is known that A allele is the disease allele and two copies of A are required for a person to be affected. However, not all persons with AA genotype are affected. This is called incomplete penetrance. It is seen that $P(\text{person is affected} \mid \text{person has genotype } AA) = 0.83$. What is the probability that person 5 will be affected?

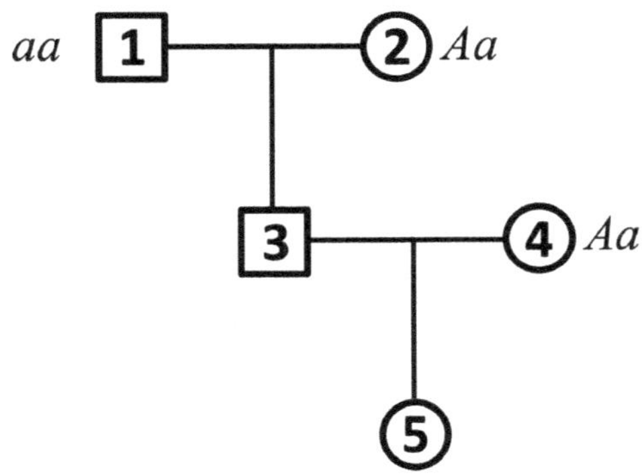

6.2 In a nuclear family, a child is carrying genotype DD at a bi-allelic locus. Assuming random mating and allele frequency of allele D in the population is p, what is the probability that her father has genotype (a) dd, and (b) Dd?

6.3 Suppose at a bi-allelic locus, HWE is satisfied. If the minor allele frequency at this locus is 0.12, what are the genotype frequencies at that locus?

6.4 Assuming HWE exists at a bi-allelic locus with two alleles M and m, if the genotype frequencies of MM and Mm are 0.04 and 0.32 respectively, what is the minor allele frequency at this locus?

6.5 At a bi-allelic locus with alleles A and a, if the genotype frequencies $P(AA)$, $P(Aa)$, and $P(aa)$ are respectively a, ar, and ar^2, for some a and r $(a, r > 0)$, what can you say about HWE at this locus?

6.6 Suppose at a locus there are three alleles A_1, A_2 and A_3. Based on genotype data n_{ij} for genotype A_iA_j, for $i = 1, 2, 3$ and $j = 1, 2, 3$, find the maximum likelihood estimates of all three allele frequencies. Can you get the same estimates if you adopt the gene counting method? Justify your answer.

6.7 Genotyping at a bi-allelic locus with alleles D and d was done for a sample from a small island in South Pacific Ocean. The observed frequencies for genotypes DD, Dd and dd are 243, 41 and 6 respectively. Check whether these genotype frequencies are in HWE.

6.8 In Question 6.7, suppose the given genotype frequencies are for females and D locus is on X -chromosome. Let the genotype frequencies for D and d at this locus among males be 280 and 30 respectively. What can you say about HWE at this locus?

6.9 Suppose allele frequencies at a bi-allelic locus L are in HWE with $P(A) = p$. After two generations, due to severe epidemic, genotype frequencies of Aa and aa have been reduced by $2pqr$ and pqr respectively, where $0 < r < 1$. How many generations will be required to stabilise the genotype frequencies? Show that genotype frequencies of AA, Aa, and aa genotypes will be $p^2(1 + 2qr)^2$, $2pq(1 + 2qr)(1 - 2pr)$ and $q^2(1 - 2pr)^2$ respectively, when they again attain HWE.

6.10 Among many Indian tribes, the allele frequency of B in the ABO blood groups is extremely low. For example, in one sample of 600 people from one location, there were 37 A and 563 O blood types. What are the best estimates of the

allele frequencies of A, B, and O in this population and what are the expected genotype frequencies assuming HWE?

6.11 At a particular bi-allelic locus, we have the allele frequencies as 0.575 and 0.425 respectively for alleles A and a. The genotype counts in a sample from a particular population are 30, 55, and 15 respectively for genotypes AA, Aa, and aa.

(a) Define the test statistic and specify its null distribution along with the degrees of freedom to test whether the genotype frequencies in this population are significantly different than what would be expected if the population is in Hardy-Weinberg equilibrium for that locus (use 5% level of significance).
(b) What is the value of the test statistic? Draw inference by comparing the value with the value obtained from Chi-square cut off at 5% level of significance.

6.12 The colour of a person's eyes is determined by a single pair of alleles at a locus. If they are both blue-eyed alleles (BB), then the person will have blue eyes; if they are both brown-eyed alleles (bb), then the person will have brown eyes; and if one of them is blue-eyed and the other a brown-eyed allele (Bb), then the person will have brown eyes. (Because of the latter fact, we say that the brown-eyed gene is dominant over the blue-eyed one.) A new-born child independently receives one eye gene from each of its parents, and the gene it receives from a parent is equally likely to be either of the two eye genes of that parent. It is found that, $P(B) = 0.08$. We have 150 brown eyed and 14 blue eyed people in a sample drawn from a population. Can we test whether the locus is at Hardy-Weinberg equilibrium? Draw statistical inference from the data.

6.13 Suppose at a tri-allelic locus data are collected on the number of people having different genotypes. Assuming all possible genotypes that can be formed by three alleles, are present in the data, how do you estimate allele frequencies? Also calculate the expectation and variance of your estimators.

6.14 Suppose the p-values for testing odds ratio in three situations are calculated based on the asymptotic distribution of logarithm of sample odds ratio. Let p-values thus obtained be 0.000031, 0.053, and 0.521. How do you interpret each case, assuming that the level of significance is taken as 0.01.

6.15 In Question 6.14, suppose these p-values are calculated for three genetic loci. What do you do to preserve the maximum type I error rate at 0.01? Interpret your result.

6.16 A case-control study on smoking status and oral cancer reveals that there are 446 cases who are smokers whereas 179 controls who are also smokers. Among non-smokers, these numbers are 13 and 29 respectively. Calculate odds ratio and comment on your result.

6.17 Later some more information becomes available with respect to the problem in Question 6.16. The data can further be classified based on alcohol use, as given below. Calculate odds ratio for each of the two groups based on alcohol

use. Comment on your result combining findings with that in Question 6.16. You may calculate other statistical quantities and/or adopt some other methods to get a comprehensive view.

	Alcohol users		non-alcohol users	
	Cases	Controls	Cases	Controls
Smokers	201	111	245	68
Non-smokers	8	9	5	20

6.18 What is a matched case-control study design and why is matching important in the context of identifying the association of a genetic marker with a disease?

6.19 For the data in Table 6.1, do the following.

(a) Assume that the disease is recessive, i.e. caused only if a person has two copies of disease alleles at a locus. Estimate the minor allele frequency at each locus based on case-control status data only.
(b) Check HWE at each locus.
(c) Test whether any locus is associated with the affectedness of people using genotype data and affection status. Do not forget to do multiple testing correction before you reach your conclusion.
(d) Using the data on cholesterol, check whether any locus is associated with cholesterol level.
(e) Suppose it is known that the cholesterol data do not follow a normal distribution. How do you test for association using cholesterol data at each locus?
(f) Divide the entire sample into two groups: those having age less than or equal to 50 years, and those above 50 years of age. Find the odds ratio for disease for the older group with respect to younger group. What is the expectation and variance of this odds ratio? Can you devise a statistical test to conclude that odds ratio is equal to 1.5? If so, please do it; otherwise, justify your answer.
(g) Address the question in 6.19(c), but in presence of covariate age. What is your conclusion?
(h) Address the same question in 6.19(d) in presence of age. What is your conclusion?
(i) Apply KBAT and SKAT tests to see whether the loci G1, G2, G3, and G4 are associated with disease, using phenotype data as case-control status.
(j) Write a detailed report on your findings obtained so far.

6.20 We are interested to see whether a quantitative trait is associated with a di-allelic locus with genotypes AA, Aa, and aa, where 'a' is the minor allele. Develop a statistical test for genetic association by writing down the model, the null and alternative hypotheses with justification, the test statistic, and critical region for the test, when we have no knowledge about the mode of inheritance of the alleles. If it is known, in this scenario, that more number of disease alleles

might increase the disease risk, how would you modify your analysis? Which one do you think to be more powerful and why?

6.21 An investigation carried out to study the toxic effects of mercury was described in the article "Comparative Responses of the Action of Different Mercury Compounds on Barley" (International Journal of Environmental Studies [1983]). Ten different concentrations of mercury were compared with respect to their effects on average dry weight (per 100 seven-day-old seedlings). The basic experiment was replicated 4 times for a total of 40 observations. The article reported an ANOVA F statistic of 1.895. Using a significance level of 0.05, test the hypothesis that the true mean dry weight is the same for all 10 concentration levels.

6.22 We have three genotypes: AA, Aa and aa. After giving a particular treatment, we measured the inflammatory response of 30 individuals divided into the above 3 genotype groups by the individual levels of a particular pro-inflammatory cytokine(pg/mL).

AA	12.5, 13.2, 17.8, 15.2, 14.1, 14.9, 14.7, 13.1, 13.5, 11.9, 12.4, 13.8
Aa	12.9, 14.9, 16.8, 14.1, 13.5, 14.2
aa	25.8, 21.6, 19.6, 14.8, 23.7, 24.9, 21.3, 19.8, 20.2, 23.8, 26.1,20.9

On the basis of the data, can we say that the treatment effect differs among these three genotype groups? Which genotype is associated with a significantly higher mean pro-inflammatory response for the treatment as compared to the other two genotypes?

References

Baker, E., Schmidt, K. M., Sims, R., O'Donovan, M. C., Williams, J., Holmans, P., Escott-Price, V., & Consortium, W. T. G. (2018). POLARIS: Polygenic LD-adjusted risk score approach for set-based analysis of GWAS data. *Genetic Epidemiology, 42*(4), 366–377.

Choi, S. W., Mak, T. S., & O'Reilly, P. F. (2020). Tutorial: A guide to performing polygenic risk score analyses. *Nature Protocols, 15*(9), 2759–2772.

Davies, R. B. (1980). Algorithm as 155: The distribution of a linear combination of χ^2 random variables. *Journal of the Royal Statistical Society, Series C, 29*(3), 323–333.

Dudbridge, F. (2013). Power and predictive accuracy of polygenic risk scores. *PLoS Genetics, 9*, e1003348.

Mukhopadhyay, I., Feingold, E., Weeks, D. E., & Thalamuthu, A. (2010). Association tests using kernel-based measures of multi-locus genotype similarity between individuals. *Genetic Epidemiology, 34*(3), 213–221.

Purcell, S. M., et al. (2009). Common polygenic variation contributes to risk of schizophrenia and bipolar disorder. *Nature, 460*, 748–752.

Wu, M. C., Lee, S., Cai, T., Li, Y., Boehnke, M., & Lin, X. (2011). Rare-variant association testing for sequencing data with the sequence kernel association test. *American Journal of Human Genetics, 89*(1): 82–93.

Chapter 7
Some Extensions of Genetic Association Study

Now that we know how to identify a genetic locus associated with a phenotype, be it qualitative or quantitative, there might arise different situations, each of which demands serious attention. This requires special treatments to search for some answers to the newly arisen questions, starting from development and/or modifying existing methods, to meeting computational challenges, and sometimes, searching for prior information to handle those issues. Each issue comes with its own special characteristics. It needs to be addressed accordingly, in a specific way in tune with the issue. Thus, one unified approach to solve all issues might seem too ambitious. If any such method exists, it needs to be evaluated carefully because there might be some compromise in performance.

As far as the genomic location is concerned, location of one marker may be very close to another marker giving rise to strong correlation between them. When we study association of a locus, we usually ignore the presence of other loci although they might be correlated with the previous one, resulting in loss of information. So it raises a major issue of incorporating this information while studying genetic association with a phenotype. Naturally, the methods that we have already learnt, may not be applicable mainly because here we have to deal with a combination of alleles from two or more different markers that might or might not be correlated. Moreover, degree of this correlation varies.

Although simple inclusion of a covariate in the model might work well to find genetic association after adjustment of covariate effects like effect of age, gender, etc., it might not be enough. We need to assess whether this really gives more reliable results. Few recent studies indicate that there might be some issues that need to be considered; otherwise the power of the test might not always be higher than the test for genetic association without this covariate. More than one covariate situation requires appropriate attention while pursuing the analysis.

In case of qualitative phenotype we usually have case-control data. This kind of data set gives only two categories of phenotype, case or control, affected or unaffected, etc. However, in many diseases like cancer, information is available for more than two phenotype categories. As for cancer, while collecting data, we have the idea of the stage of cancer that the person is suffering from. This leads to more than two

I. Mukhopadhyay and P. P. Majumder, *Statistical Methods in Human Genetics*, Indian Statistical Institute Series, https://doi.org/10.1007/978-981-99-3220-7_7

categories, like, normal, stage I, stage II, stage II, stage IV, etc. Hence it is imperative to understand this multi-category data set for phenotype and appropriate statistical methods are required to analyse this type of data.

7.1 Generating Questions

In order to appropriately anlayse genetic data, we need to understand the data and its immediate features; what types of variables are there in the data and what should be their roles; how to incorporate every available information in the genetic association study, and many other related issues. These questions would add a new dimension to data analysis.

(1) How do we conceptualise the situation that one genetic marker is related to another?
(2) How do we measure such relationship and incorporate it in the genetic association study?
(3) Can we modify existing methods to incorporate such a situation? Or do we need to develop other statistical method(s) to tackle this issue? If so, what is the solution?
(4) If phenotype data correspond to more than two classes, can we check whether the gene expressions are the same, on an average, for all classes or groups?
(5) Do the data in each group follow a normal distribution?
(6) While applying a statistical method, do the data satisfy the assumptions of that method?
(7) If yes, fine; otherwise can we find a better method?
(8) What happens when data on more than one relevant covariate are collected?
(9) How do we perform genetic association study in presence of such covariates that might influence the phenotype?
(10) How do we study genetic association when there are more than one category in the qualitative phenotype under study?
(11) Should we study all covariates that are available?
(12) How do we do genetic association test with a locus adjusting the effects of covariates?
(13) Do these methods differ depending on the nature of phenotype (qualitative/ quantitative)?
(14) Are there any other issues that are involved while adjusting covariate effect in association study?

7.2 Haplotype Association

As we have already discussed in Chap. 1, our DNA has several million nucleotide positions at which all individuals do not carry the same nucleotide. These are called DNA variants. Because chromosomes segregate from parents to offspring independently, DNA variants on different chromosomes segregate independently. It is expected that the set of DNA variants should be passed on to the offspring with the chromosomes. But this is not the case always. Even when a set of DNA variants are on the same chromosome, these may not be transmitted together from parents to their offspring. The major reason is that crossing-over events take place between a pair of homologous chromosomes that result in exchange of genetic material between the members of the pair. This exchange often leads to the fragmentation of specific arrangement of the DNA variants. Crossing-over events do not take place between two small portions of homologous chromosomes. Hence, arrangements of DNA variants within the small portions remain undisturbed and the variants get transmitted 'as a block' from a parent to offspring. The combination of DNA variants (alleles) that are transmitted together from a parent to offspring is called a haplotype. Necessarily, the variant loci underlying a haplotype must be very close on a chromosome. However, if two loci on the same chromosome are far away from each other, there is a chance that during meiosis a combination of alleles may be transmitted into offspring that are not present in parents. This can happen due to a cross-over event during meiosis (Fig. 7.1).

7.2.1 Linkage Disequilibrium

When alleles at nearby loci are transmitted together from one generation to another (i.e., co-segregate) more often than is expected by chance, the loci are said to be associated non-randomly or alternatively to be in linkage disequilibrium (LD). Co-segregation is of fundamental importance in positioning a disease locus on the genome. The positions of loci, also called marker loci, that are used in genome-wide studies of diseases are usually known. Thus we know the information that these loci are located on which chromosome, which arm, which cytogenetic band, and at which nucleotide positions. If alleles at a marker locus co-segregate with a disease, we can infer that the disease locus is in proximity to the marker locus, thus enabling the determination of the approximate position of the disease locus. Figure 7.1 provides a schematic of this concept.

Variations in the human genome arise by mutation. Since the mutant allele arises on a chromosome carrying a specific set of alleles at adjacent loci, the mutant allele co-segregates with those alleles initially after it arises. If the newly arisen mutant allele causes a disease, then the disease will co-segregate with many marker loci; hence, it will be easy to position the disease locus. But as this chromosome segregates from one generation to another, cross-over events take place and the strength of association

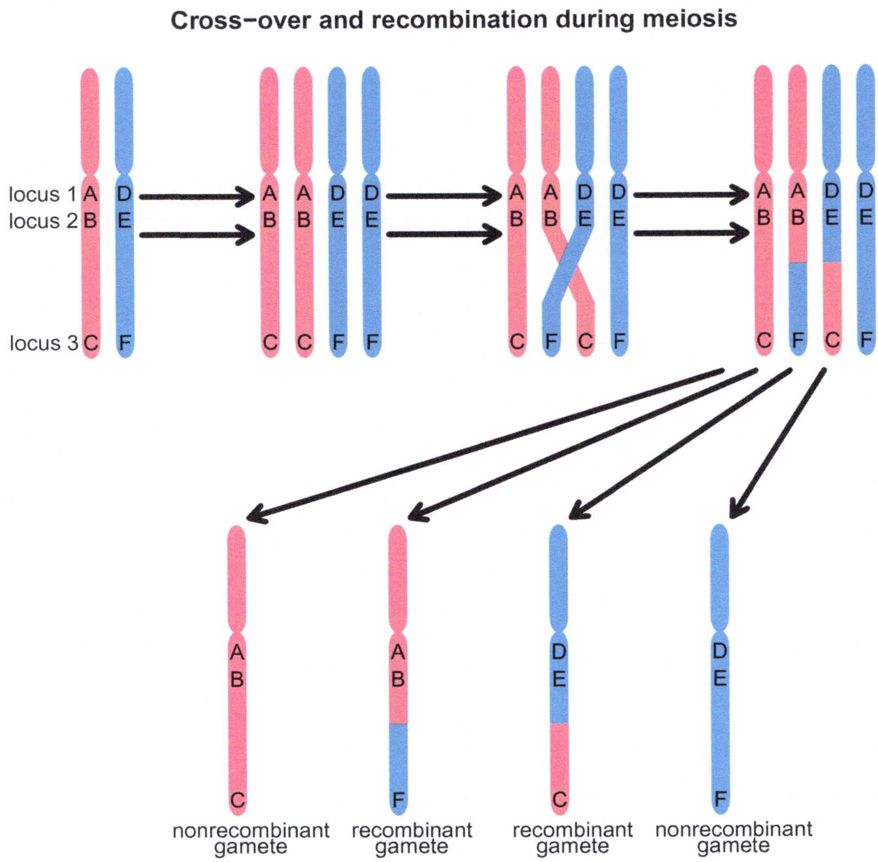

Fig. 7.1 The loci 1 and 2 are physically close on a pair of homologous chromosomes, while locus 3 is distant from these loci. Recombination events are expected to occur at higher frequency between loci 2 and 3 than between loci 1 and 2. Recombination events produce different combinations of alleles between loci 2 and 3

of the mutant locus with pre-existing variant loci around it decreases. To exemplify, suppose the frequency of recombination between two loci (with alleles A_1 and A_2 at one locus, and B_1 and B_2 at the other locus) per generation is $r = 0.1$. Then, an individual who is a double heterozygote A_1B_1/A_2B_2 is expected to produce gametes in the following proportions A_1B_1, 0.45; A_2B_2, 0.45; A_1B_2, 0.05; A_2B_1, 0.05. Note that without recombination, the gametes A_1B_2 and A_2B_1 are not expected to be produced by this individual.

After many generations of evolution, the strength of linkage disequilibrium of the mutant locus with an adjacent locus diminishes to a level such that the two loci start to behave as near-independent loci (Fig. 7.2). In humans, many disease-causing variants have arisen newly. It is essential to determine the position of a locus on the

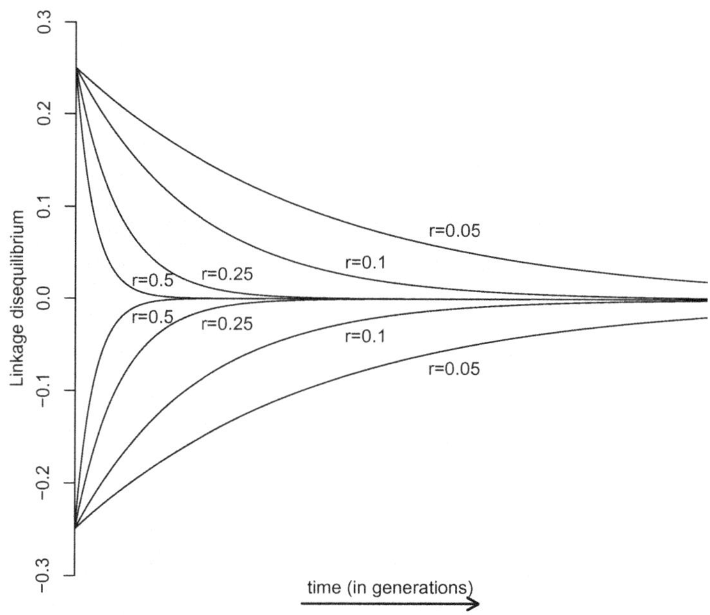

Fig. 7.2 Behaviour of linkage disequilibrium over time (generations)

genome. This is known as 'mapping'. This mapping is usually done by exploiting the association of the disease with alleles at loci whose genomic locations are known, which helps in knowing the position of the locus that causes the disease.

In fact, the concept and the statistical expectations of the behaviour of LD over generations form the basis of case-control study of a disease for mapping the disease locus. Although we describe a case-control association study in greater detail elsewhere in this book, the basis is pictorially depicted in Fig. 7.3.

Linkage disequilibrium between two loci, or the non-random association of alleles at the two loci, can be measured in various ways. Consider two loci with alleles (A, a) and (B, b). In the following table (Table 7.1), $g(-+)$ denotes the relative frequency of the gamete $(-+)$.

If $g(AB) = p \times s$, $g(Ab) = p \times t$, $g(aB) = q \times s$, and $g(ab) = q \times t$, then there is no association between the alleles at the two loci. Any significant deviation implies that there is an association. This provides a natural measure (D) of linkage disequilibrium (LD):

$$D = g(AB) - (p \times s) \text{ or } D = g(ab) - (q \times t).$$

If the two loci are not in LD, i.e. they are in *linkage equilibrium*, then $D = 0$. LD between two loci, as mentioned earlier, declines with the passage of generations since recombination events occur between the loci. If r denotes the probability of recombination between two loci, then:

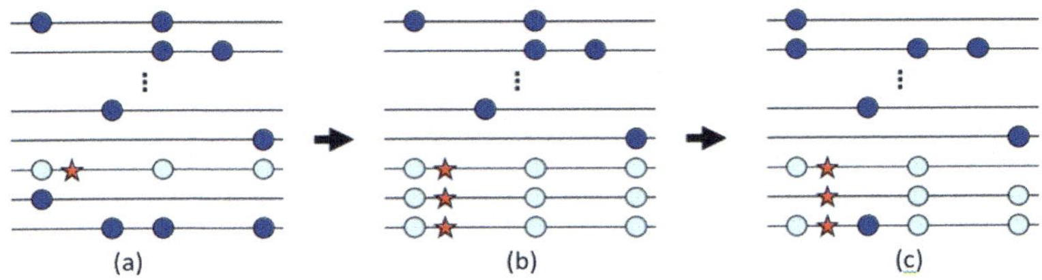

Fig. 7.3 Each line represents a chromosome and each solid circle represents a locus on that chromosome. Different colours represent different alleles. **a** The red star is a mutation that arises on a chromosome which is a specific combination of alleles at different loci. The red star mutation causes a disease and hence is found only in individuals with the disease. **b** For some generations, all individuals with the disease (cases, in a case-control association study) will have the specific combination of alleles 'the haplotype' on which the disease-causing mutation had arisen; this combination of alleles will be absent among individuals without disease (controls, in a case-control study). **c** With the passage of generations, recombination events will start taking place between loci, and hence the disease-causing mutation will also be observed on other haplotypes. In other words, the strength of association of the disease-causing allele with the haplotype on which it had arisen will weaken

Table 7.1 Alleles at two loci, combinations of alleles in gametes and their relative frequencies

	B	b	
A	$g(AB)$	$g(Ab)$	p
a	$g(aB)$	$g(ab)$	q
	s	t	1

$$D_t = (1-r)D_{t-1} = (1-r)^2 D_{t-2} = \cdots = (1-r)^t D_0,$$

where D_t is the LD between the loci in generation t.

Note that recombination events are not so frequent and the probability of recombination is much less than the probability of no recombination. Thus $r < (1-r)$ implying $0 < r < 0.5$. Also from the relation $D_t = (1-r)^t D_0$, it is clear that as t increases, LD decreases (Fig. 7.2). Thus after a large number of generations, LD between two loci becomes very small and hence two loci would be independent (nearly). The value of D obviously depends on allele frequencies at the two loci. Therefore, a normalised measure of D, D', based on the theoretical maximum and minimum, has been suggested.

When $D \geq 0$, $D' = D/D_{max}$, where D_{max} is the smaller of pt and qs.

When $D < 0$, $D' = D/D_{min}$, where D_{min} is the larger of $-pt$ and $-qs$.

Another alternative measure that has been proposed is based on the association between the two loci and is defined as

Table 7.2 Relative frequencies of alleles and haplotypes with two loci

	B	b	
A	$g(AB) = 0.6$	$g(Ab) = 0.1$	$p = 0.7$
a	$g(aB) = 0.2$	$g(ab) = 0.1$	$q = 0.3$
	$s = 0.8$	$t = 0.2$	1

Table 7.3 Observed and expected frequencies when two loci are considered together

Observed frequencies					Expected frequencies (y_{ij})				
	BB	Bb	bb	Total		BB	Bb	bb	Total
AA	n_{11}	n_{12}	n_{13}	n_{10}	AA	$g(AB) \times g(AB)$	$2 \times g(AB) \times g(Ab)$	$g(Ab) \times g(Ab)$	m_{10}
Aa	n_{21}	n_{22}	n_{23}	n_{20}	Aa	$2 \times g(AB) \times g(aB)$	$2 \times g(AB) \times g(ab) + 2 \times g(aB) \times g(aB)$	$g(Ab) \times g(ab)$	m_{20}
aa	n_{31}	n_{32}	n_{33}	n_{30}	aa	$g(aB) \times g(ab)$	$2 \times g(aB) \times g(ab)$	$g(ab) \times g(ab)$	m_{30}
Total	n_{10}	n_{20}	n_{30}	n	Total	m_{01}	m_{02}	m_{03}	n

$$r^2 = D^2/(p \times q \times s \times t),$$

the value of which ranges from 0 (no LD) to 1 (complete LD).

Ex-7.1 Consider two loci with alleles (A, a) and (B, b). Let the proportions of the various haplotypes and alleles be as provided in the Table 7.2.

The values of the various measures of LD from the above data (Table 7.2) will be

$$D = \big(g(AB) \times g(ab)\big) - \big(g(aB) \times g(Ab)\big) = 0.6 \times 0.1 - 0.1 \times 0.2 = 0.04$$
$$D = g(AB) - p \times s = 0.6 - (0.7)(0.8) = 0.60 - 0.56 = 0.04$$

Since $D > 0$, use D_{max}. Now D_{max} is the smaller of pt and qs. Noting that $pt = 0.14$ and $qs = 0.24$, we have

$$D' = D/D_{max} = 0.04/0.14 = 0.286.$$

$$\text{Finally, } r^2 = \frac{D^2}{p \times q \times s \times t} = \frac{0.04^2}{0.7 \times 0.3 \times 0.8 \times 0.2} = 0.048.$$

7.2.2 Estimation of LD and Other Parameters from Genotype Counts

When haplotypes are directly observable, i.e. linkage phase is known and hence coupling and repulsion heterozygotes can be separated, then from the equations provided in the earlier section, D can easily be estimated from gamete frequencies. However, when phase is unknown, the underlying pair of haplotypes is ambiguous for the double-heterozygotes. Therefore, one cannot directly measure LD. We describe a procedure to estimate D, when both loci A and B, with alleles (A, a) and (B, b), respectively, are codominant. Let (p, q) and (s, t) be the frequencies of (A, a) and (B, b), respectively, in a random mating population. Under this framework, there will be nine genotypes whose observed and expected frequencies are denoted, respectively, as in Table 7.3.

From Table 7.3, let $x_{11} = 2n_{11} + n_{12} + n_{21}, x_{12} = 2n_{13} + n_{12} + n_{22}, x_{21} = 2n_{31} + n_{21} + n_{31},\ x_{22} = 2n_{33} + n_{23} + n_{32},\ g(AB) = f_{11},\ g(aB) = f_{12},\ g(aB) = f_{21}$, etc.

The log-likelihood function of the observations is

$$\log(L) = \text{constant } + \sum_{i=1}^{3}\sum_{j=1}^{3} n_{ij} \log y_{ij} \tag{7.1}$$

We use Expectation-Maximisation (EM) algorithm to obtain maximum likelihood estimates of allele frequencies and LD. We provide a brief description of the EM algorithm which is used for obtaining maximum likelihood estimates of parameters when some of the data are missing or latent/unobserved. An example is with respect to ABO blood groups. At the ABO blood group locus, genotypes of individuals of A blood group are latent. An individual with A blood group can either be of the AA genotype or the AO genotype; the actual genotype cannot be known from the blood group. Similarly, the haplotype of an individual with genotypes Aa and Bb at two loci cannot be unambiguously known; the individual may be AB/ab or Ab/aB. In such cases, EM algorithm can be used to estimate allele or haplotype frequencies.

We assume that the complete data set consists of $Z = (X, Y)$, but only X is observed. The complete data log likelihood is then denoted by $L(\theta;\ X, Y)$ where θ is the unknown parameter vector for which we wish to find the Maximum Likelihood Estimate (MLE). Basically here we replace the missing quantity by its expectation and then maximise with respect to the parameter θ. This process will continue till the parameter value converges.

E-Step: The E-step of the EM algorithm computes the expected value of $L(\theta;\ X, Y)$ given the observed data, X, and the current parameter estimate, θ_C say. In particular, we define

$$Q(\theta; \theta_C) = E[L(\theta;\ X, Y)|X, \theta_C] = \int L(\theta;\ X, y)p(y|X, \theta_C)dy \tag{7.2}$$

where $p(\cdot|X, \theta_C)$ is the conditional density of Y given the observed data, X, and assuming $\theta = \theta_C$.

M-Step: The M-step consists of maximising over θ the expectation computed in (7.2). That is, we set $\theta_N = \max_\theta Q(\theta; \theta_C)$. Then we set $\theta_C = \theta_N$.

The two steps are repeated as necessary until the sequence of θ_N converges. Under very general circumstances convergence to a local maximum can be guaranteed (Dempster et al. 1977). It is always a good idea to run the EM algorithm many times, using a different starting value of θ_C in different runs. The ML estimate of θ is then taken to be the best (highest likelihood) of the set of local maxima obtained from the various runs of the EM algorithm.

The vast majority of the data that are normally generated in recent human genetic studies pertain to bi-allelic codominant loci, sometimes it is possible that one or both loci are dominant. Estimation in such cases has been described in Hill (1974). Further, if loci are multi-allelic, one may refer to the methodology developed by Schaid (2004).

7.2.3 Haplotype Block

Two seminal papers published in 2001 (Daly et al. 2001; Jeffreys et al. 2001) showed that there were some regions of the human genome in which there was very little evidence of recombination. This evidence was gleaned from the high values of LD between pairs of polymorphic loci in that region. Restriction of recombination in that region meant that the number of haplotypes observed in that region was small. Such a region is referred to as 'haplotype block'. Between a loci on one haplotype block on a chromosome and a locus on the adjacent haplotype block, LD was observed to be high. Thus, haplotype blocks seem to be separated by short regions with many inferred recombination events. Statistical methods have been suggested to identify haplotype blocks (Patil et al. 2001; Gabriel et al. 2002). The block structure of the human genome is intriguing, because it implies that there are short regions in the human genome within which recombination rates are very high, sometimes known as 'recombination hotspots'. These short regions define the boundaries between haplotype blocks. Some experimental evidence has been generated for the existence of recombination hotspots (Badge et al. 2000). These facts do not imply that among loci within a haplotype block recombination events do not take place. Indeed they do take place, but with a low probability. Figure 7.4 provides an idealised structure of haplotype blocks in a segment of the human genome.

This implies that with the passage of evolutionary time and the occurrence of recombination events, a large haplotype block can break up into two or more smaller blocks. Therefore, it is expected that in evolutionarily older ancestral human populations, the sizes of haplotype blocks will be smaller than in populations that have evolved more recent. Indeed, this has been observed in Africa, where humankind arose. Fewer and smaller multisite haplotype blocks are usually observed in African populations compared to other populations (Gu et al. 2007).

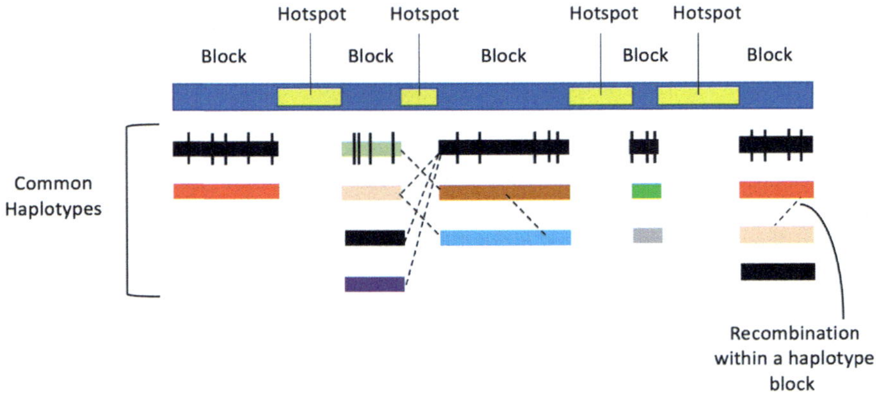

Fig. 7.4 An ideogram of block structure of a region of the human genome is shown in this diagram. Blue segments are those with little or no recombination among markers within those regions. Segments in yellow are with significantly high recombination; designated as hotspots. Physical sizes of the blocks and hotspots can be variable. The number of SNPs in a block (location of each SNP denoted by a short vertical line) can be variable and may not be correlated with block size. Within each block, there may be variable numbers of common haplotypes and a larger number of rare haplotypes. Diversity of haplotypes within each block is depicted by differences in colour. Crossovers between and within blocks are shown as dotted lines

7.2.4 Haplotype Determination

Determination of haplotypes from diploid genotype data poses a challenge. For example, consider three bi-allelic loci on a chromosome. Suppose an individual at these three loci has genotypes CG, TC, and GA. Then, the possible haplotypes can be CTG/GCA or CCG/GTA or CCA/GTG, or CTA/GCG. Thus, haplotypes may not be unique given information on genotypes at a set of loci. The number of possible haplotype combinations increases rapidly with the increase in the number of loci under consideration.

If parental genotypes are available, then the problem of haplotype determination is sometimes simplified. Suppose the genotypes of an individual are AT, AA, and CG at three loci. If the parental genotypes at these loci are available, then it may be possible to uniquely determine the haplotypes of this individual. For example, if the paternal and maternal genotypes are, respectively, (TT, AA, CG) and (TA, AA, CC), then the haplotypes of the individual are AAC/TAG. If, however, the paternal and maternal genotypes were, respectively, (AT, AA, CG) and (AT, AA, CG), then the haplotypes of the individual under consideration cannot be determined.

In general, haplotype determination is impossible. Considerable additional data on extended pedigrees can sometimes help determine haplotypes of an individual, but even that is not guaranteed. Some algorithms for haplotype determination have been proposed (Tapadar et al. 2000).

7.2.5 *Haplpotype Phasing*

Haplotype phasing is the resolution of a set of genotypes at multiple loci into the two haplotypes. There are various statistical methods for phasing. We shall outline only the maximum likelihood method via the Expectation-Maximisation (EM) algorithm (Excoffier and Slatkin, 1995).

The data are genotypes of n individuals at a set of m loci. Let the multi-loci genotypes be denoted as $1, 2, \ldots, m$ with unknown probabilities $P_1, P_2, \ldots, P_m$, and whose frequencies among the n individuals as $n_1, n_2, \ldots, n_m$, respectively. Then, the probability of observations on the sample of n individuals is

$$P(\text{sample} \mid P_1, P_2, \ldots, P_m) = \frac{n!}{n_1!n_2!\cdots n_m!} \times P_1^{n_1} \times P_2^{n_2} \times P_m^{n_m},$$

Let c_j be the number of haplotype pairings that will lead to the multi-loci genotype j. Obviously, c_j depends on the number of heterozygous loci in the multi-loci genotype j. If we assume that the observations are from individuals belonging to a random mating population, then we can write:

$$P_j = \sum_{i=1}^{c_j} P(\text{genotype } i) = \sum_{i=1}^{c_j} P(h_k h_l),$$

where $P(h_k h_l)$ is the probability that the i-th multi-loci genotype is made up of haplotypes k and l. If we denote p_k and p_l as the frequencies of the haplotypes k and l in the randomly mating population, then $P(h_k h_k) = p_k^2$ if $k = l$, and $P(h_k h_l) = 2p_k p_l$, if k and l are different. We can now write the likelihood of the observations on the multi-loci genotype counts as

$$L(p_1, p_2, \ldots, p_h) = a_1 \prod_{j=1}^{m} \left(\sum_{i=1}^{c_j} P(h_k h_l) \right)^{n_j},$$

where $p_h = 1 - p_1 - p_2 - \cdots - p_{h-1}$, and a_1 is the multinomial coefficient which is a constant.

We now use the EM algorithm to estimate the haplotype frequencies. We start with some arbitrary initial values $p_1^{(0)}, p_2^{(0)}, \ldots, p_k^{(0)}$ to estimate the genotype frequencies $P(h_k h_l)^{(0)}$. This is the E (expectation step). We use,

$$P(h_k h_l \text{ in multi-loci genotype } j)^{(0)} = P_j(h_k h_l)^{(0)} = c_j^{-1}, \text{ for all } j.$$

In the g-th iteration, we use

$$P(h_k h_l)^{(g)} = \frac{n_j}{n} . \frac{P_j(h_k h_l)^{(g)}}{P_j^{(g)}}.$$

Table 7.4 Haplotype frequency distribution among cases and controls. $H_1, \ldots, H_k$ denote the k possible haplotypes. Frequency of haplotype H_i in cases is n_{1j} and that in controls is n_{2j}, $j = 1, \ldots, k$

	H_1	H_2	...	H_k	Total
Cases	n_{11}	n_{12}	...	n_{1k}	$2n_1$
Controls	n_{21}	n_{22}	...	n_{2k}	$2n_2$

Haplotype frequencies are computed at each M (maximisation) step using,

$$\hat{P}_i^{(g+1)} = \frac{1}{2}\sum_{j=1}^{m}\sum_{i=1}^{c_j} \delta_{it} P_j (h_k h_l)^{(g)},$$

where δ_{it} is an indicator variable equal to the number of times haplotype t is present in multi-loci genotype i $(= 0, 1, 2)$. The iterative procedure is terminated when haplotype frequencies attain stable values.

There are many limitations of this EM algorithm (Excoffier and Slatkin, 1995), the most significant of which is that the performance of the algorithm rapidly diminishes with the increase in the number of polymorphic loci. The assumption of random mating, i.e. the assumption that relative frequencies of haplotypes are those expected under Hardy-Weinberg equilibrium is another serious limitation. This haplotype phasing algorithm and estimation of frequencies of inferred haplotypes have been implemented in the Arlequin software (http://cmpg.unibe.ch/software/arlequin35/).

There are many other methods that have been developed for haplotyping from genotype data on unrelated individuals. It is beyond our scope to discuss these methods and the software packages in which these methods have been implemented. A review of these methods and packages is given in Salem et al. (2005). In recent times, various computational, rather than statistical, methods are being developed for haplotyping (e.g. Tangherloni et al. 2019; and references therein). Additionally, methods have been developed for whole genome haplotyping (Kuleshov et al. 2014).

7.2.6 *Haplotype Association*

In a case-control study, when the cases and controls are unrelated to each other, a test for association is based on differences in estimated frequencies of haplotypes among cases and controls. This is performed using the chi-squared test of independence in a $2 \times k$ table (Olson & Wijsman 1994) as given in Table 7.4.

The test statistic is

$$\chi^2 = \sum_{i=1}^{k} \frac{(n_{1i} - n_{2i})^2}{(n_{1i} + n_{2i})},$$

which follows a χ^2 distribution (approximately) with $(k - 1)$ degrees of freedom.

Here, the genomic region on which the polymorphic loci are located is considered as a single multi-allelic locus, with haplotypes being considered as the alleles. While conceptually this method may be acceptable, when the number of loci increases, the number of haplotypes also rapidly increases resulting in loss of power of the chi-squared test because of a large increase in the degrees of freedom. Further, since in practice the haplotypes are inferred and not actually observed, the uncertainty in phase inference impacts on the assessment of the overall significance of the association test. When the loci are in strong LD, the impact is expected to be small.

Other approaches for measuring the association of haplotypes with a disease have been proposed. In one of these (Tzeng et al. 2003), the approach was based on the tenet that if haplotypes differ between cases and controls then it is evident that the disease-liability mutation lies in the genomic region of the haplotypes. Tzeng et al. (2003) have proposed tests based on the similarity of haplotypes among cases and controls. Score test and other more sophisticated tests have also been proposed, including simultaneous haplotype phasing and estimation of haplotype association with a disease (Schaid et al. 2002; Lin & Zeng 2006).

7.3 Studying Levels of Gene Expression at Various Stages of Cancer

Difference in gene expression levels in two independent populations, like case and control, can be studied or tested through appropriate statistical techniques. These include two-sample t-test or Wilcoxon rank sum test depending on whether the gene expression follows normal distribution or not. We can think of a generalisation to this problem.

In many diseases, especially in cancer, the severity of the disease depends on the stage of the disease. For example, stage I cancer patients usually have more survival chances after treatment compared to stage IV patients. It would be of clinical and moral importance if we can identify a few genes that are more prevalent in stage IV than that in stage I patients, in general. An overall idea about these genetic markers might be helpful during screening at the first stage. It could lead to different and more effective treatment protocols that may result in increasing the survival time of a patient.

Here data sets on gene expression consist of independent samples from patients belonging to two or more different groups. With three or more number of groups the analysis methods that we learnt, viz., t-test, Wilcoxon test, etc., cannot be directly applied. However, more groups or classes come with new information but not without other types of data analytic challenges. We discuss this issue with the help of a data set that has four groups. We assume that the individuals in a group are a random sample from the population of individuals having the same characteristics as that group. Naturally it is rational to assume that the individuals in different groups are

Table 7.5 Data of gene expression for a gene for person at different stages of cancer

Normal	Stage I	Stage II	Stage IV
1.22	−1.9	3.82	2.71
1.76	−0.49	0.83	3.34
0	−0.39	2.93	5.51
0.49	1.83	5.93	4.46
2.06	−0.1	4.15	2.47
0.22	3.41	4.85	−0.22
1.56	1.53	3.72	1.7
2.52	1.05	1.59	4.5
1.97	3.61	4.91	1.18
−0.46	−1.65	2.47	5.45
1.03	2.04	3.28	4.74
−0.18	1.14	2.34	9.13
1.61		1.97	5.07
4.01			1.28
			0.82

completely independent; features of an individual in a group do not in any way influence the features of another person in another group. Moreover, since we work with a random sample, there is also within group independence as far as sampled individuals are concerned. We start with data on gene expression of individuals distributed over four groups (Table 7.5). Here we try to mainly address the questions (4)–(7) as described in Sect. 7.1.

7.3.1 Testing Means of Multiple Samples

To begin with, first recall two-sample t-test (Sect. 4.3.6). Here we test whether the population means are equal in two groups when observations in each of the two groups follow a normal distribution with additional assumption that their variances are same. Naturally our current problem is an extension of this standard two-sample t-test, but there is a problem. Taking difference of sample means, as in case of t-test, will not work here as we have more than two means, because the number of classes is more than two. This leads to a famous technique, known as 'Analysis of Variance', abbreviated as ANOVA. This compares the means of two or more different groups of independent observations. As a passing remark, this ANOVA technique comes under a more general statistical topic called 'Linear Model'. The name 'linear model' will give an indication of the function used in computation using R.

Taking a cue from two-sample t-test, we assume that the variances of each group are the same and each observation corresponds to a random variable that follows a

normal distribution. Hence we can mathematically represent a model that describes the gene expression of an individual in a particular group. This mathematical representation is not very hard; but once we understand this, the entire downstream analytic method would be very easy. We can think that observations within a group are homogeneous. However, there are some fluctuations in data because of the presence of random error or noise that is inherent in any data set. Thus our model becomes:

$$y_{ij} = \mu_i + e_{ij};\ j = 1, \ldots, n_i;\ i = 1, \ldots, k; \tag{7.3}$$

where y_{ij} is the observation corresponding to j-th individual in the i-th group, μ_i is the average or mean expression of i-th group, and e_{ij} is the random error component associated with (i, j)-th observation, assumed to follow a normal distribution with mean 0 and variance σ^2. Needless to say that here y_{ij}s and hence, e_{ij}s are all independent.

In our example, $k = 4$ indicating that there is altogether four groups, the number of observations in i-th group being n_i, i.e. the number of observations in 1st, 2nd, 3rd, and 4th groups are n_1, n_2, n_3, and n_4 respectively.

We want to check, based on the given data, whether the group means are equal to each other, i.e. our hypotheses of interest would be

$$H_0 : \mu_1 = \mu_2 = \cdots = \mu_k \text{ against } H_1 : \text{ at least one inequality in } H_0$$

Now to test the above hypothesis, we need a test statistic. The mathematical derivation of the test statistic is slightly complicated; we try to develop it using our intuition. Just focus on what happens when H_0 is true and what happens when H_0 is not true. We can then compare these two situations that would automatically lead to a nice statistic. However, we also need to see whether the statistic developed this way would have a proper known probability distribution so that we can calculate the p-value quickly. Fortunately, this method provides a standard distribution of the test statistic.

Recall interpretations of simple mean and variance. Mean is a measure that represents the data, more precisely a central value around which the data points are clustered. If there were no fluctuations, the mean is the value that would be taken by each data point. Variance is the square of standard deviation, that gives a measure of dispersion. In other words, variance gives an idea about how the data are scattered around the mean. Higher variance indicates that the data are very much scattered from its central value whereas smaller variance ensures that observations are closely located around its mean. We just apply these simple ideas to develop a statistical testing procedure to compare the means of different groups.

First suppose that H_0 is true. That means all group means are same. We have already assumed (in our model) that variances are also same. So, we can think that all observations are supposed to come from the same distribution. Hence it is expected that the sample variances for each group are same more or less (except for sampling fluctuations) and this should be same if all groups are merged and considered as one single group. We can combine the total variance of each group to get an idea of total

variance present in the data, under the assumption that H_0 is true. This is known as 'Within Sum of Squares', as given by

$$SSW = \sum_{i=1}^{k}\sum_{j=1}^{n_i}\left(y_{ij} - \bar{y}_i\right)^2 \tag{7.4}$$

where $\bar{y}_i = \frac{1}{n_i}\sum_{j=1}^{n_i} y_{ij}$ for $i = 1, \ldots, k$. Since there are n observations but we are estimating k means, the number of independent observations involved, i.e. the degrees of freedom associated with SSW is $n - k$. We may divide SSW by $n - k$ to have an idea of average of such total variation. This measure, known as 'Mean Square Error', is defined as $MSW = SSW/(n - k)$. Sometimes it is also denoted as 'MSE' which stands for 'Mean Square Error'.

Again look at the problem from a different view point. If H_0 is true, then each of the group means would be very close to the global mean, i.e. mean of all observations in the data set. So the difference between $\bar{y}_i$ and $\bar{y}$ is very close to zero. Now this difference may be positive or negative so it is better to take the squared difference. Since there n_i observations in i-th group, we consider the total such squared difference as

$$SSB = \sum_{i=1}^{k} n_i\left(\bar{y}_i - \bar{y}\right)^2. \tag{7.5}$$

This is known as 'Between Sum of Squares'. Similar to the case of SSW, here also we define 'Mean Square due to Between groups' as $MSB = SSB/(k - 1)$, since the number of independent observations involved, i.e. degrees of freedom associated with SSB is $k - 1$.

Now let us look at SSW and SSB more carefully. Clearly, if H_0 is true, MSB should be and close to MSW. On the other hand, if H_1 is true, i.e. if H_0 is false, then the groups means are not equal to each other. So the total variation present in each group is large and hence MSB is substantially larger compared to MSW. Hence we can propose a test statistic as

$$\mathscr{F} = \frac{MSB}{MSW}.$$

It is expected that $\mathscr{F}$ is small when H_0 is true, i.e. group means are equal; but when H_0 is false, $\mathscr{F}$ is likely to be large.

Fortunately, under the assumptions of the model, it can be shown that this test statistic $\mathscr{F} \sim F_{k-1,n-k}$ under H_0. So we can easily calculate the p-value using an F distribution with degrees of freedom $k - 1$ and $n - k$. Naturally we reject H_0 if the p-value is very small compared to the chosen level of significance α. Note that we emphasise that the level of significance for the test should be chosen before data collection, keeping in mind the degree of seriousness of type I error.

7.3.1.1 Validating Assumption Underlying ANOVA

Although we have described ANOVA for testing whether average gene expressions are same, at least approximately, for all the groups, we need to see whether the data set under consideration satisfies the assumptions underlying ANOVA. We know how to check whether observations in a data set comes from a normal distribution using Kolmogorov-Smirnov test or Shapiro-Wilk test. So we can easily apply any of these tests to each group to examine whether it follows a normal distribution.

But another important assumption of ANOVA is that the variances of all groups are same. To test this, our hypotheses of interest are

$$H_0 : \sigma_1^2 = \cdots = \sigma_k^2 \text{ against } H_1 : \text{ at least one inequality in } H_0.$$

We are not going into the details of the appropriate test as it is slightly technically difficult. The test statistic for the above H_0 is

$$T = \frac{(n-k)\log(S^2) - \sum_{i=1}^{k}(n_i - 1)\log(S_i^2)}{1 + \frac{1}{3(k-1)}\left(\sum_{i=1}^{k}\frac{1}{n_1 - 1} - \frac{1}{n-k}\right)}$$

where $n = \sum_{i=1}^{k} n_i$, $S_i^2 = \frac{1}{n_i - 1}\sum_{j=1}^{n_i}\left(y_{ij} - \bar{y}_i\right)^2$, and $S^2 = \frac{1}{n-k}\sum_{i=1}^{k}(n_i - 1)S_i^2$.

S^2 is known as 'pooled variance' based on k groups. This test is known as Bartlett's test for homogeneity of variances. It can be shown that under H_0, T follows approximately (asymptotically) χ^2_{k-1}. Hence, for a chosen level of significance α, we can calculate the p-value corresponding to the test statistic T using a χ^2 distribution with degrees of freedom $k - 1$. If the p-value is large compared to α, we can conclude that there is no reason to believe that the variances are unequal at level of significance α approximately. Note that the level of significance is approximately α as this is based on an approximate test, which in this case is χ^2_{k-1}. Of course, here the total sample size n should be large enough to apply this test.

7.3.2 *Kruskal-Wallis Test*

Discussion in Sect. 7.3.1.1 raises an important point. If any of the assumptions of ANOVA are violated, what should we do? Is it still okay to go with ANOVA? Strictly speaking, the answer is 'NO'. Instead it would be better if we can find another statistical test that does not require these assumptions. Fortunately one such test exists; however, it is less powerful than ANOVA provided the assumptions of ANOVA are valid. So when they are not valid, we have only one choice; use a nonparametric test instead of standard ANOVA. The most commonly used test is

known as Kruskal-Wallis test. Although we mentioned this test earlier, only for three classes, however, this can be generalised to any number of classes, as discussed in detail below.

Kruskal-Wallis test requires only one assumption: observations are continuous in nature, i.e. they are a random sample from an absolutely continuous random variable. In our case, gene expression values are continuous and hence there is no problem in using this nonparametric test. This test is based on ranks of observations instead of their actual values.

Let R_{ij} be the rank of y_{ij} in the entire data set, where y_{ij} is the jth observation in i^{th} class; $j = 1, \ldots, n_i$; $i = 1, \ldots, k$. First we arrange the data irrespective of the groups they belong to, in increasing order with respect to their values. Define R_{ij} as rank of y_{ij} as the number of observation that are less than or equal to y_{ij} among n observations. So minimum of all observations has rank 1 whereas maximum has rank n. The test statistic compares the differences due to ranks in different groups with respect to the total variance of all ranks when the data are taken without considering any groups. Define $\bar{R}_i = \frac{1}{n_i}\sum_{j=1}^{n_i} R_{ij}$ for $i = 1, \ldots, k$ and $\bar{R} = \frac{1}{n}\sum_{i=1}^{k}\sum_{j=1}^{n_i} R_{ij}$ where $n = \sum_{i=1}^{k} n_i$. Kruskal-Wallis test statistic is given by

$$T_{KW} = (n-1)\frac{\sum_{i=1}^{k} n_i\left(\bar{R}_i - \bar{R}\right)^2}{\sum_{i=1}^{k}\sum_{j=1}^{n_i}\left(R_{ij} - \bar{R}\right)^2} \tag{7.6}$$

Although observations are continuous, in a real data set few observations might be same giving rise to confusion in ranking them. Usually if two observations are same, we replace their ranks by the average of their individual ranks if they were unequal. Suppose the ranks of two individuals with respect to their phenotype values (or variable of interest) are both 5. It indicates a tie in ranks. However if there is no tie, their ranks would have been 5 and 6. Hence, we assign an average rank $(5+6)/2 = 5.5$ to each individual so the next rank would start from 7. Thus we use tied ranks in case a few observations are equal to each other. However, if there is no tie, simplified expression of the test statistic can be obtained as

$$T_{KW} = \frac{12}{n(n+1)}\sum_{i=1}^{k} n_i\left(\bar{R}_i - \frac{n+1}{2}\right)^2 = \frac{12}{n(n+1)}\sum_{i=1}^{k} n_i\bar{R}_i^2 - 3(n+1) \tag{7.7}$$

It can be shown that when the sample size is large, T_{KW} follows approximately (asymptotically) a χ^2 with $k-1$ degrees of freedom. We can easily calculate the p-value and take decision accordingly comparing it with the level of significance α approximately.

7.4 Covariate Adjustment in a Genetic Association Study

Genetic association study involves genetic data as well as phenotype data with which we seek association of genetic variants. We are usually interested in disease phenotype, although any phenotype can be studied in the same paradigm. Phenotype might be of two types: (1) qualitative and (2) quantitative. For example, a qualitative phenotype might be the status of a disease for a particular individual under study. It indicates only whether the person is affected or unaffected. In most cases if Y denotes the disease status, we designate 0 or 1 value for Y if the person is unaffected or affected respectively. On the other hand, for traits like HDL, blood pressure, etc., it is easy to get the actual value for an individual. If we assume that 120 as a normal value for diastolic blood pressure, 140 and 160 are both indicative of high blood pressure. However, a qualitative phenotype, in this respect, can put both individuals in the same group, i.e. cases or affected individuals. Thus it is clear that quantitative phenotype is usually more informative than qualitative phenotype provided it is available. In many situations, sometimes it is difficult to get the values of quantitative phenotype or in fact, it is difficult or sometimes impossible to define an appropriate quantitative phenotype. In such cases, we have to work with the available qualitative phenotype, where phenotypic discrimination can be done unambiguously (for example, case or control).

Whatever be the situation, utmost care should be taken while collecting data, both genetic and phenotype. Genetic data are collected with a certain amount of precision that depends on the machine that generates data from the genetic material, i.e. DNA or mRNA, which is extracted from blood or other tissue from the person. However, phenotype data are external. Collection of such data does not depend on genetic material or blood or tissue. Often it is collected by external measurement or by interview. For example, we can measure blood pressure using a sphygmomanometer which is a completely non-invasive device. Cardiovascular disease can be identified, at least at a preliminary level, using ECG or other external non-invasive tests. HDL and other cholesterol levels can be measured by a simple blood test. Moreover, when a person is identified with a high possibility of cardiovascular disease, the doctor usually wants to know whether the person is a smoker, her/his food habits, age, gender, etc. These variables that might have some relation with the phenotype are known as 'covariates'. Note that in most scenarios, data on covariates are easy to collect and these are usually collected, especially when a person with some health issues meets the doctor. Few diseases are only observed when a person reaches a certain age. Thus age of onset is sometimes worth studying while studying the progression of a disease. Thus, it is clear that along with genetic and phenotype data, another set of variables can provide data that might have some influence on phenotype, in fact, on the whole analysis regarding genetic association and subsequent inference. In this section, we shall try to address questions (8)–(14) as described in Sect. 7.1.

7.4.1 Example Data Set with Covariates

In Table 7.6, we present a data set with both qualitative and quantitative phenotypes, genotype data at a few genetic loci, and three covariates, viz., age, gender, and smoking status for each individual. For gender, 'M' and 'F' denote male and female respectively where in the smoking status column, '1' means smoker and '0' means non-smoker. Age is given in years although it can be exact age. We consider age as a continuous variable whereas gender and smoking status are categorical variables. The disease status column gives information on whether a person is affected or unaffected with respect to the disease of interest. Here we use '1' to indicate that a person is affected while '0' is indicative of a normal (control) individual. For example, in the context of cardiovascular disease, '1' means that the person is suffering from cardiovascular disease and '0' means the person does not have cardiovascular disease. Quantitative phenotype provides a measurement of a variable that is in some way phenotypic marker of the disease, like low level of HDL might increase the risk of cardiovascular disease. In Table 7.6, we generate a representative data set containing these types of variables. We include genotype data at three genetic loci.

In Table 7.6 G1, G2, and G3 denote the genotype data for all individuals at three loci. If A and a are two alleles at a locus, usually genotypes are denoted by AA, Aa, aa. However, we can recode it as 0, 1, and 2 respectively. This indicates the number of minor alleles in a given genotype. Here, we observe that a is the minor allele and hence AA, Aa, aa genotypes are coded as 0, 1, and 2 respectively. The same is true for all loci under consideration. 'CC status' presents the case-control status, i.e. whether a person is affected or unaffected and 'Quant Phenotype' column gives the value of a quantitative variable of interest for each individual. Each row represents data for an individual. We assume that each individual is independent of the other because the samples are drawn randomly from the population.

7.4.2 Covariate Adjustment for Quantitative Phenotype

We first consider a quantitative phenotype. If we are interested to see any genetic association with the phenotype, we can adopt a regression model or an ANOVA model for analysing the data. Both these models belong to the class of 'linear models' that has enough flexibility to extend itself in various directions. We first explain the use of a linear regression model that can adjust for covariates effects.

Let y_i be the quantitative phenotype value for the i-th individual. Let x_i be the value of the covariate, like age for the i-th individual. Since we want to adjust the effect of this covariate, we need to include it in the model and develop an analysis protocol so that keeping or isolating the effect of x, we can search for association of genetic locus with the phenotype, which is, in this case, quantitative in nature.

Thus we propose our model as

Table 7.6 Data for case-control status, genotypes at three loci, a quantitative phenotype and age and gender as two covariates

G1	G2	G3	Age	Smoking status	Gender	CC status	Quant phenotype	G1	G2	G3	Age	Smoking status	Gender	CC status	Quant phenotype
0	2	1	36	1	0	0	49.52	0	0	2	41	0	0	0	51.63
0	2	1	50	0	0	1	50.72	0	1	2	38	0	0	0	53.37
0	1	2	39	0	0	0	48.95	0	1	0	38	0	0	0	54.57
0	2	2	39	1	0	0	48.75	0	0	0	46	0	0	1	53.87
0	2	1	41	1	0	0	50.68	0	0	0	37	1	0	0	48.3
0	1	2	40	1	0	0	48.14	0	2	1	34	1	0	0	52.85
0	1	2	47	1	0	1	52.37	0	0	2	35	1	0	0	50.4
1	0	2	46	1	1	1	57.57	0	2	0	50	0	0	1	54.37
0	2	0	52	1	0	1	47.58	2	0	0	64	1	0	1	70.66
0	1	2	41	0	0	0	52.54	0	1	2	39	0	1	0	49.05
0	1	1	38	0	0	0	49.45	0	0	1	40	0	0	0	51.99
0	0	1	32	0	0	0	48.43	0	1	0	40	1	0	0	55.18
1	1	0	56	1	0	1	53.91	0	0	2	32	1	0	0	43.27
1	0	1	51	1	0	1	54.39	0	0	1	48	1	0	1	49.17
0	1	0	42	1	0	0	47.76	1	0	0	37	0	1	1	52.48
1	1	0	47	1	0	1	54.77	0	1	0	34	1	1	0	49.48
0	2	0	41	0	0	0	53.38	0	2	0	42	1	0	0	50.5
0	1	2	47	1	0	1	48.1	0	0	1	39	0	0	0	51.22
0	2	0	35	0	0	0	45.95	0	0	1	38	0	0	0	46.53

(continued)

Table 7.6 (continued)

G1	G2	G3	Age	Smoking status	Gender	CC status	Quant phenotype	G1	G2	G3	Age	Smoking status	Gender	CC status	Quant phenotype
0	2	1	46	0	1	1	51.41	0	0	1	37	1	0	0	50.12
0	1	1	41	1	1	0	52.57	0	0	2	43	1	0	1	54.24
0	0	2	49	1	0	1	52.65	1	1	0	41	0	0	1	53.07
1	0	0	54	0	0	1	54.48	0	0	2	32	1	0	0	47.67
0	1	2	39	1	0	0	50.25	0	0	2	40	1	0	0	50.26
0	0	2	40	0	0	0	50.23	1	0	2	42	1	0	1	57.13
0	0	2	48	0	0	1	47.94	0	2	1	42	0	0	0	53.39
0	1	2	50	1	0	1	49.32	0	2	1	41	0	0	0	55.1
0	0	0	40	1	0	0	49.29	0	0	1	57	1	0	1	53.27
0	0	0	34	1	0	0	51.42	0	0	2	37	1	0	0	49.1
0	0	0	53	1	1	1	51.01	0	2	0	44	0	0	1	45.24
0	2	0	49	1	0	1	46.92	0	0	0	44	0	0	1	52.8
0	0	1	54	1	0	1	48.82	0	1	2	40	1	0	0	47.68
0	1	2	30	1	0	0	50.42	2	0	0	52	1	1	1	70.81
0	0	0	41	1	0	0	50.18	0	1	2	40	0	0	0	51.78
1	1	0	46	1	1	1	52.16	0	1	0	53	1	0	1	49.08
1	2	1	42	1	0	1	53.82	0	1	1	43	0	0	1	47.6
0	1	1	41	0	0	0	51.16	0	1	2	53	1	0	1	48.5
0	0	1	38	1	0	0	52.96	0	1	0	54	1	0	1	45.72
0	1	0	41	0	0	0	53.41	1	1	0	45	1	0	1	55.49
0	1	1	52	0	0	1	49.27	0	0	2	50	1	1	1	48.55
0	2	1	53	0	0	1	50.42	0	1	0	52	0	1	1	51.29
0	1	1	39	1	0	0	48.08	0	2	2	41	1	0	0	49.16
0	0	0	41	0	0	0	52.2	0	1	0	56	1	0	1	51.65

(continued)

Table 7.6 (continued)

G1	G2	G3	Age	Smoking status	Gender	CC status	Quant phenotype	G1	G2	G3	Age	Smoking status	Gender	CC status	Quant phenotype
1	1	2	55	0	0	1	58.24	1	0	0	46	0	0	1	53.21
0	0	0	58	0	0	1	56.34	0	1	1	40	0	0	0	51.93
0	1	2	41	0	0	0	49.38	0	1	1	37	1	0	0	50.16
0	0	1	47	0	0	1	51.51	0	1	2	38	0	0	0	49.77
0	1	2	38	1	0	0	46.32	0	1	0	47	0	0	1	48.94
0	2	1	45	1	0	1	48.98	0	0	0	55	1	0	1	53.02
0	1	1	62	1	0	1	48.66	0	1	2	42	0	0	0	50.21
0	1	0	58	1	0	1	52.65	2	0	1	54	1	1	1	70.72
0	1	0	51	1	0	1	52.4	0	1	0	55	0	0	1	53.49
0	2	0	54	1	0	1	48	1	0	0	51	0	1	1	55.73
1	0	2	37	1	1	1	55.43	0	1	0	41	0	0	0	48.84
0	1	0	42	0	0	1	52.76	1	2	2	42	0	0	1	52.58
0	1	2	57	1	0	1	46.95	0	0	1	45	1	0	1	48.19
0	0	0	51	0	0	1	48.6	0	2	0	40	1	0	0	51.13
0	0	0	40	1	0	0	46.91	0	0	1	58	1	0	1	49.75
0	0	2	42	1	0	0	51.46	0	0	2	52	1	0	1	47.76
0	1	2	40	0	0	0	49.93	0	1	0	42	0	0	0	52.99

$$y_{ij} = \mu_i + \beta(x_{ij} - \bar{x}) + e_{ij}, \quad j = 1, \ldots, n_i; i = 1, \ldots, k. \tag{7.8}$$

Here μ_i is the effect of i-th genotype and β is the effect of the covariate. We assume that the random error components, i.e. $e_{ij} \sim N(0, \sigma^2)$ for all i and j.

Note that this model can also estimate the effect of the covariate along with the association of genetic locus after adjusting for the covariate. To see whether there exists any effect of covariate, we can frame this as a hypothesis testing problem. The relevant hypotheses are

$$H_{01} : \beta = 0 \text{ against } H_{11} : \beta \neq 0.$$

Similarly, to identify any genetic association after adjusting the effect of the covariate, we can also frame another set of hypotheses as

$$H_{02} : \mu_1 = \mu_2 = \mu_3 = 0 \text{ against } H_{12} : \mu_i \neq 0 \text{ for at least one } i = 1, 2, 3..$$

We should first find the estimates of the parameters involved in the model and then only should we proceed to carry out testing. Using our knowledge of least squares method, we can derive the least squares estimates of the parameters as

$$\hat{\mu}_i = \bar{y}_i - \bar{y} - \hat{\beta}(\bar{x}_i - \bar{x}), \ i = 1, 2, 3; \ \hat{\beta} = \frac{\sum_{i=1}^{3}\sum_{j=1}^{n_i}(y_{ij} - \bar{y}_i)(x_{ij} - \bar{x}_i)}{\sum_{i=1}^{3}\sum_{j=1}^{n_i}(x_{ij} - \bar{x}_i)^2}. \tag{7.9}$$

It can be shown that

$$\hat{\beta} \sim N(\beta, \sigma_{\beta}^2) \text{ where } \sigma_{\beta}^2 = \frac{\sigma^2}{\sum_{i=1}^{3}\sum_{j=1}^{n_i}(x_{ij} - \bar{x}_i)^2}$$

Hence a test statistic for testing H_{0i1} can be given as

$$T = \frac{\hat{\beta} - 0}{\sqrt{MSE/\sum_{i=1}^{3}\sum_{j=1}^{n_i}(x_{ij} - \bar{x}_i)^2}} \sim t_{n-k-1} \text{ under } H_{01} \text{ with } k = 3. \tag{7.10}$$

where $MSE = \frac{1}{n-4}\sum_{i=1}^{3}\sum_{j=1}^{n_i}\left(y_{ij} - \hat{\mu}_i - \hat{\beta}(x_{ij} - \bar{x})\right)^2$.

Given a fixed sample size, we can easily calculate the p-value. If this p-value is significantly small compared to the level of significance, we may reject H_{01} concluding that there is a significant effect of the covariate and it is of utmost importance

that we should adjust its effect in order to study the genetic association of the locus of interest with the quantitative phenotype.

We now propose a method for testing H_{02}, i.e. for the association of genotype, following the idea of ANOVA but in presence of a covariate. This is known as Analysis of Covariance (ANOCOVA). In this situation, the test statistic proposed is

$$\mathscr{F} = \frac{SSE^* - SSE}{SSE} . \frac{n-k-1}{k-1} \sim F_{k-1,n-k-1} \text{ under } H_{02} \text{ with } k = 3, \tag{7.11}$$

where $n = n_1 + n_2 + n_3$, $SSE = \sum_{i=1}^{3}\sum_{j=1}^{n_i}(y_{ij} - \bar{y}_i)^2 - \hat{\beta}^2 \sum_{i=1}^{3}\sum_{j=1}^{n_i}(x_{ij} - \bar{x}_i)^2$, $SSE^* = \sum_{i=1}^{3}\sum_{j=1}^{n_i}(y_{ij} - \bar{y})^2 - \hat{\beta^*}^2 \sum_{i=1}^{3}\sum_{j=1}^{n_i}(x_{ij} - \bar{x})^2$,

$$\bar{y}_i = \frac{1}{n_i}\sum_{j=1}^{n_i} y_{ij}, \qquad \bar{x}_i = \frac{1}{n_i}\sum_{j=1}^{n_i} x_{ij}, \qquad \bar{y} = \frac{1}{n}\sum_{i=1}^{3}\sum_{j=1}^{n_i} y_{ij}, \qquad \bar{x} = \frac{1}{n}\sum_{i=1}^{3}\sum_{j=1}^{n_i} x_{ij},$$

$$\hat{\beta} = \frac{\sum_{i=1}^{3}\sum_{j=1}^{n_i}(y_{ij}-\bar{y}_i)(x_{ij}-\bar{x}_i)}{\sum_{i=1}^{3}\sum_{j=1}^{n_i}(x_{ij}-\bar{x}_i)^2}, \quad \hat{\beta}^* = \frac{\sum_{i=1}^{3}\sum_{j=1}^{n_i}(y_{ij}-\bar{y})(x_{ij}-\bar{x})}{\sum_{i=1}^{3}\sum_{j=1}^{n_i}(x_{ij}-\bar{x})^2}.$$

A very small p-value calculated using the F distribution with 2 and $n-4$ degrees of freedom would indicate the genetic association with the quantitative in presence of a covariate x. We can also test whether this covariate needs to be included in the model at all. In such a case we test H_{01} using the statistic as given in (7.10) and a small p-value rejects this null hypothesis in favour of including the covariate in the model. However, if the p-value is much larger than the level of significance, we can conclude that it is not necessary to include the covariate in the model and we can proceed safely with an ANOVA model as discussed in Sect. 6.5.5.

We take this as an example where we write a code that is meant for this analysis without using a given function for ANOCOVA in R. This helps in knowing how to write a code if we require a code that is not available in R. We demonstrate this using the data in Table 7.6. We assume that the data in Table 7.6 are stored in 'a' in R workspace. An user can import the data by using 'read.table()' command.

It is seen that the p-value is extremely small compared to the chosen level of significance 0.05. Hence, we can conclude that even after adjustment of the effect of age, G1 seems to be associated with the phenotype, as revealed from the given data.

```
> head(a)     ## prints first six rows of the data set
> x1<- a$Age[which(a$G1==0)]     ## extracts age values for people with G1=0
> x2<- a$Age[which(a$G1==1)]     ## extracts age values for people with G1=1
> x3<- a$Age[which(a$G1==2)]     ## extracts age values for people with G1=2
> y1<- a$Phenotype[which(a$G1==0)]     ## extracts phenotype values for
                                              people with G1=0
> y2<- a$Phenotype[which(a$G1==1)]     ## extracts phenotype values for
                                              people with G1=1
> y3<- a$Phenotype[which(a$G1==2)]     ## extracts phenotype values for
                                              people with G1=2

> n1<- length(x1)          ## number of individuals with G1=0
> n2<- length(x2)          ## number of individuals with G1=1
> n3<- length(x3)          ## number of individuals with G1=2

> x<- c(x1,x2,x3)       ## merge age values for all individuals in one variable 'x'
> y<- c(y1,y2,y3)       ## merge phenotype values for all individuals in one
                             variable 'y'

>           beta.hat<          -(sum((y1-mean(y1))*(x1-mean(x1)))+sum((y2-
mean(y2))*(x2-mean(x2)))+sum((y3-mean(y3))*(x3-mean(x3))))/(sum((x1-
mean(x1))^2)+sum((x2-mean(x2))^2)+sum((x3-mean(x3))^2))
> beta.1<- cov(x,y)/var(x)

> SSE<- sum((y1-mean(y1))^2)+sum((y2-mean(y2))^2)+sum((y3-mean(y3))^2)-
beta.hat*beta.hat*(sum((x1-mean(x1))^2)+sum((x2-mean(x2))^2)+sum((x3-
mean(x3))^2))
> SSE.1<- sum((y-mean(y))^2) - beta.1*beta.1*(sum((x-mean(x))^2))

> F.statistic<- ((SSE-SSE.1)/SSE)*((n-3-1)/(3-1))
## calculates F statistic which is used as test statistic

> pvalue<- pf(F.statistic,2,n1+n2+n3-3,lower.tail=F)        ## calculates p-value
```

Suppose in the same setup it is known that disease risk increases with an increase in the number of disease alleles. If the disease associated allele is not known beforehand, we may assume that it would be the minor allele at the locus, if it is associated with the disease. We have seen that a simple linear regression model would suffice to explore genetic association in such a case but without any covariate. However, this model can be easily extended even when covariates are present. We consider only age as a covariate. Inclusion of multiple covariates would be straightforward.

In this situation, our model becomes

$$y_i = \alpha + \beta g_i + \gamma x_i + e_i, i = 1, \ldots, n$$

where y_i, g_i and x_i denote respectively, the quantitative phenotype, genotype value (0,1, or 2 depending on the number of minor alleles at the locus), and covariate value

corresponding to i-th individual; total number of individuals in the sample being n. e_is are random error component assumed to follow $N(0, \sigma^2)$ independently.

Simple 'lm()' function in R provides all information and p-value for tesing genetic association $H_0 : \beta = 0$ in presence of a covariate. However, we might be interested to know whether there is any effect of a covariate at all, i.e. how to test $H_0 : \gamma = 0$. The p-value for this test is also available from summary of 'lm( )' function, based on which we can take our decision.

7.4.3 *Covariate Adjustment for Qualitative Phenotype*

Adjusting for a single covariate in a genetic association study is usually done by including the covariate in the right-hand side of the model that explains how phenotype is dependent on genotype. In case of quantitative phenotype, this leads to a nice statistical model in the framework of regression analysis. This is nothing but an extension of standard ANOVA technique incorporating the effect of covariate, known as ANOCOVA, or simple inclusion of covariates in a linear regression model, as discussed in the previous section. The same idea can be thought of as an appropriate one, even in case of a qualitative phenotype. We have used a logistic regression model (Sect. 6.5.4) while exploring genetic association using a qualitative phenotype. We see that although a linear model with qualitative or categorical variable as response variable is meaningless, a transformation, in this case a'logit' transformation is linear in explanatory variables. We can easily include covariates on the right-hand side as explanatory variables. Thus our model becomes

$$logit(y_i) = \alpha + \beta g_i + \gamma x_i \tag{7.12}$$

i.e. here we consider the conditional probability of y_i given values of x_i and g_i, as

$$P(Y_i = y_i | x_i, g_i) = \left(\frac{e^{\alpha+\beta g_i+\gamma x_i}}{1 + e^{\alpha+\beta g_i+\gamma x_i}}\right)^{y_i} \left(\frac{1}{1 + e^{\alpha+\beta g_i+\gamma x_i}}\right)^{1-y_i}, \quad i = 1, \ldots, n$$

where g_i and x_i are genotype and covariate value respectively for i-th individual, y_i being the affection status taking values 0 or 1 depending on whether i-th person is unaffected or affected respectively.

To test the genetic association, our hypothesis of interest would be

$$H_0 : \beta = 0 \text{ against } H_1 : \beta > 0.$$

Note that here we are interested to see whether the genetic locus under consideration increases the disease risk or somehow associated with disease adversely, i.e. whether increase in the number of alleles would increase the disease risk. However, if we are

interested to see whether the genetic locus is in any way associated with disease, be protective or causal, we should consider the alternative hypothesis as $H_1 : \beta \neq 0$.

Smaller p-value in comparison to a chosen level of significance would ensure the existence of genetic association in the presence of a covariate. Using this approach we can also check whether the inclusion of the covariate is at all required, i.e. whether there is any possibility that genetic association gets affected by the effect of a covariate. To do this, we can frame this as another test of the hypothesis problem with the following hypotheses.

$$H_0 : \gamma = 0 \text{ against } H_1 : \gamma \neq 0.$$

The R code for applying this logistic regression includes the p-values for both the hypotheses described above. Hence a look at these p-values would answer our query very easily. It is to be noted that if $H_0 : \gamma = 0$ is accepted, it indicates that the covariate has no major role to play in genetic association with the qualitative phenotype under study. In that case, we need not use any covariate in our model; rather we should use the statistical tests as described in Sect. 6.5.4 without any covariate. It is advisable that we should first check whether there is any significant effect of the covariate on the phenotype. Depending on our conclusion, we can proceed with the appropriate logistic regression model with or without the covariate effect.

Here the covariate x can be categorical (e.g. gender. etc.) or continuous (e.g. age, etc.) in nature. Logistic regression can easily accommodate any type of variable as covariate. Moreover, we can include more than one covariate. However, inclusion of each covariate would increase the number of parameters making estimation of parameters less efficient. Hence, we should not blindly increase the number of covariates; covariates should be chosen judiciously and carefully considering the problem of association. We must have a clear idea about which variables might have an effect on the qualitative phenotype and we should only include them in the model.

```
> head(a)          ## prints first six rows of the data set
> x<- a$Age          ## extracts age for all individuals
> g<- a$G2          ## extracts genotype data at locus G2 for all individuals
> y<- a$CC.Status          ## extracts case-control status for all individuals
> b<- glm(y~ x+g, family="binomial")      ## fits a logistic regression model
> summary(b)

Coefficients:
              Estimate  Std. Error  z value  Pr(> |z|)
(Intercept)   -38.0500    9.0292    -4.214   2.51e-05 ***
x               0.9078    0.2189     4.148   3.36e-05 ***
g              -0.4861    0.5000    -0.972     0.331
```

With reference to the data in Table 7.6, suppose we are interested to see whether there exists any association between G2 and the qualitative phenotype, i.e. case-control status in this case, we can adopt a logistic regression model with additional

variable age. R code for this analysis is given. As before, we assume that the entire data is stored in the variable 'a' in R workspace.

The output clearly reveals that age has a significant effect as the p-value corresponding to it is very small, i.e. 0.000034. Hence, we cannot ignore the covariate effect on the phenotype and would perform a genetic association test after adjusting for the covariate effect. However, this result shows a lack of any association signal as the p-value is 0.331, which is very large compared to the chosen level of significance of 0.05. Thus we conclude that after adjusting the effect of age, it seems that there isn't enough evidence in favour of any association signal. Interested readers can do an exercise to see what happens if we use a logistic regression model without including age as a covariate.

7.4.4 A Few Issues on Covariate Adjustment

By now we understand that we can include the effect of a covariate in our linear regression model or logistic regression model while studying genetic association using quantitative or qualitative phenotype, as the case may be. We have just discussed in the previous two sections, how one can modify the model to adjust the effect of covariates and isolate the genetic effect eliminating covariate effect. So the question naturally arises whether we can include more than one covariate in the same model. Of course, we can do, but there is a limitation. We know that in a regression setup we cannot estimate more parameters than the number of observations. Usually, in our genetic association study, the number of sample individuals is much larger than the number of covariates. So it is expected that in this context this kind of problem should not arise.

This approach of including the covariate in the model might also lead to some other problems. Conditioning on covariates does not include information on disease prevalence and non-random ascertainment. This might induce a correlation between candidate variants and known associated variants. There are different sampling schemes to collect data, especially for affected individuals. Sometimes, failure to adjust the mechanism of sampling and some other issues can cause reduction in power of the genetic association test (Kuo & Feingold 2010). In the context of logistic regression, Neuhaus and Kalbfleisch (1998) showed that the power of the test gets decreased if the covariate is correlated to the random variable that is being tested. This standard conditioning on known variants, might significantly reduce power because it induces a dependence structure between associated variants in a case-control data set (Zaitlen et al. 2012). Thus, retrospective sampling, non-availability of covariate specific prevalence information, especially for rare phenotypes, are among the main problems that occur during adjustment of covariates in genetic association using regression or logistic regression models. Moreover, it is really an interesting idea to know whether the known risk variants can be exploited, at least to some extent, in the discovery of new variants that are expected to increase disease risk. With vast amount of information available, appropriate use of prior knowledge in a relevant context

sounds promising. However, the context and relevance need to be carefully justified before using such prior information; otherwise it might induce some confounding effect and give misleading results. We discuss two main approaches to overcome these issues. This involves little more sophisticated statistical techniques and may be omitted during the first reading.

7.4.4.1 Liability Thresholding Model

Zaitlen et al. (2012) proposed a liability thresholding model in the context of logistic regression-based association study when case-control phenotypes are ascertained non-randomly. We assume that there exists a latent random variable, i.e. liability Z (say) that ascertains whether an individual will be termed as a case or a control. We label those individuals as cases if $Z \geq 0$ and controls for those $Z < 0$. This Z remains unobserved. Assuming that Z follows a normal distribution and F is the prevalence of the disease, we can consider our phenotype model as

$$z = -\mu + \varepsilon, \quad \varepsilon \sim N(0, 1), \tag{7.13}$$

where $\mu = \Phi^{-1}(1 - F)$, $\Phi(x)$ being the cumulative distribution function of a normal variable. For the sake of notational simplicity, we omit the use of suffix 'i' in each variable for i-th individual.

Let s be a SNP marker for which we are interested to study the genetic association with the phenotype, i.e. case/control status in this case. The mean adjusted genotypes, i.e. g can take values $-2p$, $1 - 2p$, and $2 - 2p$. Introducing the genotypes, our model becomes

$$z = -\mu + \beta g + \varepsilon, \text{ where } \varepsilon \sim N(0, \sqrt{1 - Var(\beta g)}). \tag{7.14}$$

This leads to our basic assumption, i.e. $Var(Z) = 1$, as assumed in the model.

Note that here the discussion is not confined to adjusting covariates like age, gender, etc., but rather an important issue is addressed. There might exist a covariate or a latent variable that gives rise to the case-control status ascertainment. Thus the main focus is to control such latent variable that happens mainly due to retrospective sampling.

Now our main job is to estimate β, which might give an idea about the possible association of the variant with the phenotype. However, we need to know the value of F, the disease prevalence. Fortunately the value of F is known from prior knowledge and available from literature. At this point, we should be careful that this prior knowledge about F would be applicable for the population under study.

Let p^+ and p^- be the minor allele frequencies for g in cases and controls respectively. Assume that we know the ascertainment probability F apriori. Under this circumstance, a better estimate of minor allele frequency (MAF) is given by

$$\hat{p} = p^+ F + p^-(1 - F).$$

Let $\sigma_\varepsilon^2 = 1 - Var(\beta g)$. If $\hat{\beta}$ is an estimate of β, the effect size for the marker under consideration, we have

$$P(Y = 1|g = 0) = \big(1 - \Phi(\mu, \hat{\beta}(-2\hat{p}), \sigma_\varepsilon^2)\big) \quad (7.15)$$

$$P(Y = 1|g = 1) = \big(1 - \Phi(\mu, \hat{\beta}(1 - 2\hat{p}), \sigma_\varepsilon^2)\big) \quad (7.16)$$

$$P(Y = 1|g = 2) = \big(1 - \Phi(\mu, \hat{\beta}(2 - 2\hat{p}), \sigma_\varepsilon^2)\big) \quad (7.17)$$

where $\Phi(x, \mu, \sigma^2)$ is the cumulative distribution function of a normal variable with mean μ and variance σ^2. Then given the phenotype values, i.e. considering the fact that $Y = 1$ for case and $Y = 0$ for control, and assuming Hardy-Weinberg equilibrium for genotype frequencies, we have the posterior probability of a genotype when the phenotype status is known, as

$$P(g = 0|Y = 1) = \frac{P(Y = 1|g = 0)P(g = 0)}{P(Y = 1)} = \frac{P(Y = 1|g = 0)(1 - \hat{p})^2}{F} \quad (7.18)$$

$$P(g = 1|Y = 1) = \frac{P(Y = 1|g = 1)2\hat{p}(1 - \hat{p})}{F} \quad (7.19)$$

$$P(g = 2|Y = 1) = \frac{P(Y = 1|g = 2)\hat{p}^2}{F} \quad (7.20)$$

Hence, the MAF among cases is given by

$$\hat{p}^+ = P(g = 1|Y = 1) + 2P(g = 2|Y = 1). \quad (7.21)$$

Proceeding similarly and considering data for controls, i.e. putting $Y = 0$, we can easily calculate the MAF at g for controls. Note that the calculation up to this point depends entirely on the initial estimate $\hat{p}$. However, there is no guarantee that it is good, in some sense of statistical theory. In order to provide a reasonably good estimate, we should use some optimisation method and ensure the goodness of our estimate. We adopt a computational approach coupled by statistical argument.

Consider the squared error between observed and expected frequencies in cases and controls. Thus we get a quantity,

$$S_e^2 = (p^+ - \hat{p}^+)^2 + (p^- - \hat{p}^-)^2. \quad (7.22)$$

Clearly S_e^2 is a function of β. Now we adopt a grid search technique and identify $\hat{\beta}$ that minimises S_e^2. This estimate of β would provide a better idea about the existing association, if any. For a detailed study, please read the paper by Zaitlen et al. (2012).

7.4.4.2 Using Profile Likelihood Function

Although the approach by Zaitlen et al. (2012) is a nice one, however, it fails to accommodate some other problems while adjusting for covariates. If the prevalence information specific to the covariate is not available, this method might not be efficient. To this end, Zhang et al. (2018) proposed a profile likelihood based approach that might be more suitable for matched case-control design. However, using profile likelihood method in a straightforward way leads to certain theoretical as well as numerical problems regarding stability of the estimates of the parameters. This problem is solved by another simple modification of the profile likelihood approach. This method also takes care of different strata present in the population.

Let the variable S denote the stratum from which frequency-matched case-control samples have been drawn. Suppose we randomly draw n_s number of cases and controls from the stratum $S = s, s = 1, \ldots, I$. Assume that X is a non-confounding covariate and G is the genotype at a bi-allelic locus coded as 0, 1, and 2, based on the number of minor alleles. Let Y denote the affection status or disease status with 0 and 1 representing control and case respectively. Then using a logistic model we can write,

$$P(Y_{si} = y_{si}|S = s, X = x_{si}, G = g_{si}) = \left(\frac{e^{\alpha+\beta g_i+\gamma_1 x_{si}+\gamma_2 s_i}}{1 + e^{\alpha+\beta g_{si}+\gamma_1 x_{si}+\gamma_2 s_i}}\right)^{y_{si}} \left(\frac{1}{1 + e^{\alpha+\beta g_{si}+\gamma_1 x_{si}+\gamma_2 s_i}}\right)^{1-y_{si}} \tag{7.23}$$

where y_{si}, x_{si}, and g_{si} are affection status, covariate value and genotype respectively for i-the individual in s-th stratum $(i = 1, \ldots, n_s; s = 1, \ldots, I)$.

Let $\pi_{si} = P(X = x_{si}|S = s)$ be the empirical distribution of X in stratum s, and $r_g(p) = P(G = g)$ be genotype frequency of genotype g. So we have $r_0(p) = (1 - p)^2, r_1(p) = 2p(1 - p)$ and $r_2(p) = p^2$ assuming that Hardy-Weinberg equilibrium distribution holds in the population of interest. We assume that

$$P(X = x_{si}, G = g|S = s) = \pi_{si} r_g(p),$$

which implies that neither X nor S confounds the association between Y and G indicating that G is independent of (X, S); however, X and S may be correlated. Moreover, stratum specific prevalence is assumed to be known, i.e. we know $f_s = P(Y = 1|S = s)$ apriori, for $s = 1, \ldots, I$.

Assuming HWE holds and independence of gene and covariate, the retrospective likelihood function is given by

$$\prod_{s=1}^{I}\prod_{i=1}^{n_s} P(X_{si} = x_{si}, G_{si} = g_{si}|S = s, Y_{si} = y_{si}) = \prod_{s=1}^{I}\prod_{i=1}^{n_s} \frac{P(X_{si} = x_{si}, G_{si} = g_{si}, Y_{si} = y_{si}|S = s)}{P(Y_{si} = y_{si}|S = s)}$$

$$= \prod_{s=1}^{I}\prod_{i=1}^{n_s} \frac{P(Y_{si} = y_{si}|X_{si} = x_{si}, G_{si} = g_{si})P(X_{si} = x_{si}, G_{si} = g_{si}|S = s)}{P(Y_{si} = y_{si}|S = s)} \tag{7.24}$$

Taking logarithm and simplifying, we have the retrospective likelihood function as

$$l(\theta) = \sum_{s=1}^{I}\sum_{i=1}^{n_s} y_{si}\log\left(\frac{e^{\alpha+\beta g_{si}+\gamma_1 x_{si}+\gamma_2 s_i}}{1+e^{\alpha+\beta g_{si}+\gamma_1 x_{si}+\gamma_2 s_i}}\right) + \sum_{s=1}^{I}\sum_{i=1}^{n_s}(1-y_{si})\left(\frac{1}{1+e^{\alpha+\beta g_{si}+\gamma_1 x_{si}+\gamma_2 s_i}}\right)$$
$$+\sum_{s=1}^{I}\sum_{i=1}^{n_s}\log \pi_{si} - \sum_{s=1}^{I}\sum_{i=1}^{n_s}\log\left\{f_s^{y_{si}}(1-f_s)^{1-y_{si}}\right\} \tag{7.25}$$

Although Eq. (7.25) looks like a simple log-likelihood function, it is subject to several restrictions making it difficult to handle mathematically. Another way to bypass this issue is to use profile likelihood function; however this leads to saddle point issue that makes it further difficult to maximise. Thus, Zhang et al. (2018) proposed a score function that can provide consistent estimates of the parameters of the likelihood function. We avoid the detailed discussion as this is mathematically more difficult; interested readers may refer to the paper by Zhang et al. (2018).

7.5 Points to Remember

So far this chapter contained some interesting discussion and statistical methods on how to tackle some issues that arise during genetic association study. Of course, when multiple loci are present, one can do multi-loci test like KBAT or SKAT. However, if the loci are close to each other, there is a chance that alleles at these loci may be transmitted together as a block from one generation to another. In such a case, along with genetic association, one should also look whether a block of DNA is being transmitted from parent to offspring that causes the disease already present in the previous generation. So not only genetic association at a single locus is important, it is worth investigating haplotype association also. However, if the number of haplotypes is large, we have to find some way to group them together or a very large sample would be required. Depending on the practical situation, we have to take a decision.

When more than one class is available, for example, in studying gene expression data for different cancer stages, we need to do ANOVA or ANOCOVA provided any covariate is absent or present, respectively. However, after the analysis, if the null hypothesis is rejected indicating that gene expressions are different over stages of cancer, we need to investigate further. It is extremely important to know which stage is actually responsible for rejection of null hypothesis. Hence, after the analysis using ANOVA or ANOCOVA technique, we have to do pairwise comparison using a two-sample t-test. To compare the gene expression between Stage II and Stage III, this is just a two-sample problem as the samples are collected independent of each other. Similarly for any pair of stages, we can proceed exactly in the same way. Care should be taken at the very beginning; we have to first test the assumptions underlying the ANOVA technique and then only do we decide which test should be adopted. If normality assumption or more strictly, equal variance assumption is

violated, we cannot do ANOVA but can do Kruskal-Wallis test easily and then do a pairwise comparison using Wilcoxon test. Blindly doing ANOVA in comparing more than two means sometimes leads to wrong conclusion.

Covariate adjustment is very important in the context of genetic association analysis. Very often we collect data on covariates without putting too much effort or resources. Hence, it would be wise to use this data when studying genetic association. As discussed, we should be careful about the pros and cons of the method used for such analysis and accordingly choose the appropriate method. Note that situation might be different depending on how we collect data, sample identification and collection mechanism, which population is under study, availability of prior knowledge about certain parameters, etc. So, there is nothing called 'best method' that would be applicable in all scenarios. This is still a fertile area of research and a few more new genetic association methods addressing several issues might come up in future.

Exercise

7.1 Suppose a genomic region has four loci, each bi-allelic in nature. Assuming Hardy-Weinberg equilibrium at each of the four loci, find out all possible haplotypes and their corresponding frequencies. Now suppose that haplotypes with at least three minor alleles are extremely rare in the population and hence can be ignored. How do you modify your answer?

7.2 Using the following table of haplotype frequencies for two loci A (with alleles A1 and A2) and B (with alleles B1 and B2), determine whether the two loci are in linkage disequilibrium.

Haplotype	Frequency
A1B1	756
A2B2	723
A2B1	123
A1B2	127

7.3 Generate Fig. 7.2 for different values of r with different starting values of D_0, where symbols have their usual meanings. To draw this figure, write an R code. This code would generate each line with a different colour that will change automatically if we want to incorporate more values of r and/or D_0. [hint: you can use appropriate loop in R code]

7.4 Data in Table 6.1 represents genotypes at four loci and case-control status.

(a) Construct the possible haplotypes from the data.
(b) Do an association test at level of significance 0.01 using these haplotype data.
(c) Write a short report on your findings.

7.5 Calculate the linkage disequilibrium for the following genotype data at two codominant loci.

	BB	Bb	bb	Total
AA	57	140	101	298
Aa	39	224	226	489
aa	3	54	156	213
Total	99	418	483	1000

7.6 Using the data in Table 7.5, perform a statistical test to see whether the variances in different stages are same? Based on your conclusion, suggest your next step to study whether this gene has anything to do with cancer stages. Suppose it is a general idea that variance of gene expression in Stage IV patients is more than that in normal people. How do you do a statistical test to validate this conjecture using data in Table 7.5?

7.7 In Table 7.6, assume that on a chromosomal region, the order of three loci is ($G1$, $G2$,G3). Assume that '1' codes for genotypes Aa and aA with probabilities 0.5 and 0.5. That means if we observe a '1', it would be Aa with probability 0.5 and aA with probability 0.5. Incorporating this uncertainty, construct a set of haplotypes and evaluate their frequencies.

7.8 Based on the result obtained in 7.7, calculate linkage disequilibrium between (i) G1 and G2, and (ii) G2 and G3. We can think of a haplotype association test as well as a multi-loci association test. So,

(a) do a haplotype association test at the level of significance 0.01, and
(b) do a multi-loci association test at the same level of significance.
(c) Interpret your results.

7.9 A research study was conducted to examine the clinical efficacy of a new antidepressant drug. Fifteen depressed patients were randomly assigned to one of three groups: a placebo group, a group that received a low dose of the drug, and a group that received a moderate dose of the drug. After four weeks of treatment, the patients completed a questionnaire. The higher the score, the more depressed the patient. The data are presented below.

Placebo	Low Dose	Moderate Dose
38	22	14
47	19	26
39	8	11
25	23	18
42	31	5

(a) What would be the null hypothesis in this study?
(b) What would be the alternate hypothesis?

(c) Is there a significant difference between the groups? [You may assume that the scores are Normally distributed, with the same variance among the three groups.]
(d) If there is a significant difference, where specifically are the differences? Interpret your answer.

7.10 Using the data in Table 7.6, do the following.

(a) Test whether gender and disease status are independent?
(b) Using case-control status as phenotype, do a statistical test to see whether a particular genetic locus is associated in presence of (i) age only, (ii) gender only, and (iii) both covariates age and gender. Do the same for all loci and give your conclusion.
(c) Using quantitative phenotype (Quant Phenotype in Table 7.6), do a statistical test to see whether a particular genetic locus is associated in presence of age and gender. You can assume that the quantitative phenotype follows a normal distribution. Do the same for all loci and give your conclusion.
(d) Based on your conclusion in (b) and (c), can you suggest any modification of your tests? Justify your answer.
(e) Suppose we are interested in genetic locus G2. What is the odds ratio in favour of disease for male and age group above 50 years compared to female and age group below 50 years? Evaluate the algebraic expression first and then estimate its value.

7.11 For the data in Table 7.6, using liability threshold model, estimate the minor allele frequency assuming (a) $F = 0.1$ and (b) $F = 0.4$. Compare these estimates with the situation when there is no ascertainment bias in sampling. Write an R code to do this.

References

Dempster, A. P., Laird, N. M., & Rubin, D. B. (1977). Maximum likelihood from incomplete data via the EM algorithm. Journal of the Royal Statistical Society. *Series B (Methodological), 39*, 1–38.

Gabriel, S. B., Schaffner, S. F., Nguyen, H., Moore, J. M., Roy, J., Blumenstiel, B., Higgins, J., DeFelice, M., Lochner, A., Faggart, M., Liu-Cordero, S. N., Rotimi, C., Adeyemo, A., Cooper, R., Ward, R., Lander, E. S., Daly, M. J., & Altshuler, D. (2002). The structure of haplotype blocks in the human genome. *Science, 296*, 2225–2229.

Badge, R. M., Yardley, J., Jeffreys, A. J., & Armour, J. A. (2000). Crossover breakpoint mapping identifies a subtelomeric hotspot for male meiotic recombination. *Human Molecular Genetics, 9*, 1239–1244.

Excoffier, L., & Slatkin, M. (1995). Maximum-likelihood estimation of molecular haplotype frequencies in a diploid population. *Molecular Biology and Evolution, 12*(5), 921–7.

Daly, M., Rioux, J. D., Schaffner, D. F., Hudson, T. J., & Lander, E. S. (2001). High-resolution haplotype structure in the human genome. *Nature Genetics, 29*, 229–232.

Gu, S., Pakstis, A., Li, H., Speed, W. C., Kidd, J. R., & Kidd, K. K. (2007). Significant variation in haplotype block structure but conservation in tagSNP patterns among global populations. *European Journal of Human Genetics, 15*, 302–312.

Hill, W. J. (1974). Estimation of linkage disequilibrium in randomly mating populations. *Heredity, 33*, 229–239.

Jeffreys, A. J., Kauppi, L., & Neumann, R. (2001). Intensely punctuated meiotic recombination in the class II region of the major histocompatibility complex. *Nature Genetics, 29*, 217–222.

Kuleshov, V., Xie, D., Chen, R., Pushkarev, D., Ma, Z., Blauwkamp, T., Kertesz, M., & Snyder, M. (2014). Whole-genome haplotyping using long reads and statistical methods. *Nature Biotechnology, 32*, 261–266.

Kuo, C. L., & Feingold, E. (2010). What's the best statistic for a simple test of genetic association in a case-control study? *Genet Epidemiol, 34*(3), 246–253.

Lin, D. Y., & Zeng, D. (2006). Likelihood-based inference on haplotype effects in genetic association studies. *Journal of the American Statistical Association, 101*, 89–104.

Neuhaus, J. M., & Kalbfleisch, J. D. (1998). Between- and within-cluster covariate effects in the analysis of clustered data. *Biometrics, 54*(2), 638–645.

Olson, J. M., & Wijsman, E. M. (1994). Design and sample size considerations in the detection of linkage disequilibrium with a disease locus. *American Journal of Human Genetics, 55*, 574–580.

Patil, N., Berno, A. J., Hinds, D. A., Barrett, W. A., Doshi, J. M., Hacker, C. R., Kautzer, C. R., Lee, D. H., Marjoribanks, C., McDonough, D. P., Nguyen, B. T., Norris, M. C., Sheehan, J. B., Shen, N., Stern, D., Stokowski, R. P., Thomas, D. J., Trulson, M. O., Vyas, K. R., & Cox, D. R. (2001). Blocks of limited haplotype diversity revealed by high-resolution scanning of human chromosome 21. *Science, 294*, 1719–1723.

Salem, R. M., Wessel, J., & Schork, N. J. (2005). A comprehensive literature review of haplotyping software and methods for use with unrelated individuals. *Human Genomics, 2*, 39–66.

Schaid, D. J. (2004). Linkage disequilibrium testing when linkage phase is unknown. *Genetics, 166*, 505–512.

Schaid, D. J., Rowland, C. M., Tines, D. E., Jacobson, R. M., & Poland, G. A. (2002). Score tests for association between traits and haplotypes when linkage phase is ambiguous. *American Journal of Human Genetics, 70*, 425–434.

Tangherloni, A., Spolaor, S., Rundo, L., Nobile, M. S., Cazzaniga, P., Mauri, G., Lio, P., Merelli, I., & Besozzi, B. (2019). GenHap: A novel computational method based on genetic algorithms for haplotype assembly. *BMC Bioinformatics, 20*, 172. https://doi.org/10.1186/s12859-019-2691-y

Tapadar, P., Ghosh, S., & Majumder, P. P. (2000). Haplotyping in pedigrees via a genetic algorithm. *Human Heredity, 50*, 43–56.

Tzeng, J. Y., Devlin, B., Wasserman, L., & Roeder, K. (2003). On the identification of disease mutations by the analysis of haplotype similarity and goodness of fit. *American Journal of Human Genetics, 72*, 891–902.

Zaitlen, N., Bogdan, P., et al. (2012). Analysis of case-control association studies with known risk variants. *Bioinformatics, 28*(13), 1729–1737.

Zhang, H., Chatterjee, N., Rader, D., & Chen, J. (2018). Adjustment of nonconfounding covariates in case-control association studies. *The Annals of Applied Statistics, 12*(1), 200–221.

Chapter 8
Exploring Multivariate Data

Moving from one gene to two genes naturally leads to the idea of existence and consideration of multiple genes together. Not only genes, this may be true for phenotypes also. If we suspect that more than two phenotypes need to be jointly considered while studying any genetic association, the same issue arises. If we have expression data for more than two genes, we might study them separately. We can use a similar idea for two or more phenotypes occurring together. Dealing with a single gene or one phenotype at a time seems okay, but there is a high possibility that we might lose the inherent correlation structure among multiple genes or multivariate phenotypes. On the other hand, a statistical way of treating them together seems more realistic; this might retain the correlation structures among genes and is more informative to decipher the genetic architecture. But how do we look at such types of data? Depending on our outlook, questions, and the problem under consideration, there are different ways and various statistical concepts and methods to deal with it.

The idea of correlation among the variables is probably the major reason to consider variables together, i.e. from a multivariate perspective. While discussing correlation, we have studied correlation and regression in the context of two variables occurring together. Similarly in the multivariate set up we might ask the question as how and to what extent a variable is related to other variables. If the variables are related, we can easily calculate pairwise correlation coefficient that gives an idea about the extent of linear relationship between any two variables. However it might so happen that two variables might seem to have a high correlation, but this correlation is only due to their relation with another third variable. This typical situation is very common in multivariate set up but does not naturally arise in case of two correlated variables. Moreover, we might also explore whether we can propose a model that explains how one variable is related to all other variables. In other words, we should explore the possibility of developing predictive models to have an idea on the value of one variable for an individual or item based on available information on other variables. This generates a plethora of questions regarding the interrelationships among variables.

I. Mukhopadhyay and P. P. Majumder, *Statistical Methods in Human Genetics*,
Indian Statistical Institute Series, https://doi.org/10.1007/978-981-99-3220-7_8

Any data analysis exercise should start with a visual presentation to get an overall idea about patterns in the data. Visualisation of two genes is possible. We can plot the expression values of two genes in two axes in a scatter plot and visualise the relationship between them. For three genes, we can get some visualisation idea as it can be plotted on a three-dimensional plane, although slightly more complicated than a scatter diagram. But what happens when there are more than three genes? There is no way we can visualise them through a diagrammatic presentation. Moreover we know how to analyse two genes together; using bivariate normal distribution or paired t-test depending on the question. But more genes trigger more questions that we try to explore here. What we said about more than two genes, same holds true for phenotypes. Although two or more genes (phenotypes) occur together, each can be considered as a single gene (phenotype). Each component is a random variable. This type of data, with two or usually more than two components (variables), is known as multivariate data. Most important feature of multivariate data in contrast to univariate data is the existence of correlations among its components. These correlations, directly or indirectly, play important roles in multivariate analysis that should not be ignored.

8.1 Generating Questions

More genes, more dimensions, more questions. We need to understand the data before planning for analysis.

(1) How do we capture the various possible correlation structure in multivariate data?
(2) How do we develop a model, may be a linear model in the line of linear regression, of one variable that might depend on other variables? If we can develop such a model, how do we justify whether the model is 'good' (in some sense)?
(3) In such a situation, where linear relation (at least to some extent) seems meaningful, can we measure such a relationship?
(4) If two variables are dependent only through a third variable, how do we identify that relation and the degree of this relationship, if it exists?
(5) Data might consist of four or more components and we have only three dimensions in our real life! Can we visualise multivariate data with more than three dimensions? If so, how?
(6) Can we apply the methods or their extensions that we learnt for univariate or bivariate data, especially in the context of genetic association study?
(7) Is there any pattern hidden in the data that we need to understand and extract even if immediate raw data visualisation is not possible?
(8) Data with four components (dimensions) are difficult to handle and impossible to visualise directly. Can we make our life simple by reducing the data to a lower dimension?

We try to explain and address these questions through our intuition, statistical knowledge, and of course, common sense. In multivariate data, we usually have $p\,(> 2)$ variables denoted by a vector $\boldsymbol{X}$ with p components or variables $\{X_1, X_2, \ldots, X_p\}$. We write it using a vector notation as

$$\boldsymbol{X} = \begin{pmatrix} X_1 \\ X_2 \\ \vdots \\ X_p \end{pmatrix}.$$

If r_{ij} is the correlation coefficient between X_i and X_j for $i \neq j = 1, \ldots, p$, we can represent the correlation coefficients in the form of a $p \times p$ matrix as

$$R = \begin{pmatrix} r_{11} & r_{12} & r_{13} & \cdots & \cdots & r_{1p} \\ r_{21} & r_{22} & r_{23} & \cdots & \cdots & r_{2p} \\ r_{31} & r_{32} & r_{33} & \cdots & \cdots & r_{3p} \\ \vdots & \vdots & \vdots & \vdots & \vdots & \vdots \\ r_{p1} & r_{p2} & r_{p3} & \cdots & \cdots & r_{pp} \end{pmatrix} = \begin{pmatrix} 1 & r_{12} & r_{13} & \cdots & \cdots & r_{1p} \\ r_{21} & 1 & r_{23} & \cdots & \cdots & r_{2p} \\ r_{31} & r_{32} & 1 & \cdots & \cdots & r_{3p} \\ \vdots & \vdots & \vdots & \vdots & \vdots & \vdots \\ r_{p1} & r_{p2} & r_{p3} & \cdots & \cdots & 1 \end{pmatrix} \text{ since } r_{ii} = 1 \text{ for all } i = 1, \ldots, p.$$

Note that the correlation matrix R is a symmetric matrix as $r_{ij} = r_{ji}$ for all $i, j = 1, \ldots, p$. Naturally $\boldsymbol{x}$ is the observed realisation of $\boldsymbol{X}$ where

$$\boldsymbol{x} = \begin{pmatrix} x_1 \\ x_2 \\ \vdots \\ x_p \end{pmatrix}.$$

8.2 A Multivariate Data Set

We first present a data set (Table 8.1) that contains four variables viz. HDL, LDL, VLDL, and TGL, measures of cholesterol levels in a person. We explain different topics in multivariate analysis using this data set. Although this is a simulated data set, we have kept our attention to standard values that one might observe in a sample from a population. In other words, we have generated these observations mimicking real-life situations. We have mentioned that more than two genes may be coexpressed, i.e. we need to study them together. The same is true for multivariate phenotypes. Here we consider only a multivariate data set that has four-variate phenotype data and a single variable for genotype. Our main focus is to understand different multivariate techniques that might be required while doing analysis when we encounter any such multivariate data. In Table 8.1, although each cholesterol indicator shows integer value, without any loss of generality, we can consider each of them as a continuous

Table 8.1 Data on different parameters measuring cholesterol level in a few individuals

HDL	LDL	VLDL	TGL	Genotype	HDL	LDL	VLDL	TGL	Genotype
49	95	23	146	0	46	107	24	157	0
45	103	22	150	0	44	107	26	168	1
44	100	21	155	0	66	58	19	100	0
63	72	20	120	0	68	65	18	104	1
50	90	23	148	1	64	66	19	119	0
47	106	23	148	0	43	105	25	153	1
66	62	19	110	0	44	106	26	148	1
63	73	21	109	1	60	75	18	110	2
62	82	16	117	0	48	106	26	158	0
68	64	17	107	1	64	60	16	106	0
65	75	18	102	0	50	113	22	147	0
68	60	20	117	0	45	92	26	153	0
45	114	28	162	0	53	98	25	156	1
64	69	16	100	0	53	96	23	146	1
52	98	17	151	1	63	79	23	120	0
61	78	20	130	1	55	91	21	144	0
65	55	14	93	1	68	55	18	105	1
47	109	22	161	2	63	68	20	112	1
61	68	20	115	0	44	101	25	161	1
65	74	14	92	1	53	96	18	135	0

variable, due to the limitation of measuring instruments. However, genotype is a discrete variable indicating the number of minor alleles at a locus of interest.

8.3 Multiple Regression

Like linear regression in the case of bivariate data, we might be interested to know how one variable depends on the other variables. Here we mainly focus our attention only to linear regression model. The idea behind such a model is to predict the value of one variable using the values of other variables for an item or individual.

Suppose we want to fit a linear regression model to explain how x_1 depends on other variables $x_2, x_3, \ldots, x_p$. Let such a linear regression model be

$$x_{1l} = \beta_0 + \beta_2 x_{2l} + \beta_3 x_{3l} + \cdots + \beta_p x_{pl} + e_l, \quad l = 1, 2, \ldots, n \tag{8.1}$$

where $e_1, e_2, \ldots, e_n$ are errors associated with observations corresponding to n individuals.

We can apply the least squares method to get estimates of the coefficients as $\hat{\beta}_0, \hat{\beta}_2, \ldots, \hat{\beta}_p$. Now using these estimated coefficients, we have the fitted value of x_1, denoted by $X_{1\bullet 23\ldots p}$ which is

$$X_{1\bullet 23\ldots,p;l} = \hat{\beta}_0 + \hat{\beta}_2 x_{2l} + \hat{\beta}_3 x_{3l} + \cdots + \hat{\beta}_p x_{pl}, \quad l = 1, 2, \ldots, n \tag{8.2}$$

This is known as multiple regression of x_1 on $x_2, x_3, \ldots, x_p$ and $X_{1\bullet 23\ldots,p;l}$ is the predicted (or fitted) value of x_1 for l^{th} observation when $x_{2l}, \ldots, x_{pl}$ are known. Here x_1 is our study variable or response and $x_2, \ldots, x_p$ are regressors or explanatory variables.

We do not discuss in detail how to derive estimates of the coefficients $\beta_0, \beta_2, \ldots, \beta_p$, because it requires knowledge of matrix algebra and some rigorous mathematical calculations. However, an R code can easily evaluate them. In fact from that R function, we can get fitted values as explained by regressor variables $x_2, \ldots, x_p$. With reference to Table 8.1, if we want to fit TGL on HDL, LDL, and VLDL, i.e. we want to see how TGL depends on other cholesterol parameters through a linear regression model, we fit this directly using in-built R function.

```
> a<-lm(b$TGL~b$HDL+b$LDL+b$VLDL)  ## fitting regression model
                                      using data stored in 'b'
                                    ## output is stored in 'a'
> summary(a)                        ## gives estimates and p-value
                                      for significant coefficients
```

'summary(a)' gives the output of the multiple regression fitted to establish the linear dependence of TGL on HDL, LDL, and VLDL using the data (Table 8.1). The multiple R^2, which is square of the multiple correlation coefficient (to be discussed in next section) is 0.9145 indicating that there is a strong effect of HDL, LDL, and VlDL when we study the linear dependence of TGL on these variables. Moreover the extremely small p-value ($< 2.2^{-16}$) is indicative of the presence of a multiple linear relationship of x_1 on $x_2, x_3, \ldots, x_p$.

8.4 Multiple Correlation Coefficient

Multiple regression provides a linear relation through which x_1 is dependent on other variables. However, if we are interested to know how good the fit is, we have to find a measure that indicates the degree of this relationship, i.e. it should measure to what extent x_1 is being explained by other variables through the fitted linear regression equation.

$X_{1\bullet 23\ldots p}$ is the fitted value of x_1 through a linear regression equation of x_1 on $x_2, \ldots, x_p$.This quantity contains the effect of $x_2, \ldots, x_p$ on x_1 through the least squares linear regression equation. On the other hand, x_1 has the effect of all other

variables along with its own effect. Hence, we propose a measure, known as 'multiple correlation coefficient', as a product moment correlation coefficient between x_1 and $X_{1\bullet 23\ldots p}$ is, and is defined as

$$r_{1\bullet 23\ldots p} = \text{correlation coefficient } (x_1, X_{1\bullet 23\ldots p}) = \frac{\text{Cov}(x_1, X_{1\bullet 23\ldots p})}{\sqrt{V(x_1)V(X_{1\bullet 23\ldots p})}}. \quad (8.3)$$

We should note one very interesting phenomenon. Let $\hat{e}_l$ be the value of error that has been associated with $X_{1\bullet 23\ldots p;l}$ and x_{1l} for each $l = 1, \ldots, n$. Note that $\hat{e}$ is uncorrelated with $X_{1\bullet 23\ldots p}$, i.e. $Cov(X_{1\bullet 23\ldots p}, \hat{e}) = 0$. Then, we have

$$x_1 = X_{1\bullet 23\ldots p} + \hat{e} \text{ and hence,}$$

$$Cov(x_1, X_{1\bullet 23\ldots p}) = Cov(X_{1\bullet 23\ldots p} + \hat{e}, X_{1\bullet 23\ldots p}) = V(X_{1\bullet 23\ldots p}) + Cov(X_{1\bullet 23\ldots p}, \hat{e}) = V(X_{1\bullet 23\ldots p})$$

$$\text{Thus, } r_{1\bullet 23\ldots p} = \frac{\text{Cov}(x_1, X_{1\bullet 23\ldots p})}{\sqrt{V(x_1)V(X_{1\bullet 23\ldots p})}} = \frac{V(X_{1\bullet 23\ldots p})}{\sqrt{V(x_1)V(X_{1\bullet 23\ldots p})}} = \sqrt{\frac{V(X_{1\bullet 23\ldots p})}{V(x_1)}}. \quad (8.4)$$

Thus multiple correlation coefficient, being the square root of ratio of two variances, is always positive. Moreover, being a product moment correlation coefficient, it cannot be greater than one. Hence, we have, $0 \leq r_{1\bullet 23\ldots p} \leq 1$. If $X_{1\bullet 23\ldots p}$ can really explain a substantial portion of x_1, this implies that x_1 and $X_{1\bullet 23\ldots p}$ have much in common. Hence $r_{1\bullet 23\ldots p}$ should be large. On the other hand, if $X_{1\bullet 23\ldots p}$ is unable to explain or capture a substantial portion of x_1, this correlation would be small. Using the correlation matrix R, a little bit of algebraic manipulation gives an expression for $r_{1\bullet 23\ldots p}$ as

$$r_{1\bullet 23\ldots p} = \left(1 - \frac{|R|}{R_{11}}\right)^{1/2} \quad (8.5)$$

where R_{11} is the cofactor of $(1, 1)th$ element of the correlation matrix R and $|R|$ is the determinant of R. Note that cofactor of $(i, j)th$ element of a matrix A is defined as $(-1)^{i+j}|M_{ij}|$ where M_{ij} is the matrix obtained by deleting i^{th} row and j^{th} column of the matrix A.

We can easily write an R code using the formula (8.5) for calculating multiple correlation coefficients. In R programming, the function 'cor( )' gives the correlation coefficient between two lists of observations from the same group of individuals. But if we input a matrix of observations on multiple variables, the same function would calculate correlations among all possible pairs of variables and arrange them in a matrix form. Thus we directly get the correlation matrix using R code.

```
> R<- cor(b[,-5])          ## correlation matrix obtained from data stored
                             in 'b'; we remove 5th column as
                           ## it contains genotype data
> R.11<- R[-1,-1]          ## matrix obtained from R deleting first row and
                             first column
> multiple.correlation<- sqrt(1 - (det(R)/det(R.11)))     ## r1•23...p as
                             given in (8.5); det() calculates
                                        ## determinant of a matrix
```

The output correlation matrix is

$$R = \begin{pmatrix} 1.000 & -0.935 & -0.772 & -0.934 \\ -0.935 & 1.000 & 0.735 & 0.920 \\ -0.772 & 0.735 & 1.000 & 0.825 \\ -0.934 & 0.920 & 0.825 & 1.000 \end{pmatrix}$$

This R matrix provides an idea about all possible pairwise correlation coefficients among p variables. Clearly as HDL increases, LDL decreases because their correlation coefficient is -0.935. The multiple correlation coefficient is 0.954 indicating that there seems to exist a strong linear dependence of LDL, VLDL, and TGL on HDL as explained by the multiple linear regression equation.

Multiple correlation coefficient reduces to absolute value of Pearson correlation coefficient for bivariate data. It indicates how much of x_1 can be explained through a linear regression model based on other variables. If $r_{1\bullet 23\ldots p}$ is very high, i.e. close to 1, we can say that the multiple regression equation is successful in explaining the response or study variable by other variables through a linear model (equation). However, a small value is indicative of its failure for such an explanation through a linear regression model. Thus, $r_{1\bullet 23\ldots p}$ provides a measure of degree of linear relationship of response with other explanatory variables.

8.5 Partial Correlation Coefficient

Multiple regression provides an idea on the dependence of one variable on other variables through a linear relation whereas multiple correlation coefficient measures the extent or degree of the influence of other variables on one variable (here it is x_1) as explained through linear regression equation. But one might be interested to know how only two variables are related. Is there any measure that can give an idea about it? This question is in relation to bivariate data where correlation coefficient measures the degree of linear relationship between two variables. This is still a good measure but is unable to take care of the effects of other variables on these two variables. It might so happen that two variables might have a very high correlation coefficient between them, but this correlation is due to the effect of another variable.

In this case, in order to get an idea of the actual relation, we need to eliminate the effect of the third variable.

In mathematical notation, suppose we want to find any possible degree of linear relationship between x_1 and x_2 but there exist variables $x_3, \ldots, x_p$ that might have some effect on x_1 and x_2. Especially, a situation might arise where both x_1 and x_2 are actually uncorrelated but they show a high degree of correlation only because of the fact that both of them are highly influenced by x_3. Naturally in such cases it is advisable to first eliminate the effect of the influencing variables, i.e. $x_3, \ldots, x_p$ from each of x_1 and x_2 and then measure the correlation between them. This correlation measure thus obtained is known as 'partial correlation coefficient' between x_1 and x_2 eliminating the effects of other variables.

But how do we eliminate the effects of the other variables? Well, we know that multiple regression captures the dependence of one variable on other variables. If we subtract the fitted value as explained by multiple regression equation of x_1 on $x_3, \ldots, x_p$ from x_1, it contains the effect of x_2 and its own effect. Let this be denoted by $e_{1\bullet 34\ldots p} = x_1 - X_{1\bullet 34\ldots p}$ where $X_{1\bullet 34\ldots p}$ is the fitted value of x_1, i.e. the portion of x_1 that is being explained by $x_3, \ldots, x_p$. Similarly $e_{2\bullet 34\ldots p} = x_2 - X_{2\bullet 34\ldots p}$ contains the effect of x_2 on x_1 eliminating the effects of other variables from x_2. Note that these variables $e_{1\bullet 34\ldots p}$ and $e_{2\bullet 34\ldots p}$ are free from the effect of $x_3, x_4, \ldots, x_p$ as captured through a linear regression equation. Hence, the product moment correlation coefficient (denoted by $r_{12\bullet 3\ldots p}$) between $e_{1\bullet 34\ldots p}$ and $e_{2\bullet 34\ldots p}$ provides a measure of linear relationship between x_1 and x_2 eliminating effects of $x_3, \ldots, x_p$ as captured by respective linear regression equation and is given by

$$r_{12\bullet 3\ldots p} = \frac{Cov(e_{1\bullet 34\ldots p}, e_{2\bullet 34\ldots p})}{\sqrt{V(e_{1\bullet 34\ldots p})V(e_{2\bullet 34\ldots p})}}. \tag{8.6}$$

Since we already have a correlation matrix, i.e. R, we should be able to find an easy expression for partial correlation coefficient that will help us to calculate this quantity directly without evaluating fitted values through multiple regression models. It can be shown that partial correlation coefficient can be expressed as

$$r_{12\bullet 3\ldots p} = -\sqrt{\frac{R_{12}}{R_{11}R_{22}}} \tag{8.7}$$

where R_{12}, R_{11} and R_{22} are the cofactors of $(1, 2)th$, $(1, 1)th$ and $(2, 2)th$ elements respectively in the correlation matrix R. Partial correlation coefficient being a product moment correlation coefficient always lies between -1 and $+1$, i.e. $-1 \le r_{12\bullet 3\ldots p} \le 1$. The R code for calculating this is given below.

```
> R.11<- R[-1,-1]        ## matrix obtained from R deleting first row and
                            first column
> R.22<- R[-2,-2]        ## matrix obtained from R deleting second row
                            and second column
> R.12<- R[-1,-2]        ## matrix obtained from R deleting first row and
                            second column
> partial.correlation<- -(-det(R.12))/sqrt(det(R.11) * det(R.22))
                         ## r1•23...p as given in Eq. 8.7.
```

The value of $r_{12\bullet 34\ldots p}$ is -0.5445 indicating that there exists a moderately strong linear relation between HDL and LDL levels eliminating the effects of VLDL and TGL based on the given data (Table 8.1).

8.6 Principal Component Analysis

The first typical feature of multivariate data is its dimensionality. If there are more than three dimensions, there is no way that we can see the data through simple plots, because we cannot plot anything beyond three dimensions. So it seems that a few standard and unified ideas should be built in order to visualise data and do appropriate data analysis. Two important things to remember: first, it would be great if we can somehow see the data; second, it is always easy if we can analyse uncorrelated or independent variables rather than correlated variables. But multivariate data are usually correlated and they are of higher dimensions preventing us from plotting them on a two-dimensional plane if the dimension of each data point is more than two. However, an intelligent solution exists to solve these issues.

Suppose a data set has p components where $p \geq 3$. It is impossible to plot them if $p > 3$ and even if we plot 3-dimensional data, not much information can be extracted immediately. This triggered an idea: can we do 'something' so that we can reduce the data to a lower dimension, say 2-dimension? If we can, it would be very easy to plot the data immediately. Moreover, while undertaking such a process, can we make the variables free from correlations? Fortunately, in both cases the answer is 'yes'. However, there is one shadow of doubt: is it possible to retain the 'originality' of the data as it was collected from the experiment? Well, not every answer can be 'yes'; we can achieve something but may have to compromise. But first we should try to develop a method that can reduce the data to a lower dimension, at least approximately, in some sense. The most commonly used method that reduces the dimensions of a multivariate data is known as 'Principal Component Analysis' (PCA).

Let's use a few mathematical notations for the sake of simplicity and clarity. Let $X_1, \ldots, X_p$ be the multivariate data with means $\mu_i = E(X_i)$, variances $\sigma_i^2 = V(X_i)$, and covariances $\sigma_{ij} = Cov(X_i, X_j)$ for $i, j = 1, \ldots, p$. Note the correlation coefficient between X_i and X_j is $\rho_{ij} = \sigma_{ij}/\sigma_i\sigma_j$.

Let's represent $X_1, \ldots, X_p$ as a column vector, i.e.

$$\boldsymbol{X} = \begin{pmatrix} X_1 \\ X_2 \\ \vdots \\ X_p \end{pmatrix}$$

Then all possible variances and covariances of the components of $\boldsymbol{X}$ can be represented in a matrix form as

$$\Sigma = V(\boldsymbol{X}) = \begin{pmatrix} \sigma_{11} & \sigma_{12} & \cdots\cdots & \sigma_{1p} \\ \sigma_{21} & \sigma_{22} & \cdots\cdots & \sigma_{2p} \\ \vdots & \vdots & \vdots \quad \vdots & \vdots \\ \sigma_{p1} & \sigma_{p2} & \cdots\cdots & \sigma_{pp} \end{pmatrix}$$

Our objective is to reduce $X_1, \ldots, X_p$ to lower dimension variables, say, $Y_1, \ldots, Y_q$ with $q < p$. Moreover, while doing so, it would be great if we can find $Y_1, \ldots, Y_q$ such that correlation coefficient between Y_i and Y_j is 0, for $i \neq j$, although there exists a correlation between X_i and X_j. Looks surprising but possible!

Since there are altogether p variables, consider p linear combinations (functions) of $X_1, \ldots, X_p$ as

$$\begin{aligned} Y_1 &= a_{11}X_1 + a_{12}X_2 + \cdots + a_{1p}X_p \\ Y_2 &= a_{21}X_1 + a_{22}X_2 + \cdots + a_{2p}X_p \\ &\vdots \\ Y_p &= a_{p1}X_1 + a_{p2}X_2 + \cdots + a_{pp}X_p \end{aligned} \tag{8.8}$$

where a_{ij}, $i, j = 1, \ldots, p$ are constants such that they have two very interesting properties:

(1) Correlation coefficient Y_i and Y_j is 0 for $i \neq j$, and
(2) $V(Y_1) \geq V(Y_2) \geq \cdots \geq V(Y_p)$

Note that, we can evaluate $V(Y_i)$ for $i = 1, \ldots, p$ and $Cov(Y_i, Y_j)$ for $i \neq j = 1, \ldots, p$ as

$$V(Y_i) = V\Big(a_{i1}X_1 + a_{i2}X_2 + \cdots + a_{ip}X_p\Big) = \sum_{k=1}^{p}\sum_{r=1}^{p} a_{ik}a_{ir}\sigma_{kr}, \tag{8.9}$$

$$\text{and } Cov(Y_i, Y_j) = Cov\Big(\sum_{k=1}^{p} a_{ik}X_k, \sum_{r=1}^{p} a_{jk}X_r\Big) = \sum_{k=1}^{p}\sum_{r=1}^{p} a_{ik}a_{jr}\sigma_{kr}. \tag{8.10}$$

We are not going into mathematical details of derivation and results related to PCA; we would rather prefer to discuss this entire issue more conceptually that might help in actual data analysis using any statistical package or R. First note that each Y_j is a linear combination, i.e. a linear function of $X_1, \ldots, X_p$, where the coefficients a_{ij}s are estimated on the basis of the given data. So, although we know the meaning of each X variable, after applying PCA, we might not get a clear idea about the interpretation and meaning of each Y_j. This definitely looks like a disadvantage! Don't worry; we have many more advantages that make this method very popular and effective.

It is clear that we have to find the coefficients $a_{11}, a_{12}, \cdots, a_{1p}$ such that variance of the first linear combination would be maximum, i.e. we have to choose $a_{11}, a_{12}, \cdots, a_{1p}$ so that $V(Y_1)$ would be maximum. It can be shown that $V(Y_1) = \lambda_1$ where λ_1 is the maximum eigen value of Σ, the dispersion matrix of $(X_1, X_2, \cdots, X_p)$.

Then, we proceed to find the coefficients $a_{21}, a_{22}, \cdots, a_{2p}$ such that $V(Y_2)$ is maximised subject to the restriction $Cov(Y_1, Y_2) = 0$. Fortunately, it can again be shown that $V(Y_2) = \lambda_2$ where λ_2 is the second largest eigen value of Σ. Thus, it is possible to derive two uncorrelated variables Y_1 and Y_2 even when there may exist some correlation between the components of X. Continue this way, i.e. maximise $V(Y_i)$ subject to the fact that it remains uncorrelated with all previous $Y_1, Y_2, \ldots, Y_{i-1}$ (for $i = 2, \ldots, p$). This can easily be done. So we have

$$V(Y_i) = \lambda_i, \quad i = 1, \ldots, p \text{ and } Cov(Y_i, Y_j) = 0 \text{ for } i \neq j = 1, \ldots, p.$$

Hence, the dispersion matrix or variance-covariance matrix of $\boldsymbol{Y}$, is given as

$$\Sigma^* = \mathrm{V}(\mathbf{X}) = \begin{pmatrix} \lambda_1 & 0 & \cdots\cdots & 0 \\ 0 & \lambda_2 & \cdots\cdots & 0 \\ \vdots & \vdots & \vdots \; \vdots & \vdots \\ 0 & 0 & \cdots\cdots & \lambda_p \end{pmatrix}$$

First note that although X_i and X_j are correlated, the transformed variables Y_i and Y_j are uncorrelated for $i \neq j, \; i, j = 1, \ldots, p$. To understand what happens inside this mechanism, when we convert X to Y, assume for the sake simplicity, that $p = 2$, i.e. we are dealing with a bivariate situation. If we plot x_1 and x_2 values in a scatter diagram, this shows the existence of correlation between them. But if we plot y_1 and y_2, any correlation is clearly absent. y_1 and y_2 are derived from the original data for each individual. So after PCA we get the transformed variables to be uncorrelated in terms of a new set of variables which are functions of x-variables. Here $\{Y_1, \ldots, Y_p\}$ retains the information that is contained in $\{X_1, \ldots, X_p\}$ with additional advantage that they are uncorrelated. If the X-variables follow normal distributions, then Y-variables become independent also. Thus in many cases, we would be able to work with independent variables derived from originally correlated normal variables. However, PCA works irrespective of whether $X_1, \ldots, X_p$ follow normal distribution or not.

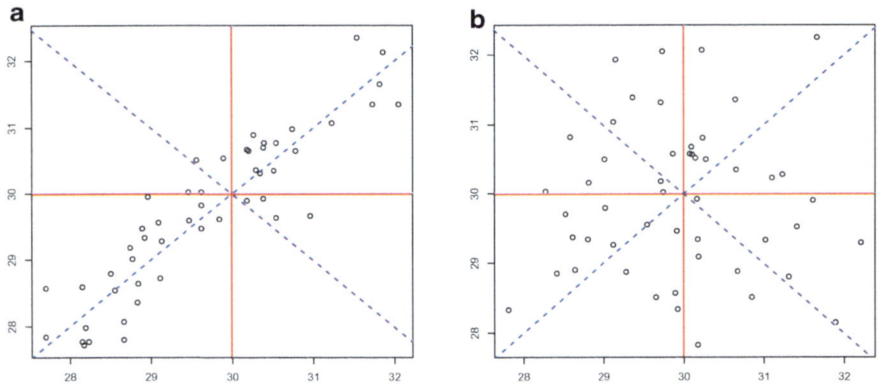

Fig. 8.1 Geometry of principal component analysis. Red lines, the original axes for X-variables fail to explain substantial variability along each axis. But the dotted lines, the axes for principal components can very well explain variability along it axes. Figure 8.1A: Blue line, indicating PC1, explains much of the variation present in the data whereas the purple line, representing PC2 explains very small amount of variation. This happens because the variables are highly correlated and PC1 explains successfully most of the variations. Figure 8.1B: Blue line and purple lines explain almost the same amount of variation because the original variables are uncorrelated and hence PCA has no advantage in this situation

There is another way to look at the principal components. Again, for ease of understanding, we consider bivariate case and draw a scatter diagram of two variables x_1 and x_2. If we rotate the axes such that the variation along Y_1 axis is much more than that along Y_2 axis, although the total variance remains same, i.e. $V(X_1) + V(X_2) = V(Y_1) + V(Y_2)$. Now if we see that $V(Y_1)$ covers almost $80 - 90\%$ of total variance, we can easily work with only Y_1 instead of working with two variables without sacrificing much information (variation). This is the biggest advantage of PCA. Thus in a high dimension scenario, if the first few principal components explain a very high amount of variance, we can work with these new variables which are very small in number compared to the original set of variables but contain a very large portion of information or substantial amount of total variability. Thus PCA reduces higher dimension variables into a lower dimension. Rest of the downstream analysis may depend on these few principal components that explain majority of variance.

The geometry of PCA can be explained by a diagram using two variables. Figure 8.1 describes how the scatter looks like when we plot the original variables and the principal components, on the same graph. As explained by Fig. 8.1, PCA should be applied only when the variables provide information through their correlations, but it is not a successful technique when the variables are uncorrelated or independent. The main reason is that when $X_1, \ldots, X_p$ are independent, the derived principal components $Y_1, \ldots, Y_p$ are also independent. Hence, it is impossible to reduce dimensions, i.e. we are unable to find a lesser number of variables that contain majority of variance.

Hence, after doing PCA, we find how many of the principal components explain a large amount (percentage) of variation present in the data as given by $\sum_{i=1}^{p} V(X_i)$. By 'large' we usually take 80% or 90% or any other large value satisfying the user (but not 100%). We take first k principal components such that

$$\frac{V(Y_1) + \cdots + V(Y_k)}{V(Y_1) + \cdots + V(Y_p)} \approx 0.80 \text{ (or 0.90)} \tag{8.11}$$

If we plot the variances, known as 'scree plot', it would give us an idea of the number of principal components that covers a substantially high amount of variation present in the data.

If we want to work with variables that explain the entire variation, i.e. 100% of total variability, we have no option but to work with all principal components, and hence, all actual variables. It is possible to capture the entire variance only if we work with all variables. Thus, in such a situation, PCA has no role to play.

We now present the data analysis of the multivariate data on cholesterol level for 40 individuals as given in Table 8.1. We need not write any rigorous code for doing this. There is an in-built R function that will produce all results of PCA.

```
> a<-read.csv("cholesterol.txt",header=T)        ## reading the datafile
> b<- princomp(a)          ## doing PCA on the data
> summary(b)          ## summary of analysis
Importance of components:
                  Comp.1      Comp.2      Comp.3      Comp.4
Standard
deviation      30.299694  5.80204682  2.53662405  1.940120798
Proportion of
Variance        0.954402  0.03499591  0.00668909  0.003913023
Cumulative
Proportion      0.954402  0.98939789  0.99608698  1.000000000

> screeplot(b)      ## a graphical presentation showing proportion
                    of variance explained.
```

The R command 'screeplot( )' produces Fig. 8.2. It represents vertical bar diagrams showing the contribution of each principal component. Figure 8.2 clearly shows that it is enough to work with only one, i.e. first principal component because it captures more than 95% of the total variance present in the data.

Here the proportion of variance explained by the principal component is

$$\frac{V(Y_1)}{V(Y_1) + V(Y_2) + V(Y_3) + V(Y_4)} = 0.95442 \text{ i.e more than } 95\%.$$

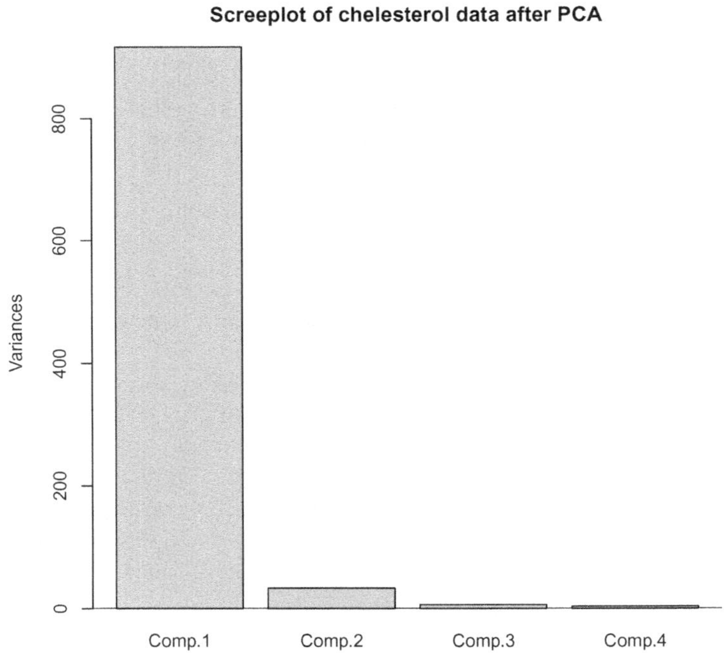

Fig. 8.2 Scree plot showing proportion of variance explained by different principal components

Hence it is enough to work only with first the principal component for further downstream analysis. This means that any further analysis is nothing but analysis of univariate data considering that information from the data are well contained in the first principal component. In most situations, where variables in multivariate set up are really correlated, we have seen that two (or sometimes three) components can explain more that 80% of total variation. Hence downstream statistical analysis may be done using bivariate (or trivariate) distributions. However, extra care should be taken while interpreting the result obtained through analysis of principal components instead of original variables. Sometimes, it becomes difficult to interpret Y-variables, i.e. principal components because they are linear combinations of X-components.

Example: Investigation of population stratification using PCA

The problem of population stratification in genome-wide association analysis was introduced in Sect. 6.6.2. Existence of cryptic population stratification can adversely impact on the inferences of an association study. It is desirable that existence of population stratification be investigated prior to conducting an association analysis so that the effect of stratification can be controlled to a certain extent. A simple method to detect population stratification is to use genotype data at a large number of polymorphic loci in the set of individuals recruited into the study and extract the first two principal components. Then, based on the first two principal components,

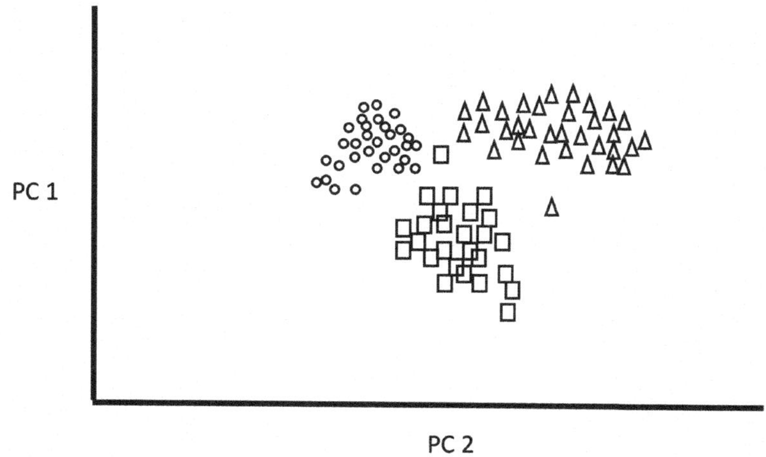

Fig. 8.3 Scatter diagram of first two PCs for individuals based on genotype data of individuals recruited into a genome-wide association study

we plot the scatter diagram of the data on the individuals. Such a scatter diagram is provided in Fig. 8.3. This figure reveals the existence of a cryptic population stratification; individuals recruited into the study were possibly from three distinct populations.

Price et al. (2006) proposed EIGENSTRAT, which computes principal components for variants across the genome to identify population stratification. If population stratification is detected, then its effect is controlled by including the first few principal components as covariates in the regression framework that relates the trait of interest and the genomic variants (Fig. 8.3).

8.7 Cluster Analysis

Although the cholesterol data set has four variables, we can think of, at least to start with, a scatter diagram for a pair of variables. A scatter diagram (Fig. 8.4) of VLDL against HDL reveals some interesting feature.

It clearly reveals that the data points form two or three groups (or clusters), although they may not be very distinct. This idea of grouping or tendency to form different groups in a single data set is the key idea behind cluster analysis. Well, although it looks easy through a scatter diagram, there are some issues that make this an active area of research. We are unable to 'see' data points beyond three dimensions. We need an appropriate method or algorithm, with justification, that may provide these groups or clusters from the data set. This method should be based on sound statistical logic or intuition along with some optimised measure. At the same time, it should retain the characteristics of the data while putting them in respective groups. Note

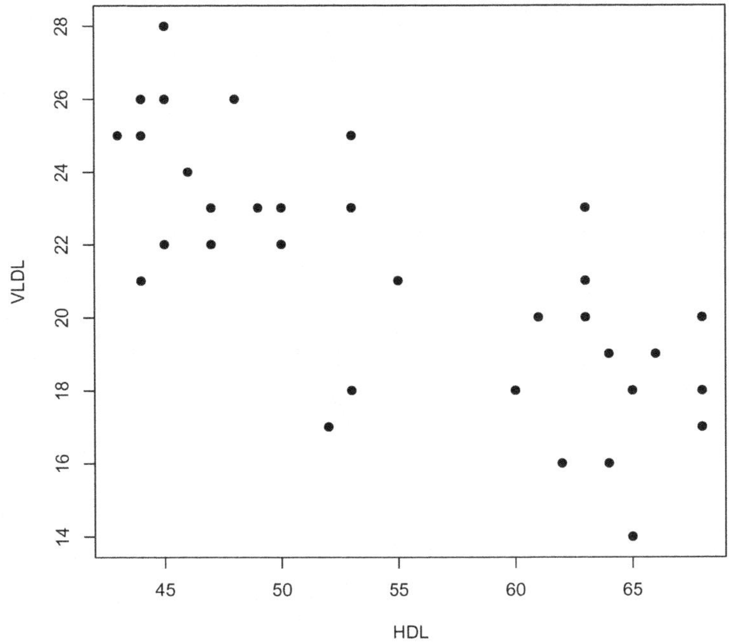

Fig. 8.4 Scatterplot of VLDL versus HDL reveals two clusters

that this group membership is not known a priori; we have to find it from the data. We also have no idea about the number of such groups, commonly called 'clusters'. We only have multivariate data.

The basic idea is that we should start assuming something logical, but that is not final. Studying the inherent features of data through statistical ideas and methods might reveal these groups. There is no prior knowledge that is being used while identifying these latent groups. Using some logic, we could first assign cluster membership to each data point, i.e. to which group a data point belongs. We have to then continuously update the process of assigning cluster membership to a point until it gives a reasonably good solution. But what is a 'good solution'? Well, a solution is good only with respect to certain criterion. So given an optimum criterion, we have to continuously update our algorithm until it is satisfied, i.e. optimised according to our criterion. This optimum criterion, may be defined in various ways. Thus, it is natural that one technique might give different solutions depending on the choice of the optimum criterion.

Nowadays, with large data it is not possible to do clustering manually. We have to use a computer. Moreover, if the number of variables is more than three, we cannot plot them and any way of identifying clusters just by looking at the graphs is out of question. Note that at this point we have no idea about how the clusters were originally there in the population or even in a sample as we cannot see the data.

Hence we need an efficient algorithm supported by statistical theory or intuition or logic to separate the clusters from the data cloud. Thus cluster analysis is sometimes also known as unsupervised machine learning. The word ‘unsupervised’ indicates lack on any prior knowledge or information about the data.

Clustering method can also be looked as a multi-dimensional optimisation because some optimum criteria must be satisfied in order to cluster the data points, each point having a number of components. Multi-dimensional optimisation may not always lead to global optimisation. However, we shall never know whether it reaches a global optimum. So, the underlying principle of the clustering method is essential and it needs to be sound with respect to the problem concerned.

The basic idea of clustering is intuitively very simple. If two points are very close to each other, they are expected to belong to the same group. On the other hand, if they are not so close or far apart, they should belong to different groups or clusters. Points within a group are homogeneous whereas there is heterogeneity between groups. Thus, the concept of distance is at the heart of cluster analysis. It represents how far one data point is with respect to the other. For a given set of n observations, we can have pairwise distances for all possible $n(n-1)/2$ pairs of data points. We can also think of the distance of a point from a chosen point that may not lie in the data set but might be represented in some sense. So while doing cluster analysis or choosing an appropriate clustering method, we need to be careful about our purpose or objective and choose a method wisely. Cluster analysis or unsupervised machine learning literature is very rich and research is still ongoing. Here we discuss a few of the most popular and recently developed methods that seem to be appropriate in genetic data analysis.

8.7.1 Hierarchical Clustering

This is one of the oldest clustering methods but its use is ever growing. Based on a distance measure between two points or two objects, first it searches for two closest data points and connects them. Then, considering these two points as a group, it continues its search to find the next two pairs of points. Here the previously grouped two points are treated as one single point. It continues until all points are connected. We use five two-dimensional points as in Fig. 8.5 and demonstrate the algorithm of this clustering method.

Points 3 and points 4 are closest; so we connect them by a branch and consider it as a group. Next point 1 and point 5 are closest; join them similarly. However, now the distance between point 1 and point 5 as a group and point 2 is the next smallest; so we join them. At the end, we connect the group consisting of points 1, 2, 5 and the group with points 3 and 4 to get the final structure. It is clear that the length of line joining two points of groups represents the distance between them. The entire set of points are connected in a sequential manner and there is an underlying hierarchy, a tree-like structure. This is known as a hierarchical clustering method.

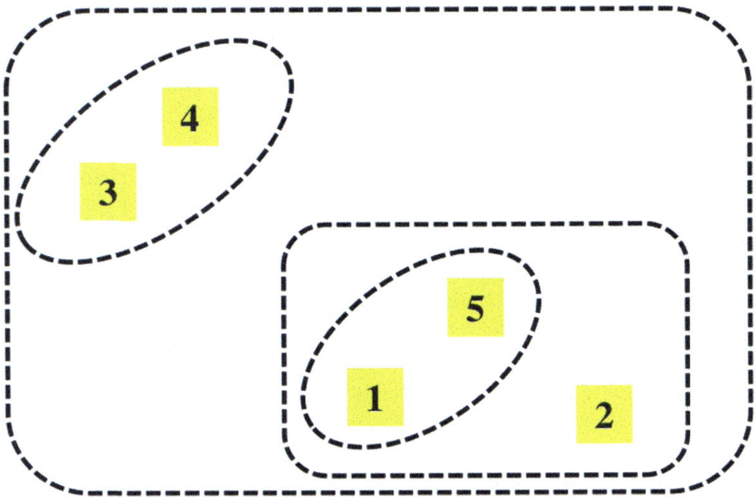

Fig. 8.5 A schematic view of hierarchical clustering

Mathematically this can be represented easily. Let $d(\{P_1, \ldots, P_m\}, \{Q_1, \ldots, Q_n\})$ be the distance between two groups of points $\{P_1, \ldots, P_m\}$ and $\{Q_1, \ldots, Q_n\}$. If $m = 1$ or $n = 1$, the corresponding group consists of only one point. According to Fig. 8.5, we have

$$
\begin{aligned}
& d(\{P_3\}, \{P_4\}) \leq d(\{P_i\}, \{P_j\}) \text{ for all } i \neq j \\
& \qquad \Longrightarrow \text{ join } P_3 \text{ and } P_4 \\
& d(\{P_1\}, \{P_5\}) \leq d(\{P_1\}, \{P_2\}) \\
& d(\{P_1\}, \{P_5\}) \leq d(\{P_2\}, \{P_5\}) \\
& d(\{P_1\}, \{P_5\}) \leq d(\{P_3, P_4\}, \{P_j\}) j = 1, 2, 5 \\
& \qquad \Longrightarrow \text{ join } P_1 \text{ and } P_5 \\
& d(\{P_1, P_5\}, \{P_2\}) < d(\{P_1, P_5\}, \{P_3, P_4\}) \\
& d(\{P_1, P_5\}, \{P_2\}) < d(\{P_2\}, \{P_3, P_4\}) \\
& \qquad \Longrightarrow \text{ join } \{P_1, P_5\} \text{ and } \{P_2\} \\
& \text{Finally join } \{P_1, P_2, P_5\} \text{ and } \{P_3, P_4\}
\end{aligned}
$$

In this type of clustering (Fig. 8.5), it does not give an output of groups of homogeneous points. All points are connected to each other, directly or indirectly through other points. However, we can use a horizontal line to cut this tree in order to get the desired number of groups. Clearly, the number of groups thus obtained is subjective; it depends on the question and the most plausible subjective representation and interpretation.

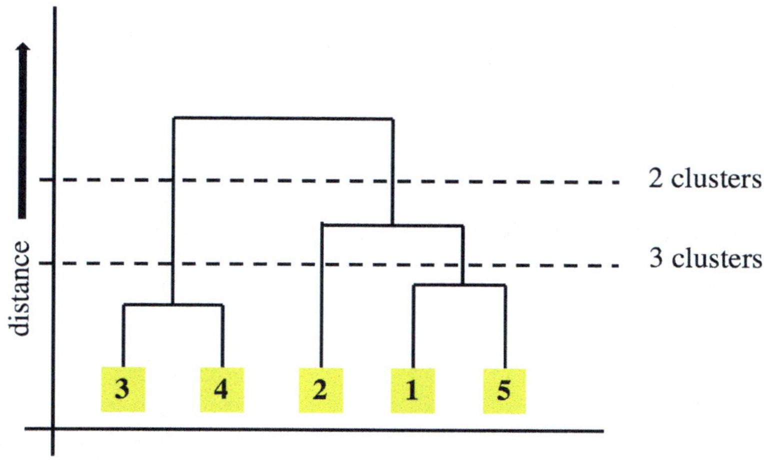

Fig. 8.6 Hierarchical clustering

One can also start from the pair of points having maximum distance and putting them farthest apart and continue the same algorithm; only here we choose the pairs of groups of points having maximum distance. So it would be a top-down approach whereas the previous one is a bottom-to-top approach. Another interesting issue about hierarchical clustering is that, conceptually, although it identifies the closest member of a point like other clustering algorithms, however, here all points are connected to each other, either closely or remotely, unless, of course, we cut the tree to get distinct clusters (Fig. 8.6).

However, if there is any error at any stage, it will continue to propagate till all points are connected. Thus, an early stage error might lead to an erroneous output and there is no way to rectify it after we get the result. This is a major disadvantage of hierarchical clustering method. On the other hand, the output can be presented using a nice diagram, known as 'dendrogram' that gives a visual impression of connections among multi-dimensional data points. Dendrogram is very useful in interpreting results in connection to the problem.

In R, 'hclust()' function does the job of constructing a hierarchical clustering. However, it requires a dissimilarity matrix as its input. We have to use the function 'daisy()' already available in the package '{cluster}'. First we install the package '{cluster}' using the command 'install.packages("cluster")', choose any mirror that will pop up automatically if the computer is connected to the internet, and then write 'library("cluster")' to make it active in the current R workspace. So hierarchical clustering is done using the command 'daisy()' followed by 'hclust()'. The R function 'daisy()' calculates the pairwise similarity measure indicating the distance between all possible pairs of points. The function 'hclust()' acts upon the output of 'daisy()' and constructs the lineage connecting all points. Note that after the construction of hierarchical clustering, in order to plot the dendrogram, we need to use

'as.dendrogram()' function inside 'plot()' function. If we want to extract k clusters, we shall use 'cutree()' command. Fig. 8.6 presents a schematic dendrogram with 2 and 3 clusters. Comprehensive R codes are given below.

```
> a<-read.txt("cholesterol.txt",header=T)      ## reading the datafile
                                               stored in "cholesterol.txt"
> d<-daisy(a,metric="euclidean")               ## calculating dissimilarity
                                               matrix using data set stored in 'a'
> b<-hclust(d)                                 ## doing hierarchical
                                               clustering
> plot(as.dendrogram(b))                       ## plotting hierarchical
                                               clustering tree
> cutree(b,k=2)                                ## provides cluster
                                               membership of each data ppoint
```

A dendrogram based on cholesterol data (Table 8.1) is given in Fig. 8.7. We did not use any 'cutree' command. This is left to the user to choose the number of clusters and accordingly create clusters.

8.7.2 k-means Clustering

One of the most popular clustering methods is 'k-means clustering'. Before executing the algorithm we assume that there are k clusters. So the value of k must be given as an input parameter. Then, the k-means clustering algorithm splits the data into k groups so that the data points within each group are homogeneous whereas they are heterogeneous between groups. That means the distance between two points is very small if they belong to the same group, but they are far apart if they belong to different groups. This algorithm works iteratively until it satisfies a convergence criterion.

First we randomly choose k data points. These are known as 'centres'. Let these centres be $c_1, c_2, \ldots, c_k$. We construct k clusters, $\mathscr{C}_1, \ldots, \mathscr{C}_k$ (say), around each centre. Calculate the distance of a data point x_i (say) from each of these k centres and assign x_i to the cluster $\mathscr{C}_j$ if its distance from c_j is minimum. Let $d(x, y)$ be the distance between two points x and y. Then we assign x_i to $\mathscr{C}_j$ if

$$d(x_i, c_j) \leq \min\left\{d(x_i, c_1), \ldots, d(x_i, c_k)\right\} \tag{8.12}$$

Once we complete assigning all points to some cluster or other, we take new centres as the mean of each cluster, i.e. $c_j = \frac{1}{n_j} \sum_{i:x_i \in \mathscr{C}_j} x_i$, where n_j is the number of point in cluster $\mathscr{C}_j$ after the first iteration. Note that, now the cluster centres are new, not the same as the one with which we started with. Moreover, after the first iteration, a cluster centre may not be a datapoint in the data set. Repeat the same

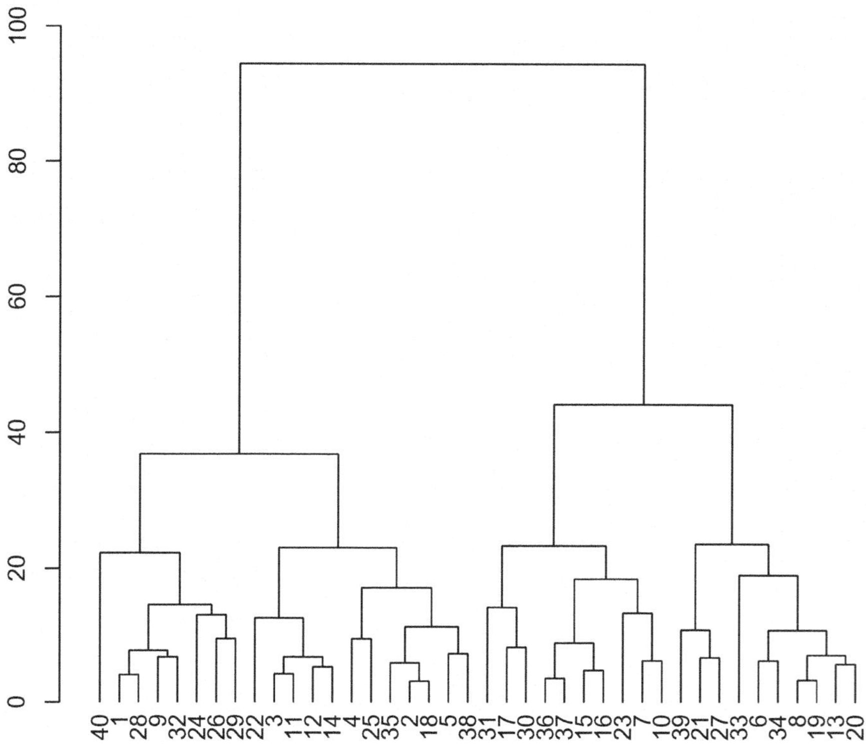

Fig. 8.7 Hierarchical clustering of cholesterol data

procedure and assign each point to a cluster satisfying the condition (8.12). Continue until two consecutive iterations give the same set of cluster centres. We then say that the algorithm is converged and it provides the final clustering scheme for all data points.

Few important points to note. Here k is known and would be given as input parameter. Each data point belongs to one and exactly one cluster. There is no data point that remains outside any cluster. Usually in k-means clustering we take Euclidean distance between two points. Let $\boldsymbol{x} = (x_1, \ldots, x_p)'$ and $\boldsymbol{y} = (y_1, \ldots, y_p)'$ be two data points each having p components. Then Euclidean distance between $\boldsymbol{x}$ and $\boldsymbol{y}$ is defined as

$$d(\boldsymbol{x}, \boldsymbol{y}) = ||\boldsymbol{x} - \boldsymbol{y}|| = \sqrt{\sum_{i=1}^{p}(x_i - y_i)^2}.$$

k-means clustering is a very popular, effective, and nice method, aligning with our intuition of distance between and among the points. It is clear that output of k-means clustering algorithm depends on the random selection of k centres. Theoretically, it

does not matter because we intend to get a global minimum through this algorithm. But the practical situation might be different.

One major drawback of k-means clustering is that it depends on calculation of mean, a measure of central tendency that is highly affected by extreme values or outliers. In such situations, if we somehow know the presence of outliers, this might not be a good clustering algorithm. A variation recommended to tackle such situation is to take $d(x, y) = |x - y|$, which is the absolute value of the difference between two points x and y. This method is known as 'Partitioning Around Medoids' (PAM). PAM is expected to be more robust than k-means clustering method.

R code for k-means clustering can be applied easily to find clusters in a data set. The input parameters are the data set itself and the number of clusters or cluster centres. We apply this algorithm on the cholesterol data set. First we present the output of R code for k-means algorithm. After that we shall explain the output.

```
> kmeans(a,2)          ## k-means algorithm

K-means clustering with 2 clusters of sizes 20, 20
Cluster means:
   HDL    LDL   VLDL  TGL
1  64.35  67.90  18.3  109.40
2  47.85 101.65  23.3  152.35
Clustering vector:
[1] 2 2 2 1 2 2 1 1 1 1 1 2 1 2 1 1 2 1 1 2 2 1 1 1 2 2 1 2 1 2 2 2 2 1 2 1 1 2 2
Within cluster sum of squares by cluster:
[1] 3203.35   2463.85
(between_SS / total_SS = 85.3%)
```

Here it is clear that clustering scheme with two centres seems reasonable as it explains 85.3% of the total variation present in the data. Increasing the number of clusters would increase the percentage of explained variation, but we have to stop at some point where this increase is not substantial. There is no hard and fast rule for choosing the number of clusters. Few authors recommended some measures, but those suggestions may not produce a deterministic result in all data sets. Cluster means are vectors of means for four variables; each row indicating centre of the corresponding cluster.

Note that k-means algorithm is basically an optimisation problem in a multi-dimensional space. Naturally, this may not lead to global optimum always. Thus solutions of k-means algorithm is a local optimum, which may or may not be the global optimum. However, even if multiple applications of k-means algorithm with different cluster centres produce different clustering schemes, they do not differ widely in most cases. So we should be happy with the result of a single application of k-means algorithm. However, if we want to be more accurate, we have to apply k-means algorithm several times with different cluster centres and a final clustering may be obtained through consensus, although there is no guarantee that the final clustering would be the best.

8.7.3 Tight Clustering

The above clustering methods are very useful and popular. Besides these, a few other clustering methods are available, among which, PAM, model-based clustering are used in many applications. However, the surge in data science research and technology has triggered the generation of different types of genomic data. These genomic data vary with respect to their characteristics and demand modification of existing clustering methods. Not only the modification of existing methods, sometimes it demands some changes in the clustering philosophy too.

Till the end of last century, clustering usually meant that each data point has a cluster membership, i.e. each data point belonged to one cluster or another. However, Tseng and Wong (2005) proposed a new clustering philosophy that seems pertinent in a few special data types like gene expression, medical data, etc.

Gene expression data are very noisy, a result of influence of many factors, some known and some unknown. Myriads of genes interact together to give the gene expression profiles. The basic philosophy or assumption behind clustering that each point must belong to a cluster seems too stringent in view of gene expression data. Tseng and Wong (2005) proposed a breakthrough philosophy that, it is not necessary for each gene to belong to a particular cluster. There might be some genes that remain scattered all over the data cloud, with their own characteristics, and do not belong to one cluster or another. Thus they have no cluster membership. This looks plausible and valid in view of the biological knowledge and background noise during data generation.

Tight clustering algorithm (Tseng & Wong, 2005) honours the above philosophy. Instead of trying to assign each data point to a cluster, the algorithm first extracts the tight-most cluster, i.e. a group of points that are homogeneous within themselves to a high degree. In order to find tight-most cluster, tight clustering algorithm uses resampling techniques. It randomly samples a substantial portion of (say 70%) of the data, applies a k-means clustering engine, and assigns membership of remaining data points. Repetition of this resampling technique for a large number times generates a consensus matrix that shows the number of times a pair of points belong to same cluster. Based on a threshold, it then extracts points that are very close to each other, i.e. it extracts the tight-most cluster at this stage or iteration. After keeping these points separate, it moves on to find a second tight-most cluster using the same approach and so on. It is expected that although subsequent iterations produce tight or homogeneous clusters, however, the degree of tightness decreases with increase in iteration number. At some point, the members within a cluster might not be really homogeneous. Thus, the sequential extraction of clusters with decreasing degree of tightness continues until the within group homogeneity, i.e. tightness of a cluster falls below a threshold. Each iteration or stage provides a cluster. When the tightness of the cluster falls below a threshold, we stop at that stage. The remaining points are left scattered over the entire space of points. These points that remained unassigned to any cluster are sometimes called noise points, or scattered genes in the context of gene expression data.

This clustering algorithm has another advantage. In k-means or PAM, the number of clusters should be given as input and the algorithm finds exactly that number of clusters. Hierarchical clustering does not require the number of clusters as input, but it connects all points and it is up to the user in extracting the number of clusters by cutting the tree representing the lineage. However, in tight clustering, it is not necessary to externally provide the number of clusters. Only an idea on the maximum number of clusters would suffice. Tight cluster algorithm will stop if the within cluster tightness does not provide sufficient evidence for being termed as a cluster. Hence it is robust to the choice of the number of clusters.

R command 'tight.clust()' in the package 'tightClust' generates output of this algorithm where the only input is the data matrix and the number of clusters that the user wants to find. Sometimes, if the number of clusters given by the user is large, tight clustering algorithm stops automatically as soon as the tightness of a group of points fall below a threshold, even if the desired number of clusters is not achieved. Thus, in some sense, this algorithm is intelligent while producing results. Thalamuthu et al. (2006) performed an extensive simulation study and real data analysis of gene expression data. It has shown the superiority of tight clustering algorithm over other common clustering methods that are applied to microarray gene expression data.

However, there is one limitation of this tight clustering algorithm. The number of genes in a standard gene expression data set is usually very large, around 20000 to 30000. Unfortunately the tight clustering algorithm proposed by Tseng and Wong (2005) is unable to handle such huge number of genes; it can run only up to ~ 3000 genes. Karmakar et al. (2019) proposed another version of tight clustering that removes this drawback. This algorithm is made specifically for handling large data. It is efficient and provides robust performance while dealing with large data sets. It can even run for 100000 data points seamlessly.

The goal of this extended version is to provide a stable algorithm that allows the use of tight clustering in large data sets. The proposed algorithm follows big data analytics approach (Karmakar & Mukhopadhyay, 2020). The broader theme of this approach is splitting a problem into smaller problems and then combining the results of those smaller problems to arrive at a solution to the original problem. This algorithm takes advantage of the various nice aspects of the original tight clustering algorithm. Since the mathematical treatment of this algorithm is little difficult, we shall only describe an outline without any mathematical details.

First we randomly partition n data points into a number of disjoint subsets such that each subset would be of a manageable size so as to run the original tight clustering algorithm. Let the number of such subsets of data be l. To reduce errors due to partitioning, this procedure is repeated r times. Hence the total number of such manageable sets becomes $M_0 = l \times r$. We extract the tight-most cluster from each of these M_0 data sets and store them in an empty bucket. Note that elements in this bucket may not necessarily be disjoint and they may not contain all data points. An element of the bucket simply tells us that the data points therein are of a similar pattern. The next step of the algorithm performs some operations on this bucket. In this step, we redefine members of the bucket so that the data points belonging to

each element will now have a similar profile and gene expressions from different members will likely have significantly different profiles.

Now we should need to combine the elements in the bucket that will lead us towards the final clusters. Consider the first member of the bucket. Using a predefined decision rule, we combine all members of the bucket with a pattern similar to that of the first element and redefine it as a new collection. We again do this process with another element left in the bucket. Continue until all elements in the bucket are put in redefined groups.

Finally to obtain the tight-most cluster from the possible choices, i.e. the redefined members of the bucket, we use an appropriate measure of tightness of a set of data points that tells us how disperse the set of data points are, within an element in the bucket or a collection of points. Using this measure we have the first tight-most cluster. The tightness measure ensures a stable cluster.

In the next iteration, after we obtain the tight-most cluster from the data, we need to find another, i.e. the second tight-most cluster. For this, we remove all data points of the already obtained cluster from the whole data set as well as from members of the bucket. Now, consider the remaining data as a new data set and repeat the process starting from partitioning and combining to extracting another tight-most cluster and so on. Once we get k desired clusters, we stop and the remaining points are treated as noise points or scattered points that might span over any space in the main data cloud. Of course, it is not necessary to extract all k clusters. If the data points in a cluster do not satisfy the tightness criterion, it will not be termed as a cluster and will remain as scattered points. Thus, although an idea about the number of clusters is required as an input, it is a very weak notion; tight cluster algorithm, even for large data sets will form clusters until a tightness criterion is satisfied.

Note that the choice of decision function and threshold depends on the user. However, different choices might lead to different clusters although there should not be much difference in the clusters as this algorithm heavily depends on the substantial number of random sampling. However, the tightness of clusters will depend on the choice of measure of tightness. This algorithm removes the limitation of original tight clustering; moreover, it is scalable to time and robust in its performance. For a detailed discussion with simulated and real data sets readers are advised to read the original paper by Karmakar et al. (2019).

8.8 Association with Multivariate Phenotypes

We have learnt different statistical methods to detect genetic association of a locus with phenotype, be it qualitative or quantitative. Discussion with multivariate data reveals another interesting and important question. We know that any clinical test for measuring cholesterol level produces results on four phenotypes, viz, HDL, LDL, VLDL, and triglyceride (TGL) levels. Similarly status or indication of hypertension is measured by two variables, systolic and diastolic blood pressures. Thus occurrence of multivariate phenotypes is natural mainly because these variables are correlated

with each other and they usually occur together. Moreover, since all these measures are indicators of a particular broad phenotype like cholesterol level (may be a trigger for cardiovascular disease) or blood pressure, we may think that genetic variant(s) responsible for the disease might be associated with all these phenotypes. For example, four phenotypes HDL, LDL, VLDL, and TGL might all be somehow associated with the same set of genetic variants or loci.

It seems that if we study multivariate phenotypes together instead of univariate or a single variable or phenotype separately, power of the study would increase as we can exploit extra information through correlations prevailing among different variables. Hence there is a need for statistical methods for studying genetic association using multivariate phenotypes.

However, extension of univariate to multivariate phenotype in the context of association study with a genetic locus is not straightforward. For example, in a regression set up, each observation is a vector (ordered set of more than one variable). The variables in a vector may be correlated. Because of this correlation structure among the variables occurring together, it demands a fresh look at all methods of genetic association. We discuss a few methods that may be used to study genetic association using multivariate phenotypes.

Till today, there are plenty of genetic association studies that identify numerous loci that are associated with different traits. However, single marker analysis is unable to explain a substantial portion of genetic heritability. One reason might be the loss of information by using a single marker instead of multiple loci. This problem has been resolved, at least to some extent, by proposing multi-loci method (Mukhopadhyay et al. 2010; Wu et al. 2011; Thalamuthu et al. 2011). Another reason may be argued as the inability to accommodate multivariate phenotypes in a single model thus failing to exploit the pleiotropy (one genetic variant affecting multiple phenotypes simultaneously) or inter-relations among different phenotypes. However, literature survey reveals a few methods that have been developed to meet these challenges. We see that existing statistical methods addressing the analysis through multivariate phenotypes can be broadly classified into two classes: (1) use of summary statistics that result from previous genetic association studies using individual variables, and (2) direct use of multivariate phenotypes to study association with a single locus. We discuss below, a few methods for genetic association at single marker using multivariate phenotypes.

8.8.1 Linear Mixed Effects Model

Linear mixed effects model looks like a regression type model but with some more extensions and assumptions. This consists of a fixed genetic effect due to the locus under study and a person specific effect that occurs due to the availability of multivariate phenotypes for each person in the sample. For the sake of simplicity, we discuss only multivariate phenotypes where all components are quantitative in nature, more specifically continuous variables.

A linear mixed effects model can be written as

$$y_{ij} = \alpha + \beta_j X(g_i) + \eta_{ij} + e_{ij},\ j = 1, \ldots, J;\ i = 1, \ldots, n \tag{8.13}$$

where y_{ij} is the phenotype value for j-th component in i-th individual, α is the general effect, $X(g_i)$ is the genetic effect for the i-th individual with β_j being the corresponding coefficient indicating the contribution of genetic effect, η_{ij} is the person specific random effect for j-th component and e_{ij} is the random error component. We assume that e_{ij}s are independently and identically distributed (i.i.d.) random variables following $N(0, \sigma_e^2)$ for all $j = 1, \ldots, J;\ i = 1, \ldots, n$. Clearly, the correlation between two univariate phenotypes is introduced through covariance among η_{ij}s, i.e.

$$Cov(y_{ij}, y_{ij'}) = Cov(\alpha + \beta_j X(g_i) + \eta_{ij} + e_{ij}, \alpha + \beta_{j'} X(g_i) + \eta_{ij'} + e_{ij'}) = \sigma_{jj'},\ \text{(say)},$$

because of independence among e_{ij}s and between e_{ij} and η_{ij}. Multivariate phenotypes occur in the model more vividly and we write the above model using vector of phenotype observations as

$$\boldsymbol{y}_i = \alpha \mathbf{1} + X(g_i)\boldsymbol{\beta} + \boldsymbol{\eta}_i + \boldsymbol{e}_i,\quad i = 1, \ldots, n \tag{8.14}$$

where

$$\boldsymbol{y}_i = \begin{pmatrix} y_{i1} \\ y_{i2} \\ \vdots \\ y_{iJ} \end{pmatrix}, \mathbf{1} = \begin{pmatrix} 1 \\ 1 \\ \vdots \\ 1 \end{pmatrix}, \boldsymbol{\beta} = \begin{pmatrix} \beta_1 \\ \beta_2 \\ \vdots \\ \beta_J \end{pmatrix}, \boldsymbol{\eta}_i = \begin{pmatrix} \eta_{i1} \\ \eta_{i2} \\ \vdots \\ \eta_{iJ} \end{pmatrix} \text{ and } \boldsymbol{e}_i = \begin{pmatrix} e_{i1} \\ e_{i2} \\ \vdots \\ e_{iJ} \end{pmatrix}$$

Now, if we want to test whether the genetic locus has no effect on any of the J phenotypes, we can frame this as a problem of testing statistical hypothesis as

$$H_0 : \beta_1 = \beta_2 = \cdots = \beta_J = 0 \text{ against } H_1 : \text{ at least one } \beta_j \neq 0.$$

During estimation, variance parameters in Σ matrix and σ_e^2 can be estimated by a more sophisticated method called 'restricted maximum likelihood method' (REML). Using likelihood ratio testing method we can test the above hypothesis and find the p-value based on asymptotic distribution of the test statistic. However, one can also employ another test statistic, known as Wald's χ^2 statistic, i.e.

$$\hat{\boldsymbol{\beta}}' \widehat{\Sigma_{\boldsymbol{\beta}}^{-1}} \hat{\boldsymbol{\beta}} \sim \chi_J^2 \text{ asymptotically, for large } n,$$

where $\hat{\boldsymbol{\beta}}$ and $\widehat{\Sigma_{\boldsymbol{\beta}}^{-1}}$ are the estimates of $\boldsymbol{\beta}$ and inverse of variance-covariance matrix Σ respectively. The p-value can easily be obtained and we can take a decision accordingly.

This procedure is an extension of the standard regression model with an additional random component η_{ij} making its analysis slightly different. This kind of model is known as 'linear mixed model', because β_js are fixed parameters whereas η_{ij}s are unknown but they are considered as a random variable with unknown parameters. However, this provides an easy way of incorporating information from all phenotypes along with their correlations in a single model. It can also be extended to include more than one genetic locus.

8.8.2 Variable Reduction Method Using PCA

To collapse information included as a whole in multivariate phenotypes, we can use some dimensional reduction method as discussed previously. Principal component analysis is probably the most popular method for such reduction of multiple variables. If the phenotype vector consists of J univariate components, we can use PCA to reduce its dimension. Now only using the first principal component (PC1) we can easily do our genetic association study by adopting any method for a single phenotype variable. However, although we use PC1 as our phenotype of interest, it might have no physical meaning as it is a linear combination of the components of multivariate phenotype vector.

In this case also, we assume that all components of phenotype vector are continuous and hence PC1 would act as a quantitative univariate quantity. PC1 contains majority of information embedded in multivariate phenotype. Hence any significant genetic association with PC1 is expected to show some association with multivariate phenotype. Based on this idea, to test any possible association with genetic locus, we can adopt any regression-based approach or any other approach suitable for association using a single variable. If, along with multiple variables and genotype data, we have covariate information for all individuals, we can easily adjust the effect of covariate and identify association signal, if any, using any method discussed previously.

If the p-value of the test statistic is extremely small, we can say that this multivariate phenotype is probably associated with the genetic locus, although some difficulty may arise while interpreting the results. If the PC1 explains more than 80% of total variation, only then should we apply this method. Otherwise, a small p-value may not carry any significance.

For the data in Table 8.1, we see that principal component analysis indicates that first principal component (PC1) explains more than 85% variance. So it is enough to work with only PC1. To find association with genotype as given in Table 8.1, we adopt a linear regression method with a number of alleles as an independent or explanatory variable. Thus our model becomes

$$PC1_i = \mu + \beta g_i + e_i, \quad i = 1, \ldots, n$$

where $PC1_i$ is the value of first principal component corresponding to i^{th} individual, g_i is the genotype of the same individual with e_i as random error component with usual assumptions as in simple regression analysis. Here, the null hypothesis, H_0 : $\beta = 0$ indicates the absence of any association of PC1 with the number of alleles at the genetic locus of interest.

Fitting a simple linear regression reveals the p-value as 0.9345 which is extremely large compared to a chosen level of significance 0.05 (say). Hence we can say that there is no evidence to declare any association of phenotype vector with the genotype of interest. Note that association of phenotype vector only implies that we assume that PC1 contains enough information to reveal the association with multivariate phenotypes.

Interpretation of result needs careful attention. If there is any association signal, it is difficult to interpret in terms of individual components, because from PC1 we may not get a clear idea about which component is really associated. Sometimes, the phenotype vector, as a whole, is associated due to synergistic effect of all components. One gene might affect multiple phenotypes with varying degrees. In such cases, it is not wise to go for interpretation in terms of individual components; rather any comment or conclusion on association with the phenotype vector as a whole seems more meaningful.

8.8.3 O'Brien's Method of Combining Univariate Test Statistics

Various genetic studies have made available the summary statistics from testing association between a univariate phenotype and a genetic locus of interest. So the question naturally arises, whether it is possible to combine these results obtained through univariate analysis in order to get some idea about pleiotropy, i.e. whether a single marker is associated with multiple phenotypes. O'Brien (1984) proposed a method that combines univariate statistics. However, here the basic demand is how to use the available pairwise correlations among the components of the vector of phenotypes.

Let $\boldsymbol{T} = (T_1, T_2, \ldots, T_K)'$ be a multivariate phenotype vector of order K, i.e. it has K univariate components. We have seen in univariate genetic association study that most of the test statistics follow normal distribution either exactly or asymptotically. For example, even when the phenotype is qualitative, the test statistic, which is an odds ratio, follows an asymptotic normal distribution. Thus, without any loss of generality, we can assume that

$$\boldsymbol{T} \sim N_K(\boldsymbol{\beta}, \Sigma),$$

where $\boldsymbol{\beta} = (\beta_1, \beta_2, \ldots, \beta_K)'$ is the vector of parameters that is to be tested and Σ is the variance-covariance matrix. Now $H_0 : \beta_1 = \cdots = \beta_K = 0$ provides a null

hypothesis which when true indicates the absence of genetic association with the entire phenotype vector, i.e. with all components of multivariate phenotype vector as a whole.

Let for the phenotype vector $\boldsymbol{T}$, Σ and Σ^{-1} be variance-covariance matrix and its inverse respectively, as given by

$$\Sigma = \begin{pmatrix} \sigma_{11} & \sigma_{12} & \cdots & \sigma_{1K} \\ \sigma_{21} & \sigma_{22} & \cdots & \sigma_{2K} \\ \vdots & \vdots & \cdots & \vdots \\ \sigma_{K1} & \sigma_{K2} & \cdots & \sigma_{KK} \end{pmatrix} \quad \text{and} \quad \Sigma^{-1} = \begin{pmatrix} \sigma^{11} & \sigma^{12} & \cdots & \sigma^{1K} \\ \sigma^{21} & \sigma^{22} & \cdots & \sigma^{2K} \\ \vdots & \vdots & \cdots & \vdots \\ \sigma^{K1} & \sigma^{K2} & \cdots & \sigma^{KK} \end{pmatrix},$$

where $\sigma_{ij} = cov(T_i, T_j)$, for $i, j = 1, \ldots, K$.

Taking $\mathbf{1} = (1, 1, \ldots, 1)'$ as a vector with all components as unity, O'Brien proposed a test statistic as

$$V = \mathbf{1}'\hat{\Sigma}^{-1}\boldsymbol{T} = \sum_{i=1}^{K}\sum_{j=1}^{K} \hat{\sigma}^{ij} T_i, \tag{8.15}$$

where $\hat{\sigma}^{ij}$ is the (i, j)-th element of the estimated inverse of Σ.

It can be shown that, V follows asymptotic univariate normal distribution with mean $\mathbf{0} = (0, 0, \ldots, 0)'$ and variance $\mathbf{1}'\hat{\Sigma}^{-1}\mathbf{1} = \sum_{i=1}^{k}\sum_{j=1}^{k} \hat{\sigma}^{ij}$, under H_0.

This test does not require the entire data on all individuals. It only requires the summary statistics or the statistics that have been used in identifying genetic association of each variable with a genetic marker. Thus in this sense it can handle missing data, if any, because missingness has already been taken care of while evaluating test with respect to univariate phenotype.

We need to be careful about some issues while applying O'Brien's test. The most important issue is the estimation of Σ. From the given data, we can have only one set of statistics for phenotypes and it is not possible to get such an estimate from the given data. Usually some prior information about possible correlation might be helpful in this regard. Summary statistics are obtained from previous studies. While doing any genetic association study, background information about the population under study, sampling schemes, etc., are very important. Thus while applying such a test, we should be very careful whether information through summary statistics is really applicable to the population under study. However, the recent richness in databases has become extremely helpful in eliminating such issues, if not fully, to a great extent.

8.8.4 *Methods with Heterogeneous Genetic Effects*

One limitation of O'Brien's method is that although this is the most powerful test among the class of tests that are linear combinations of univariate test statistics when $\beta_1 = \cdots = \beta_K > 0$, provided Σ is known; however, there is a drop in power when $\beta_1, \ldots, \beta_K$ are not homogeneous. Another drawback of this test is that it requires the knowledge of estimates of covariances or correlations present between two univariate test statistics. In most cases, we do these kinds of tests using odds ratio. Hence, the summary statistics used are odds ratio obtained previously using single marker association test. Thus, as an approximation, we can use the pairwise correlation between two SNP markers.

To remove this limitation, Yang et al. (2010) proposed some modifications. They proposed the use of non-uniform weights to incorporate the potential heterogeneity of $\beta_1, \beta_2, \ldots, \beta_K$. The strategy is to split the entire data set into two random subsets. Use one data set to estimate the weights of $\beta_1, \ldots, \beta_K$. Then, we can construct the test statistic as given in (8.15) based on the second subset. Let the test statistic obtained using the estimation set be $\hat{\boldsymbol{T}}_w$ and that using the other set be $\boldsymbol{T}$. Then the proposed test statistic is

$$S = \hat{\boldsymbol{T}}_w' \hat{\Sigma}^{-1} \boldsymbol{T}. \tag{8.16}$$

It can be shown that S follows univariate normal distribution asymptotically with variance $\hat{\boldsymbol{T}}_w' \Sigma^{-1} \hat{\boldsymbol{T}}_w$, conditionally on the splitting of data sets.

The only assumption here is that the sample size should be large so that even after randomly splitting the data set into two subsets, we should have a good estimate of weights. Also, the sample size for the second subset of data should be large enough so that asymptotic normality of the distribution will hold. This method requires the entire data set.

The only limitation is that the performance of the test statistic may depend on the actual splitting of the data. Yang et al. (2010) proposed a modification of this method using cross-validation method where p-value can be calculated using permutation-based technique. However, splitting of the original data set may sometimes lead to loss of information, because we are working with a portion of data, not the entire data set for a single operation.

Another class of test statistic is based on a quadratic form of the components of multivariate phenotypes, of course including the pairwise correlations among univariate test statistics. Xu et al. (2003) proposed a Wald's chi-square type test statistic as

$$S = \boldsymbol{T}' \hat{\Sigma}^{-1} \boldsymbol{T}. \tag{8.17}$$

Here we replace the vector $\mathbf{1} = (1, 1, \ldots, 1)'$ as in (8.15) by $\boldsymbol{T} = (T_1, T_2, \ldots, T_K)'$, the vector of univariate test statistics. It can be shown that for large sample size, the test statistic S asymptotically follows a mixture of chi-square variables; however, this

distribution can be approximated by a linear function of another chi-square variable. Thus we have, in a precise form,

$$S \overset{D}{\sim} \sum_{i=1}^{K} c_i \chi_1^2 \approx a\chi_d^2 + b \text{ where } d = \frac{\sum_{i=1}^{K} c_i^3}{\sum_{i=1}^{K} c_i^2},\ b = \sum_{i=1}^{K} c_i - \frac{\left(\sum_{i=1}^{K} c_i^2\right)^2}{\sum_{i=1}^{K} c_i^3},\ \text{and } a = \frac{\left(\sum_{i=1}^{K} c_i^2\right)^3}{\left(\sum_{i=1}^{K} c_i^3\right)^2}$$

where $X \overset{D}{\sim} Y$ indicates that X and Y have the same probability distribution for large sample size.

Let s be the observed value of the test statistic as given in (8.17). Based on the above approximate distribution, we can easily calculate the p-value approximately as

$$\text{p-value} \approx P_{H_0}(\chi_d^2 > (s - b)/a),$$

and take a decision accordingly. Note that whatever decision we take, we have performed this test at level of significance α $(0 < \alpha < 1)$ approximately.

Exercise

First we provide a table on measurements of diastolic and systolic blood pressure, and genotypes at three SNP loci for a number of individuals. A few problems in this exercise are based on this data set. Each genotype is coded as 0, 1, or 2 according to the number of minor alleles.

8.1 Distinguish, with an example, among simple correlation coefficient, multiple and partial correlation coefficients.

8.2 Product moment correlation coefficient may be negative; however, multiple correlation coefficient is always positive. Justify and interpret this comment.

8.3 For the correlation matrix as given in Sect. 8.4, calculate all possible multiple and partial correlation coefficients.

8.4 For four variables, the correlation matrix R is given by

$$R = \begin{pmatrix} 1.000 & 0.245 & 0.765 & -0.526 \\ 0.245 & 1.000 & 0.707 & 0.384 \\ 0.765 & 0.707 & 1.000 & 0.074 \\ -0.526 & 0.384 & 0.074 & 1.000 \end{pmatrix}$$

Calculate multiple correlation coefficients $r_{1\bullet 234}$, $r_{2\bullet 134}$, and partial correlation coefficients $r_{12\bullet 34}$, $r_{13\bullet 24}$ and $r_{14\bullet 23}$. Comment on your findings.

8.5 Let there be three variables x_1, x_2, and x_3 such that

$$x_1 = 2x_3 + \epsilon_1$$
$$x_2 = 3x_3 + \epsilon_2$$

where $\epsilon_1, \epsilon_2 \sim N(0, \sigma_1^2)$ are independently and identically distributed, and $v(x_3) = \sigma_3^2$. Find r_{12} and $r_{12\bullet 3}$. Comment on your findings.

8.6 Table 8.2 contains data on bivariate phenotype on blood pressure, i.e. systolic and diastolic measurements from the same group of individuals.

(a) Perform a principal components analysis on the bivariate phenotype.
(b) Do a scree plot and determine the percentage of variation explained by the first principal component.
(c) Check whether the total variation of two principal components is same as the total variance of the original variables. Whatever be the answer, justify.
(d) Plot systolic measurements against diastolic measurements for all individuals and also plot one principal component against the other one. Comment on your findings using these plots.

8.7 We want to know whether there exist different groups in the population with respect to cholesterol levels, using the phenotype data in Table 8.1.

(a) Perform a cluster analysis using hierarchical clustering. Identify the groups using two different cutrees, one giving two groups and another giving three groups. Which one you would prefer and why?
(b) Perform a cluster analysis using k-means clustering method and identify the group label for each person and present it in a table. Compare the groups found by using k-means clustering method with that obtained by hierarchical clustering method.

8.8 The objective of a study is to understand the factors that are associated with systolic blood pressure in infants. Systolic blood pressure (mmHg), weight (grams) and age (days) are measured in 100 infants. A multiple linear regression is performed to predict blood pressure (mm Hg) from age and weight. The following results were obtained.

Regression coefficient	Estimate	SD(estimate)
Intercept (β_0)	50.0	4.0
Weight (gm) (β_1)	0.1	0.3
Age (days) (β_2)	4.0	0.6

How much higher would you expect the blood pressure to be of an infant who weighed 3400 grams compared to an infant who weighed 2550 grams if both infants were of exactly the same age?

8.9 You can think that systolic and diastolic measurements as in Table 8.2 are related to blood pressure level of an individual as a bivariate phenotype.

Table 8.2 Data on systolic and diastolic blood pressure for a group of individuals

Diastolic	Systolic	G1	G2	G3	Diastolic	Systolic	G1	G2	G3	Diastolic	Systolic	G1	G2	G3
88	148	0	1	2	78	121	0	0	0	81	123	0	0	0
107	144	0	0	1	82	117	0	0	0	91	116	0	0	1
83	112	0	0	1	71	112	0	1	0	87	114	0	0	1
85	115	1	0	0	80	128	0	0	0	72	120	0	1	0
72	127	0	1	0	86	132	0	2	2	79	124	0	0	0
81	122	1	0	0	76	119	0	2	0	78	116	1	0	0
75	114	0	0	0	87	133	0	1	2	75	115	0	1	0
86	122	0	2	1	71	116	1	0	1	72	118	0	1	0
72	118	0	1	0	91	135	0	2	1	88	124	1	2	0
81	115	0	2	1	77	114	0	2	0	89	122	0	0	0
95	131	1	0	1	80	123	0	0	0	78	122	0	0	0
93	132	0	1	2	72	116	1	0	0	90	126	0	1	0
84	125	0	0	0	76	112	1	0	1	97	141	1	1	2
82	118	1	0	0	97	133	0	0	2	90	125	0	2	0
88	128	1	0	0	74	111	0	0	0	87	117	0	0	1
87	126	0	0	2	87	127	1	1	0	94	133	1	0	0
88	129	0	0	0	97	137	0	0	1	88	140	0	0	1
72	122	2	0	0	75	117	1	0	1	106	131	0	0	2
93	119	0	1	0	94	131	0	2	2	82	129	0	0	0
85	124	0	0	0	80	112	0	0	0	77	120	0	0	0
88	138	0	1	2	78	112	0	2	0	84	122	1	0	0
84	130	1	0	0	95	125	0	0	0	84	117	1	0	0
79	112	0	0	0	99	131	0	2	2	73	118	0	0	0
84	141	0	1	0	86	113	0	0	0	93	126	1	0	1

(continued)

Table 8.2 (continued)

Diastolic	Systolic	G1	G2	G3	Diastolic	Systolic	G1	G2	G3	Diastolic	Systolic	G1	G2	G3
94	128	0	1	1	82	123	1	0	1	71	115	0	0	0
71	118	0	2	0	71	117	0	0	0	85	121	1	0	0
88	113	0	1	0	85	134	0	0	0	85	125	1	0	1
83	120	0	1	0	78	115	0	0	0	73	124	0	0	0
73	127	0	0	0	91	122	0	1	0	88	131	0	2	2
90	116	1	0	0	94	118	1	0	0	79	125	0	0	1
100	120	1	2	0	93	117	0	0	2	76	114	0	0	0
83	128	0	0	0	87	126	0	1	2	79	119	0	0	0
85	134	1	0	0	76	120	0	0	1	85	123	0	1	0
74	128	0	0	0	82	129	2	1	0	80	126	0	0	0
88	124	0	2	0	84	133	0	2	0	85	117	0	1	0
80	116	0	0	0	74	118	1	1	0	98	124	0	0	2
99	131	0	0	2	85	122	0	1	0	75	112	0	0	0
77	117	0	0	0	78	132	1	1	0	71	119	0	0	0
85	118	0	2	0	73	118	0	0	0	80	113	0	0	0
91	120	0	2	0	79	119	0	1	0	93	123	0	0	0
83	127	1	1	0	86	122	0	0	0	85	131	0	1	1
91	138	0	2	2	88	123	0	2	0	82	128	0	0	0
78	116	0	0	0	101	132	1	2	2	74	122	0	0	0
78	114	0	1	0	71	113	0	1	0	89	122	0	0	0
74	115	0	2	0	84	131	0	2	0	83	124	0	0	0
86	112	0	0	0	86	136	0	2	2	81	122	0	2	1
77	116	0	1	0	75	117	1	0	0	74	116	0	0	0

(a) Perform an association analysis using bivariate phenotype for each locus.
(b) Also perform association analysis using each phenotype, i.e. univariate phenotypes for each locus.
(c) Compare the results obtained in (a) and (b). Comment on your results.
(d) Can you find any heterogeneity in the sample using data in Table 8.2? Justify your answer.

8.10 Consider the observations in Table 8.1.

(a) Draw histograms for each variable on a single graph.
(b) Draw scatter diagrams considering all possible pairs of variables and present them on a single graph.
(c) Find multiple correlation coefficients of TGL on HDL, LDL, and VLDL.
(d) Evaluate partial correlation coefficients of (i) HDL and (ii) LDL eliminating the effects of other variables.

8.11 For the data in Table 8.1, first divide the sample into two groups: one having at least one minor allele at the genetic locus, and the remaining persons as another group. Let these be Group 1 and Group 2 respectively.

(a) What is the multiple regression equation of HDL on LDL, VLDL, and TGL for (i) the entire group, (ii) Group 1, and (iii) Group 2?
(b) For each Group calculate multiple correlation coefficients of TGL on HDL, LDL, and VLDL. Comment on your results.
(c) For each Group calculate partial correlation coefficient between HDL and LDL eliminating the effects of TGL and VLDL. Comment on your results.

8.12 For the data in Table 8.1, do the following.

(a) Identify how many principal components explain at least (i) 70%, (ii) 80%, and (iii) 90% of total variance present in the data. Show your justification graphically and comment.
(b) Use cluster analysis to see whether there are any distinct groups in which the individuals can be grouped with respect to their phenotypes.
(c) Use an appropriate cluster analysis method to see whether any individual has any special characteristics so that s/he can be left out of any cluster.
(d) Use principal component regression to see whether there is any genetic association with any of the loci.
(e) Try to write your own R code to do association study with all cholesterol indicators using linear mixed model approach, taking one locus at a time.

References

Karmakar, B., & Mukhopadhyay, I. (2020). Statistical validity and consistency of big data analytics: A general framework. *Statistics and Applications, 18*(2), 369–381.

Karmakar, B., Das, S., Bhattacharya, S., Sarkar, R., & Mukhopadhyay, I. (2019). Tight clustering for large datasets with an application to gene expression data. *Scientific reports, 9*(1), 3053.

Mukhopadhyay, I., Feingold, E., Weeks, D. E., & Thalamuthu, A. (2010). Association tests using kernel-based measures of multi-locus genotype similarity between individuals. *Genetic Epidemiology, 34*(3), 213–221.

O'Brien, P. C. (1984). Procedures for comparing samples with multiple endpoints. *Biometrics, 40*, 1079–1087.

Price, A. L., Patterson, N. J., Plenge, R. M., Weinblatt, M. E., Shadick, N. A., & Reich, D. (2006). Principal components analysis corrects for stratification in genome-wide association studies. *Nature Genetics, 38*, 904–909.

Thalamuthu, A., Mukhopadhyay, I., Zheng, X., & Tseng, G. C. (2006). Evaluation and comparison of gene clustering methods in microarray analysis. *Bioinformatics, 22*(19), 2405–2412.

Thalamuthu, A., Zhao, J., Keong, G. T., Kondragunta, V., & Mukhopadhyay, I. (2011). Association tests for rare and common variants based on genotypic and phenotypic measures of similarity between individuals. *BMC proceedings, 5*(Suppl 9), S89.

Tseng, G. C., & Wong, W. H. (2005). Tight clustering: A resampling-based approach for identifying stable and tight patterns in data. *Biometrics, 61*(1), 10–16.

Wu, M. C., Lee, S., Cai, T., Li, Y., Boehnke, M., & Lin, X. (2011). Rare-variant association testing for sequencing data with the sequence kernel association test. *American Journal of Human Genetics, 89*(1), 82–93.

Xu, X., Tian, L., & Wei, L. J. (2003). Combining dependent tests for linkage or association across multiple phenotypic traits. *Biostatistics, 4*, 223–229.

Yang, Q., Wu, H., Guo, C. Y., & Fox, C. S. (2010). Analyze multivariate phenotypes in genetic association studies by combining univariate association tests. *Genetic Epidemiology, 34*, 444–454.

Appendix A

What is R

Before starting to describe the salient features of R, first note that we have to change our mindset in terms of using a computer. We are habituated with mouse clicking most of the times; now change that habit to write a few 'commands' in the terminal. 'Commands' are nothing but a set of instructions given to the computer so that it can understand our need and do the computing job for us. This is easy as we always type a letter or document using any word processor. Likewise, here we have to write the code in R or the 'R workspace'.

But first THREE QUESTIONS that would immediately come to mind:

(1) What is R?
(2) How do we get it?
(3) How do we learn it?

The brief but full answers to these THREE QUESTIONS are:

(1) R is a programming language, but can also be viewed as a software if we stick the objective of a software, that it is able to do lots of things using some inbuilt 'functions'. R has many functions already built-in. At the same time it is developed such that we can use R as a language in order to write our own code, using the available functions. Like all other programming language, it has the flexibility to use loops of many types, logical statements, reading and writing files, plotting high quality graphs, etc. Practically it has flexibility of a programming language and strength of a computer software. Moreover, we can also build a comprehensive code or a 'package' for doing a specific job. A package is a set of commands that can be used later for a more general purpose or for working with other data sets. Since R is freely available, one can write a package and upload it so that people from other corners of the world can use it; thus spreading scientific knowledge across the world. Note that, in R, similar types of functions or add ons for them are grouped into a 'package'. In order to do a particular

I. Mukhopadhyay and P. P. Majumder, *Statistical Methods in Human Genetics*, Indian Statistical Institute Series, https://doi.org/10.1007/978-981-99-3220-7

work, we first have to install the package that contains the required functions. For example, if we want to draw graphs in R, we have to install the package 'graphics' and use the 'plot' function to draw a graph. However, packages that contain commonly used functions would automatically be installed as default while installing R on the computer.

(2) R can be downloaded from the website 'http://cran.r-project.org'. It is available for many computing platforms and operating systems like Windows, Mac, Linux, and Unix. Every now and then, the R version is upgraded and so we have to upgrade it time to time in order to use the full power of R. Note that while installing R in the machine, it automatically installs some base packages, which includes thousands of commonly used functions. If we want to do some sophisticated calculation or operation, we need to find out the function and the corresponding package and install that package, again free of cost. Note that you can download basic R or R Studio, which seems to be a more user-friendly version of R. The same R will run; only in R studio, there are four windows (or terminals) each of which is reserved for a specific job. Like, if we want any help in R command, we can see it on one window.

(3) But the most important issue is how one can learn R. The best way to learn R is **self learning**. However, there are plenty of online materials, books, tutorials, etc. available in the internet. One such book is 'Introductory Statistics with R' by Peter Dalgaard. Just go through it for the first few pages at least, write the commands and codes as given in the book, understand each and every command, and one will acquire proficiency in using R. However, one may face a few problems while using R. The major problem is that, if we want do a particular work, how do we know the function that does this work? One option is to guess and in most of the cases the guess would be right. For example, if we want to calculate the arithmetic mean of a set of numbers already input in R, we can guess a command that would do this job. The most plausible guesses can be 'average', 'avg', 'mean', 'am,' etc. Surprisingly one of them is correct and for calculating arithmetic mean, 'mean' is the command in R. We have done one experiment as ask to guess R command for a job; every time the correct command emerges within first three guesses. However, one can always search in the internet using any search engine and find the command. In the above example, if we write 'calculate mean in R' in the search engine, a page will be opened that informs the actual command, the package in which it belongs, how to use that function, and all other relevant information about the command. Note that it will give the package name and we have to install it if it is not already available in the R workspace.

Basic operations in R

We now describe some basic operations in R in order to give confidence to the readers, who are still hesitating to use it. First we double click the icon of R (or R studio)

and a terminal will be opened. It is always advisable to change the directory from the menu and go to the directory where the data set for the analysis is stored. Now that we are in the same folder, whatever analysis we do using R, the resulting files (we have to save them) will be available in that directory only. If we set the directory (folder) at the very beginning, we don't need to set the path while saving any result because the result file would be saved automatically in that directory.

Variables

First note that R can be looked upon as a calculator. If we type $2 + 3$ and press 'enter', it will give 5. $sqrt()$ is the command for calculating square root of a number. Now $sqrt(49)$ gives 7, but $sqrt(\text{"}statistics\text{"})$ will give error, because 'statistics' is not a numeric value for which we can calculate square root. Note that '$< -$' or '$=$' is used to assign value to a variable. If we write '$a < -49$', press 'enter', the value 49 is stored in 'a'.

```
> 2+3
[1] 5
> sqrt(49)
[1] 7
> a< −49
> a
[1] 49
> srqt("statistics")
Error in srqt("statistics") : could not find function "srqt"
```

The above discussion leads us to define three types of variables used in R. They are:

(1) Numeric variable: $a < -2.53$; a is a numeric variable. We can do arithmetic operations using a.
(2) Character variable: $a < -$"*Knowledge of Statistics is essential for data analysis*"; a is a character variable and we cannot do arithmetic operations using a, but can do other things. If we write *sub*('*Knowledge*', '*Knowledge and practice*', a), we will get '*Knowledge and practice of Statistics is essential for data analysis*'. Here the word '*Knowledge*' in the variable a is being substituted by '*Knowledge and practice*'.
(3) Logical variable: $a < -(2 + 2 == 5)$; a is a logical variable giving the output *FALSE*, because $2 + 2$ can never be 5!

Data entry and simple operations

To input data into R, we give the data as a vector or a list of observations. After each command we have to press enter. The command is

```
x<-c(1,2,4,7,3.4)
```

Similarly if we write $x < -seq(-3, 3, 0.1)$, we input all the values starting at -3 and ending at 3 with an increment 0.1. Thus we enter 61 observations following a rule, i.e. they are equispaced between two known values. Now we can do some simple operations with the values in x.

```
y<-x*x        ## makes the square of each observation in x and store it in the same position
              ## in another variable y
```

We provide a few examples of simple R code that are very useful when we work with any data.

```
x<- c(2,5,3,7,1,9,6)        ## these values are stored as a list (vector) of values in the variable x
length(x)          ## finding the number of elements (numbers) stored in x
x[1]          ## first element of x
x[length(x)]          ## last element of x without knowing the number of elements in x
min(x)          ## finding minimum value
max(x)          ## finding maximum value
c(min(x), max(x))          ## minimum and maximum values together as a list (vector)
```

Few inbuilt functions in R help in typical data input. We shall discuss two such functions '*rep()*' and '*seq()*'. 'rep' is used when we want to repeat any number and 'seq' helps in writing consecutive numbers, not necessarily integers.

```
rep(2,4)          ## the number 2 is being repeated 4 times
c(rep(4,4),rep(3,3),rep(5,3))          ## creates a list 4, 4, 4, 4, 3, 3, 3, 5, 5, 5
rep(c(4,3,5),c(4,3,3))          ## easier way to create the previous list

seq(3,7,by=0.5)          ## all numbers starting with 3 and ending with 7 with an increment of 0.05
x<- seq(3,7,by=0.5)          ## store the above numbers
```

Some simple statistical measures

Simple statistical measures like mean, median, standard deviation, etc. can easily be computed for a list of observations using a single command, in most cases. Sometimes we can merge a few functions to get slightly more complicated or non-standard measures.

```
x<-c(8.60,11.68,9.58,9.17,9.11,8.22,10.74,8.25, 9.84,12.14,9.58,10.80,12.17,7.16,13.32)  ## data entry
y<-c(3,2,4,1,5,7,2,9,1,4,7,3,6,1,9)      ## list of other values stored in y
n<- length(x)        ## finds the number of observations in x
sum(x)        ## finds the sum of observations in x
mean(x)        ## calculates arithmetic mean of values in x
median(x)        ## calculates median of values in x
sd(x)        ## calculates standard deviation of values in x, divisor is n − 1
cor(x,y)        ## calculates correlation coefficient between x and y
```

Using these simple functions, we can easily compute a few other measures. For example, suppose we want to calculate a measure, known as mean deviation about median for values in y. This is defined as

$$MD = \frac{1}{n}\sum_{i=1}^{n}|y_i - \tilde{y}| \text{ where } \tilde{y} \text{ is the median of } y.$$

An R code for mean deviation about median is written as

```
MD<- mean(abs(y-median(y)))
```

Dataframe

Sometimes, we have data on multiple variables from the same individual. Data set given in Table 8.1 is an example of such data. We need to understand how to create or present data in a tabular format using R. Here we explain only using a small portion of the data in Table 8.1. Suppose we have data in four variables HDL, LDL, VLDL, and TGL, as follows.

```
HDL<- c(49,45,44,63,50,47,66,63,62,68)        ## data input for HDL
LDL<- c(95,103,100,72,90,106,62,73,82,64)        ## data input for LDL
VLDL<- c(23,22,21,20,23,23,19,21,16,17)        ## data input for VLDL
TGL<- c(146,150,155,120,148,148,110,109,117,107)        ## data input for TGL
```

First we should check whether each variable has the same number of values. It can be done in many ways. We use ‘length( )’ command in a logical mode to check this.

```
length(HDL)        ## gives the number of values in variable ‘HDL’
length(HDL)==length(LDL)        ## checks whether HDL and LDL have
                                    same number of values.;
        ## its output is “TRUE” or “FALSE”
```

Similarly we can check whether all variables are of same length. This is essential in order to present multivariate data. Now to present them in a tabular format, we use the function ‘data.frame( )’.

```
d<- data.frame(HDL, LDL, VLDL, TGL)          ## stores data in a table in this order
d                                             ## prints the output on the screen, as follows:

   HDL  LDL  VLDL  TGL
1   49   95    23  146
2   45  103    22  150
3   44  100    21  155
4   63   72    20  120
5   50   90    23  148
6   47  106    23  148
7   66   62    19  110
8   63   73    21  109
9   62   82    16  117
10  68   64    17  107
```

Given this dataframe or data set, we might want to see the values only for HDL, or calculate the mean of each column using one command or function in R. We might also add another column, say, age, on the right of this dataframe. All these can be done very easily.

```
apply(d,2,mean)          ##apply 'mean' function to each column (2 indicates column, 1 indicates row)

HDL    LDL    VLDL   TGL                 ##Output of 'apply' function applied on 'd'
55.7   84.7   20.5   131.0

age<- c(37, 41, 47, 51, 37, 44, 37, 36, 59, 53)          ## stores age data in 'age'
d<-cbind(d,age)                ## creates another column 'age' as 5-th column in 'd'
d                     ## gives the output of 'd'

d$age<- c(37, 41, 47, 51, 37, 44, 37, 36, 59, 53) ## directly creates 5-th column as age; check output of 'd'
```

Reading data files

Another issue without which this discussion would remain incomplete is that we need to know how to input data through a datafile. Usually most data files are available in *text* format (*.txt*) or *comma separated value* format (*.csv*). In order to directly import a datafile into the R workspace, the commands are:

(1) $read.table(\text{“}filename.txt\text{”}, header = T, sep = \text{“}\backslash t\text{”})$: to read .txt file, if there are column names, write ‘header = T’, otherwise ‘header = F’ and mention the separator, if any.
(2) $read.csv(\text{“}filename.csv\text{”}, header = T)$: to read .txt file, if there are column names, write ‘header = T’, otherwise ‘header = F’.

We can do some operations after reading a table of data. These operations would give an idea about some features of the table. Here also we use ‘apply’ function of R to get accustomed to working with tables after reading them directly from the directory.

```
a<- read.table("filename.txt", header=T,sep="\ t")        ##reads a table and stores in a
dim(a)          ## dimension i.e. number of rows and columns in a
head(a)         ## first six rows of a
tail(a)         ## last six rows of a
a[1:5, 1:7]         ## first five rows and first seven columns of a; note that within square bracket, before
                    ## comma (',') put the row indicators and after comma, put column indicators
a[c(1,3,4),c(2,3,7,6)]          ## values corresponding to row number 1, 3 and 4, and column
                                ## number 2, 3, 7, and 6, in this order, from the table stored in a
mean(a[,2])         ## calculates mean of elements in second column of a; note that within square bracket,
                    ## before comma, there is nothing, indicating all rows have been considered
sd(a[5,])           ## calculates standard deviation of elements in fifth row of a; note that within square
                    ## bracket, after comma, there is nothing, indicating all columns i.e. all elements in the
                    ## entire fifth row have been considered
apply(a,1,mean)         ## a nice function, calculates means of each row of a; the function 'apply' is
                        ## applying 'mean' function on the data stored in 'a' on each row ('1' indicates
                        ## row, '2' indicates column)
apply(a,2,sd)       ## calculates standard deviations of each column of a; the function 'apply' is applying
                    ## 'sd' function on the data stored in 'a' on each column ('2' indicates column)
```

Loop in R

There are mainly two loops, as common in all programming languages, in R. One is 'for' loop and another is 'while' loop. We shall describe how to write 'for' loop in R. We leave it to the reader to explore and find how to write a loop using 'while' command.

Loops are used to do a job repeatedly. In the calculation of arithmetic mean, we have to first add second number to the first, then add third number to the previous sum, and so on. Thus in each step we have to collect one number to be added to the result obtained at the previous step. It ends as soon as we add the last number. So it seems that we need three quantities: the initial number (initiation), an increment to get the next number, and an identity of the last number. We shall write an R code for calculating arithmetic mean of a set of numbers stored in x.

```
x<- 11:20           ## the consecutive numbers from 11 to 20 are stored in x
n<- length(x)           ## number of numbers stored in x
s<- 0           ## initialisation of a variable where the sum of numbers will be stored
for (i in 1:n)          ## 'for' loop starts with taking the first value, it will take increment of one to get the
                        ## next number i.e. second number and so on; loop continues until it reaches the last
                        ## number (as given by 'n')
{           ## beginning of the operations to be done inside the loop
s<- s+x[i]      ## first number of x is being added to s. It continues as long i takes consecutive values up to n
}           ## end of operation inside the loop; it then searches for the next value of i; loop ends when i = n
m<- s/n         ## after the loop ends, s contains the sum of all numbers in x; mean is then just s/n
cat("mean is ", round(m,2))         ## prints as "mean is ", and then the value of mean rounded up to two
                                    ## places of decimal
```

We can do multiple jobs inside the loop, i.e. inside {...} and get all results after the loop ends. There is no need to write the loop multiple times for multiples jobs over the same set of numbers.

Writing data files

Similarly we can save output files, provided it is described in rows and columns. Suppose '*a*' contains the output datafile and we want to save it in file named *output.txt* with no row heading but same column heading; elements of a row are separated by *tab*. The command for that would be
$write.table(a, \text{``}output.txt\text{''}, row.names = F, col.names = T, quote = F, sep = \text{``}\backslash t\text{''})$.
We write 'quote = F' because we don't want the numerical figures to be enclosed within double quotes. It is also possible to save a file which are not represented in rows and columns.

If we want to save the table, which is the output file, in .csv format, the command would be
$write.csv(d, \text{``}output.csv\text{''}, row.names = F, quote = F)$
There is no to need write 'col.names = F' as it always keeps column heads while saving in .csv format. While reading back this file, we can use either 'read.csv()' or 'read.table()' with *sep* inside the function. Check what happens if you don't use *sep* inside 'read.table()'!

Plots

If we want to plot y against x, just type $plot(x, y)$, we see the graph, with all points of y plotted against the corresponding value of x. Note that the points are 'empty circles'. If we want a smooth curve, just type $plot(x, y, ty = \text{``}l\text{''})$. Here the consecutive points are joined by straight line ('l' for line); if the points are very close to each other, the curve looks smooth. We can play around with this graph in many ways.

One important lesson: if we want to know the usage of a known function, say for the 'plot' function, in the terminal type '$?plot()$'; a full page of instructions will be opened, indicating the input variable to the function, output variables, the interpretation of each parameter that one can pass into the function along with some nice examples. Many examples of application of 'plot()' function will be available throughout this book and will serve as a good learning exercise for the reader. Moreover, there are options for other related commands that can be used for more sophistication. Thus R itself has many ways and tools that may be extremely helpful for self learning.

We shall only demonstrate scatter plot for two variables using the cholesterol indicators data stored in dataframe 'd'. We want to draw six scatter diagrams taking one pair of variables at a time. We have four variables; the number of possible pairs of variables is six, viz., (HDL, LDL), (HDL, VLDL), ..., (VLDL, TGL). We also want to save the graph in 'pdf' format. R can do that easily!

```
pdf("Cholesterol.pdf")          ## save the output plot in file named 'Cholesterol.pdf' in the same directory
we are working. Drop this line if we don't want to save this.

par(mfrow=c(2,3))          ##splits one page into six cell, arranged in 2 rows and 3 columns
plot(d$HDL,d$LDL,xlab="HDL",ylab="LDL")          ## simple scatter plot with default setting
plot(d$HDL,d$VLDL,xlab="HDL",ylab="VLDL",pch=19)    ## scatter plot another point character
                                                     # with default colour (black)
plot(d$HDL,d$TGL,xlab="HDL",ylab="TGL",pch=8,col="red")    ## scatter plot with another point
                                                            #character in red colour
plot(d$LDL,d$VLDL,xlab="LDL",ylab="VLDL",pch=7,col="blue")    ## scatter plot with another point
                                                               # character in blue colour
plot(d$LDL,d$TGL,xlab="LDL",ylab="TGL",pch=17,col=10)    ## scatter plot with another point
                          # character in colour number 10 in R plot colour library
plot(d$VLDL,d$TGL,xlab="VLDL",ylab="TGL",pch=15,col=6)    ## scatter plot with another point
                                                           # character in another colour
dev.off()          ## if we save the plot in a file, we have to write it; otherwise don't write.
```

Note that we can play with the type of points, colour, font size, etc. within a 'plot' function. Moreover, we can change the labels of x-axis and y-axis by changing the text inside 'xlab' and 'ylab' in a 'plot' function. We present the output of the above code, which is written in a very simple manner, for ease of understanding. However, we see that there are some spaces or gaps between plots in each column (and/or row). This can be adjusted by controlling other parameters that define the margins. Typing '?par()' at R console, a page will be opened with many options for everything that we want to change. Play with that to produce beautiful graphs. Of course, we can write the main title and subtitle in this entire plot.

In other words, think of anything that you want to put in a plot to present your data and results in the best possible way. In fact, you can draw nice figures. The entire Fig. A.1 is drawn using R code!

Writing a new function

We know R works through functions. For example, mean, sd, cor, etc. are all functions that are inbuilt in R. However, in some cases we might require to do a job repeatedly in our program. Writing every time would make the program unnecessarily long. This can be made more precise and faster if we write a small R code for that job. This is called writing a function for a specific purpose relevant to our study or problem. Let's demonstrate this idea using a simple example.

Suppose we want to write our R code for calculating the correlation coefficient based on data stored in x and y variables. Careful attention to the formula for correlation coefficient reveals that we need $\sum_{i=1}^{n} x_i$, $\sum_{i=1}^{n} y_i$, $\sum_{i=1}^{n} x_i^2$, $\sum_{i=1}^{n} y_i^2$, and $\sum_{i=1}^{n} x_i y_i$. Thus, we need to write a function that calculates the sum of values in a variable. Suppose 'a' contains a list of observations. Let the name of our function be $func$, which will calculate the sum. We write the code for this function and apply it to calculate the correlation coefficient.

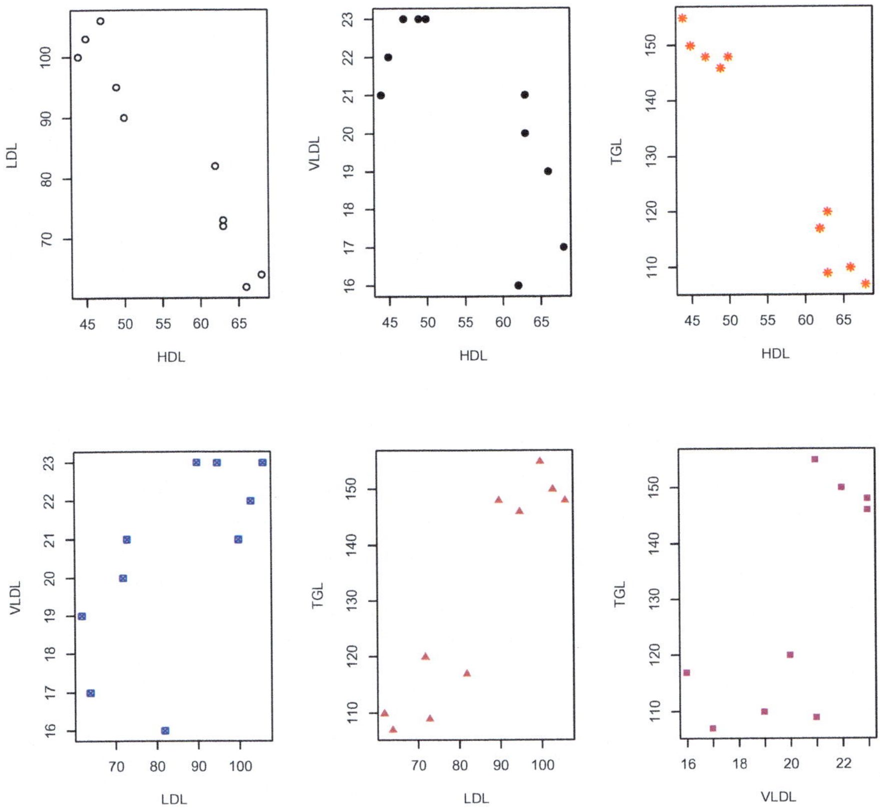

Fig. A.1 A demonstration of R plots

```
func<- function(a)          ## declaration of function named 'func' with input 'a'
{                            ## function starts
   out<- 0                   ## initiation of a variable where output will be stored
   n<- length(a)             ## number of observations in input variable
   for (i in 1:n)            ## 'for loop' starts
      out<- out+a[i]         ## number from 'a' is being added to previous result in 'out', one at a time
   return(out)               ## returns the sum of numbers in 'a'
}                            ## function ends
mean.x<- func(x)/length(x)             ## using 'func', mean calculation of numbers in 'x'
mean.y<- func(y)/length(y)             ## using 'func', mean calculation of numbers in 'y'
sd.x<- sqrt(func(x*x)/length(x)-mean.x^2)         ## using 'func', sd calculation of numbers in 'x'
sd.y<- sqrt(func((y-mean.y)^2)/length(y))         ## using 'func', sd calculation of numbers in 'y'
cov.xy<- func(x*y)/length(x)-mean.x*mean.y          ## using 'func', covariance calculation for 'x' and
'y'
corr.xy<- cov.xy/(sd.x*sd.y)  ## calculation of correlation coefficient between 'x' and 'y'
corr                               ## print correlation coefficient on the screen
```

We demonstrate another function that takes a DNA sequence and a codon as input. It calculates the frequency of that codon in the sequence. Using this function we can prepare a frequency distribution of codons of interest. We then generate a DNA sequence of length 200 and search for a few codons.

Note that, we can of course write more than one R function inside one program, i.e. a program might require multiple user-defined functions.

```
codonSearch<-function(sequence,codon)            ## declaration of function with input 'sequence' of DNA
                                                      and a 'codon'
  {
 x<-unlist(strsplit(sequence,split=""))          ## splits input DNA sequence into a sequence of nucleotides
 y<-unlist(strsplit(codon,split=""))           ## splits input codon sequence into a sequence of 3 nucleotides
   id1<-which(x==y[1])           ## identifies position of first nucleotide of the codon within DNA sequence
   if (length(id1>0))          ## checks whether any nucleotide matches with first codon nucleotide
     {
     id2<-which(x[id1+1]==y[2])  ## stores positions of sequence matching with second codon nucleotide
     if (length(id2>0))          ## checks whether any nucleotide matches with second codon nucleotide
       {
       id<-id1[id2]          ## stores positions from which two consecutive nucleotides match with first
                             # two nucleotides of the codon
       id3<-which(x[id+2]==y[3])          ## stores positions from which three consecutive nucleotides
                                          # match with the codon
       k<-length(id[id3])          ## number of such positions i.e. frequency of codon in the sequence
       }
     }
return(k)                       ## returns output which frequency of codon in input DNA sequence
  }
```

Suppose we randomly generate a DNA sequence of length 10000 and try to calculate the frequency distribution of a few codons, viz., $AGC, ATA, GTT, CCC, CTC, CGA, CGT, TGA, TCG$, and TTG. The R code using the function 'codonSearch' can be written as

```
DNAsequence<-sample(c("A","T","G","C"),1000,replace=T)
codonChoice<-c("AGC","ATA","GTT","CCC","CTC", "CGA","CGT", "TGA", "TCG", "TTG")
n<-length(codonChoice)          ## number of codons to be studied
freq<-rep(0,n)          ## initiation of 'freq' variable to store frequency of each codon
for (i in 1:n)          ## 'for loop' starts
{
freq[i]<-codonSearch(DNAsequence,codonChoice[i])          ## frequency calculation for ith codon from
                                                          #'codonChoice' list
}
freq.table<-data.frame(codonChoice,freq)          ## gives frequency distribution of codons
```

Definitely you can do a lot using this function. Similar other functions can be written to facilitate our computation in R.

Exercise

A.1 Enter the data {2, 5, 3, 7, 1, 9, 6} in a variable x. Then type x and press ‘enter’.

(a) Find the number of elements in x, i.e. in the data list.
(b) Find the last element of x.
(c) Find the 3rd element in the data list.
(d) Find 3rd to 5th element in the data list.
(e) Find 2nd, 5th, and 6th element from the list.
(f) Find the minimum element of x.
(g) Print the elements in x in reverse order without entering data again.
(h) Arrange the data in increasing and decreasing order.

A.2 Type the command $rep(2, 4)$ and see what it gives.

(a) Create a data list (4, 4, 4, 4, 3, 3, 3, 5, 5, 5) using *rep* function.
(b) Create a list (4, 6, 3, 4, 6, 3, ..., 4, 6, 3) where there are 10 occurrences of 4, 6, and 3 in the given order.
(c) Create a list (3, 1, 5, 3, 2, 3, 4, 5, 7, 7, 7, 7, 7, 8, 7, 6, 5, 4, 3, 2, 1, 34, 21, 54) using a one-line command.
(d) First create a list (2, 1, 3, 4). Then append this list at the end with another list (5, 7, 12, 6, −8). Check whether the number of elements in the augmented list is 11.
(e) Find the numbers in odd positions.

A.3 Use of *seq* function.

(a) Print all numbers starting with 3 and ending with 7 with an increment of 0.5. Store these numbers in x.
(b) Print all 17 equispaced numbers between 12 and 37, both inclusive.
(c) Print all even numbers between 2 and 14 (both inclusive).
(d) Type $2 * x$ and see what you get.
(e) Type $x * x$ and see what you get.
(f) Type $x/10$ and see what you get.
(g) If y contains first n positive integers where n is the length of x, what do you get if you write $x + y$, $x * y$ and x/y?

A.4 Enter the data {1, 2, ..., 19, 20} in a variable x.

(a) Find the sum of numbers positioned at odd numbered places.
(b) Find the sum of numbers at 2nd, 5th, 13th, and 18th positions in the list.
(c) Find mean, median, and standard deviation as a list of numbers.
(d) Find the sum of squares of the values in x.
(e) Calculate mean deviation about mean and mean deviation about median and print them together, as a list.
(f) Check whether both mean deviation about mean and mean deviation about median are less than the standard deviation.

A.5 R has a few in built data sets. One of them is 'iris' data set. Just type '$a < -iris$'. The entire datafile is stored in 'a'.

(a) Just type 'a' and see what you get.
(b) How many flowers constitute this data set?
(c) How many variables are there in 'iris' data set?
(d) Print first six rows of a.
(e) Print ten rows of a.
(f) Print first six rows and first four columns of a.
(g) Extract the data only for 'setosa' species.
(h) Draw histogram for petal length for setosa species.
(i) For setosa species data, draw four histograms on the same graph paper for petal length, petal width, sepal length, and sepal width.
(j) Store the data for setosa species in a separate file, dropping the last column.
(k) Find the correlation coefficient between petal length and sepal length for the entire data.
(l) Find a linear regression of petal width on petal length for the entire data.
(m) Drawn a scatter diagram for petal length and petal width and also plot the regression line for the entire data on this diagram.
(n) Fit a linear regression of petal width on petal length, sepal length, and sepal width for the entire data.
(o) Find all possible pairwise correlation coefficients for sepal length, sepal width, petal length, and petal width.
(p) Extract the data only for virginica.
(q) Fit regression equations of petal width on petal length for setosa species and for virginica species separately.
(r) For each of setosa and virginica species, draw scatter diagrams of petal length and petal width separately; superimpose the respective regression lines on the same plots.
(s) Merge three plots for setosa and virginica species that you have done in (m) and (r).
(t) Calculate the column means and column standard deviations for each column except the fifth column. Augment the full data set by adding one row for mean and one row for standard deviation. Note that in each newly added row, you have to keep the fifth position as blank. Store this augmented file in a new file using 'write.table' command in R.
(u) Do a two-sample t-test to see whether means of petal lengths are same in versicolor and virginica species.
(v) Do a statistical test to see whether means of petal lengths are the same in all three species.
(w) Calculate multiple correlation coefficients of petal width on petal length, sepal length, and sepal width for (i) the entire data, and (ii) for each species separately.

(x) Calculate partial correlation coefficient between petal length and petal width, eliminating the effects of sepal length and sepal width for (i) the entire data, and (ii) for each species separately.
(y) Write a program to calculate the mean of petal length for all flowers using a loop in R. Check whether your result matches with the result obtained through 'mean' function of R.
(z) Using only one loop, write a program to calculate mean, median, and standard deviation of sepal length for all flowers in the data. Present them as a list of three values indicating the measures.
[Note: For questions (u) - (x), you can do it after reading the book.]

A.6 Using the iris data set, do the following.

(a) Draw a histogram for petal length for all flowers.
(b) Draw boxplots for sepal length for (i) all flowers, (ii) setosa only, (iii) versicolor only, and (iv) virginica only, on the same graph using 'par()' command in R.
(c) Draw a scatter diagram for petal width against petal length for all flowers. However, each species should have its own colour different from others so as to identify them properly. Also put the legend for colours used inside the graph.
(d) Draw Q-Q plots for (i) setosa only, (ii) virginica only, (iii) versicolor only, and (iv) the whole data set. Comment on your findings in comparison to a normal distribution.
(e) Do a principal component analysis with the entire data set. Note that you have to remove the last column, i.e. 'Species' from the data before doing this analysis. What is the percentage of variance explained by the first two principal components? Plot the amount of variance explained against number of principal components and comment on how you would choose the number of such components to work with.
(f) Plot first and second principal components against each other and comment on the plot.
(g) Do a cluster analysis using (i) hierarchical clustering and (ii) k-means clustering methods. Compare the results obtained by these two methods and interpret your result.

Appendix B

We have drawn Fig. 2.1 to explain histogram and its components. The R command 'hist( )' will give only histogram. R code that produces Fig. 2.1 involves little bit of coding using different R commands. The entire code is given below. It can be made more precise; but we intentionally keep it as this, because you can try it yourself. Just copy each line one by one, paste in R terminal, and press enter; see what happens and how this entire graph is emerged at the end. Once you get Fig. 2.1, you may try to make it shorter and add different colours to different components.

Similarly we also provide below the R code for reproducing Fig. 2.3 which represents ogives, both less than and greater than types with some additional parts required for explanation. As in the previous case, use this R code to reproduce Fig. 2.3 and then try to add some changes as you wish.

I. Mukhopadhyay and P. P. Majumder, *Statistical Methods in Human Genetics*, Indian Statistical Institute Series, https://doi.org/10.1007/978-981-99-3220-7

```
> br <- round(seq(0,10,length = 12),3)
> hb <- hist(x,breaks = br)
> d <- data.frame(br[−1],cumsum(hb$counts), br[−length(br)],
cumsum(hb$counts[length(hb$counts):1]))
> d <- rbind(c(0,0,0,90),d)
> d <- rbind(c(−0.909,0,−0.909,90),d)
> d <- rbind(d,c(10.909,90,10.909,0))
> d <- rbind(d,c(10.909,90,10.909,0))
> plot(d[,1],d[,2],pch=20,xaxt= 'n',xlab="Class boundary",
ylab="Cumulative frequency")
> lines(d[,1],d[,2])]
> axis(side = 1,at = d[,1],labels = d[,1],las = 2)
> points(d[,3],d[,4], pch = 20)
> lines(d[,3],d[,4])
> lines(c(2.927,2.927),c(−4,45))
> text(6,30,"Median")
> arrows(5.32,30,2.93,−3.5,length = 0.15,angle = 25,code = 2)
> text (8.2, 50, "Less-than type ogive")
> arrows(6.4,50,3.2,38,length = 0.15,angle = 25,code = 2)
> text (8.2, 70, "Greater-than type ogive")
> arrows(6.2,70,1.3,76,length = 0.15,angle = 25,code = 2)
> text(3.2,45,"O")
> lines(c(−1.4,2.927),c(45,45))
> axis(side = 1, at = 2.927, labels = "M", las = 1, line = −0.9)
> axis(side = 2, at = 45, labels = "45", las = 1, line = −0.9)
```

R code for boxplot (Fig. 2.4) is given below that contains codes for drawing arrows, inserting texts in any position of the graph, etc. You may try to make this code more precise, use other fonts and colours, etc.

```
> boxplot(b,lwd = 1.5,pch = 20)
> text(1.4,6,"Median")
> arrows(1.34,5.9,1.1,3,length = 0.15,angle = 25,code = 2)
> text(x = 1.03, y = 9.47, '{', srt = 180, cex = 3.48,family = 'CenturySch')
> text(1.3,9.5,"Outliers")
> arrows(1.235,9.5,1.05,9.5,length = 0.15,angle = 25,code = 2)
> text(x = 1.24, y = 2.935, '{', srt = 180, cex = 7,family = 'CenturySch')
> text(1.37,3.25,"Middle 50%")
> text(1.37,2.8,"observations")
> text(0.56,2.8,"Whiskers")
> arrows(0.98,6,0.56,3,length = 0.15,angle = 25,code = 2)
> arrows(0.98,0.8,0.56,2.5,length = 0.15,angle = 25,code = 2)
```

Appendix C

Theorem 18 $P(A \cup B) = P(A) + P(B) - P(A \cap B)$.

Proof From Fig. 3.2 it is clear that the set B can be written as an union of two mutually exclusive sets, $A \cap B$ and $A^c \cap B$. Noting that the intersection of these two sets is a null set, we have by the definition of probability,

$$\begin{aligned} & B = (A \cap B) \cup (A^c \cap B) \\ \Longrightarrow\ & P(B) = P\Big[(A \cap B) \cup (A^c \cap B)\Big] = P(A \cap B) + P(A^c \cap B) \\ \Longrightarrow\ & P(A^c \cap B) = P(B) - P(A \cap B) \end{aligned}$$

Again, from Fig. 3.2, it is clear that $A \cup B$ can be written as an union of two mutually exclusive sets, A and $A^c \cap B$. Hence by definition of probability, we have

$$P(A \cup B) = P\Big[A \cup (A^c \cap B)\Big] = P(A) + P(A^c \cap B) = P(A) + P(B) - P(A \cap B).$$

Theorem 19 $P(\phi) = 0$.

Proof Clearly, $\Omega = \Omega \cup \phi$. Noting that $P(\Omega) = 1$, we have,

$$1 = P(\Omega) = P(\Omega \cup \phi) = P(\Omega) + P(\phi) = 1 + P(\phi) \implies P(\phi) = 0.$$

Theorem 20 *If A and B are two mutually exclusive sets,* $P(A \cap B) = 0$.

Proof For two mutually exclusive events A and B, we have $A \cap B = \phi$. Hence, by Theorem 19, $P(A \cap B) = P(\phi) = 0$.

Theorem 21 *For any A and B such that $A \subset B$, $P(A) \leq P(B)$.*

Proof Using the techniques in previous theorems and from first graph of Fig. 3.3, it is easy to see that $B = A \cup (A^c \cap B)$. Moreover A and $(A^c \cap B)$ are mutually exclusive events so that $P\big[A \cap (A^c \cap B)\big] = 0$, by Theorem 20. Hence applying the definition of probability and using Theorem 18, we have

I. Mukhopadhyay and P. P. Majumder, *Statistical Methods in Human Genetics*, Indian Statistical Institute Series, https://doi.org/10.1007/978-981-99-3220-7

$$P(B) = P\Big[A \cup (A^c \cap B)\Big] = P(A) + P(A^c \cap B) \geq P(A) \qquad \text{[since, } P(A^c \cap B) \geq 0\text{]}$$

Theorem 22 $0 \leq P(A) \leq 1$ *for any* $A \in \mathscr{A}$.

Proof Since $A \in \mathscr{A}$, we have by definition $P(A) \geq 0$. Again Ω being the largest set, $A \subset \Omega$. Using the fact that $P(\Omega) = 1$ by definition and using Theorem 21, we have, $0 \leq P(A) \leq P(\Omega) = 1$, i.e. $0 \leq P(A) \leq 1$.

Theorem 23 $P(A^c) = 1 - P(A)$.

Proof Clearly, from Fig. 3.2, we can write, $\Omega = A \cup A^c$. Moreover $A \cap A^c = \phi$. Hence, by definition and using Theorem 18, we have,

$$1 = P(\Omega) = P(A \cup A^c) = P(A) + P(A^c) \implies P(A^c) = 1 - P(A).$$

Appendix D

Proof of Hardy-Weinberg equilibrium

First we shall prove Hardy-Weinberg Equilibrium (HWE) in the context of bi-allelic locus.

Proof To prove HWE, first we have to calculate offspring genotype probability conditional on parental genotypes. Suppose, a male with genotype AA undergoes random mating with a female having genotype AA. The offspring genotype due to this random mating will always be AA, because female will transmit A allele with probability 1 and similarly the male will transmit A allele with probability 1. Moreover, transmission of an allele by mother is completely independent of the transmission of allele by the father. Hence,

$$\begin{aligned} P(\text{Offspring genotype is } AA|AA \times AA) &= 1, \\ P(\text{Offspring genotype is } Aa|AA \times AA) &= 0, \\ \text{and } P(\text{Offspring genotype is } aa|AA \times AA) &= 0. \end{aligned}$$

To write this precisely, we use a few notations. Let O_g, F_g and M_g be the genotype of offspring, female and male individuals respectively and $F_g \times M_g$ denote random mating between the male and the female. Hence, using these notations, we can write

$$P(O_g = x|M_g = AA, F_g = AA) = \begin{cases} 1 & \text{if } x = AA \\ 0 & \text{if } x = Aa \text{ or } aa \end{cases}$$

where x denotes the genotype of the offspring under random mating between F_g and M_g. Now suppose that $F_g = Aa$ and $M_g = Aa$. This male can transmit A allele with probability 0.5 and the female can transmit A allele with probability 0.5. These two events are also independent of each other. So,

$$P(O_g = AA|F_g = Aa, M_g = Aa) = 0.5 \times 0.5 = 0.25.$$

I. Mukhopadhyay and P. P. Majumder, *Statistical Methods in Human Genetics*, Indian Statistical Institute Series, https://doi.org/10.1007/978-981-99-3220-7

To calculate $P(O_g = Aa|F_g = Aa, M_g = Aa)$, note that male and female transmit A allele and a alleles respectively each with probability 0.5 resulting in offspring genotype as Aa whereas male and female transmit a allele and A alleles respectively each with probability 0.5 resulting in the same offspring genotype i.e Aa. Also note that transmission of allele by mother (female) and transmission of allele by father (male) are independent, we have

$$\begin{aligned} P(O_g = Aa|F_g = Aa, M_g = Aa) &= P(M_g \to A, F_g \to a) + P(M_g \to a, F_g \to A) \\ &= P(M_g \to A)P(F_g \to a) + P(M_g \to a)P(F_g \to A) \\ &= 0.5 \times 0.5 + 0.5 \times 0.5 = 0.5 \end{aligned}$$

Hence, we have

$$P(O_g = x|M_g = Aa, F_g = Aa) = \begin{cases} 0.25 & \text{if } x = AA \\ 0.5 & \text{if } x = Aa \\ 0.25 & \text{if } x = aa \end{cases}$$

Proceeding similarly, we have the offspring genotype probabilities for all possible genotypes due to random mating and is given in Table A.1.

Now assume that genotype frequencies (probabilities) in the population in Generation 0 for three possible genotypes are $Q = P(AA)$, $R = P(Aa)$, and $H = P(aa)$. Table A.1 presents the conditional probability of offspring genotype given parental genotype combinations. So, in the next generation, i.e. Generation 1, after random mating at Generation 0, the offspring genotype probabilities become

Table A.1 Offspring probability conditional on parental mating

Mating type	Offspring genotype probability		
$F_g \times M_g$	AA	Aa	aa
$AA \times AA$	1	0	0
$AA \times Aa$	0.5	0.5	0
$AA \times aa$	0	1	0
$Aa \times AA$	0.5	0.5	0
$Aa \times Aa$	0.25	0.5	0.25
$Aa \times aa$	0	0.5	0.5
$aa \times AA$	0	1	0
$aa \times Aa$	0	0.5	0.5
$aa \times aa$	0	0	1

$$
\begin{aligned}
P(O_g = AA) &= P(O_g = AA|AA \times AA)P(AA \times AA) + P(O_g = AA|AA \times Aa)P(AA \times Aa) + \cdots \\
&= 1 \times Q^2 + 0.5 \times QR + 0 \times QH + 0.5 \times RQ + 0.25 \times R^2 + 0 \times RH + 0 \times HQ + 0 \times HR + 0 \times H^2 \\
&= Q^2 + \frac{1}{2}QR + \frac{1}{2}QR + \frac{1}{4}R^2 = (Q + \frac{R}{2})^2 = Q_1 \text{ (say)}, \\
P(O_g = Aa) &= P(O_g = Aa|AA \times AA)P(AA \times AA) + P(O_g = Aa|AA \times Aa)P(AA \times Aa) + \cdots \\
&= 0 \times Q^2 + 0.5 \times QR + 1 \times QH + 0.5 \times RQ + 0.5 \times R^2 + 0.5 \times RH + 1 \times HQ + 0.5 \times HR + 0 \times H^2 \\
&= \frac{1}{2}QR + QH + \frac{1}{2}QR + \frac{1}{2}R^2 + \frac{1}{2}RH + QH + \frac{1}{2}RH \\
&= 2(Q + \frac{R}{2})(H + \frac{R}{2}) = R_1 \text{ (say), and} \\
P(O_g = aa) &= (H + \frac{R}{2})^2 = H_1 \text{ (say)}.
\end{aligned}
$$

If HWE is true, we can claim that these genotype frequencies, i.e. $Q_1 = P(AA)$, $R_1 = P(Aa)$, and $H_1 = P(aa)$ in Generation 1 becomes stable and will continue to remain the same in subsequent generation as long as only random mating continues. To show whether these become stable, we have to consider another generation of random mating i.e. in Generation 1. Noting that $Q + R + H = 1$ and proceeding as explained above, but with genotype frequencies Q_1, R_1, and H_1 for genotypes AA, Aa, and aa respectively, we have the genotype frequencies in Generation 2 as

$$
\begin{aligned}
P(O_g = AA) &= (Q_1 + \frac{R_1}{2})^2 = \Big\{(Q + \frac{R}{2})^2 + \frac{1}{2} \times 2(Q + \frac{R}{2})(H + \frac{R}{2})\Big\}^2 \\
&= (Q + \frac{R}{2})^2(Q + \frac{R}{2} + H + \frac{R}{2})^2 = (Q + \frac{R}{2})^2(Q + R + H)^2 = (Q + \frac{R}{2})^2 = Q_1, \\
P(O_g = Aa) &= 2(Q_1 + \frac{R_1}{2})(H_1 + \frac{R_1}{2}) \\
&= 2\Big\{(Q + \frac{R}{2})^2 + \frac{1}{2} \times 2(Q + \frac{R}{2})(H + \frac{R}{2})\Big\}\Big\{(H + \frac{R}{2})^2 + \frac{1}{2} \times 2(Q + \frac{R}{2})(H + \frac{R}{2})\Big\} \\
&= 2\Big\{(Q + \frac{R}{2})(Q + R + H)\Big\}\Big\{(H + \frac{R}{2})(Q + R + H)\Big\} \\
&= 2(Q + \frac{R}{2})(H + \frac{R}{2}) = R_1, \text{ and} \\
P(O_g = aa) &= (H_1 + \frac{R_1}{2})^2 = \Big\{(H + \frac{R}{2})^2 + \frac{1}{2} \times 2(Q + \frac{R}{2})(H + \frac{R}{2})\Big\}^2 \\
&= (H + \frac{R}{2})^2(Q + \frac{R}{2} + H + \frac{R}{2})^2 = (H + \frac{R}{2})^2(Q + R + H)^2 = (H + \frac{R}{2})^2 = H_1.
\end{aligned}
$$

Hence, the genotype frequencies remain the same in Generation 2 and subsequent generations, provided random mating continues in every generation and there is no external force acting in any generation.

Properties of allele frequency estimate

We have seen that $\hat{p} = \frac{2n_{AA}+n_{Aa}}{2n}$. Now if we consider n_{AA} as the number of individuals having genotype AA and the rest have genotype either Aa or aa, then we can say that, $n_{AA} \sim Bin(n, p^2)$. Similarly $n_{Aa} \sim Bin(n, 2pq)$. Note that if X is a random variable

that follows a binomial distribution with parameters n and p, we have $E(X) = np$ and $V(X) = npq$ where $q = 1 - p$. Thus we have

$$E(\hat{p}) = E\left(\frac{2n_{AA} + n_{Aa}}{2n}\right) = \frac{2E(n_{AA}) + E(n_{Aa})}{2n} = \frac{2n.p^2 + n.2pq}{2n} = p \qquad \text{(since, } p + q = 1\text{)}$$

To evaluate V(Y), we consider $Y = n_{AA} + n_{Aa}$ as a random variable that denotes the number of individuals having genotypes either AA or Aa and so the rest have genotype aa. Thus, $Y \sim Bin(n, p^2 + 2pq)$. So we have

$$\begin{aligned} &V(Y) = V(n_{AA} + n_{Aa}) = V(n_{AA}) + V(n_{Aa}) + 2Cov(n_{AA}, n_{Aa}) \\ \Longrightarrow\ &n(p^2 + 2pq)(1 - p^2 - 2pq) = np^2(1 - p^2) + 2npq(1 - 2npq) + 2Cov(n_{AA}, n_{Aa}) \\ \Longrightarrow\ &Cov(n_{AA}, n_{Aa}) = -2np^3q \end{aligned}$$

$$\begin{aligned} \text{Hence, } V(\hat{p}) = V\left(\frac{2n_{AA} + n_{Aa}}{2n}\right) &= \frac{1}{2n^2}.\left[4V(n_{AA}) + V(n_{Aa}) + 4.cov(n_{AA}, n_{Aa})\right] \\ &= \frac{1}{2n^2}\left[4np^2(1 - p^2) + 2npq(1 - 2pq) - 8np^3q\right] = \frac{pq}{n}. \end{aligned}$$

The manufacturer's authorised representative in the EU is Springer Nature Customer Service Centre GmbH, Europaplatz 3, 69115 Heidelberg, Germany. If you have any concerns regarding our products, please contact ProductSafety@springernature.com

Printed and bound by CPI Group (UK) Ltd, Croydon, CR0 4YY

05/10/2026

02228132-0001

The manufacturer's authorised representative in the EU is Springer Nature Customer Service Centre GmbH, Europaplatz 3, 69115 Heidelberg, Germany. If you have any concerns regarding our products, please contact ProductSafety@springernature.com

Printed and bound by CPI Group (UK) Ltd, Croydon, CR0 4YY
01/10/2026
02225900-0010